Human and Environmental Risks of Chlorinated Dioxins and Related Compounds

ENVIRONMENTAL SCIENCE RESEARCH

Recent Volumes in this Series

A Continuation Order Plan is available for this series. A continuation order will bring
delivery of each new volume immediately upon publication. Volumes are billed only upon
actual shipment. For further information please contact the publisher.

Human and Environmental Risks of Chlorinated Dioxins and Related Compounds

Edited by
Richard E. Tucker

Dynamac Corp Enviro Control Division
Rockville, Maryland

Alvin L. Young

U.S. Veterans Administration
Washington, D.C.

and
Allan P. Gray

Dynamac Corp Enviro Control Division
Rockville, Maryland

PLENUM PRESS ● NEW YORK AND LONDON

Library of Congress Cataloging in Publication Data

Main entry under title:

Human and environmental risks of chlorinated dioxins and related compounds.

(Environmental science research; v. 26)
Proceedings of an international symposium on Chlorinated Dioxins and Related Compounds, held October 25–29, 1981 in Arlington, Va.
Includes bibliographical references and index.
1.Tetrachlorodibenzodioxin—Environmental aspects—Congresses. 2. Tetrachloro-dibenzodioxin—Toxicology—Congresses. I. Tucker, Richard E. II. Young, Alvin L. III Gray, Allan P. IV. Series.
QH545.P4H85 1982   661′.8915   82-18121

ISBN-13: 978-1-4613-3601-3   e-ISBN-13: 978-1-4613-3599-3
DOI: 10.1007/978-1-4613-3599-3

Proceedings of an international symposium on Chlorinated Dioxins and Related Compounds, held October 25–29, 1981, in Arlington, Virginia

© 1983 Plenum Press, New York
Softcover reprint of the hardcover 1st edition 1983

A Division of Plenum Publishing Corporation
233 Spring Street, New York, N.Y. 10013

PREFACE

Increasing international concern is being expressed regarding the contamination of the environment with polychlorinated dibenzo-p-dioxins (PCDDs) and polychlorinated dibenzofurans because certain of these chemicals have been shown to be highly toxic to animals and are ubiquitous in the environment. They are known to be distributed as contaminants of commercial products and as by-products from combustion processes.

A considerable volume of information has accumulated on these chemicals in the past two decades, particularly for the most toxic of them, 2,3,7,8-tetrachlorodibenzo-p-dioxin (2,3,7,8-TCDD). However, this body of knowledge has not succeeded in resolving genuine judgmental differences among experts in the field as to the degree of hazard to human health and the environment. In light of the widespread public concern, it is clearly imperative to come to grips with the continuing scientific controversy, to review the data, assess the issues, to see where areas of agreement exist, and where further research is needed to resolve remaining areas of disagreement.

This volume represents an effort to contribute to these goals. The volume contains the Proceedings from an International Symposium on Chlorinated Dioxins and Related Compounds which was held October 25 to 29, 1981, in Arlington, Virginia. The objectives of the meeting were to bring together scientists from a wide range of disciplines, all of whom were directly concerned with one aspect or another of the dioxin problem, to review the existing information, evaluate controversial data, present new data, identify areas of agreement, and indicate directions for future research. The 56 papers and panel reports that comprise this volume testify to the success of the Symposium.

The meeting was divided into nine sections:

Definition of the Problem
Analytical Chemistry
Environmental Chemistry
Animal Toxicology ·

        Biochemistry and Metabolism
        Environmental Toxicology
        Human Observations
        Risk Assessment
        Laboratory Safety and Waste Management

Panel sessions were organized for the major sections and the conclusions arrived at in these discussions are reported at the end of the volume.

The intense interest in the subject area can be appreciated by examining the diverse backgrounds of the symposium participants. Individuals from 20 different countries attended. Not only did scientists representing the various disciplines covered in the program attend, but also lawyers, engineers, economists, and representatives of the media. It is especially interesting to note that over 70 physicians attended the four-day Symposium.

We thank all who made this meeting possible, particularly the participants and authors. We are especially grateful to Ida McDonald Sandy Hugus, Barbara Brandau, and Dr. Janice Longstreth for their help with the preparation of this volume.

CONTENTS

DEFINITION OF THE PROBLEM

Chairman: Barclay M. Shepard
               Veterans Administration
               Washington, D. C. USA

DIOXINS AS CONTAMINANTS OF HERBICIDES: THE U.S. PERSPECTIVE

B. M. Shepard and A. L. Young

Office of Environmental Medicine
Veterans Administration
Washington, DC, USA

ABSTRACT

For almost two decades, the United States has been involved in controversy over the tactical use of herbicides in the Vietnam Conflict. Few environmental or occupational health issues have received the sustained national attention that has been focused on "Agent Orange". The Veterans Administration in cooperation with other Federal agencies in the United States Government has initiated extensive health studies of veterans exposed to Agent Orange and its dioxin contaminant, 2,3,7,8-tetrachlorodibenzo-p-dioxin (TCDD) during the Vietnam Conflict. The basis for resolving the Agent Orange controversy must in large measure stem from the results of scientific inquiry.

DISCUSSION

The use of chemical (herbicides) to control vegetation has been one of the most controversial subjects arising from the Vietnam Conflict. The U.S. Air Force applied most of these herbicides in dense jungle areas to uncover and expose hidden enemy staging areas, clear vegetation from the vicinity of military bases and camps, and along lines of communication. The objective was to provide defoliated zones that would reduce ambushes and disrupt enemy tacts. The most commonly used "defoliant" was "Agent Orange," a mixture of two commercial herbicides widely employed for a number of years in brush control programs throughout the United States and for weed control in agriculture.

During a 5-year period from 1965 to 1970, the U.S. Air Force applied more than 10 million gallons of Agent Orange in South Vietnam. Some two million American military personnel served 1-year tours during the same period. Recently, many veterans of that era have reported medical problems that possibly stem from exposure to Agent Orange during their military assignments. Their complaints have ranged from tingling in the extremities to rare forms of cancer. Some veterans have fathered children with birth defects and have suggested that Agent Orange is the culprit. But the majority of scientific data on the toxicology of the chemical components in Agent Orange do not substantiate these claims. Nevertheless, the media has given intense sympathetic coverage to the veterans and their medical complaints. In the meantime, the Veterans Administration has been directed by the United States Congress to conduct an extensive health study of veterans who were exposed to herbicides in South Vietnam. The issue is whether actual or perceived health problems stem from herbicide exposure during the Vietnam conflict or whether other factors drive the controversy.

A key question that must be considered in reviewing present concerns over Agent Orange is "why has the Agent Orange issue surfaced so long after the herbicides were used in Vietnam?" One answer may be that serious concerns about potential health effects from exposure to the herbicide have just recently appeared. Another possible answer is that the general public and Congress have become more sympathetic to the concerns of Vietnam veterans, and Agent Orange is a vehicle on which to focus those concerns. Certainly, the acrimony and bitterness over U.S. involvement in Vietnam caused many Americans to repress memories of that war. As a result, they have tended either to ignore veterans of the Vietnam era or to relegate them to a lesser status than veterans of other wars. Recent gains in respect for Vietnam veterans have coincided with increasing American interest in health and environmental issues. Thus, the controversy surrounding Agent Orange has surfaced primarily because it involves the veterans and herbicides, both of which have been the center of controversy since the Vietnam conflict.

Health concerns involving Agent Orange, its component herbicides and the toxic dioxin contaminant 2,3,7,8-tetrachloro-dibenzo-p-dioxin (TCDD) date from 1970. Current interest is merely an extension and popularization of issues first publicized in 1970 and again in 1974. A large volume of toxicological data on 2,4,5-trichlorophenoxyacetic acid (2,4,5-T) and 2,4-dichloro-phenoxyacetic acid (2,4-D), the two herbicides in Agent Orange, were available during the final years of U.S. involvement in Vietnam, but the serious inadequacy of toxicological and environmental data on TCDD precluded any rapid or easy resolution

of the issue. Although scientists recognized that TCDD was
acutely toxic and teratogenic (capable of producing birth
defects) in laboratory animals, no studies were available which
had measured the effects of chronic long-term low-level exposures
on lower mammalian species. Furthermore, numerous occupational
exposures to TCDD were reported during the industrial production
of trichlorophenol but with few exceptions, extensive long-term
follow-up human epidemiologic studies are not available despite
documented exposures as early as 1949. Thus, in order to resolve
the present controversy, scientists must determine whether they
can assess the long-term effects of exposure to TCDD on the basis
of existing data or whether the resolution must await results from
new research initiatives.

Most of the Agent Orange was sprayed from fixed-winged
aircraft, C-123's, in what was then called "Operation RANCH HAND",
a code name for tactical spraying operations. A relatively small
amount of Agent Orange was sprayed from helicopters, from ground
vehicle-mounted containers and hand-operated portable equipment.
In April 1970 the Secretaries of Agriculture; Health, Education,
and Welfare; and the Interior suspended most domestic uses of
2,4,5-T. This suspension resulted from studies which revealed
that 2,4,5-T had toxic effects on animals. This toxicity was
subsequently found to be the effect of the TCDD contaminant. The
Department of Defense at the same time suspended the use of Agent
Orange in Vietnam.

The health issues surrounding Agent Orange have proven to be
very complex. There is also the confounding factor that, with the
exception of the Air Force records of "Operation RANCH HAND",
there is little or no systematic documentation of ground troops or
military units exposed to herbicides in Vietnam. So we have only
limited exposure data. In addition, there are no known laboratory
tests for previous exposure, so it is impossible at this time to
examine an individual and determine what his level of exposure may
have been during that period of service. Furthermore, veterans
have often complained of a wide range of symptoms, none of which
really leads to a diagnosis or a group of diseases. In addition a
great deal of conflicting scientific evidence has emerged both
from Europe and from Vietnam itself concerning the health effects
of exposure to herbicides.

What has been the Veterans Administration (VA) response to
the critical issues raised by Agent Orange? Early action by the
VA included the development of a communication process for
obtaining scientific and medical information which would bear
directly on the problem. In addition, early in 1978, the VA
established the Agent Orange Registry in order to accomplish four
objectives:

(1) To identify all Vietnam veterans expressing a concern
about the possible adverse health effects of their exposure
to Agent Orange.

(2) To provide a mechanism for Vietnam veterans to voice
their concerns to a physician, receive a physical examination
and obtain responsible answers to some of their questions.

(3) To serve as a mechanism for follow-up of these
individuals should significant information develop as a
result of the various research efforts underway or being
planned.

(4) To obtain, as a by-product, some preliminary information
on the current health status of the veterans who have
participated in the registry.

The Agent Orange Registry provides us essentially with a
listing of names and addresses, together with some background data
on a self-reported group of veterans with service in Vietnam. A
complete physical examination and a group of baseline laboratory
tests are provided to each veteran. The information from this
effort is currently being placed into a computer data bank. There
are now in excess of 80,000 Vietnam veterans who have been
examined by the VA. Of these, approximately 60,000 have had the
information from their examinations entered into our computer data
bank. We are in the process of collating the data to determine
whether or not there are any significant health trends in these
individuals which might suggest areas for further investigation.

We are also engaged either directly or cooperatively in a
number of research activities. Let me briefly describe six of
these studies.

TCDD Assay of Human Adipose Tissue

Since TCDD is known to accumulate preferentially in the
adipose tissue of certain species of laboratory animals, it was
suggested, early in the history of the Agent Orange issue, that
the analysis of human fat for TCDD might provide a way of
determining Agent Orange exposure. In addition, if detectable
levels of this substance were present, there might be some
relationship to health problems. Although methods for TCDD
analysis have improved in recent years, no such study had been
carried out in humans with known exposure to herbicides containing
this toxic contaminant. Consequently the VA embarked on a small
feasibility study to test the methodology and see what conclusions
might be drawn regarding the significance of the results. The
study was carried out in three groups of adult males as follows:

a) Twenty Vietnam veterans who claimed health problems
related to Agent Orange exposure and who volunteered for the
fat biopsy.

b) Three U.S. Air Force officers with known heavy and
relatively recent exposure in connection with herbicide
testing, but who did not serve in Vietnam.

c) Ten veterans with no service in Vietnam and no known
exposure to herbicides who were undergoing elective abdominal
surgery and volunteered to serve as controls.

The procedure called for the removal of 20-30 grams of
subcutaneous adipose tissue from the abdominal wall. The
University of Nebraska, Lincoln was awarded the contract for the
analysis since its laboratory was one of the few in this country
with recognized experience in TCDD analysis in minute
concentrations using high resolution mass spectrometry and gas
chromatography.

The results of this very complex and technically difficult
analysis indicated that very low levels of TCDD, believed to be
the 2,3,7,8 isomer, was detected in the range of a few parts-per-
trillion in many of the veterans, but the levels did not correlate
well with known exposure and non-exposure, and that there was no
correlation with health status. Nevertheless, the results
indicated that the assay method was feasible, but would serve no
clinically or administratively useful purpose at the present
time.

Review and Analysis of Literature

Public Law 96-151, enacted by Congress in December 1979,
directed the Administrator of Veterans Affairs to "conduct a
comprehensive review and analysis" of the worldwide literature on
Agent Orange and other phenoxy herbicides. The goal of this
report was to present a balanced and critical review of the
current state of published scientific literature relevant to the
problem of exposure to herbicides, particularly, phenoxy
herbicides used in Vietnam. The contract for this review was
awarded to JRB Associates, Inc., of McLean, Virginia, and
submitted in final form to the VA in October 1981. For the
purpose of this review, the contractor used an extensive search
strategy and developed careful parameters for scientific
relevance. Over 1,400 published articles were acquired and
examined. Aproximately 1,200 were judged to meet the relevancy
criteria. These references underwent an extensive review and
analysis by a panel of recognized experts. The report was
prepared in two volumes: the first is a critical analysis of the
relevant citations and the second is an annotated bibliography.

Vietnam Veteran Mortality Studies

The VA is in the process of designing two large mortality studies which will analyze and compare death rates and cause-of-death profiles between veterans with service in Vietnam and comparable veterans with no service in Vietnam. It is estimated that approximately 300,000 Vietnam and Vietnam-Era veterans have died since the start of the Vietnam Conflict. This number includes approximately 52,000 combat deaths. The studies will use existing computer records to assemble cohorts of veterans and determine their mortality experience. The studies also include independent validation of the computer data bases used in the studies. By using automated records, a large subsample of veterans may be studied and results may be obtained relatively quickly and inexpensively. It should be noted, however, that the mortality studies are not intended to provide definitive answers, but will instead provide mortality information which may prove useful primarily in suggesting areas for further scientific study. In addition, it is important to note that the mortality studies will not relate specifically to phenoxy herbicide exposure, but rather to service in Vietnam.

Birth Defects and Military Service in Vietnam Study

In 1981 the Centers for Disease Control (CDC) initiated a study designed to determine if Vietnam veterans are at increased risk of having children with birth defects. Since 1968 CDC has maintained a registry of all babies born with defects in the greater metropolitan Atlanta area. Of the total 15,000 children in this registry, approximately 7,500 had significant anatomical defects at birth. The investigators will attempt to locate and interview the parents of all 7,500 of the children in this group. In addition, the parents of 3,000 matched control normal babies born during the same time period will be interviewed. Since the major objective of this study will be to determine whether an unusually high proportion of fathers of babies born with defects served in Vietnam, information will be gathered about Vietnam service as well as other factors which may be associated with the occurrence of birth defects. If the study demonstrates that a Vietnam veteran has an increased risk of fathering a child with a defect it may be desirable to attempt to determine if the increase is associated with Agent Orange exposure or with some other factor(s). The study is scheduled to be completed by September, 1983.

The Air Force Health Study

In 1979 the Air Force initiated the protocol for a comprehensive epidemiologic study of the RANCH HAND personnel, a group of approximately 1,250 men who conducted the aerial,

fixed-wing herbicide spraying missions in Vietnam from 1962
through 1971. The study is designed around the question, "Have
there been, are there now, or will there be in the reasonably
foreseeable future, any adverse health effects among RANCH HAND
personnel caused by repeated exposure to Herbicide Orange?"

The investigation is composed of three integrated elements --
a mortality study of those individuals who have died since their
exposure, a morbidity study to examine the current health status
of the study subjects, and a follow-up study to look for delayed
effects over the next 20 years. The mortality and morbidity study
elements are being conducted simultaneously on the RANCH HAND
personnel together with a carefully matched control group by the
use of personnel tracking procedures coupled with an extensive
review of military medical and personnel records; detailed
face-to-face questionnaires to ascertain current and past health
events as well as occupational and family data; and comprehensive
physical examinations, psychological testing and diagnostic
laboratory studies to determine exact health status and to provide
baseline data with which to correlate symptoms. Additional
questionnaires and physical examinations are to be administered
periodically during the follow-up phase. Results from the
mortality analysis are to be reported in late 1982 and the
results of the initial questionnaire and current health status
analysis are to be released in 1983.

VA Epidemiologic Study of Ground Troops Exposed to Agent Orange

Public Law 96-151 also charged the Veterans Administration to
"design a protocol for and conduct an epidemiological study of
persons, who while serving in the Armed Forces of the United
States during the period of the Vietnam conflict, were exposed to
any of the class of chemicals known as 'the dioxins' produced
during the manufacture of the various phenoxy herbicides
(including the herbicide known as 'Agent Orange') to determine if
there may be long-term adverse health effects in such persons from
such exposure."

On May 1, 1981, the VA contracted with the School of Public
Health of the University of California at Los Angeles, to design
the mandated study. Although offering information and assistance,
the VA has carefully avoided any attempts to influence the design.
The study protocol will be sent to four groups for review, namely
the Office of Technology Assessment, the Science Panel of the
Agent Orange Working Group, the VA Advisory Committee on
Health-Related Effects of Herbicides and the National Academy of
Sciences. The VA Advisory Committee has members from veterans
organizations and thus provides a direct means to solicit comments
from the veteran community. The recommended study protocol uses
an historical cohort design comparing the health status of

presumed highly and minimally exposed cohorts. The study cohorts
are being defined through the use of Army and Marine Corps records
for the period 1967-1969. All members of the final cohorts will
be traced as necessary to determine current vital status, and will
be examined by a standard protocol including an extensive
questionnaire, physical examination, and laboratory testing.
Final results from this study are not to be expected for at least
4 to 5 years.

CONCLUSION

Neither the government nor the scientific community has
resolved the numerous controversies (environmental, medical, or
political) involving the use of Agent Orange in Vietnam from 1962
to 1970. A report by the National Academy of Sciences in 1974
documented some of the environmental impacts of Agent Orange, but,
unfortunately, the arrangements that terminated the conflict
precluded additional scientific studies in that area. Such
studies might have minimized current medical concerns about
herbicide exposure.

The controversy surrounding the military and civilian uses of
herbicides only added fuel to emotional issues related to U.S.
involvement in Vietnam. Any answer to the question of whether the
use of herbicides was "right" or "wrong" depends on personal
perspectives of the conflict. There will never be accurate
figures reflecting the number of American lives saved because of
the effective use of herbicides in preventing many ambushes or
limiting the enemy's combat capabilities. Conversely, the impact
of using Agent Orange will be viewed in a different light if the
herbicides, in fact, caused health problems for veterans of that
conflict. In either case, resolution of the controversy must come
through a process that separates factual, scientific elements from
political considerations. Once the science is clearly defined,
the issue then turns to resolution of critical differences in
value systems that too frequently place scientists, government
officials, and individual private citizens in adversary
relationships.

The scientific community must conduct valid research on
controversial environmental and health-related issues to provide
reliable data for use in appropriate decision making. The
Veterans Administration stands firmly committed to working closely
with other agencies of the Federal Government as well as with the
private sector to obtain as many answers as quickly as possible
consistent with sound scientific principles in order to resolve
this perplexing issue. The resolution of the controversy,
however, will be achieved only following the public's acceptance
of the outcome of scientific investigations. To that end

scientists must accept the responsibility of not only conducting
quality research, but also to translate the results of their
efforts to legislators, to the courts, to the media and
ultimately to the public at large.

DIOXINS AS CONTAMINANTS OF HERBICIDES: AUSTRALIAN PERSPECTIVE

A. L. Black

Commonwealth Department of Health
Woden
A.C.T. Australia

ABSTRACT

In Australia the TCDD contaminant of 2,4,5-T herbicidal formulations is maintained at or below 0.1ppm and the use and safety of
these materials is subject to continuing review. The Australian
population has experienced no known adverse health affects from the
regulated use of chlorophenoxyacetic acid herbicides. Claims of ill
effects by Vietnam veterans resulting from exposure to dioxin contaminated defoliants in Vietnam have proved a powerful influence on
public debate on the merits of the continued agricultural usage of
these herbicides.

Australia is in a unique situation with regard to the international controversy over chlorinated dioxins for a number of reasons.
Despite its relatively small population, Australia is a significant
agricultural producer on the world scene. On the basis of its experience over many years, it is reluctant to forego the continued use of
chlorinated phenoxyacetic acid herbicides without adequate justification. Nevertheless, the Australian population rightly expects protection from the potential hazards of dioxin contaminants inevitably
present in these materials. In addition it should not be forgotten
that Australia was one of the few western countries which participated
in the Vietnamese war. Subsequent claims by Vietnam veterans of
adverse health affects from exposure to chemical defoliants in
Vietnam have significantly influenced consideration of these issues
within the Australian society.

Unlike other federal systems, Australia is a federation of
sovereign states which have retained a considerable degree of autonomy

13

under the Australian constitution, adopted in 1901. In fact, the
national or Commonwealth Government has only limited powers to legis-
late with respect to matters such as health and agriculture, the
states having largely retained primacy in these areas.

In the 80 years since Federation, eminent advisory bodies such
as the National Health and Medical Research Council (NH&MRC) and the
Australian Agricultural Council (AAC) have evolved to advise on and
coordinate activity by Government in these fields. The development
of uniform national policies in health related matters has been
largely achieved through the NH&MRC. The functions of this Council,
established in 1926, can be briefly summarized as advising the
Commonwealth and the states on public health legislation and administration, and more generally on matters relating to medical and
dental health and on medical research. Council's membership is composed of representatives of Commonwealth and state Departments of
Health, the various medical and dental professional colleges, eminent
academics and lay people and consumer organizations.

The credibility of NH&MRC's recommendations within the
Australian community is based upon the integrity of the independent
scientific assessments undertaken by its various expert committees.
In matters of public health, Council's deliberations are guided by
the Public Health Advisory Committee which, in turn, is advised by
expert committees such as the Pesticides and Agricultural Chemicals
(Standing) Committee and Poisons Schedule (Standing) Committee.
Although NH&MRC's recommendations have no statutory authority, they
often provide the basis for legislative action by Australian states
and have been instrumental in achieving a fair degree of uniformity.

It is not surprising therefore that Australia has approached the
problem of chlorinated dioxins and related materials through its
well established system of pesticide regulation. Deliberate efforts
have been made to contain the assessment of potential hazards of
these materials within the realms of science and to eschew unwarranted speculation or recourse to legal debate on matters which are
properly the domain of scientists.

In Australia, appropriate poisons classification and residue
limits have long been recommended and implemented for the phenoxyacetic acid herbicides so as to limit but not inhibit their use in
agriculture or domestically. Following recognition of the teratogenicity of certain 2,4,5-T products containing 2,3,7,8-tetrachloro-
dibenzo-p-dioxin (TCDD) in 1969, NH&MRC investigated 2,4,5-T
formulations and their safety for use in Australia. Council subsequently recommended that 2,4,5-T containing more than 0.1ppm TCDD
should not be permitted for use as a herbicide in Australia. This
recommendation was incorporated into the Australian Standard for
2,4,5-T and has formed the basis for relevant state legislation.

The states have undertaken regular monitoring by chemical analysis to
insure that 2,4,5-T products used in Australia conform with or are
beneath this level of TCDD contamination. Recently, state authorities
have been reviewing the current use of these herbicides independently
of NH&MRC. No reputable scientific evidence has emerged to date to
indicate that the continued use of 2,4,5-T and related products which
conform with NH&MRC's recommendations have produced deleterious
effects to human health or the environment in Australia. Consequently NH&MRC believes that, despite continuing public debate, there
is no scientific and technical evidence as yet to justify alterations to its recommendations.

A review of the evidence produced in the continuing international
debate on 2,4,5-T and chlorinated dioxins has also been maintained.
Certain issues which have assumed special significance in Australia
warrant special comment. The alleged association between human congenital abnormalities and exposure to chlorinated dioxins that could
result from the use of herbicidal products containing TCDD has been a
subject of considerable concern in Australia. NH&MRC has reviewed
the use of these materials in Australia and has concluded that, even
though TCDD has been shown to have teratogenic effects in experimental animals, there is as yet no evidence that the use of herbicidal formulations containing recommended levels of TCDD are of
significance in production of human birth defects. In order to
facilitate the on-going assessment of this and related problems,
Council encouraged the recent establishment of a National Perinatal
Statistics Unit.

Following the action of the United States Environmental
Protection Agency in restricting the use of 2,4,5-T in 1979, NH&MRC
analyzed the report of the Six Year Spontaneous Abortion Rate Study in
Oregon (Alsea Study). Its Working Party on the use and safety of
2,4,5-T concluded that the Alsea report neither substantiated the
conclusions it reached nor provided a basis for concluding whether
2,4,5-T causes or does not cause an increase in spontaneous abortion.
No additional restrictions on the use of 2,4,5-T in Australia were
recommended after this review.

More recently the potential importance of the well-known series
of Swedish epidemiological studies relating to the use of 2,4,5-T,
chlorophenols and other materials, to a variety of neoplasms has been
recognized. These studies are under active review at this time.

In recent times the dioxin controversy in Australia has received
considerable impetus and an added degree of urgency because of stren-
uous claims by some of Australia's Vietnam veterans that their
physical and mental health has suffered as a result of exposure in
Vietnam to dioxin-contaminated defoliants. Although Australian
forces are not known to have used 2,4,5-T containing materials in

Vietnam, there have been repeated claims of various conditions such
as skin rashes, peripheral neuropathies and psychiatric conditions.
Particularly distressing for Veterans and their families are allegations of increased incidences of suicide and neoplasia amongst the
veterans and claims of an increased incidence of congenital malformations in their offspring. While Australia has reason to be proud
of the equitable and comprehensive care it has traditionally provided
for its war veterans, these novel claims present a new challenge for
the Veterans Affairs Administration.

Unfortunately, at this time, there are insufficient data to
determine the validity or otherwise of these claims. Accordingly,
the Australian Government has initiated independent epidemiological
investigations to document the health experiences of Australian's
Vietnam veterans and their families. It is believed that only in this
way will adequate scientific data be obtained to enable satisfactory
resolution of these allegations.

In conclusion, Australia's public health and pesticide regulatory
authorities will, no doubt, continue to review the biological significance of chlorodioxins as contaminants of herbicides and act accordingly. Data presented and conclusions drawn at forums such as this
international symposium will undoubtedly influence future debate and
actions in Australia in this controversial domain.

THE USE OF SCIENTIFIC INFORMATION

IN THE DEVELOPMENT OF PUBLIC POLICIES

Jarrell E. Southall

Arkansas Department of Pollution Control and Ecology
Little Rock, Arkansas, USA

ABSTRACT

This paper explores the problems related to developing and implementing public policies with insufficient scientific information. The four recommendations to improve the quality and exchange of scientific information include: 1) establish the institutional structures which would support the development of scientific information on a priority basis; 2) develop procedures by which the development of public policies could be more closely synchronized with the development of scientific information; 3) establish procedures to permit the refinement of public policies as scientific information is further developed; and 4) improve the opportunity for the public to understand the process by which scientific information is incorporated into public policies. To facilitate these recommendations, a transfer of standard-setting responsibilities to a national laboratory operating under a central control is proposed.

DISCUSSION

In Jacksonville, Arkansas, a 2,3,7,8-TCDD (dioxin) contamination problem originated from the manufacture of 2,4,5-T. Although the problem is fairly well contained, a number of problems remain; e.g., the extent of the cleanup required, the methods by which the cleanup will be achieved, and the long-term consequences of the environmental contamination. Environmental and health-effects data presented in these proceedings will be important to the resolution of those questions.

Many specific questions concerning the Jacksonville situation
are involved in a federal suit which has yet to be settled. Those
questions suggest problems related to the broader issue of how
scientific findings are incorporated into the process of
developing and implementing public policies, including specific
standards and regulations. These problems apply not only to
dioxin but to a host of other environmental contaminants.

The first problem is the lack of synchronization between the
development of public policies and the development of scientific
information. I am sure that we can all think of situations where
public policy had been formulated before there was sufficient
information to adequately justify those policies. Although
situations do emerge where some type of action is indicated prior
to the development of a sound scientific base for decision making,
I think we would all agree that the goal should be to have public
policies based upon sound scientific information. If that is our
goal in the field of environmental regulation, I think we should
work to achieve four supporting objectives. They are:

(1) Establish the institutional structures which would
support the development of scientific information on a
priority basis and with the maximum efficiency possible.

(2) Develop procedures by which the development of public
policies could be more closely synchronized with the
development of scientific information.

(3) Establish procedures which will not only permit—but will
encourage, if not require—the refinement of public policies
as scientific information is further developed.

(4) Improve the opportunity for the public to understand the
process by which scientific information is, has been, and
will be, incorporated into public policies.

Perhaps the fourth stated objective is the most important of
all if science and technology are to be properly used for the
public benefit. At this point I note the erosion of the public's
confidence in science and technology as well as in government in
general. Confidence can be restored if we improve the processes
of policy development and decision making, and if the public has
understanding of those processes and their weaknesses.

Unfortunately, there appears to be no mechanism to encourage
government to move on its own initiative to refine its policies as
new information becomes available. This lack of encouragement is
especially evident in those situations when new information
indicates that the existing policies are unnecessarily stringent.
However, when government moves to relax its policies it is met

with charges that it is sacrificing public health in the interest
of economics or politics.

What is most disturbing about such charges is that few
individuals stop to realize that the government may have acted in
the public interest in the first instance by setting a standard
with a high margin of safety (in view of uncertain or limited
health-effects data) and that the government, in the second
instance, again was acting in the public interest by adjusting the
standard to fit improved data. The point is that there is little
incentive for the professionals in government to improve earlier
decisions--decisions which were based upon preliminary information
at best.

When new or improved information indicates that public
policies are misdirected or unnecessarily stringent, is it in the
public interest to change such policies? The answer must be in
the affirmative if we realize that such policies place costs upon
the affected parties in excess of the benefits received by the
public. As a consequence, opportunities for achieving greater
public benefits are lost. The lost opportunities become evident
when we think of the depletion of natural resources which results
from misdirected or wasteful activities.

In addition, such policies divert capital which could be used
for more productive purposes. Such policies increase regulatory
confusion which further increase the dimensions of such costs.
Let me explain what I mean by the term "regulatory confusion." If
a policy is allowed to stand without adequate factual basis, what
is to prevent similar policies with an equally inadequate basis
from being applied in other areas? I think we all know the answer
to that question, and I think we also see the logical consequence
of such activities--increased confusion on the part of the public,
erosion of the concept of accountability, and additional
opportunity for single-interest groups to acquire gains at the
expense of the public.

Although many laws and regulations were enacted, and programs
were established over the last decade at the Federal level, we
still do not have a centralized process to prioritize research
needs, to support those needs, and to assure that public policies
which are needed before a sound scientific base is available do
not subsequently frustrate related research or distort the
findings of that research.

We now have a collection of regulatory programs, each program
having the responsibility to adopt standards, design, and
implement regulations. Too often, public demands or legislative
madates require that greater attention be given to the regulations
than to the standards upon which the regulations are based. Such

a situation leads not only to "regulatory confusion" but, it
minimizes the opportunity for research related to contaminants
which fall outside the regulatory programs. The fact that many
serious questions about dioxin exist a decade or so after they
were first raised suggests that the system is not working as well
as it should.

How could the system be improved? I would support the
concept of a national laboratory which would direct the research
for setting standards and which would operate under a central
control separate from the regulatory agencies. Regulatory
agencies would not longer have standard-setting responsibilities
under such an arrangement. They would, however, have the
responsibilities of (1) forwarding proposed agendas for standard
setting to the national laboratory and (2) developing and
implementing regulations to meet the standards set by the national
laboratory.

The purpose of transferring the standard-setting
responsibilities of the regulatory agencies to an independent
laboratory is not only to achieve a more efficient process for
setting standards, it is to improve the effectiveness and
credibility of both the standards and the regulatory agencies.
Such a transfer would eliminate the conflicts, or potential
conflicts, within regulatory agencies which are currently charged
with the dual responsibility of setting standards and enforcing
standards.

The potential for conflict exists in many forms. The basic
conflict, however, is between the need to establish sound
standards and the need for prompt responses to particular
pressures, whether they take the form of statutory mandates or
thrusts by single-interest groups. Because of that conflict, the
regulatory agencies face accusations that inappropriate
standard-setting actions occurred in order to increase the
workload and authority of the agency, or perhaps, in order to
justify an earlier political decision of the agency. Alternatively, they face accusations that inappropriate actions occurred
because the agency was fearful of the consequences of adverse
economic and social impacts. I do not believe that such a
situation is fair to the public or to the regulatory agencies.

The transfer of standard-setting responsibilities would not
eliminate the authority of the regulatory agencies to (1) operate
under existing standards while they are reviewed by the national
laboratory or (2) establish provisional standards (action levels)
in the absence of standards or guidance by the national laboratory
where some form of emergency action is indicated. The provisional
standards would remain in effect until changed by the national
laboratory. The authority to set provisional standards would

provide the agency with the means of dealing with problems on a
case by case basis and would limit the potential for
misapplication.

Although the concept of provisional standards or action
levels has potential application throughout the field of
environmental regulation, it is most applicable in situations
involving contaminants such as dioxin. In such situations,
personnel are now forced to operate with little authoritative
guidance in an area which is bounded by the popular desire of
zero-exposure, zero-risk, and the reality of limited resources.
The action levels would guide those personnel between what is
desirable and what is prudently necessary, thereby enabling them
to concentrate on attaining the latter before attempting the
former.

The Jacksonville problem could have been handled much more
expeditiously had action levels been available for those in charge
of the cleanup. Such levels not only would have permitted the
necessary cleanup activities to proceed with some rational basis
priority, they would have indicated where one phase of the
cleanup could have stopped and another commenced. The efficiency
inherent in such an approach becomes critically important because
time and money are two of the most important factors of a cleanup
process.

From a broader perspective, the possibility that the action
levels could fall short of some final standard, if and when a
final standard is developed, cannot be minimized. The alternative,
however, is an all or nothing approach which either stresses the
environment and public health or our ability to pay for other
improvements. The ability to while we continually seek to move
forward improve the basis for future actions is a far better
alternative than the all or nothing approach that has frustrated
our past efforts to cleanup the environment and which has
contributed to the erosion of public confidence in science,
technology, and government.

The institutions and processes for developing public policy
must be changed to enable and encourage the objective and timely
use of scientific information. Otherwise, neither the public nor
the scientific community will be able to realize the benefits
which our society is capable of producing.

REGULATORY ACTIONS ON DIOXINS AND RELATED COMPOUNDS

D. G. Barnes

Chlorinated Dioxins Work Group
U.S. Environmental Protection Agency
Washington, D.C. USA

ABSTRACT

The Environmental Protection Agency (EPA) has been involved
with polychlorinated dibenzo-p-dioxins (PCDDs) for more than 10
years, and, more recently, with polychlorinated dibenzofurans
(PCDFs). This paper briefly tracks the regulatory activities in
the area of herbicides, industrial wastes, abandoned dumpsites,
and controlled combustion.

INTRODUCTION

Over the past 15 to 20 years the word "dioxin" has grown from
a precise term used by the practitioners of the chemical arts to
a multifaceted term which carries different meanings to different
people. This change is a consequence of scientific discoveries
in the laboratory, imaginative investigations in the field,
military action halfway around the world, and a host of
fascinating questions, many of which remain unanswered.

A certain portion of our current "dioxin problem" is related
to imprecise language. The title of this paper—and even the
title of this Conference, I submit—are misnomers. There is a
strong likelihood that not a single speaker this week will refer
to a molecule called dioxin. Rather, we are here discussing polychlorinated dibenzo-p-dioxins (PCDDs), in general, tetrachlorodibenzo-p-dioxins (TCDDs), as a class, and 2,3,7,8-TCDD, in
particular, with some discussion of the polychlorinated dibenzofuran (PCDF) analogues. If we as scientists persist in our short-
hand reference to the "dioxin problem" when we actually mean the

vaguely defined PCDD problem or the more sharply focused TCDD or
2,3,7,8-TCDD problem, then we will surely reap the whirlwind of
media report that fail to distinguish the 2,3,7,8-isomer from its
less toxic or less well-studied cousins and, consequently,
declare such things as "dioxins are the most toxic materials made
by man" and "dioxins are mutagenic, teratogenic, and carcinogenic."
For us as scientists to be linguistically vague or imprecise
opens the door to the spreading of misinformation among those in
the public who are scientifically unsophisticated, but
emotionally and politically involved. Thus, while the well-known
2,3,7,8-TCDD problem has it own special niche, the more general
more complex problem of PCDDs and PCDFs in the environment is
showing itself to be a true and worthy challenge to our best
efforts--and consequently deserves to be called by its proper
name.

The United States Environmental Protection Agency has been
active in the investigation and control of PCDDs for many years.
During this period the Agency has dealt with PCDD-related
situations under different statutes. In each case, however, the
Agency's goal has been to avoid unacceptable risks, which
generally involves a balancing of risks and benefits. Today I
would like to review some of this history in order to illustrate
ways the Agency has dealt with the presence of PCDDs and PCDFs in
the environment.

ENCOUNTERS WITH PCDDs AND PCDFs

Herbicides

In the late 1960's and early 1970's, laboratory studies
revealed toxic properties associated with 2,3,7,8-TCDD, a contaminant of certain pesticide products, including the herbicide,
2,4,5-T. Following the uncontested cancellation of some registrations of 2,4,5-T, including home and garden uses, the Agency,
acting under the Federal Insecticide, Fungicide and Rodenticide
Act (FIFRA), took steps to cancel registrations for all remaining
uses. The final decision turned on whether or not the material
could be used without "unreasonable adverse effects to health or
the environment." The ensuring administrative law hearing was
terminated before any evidence was presented, due to the lack of
information on the levels of 2,3,7,8-TCDD residues in the environment. In an effort to fill this data gap, the major parties
involved in the hearing joined to form the Dioxin Implementation
Plan (DIP) for the primary purpose of developing a sensitive,
validated method of analysis for TCDDs in environmental samples.
In those early stages, there was no particular thrust to develop
an isomer-specific method. It was that pioneering effort, however,
that has contributed to making today's meeting possible.

In the late 1970's, 2,4,5-T was reviewed under the Rebuttable
Presumption Against Registration (RPAR) program. In 1978 the
Agency concluded that a rebuttable presumption had arisen
against the continued registration of 2,4,5-T. Registrants and
other interested parties submitted over 2,000 responses to the
RPAR notice. The Agency published its findings and responses to
the comments in a series of position documents and Federal
Register notices (43 FR 17116 April 21, 1978; 44 FR 41531 July
17, 1979; and 44 FR 72316 December 13, 1979).

The accumulating laboratory results from animal studies on
the effects of 2,4,5-T and/or 2,3,7,8-TCDD, coupled with
epidemiological evidence from forest spray areas in the
Northwest, prompted the Agency in 1979 to issue an emergency
suspension against certain uses of 2,4,5-T and Silvex, a compound
which is chemically and functionally similar to 2,4,5-T and which
is also contaminated with 2,3,7,8-TCDD. In 1980 and administrative
law proceeding, In re: Dow Chemical Company et al, was begun to
hear evidence on the Agency's notices of intent to cancel the
registrations of the suspended uses of 2,4,5-T and Silvex and of
intent to determine whether or not the other uses should be
cancelled. After more than a year of gathering evidence,
including more than 50,000 pages of testimony, exhibits, and
cross-examination, the major parties asked the judge for a recess
in the formal proceedings to attempt to negotiate a settlement.
Since these negotiations are currently under way, it would not be
appropriate for me to say more about this situation at this time.

Industrial Wastes

In the process of manufacturing 2,4,5-trichlorophenol (2,4,5-
TCP) used in the manufacture of certain herbicides and pesticides,
waste products contaminated with 2,3,7,8-TCDD accumulate. Before
we were fully aware of the hazards associated with such
materials, some manufacturers incinerated these wastes; others
stored them; and still others disposed of them in landfills of
varying degrees of sophistication.

In 1980 EPA learned that the Vertac Chemical Company of
Jacksonville, Arkansas, intended to move some of its drummed
2,4,5-TCP wastes containing parts-per-million levels of
2,3,7,8-TCDD to a commercial land fill out of the state. Acting
under the authority of Section 6(d) of the Toxic Substances
Control Act (TSCA), the Agency issued an "immediately effective
rule," blocking this action, based on the finding that the
proposed move would present an unreasonable risk when compared to
retaining and maintaining the drums on-site (44 FR 15592 March
11, 1980). In addition the rule requires that anyone who plans
to move waste containing 2,3,7,8-TCDD derived from the manufacture

of 2,4,5-TCP notify EPA at least 60 days prior to such a move.
During the 60 days, the Agency has the opportunity to take action
if it determines that an unreasonable risk would result. To
date, the Agency has received three notices.

When such industrial wastes are in liquid form, methods do
exist for treating the problem more directly. In Verona,
Missouri, Syntex Agribusiness Inc. found itself with liquid
2,3,7,8-TCDD-contaminated wastes on its hands which has been
produced by a previous owner. With the Agency maintaining a
watchful eye on activities, the corporation developed, installed
and operated a method for photodegrading much of the 2,3,7,8-TCDD
(Waste Age, 1980). While some problems remain, this effort
represents successful cooperation between industry and EPA in
mitigating a 2,3,7,8-TCDD problem, outside of the formal
regulatory framework.

Abandoned Dumpsites

In the same area of Missouri, our Regional office is
gathering evidence of sites where 2,4,5-TCP wastes were dumped by
producers who have long since left the area. There is now a
concern about the integrity of the drums which were buried and
abandoned at these sites years ago and about the ultimate fate of
2,3,7,8-TCDD they contain.

Syntex Agribusiness Inc. developed a plan to uncover one such
location, remove any intact drums, process liquid wastes at its
Verona facility, and excavate contaminated soil. The plan
provides for experiments with microorganisms in the hopes of
finding microbe(s) which will decompose 2,3,7,8-TCDD in the soil.
This ambitious initiative is now in its first phase. The
corporation will provide its findings to the Agency, and together
they will formulate plans for the next phase of the cleanup
operation.

Because the Agency is making decisions about TCDD in
dumpsites based upon a weighing of risks and benefits, each
situation is evaluated on its own merits. An action taken at one
location may not be appropriate at another. For example, while
excavation was recommended at the Missouri site--a remote spot in
a heavily wooded area, far from any dwellings--the Agency
recommended against excavation at the Bloody Run site near the
Niagara River in New York. In the latter case, the contaminated
area is close to occupied dwellings and, therefore, excavation
would appear to present an unreasonable risk. Instead, a
combination of capping of the site, coupled with stream diversion
around the area and installation of monitoring wells, is the
favored method of treatment. The Agency is presently negotiating
a settlement of this New York site.

Controlled Combustion

In the late 1970's, researchers in this country and Europe
reported finding PCDDs and PCDFs associated with products of
various kinds of combustion. In the United States, the Dow
Chemical Company generated much of these early data and used them
as a basis for the so-called "trace chemistries of fire"
hypothesis. Later in this Conference we will be treated to
further findings--both empirical and theoretical--by capable and
imaginative workers at Dow and elsewhere. Suffice to say that
while some of the concepts presented in this hypothesis have
gained support through the continuing work of researchers,
including those at EPA, some of the broad generalizations have
not been verified by additional data. Be that as it may, the
findings to date have stimulated many workers to conduct new
investigations and sensitized all of us to the possibility of
PCDD- and PCDF-formation/emission through combustion.

Studies in the last few years (Morita et al, 1977; Buser et
al, 1978) have associated the formation of PCDFs with the
incineration of polychlorinated biphenyls (PCBs). Consequently,
in 1980 when two commerical incinerators applied for permits to
destroy PCBs through thermal means, the Agency was concerned
about the possible risks from highly toxic combustion products.
Therefore, the Agency initiated extensive sampling and analytical
efforts to check for the emission of TCDDs and TCDFs, especially
2,3,7,8-TCDF. Using a modified EPA Method 5 stack sampler,
designed to trap both particulate material and organic vapors, a
contractor obtained the critical samples in duplicate. The
results of extraction and analysis are shown in Table 1, which
displays the amounts of TCDDs, and TCDFs, and 2,3,7,8-TCDF found
in various parts of the sampling train operated during the
combustion of routine wastes and a carefully metered amount of
PCBs. Most of the TCDDs and TCDFs appeared in or on the probe
and the first impinger of the sampler, a situation which is
consistent with these materials being associated with particulate
matter. The feed material to the flame was not analyzed for
TCDDs or TCDFs, so while the origin of the components might be
the subject of speculation, it is clear that TCDDs and TCDFs were
emitted during the combustion process, albeit in very small
amounts.

In order for the Agency to estimate the magnitude of the risk
the incineration of PCBs might represent, information was needed
about a) the toxic hazard of all these materials, and b) the concentrations to which people might conceivably be exposed. While
2,3,7,8-TCDD has been shown to have a variety of toxic effects,
few experimental data have been collected on the chronic effects
of other TCDD isomers or any of the TCDFs. Therefore, the Agency
had to make some assumptions about the possible effects of these

Table 1. Levels of TCDD and TCDF Nanograms Found in Various Parts of a Sampling Train Operated During Combustion of PCBs

Sample	Total TCDD	Percent Recovery	Total TCDF	2,3,7,8-TCDF	Percent Recovery (PCBB)
			Site A		
Impinger	0.48	41	4.0(.05)*	1.0	100
Prove wash	ND(.25)	97	0.5(.09)	0.2	75
XAD-2	ND(.38)	134	1.5(.05)	0.3	100+
Florisil	ND(.66)	51	ND(.08)	ND (0.8)	100
	0.48+(1.29)		6+(.08)	1.5+(0.8)	
			Site B		
Impinger	ND(.27)	75	6.0(.01)	0.8	72
Probe wash	1.42(.27)	129	14.0(.01)	1.4	122
XAD-2A/XAD-2B**	ND(.38)	107	2.0(.07)	0.25	100
Florisil	ND(.48)	79	ND(.09)	ND (0.9)	85
	1.42+(1.29)		22.0+(.09)	2.4+(0.9)	

* Detection limits in parentheses
** Composited Sample

emissions. Similarly, the need for exposure data could only be
met by making a series of assumptions about the doses people
might receive, based on the emission values seen in the stack.
Table 2 lists the assumptions which were made. These generally
"worst case" assumptions, coupled with data from animal tests--
specifically carcinogenicity (Kociba et al, 1978) and
reproductive effects (Murray et al, 1979)--made it possible to
estimate the higher range of risks associated with the
incineration of PCBs.

To make an "unreasonable risk" determination, this risk information had to be compared to the benefits--removal of PCBs from
the environment. This analysis of benefits also included consideration of alternatives for handling PCBs; e.g., landfilling and
chemical means of detoxification.

Table 2. EPA Assumptions About the Possible Effects of Emissions
          from the Incineration of PCBs.

---

1. Carcinogenicity of 2,3,7,8-TCDF equal to that of 2,3,7,8-TCDD.
2. All TCDDs and TCDFs have the same carcinogenicity potential
    as 2,3,7,8-TCDD.
3. All emissions from the stack are respirable and are retained
    in the body.
4. All the TCDDs and TCDFs that are respired are retained and are
    biologically available to the organism even though some
    significant fraction is tenaciously bound in or on the
    emission particles.
5. The composition of the emission products found in the stack
    is identical to the composition (but not the concentration)
    found at the ground level.
6. The combustion of the PCBs will continue for 24 hours a day
    for a 70-year lifetime.
7. The air dispersion modeling is representtive of what happens
    at the two sites.
8. The entire population is subjected to the maximum average
    annual concentration found at the point of highest
    concentration.
9. The linearized, multi-stage extrapolation method of cancer
    risk assessment is valid.
10. Laboratory animals are faithful surrogates for humans.
11. The other assumptions for cancer risk assessment as found in
    FR 44 131, July 6, 1979 are valid.
12. The assumptions made by the cancer Assessment Group in
    determining unit risk assessments for the Air Programs are
    valid.

When this information was compiled and placed before the Regional Administrator, who had permitting authority in this case, she judged that the estimated risk involved in burning the PCBs did not outweigh the benefits. Therefore, she granted the permits, which contained a number of stringent conditions. A full report of the sampling, analysis, and decision will soon be available from the National Technical Information Service.

A second example of PCDD and PCDF emission from controlled combustion sources is municipal waste combustion. Because our country is learning the value of reclaiming resources imbedded in our urban refuse, what was once viewed as a jumble of junk to be burned (incinerated) simply to reduce the volume requiring landfilling, is now seen as a resource ripe for harvesting. For example, systems designed for the separation and recovery of glass, metal, newpaper, etc., followed by controlled combustion to recover heat energy for use in generating electricity and operating industrial equipment, are becoming more common. Such facilities are varyingly referred to as resource recovery facilities or municipal waste combustors. Although the Agency encourages this type of recycling of resources, the Agency also has a strict obligation to see to it that these goals of resource recovery can be achieved without unacceptable risks.

In 1979 the Agency received indications that at least one municipal waste combustor was emitting TCDDs. These early data, however, would not support a quantitative determination. Shortly thereafter data produced outside the United States appeared showing that PCDDs and PCDFs were commonly found in bottom ash of facilities in those countries. During the same period, the Agency was conducting its own series of studies to assess the qualitative and quantitative emissions, including PCDDs and PCDFs from combustors in this country. Data are now available from the first three plants in these studies, two of which will be reported on later in this Conference.

This information is being collected and analyzed in a manner similar to, but not identical with, the analysis done on the PCB incineration question. The EPA Administrator is currently reviewing this information in order to address the considerable public and governmental concern about the problem of TCDDs from municipal waste combustors. The Agency's interim assessment of the data will be supplemented by additional data that we plan to collect on such emissions in the future.

CONCLUSION

EPA has encountered PCDDs and PCDFs in other situations, as well as those mentioned in this paper. For example, PCB/chloro-

benzene-filled transformers involved in fires have yielded PCDDs
and PCDFs as combustion products. In addition, 2,3,7,8-TCDD has
been found in fish and herring gull eggs from the Great Lakes
area. In these instances the Agency has served in an advisory
capacity to other authorities in their efforts to deal with a
PCDD/PCDF problem, and these cooperative efforts continue.

The lack of information in some areas (e.g., toxicity of the
various kinds and isomers of PCDDs and PCDFs, background levels
in the environment, and bioavailability of the materials when
absorbed to particulate matter) makes the task of dealing with
PCDDs and PCDFs in a regulatory framework more difficult. We
look forward to the development of this important information in
the future so that hard data can replace some of the assumptions
we now have to make. In the interim the Agency will continue to
use the best available information in making its decisions on how
best to protect man and the environment from unreasonable risks
associated with PCDDs and PCDFs.

REFERENCES

Buser, H. R., Bosshardt, H.R. and Rappe, C., 1978, Formation of
    polychlorinated dibenzofurans (PCDFs) from pyrolysis of PCBs,
    Chemosphere, 1:109-119.
Kociba, R. J., Keyes, D. G., Beyer, J. E., Carreon, R. M., Wade,
    C. E., Dittenber, D. A., Kalninis, R. P., Frauson, L., Park,
    C. N., Barnard, S. D., Hummell, R. A., and Humiston, G. C.
    1977, Results of a two-year chronic toxicity and oncogencity
    study of 2,3,7,8-TCDD in rats, Toxicol. Appl. Pharmacol.,
    46:279-303.
Morita, M., Nakagawa, J., and Rappe, C., 1977, Polychlorinated
    dibenzofuran (PCDF) formation from PCB mixture by heat and
    oxygen, Bull. Environ. Contam. Toxicol., 19:665-670.
Murray, F. J., Smith, F. A., Nitschuke, K. D., Humiston, C.G.,
    Kociba, R. J., and Schwetz, B. A., 1979, Three generation
    reproduction study of rats given 2,3,7,8-tetrachlorodibenzo-
    p-dioxin (TCDD) in the diet, Toxicol. and Appl. Pharmacol.,
    50:241-252.
Waste Age, 1980, Destroying dioxin: a unique approach, October
    60-63.

DIOXINS AND RELATED COMPOUNDS AS ISSUES OF INTERNATIONAL CONCERN

E. Somers and V.M. Douglas

Environmental Health Directorate
Health Protection Branch, Department of National
    Health and Welfare
Ottawa, Canada

In recent years, there has been ever-increasing concern for the real and potential hazards posed to man and the environment by polychlorinated dibenzo-p-dioxins (PCDDs) a class of organochlorine contaminants comprising 75 congeners, the most notorious of which is 2,3,7,8-TCDD. International attention on the dioxins was focused in 1973 by the National Institute of Environmental Health Sciences conference (Huff and Wassom, 1973). Intensive investigations have followed. Physical and chemical characteristics of the dioxins that influence their distribution and persistence in the environment include:

1. Relatively insoluble in water, but soluble in lipids
2. Easily absorbed onto soil
3. Environmentally stable
4. Metabolized slowly

This paper will address--from the Canadian experience--the issues of:

1. The transport of dioxins in air
2. The transport of dioxins in water
3. Occurrence of dioxins in foods subject to international trade
4. International trade of herbicides, wood preservatives and other chemicals which may contain dioxins

It is important to recognize that there are wide ranging differences in physical, chemical, and biological properties between the dioxin congeners. In fact, isomers of the same chlorinated

33

dibenzo-p-dioxin can exhibit markedly different toxicities, hence
making their individual, unequivocal identification vital. It is
known that 2,7-dichloro-, and octachloro-dibenzodioxin have low
acute toxicities, while certain hexachlorodibenzodioxins and tetra-
chlorodibenzodioxins are very toxic (Fishbein, 1977): the acute
toxicity of 2,3,7,8-TCDD in fact is up to 3000 times higher than
that of the di- or the octachlorinated homologs. The toxicity of
the individual chlorinated dioxin is dependent on its structure, high
toxicity requiring chlorine atoms in the 2,3,7 and 8 positions
(Poland and Glover, 1973), while the bioaccumulation potential of
chlorinated organics is dependent on lipid solubility, which
obviously varies with degree of chlorine content.

     Bioaccumulation of a contaminant in a real exosystem is depen-
dent on other factors as well, such as growth of an organism, food
conversion efficiency and metabolism and excretion efficiencies
(Norstrom et al, 1976). Although bioaccumulation of dioxins in
both animal and plant tissues has been demonstrated (IJC, 1980;
Esposito et al, 1980) their full bioconcentration and bioaccumulation potential has not been fully elucidated.

     These factors inevitably have implications in the international
assessment of the effects of dioxins and it is important to realize
that there are at least three major mechanisms available to governments to control the health or environmental hazard from a chemical:
economic--fining the polluter or subsidizing safety activities;
educational--discouraging hazardous activities or increasing public
awareness of the health effects of the chemical; or regulatory.
With the regulatory option a wide range of strategies is available
ranging from the outright banning--to setting national emission
standards or to establishing maximum allowable concentrations.
Even less stringent is to recommend guidelines. In diminishing
scale of authority, national governments can avail themselves of
these control techniques.

TRANSPORT IN AIR

     The long range aerial transport of chemicals has been recognized
since 1960. There is evidence for example, that DDT has travelled
great distances from its origin of application to the remote
regions of Antarctica (WHO, 1979).

     Air pollutants may be in the form of gases, aerosols or parti-
culates. Some of these diffuse upward to levels where they may
be degraded by ultraviolet radiation; (some TCDDs for example are
easily photodegraded), others are absorbed onto suspended particulates, and yet others are dissolved in water droplets and returned
to earth in rainfall. In some cases, global atmospheric processes
may include cycles of elimination and reintroduction of a substance.
Since these processes are not well understood, rates of transfer

are difficult to predict. It is well known that aerosols, such as those of sulfur oxides can remain in the atmosphere for several days and be transported appreciable distances from the source. For example, sulfur oxides found in Sweden and Norway were traced to sources as far away as the United Kingdom and France, and aerosols in the Caribbean area have been associated with the arid regions in Africa (Rodhe et al, 1972; Prospero et al, 1970). Studies by Hamilton (1979) have indicated a strong correlation between sulfate levels in the North Eastern United States and excess mortality, and morbidity. It must be emphasized that it has not been demonstrated that sulfate is the toxic agent but it serves as a surrogate for the causative factor.

The long range transport of dioxins has been debated for a number of years. Dioxins are known to be widely distributed in the environment. They have been reported to be present in at least ten pesticides and pesticide mixtures, in food, air, and water. Recently, emissions of dioxins and similar or related polycyclic aromatic compounds from combustion sources have been reported by Dow Chemical Co. (1978) and other investigators (Buser et al, 1978; Olie et al, 1977; Karasek, 1980).

The "trace chemistries of fire" hypothesis of Dow Chemical Co. has been proposed to explain the ubiquitous presence of dioxins and related compounds in a variety of media such as soil, dust, cigarette smoke, municipal incinerators. It is suggested that dioxins and other chlorinated substances may form during combustion processes: research and theoretical studies support the hypothesis (Hutzinger et al, 1981; Harris et al, 1981).

Analyses to date are not unequivocal but do indicate the presence of dioxins including the 2,3,7,8-TCDD isomer, as well as polychlorodibenzofurans and PCBs, in fly ash and flue gases from a number of sources (U.S. EPA, 1980). The emission rates of dioxins from incineration are highly variable, depending on fuel type and combustion methodology. It has been shown that the formation of chlorinated organics are favoured by high fuel-chlorine content, low temperatures, inefficient mixing and especially by the presence of precursors in the fuel source. Known precursors of dioxins and dibenzofurans include chlorophenols, chlorophenoxy herbicides and chlorobenzenes. PCBs are also known precursors to polychlorinated dibenzofurans (PCDFs). Inefficient combustion of municipal solid wastes in small incinerator units with inadequate emission control is the major source of such pollutants in the atmosphere. The data indicate that emissions of chlorinated dioxins and chlorinated dibenzofurans are "general phenomena related to all combustion processes" (Shih et al, 1980). The prevailing pattern of air movement on this continent would suggest that, if the long range transport of dioxins does occur, Canada will be the recipient of pollution from the northeastern U.S.A., particularly the Ohio Valley.

One of the most difficult questions to answer concerns the total amount of PCDDs which may enter the environment annually. The total annual amount of dioxins emitted by incineration processes in the U.S.A. is estimated to be about 1000 kg (Crummett, 1981). The estimate for dioxin emissions from municipal inciner-ators in Canada is about 13.4 kg/yr (NRC Canada, 1981). Since input into the environment from combustion may be a significant problem, particularly for long range pollution, it is important to establish criteria for acceptable dioxin and dibenzofuran emission levels. The total gross estimate of polychlorinated dibenzodioxins released into the Canadian environment is 1.5 tons per annum (NRC Canada, 1981), which was estimated based on contaminant levels in commercial products commonly used in Canada. The major source of dioxins is the chemical industry, particularly pentachlorophenol producers and users. In other countries, total estimates of dioxins may be based on patterns of use of commercial products and manufacturing processes. However, it should be pointed out that the total amount is less critical than the degree of availability and the concentrations in various matrices in assessing risk. The estimate of total dioxins indicates a potential threat only since much of the dioxins may be bound in landfill sites or may have been biodegraded or chemically transformed.

TRANSPORT IN WATER

Contamination of water bodies by dioxins can occur following the spraying of chlorophenoxy herbicides on forests. Possible routes of water contamination from spraying are direct application, drift of the spray and overland transport after heavy rains (Miller et al, 1973). The transport of dioxin-contaminated soil into lakes or streams by erosion constitutes another route of contamination. This is shown by the detection of 2,3,7,8-TCDD in water samples from a Florida pond adjacent to a highly-contaminated land area (Bartleson et al, 1975).

Such findings are consistent with recent reports that TCDDs can migrate to nearby water bodies from industrial chlorophenol wastes, buried or stored in various landfills. At Niagara Falls, New York, for example 1.5 ppb TCDDs have been detected at an onsite lagoon at the Hyde Park dump, where 3300 tons of 2,4,5-trichloro-phenol wastes are buried (Chemical Week, 1979; Wright State Univ., 1979). Sediment from a creek adjacent to the Hyde Park fill is also contaminated with ppb levels of dioxins (Chemcial Week, 1979). There is evidence that TCDDs have migrated from process waste containers in the landfill of a former 2,4,5-T production site. Dioxins have been found both in surface water on the site (at 500 ppb levels) and downstream of the facility in the local sewage treatment plant, in bayou botton sediments and in mussels and fish (Richards, 1979; Fadiman, 1979; Tiernan et al, 1980). TCDDs are apparently also leaching into surface and ground waters from an 880-acre dump site of the Hooker Chemical Company at Montague, Michigan (Chemical Week, 1979; Chemical Regulation Reporter, 1979). Dioxin levels at the Hooker site were approaching 800 ppt.

Water contamination can inevitably cross national boundaries and international focus was given to this problem in 1980 by the reported presence of 2,3,7,8-TCDD in Lake Ontario herring gull eggs (Ogilvie, 1981; Norstrom, 1981). Retrospective analysis showed a decrease of TCDD levels has occurred since 1971. It should be noted, however, that regular monitoring of Canadian drinking water supplies from municipalities in the Niagara area has failed to show any dioxins in the potable water. Nevertheless, the finding of traces of dioxins in the Great Lakes has led to a recommendation to the Great Lakes Science Advisory Board of the International Joint Commission (IJC, 1980) that TCDD should be absent from all compartments of the ecosystem, including air, land, water, sediment, and biota, where absent means not detectable by the best available technology (at present 0.01 µg/kg in tissue and sediment, and 0.00001 µg/L in water). This proposal, if accepted, should be viewed as an ecological objective to safeguard the future of the Great Lakes.

OCCURRENCE IN FOOD

Analyses of fish from Lake Ontario in 1979 showed measureable levels of TCDD in a number of species (New York State Dept. Health, 1979). At a meeting in July 1981, a comprehensive review of dioxin analyses on Great Lakes fish was carried out by Canadian and U.S. scientists. The conclusions were that:

1. Dioxin residues were confined mainly to Lake Ontario fish
2. Commercial fish with the highest levels were eels and smelt (>20 ppt)
3. Sport fish with the highest levels were the salmonids (20-25 ppt)
4. Dioxin content positively correlated with fat content of the fish

The hazard to the public depends, of course, on the extent of fish consumption. Canadians consume about 13 lbs of fish/person/year or about 16g/person/day. U.S. estimates of fish consumption range up to twice the Canadian estimates.

As a result of the meeting, the Minister of National Health and Welfare announced the establishment of a guideline of 20 ppt of 2,3,7,8-TCDD in commercial fish as part of the federal program to control health hazards from environmental contaminants (Health and Welfare Canada, 1981).

In Canada, we have based our estimates on the rodent studies carried out by Dow Chemical. If the level of dioxin in fish is 20 ppt, then the safety margin for a Canadian ingesting 16g of fish daily is over 2000, based on rat reproductive studies. In other words, this dioxin concentration represents a 2000 times lower level than the first measurable "effect level" in rat reproductive studies. The safety margin for carcinogenicity is of the same order of magnitude: 2000 for 20 ppt dioxin in fish (Grant, 1981).

Thus, an administrative guideline has been established for TCDD in fish. More stringent regulations have been established under the Food and Drugs Act and Regulations which controls the sale of food in Canada. As early as 1964 food products could not be sold if they contained "fatty acids and their salts containing chick-edema factor or other toxic factors": a remarkable example of bureaucratic pre-cognition. Similarly, from 1980, foods cannot be sold if they contain chlorinated dibenzo-p-dioxins. The key to implementation is of course the concentration of dioxin as detected by a recognized method of analysis.

The Food and Drugs Act only applies to foods intended for consumption in Canada. Overprinting the package with the word 'Export' removes the requirement to comply as long as a certificate that the package and contents do not contravene any know requirement of the country to which it is to be consigned has been issued. The sensitive political question of exporting foods to countries with less stringent regulations is one that has many conflicting components—not only the moral issue but also nutritional, social, and cultural concerns.

TRADE IN DIOXIN-CONTAMINATED PRODUCTS

In Canada, we import pesticides and wood preserving agents that may contain dioxions: we export pentachlorophenol and wood products treated with this chemical. For use in Canada, controls have been established, through an interim standard of the Pest Control Products Act, of a maximum concentration of 0.1 ppm of TCDD in 2,4,5-T and Silvex. For other products, decisions are made on a case-by-case basis with the objective of reducing contamination to the lowest practival level. Thus for 2,4-D products, that can contain isomers of mono-, di-, tri-, and tetra-chlorodioxins, the Department of Agriculture (Agriculture Canada, 1981) has declared that at present 2,4-D must contain less than 10 ppb of a specific dioxin isomer. We have no guidelines at present for pentachlorophenol.

These controls, albeit limited, are for one country. Consistency of approach is certainly required among all trading nations. Probably the most serious international concern with dioxin contamination at present relates to the disposal, incineration and uncontrolled emission. However we deal with these problems in our own countries, international accord, inevitably mediated through such UN agencies as WHO and UNEP, is the only route we have to achieve their resolution.

REFERENCES

Agriculture Canada. Trade Memorandum T-1-223, Food Products and Inspection Branch, Pesticide Section. Aug. 28, 1981.
Bartleson, F.D., Jr., Harrison, D.D., Morgan, J.B. Field Studies of Wildlife Exposed to TCDD Contaminated Soils. Air Force Armament Lab. Eglin A.F. Base, Florida, 1975.
Buser, H.R., Bosshardt, H., and Rappe, C. Identification of poly-chlorinated dibenzo-p-dioxin isomers found in flyash. Chemosphere, 7:165-172, 1978.
Chemical Regulation Reporter. Hooker Chemicals installing toxic controls at Monaque, Michigan Site, pp. 457-458, June 22, 1979.
Chemical Week. Hooker Dumpsites may pose dioxin threat. 124(1):16, Jan. 3, 1979.
Chemical Week. More Agent Orange Suits Filed in Chicago; Still Others Will Follow. 124(9), Feb. 28, 1979.
Crummett, W.B., Bumb, R.R., Lamparski, L.L., Mahle, N.H., Nestrick, T.J. and Whiting, L.W. Environmental chlorinated dioxins from combustion - The trace chemistries of fire hypothesis In Proc. Workshop. Impact of Chlorinated Dioxins and Related Compounds on the Environment. Rome. Pergamon Press. 1980.
Dow Chemical Co. The trace chemistries of fire - a source of and routes for entry of chlorinated dioxins into the Environment. The Chlorinated Dioxin Task Force. Dow Chemicals Co. Midland, Michigan. 46 pp. plus appendices, 1978.
Esposito, M.P., Tiernan, T.O., and Dryden, F.E. Dioxins. Contract Nos. 68-03-2577, 68-03-2659, 68-03-2579. Industrial Environmental Research Laboratory, U.S. EPA, Cincinnati, Ohio, 45268. 351 pp. EPA 600/2-80-197, Nov. 1980.
Fadiman, A. A Poisoned Town. Life Magazine, pp. 43-46, 49. Sept., 1979.
Fishbein, L. Trace organic contaminants: 1. Chlorinated dioxins and dibenzo-furans. Int. J. Ecol. Environ. Sci. 2(2-3): 69-82, 1977.
Grant, D. Personal communication. Health and Welfare Canada, Ottawa, Canada. 1981.
Hamilton, L.D. Health effects of electricity generation. Conf. on Health Effects of Energy Production, Chalk River, Ontario. Sept., 1979.
Harris, J.C., Anderson, R.C., Goodwin, B.E. and Rechsteiner. Dioxin emissions from combustion sources: A review of the current state of knowledge. A.D. Little Inc. Rpt. to Res. Committee on Indus. Municipal Wastes. Am. Soc. Mechan. Eng. New York. 1981.

Health and Welfare Canada. Dioxin guideline announced for com-
    mercial fish. News release. July 17, 1981.
Huff, J.E. and Wassom, J.S. Chlorinated dibenzodioxins and
    dibenzofurans. Environ. Health Perspect. 5:283-312, 1973.
Hutzinger, O., Choudhry, G.G. and Olie, K. Mechanisms in the thermal
    formation of polychlorinated dibenzo-p-dioxins and related
    compounds. In Proc. Workshop. Impact of chlorinated dioxins
    and related compounds in the environment. Rome, 1980.
International Joint Commission Great Lakes Science Advisory Board.
    Report of the aquatic ecosystem objectives committee, pp.
    26-40, Nov., 1980.
Karasek, F.W. Dioxins from garbage: previously unknown source of
    toxic compounds is being uncovered using advanced analytical
    instrumentation. Canadian Research. 1316:50, 52, 54, 56.
    1980.
Miller, R.A., Norris, L.A., and Hawkes, C.L. Toxicity of 2,3,7,8-
    TCDD in Aquatic Organisms. Environ. Health Perspect. 5:177-
    187, Sept., 1973.
National Research Council of Canada. Polychlorinated dibenzo-p-
    dioxins: Criteria for their effects on Man and his environment.
    Associate Committee on Scientific Criteria for Environmental
    Quality. Draft publication NRCC No. 18574, 1981.
New York State Department of Health. News Release, April 24, 1979.
Norstrom, R. Personal communication. Canada Wildlife Service.
    Environment Canada, 1981.
Norstrom, R.J., McKinnon, A.E., deFreitas, A.S.W. A bioenergetics -
    based model for pollutant accumulation by fish. Simulation of
    PCB and Methylmercury residue levels in Ottawa River Yellow
    Perch (Perca flavescens). J. Fish. Res. Board Can., 33:248-267,
    1976.
Ogilvie, D. Dioxin Found in the Great Lakes Basin. Ambio, 10(1):
    38-39, 1981.
Olie, K., Vermeulen, P.L., and Hutzinger, O. Chlorodibenzo-p-dioxins
    and chlorodibenzofurans are trace components of flyash and
    flue gas of some municipal incinerators in the Netherlands.
    Chemosphere 6:455-459, 1977.
Poland, A. and Glover, E. Studies on the mechanism of toxicity of
    the chlorinated dibenzo-p-dioxins. Environ. Health Persp.,
    5:245-251, 1973.
Prospero, J.M. Dust in the Caribbean atmosphere traced to an African
    dust storm. Earth Planet Sci. Lett., 9:287-293, 1970.
Richards, B. Arkansas Site May Hold Clue. The Washington Post,
    July 25, 1979.
Rodhe, H., Persson, C., Akesson, O. An investigation into regional
    transport of soot and sulfate aerosols. Atmospheric Env.,
    6:675-693, 1972.

Shih, C., Ackerman, D., Scinto, L. and Johnson, B. Emissions of
    polychlorinated dibenzo-p-dioxins (PCDDs) and dibenzofurans
    (PCDFs) from the combustion of fossil fuels, wood, and coal
    refuse. U.S. Environmental Protection Agency Contract 68-02-
    3138. Draft, Dec., 1980.
Tiernan, T.O. Analyses of Industrial Samples for tetrachloro-
    dibenzo-p-dioxins (TCDDs). Final Report U.S. EPA Contract No.
    68-03-2830 and Order Nos. 9T-1501-NTEX and OT-0267-NAEX.
    April 1, 1980.
U.S. EPA. Shih, C., Ackerman, D., Scinto, L., Moon, E., Fishman, E.
    POM Emissions from stationary conventional combustion processes
    with emphasis on polychlorinated compounds of dibenzo-p-dioxins
    (PCDDs) biphenyl (PCBs) dibenzofurans (PCDFs), OCEA issue
    paper. Prepared by TRW Inc. for U.S. EPA-IERL. 118 pp.,
    Jan., 1980.
World Health Organization. Environmental Health Criteria No. 9.
    DDT and its derivatives. Published under the joint sponsorship
    of the United Nations Environment Programme and the World
    Health Organization. Geneva. 194 pp., 1979.
Wright State University. Report on Analyses of Love Canal Samples
    for 2,3,7,8-Tetrachlorodibenzo-p-dioxin (TCDD). New York
    State Department of Health Purchase Order No. 5975. Jan. 21
    and April 20, 1979.

ANALYTICAL CHEMISTRY

Chairman: Warren B. Crummett
Dow Chemical Company
Midland, Michigan USA

ANALYTICAL METHODOLOGY FOR THE DETERMINATION OF PCDDs AND PCDFs IN

PRODUCTS AND ENVIRONMENTAL SAMPLES: AN OVERVIEW AND CRITIQUE

Warren B. Crummett

Dow Chemical, USA
Analytical Laboratories
Midland, Michigan   USA

Progress in the development of analytical methodology for the determination of chlorinated dibenzodioxins (CDDs) and dibenzofurans (CDFs) in products and environmental samples has been extensive and dramatic during the last decade. Thus, the limit of detection for the tetrachlorodibenzo-p-dioxins (TCDD) in products has been lowered from one part per million in 1969 to 1 part per billion in 1980. Similarly the limit of detection for 2,3,7,8-TCDD in environmental samples has developed to a part per trillion in 1978 from 50 parts per billion in 1970. Furthermore, the ability to separate a specific isomer in a particular isomer group from all of its isomers and other congeners has advanced from the ability to separate 2,3,7,8-TCDD from only two of its isomers in 1974 to an ability to separate all of the 22 TCDD isomers in 1978. Likewise all 10 isomers of H_6CDD have been separated as have the two isomers of H_7CDD.

Such rapid development of highly sensitive methodology sutiable for the determination of specific compounds among large numbers of isomers in a series of homologous compounds, as well as a vast number of other related compounds, sets new standards for progress in analytical science. It was achieved by the continuous investment in the finest manpower and eouipment, both operating near their optimum potential. Leadership and cooperation by industry, academic, and government agencies were required to accomplish the goal.

As progress was made, time demands on the analytical chemist and his tools became more acute. This resulted in the inevitable inability of many to generate data sufficient to establish the certainty of results obtained by a particular method. Thus sampling data, validation data, and collaborative data are often lacking.

In this overview we examine the sophistication and variation of the methodology, evaluate the results obtained, and develop a philosophy which we hope will assure the collection of more meaningful data in the future. To accomplish this, we will rely heavily on two excellent recent reviews - one by Mahle and Shadoff (1981) and the other by a National Research Council of Canada panel chaired by H. McLeod (1981) for technical information and interpretation. Evaluation of the state of the methodology will be made by use of the American Chemical Society "Guidelines For Data Acquisition and Data Quality Evaluation in Environmental Chemistry" (1980).

The ACS guidelines identify the components of the analytical measurement as consisting of planning, quality assurance, sampling, calibration and standardization, measurement, calculations, verification, and documentation. The degree of certainty attained by a particular method is measured by the level at which the analytical scientist was able to meet the criteria outlined in these guidelines. Although planning, quality assurance, sampling, and documentation are very important parts of the analytical process and are generally recognized as being inadequately conducted, the discussion which follows will largely ignore them and concentrate on measurement and verification.

ANALYTICAL MEASUREMENT

From a mass spectroscopist's viewpoint, the analytical methodology for the determination of polychlorodibenzodioxins (PCDDs) and polychlorodibenzofurans (PCDFs) consists of three stages: 1) sample preparation, 2) sample introduction into the mass spectrometer, and 3) mass spectrometry.

SAMPLE PREPARATION

Sample preparation consists of 1) initial sample treatment, 2) liquid-liquid partition, 3) column chromatography, and 4) high pressure liquid chromatography (HPLC). These are used in numerous combinations to isolate the PCDDs from most of the interferences. The extent of this use is summarized in the following:

STAGE I   SAMPLE PREPARATION

INITIAL SAMPLE TREATMENT	NUMBER OF LITERATURE REFERENCES	EARLY REFERENCE
DISSOLUTION	10	
DIGESTION		RESS et al (1970)
IN BASE	3	BAUGHMAN & MESELSON (1973 a&b)
IN ACID	2	LANGHORST & SHADOFF (1980)
		ZABIK & ZABIK (1980)

INITIAL SAMPLE TREATMENT	NUMBER OF LITERATURE REFERENCES	EARLY REFERENCE
DIRECT EXTRACTION	18	CRUMMETT & STEHL (1973)
LIQUID-LIQUID PARTITION AGAINST CONC. H_2SO_4	12	RESS et al (1970)
AGAINST AQUEOUS BASE	9	CRUMMETT & STEHL (1973)
COLUMN CHROMATOGRAPHY ALUMINA	30	RESS et al (1970)
SILICA GEL	10	CRUMMETT & STEHL (1973)
FLORISIL	10	FUKUHARA et al (1975)
ION EXCHANGE RESIN	3	CRUMMETT & STEHL (1973)
COLUMN CHROMATOGRAPHY CHARCOAL	4	STALLING et al (1975)
GEL PERMEATION	7	STALLING et al (1975)
REAGENT MODIFIED SILICA	6	BUSER (1977)
HIGH PRESSURE LIQUID CHROMATOGRAPHY NORMAL PHASE	4	NESTRICK et al (1979)
REVERSED PHASE	8	PFEIFFER (1976)
REVERSED + NORMAL PHASE	4	NESTRICK et al (1979)

These sample preparation techniques, in various combinations, have been used successfully to separate PCDDs and PCDFs from a wide variety of matrices including plant and animal tissues and products, chemical products, sediment, soil, combustion products, and waste materials. A great deal is known about the techniques for doing this, thus assuring that measurable signals are obtained. However, little is known about the relative effectiveness of these techniques for 1) the quantitative removal of PCDDs and PCDFs from the sample, 2) the removal of interferences, and 3) the reproducibility of the process. More studies such as those of Lustenhouwer (1980) and Cutie (1981) on extraction recovery efficiencies are needed to optimize conditions.

High pressure high efficiency liquid chromatography is especially useful to seqarate isomers. Ten reports in the literature have used this approach. The power of this technique is shown by the separation of all 22 TCDDs and 10 H_6CCDs by using reversed phase and normal phase HPLC consecutively. Unfortunately, these valuable techniques have not been collaboratively studied.

SAMPLE INTRODUCTION

Introduction of the sample into the mass spectrometer is accomplished by 1) direct probe, 2) packed column gas chromatography, or 3) capillary gas chromatography. The various types of techniques are summarized in the following:

STAGE II MASS SPECTROMETER SAMPLE INTRODUCTION

	NUMBER OF REFERENCES	EARLY REFERENCE
DIRECT PROBE NO GAS CHROMATOGRAPHY		BAUGHMAN & MESELSON (1973 a&b)
PACKED COLUMN GAS CHROMATOGRAPHY		
SE-30	4	RAPPE & NILSSON (1972)
SE-52	1	RESS, et al (1970)
XE-60	3	HUSTON (1972) WILLIAMS & BLANCHFIELD (1972)
OV-1	3	VILLANUEVA et al (1975)
OV-101	8	FIRESTONE et al (1972) METCALFE (1972)
OV-3	9	SHADOFF et al (1977)
OV-7	2	HUCKINS et al (1978)
OV-17	4	CRUMMETT & STEHL (1973)
OV-105	2	SHADOFF et al (1977)
OV-210	3	CRUMMETT & STEHL (1973)
OV-225	2	BUSER & BOSSHART (1974)
OV-17/POLY S-179	8	PFEIFFER et al (1978)
BMBT LIQUID CRYSTAL	1	SHADOFF (1980)
DEGS	1	GRAY et al (1976)

PACKED COLUMN GAS CHROMATOGRAPHY	NUMBER OF REFERENCES	EARLY REFERENCE
CARBOWAX 20M	4	OLIE et al (1977)
SP-2100	2	FIRESTONE et al (1979)

CAPILLARY COLUMN GAS CHROMATOGRAPHY		
DEXIL 300	3	GRAY et al (1976)
OV-17 WCOT	2	BUSER et al (1978)
OV-61 WCOT	1	BUSER (1975)
OV-101 WCOT	2	BUSER (1976a)
OV-17/POLY S-179	1	NESTRICK et al (1980)
SILAR 10C	2	BUSER (1976a)
SE-30 WCOT	2	HARLESS et al (1980)
SP-2100 SCOT	1	MITCHUM et al (1980)
METHYL SILICONE WCOT	1	NORSTROM et al (1981)

A wide variety of columns has been tried. Some have been used extensively enough by different laboratories to command considerable confidence in their separating capabilities. Others, used only once or twice by a single laboratory, can be said to show great promise but their performance is not yet established. Unfortunately, most of the work reported on the use of capillary columns is in the latter category. This is regrettable since they have much greater power to resolve isomers. In fact, the use of three capillary gas chromatographic columns has allowed the separation of 2,3,7,8-TCDD from all of its isomers. This is summarized in the following table along with the use of both reversed phase and normal phase high pressure liquid chromatography to achieve separation of all 22 TCDD isomers.

SEPARATIONS

22 ISOMERS OF TCDD

TECHNIQUE	COLUMN 1	COLUMN 2	COLUMN 3	REFERENCE
HPLC	RP ZORBAX ODS	NP ZORBAX SIL	--	NESTRICK et al (1979)

SEPARATIONS (cont.)

2,3,7,8-TCDD FROM ITS 21 ISOMERS

	COLUMN			
TECHNIQUE	1	2	3	REFERENCE
CAPILLARY COLUMN GC	SILAR 10C	OV-17	OV-101	BUSER & RAPPE (1980)

MEASUREMENT BY MASS SPECTROMETRY

Mass spectrometry (MS) in the electron impact mode was first reported for the identification of PCDDs by Ress et al (1970). Since then it has had extensive development. As shown in the table below, at least a dozen models operating at low resolution ($M/\Delta M < 3000$) have been used to both identify and determine PCDDs after separation in Stages I and II.

STAGE III MASS SPECTROMETRY
(LOW RESOLUTION ELECTRON IMPACT
$M/\Delta M \leqslant 2000$

MS MODEL	TCDD LIMIT OF DETECTION pg	$M/\Delta M$	REFERENCE
LKB-9000	6	600	CRUMMETT & STEHL (1973)
	5-10	400	HUMMEL (1977)
	2-10	400	SHADOFF & HUMMEL (1978)
	50	---	ADAMOLI et al (1978)
LKB-9000S	20	---	DI DOMENICO et al (1979)
LKB-2091	250	400	FANELLI et al (1980)
FINNIGAN 3000	50	UNIT	ADAMOLI et al (1978)
FINNIGAN 3200	20	UNIT	DI DOMENICO et al (1979)
	40-80	UNIT	CAVALLARO et al (1980)
HEWP-5984A	20	UNIT	DI DOMENICO et al (1979)
-5992A	40-60	UNIT	LAMPARSKI & NESTRICK (1980)
-5992	5-10	UNIT	NORSTROM et al (1981)
-	1-5	UNIT	REES et al (1981)
EXTRA-NUC.	1000	UNIT	TIERNAN et al (1980)
KRATOS 30	5	1000	LANGHORST & SHADOFF (1980)
VARIAN - 311A	10	1000	RYAN & RILON (1980)

Very little work has been done at medium resolution (M/ΔM = 2500 to 9000) as shown below. This approach would be expected to have some advantage in sensitivity over that of the same instrument operated at higher resolution.

STAGE III   MASS SPECTROMETRY

MEDIUM RESOLUTION ELECTRON IMPACT

MS MODEL	TCDD LIMIT OF DETECTION pg	M/ΔM	REFERENCE
KRATOS-30	5-10	3000	HUMMEL (1977)
KRATOS-50	50	2000	CHESS & GROSS (1980)

Although high resolution mass spectrometry (M/ΔM ⩽ 900) was used in the early work of Baughman and Meselson (1973), it has only recently been emphasized by some mass spectroscopists as being important. This is primarily due to the work of Harless et al (1980). The experience with this approach is summarized below:

STAGE III   MASS SPECTROMETRY

HIGH RESOLUTION ELECTRON IMPACT

MS MODEL	TCDD LIMIT OF DETECTION pg	M/ΔM	REFERENCE
VARIAN 311A	5-10	9000	HARLESS et al (1980)
AEI MS-9	5	10000	BAUGHMAN & MESELSON (1973 a&b)
KRATOS-30	100	12000	TIERNAN & et al (1980)
AEI MS-30	2-10	9000	SHADOFF & HUMMEL (1978)

Chemical ionization mass spectrometry has been under investigation as a possible detector and monitor for PCDDs. The quantitative results of this work are summarized below:

STAGE III   MASS SPECTROMETRY

CHEMICAL IONIZATION

MS MODEL	TCDD LIMIT OF DETECTION pg	M/ΔM	REFERENCE
FINN. 3300	50-500	UNIT	HAAS et al (1978)
EXTRA-NUC.	10-30	UNIT	MITCHUM et al (1980)
			CAIRNS et al (1980)

It is evident from the tables that the limit of detection for TCDD is very much the same on high resolution instruments as on low resolution ones. The advantage of high resolution lies in its ability to separate interferences in the mass spectrometer.

Criteria for the confirmation of the homologue and specific isomer is essential to any analysis. Harless et al (1980) have proposed the following:

HARLESS CRITERIA FOR HOMOLOGUE AND SPECIFIC ISOMER CONFIRMATION

1.   Correct high resolution gas chromatography-high resolution mass spectrometry (HRGC-HRMS) retention time for 2,3,7,8-TCDD

2.   Correct HRGC-HRMS multiple ion response for ^{37}Cl-TCDD and TCDD masses (simultaneous response for elemental composition of m/z 320, m/z 322, m/z 328).

3.   Correct chlorine isotope ratio for the molecular ions (m/z 320 and m/z 322).

4.   Correct responses for the co-injection of sample fortified with ^{37}Cl-TCDD and TCDD standard.

5.   Response of the m/z 320 and m/z 322 must be greater than 2.5 times the noise level.

SUPPLEMENTAL CRITERIA PROPOSED BY HARLESS

(A)   COCl loss indicative of TCDD structure, and

(B)   HRGC-HRMS peak matching analysis of m/z 320 and m/z 322 in real time to confirm the TCDD elemental composition.

These criteria are useful and are agreeable to almost all analytical scientists. However, it is not necessary to specify capillary
gas chromatography (HRGC) in either criterion 1 or 2 since high resolution liquid chromatography eliminates the need to use it. Also,
the limit of detection in criteria 5 would be better defined as 3
times the noise level. The supplemental criteria are nice to have,
but usually cannot be obtained at concentration levels near the limit
of detection where the information is most needed.

THE ANALYTICAL PROCESS

All three stages of the analytical process may be thought of as
consisting of three categories of resolving power: low (L), medium
(M), and high (H). The categories of sample preparation are described
as follows:

STAGE I   SAMPLE PREPARATION

RESOLVING POWER	DESCRIPTION
L	CHEMICAL TREATMENT AND/OR EXTRACTION NO CHROMATOGRAPHY
M	L + COLUMN CHROMATOGRAPHY
H	M + HIGH PRESSURE LIQUID CHROMATOGRAPHY

The sample introduction into the mass spectrometer may be described as follows:

STAGE II   MS SAMPLE INTRODUCTION

RESOLVING POWER	DESCRIPTION
L	DIRECT PROBE (NO GAS CHROMATOGRAPHY)
M	PACKED COLUMN GC
H	CAPILLARY COLUMN GC

Likewise, the operation of the mass spectrometer may be assigned
rankings as follows:

STAGE III   MASS SPECTROMETRY

RESOLVING POWER	DESCRIPTION
L	LOW RESOLUTION (M/ΔM ≤ 2500)
M	MEDIUM RESOLUTION (M/ΔM = 2500 to 9000)
H	HIGH RESOLUTION (M/ΔM ≥ 9000)

Obviously, these can be put together in all various possible combinations to create a method. Thus, methods could be developed which could be described by these acronyms.

POSSIBLE ANALYTICAL METHODS FOR THE DETERMINATION OF PCDDs AND PCDFs

LLL	MMM	HHH
LLM	MLL	HLL
LLH	MLM	HLM
LMH	MLH	HLH
LHH	MMH	HMH
LML	MHH	HMM
LMM	MML	HHM
LHL	MHL	HML
LHM	MHM	HHL

Remembering that each of the L, M, and H categories have many variations, the possible number of methods for the determination of PCDDs and PCDFs becomes very large. Methods for the determination of PCDDs in chlorophenol and derivatives are categorized as follows:

METHODOLOGY FOR THE DETERMINATION OF CHLORINATED DIOXINS IN CHLOROPHENOLS AND DERIVATIVES

METHOD	DEGREE OF CHLORINATION	NUMBER OF PUBLICATIONS
LML	4	1
LHL	4–8	1
MML	4–8	8
MHL	4–8	1

Likewise, methods for the environmental/biological samples consist of the following:

METHODOLOGY FOR THE DETERMINATION OF CHLORINATED DIOXINS
OF
ENVIRONMENTAL/BIOLOGICAL SAMPLES

METHOD	DEGREE OF CHLORINATION	NUMBER OF PUBLICATIONS
MLL	4	1
MLH	4	4
MML	4	24
MMM	6,8	1
MMH	4	5
MHL	4-8	8
MHM	4	1
MHH	4	1
HML	4-8	5
HHL	4	2

A compilation of these methods with their approximate perform-ance parameters is tabulated in Table 1.

Consideration of this data makes it doubtful that insistence on the most sophisticated methodology resulting in HHH should be made and, indeed, use of such methodology has not yet been reported. From the table HHH would not be expected to give significantly better data than HML or HHL. The choice of methodology thus becomes a matter of personal preference. Without appropriate validation, data collected by capillary column gas chromatography high resolution mass spectro-metry is meaningless as is unvalidated data collected by other methodologies, It is thus as inappropriate for some to insist that LHH be used as it is for others to insist that HML be used. Of course it is appropriate that a scientist be proud of his own work.

USE OF ACS GUIDELINES

Using the ACS Guidelines, the Canadian Panel developed criteria for rating publications. These are given verbatim below:

Table 1. Performance of Analytical Methods for the Determination of Chlorinated Dioxins in Environmental Samples

METHOD	NUMBER OF REFERENCES	RELATIVE CLEAN-UP EFFICIENCIES DDE, PCBs, AND HOMOLOGUES	BENEFITS	
			NUMBER OF TCDD ISOMERS SEPARATED	TYPICAL LIMIT OF DETECTIONS FOR 2,3,7,8-TCDD pg./g.
MLL	1	1	NONE	5
MLH	2	5	NONE	1
MML	24	50	4-12	10-50
MMM	1	50	4-12	----
MMH	5	50	4-12	1-50
MHL	8	50	10-15	10-50
MHM	1	500	15-22	1-7
MHH	1	500	15-22	1-7
HML	5	50,000	15-22	1-50
HHL	2	200,000	22	1-50
HHH	0	500,000	22	1-50

CRITERIA FOR RATING PUBLICATIONS ON PCDD ANALYTICAL METHODS

POINT RATING	ESSENTIAL ELEMENTS	NUMBER OF PUBLICATIONS
1 (highest)	Complete quality assurance as described by ACS (1980). An ideally developed, evaluated method including collaborative studies.	0
2	Isomer specific, extensive recovery studies, interferences removed and separation achieved through extensive chemical workup, but lacks collaborative evaluation and assumes confirmation.	2
3	Incompletely isomer specific, some recovery studies, interferences partially removed and partial separation achieved through chemical workup, lacks collaborative evaluation and assumes confirmation.	10
4	Essentially a screening method for most homologues, interferences partially removed and partial separation through limited chemical workup, lacks collaborative evaluation and assumes comfirmation.	21
5	Same as 4, except inadequately documented for recovery, clean-up, etc.	20
6	Insufficient for the present "state of the art"	10

Applying these criteria to publications they arrived at the following ratings.

NATIONAL RESEARCH COUNCIL OF CANADA
RATINGS

	RATING					
YEAR	1	2	3	4	5	6
1971					2	1
1972				1	1	3
1973					2	3
1974						
1975				1		
1976				2	2	2
1977			1	3	3	
1978			2	5	2	
1979			1	3	4	
1980		2	4	5	2	1
1981			2	1	2	

It is interesting to note that the ratings are better in recent years. However, some of the early work rated 6 was state of the art at the time it was published. It is understandable that no dioxin work has a rating of 1, but it would appear that there should have been more work rated 2 or 3. In my view, many of the low ratings are justified. The reasons are many. Among them are:

o Lack documentation of the extent of validation data

o Lack of analytical standards

o Failure to show appreciation for the role of possible interferences

o Detection limits inappropriate for requirements

o Limits of detection and quantitation confused

o Sampling protocol not documented

o Inappropriate extrapolation of data.

The problems of measurement near the limit of detection have been discussed by Crummett (1979 a+b). These are the problems of sample selection, sample contamination, sample degradation, background noise, interferences, signal detection, signal measurement, identification, confirmation, corroboration, and contamination source.

Although solutions to these problems are not always apparent, much useful information can be gleaned from the published work. It is appropriate, for example, to conclude that PCDEs or PCDFs are not present in a sample at a concentration level of 10 times noise if they were not detected at a level of 3 times noise. One can thus get information about the concentration level above which dioxins cannot possibly be present. Such information is often very important.

Extensive validation data and collaborative studies are necessary for regulatory purposes if the regulation is based on the limit of detection (3 x noise). Realistically, however, regulations should be based on concentration levels at or above the limit of quantitation thus adding greatly to the certainty of the data without extensive validation and collaborative studies.

COLLABORATION

Recently IUPAC held a conference in Helsinki, Finland on "Harmonization of Collaborative Analytical Studies". Many speakers emphasized that collaborative studies should insist that participants use exactly the same techniques. They would not be satisfied that all laboratories use a MHH mehtod but that all separation and measurement devices be identical. Others expressed the need to study methods said to be equivalent but not identical. Of course it is apparent that the determination of PCDDs and PCDFs falls into the latter category and we must be prepared to do collaborative studies in this mode. In my view, agreement in results on environmental samples obtained from a variety of analytical techniques may be regarded as excellent confirmation.

BIOASSAY

Bioassay methods are listed in the following table. They represent alternate screening techniques for the presence of PCDDs based on biological and biochemical properties. They are all in an early development stage and are non-specific for any isomer. They are less sensitive than current analytical methodologies. If made reliable, however, they would have the advantages of cheapness and rapidity and offer hope for large volume screening. Bioassay also offers the possibility of presenting a toxicologically significant result.

BIOASSAY METHODS

METHOD	STAGE	LIMIT OF DETECTION pg	REFERENCE
RADIOIMMUNOASSAY	EARLY DEVELOPMENT	100	MCKINNEY et al. (1980)
ARYL HYDROCARBON HYDROXYLASE-(AHH)- INDUCTION ASSAY	EARLY DEVELOPMENT	UNKNOWN	BRADLAW & CASTERLINE (1979)
CYTOSOL-RECEPTOR- ASSAY	EARLY DEVELOPMENT	100-200	HUTZINGER et al. (1981)

SUMMARY AND CONCLUSIONS

Many techniques are available to dissolve or extract samples containing chlorinated dioxins, to remove the bulk of the interferences, to separate homologues and isomers, and to measure isomers either specifically or in combination. These techniques can be combined in an almost infinite number of ways to yield procedures which appear to produce reliable data. Because these procedures are tedious and require considerable time to carry out, true measures of the certainty of results are often neglected. Thus, validation and collaborative studies are seldom done to the satisfaction of the scientific community.

Enough is known, however, to allow analytical scientists to write protocols which are based on good estimates of the sensitivity and specificity of the total analytical scheme. This approach is much preferred over the use of the "best available analytical methodology" often specified by regulators.

More validation and collaborative studies are needed if regulation is indicated. These should not require that all participants use identical procedures but that the methodologies which are used are equivalent. Both intra- and inter-laboratory studies are required. However, regulations should be based on inter-laboratory studies.

If regulations are to be promulgated for a number of PCDDs and PCDFs, attention must be given to the availability of appropriate analytical standards. There is an urgent need to know the concentration level of regulatory concern and to devise methods which have limits of detection well below the level at which decisions will be made.

Presumably, the session on the analytical chemistry of chlorinated dioxins will provide us with more validation and collaboration information as well as better analytical techniques. I, for one, will be pleased if this symposium shows that some of my comments are incorrect or even unnecessary. I am well aware that it is easier to be critical than correct.

REFERENCES

1. Adamoli, P., et al. Ecol Bull, 27:31, 1978.
2. American Chemical Society Committee on Environmental Improvement.
       Anal Chem, 52:2242, 1980.
3. Baughman, R., and Meselson, M., Adv Chem Ser, 120:92, 1973a.
4. Baughman, R., and Meselson, M., Environ Health Persp, 5:27, 1973b.
5. Bradlaw, J.A., and Casterline, J.L., J. Assoc Off Anal Chem,
       62:904.
6. Buser, H.R., Anal Chem, 48:1553, 1976a.
7. Buser, H.R., J Chromatogr, 107:295, 1975a.
8. Buser, H.R., J Chromatogr, 114:95, 1975b.
9. Buser, H.R., J Chromatogr, 129:303, 1976b.
10. Buser, H.R., Anal Chem, 49:918, 1977.
11. Buser, H.R., and Bosshardt, H.P., J Chromatogr, 90:71, 1974.
12. Buser, H.R., and Bosshardt, H.P., J Assoc Off Anal Chem, 59:562,
       1976.
13. Buser, H.R., Bosshardt, H.P., and Rappe, C., Chemosphere, 7:165,
       1978.
14. Buser, H.R., and Rappe, C., Anal Chem, 52:2257, 1980.
15. Buser, H.R., and Rappe, C., Chemosphere, 7:199, 1978.
16. Carins, T., Fishbein, L., and Mitchum, R.K., Biomed Mass Spectrom,
       7:484, 1980.
17. Cavallaro, A., Bandi, G., Invernizzi, G., Luciani, L., Mongini,E.,
       and Gorni, A., Chemosphere, 9:611, 1980.
18. Chess, E.K., and Gross, M.L., Anal Chem, 52:2057, 1980.
19. Crummett, W.B., Ann New York Acad Sci, 320:43, 1979.
20. Curmmett, W.B., Toxicol Environ Chem Rev, 3:61, 1979.
21. Curmmett, W.B., and Stehl, R.H., Environ Health Persp, 5:15,
       1973.
22. diDomenico, A., Merli, F., Boniforti, L., Camoni, I., DiMuccio,
       A., Taggi, F., Vergori, L., Colli, G., Elli, G., Gorni, A.,
       Grassi, P., Invernizzi, G., Jemma, A., Luciani, L., Cattabeni,
       F., DeAngelis, L., Galli, G., Chiabrando, C., and Fanelli,
       R., Anal Chem, 51:735, 1979.

23. Fanelli, R., Castelli, M.G., Martelli, G.P., Noseda, A., and Garattini, S. Bull Environ Contam Toxicol, 24:460, 1980.

24. Firestone, D., Clower, M., Borsetti, A.P., Teske, R.H., and Long, P.E. J Agric Food Chem, 27:1171, 1979.

25. Firestone, D., Ress, J., Brown, N.L., Barron, R.P., and Damico, J.N. J Assoc Off Anal Chem, 55:85, 1972.

26. Fukuhara, K., Takeda, M., Uchiyama, M., and Tanabe, H. Eisei Kagaku, 21:318, 1975.

27. Gray, A.P., Cepa. S.P., Solomn, ILJ., and Aniline, O. J Org Chem, 41:2435, 1976.

28. Harless, R.L., Oswald, E.O., Wilkinson, M.K., Dupuy, Jr., A.E., McDaniel, D.D., and Tai, H. Anal Chem, 52:1239, 1980.

29. Hass, J.R., Friesen, M.D., Harvan, D.J., and Parker, C.E. Anal Chem, 50:1474, 1978.

30. Huckins, J.N., Stalling, D.L., and Smith, W.A. J Assoc Off Anal Chem, 61:32, 1978.

31. Hummel, R.A. J Agric Food Chem, 25:1049, 1977.

32. Hummel, R.A., and Shadoff, L.A. Anal Chem, 52:191, 1980.

33. Huston, B.L. J Agric Food Chem, 20:724, 1972.

34. Hutzinger, O., Frev, R.W., Merian, E. and Pocchiari, F. Rome Conference, 1981.

35. International Union of Pure and Applied Chemistry, Helsinki Conference on "Harmonization of Collaborative Analytical Studies", Proceedings to be published, Pergamon Press, 1981.

36. Lamparski, L.L., and Nestrick, T.J. Anal Chem, 52:2045, 1980.

37. Langhorst, M.L., and Shadoff, L.A. Anal Chem, 52:2040, 1980.

38. Mahle, N.H., and Shadoff, L.A. Review to be published, Biomed Mass Spectrom, 1982.

39. McLeod, H1, et al. National Research Council Canada, Publ NRCC NO. 18576 of the Environmental Secretariat, 1981.

40. McKinney, J., Albro, P., Luster, M., Corbett, B., Schroeder, J., and Lawson, L. Rome, Oct. 1980.

41. Metcalfe, L.D. J Assoc Off Anal Chem, 55:542, 1972.

42. Mitchum, R.K., Moler, G.F., and Krofmacher, W.A. Anal Chem, 52:2278, 1980.

43. Nestrick, T.J., Lamparski, L.L., and Stehl, R.H. Anal Chem, 51:2273, 1979.

44. Nestrick, T.J., Lamparski, L.L., and Townsend, D.I. Anal Chem, 52:1865, 1980.

45. Norstrom, R.J., Hallett, D.J., Simon, M., and Mulinhill, M.J. Private communication, 1981.

46. Olie, K., Vermeulen, P.L., and Hutzinger, O. Chemosphere, 6:455, 1977.

47. Pfeiffer, C.D., Nestrick, T.J., Kocher, C.W. Anal Chem, 50:800, 1978.

48. Pfeiffer, C.D. J Chromatogr Sci, 14:386, 1976.

49. Rappe, C., and Nilsson, C.A. J Chromatogr, 67:247, 1972.

50. Rees, G., Smillie, R.D., and Tosine, H. National Research Council of Canada Review, Draft, 1981.

51. Ress, J., Higginbotham, G.R., and Firestone, D. J Assoc Off
        Anal Chem, 53:628, 1970.
52. Ryan, J.J., and Pilon, J.C. J Chromatogr, 197:171, 1980.
53. Shadoff, L.A. ACS Symposium Series, 136:277, 1980.
54. Shadoff, L.A., and Hummel, R.A. Biomed Mass Spectrom, 5:7, 1978.
55. Shadoff, L.A., Hummel, R.A., Jensen, D.J., and Mahle, N.H.
        Ann Chim (Rome), 67:583, 1977a.
56. Shadoff, L.A., Hummel, R.A., Lamparski, L.L., and Davidson, J.L.
        Bull Environ Contam Toxicol, 18:478, 1977b.
57. Stalling, D.L., Johnson, J., and Huckins, J.N. Environmental
        Quality and Safety. Vol. 3, Coulston, F., and Korte, F.,
        Ed., G. Thieme Publ., Stuttgart, Germany. 1975, pp. 12-18.
58. Tiernan, T.O., Taylor, M.L., Erk, S.D., Solch. J.G., VanNess, G.,
        and Dryden, J. USEPA 600/2-80-57, 1980.
59. Villanueve, E.C., Jannings, R.Q., Burse, V.W., and Kimbrough,
        R.D. J Agric Food Chem, 23:1089, 1975.
60. Williams, D.T., and Blachfield, B.J. J Assoc Off Anal Chem,
        55:93, 1972.
61. Zabik, M.E., and Zabik, M.J. Bull Environ Contam Toxicol,
        24:344, 1980.

HIGH RESOLUTION MASS SPECTROMETRY METHODS OF ANALYSIS FOR

CHLORINATED DIBENZO-P-DIOXINS AND DIBENZOFURANS

R. L. Harless,* A. E. Dupuy,** and D. D. McDaniel**

*U.S. Environmental Protection Agency
 Environmental Monitoring Systems Laboratory
 Research Triangle Park, North Carolina   USA
**Toxicant Analysis Center
 Bay St. Louis, Mississippi   USA

ABSTRACT

Methods of analysis for chlorinated dibenzo-p-dioxins (CDDs) and dibenzofurans (CDFs) based on high resolution mass spectrometry detection techniques have been developed and applied. The application of these methods for identification and quantification of CDDs and CDFs in samples collected from accidental fires involving electrical transformers and incineration processes are described. Analytical results for specific incidents, criteria used for confirmation of CDDs and CDFs and the problems caused by the lack of sufficient reference standards are discussed.

INTRODUCTION

Considerable work has been directed toward development of reliable procedures for the quantitative determination of chlorinated dibenzo-p-dioxins (CDDs) and dibenzofurans (CDFs) during the past decade. Most of this effort can be attributed to concern over 2,3,7,8-tetrachlorodibenzo-p-dioxin (2,3,7,8-TCDD), a contaminant in 2,4,5-trichlorophenoxyacetic acid (2,4,5-T). Public disclosure that 2,3,7,8-TCDD was a ppb- to ppm level contaminant in the widely used herbicide, 2,4,5-T, triggered intensive research efforts and monitoring activities in the early 1970s. The first procedures developed specifically for determination of pg/g (ppt) levels of 2,3,7,8-TCDD were reported by Baughman and Meselson (1973). The evolution of EPA methodology, which relies on high resolution mass spectrometry detection techniques for determination of ppt to ppm levels of 2,3,7,8-TCDD in various matrices, is shown in Table 1.

65

Table 1.   Evolution of EPA Methodology for Determination of
           2,3,7,8-TCDD

Technique	Application	Minimum Limit of Detection (ppt range)	Specificity
Direct probe HRMA	1972-74	10-50	None
LRGC-HRMS	1974-76	10-30	Minimum
HRGC-HRMS	1976-present	1-10	Maximum

     The methods used by EPA during 1972-76 were not specific for
2,3,7,8-TCDD but it is emphasized that this does not necessarily mean
that the analytical technique and/or results were invalid.   During
this time frame monitoring studies were directed at specific areas,
animals, etc., that had been exposed to phenoxy-herbicides known to
contain ppm to ppb levels of 2,3,7,8-TCDD.   The early findings (Ryan,
Biros and Harless, 1974) showed that 2,3,7,8-TCDD could be detected
in samples of specific species of wildlife and that it could be quantitatively determined in adipose tissues and livers of domestic
animals intentionally exposed (control studies) to 2,4,5-T-treated
feed and rangeland.   However, these findings were not considered conclusive because of the presumed lack of specificity of the analytical
method employed.   In 1975-76 four basic principles were used to validate the analytical methodology used by EPA and collaborating
laboratories: Validation studies; criteria for confirmation of 2,3,7,
8-TCDD; quality assurance (QA) programs; and multiple laboratory participation.   The methodology was validated down to about 10 ppt and
small scale monitoring activities were then conducted by collaborating
laboratories.

     Up to this point, 2,3,7,8-TCDD entry into the environment was
assumed to involve only the synthetic chemical industry.   Suddenly
an unexpected route for CDDs and CDFs to enter the environment was
reported (Ollie et al. 1977; Buser et al. 1978; Dow Chemical
Company, 1978).   These reports indicated CDDs and TCDD isomers, including 2,3,7,8-TCDD, were being emitted into the environment from
specific combustion sources.   It was then obvious that valid analytical procedures must incorporate appropriate specificity for
determination of 2,3,7,8-TCDD alone or in the presence of other
isomers in all monitoring activities.   Specificity can be incorporated in the sample preparation procedures or in the detection
technique.   This laboratory had coupled high resolution gas chromatography (HRGC) to a high resolution mass spectrometer (HRMS) in 1976
for purposes other than resolution of TCDD isomers.   Fortunately, the
HRGC-HRMS technique also provided the required specificity for 2,3,7,
8-TCDD determination.   This technique has been used to identify and

quantify ppt to ppm levels of 2,3,7,8-TCDD in a variety of sample matrices since 1976 (Harless and Lewis, 1980), i.e. water from contaminated sources, soil, sediment, human adipose tissue, fish, deer, elk, herring gull, chemical products, chemical disposal sites, municipal incinerators, accidental fires, etc.

Controversial issues, intense scrutiny by the scientific community and technological advances, have served to encourage continual improvements in analytical procedures and involvement by many laboratories since 1971. Today, many laboratories, including this one have available the 22 TCDD isomers and can make specific isomer assignments. The analytical procedures developed for 2,3,7,8-TCDD are also applicable to other CDDs and CDFs. However, the lack of sufficient references standards, validated analytical methodology and sampling devices seriously detracts from the scientific value of the analyses and limit their usefulness for meaningful assessment purposes.

In this paper, the analytical methodology, the deficiencies, and the findings of CDDs and CDFs in samples collected from accidental fires and a municipal incinerator are discussed. It should be made clear that this is not an official EPA report, but rather, a brief description and summary of work performed by EPA scientists.

EXPERIMENTAL

Analytical Methodology

The sample preparation procedures, HRGC-HRMS methods of analysis and criteria were developed specifically for 2,3,7,8-TCDD (Harless et al., 1980) but the methodology was determined also to be suitable for determination of other CDDs and CDFs. Briefly, sample preparation involved: Fortification of sample, milligrams to 10g, with 5 or 10ng $^{37}Cl_4$-TCDD and 10ng $^{37}CL_8$-OCDD for determination of sample preparation efficiency; saponification with hot alkali followed by extraction with hexane; treatment with concentrated sulfuric acid; chromatographic cleanup on alumina; and concentration of the alumina column eluent to 60ul for HRGC-HRMS analysis.

A Varian MAT 311A MS, directly coupled to a Varian Model 2700 GC, was used for these analyses. The GC was equipped with 60 and 30-m x 0.25-mm i.d.SE-30, OV-101 or OV-17 WCOT glass columns. The MS was operated in the electron impact (EI) mode and utilized an eight-channel hardware multiple ion selection (MIS) device. The HRGC-HRMS parameters were: injection port temperature, 265° C; oven program, 6 mins at 80° C then programmed to 270°C at 34° C/min, or, alternatively, programmed to 150° C at 34° C/min and then programmed to 270°C at 4° C/min; GC transfer line into MS ion source, 255° C; ion source temperature, 240° C; variable acceleration voltage, 3 kV maximum;

electron energy 70 eV; filament emission, 1 mA; mass resolution, 8000-10,000; multiplier gain greater than 10^6.

The exact masses corresponding to M+ and M+2 of respective CDDs and CDFs were monitored in the multiple ion monitoring mode utilizing exact masses of perfluorokersene as the reference. The criteria used for confirmation of CDDs and CDFs were: multiple ion monitoring response for M+ and M+2 chlorine isotope ratio; HRGC-HRMS confirmation of exact masses of M+ and M+2 corresponding to elemental compositions of respective CDDs or CDFs. Quantification of CDDs and CDFs was based on several reference standards, i.e. 2,3,7,8-TCDD and TCDF, 1,2,3,4,7,8-hexa-CDD, 1,2,3,4,6,7,8-hepta-CDD, OCDD and OCDF.

DISCUSSION

The identification and quantitative determination of CDDs and CDFs in samples from combustion sources has recently become more important because of the need to assess accurately this route of entry into the environment. The EPA analytical methodology was determined to be adequate for the isolation and quantitative determination of specific CDDs and CDFs. However, the deficiencies shown in Table 2 prohibit full utilization of its capabilities. With the exception of TCDDs, it is almost impossible to make valid isomer assignments for most of the other di- through hepta- CDDs and CDFs. Also, the final analytical results can only indicate minimum amounts of CDDs and CDFs present in the extract and/or emitted from the process because the analytical methodology and sampling device efficiency are unknown.

Table 2.   Deficiencies in Methodology for Determination of CDDs and CDFs

---

   ° Insufficient number of CDD (75) and CDF (135) reference standards available for identification purposes.

   ° Insufficient number of labeled CDDs and CDFs available for use as internal standards to monitor analytical methodology efficiency.

   ° Analytical methodology efficiency for quantitative determination of mono- through octa-CDDs and CDFs has not been determined and validated.

   ° Sampling device efficiency (collected and retention) for CDDs and CDFs has not been determined and validated.

The elimation of these deficiences is costly and time consuming. Progress is being made, i.e. both "native" and "labeled" reference standards are slowly becoming available and sampling device efficiencies are being determined.

Accidental Fires

The possibility of human exposure to CDDs and CDFs that may be formed in accidental fire has caused much concern since 1978. In one incident, specific fire department equipment was quarantined until CDD and CDF analyses were completed. Because of the lack of sufficient reference standards, primary emphasis has been placed upon determinations of TCDDs and TCDFs in these limited scope investigations. As shown in Table 3, picogram to nanogram amounts of CDDs and CDFs have been detected in samples associated with accidental fires involving electrical transformers. CDF levels were higher than CDD levels in all samples collected from these types of sources.

Table 3. Identification of TCDDs and CDFs in Samples Associated with Accidental Fires Involving Electrical Transformers

Incident and Identification	Amount of TCDDs Including 2,3,7,8-TCDD Detected	Estimated Amounts of CDFs Detected
A Wipe samples from capacitors, roofing material and building walls	30 to 500 pg	Tri- 1-600 ng Tetra- 20-400 ng Penta- 100-500 ng
B Wipe samples	129 to 260 pg	Tri- 80 ng Tetra- 100 ng Penta- 60 ng Hexa- 40 ng
C Soot Sample[a]	1.2 ng 2,3,7,8-TCDF	

[a]Subjected to New York State Department of Health specific sample preparation procedures.

The findings show that CDDs and CDFs can be detected in samples collected from accident fires and that human exposure to these compounds is possible. CDDs and CDFs are known contaminants of certain chemical products and they may be formed in combustion processes. However, it is emphasized that these limited scope investigations were not designed to answer the many questions that arise.

Incinerator

Determinations of specific CDDs and CDFs were recently performed on the stack gas effluent from a municipal incinerator. The results, Table 4, show that nanogram amounts of TCDDs, TCDFs and hexa-CDDs were detected. 2,3,7,8-TCDD, one of 11 resolved isomers, was a minor component (less than 2 percent). 1,3,6,8-TCDD was one of several major isomers detected. Sixteen partially resolved TCDF isomers were detected. One isomer exhibited the same retention time as 2,3,7,8-TCDF, Figure 1, but conclusive isomer assignment can not be made because only one isomer, 2,3,7,8-TCDF, is available in this laboratory. Five hexa-CDD isomers were detected but none exhibited a retention time identical to that of the 1,2,3,4,7,8-hexa-CDD standard.

Table 4. Analytical Results for CDDs and TCDFs in Gas Phase Emissions from a Municipal Incinerator

ID	Minimum Number of Isomers	Estimated Amounts Detected in Extracts (ng)
2,3,7,8-TCDD		3 to 5
2,3,7,8-TCDF		90
TCDD isomers	11	329 to 466
TCDF isomers	16	820
Hexa-CDDs	5	141

SUMMARY

The intense interest focused on 2,3,7,8-TCDD during the past decade led to the rapid development of state-of-the-art analytical procedures and the discovery that combustion processes represent another major route for 2,3,7,8-TCDD, other CDDs and CDFs, to enter the environment. Analysis of extracts from these sources presents challenging problems, but the many analytical procedures in use today are capable of characterizing the extracts and assessing the processes if sufficient reference standards are made available and sampling

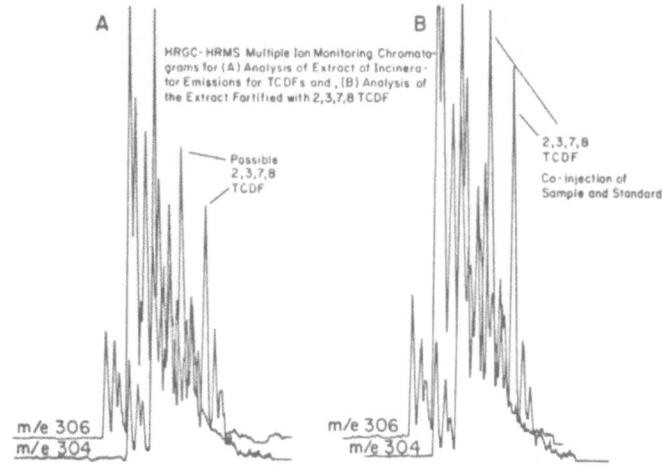

Figure 1. HRGC–HRMS Mass Chromatograms Obtained in Analysis of
Effluents from a Municipal Incinerator for TCDFs
(EPA/EMSL-RIP).

device efficiencies are documented. The findings of CDDs and CDFs in
the gas phase effluent from a raw-refuse-fired municipal incinerator
are similiar to those previously reported by this laboratory and many
other laboratories. Several investigations, "Trace Chemistries of
Fire", and unpublished reports indicate that CDD and CDF emissions can
be reduced and kept at or below present detectable limits if supplemental fuel is used in the combustion process. Preliminary findings
that will be useful in assessing combustion processes are slowly un-
folding. The findings also confirm that CDDs and CDFs can be detected
in samples collected from accidental fires. Therefore, human exposure
to these compounds is possible. In-depth studies need to be performed
in order to assess the risks properly and to develop adequate safety
measures for personnel.

REFERENCES

Baughman, R., and Meselson, M., Environmental Health Perspectives,
5: 27-35, 1973.
Buser, H.R., Bosshardt, H.P., and Rappe, C., Chemosphere, No. 2,
165-172, 1978.
Dow Chemical Co., The Chlorinated Dioxin Task Force. The Trace
Chemistries of Fire--A Source of and Routes for the Entry of
Chlorinated Dioxins into the Environment. Michigan Division
Dow Chemical Company, Midland, Michigan, 1978.
Harless, R.L., Oswald, E.O., Wilkerson, M.K., Dupuy, A.E.,
McDaniel, D.D., and Tai, Han. Analytical Chemistry, 52:1239-
1245, 1980.

Harless, R.L., and Lewis, R.G. Symposium-Impact of Chlorinated
    Dioxins and Related Compounds on the Environment, Instituto
    Superiore di Sanita, Rome, Italy, October 22-24, 1980.
Ollie, K., Vermeulen, P.L., and Hutzinger, O. Chemosphere, No. 8,
    455-459, 1977.
Ryan, J.F., Biros, F.J., and Harless, R.L. Proceedings 22nd
    Annual Conference on Mass Spectrometry and Allied Topics,
    Philadelphia, Pa., May 19-24, 1974.

DETERMINATION OF TETRACHLORODIBENZO-P-DIOXINS AND TETRACHLORO-DIBENZOFURANS IN ENVIRONMENTAL SAMPLES BY HIGH PERFORMANCE LIQUID CHROMATOGRAPHY, CAPILLARY GAS CHROMATOGRAPHY AND HIGH RESOLUTION MASS SPECTROMETRY

R.M. Smith, D.R. Hilker, P.W. O'Keefe, K.M. Aldous,
C.M. Meyer, S.N. Kumar and B.M. Jelus-Tyror

Toxicology Institute
New York State Department of Health
Albany, New York   USA

INTRODUCTION

Regulatory agencies have recently been faced with the formidable problem of possible human exposure to complex mixtures of organic compounds such as those found in chemical wastes. For many of these compounds, toxicity data is either limited or completely unknown and standards may be impure or completely unavailable for testing. The problem is often compounded by the lack of a standard analytical method that provides adequate sensitivity (e.g., part-per-trillion) and accuracy.

An approach that we have taken to the analysis of the polychlorinated dibenzo-p-dioxins (PCDDs) and polychlorinated dibenzofurans (PCDFs) in environmental samples is first to accurately determine only the highly toxic 2,3,7,8-tetrachlorodibenzo-p-dioxin (2,3,7,8-TCDD) with the aid of an improved version of a chromatographic separation scheme used by Nestrick et al, (1979). Unlike most of the other PCDDs and PCDFs, many literature references are available concerning the toxicity and analytical measurement of this compound. In addition, from a quantitative viewpoint, ^{37}Cl- and ^{13}C-labeled 2,3,7,8-TCDD standards can be obtained for accurate and precise determination using ion monitoring mass spectrometry (MS) techniques. A rigorous sequence of high-performance liquid chromatographic (HPLC) sample clean-up, high-resolution capillary gas chromatography (capillary GC) and high-resolution mass spectrometric analysis (HRMS) is used to eliminate interfering compounds (e.g. PCB, DDE) and separate 2,3,7,8-TCDD from the 21 other TCDD isomers. The isomer-specific results that we obtain provide a basis for comparison with simpler

analytical methodologies which are designed to investigate all PCDDs
and/or PCDFs. Future developments, in particular the synthesis of
different labeled internal standards, should make it possible to
obtain rigorous quantitative data using these methodologies.

ANALYTICAL METHOD

Test and control samples are carefully chosen and dried, $(U-^{13}C)-$
2,3,7,8-TCDD (or other labeled internal standard) is added, and the
samples are then extracted using one of several solvent extraction
techniques. Each sample is then processed through a "preliminary
clean-up" procedure that has been specifically designed to remove the
problem compounds for a given sample matrix (see Figure 1).

High Performance Liquid Chromatography

For an isomer-specific analysis, the extract from the "preliminary clean-up"is concentrated to 50 µl and injected onto the first
of two HPLC columns. Reversed-phase Zorbax ODS chromatography provides clean-up and partial TCDD isomer separation using methanol
solvent (2 ml/min) at 40°C. We have found it to be easy to use and
to give reproducible retention times. The proper fraction is collected at the retention time of a standard. It is then concentrated
and injected onto a normal-phase Zorbax Sil liquid chromatographic
column. This column was found to yield optimum TCDD isomer separation
with minimal retention time drift when the silica gel packing is
partially deactivated with water as suggested by Krikland (1973)
(see Figure 2).

We routinely activate the packing with CH_2Cl_2, and partially
deactivate with water-saturated hexane. For additional resolution,
0.4% toluene (water-saturated) in hexane (2ml/min) is used as a
chromatographic eluent (O'Keefe et al, submitted 1981). The sample
fraction is collected at the retention time of the standard, concentrated and injected onto the capillary GC.

High Resolution Gas Chromatography

The GC capillary column that provides the best selectivity for
TCDD isomers is a 0.35 mm x 40 m HCL (gas) etched soda-glass column
that has been carbowax 20-M deactivated and coated with the dicyanoal-
lyl liquid phase, OV-275, using a Hg plug technique developed by
Schomburg (1975). Individual TCDD isomers (or isomer pairs) have been
injected onto this column and the elution order relative to a $(U-^{13}C)-$
2,3,7,8-TCDD internal standard is shown in the composite chromatogram
(Figure 3).

Most of the 22 TCDD isomers appear to be separated by the capillary GC step alone. It has been shown that 2,3,7,8-TCDD can be

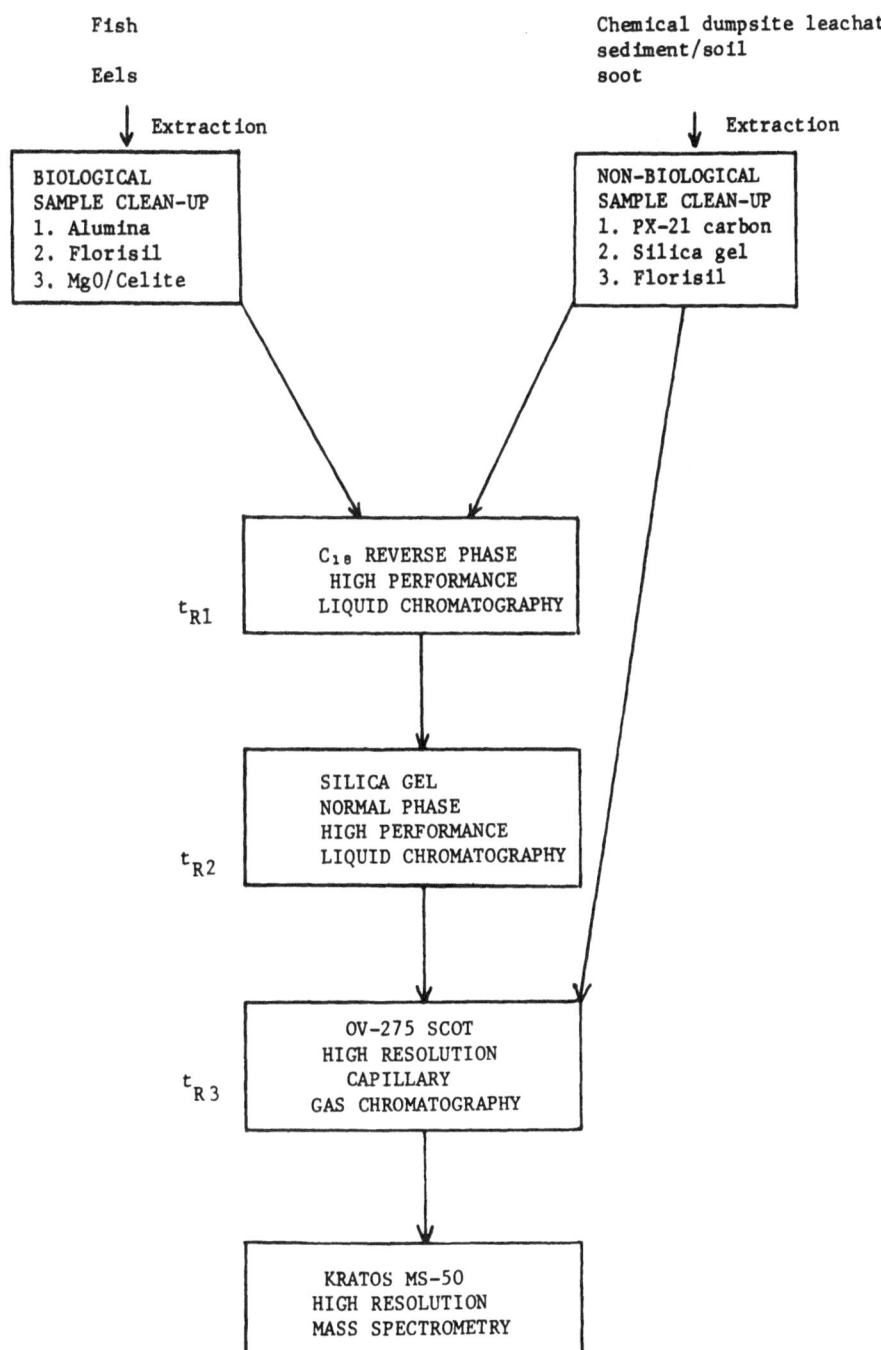

Figure 1.   Schematic of the Isomer-Specific TCDD Method

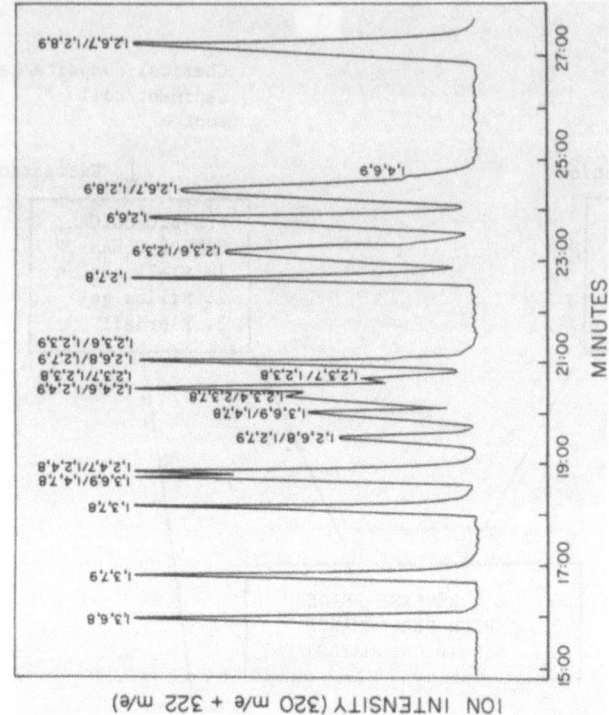

Figure 3. Composite Capillary Gas Chromatogram
of 22 TCDD Isomers.

Column: OV-275 SCOT, 0.35 mm, 45 m.,
He carrier, Temp: 80°C for 1 min.
then 10°/min to 180°C.

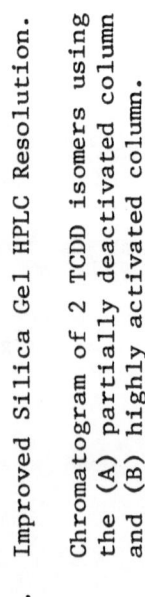

Figure 2. Improved Silica Gel HPLC Resolution.

Chromatogram of 2 TCDD isomers using
the (A) partially deactivated column
and (B) highly activated column.

separated from the other 21 TCDD isomers using a capillary column
coated with a related cyanophase, Silar 10C, although a complete separation of all 22 TCDD isomers has not yet been achieved (Buser and
Rappe, 1980). As better capillary columns are developed is is expected that capillary GC will be used as the only means of isomer
separation.

Isomer Specificity

The TCDD isomers that are not easily separated on the OV-275
capillary column can be separated by HPLC-trapping as may be seen
from the relative retention times given in Table 1. When a standard
mixture containing all 22 TCDD isomers (21 + ^{37}Cl-2,3,7,8-TCDD)
(Figure 4A) is analyzed by this method, only the ^{37}Cl-2,3,7,8-TCDD
is identified (Figure 4B). The isomer-specific method has also
been applied to 2,3,7,8-TCDF.

Non-Specific Clean-up Method

A low pressure liquid chromatographic clean-up involving PX-21
carbon (Stalling et al, 1980), silica gel and Florisil followed by
capillary GC/HRMS is presently being used to identify other TCDD
and TCDF isomers in environmental samples.

INSTRUMENTATION

Our instrumentation for analysis of PCDDs and PCDFs (Figure 5)
consists of a Kratos DS-55 data system interfaced to a Kratos MS-50
high resolution mass spectrometer. The mass spectrometer is in turn
coupled via a glass jet separator to a Carlo Erba Fractovap Series
460 capillary gas chromatograph with an on-column injector. For
optimum transfer of organic compounds from a capillary column into
the mass spectrometer it is necessary to introduce a make-up gas on
the high resolution (10,000) ion monitoring mode. Data acquisition
is controlled through a software program called HRMPM (Chapman et al,
1980).

The most important feature of the program is that we can obtain
real-time chromatographic data in addition to post-acquisition mass
profiles. This accomplished in the following manner: The total
signal in individual mass scans for selected ions are summed in real-
time to give mass chromatograms. From one to four ions can be scanned
over defined mass windows in any single experiment. For display purposes during an experiment the ion mass chromatograms are added together to give a "TIC" chromatogram.

After completion of an experiment, any desired number of scans
in the region of a signal on a mass chromatogram can be used to form

Table 1.   Relative Retention Times for 22 TCDD
Isomers on HPLC and Capillary GC

TCDD Isomers	Relative Retention Time		
	RP-HPLC[a]	Silica-HPLC[a]	GC-capillary OV275[b]
1267/1289	0.853	1.471	1.208
	0.853	1.689	1.365
1268/1279	0.953	1.191	1.037
	0.997	1.240	0.963
1269	0.833	1.670	1.170
1278	0.956	1.240	1.118
1368	1.100	0.963	0.795
1369/1478	0.950	1.294	0.992
	0.963	1.183	0.923
1378	1.068	1.001	0.896
1379	1.100	0.933	0.838
1469	0.833	1.582	1.212
2378	1.000	1.000	1.004
1236/1239	1.027	1.331	1.036
	1.016	1.331	1.130
1237/1238	1.007	1.086	1.013
	1.007	1.100	1.021
1246/1249	0.983	1.289	1.009
	0.983	1.367	1.014
1247/1248	1.013	1.114	0.928
	1.013	1.159	0.930
1234	1.132	1.238	1.002

[a] Retention time relative to absolute retention time of 2,3,7,8-TCDD.

[b] Retention time relative to absolute retention time of ^{13}C-2378TCDD.

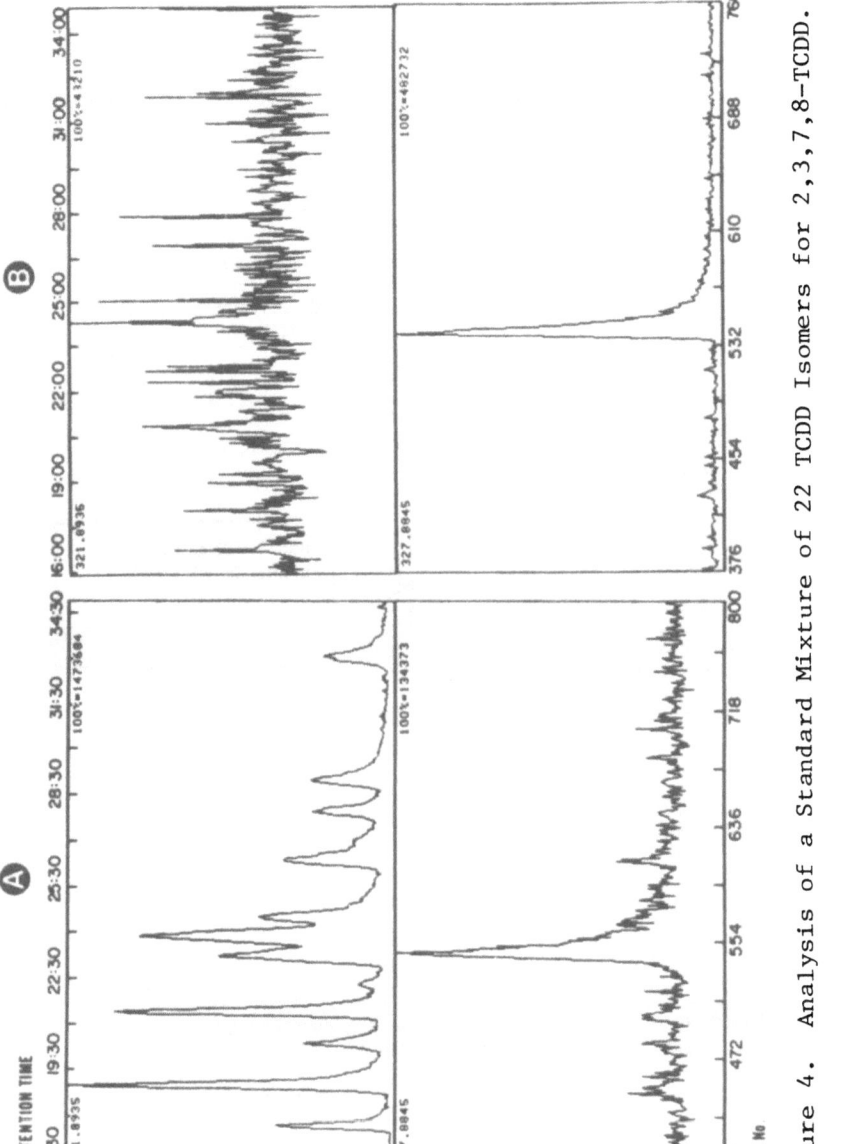

Figure 4. Analysis of a Standard Mixture of 22 TCDD Isomers for 2,3,7,8-TCDD.

(A) Single ion mass chromatograms of starting material: 21 TCDD isomers and ^{37}Cl labeled 2,3,7,8-TCDD. (B) Mass chromatograms after isomer-specific method (expanded scale m/e 322).

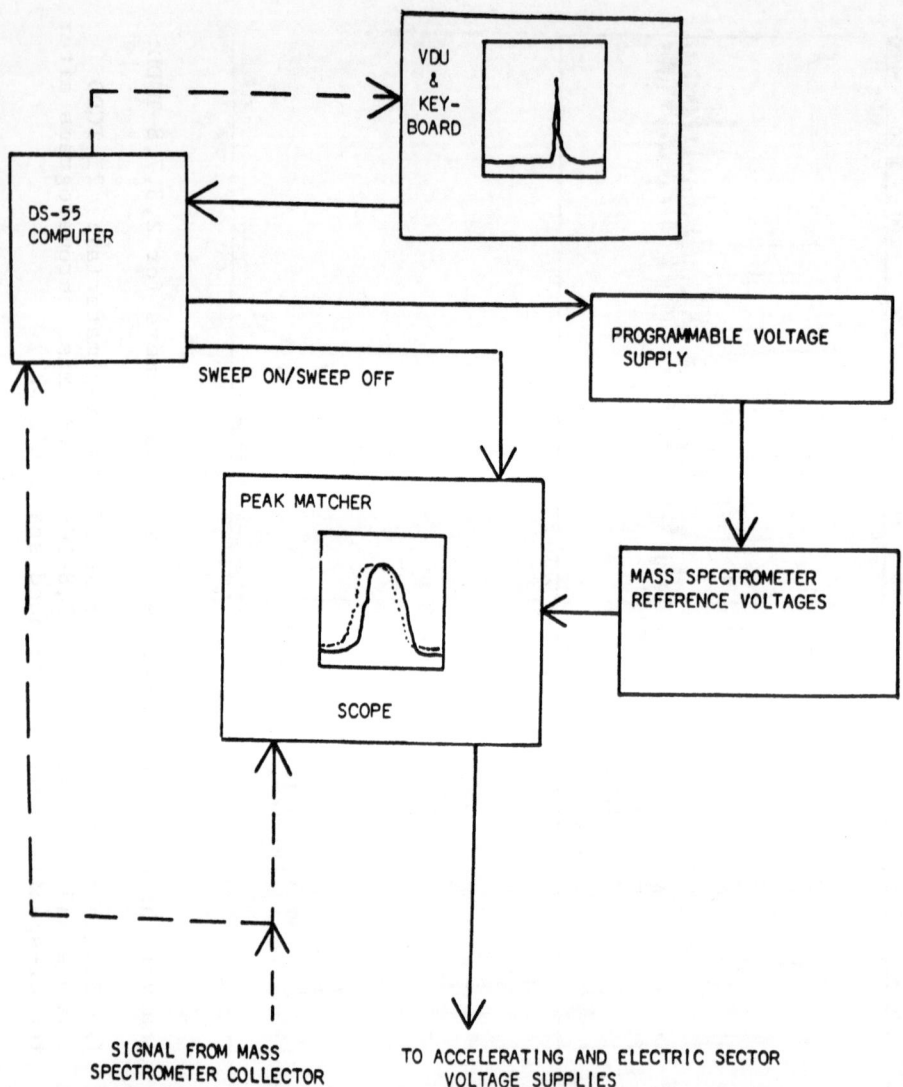

Figure 5.   Schematic of the Ion Monitoring System

‒ ‒ ‒ ‒ ‒ ‒ Data Signals
——————— Control Signals

averaged mass profiles. The averaging process results in a significant improvement in the signal-to-noise ratio which is important
since the mass profile areas are used for quantitative calculations
as described below. Each mass profile also enables us to determine
the presence of interfering compounds which occur within the narrow
mass region (generally 0.1 mass units). Data acquired in a HRMPM
analysis are illustrated in schematic outline in Figure 6.

   Experimental parameters are entered into the program through the
video display unit. The important parameters used with the current
software consist of the number of ions monitored, the exact mass of
each ion, the mass sweep width for each ion, scan time for each ion
(0.3, /1 or 3 seconds). Based on the exact mass of each ion entered
into the program, the computer calculates the value of a voltage sup-
plied to a programmable power supply which in turn generates the
reference voltages for the electric sector and the accelerating
voltage. The calculation is carried out using data previously acquired from an electric sector scan of perfluorokerosene.

ION MONITORING OF ENVIRONMENTAL SAMPLES

   In the analysis of extracts from environmental samples for
2,3,7,8-TCDD, m/z 319.8965, m/z 321.8936 and m/z 333.9338 or m/z
327.8847 are selected for ion monitoring. These ions respectively
represent the two most intense ions for 2,3,7,8-TCDD and the most
intense for one of the labeled internal standards (U-^{13}C)-2,3,7,8-TCDD
or (U-^{37}Cl)-2,3,7,8-TCDD. The data obtained from a typical 2,3,7,8-
TCDD determination can be illustrated by results shown in Figure 7
from the isomer-specific analysis of a small-mouth bass from Lake
Ontario.

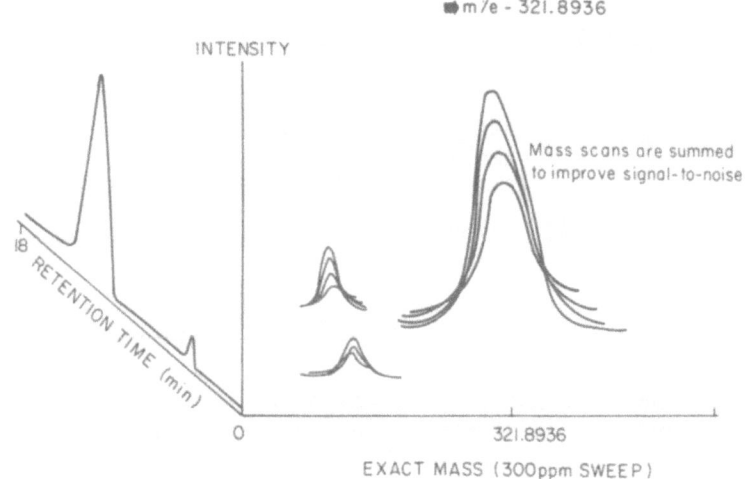

Figure 6. Typical Post-Acquisition Information

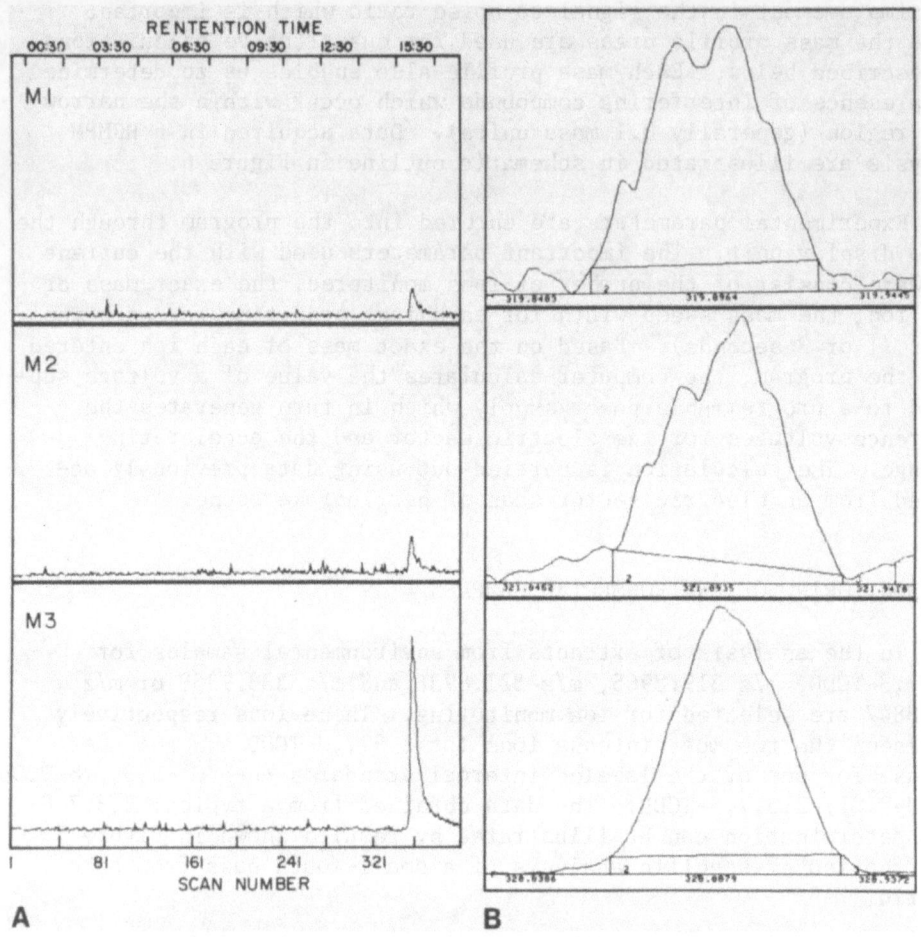

Figure 7.   5 PPT Small-Mouth Bass Sample

(A)   Exact mass chromatograms of
Lake Ontario small-mouth bass
sample.
(B)   Mass profiles of same sample.

    The retention time for (U-^{37}Cl)-2,3,7,8-TCDD as determined from
the mass chromatogram at m/z 327.8847 is used as a basis for selecting scans for the averaged mass prodiles in Figure 7B.   Each mass
profile has been divided into three areas.   Areas 1 and 3 are averaged to obtain the background noise.   Area 2 is then skimmed to delineate the signal from the appropriate TCDD ions.   This data for
Figure 7B, is shown in tabular form in Figure 8.

```
                    DOSS HIGH RESOLUTION MPM

                    PEAK SUMMATION REPORT


          RUNNAME TUDH10   DATE   4/21/81    TIME 16:37

          MASS            321.8935
          SCAN WIDTH       300 PPM
          SCAN TIME        0.3 SECS
          SCAN NUMBERS    355- 386
          STANDARD         0.0000
          FACTOR            0


 LAKE ONTARIO SMALL MOUTH BASS P0186 011120 2.5UL OF 5.2UL

 MASS      ITEM          AREA    BASELINE    BASELINE   %TOTAL   RELATIVE
 CENTROID                        SUBTRACTED  SKIMMED    AREA     TO STANDARD
 321.8972  TOTAL       753485.     YES        NO        76.09    0.00
 321.8616    1          55999.     YES        NO         5.65    0.00
 321.9004    2         583097.     YES        YES       58.88    0.00
 321.9351    3          11764.     YES        NO         1.19    0.00


                    DOSS HIGH RESOLUTION MPM

                    PEAK SUMMATION REPORT


          RUNNAME TUDH10   DATE   4 21 81    TIME 16:37

          MASS            328.8879
          SCAN WIDTH       300 PPM
          SCAN TIME        0.3 SECS
          SCAN NUMBERS    355- 386
          STANDARD         0.0000
          FACTOR            0


 LAKE ONTARIO SMALL MOUTH BASS P0186 011120 2.5UL OF 5.2UL

 MASS      ITEM          AREA    BASELINE    BASELINE   %TOTAL   RELATIVE
 CENTROID                        SUBTRACTED  SKIMMED    AREA     TO STANDARD
 328.8943  TOTAL      3571674.     YES        NO        92.44    0.00
 328.8518    1          32492.     YES        NO         0.58    0.00
 328.8950    2        3275570.     YES        YES       84.78    0.00
 328.9295    3          22503.     YES        NO         0.58    0.00
```

Figure 8.   Tabular Report of Mass Profile Data of 5 PPT
            Lake Ontario Small-Mouth Bass Sample

The value of this area along with the corresponding value for
the internal standard ion is used in the following ratio calculation
to determine the concentration of 2,3,7,8–TCDD in a sample:

$$C = (X/Y)RC_1$$

Where C = concentration of native TCDD in sample

C_1 = concentration of isotopically labeled internal
standard added prior to cleanup

R = response factor of labeled internal standard
relative to native TCDD at m/z 321.8936

X = area designated in the tabular report at m/z 321.8936

Y = area designated in the tabular report at the m/z of
the internal standard ions

This particular fish sample was calculated to contain 5 ppt TCDD with a limit of detection of less than 1 ppt using a 3 to 1 signal to noise criteria. Figure 9 shows results from a Lake Erie fish sample which was blank at the same detection limit.

High resolution ion monitoring mass spectrometry has one very distinct advantage over its low resolution counterpart, i.e., the capability of distinguishing compounds of the same nominal mass but differing in exact mass. Figure 10A is the exact mass chromatogram of a storm sewer sediment sample. This extract had been generated using HPLC along with other extraction and chromatographic procedures.

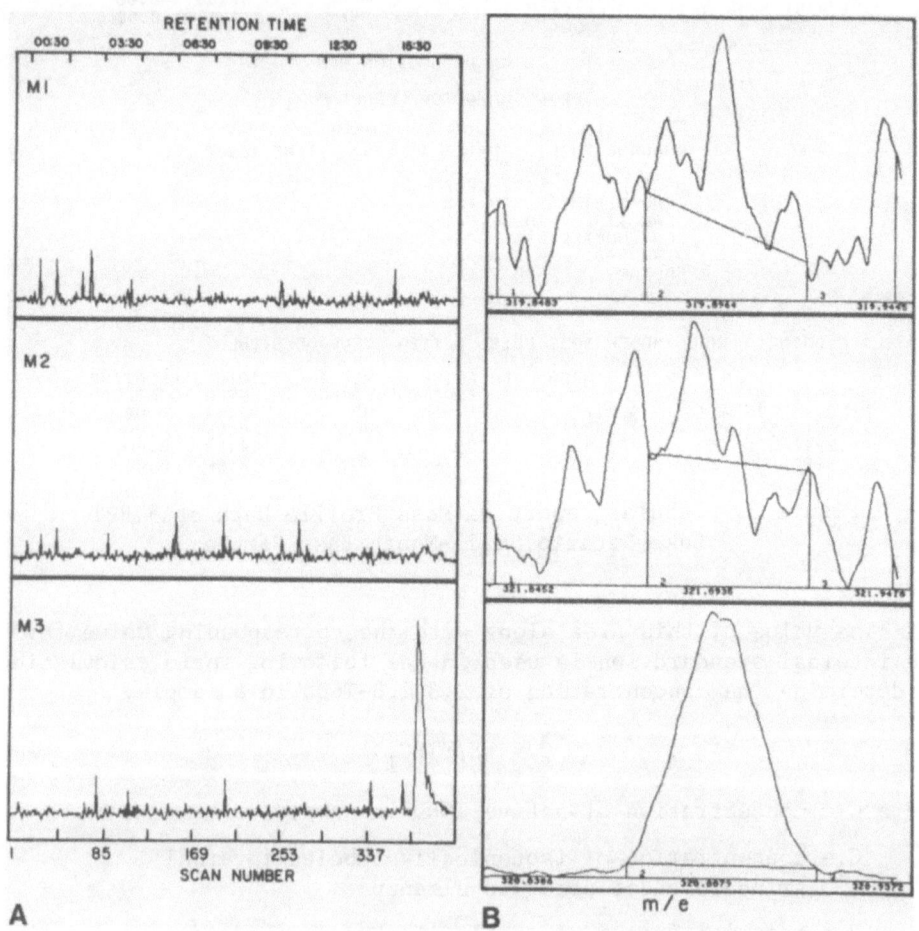

Figure 9.   Lake Erie Blank Fish Sample.

(A) Exact mass chromatogram of
a Lake Erie fish sample.
(B) Mass profile of same sample.

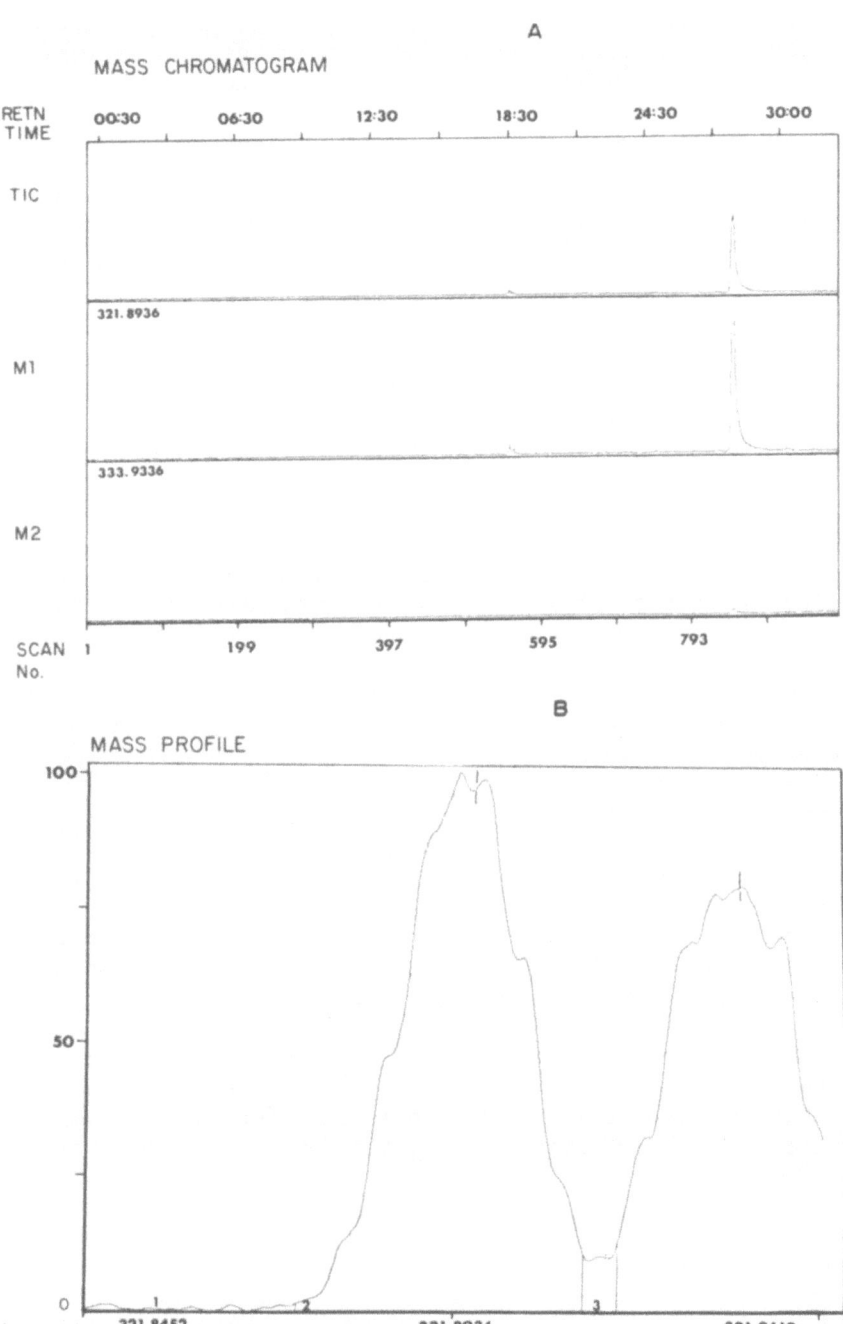

Figure 10. Storm Sewer Sediment Sample.

(A) Exact mass chromatogram of storm sewer sediment sample.
(B) Mass profile data of same; scans 829-883.

It had been expected to contain only two substances, 2,3,7,8-TCDD and added (U-^{13}C)-2,3,7,8-TCDD standard. However the mass data shown in Figure 10B confirms the presence of another substance in the extract which would have gone unnoticed and might have been quantitated as native TCDD if only low resolution ion monitoring data had been acquired.

FISH ANALYSIS

A study of concentrations of 2,3,7,8-TCDD in fish from the Great Lakes has been undertaken in order to determine whether industrial pollution has released significant amounts of this toxic isomer into the environment (O'Keefe et al, submitted, 1982). It is often necessary to determine ppt concentrations of 2,3,7,8-TCDD in the presence of large amounts of PCB-containing lipid. In order to determine the accuracy of analysis, samples were fortified with native 2,3,7,8-TCDD, cleaned up by a combination of alumina, Florisil and MgO/Celite columns and analyzed by the isomer-specific method. Samples containing 2.5 to 100 ppt, blanks, and control fish were analyzed. Results in the 5 to 100 ppt range were found to be within + 15 percent of the true value as shown in Table 2. TCDD could be detected at less than 5 ppt with reduced accuracy. The overall recovery is typically low due to the many separation and concentration steps; this makes the internal standard necessary for accurate quantitation. PCB interference has only been observed in fish samples which were analyzed by non-isomer-specific methods.

Fish collection stations where fish containing $\leq$ 3 ppt 2,3,7,8-TCDD were found are shown on the map in Figure 11. All sixty-four fish of various fish species collected in the Lake Ontario-Niagara River aquatic system and from Lake Huron between 1978 and 1980 were found to contain 2,3,7,8-TCDD in concentrations ranging from 2 to 162 ppt, while the 29 fish from Lake Superior, Lake Michigan and Lake Erie had levels usually below the detection limit.

Seventeen of these fish were also analyzed by several leading TCDD laboratories (Table 3) using different analytical methods. Although there are discrepancies in the results for certain samples, in general there is considerable agreement among the four different laboratories.

HAZARDOUS CHEMICAL WASTES

The analysis of complex chemical waste mixtures and soils or sediments for trace organics requires highly selective analytical techniques. The isomer-specific analytical method shown in Figure 1 has been used to determine 2,3,7,8-TCDD in sediment samples taken

Table 2. Analysis of Fortified and Blank
Fish for 2,3,7,8–TCDD

Sample (Replicates)	Fortification Level	Analytical[1] Result	Limit of[2] Detection	320/322 Ratio	Recovery[3] (%)
		TCDD Concentration (ppt ± s.d.)			
Procedural Blanks (5)	-	ND	2 ± 1.8		26 ± 20
Coho Salmon (Lake Erie)	-	1.6,1.3	0.1,0.4	0.74,0.56	29,36
American Shad (8) (Delaware River)	-	ND	2.8 ± 1.4	-	19 ± 11
Rainbow Trout (Adirondack Mtns.)	-	ND	3.0	-	20
Bluefish (Boston Bay)	-	ND	3.3	-	8
	2.5	ND	6.3	-	4
	2.5	3.1,4.6	0.5,4.4	0.63	42,39
Coho (4)	5.0	5.4 ± 0.5	1.1 ± 0.5	0.69 ± 0.07	41 ± 16
American Shad	10	9.4	2.1	0.87	25
Coho (3), American Shad	40	39 ± 3	1.2 ± 0.2	0.79 ± 0.02	31 ± 5
Coho	100	115	3.1	0.78	20

[1]Determined from an isotope ratio calculation (see Ion Monitoring). Replicates were separate portions of tissue homogenate from the sample.

[2]Based on 3 times the average of the noise area each side of the m/e 322 TCDD in the 300 ppm mass profile (see Figures 7 and 9).

[3]Determined by comparison of the [37]Cl-TCDD signal to the signal from an external [37]Cl-TCDD standard. Not used for concentration calculations.

from the storm sewer systems near the Love Canal chemical dumpsite, Niagra Falls, New York. In addition to conventional adsorbents (silica gel and Florisil) the clean-up involved the use of a graphitized carbon material which has been shown to have specific adsorption properties for planar molecules such as PCDDs and PCDFs. The analytical results are shown on the map of the storm sewers near the Canal (Figure 12).

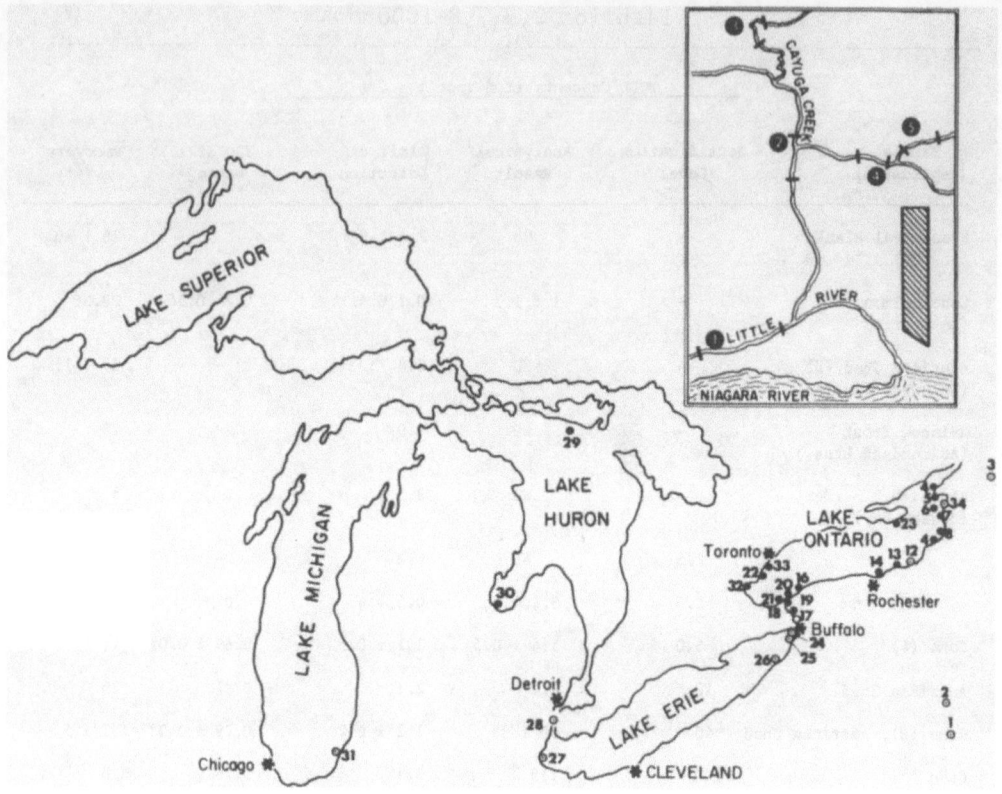

Figure 11.   Fish Collection Stations for 2,3,7,8-TCDD Survey of
             the Great Lakes.

             0 = Stations where concentrations of TCDD in fish
                 was $\geq$ 3 ppt.

             θ = Stations where concentrations of TCDD in fish
                 was < 3 ppt.

     Concentrations of up to 312 ppb 2,3,7,8-TCDD were found in the
storm sewer sediments close to the Canal (Smith et al, submitted,
1981).   The TCDD found is presumed to have come from trichlorophenol
wastes that were dumped in the Lover Canal.

COMBUSTION PRODUCTS

     An explosion of a transformer filled with PCBs and trichloro-
benzenes in the basement of a state office building in Binghamton,
New York, spread a soot-like material throughout the eighteen floors

Tabel 3.   Interlaboratory Comparison of TCDD Concentrations in Fish

		TCDD Level (ppt)	
Species	Lake	NYS - DOH	Other Laboratories[1]
Chinook Salmon	Ontario	39	64(A)
Coho Salmon	Ontario	20	29(B); 18(C)
Coho Salmon	Ontario	19	29(B); 95(C)
Coho Salmon	Ontario	19	31(A)
Lake Trout	Ontario	42	30(D)
Lake Trout	Ontario	41	36(D)
Brown Trout	Ontario	162	170(B); ND(C)
Brown Trout	Ontario	11,11	12(A)
Smallmouth Bass	Ontario	4.8	24(A)
Brown Bullhead	Ontario	3.8	3.6(A)
Lake Trout	Hudson	21,25	ND(D)
Carp	Hudson	22	22(A)
Carp	Hudson	29	61(A)
Catfish	Hudson	17	20(A)
Catfish	Hudson	22	40(A)
White Sucker	Hudson	2.5	2.1(A)
Perch	Hudson	ND	ND(A)

[1]A = Acid/Base Digestion, gel permeation HPLC, reverse-phase HPLC, capillary
    GC/low res. MS, No Internal Standard

 B = Ambient Acid Digestion, Silica, Chemically Treated Silica, Alumina
     reverse-phase HPLC, silica gel HPLC, GC/low res. MS

 C = Acid/Base Digestion, Adsorption Chromatography, GC/high res. MS

 D = As for B without silica gel HPLC but with capillary GC

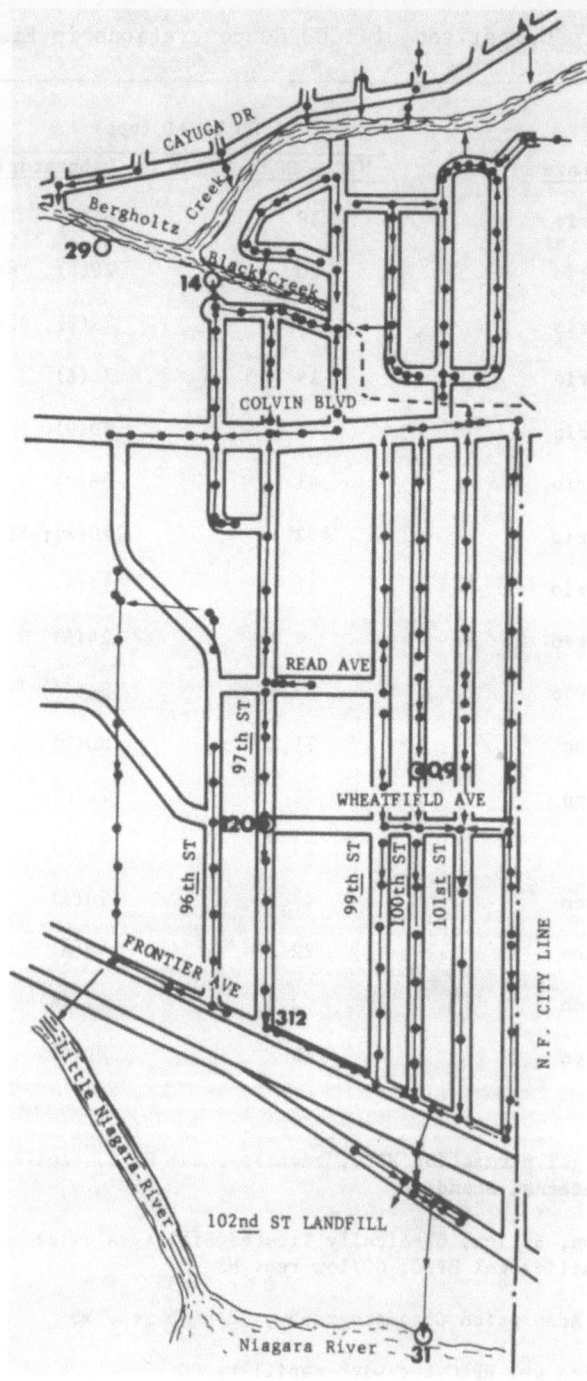

Figure 12. Map of the Love Canal Area Showing Concentrations of 2,3,7,8-TCDD Found in Storm Sewer Sediments Concentrations shown in ppb.

O = sampling site
O = manhole
——> = indicates direction of water flow.

of the building. In view of the literature reports demonstrating
that PCBs can be converted to PCDFs when heated to temperatures from
200 to 600°C (Buser et al, 1978; Buser and Rappe, 1979), the combustion products were investigated in our laboratory. Analysis of fortified fireplace soot samples (controls) for 2,3,7,8-TCDD and 2,3,7,8-
TCDF using a 16 hr benzene Soxhlet extraction and the isomer-specific
method produced the results given in Table 4.

Table 4. Analysis of Fortified Soot Samples
for 2,3,7,8-TCDD and 2,3,7,8-TCDF

	TCDD			TCDF		
	Concentration (ppm)		^{13}C TCDD	Concentration (ppm)		
Sample	Fortified	Found[1]	Recovery (%)	Fortified	Found[2]	Recovery (%)
1	8.4	9.1	5.6	15	20	17.7
2	8.4×10^{-2}	8.8×10^{-2}	4.0	15×10^{-2}	6.3×10^{-2}	5.5

[1] Internal standard method of calculation

[2] External standard method of calculation using 13% recovery (average value for
TCDD and TCDF, Tables 1 and 2)

     Quantitation of 2,3,7,8-TCDF is accomplished using an external
standard because an isotopically-labeled standard was unavailable.
Therefore, results for 2,3,7,8-TCDF are expected to be less accurate
than those for TCDD. The results of a duplicate analysis of a black
soot sample taken from the 7th floor of the building just after the
fire are shown in Table 5.

     The average concentrations of 2,3,7,8-TCDF and 2,3,7,8-TCDD were
found to be about 200 ppm and 2.9 ppm, respectively. Capillary
GC/HRMS analyses of extracts from one of these samples were carried
out by another government laboratory which reported that the sample
contained 89 percent of the 2,3,7,8-TCDD and 59 percent of the
2,3,7,8-TCDF reported by our laboratory. Additional analysis of the
sample were obtained by two laboratories using, respectively, capillary GC/electron impact low resolution mass spectrometry (EILRMS)
and capillary GC/negative chemical ionization low resolution mass
spectrometry (NCILRMS). The extract was prepared by a mild extraction
technique (room temperature equilibration with toluene) and then
cleaned up by a carbon column method analogous to that described above.
By capillary GC/EILRMS the sample was indicated to contain 12 ppm

Table 5.   Analysis of Transformer Fire Soot Samples
for 2,3,7,8-TCDD and 2,3,7,8-TCDF

| | Analytical Result (ppm) | | Recovery(%) | Molecular Ion Isotope Ratios[1] | | |
| | | | | TCDD | TCDF | |
Sample	TCDD	TCDF	[13]C TCDD	320/322	304/306	308/306
Binghampton Soot 1	2.8	270	26.8	0.73	0.77	0.47
Binghampton Soot 2	2.9	120	16	0.85	0.79	0.43
Solvent Blank 1	ND($<1.2 \times 10^{-2}$)[2]	ND($<8.5 \times 10^{-2}$)[2]	13.6	-	-	-
Fireplace Soot	2.7×10^{-3}	7.2×10^{-3}	-[3]	1.2	0.78	0.41

[1]   Theoretical isotope ratios: 320/322 = 0.77, 304/306 = 0.77, 308/306 = 0.49

[2]   Based on a theoretical sample weight of 3.5 mgs.

[3]   Not calculated since these were an error in sample trapping from one HPLC step.

TCDF and 0.6 ppm TCDD together with a considerable number of other PCDDs and PCDFs. Due to sensitivity limitations, PCDDs were not detected in the capillary GC/NCILRMS analysis with which the 2,3,7,8-TCDF level was reported to be 3.7 ppm. Since these comparative results were obtained with two separate extracts from only one sample it is not possible at this time to explain the differences between the HRMS and the LRMS results.

A high-volume air sample taken on the 7th floor and a homogenized soot sample currently being used in animal test studies were analyzed for total TCDF and TCDD using the non-isomer-specific method indicated in Figure 1. Both samples contained a mixture of TCDFs as shown in Figure 13. The concentrations of TCDFs and TCDDs were found to be slightly lower in the air particulates than in the homogenized soot.

The data indicate that the TCDF and TCDD concentrations found in these samples may be a result of dilution of the soot material with normal "dust" particles.

Work is presently in progress to quantitate the higher chlorinated PCDFs and PCDDs.

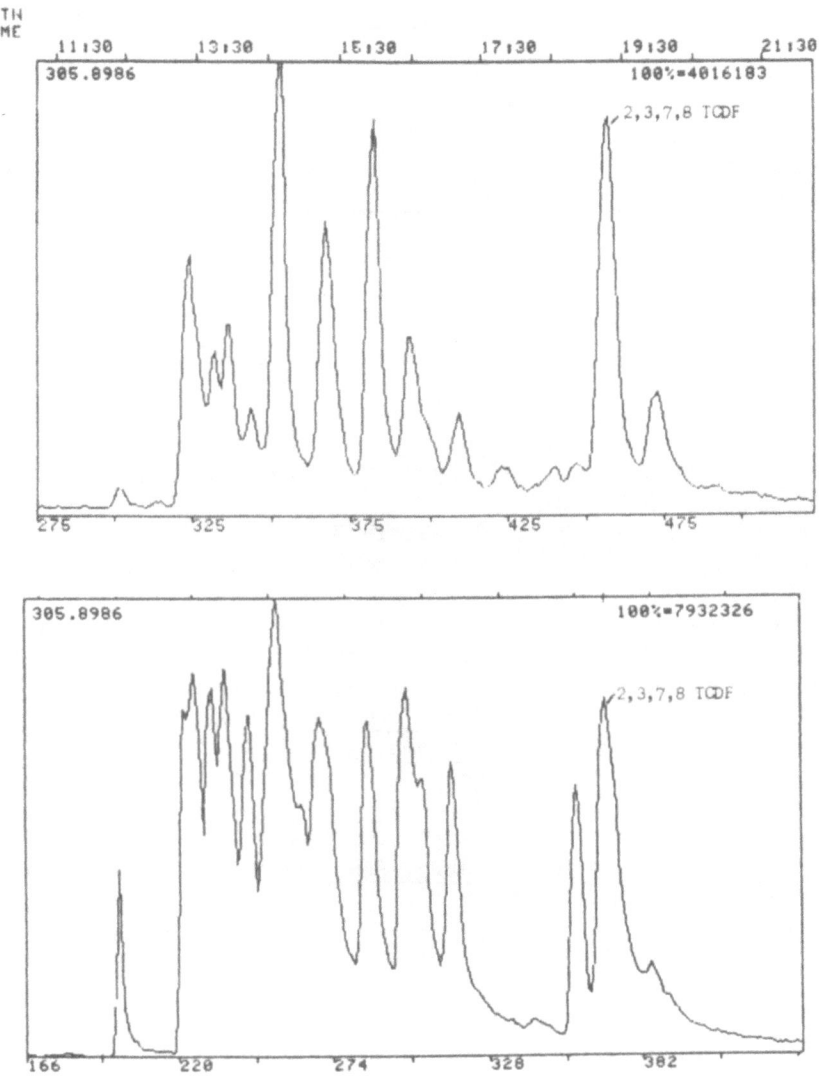

Figure 13.   Single Ion High Resolution Mass Chromatograms of
             Samples from a Contaminated State Office Building.

             (Top) High volume air particulates
             (Bottom) Homogenized soot.

Table 6. Concentrations of TCDFs and TCDDs in Samples from the
          Binghamton State Office Building (ppm)

	TCDF		TCDD	
	Total	2,3,7,8	Total	2,3,7,8
Soot	597	48	1.8	1.2
Air Particulates	111	21	0.5	0.3

REFERENCES

Buser, H.R., Bosshart, H.P. and Rappe, C., Chemosphere 7: 109–119, 1978.
Buser, H.R. and Rappe, C., Anal. Chem., 52:2259, 1980.
Buser, H.R. and Rappe, C., Chemosphere 8:157–174, 1979.
Chapman, J.R., Warburton, G.A., Ryan, P.A. and Maselby, D. Bio. Mass. Spect., 7:597–599, 1980.
Kirkland, J.J. Chromatogr., 83:149, 1973.
Nestrick, T.J., Lamparski, L.L. and Stehl. Anal. Chem., 51: 2273, 1979.
O'Keefe, P., Meyer, C., Hilker, D., Aldous, K., Jelus-Tyror, B., Dillon, K. and Donnelly, R. Submitted to Anal. Chem., 1982.
O'Keefe, P., Smith, R., Meyer, C., Hilker, D., Aldous, K. and Jelus-Tyror, B. Submitted to J. Chrom., 1981.
Schomburg, G. and Husmann, H. Chromatographia, 8:517, 1975.
Smith, R.M., O'Keefe, P.W., Aldous, K.M., Hilker, D.R. and O'Brien, J.E. Manuscript submitted to Env. Sci. and Tech., 1981.
Stalling, D.L., Smith, L.M., Petty, J.D. In McKinney ed., Environmental Health Chemistry, Ann Arbor: Ann Arbor Science Publishers, p. 177, 1980.

METHODOLOGY AND PRELIMINARY RESULTS FOR THE ISOMER-SPECIFIC
DETERMINATION OF TCDDs AND HIGHER CHLORINATED DIBENZO-P-DIOXINS
IN CHIMNEY PARTICULATES FROM WOOD-FUELED DOMESTIC FURNACES
LOCATED IN EASTERN, CENTRAL, AND WESTERN REGIONS OF THE
UNITED STATES

T.J. Nestrick, L.L. Lamparski, L.A. Shadoff, and
T.L. Peters

The Dow Chemical Company
Michigan Division Analytical Laboratories
Midland, Michigan USA

ABSTRACT

The Dow Chemical Company has recently conducted a survey of
domestic wood-fueled furnaces located in three different regions of
the US for chlorinated dibenzo-p-dioxin (CDD) emissions. This study
was undertaken to determine if CDDs are produced in significant quantities during the common combustion of a naturally occurring fuel.
Particulate matter was collected from the chimneys of domestic wood
furnaces that were rurally located and were fueled with natural wood
obtained from areas where contamination from pesticides and herbicides
was expected to be minimal. The development and application of appropriate analytical methodology have provided preliminary evidence
that significant concentrations of CDDs are produced under these
circumstances.

INTRODUCTION

    In 1978 The Dow Chemical Company published the results of a study
entitled "The Trace Chemistries of Fire - A Source of and Routes for
the Entry of Chlorinated Dioxins into the Environment" (Chlorinated
Dioxin Task Force 1978). Based upon data from this investigation, a
hypothesis was proposed that CDDs are ubiquitous in the environment
as the result of their being byproducts of common combustion and
therefore have been present since the advent of fire. Since 1978,
numerous scientific studies have been conducted to evaluate this
hypothesis, many of which are cited in a recent review (Lustenhouwer
et al, 1980). Although many of these studies demonstrate the presence

95

of CDDs in effluent from a variety of modern combustion sources,
little is known about the mechanism of formation in "real world"
fires. Consequently, most of this inforamtion neither contradicts
the hypothesis of the Trace Chemistries of Fire (TCOF) nor for that
matter, directly supports it. Similarly, the survey analytical data
published by Dow in 1978, upon which the TCOF hypothesis was based,
involved such combustion-related sources as industrial incinerators,
municipal incinerators, automobile mufflers, diesel mufflers,
cigarette smoke, and charcoal-broiled meat. Taking advantage of
hindsight, we have to admit none of these combustion sources truly
demonstrate situations representative of "common combustion" on a
time-scale which includes the "advent of fire." For this reason we
have undertaken a more fundamental study involving the combustion of
naturally occurring wood.

        The selection of wood as a combustion fuel for this survey study
of CDD emissions was based upon several considerations. Although the
TCOF predicts that CDDs are ubiquitous in the environment, it does
not define expected concentrations for these species. In fact, data
presented in the original 1978 TCOF publication indicated that CDD
concentrations present in effluents collected in the immediate vicinity of relatively large, modern combustion sources were very often in
the part-per-billion (ppb) to part-per-trillion (ppt) range. Because
low ppt concentrations of CDDs in particulate combustion effluents
are relatively close to the current limit of analytical capability
for their detection, it is unreasonable to expect that they can be
directly observed in the general environment following the extreme
dilution which occurs as they are distributed away from the source.
Therefore we have chosen to investigate whether the combustion of one
of the world's most common renewable fuel sources, wood, produces
significant levels of CDDs when examined within the combustion source.
In view of the magnitude of controlled and uncontrolled wood combustion in the world today, if CDDs are emitted by this process their
ubiquity in the environment can no longer be ignored, whether or not
they can be observed by direct analyses. In addition, the detection
of CDDs from the burning of wood would provide better support for the
remaining portions of the TCOF hypothesis than from other fuel sources
previously examined. First, wood represents a fuel of completely
natural composition. When it is obtained and burned in rural locations expected to have minimum contact with herbicides and/or pesticides, its combustion products should not be influenced by contamination from man-made chemical precursors cited by others (Rappe et al,
1979, Lustenhouwer et al, 1980) as being required for the production
of CDDs in combustion processes. Secondly, since wood is a constantly
renewed fuel source of rather ancient origin on earth, if CDDs can be
detected from its common combustion today then it becomes reasonable
to assume that CDDs have been present in the environment from at
least as far back as the advent of fire.

EXPERIMENTAL

Reagents, Solvents, Adsorbents, and Dioxin Standards

A majority of these materials have been previously described
(Lamparski and Nestrick, 1980).

Apparatus

The high performance liquid chromatography (HPLC) systems employed for CDD residue fractionation have been described (Lamparski
and Nestrick, 1980). Modification of the reverse phase-HPLC (RP-
HPLC) to substitute a 9.4 x 250mm Zorbax® ODS front column and an
increase in eluent flowrate to 3.3mL/min should be noted. These
changes permit the injection of CDDs in as much as ∿55µL of chloroform without significant loss of performance.

High resolution gas chromatography-low resolution mass spectrometry (HRGC-LRMS) used for CDD identification and quantitation was
accomplished on a significantly modified Kratos MS-80 equipped with
a DS-55 data system operating in the multiple peak monitoring (MPM)
mode: column, 0.32mm I.D. x 30m -J&W Fused Silica DB-5 (0.25µm film
thickness): mass resolution, 1000. TCDD analysis conditions:
isothermal column temperature, 210°C; ions monitored, native TCDDs
at m/z 319.896 and 321.893, and ^{13}C-2,3,7,8-TCDD internal standard at
m/z 335.930. Hexachlorodibenzo-p-dioxin (HCDD) analysis conditions:
isothermal column temperature, 240°C; native HCDDs at m/z 387.818
and 389.815, and ^{13}C-1,2,3,4,7,8-HCDD internal standard at m/z
397.847. Heptachlorodibenzo-p-dioxin (H7CDD) analysis conditions:
isothermal column temperature, 280°C; native H7CDDs at m/z 423.776
and 425.773. Octachlorodibenzo-p-dioxin (OCDD) analysis conditions:
isothermal column temperature, 280°C; native OCDD at m/z 457.737 and
459.734, and ^{13}C-OCDD internal standard at m/z 469.768.

HRGC-high resolution mass spectrometry (HRGC-HRMS) used for TCDD
confirmation was accomplished on a Kratos MS-50 equipped with a DS-55
data system operating in the MPM mode: column, 0.32mm ID x 60m -
J&W Fused Silica SE-54 (0.25µm film thickness); mass resolution,
9000. TCDD analysis conditions: programmed column temperature,
95°C (1 min) then ballistic to 240°C (hold); TCDD ions monitored same
as HRGC-LRMS. Trichlorodibenzofuran (T3CDF) and tetrachlorodibenzo-
furan (TCDF) analysis conditions: programmed column temperature,
175°C to 255°C at 1.5°C/min; mass resolution, ∿11000; native T3CDFs,
m/z 269.940; native TCDFs, m/z 305.898.

HRGC-LRMS used for chlorinated dibenzofuran (CDFs) identification and quantitative estimation (TCDFs only) was accomplished on a
Hewlett-Packard 5985B GC-MS-DS operating in the selected ion monitoring (SIM) mode: column, 0.32mm ID x 30m - Fused Silica DC-410
(0.80µm film thickness); mass resolution, unit. CDF analysis

conditions: programmed column temperature, 100°C to 280°C at 5°C/min;
monochlorodibenzofurans (MCDFs) at m/z 202 and 139; dichlorodibenzo-
furans (DCDFs) at m/z 236 and 173; trichlorodibenzofurans (T3CDFs)
at m/z 270 and 135 (pentachlorodiphenylethers at m/z 340); tetrachlorodibenzofurans (TCDFs) at m/z 304 and 241 (hexachlorodiphenyl-
ethers at m/z 374); pentachlorodibenzofurans (PCDFs) at m/z 338 and
275 (heptachlorodiphenylethers at m/z 408).

Experimental Combustion Effluent Particulate Samples

All combustion effluent particulate samples were obtained from
domestic heating systems. Particulate samples ranging between ∿25
and 65g were transferred, using a clean kitchen spoon obtained at
each site, from the source to thoroughly cleaned 4-oz. glass bottles
equipped with Poly-Seal® caps and Teflon® tape-wrapped threads.
Preliminary evaluation samples were obtained from the flues of an
exclusive oil heating system and an exclusive wood heating system
located in Cape Cod, MA, as well as from a primary wood heating system
located ∿15 miles west of Midland, MI. For the actual survey experiment, six individual samples of chimney particulates were collected
from each of three rural regions of the US. The eastern region samples (Eastern) were obtained from residents in the vicinity of
Farmington, NH, an agricultural community. Wood combustion was the
primary heat source for each residence, and in all cases, modern
firebrick and metal furnaces vented through uninsulated metal flues
to masonry chimneys were used. All Eastern samples were particulate
matter collected from the base of the chimney. Four central region
samples (Central) were collected from residences ∿20 miles north of
Grand Marais, MN, an area characterized by tourism and logging. Although wood combustion in modern firebrick and metal furnaces was the
primary heat source, two different types of chimneys were encountered.
One sample came from a system having a masonry chimney, and therefore
it was collected from the chimney base. The others came from systems
in which an uninsulated flue pipe connected directly to the top of
the furnace was used as a chimney. Particulate effluent samples were
collected from the spark-arrester baffle at the top of the chimney
for these cases. The remaining two Central samples were collected
from residences having systems with masonry chimneys similar to those
of the Eastern region; samples were taken at the base. One of these
residences was located ∿30 miles northeast of Duluth, MN, on the site
of an old farm. The other was located in a rural area near Garden,
MI, in the upper peninsula ∿25 miles east of Escanaba. The western
region samples (Western) were collected from residences in a rural
area near Creswell, OR. In all cases, primary wood heating was
delivered via modern firebrick and metal furnaces connected directly
to metal flue chimneys. Particulate effluent samples were taken from
the spark-arrester baffle located on the top of the chimney.

Sample Preparation

Because this study is preliminary in nature and not yet complete, a full disclosure of methodology will be presented in a future publication. The procedure employed is fundamentally an adaptation of our published technique for determining ppt concentrations of CDDs in particulate materials (Lamparski and Nestrick, 1980).

Quantitative Nomenclature

The ^{13}C-enriched CDD internal standards are incorporated into every sample and reagent blank processed, for quality control purposes only. Observed native CDD concentrations have not been corrected for recoveries. Reagent blanks are routinely examined; however, for purposes of brevity these data will not be reported here. In all cases associated with reported data, the reagent blank yielded a value of not detected (ND) or was insignificant (<~5 percent) relative to observed species; only OCDD was occasionally observed in reagent blanks.

The limit of detection (LoD) for all species is defined as 2.5 times peak-to-valley moise in an adjacent region of the mass chromatogram. The nomenclature for observed concentrations is as follows: species having concentrations ranging from ND to 10 x LoD will be followed by a number in parentheses which is the LoD for that determination, and species observed at concentrations greater than 10 x LoD will be reported without presenting the LoD.

DISCUSSION

During the developmental stages of this project, several important considerations resulted in our choice of residential wood combustion as a source of effluent material for studying CDDs. Foremost of these involved the rather extreme range of combustion conditions associated with reasonably large pieces of burning natural wood; it was necessary to include the time period over which reactions can occur, often in excess of ten hours. Although samples obtained from uncontrolled open-burning situations (i.e., forest fires) were anticipated to provide the best overall data with respect to environmentally significant CDD emissions, problems associated with obtaining a representative sample of all stages of the combustion process, determination of exactly what was being burned, and basic physical danger precluded initial use of this type of experiment. Limited open-burning experiments were also deemed unsatisfactory since the combustion of only a few kilograms of wood would necessitate hours of continuous sampling, would require multiple analyses to establish the total emissions picture, and ultimately would represent only a single fire which consumed a relatively small amount of fuel. Because of these problems, it was decided that wood-fueled stoves

represented the most reasonable experimental compromise for an initial
evaluation of the possible significance of CDD emissions arising from
an appropriate location within this source-type, it was anticipated
that CDD concentrations would be representative of the time-averaged
combustion of several tons of natural wood that would have occured
under a variety of different combustion conditions including those
associated with open-burning. Since wood-type, furnace construction
and operator peculiarities were expected variables for these systems,
a minimum of six different samples collected from three distant geographic locations were considered necessary to provide normalization
and representative CDD concentration ranges for "typical" wood combustion. In order to provide additional assurance that our data
would not be biased, several precautions were taken during sample
selection and collection. All samples were collected from rural areas
where locally obtained wood would be expected to have minimal exposure
to pesticides and/or herbicides, and where there would be no nearby
large industrial or municipal incinerators. Only wood-fueled systems
used as a primary heat source were selected. No samples were obtained
from systems where any type of treated or processed wood was burned.
Also, no samples were obtained where, to the best of the operators'
knowledge, either the fuel or the residence had ever been exposed to
2,4-D, 2,4,5-T, or pentachlorophenol.

Prior to collection of the regional samples, two preliminary
experiments were conducted. The first of these was designed to demonstrate that auxiliary gas or oil heating equipment would not significantly influence CDD levels when in the vicinity of, or directly
attached to, a primary heat source fueled with wood. For this purpose
two different flue particulate samples were collected from residences
located near Cape Cod, MA. One was taken from an exclusive oil-burn-
ing system and the other from an exclusive wood-burning system.
Analysis of these materials provided the observed CDD concentration
data given in Table 1. As indicated, the exclusive wood-burning
system yielded easily detectable quantities of tetra- through octa-
chlorinated dioxins, whereas the oil-fueled system produced non-
detectable concentrations of TCDDs and HCDDs (LoD 1 ppt and 3 ppt,
respectively), and reduced amounts of H7CDDs and OCDD relative to
the wood system. Based upon these data, it is anticipated that such
auxiliary heating equipment should not significantly contribute to
CDD levels observed when wood is the primary fuel consumed.

The second preliminary experiment was designed to indicate the
optimum sampling location with respect to maximizing CDD concentrations in particulate matter within a "typical" wood-fueled furnace.
For this purpose a heating system (owned and operated by one of the
authors) that is located ∿15 miles west of Midland, MI was selected.
As shown in Figure 1, the component construction of this firebrick
and metal furnace, metal heat exchanger and uninsulated flue, and
masonry chimney are reasonable typical of modern equipment. Samples
of particulate material were collected from the base of the chimney

Table 1.   CDD Concentrations Observed in Flue Particulates from
           Exclusive Oil- and Wood-Fueled Systems from Cape Cod, MA.

	Concentration in ppt (pg/g)		
	Reagent Blank	Oil	Wood
Total TCDDs	ND (1)	ND (1)	170
Total HCDDs	ND (2)	ND (3)	260
Total H7CDDs	ND (2)	18 (2)	330
OCDD	ND (11)	92 (14)	210

(Chimney), the rear portion of the flue after the draft control (Flue-
back),. the front portion of the flue after the heat exchanger (Flue-
front), and from the ash pit (Ash).   Analysis of these samples provided the isomer-specific TCDD data and the total concentrations for
higher chlorinated dioxins shown in Table 2.   These data establish a
well defined concentration profile within the source, which indicates
that chimney particlates would be expected to contain the highest
concentrations fo the denoted CDDs.   Although it is tempting to draw
certain mechanistic and physical property conclusions from this information, we will limit ourselves to one which appears to have some
bearing on data to be presented for certain Central and all Western
region chimney particulate samples.   Based on the data in Table 2,
it is reasonable to conclude that wide variations in particulate
matter CDD content can be observed within a given heating system, and

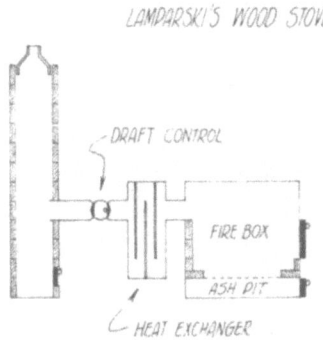

Figure 1.   Furnace Design Used for the Preliminary Experimental
            Determination of Optimum Sampling Point Within a
            Modern Residential Wood Combustion Unit.   Source
            Location ∿15 Miles West of Midland, Michigan.

Table 2. CDD Concentration Profile for Midland, Michigan,
          Test Furnace.

CDD Isomers	Concentration in ppt (pg.g)			
	Chimney	Flue-back	Flue-front	Ash Pit
^{13}C-2378 Recovery	82%	63%	59%	57%
2378-TCDD	26	ND (0.3)	0.3 (0.2)	ND (0.3)
1469-TCDD	5.0 (1.2)	ND (0.3)	ND (0.3)	ND (0.4)
1269-TCDD	20	ND (0.3)	ND (0.3)	ND (0.4)
1267-TCDD	14	ND (0.3)	ND (0.3)	ND (0.4)
1289-TCDD	12	ND (0.3)	ND (0.3)	ND (0.4)
1369-TCDD	33	ND (0.3)	ND (0.3)	ND (0.4)
1247+1248TCDDs	60	0.2 (0.2)	ND (0.3)	ND (0.4)
1278-TCDD	37	ND (0.2)	ND (0.4)	ND (0.4)
1268-TCDD	35	Nd (0.3)	ND (0.4)	
1237+1238-TCDDs	50	0.3 (0.2)	ND (0.3)	ND (0.4)
1279-TCDD	45	ND (0.3)	ND (0.3)	ND (0.4)
1246-TCDD	28	ND (0.2)	ND (0.4)	ND (0.4)
1478--TCDD	14 (1.9)	ND (0.3)	ND (0.3)	ND (0.4)
1236-TCDD	20	ND (0.2)	ND (0.4)	ND (0.4)
1239-TCDD	24	ND (0.2)	ND (0.4)	ND (0.4)
1249-TCDD	10 (3.0)	ND (0.3)	ND (0.3)	ND (0.4)
1368-TCDD	145	2.8 (0.4)	0.8 (0.3)	0.9 (0.4)
1379-TCDD	100	1.4 (0.4)	0.6 (0.3)	ND (0.4)
1378-TCDD	84	ND (0.4)	ND (0.3)	ND (0.4)
1234-TCDD	15	ND (0.4)	ND (0.3)	ND (0.4)
Total TCDDs	777	4.7	1.7	0.9
^{13}C-Hcdd Recovery	90%	90%	74%	81%
Total HCDDs	3100	15 (1.8)	12 (1.5)	ND (1.6)
Total H7CDDs	7200	68	57	22
^{13}C-OCDD Recovery	88%	61%	66%	64%
OCDD	10600	150	160	77

1,2,4,6-TCDD : chromatographic characteristics permit separation,
but absolute identity not established, may be 1,2,4,9-TCDD.

therefore differences in component design, which impose constraints
upon the sample collection point, may significantly influence apparent CDD emission levels reported. With respect to this point, it is
important to note that three Central and all Western samples were
obtained from the top of metal flue chimneys whereas all Eastern and

three Central samples came from the base of masonry chimneys. Other
characteristics for the regional chimney particulate samples are
given in Table 3.

Table 3.   General Characteristics for Regional
Chimney Particulate Samples.

Eastern Region

Combustion Unit: modern construction, firebrick and metal
Chimney: masonry
Sampling Point: chimney base
Primary Wood Fuel: Red Oak
Typical Operation Mode: high air restriction

Central Region

Combustion Unit: modern construction, firebrick and metal
Chimney: 3 masonry, 3 metal (non-insulated)
Sampling Point: masonry = chimney base, metal = chimney top
Primary Wood Fuel: Birch
Typical Operation Mode: low air restriction

Western Region

Combustion Unit: modern construction, firebrick and metal
Chimney: metal (insulated)
Sampling Point: chimney top
Primary Wood Fuel: Fir
Typical Operation Mode: high air restriction

Following completion of the preliminary experiments and collection of the regional samples we discovered that our publisher procedure (Lamparski and Nestrick, 1980), which had been used successfully
thus far, could not provide adequate CDD detectio capabilities for
many of the particulates. Subsequent modifications alleviated this
problem. The degree of success achieved is exemplified by the HRGC-
LRMS isomer-specific TCDD mass chromatograms for Western sample #10
shown in Figure 2. Each of the TCDDs identified in this sample have
been confirmed by HRGC-HRMS at a mass resolution of ∿9000. Note that
each of the TCDD mass chromatograms monitored at m/z 322 (mass resolution ∿1000) shown in Figure 2 are completely free of extraneous
peaks other than the expected TCDD isomers, even though their concentrations are very close to the LoD. This is also true for m/z
320 and 336. For 9 out of 10 samples examined to date, we have found

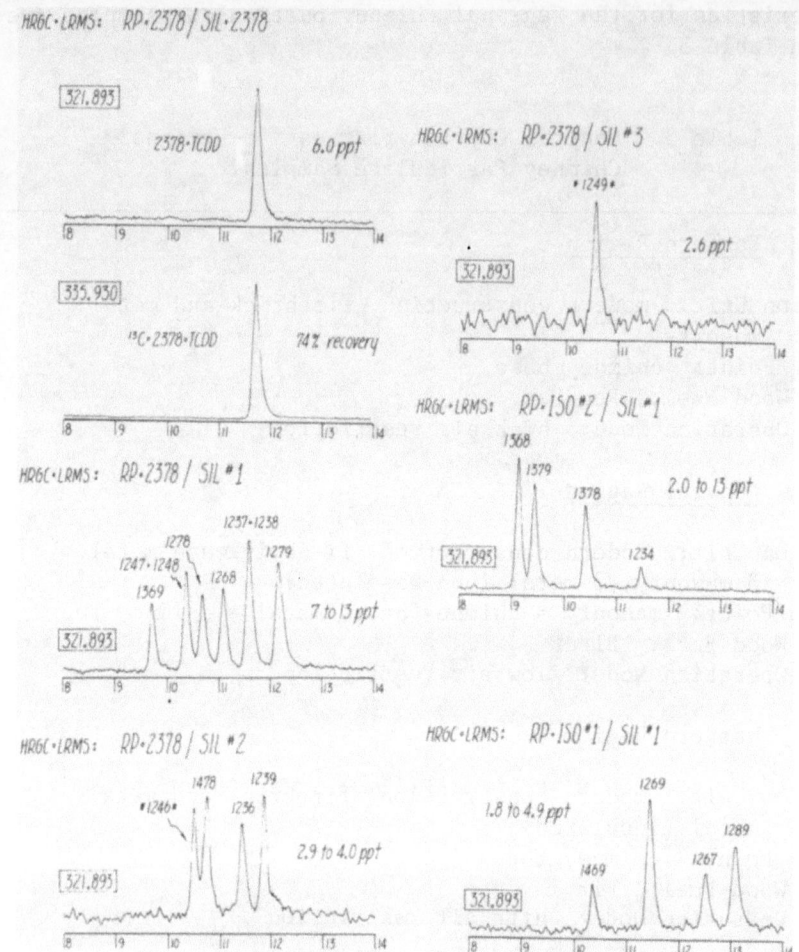

Figure 2.   HRGC-LRMS isomer-specific TCDD mass chromatograms for
            Western #10 chimney particulates (each run is equiv-
            alent to 2g of particulates injected).

that the LoD for TCDDs is totally dependent upon instrumental noise
when a 10g particulate sample is subjected to our clean-up procedures,
which incidentally also is true for higher chlorinated dioxins.

     To date, we have completed the analysis of 10 of the 18 regional
samples collected. Because this survey study is therefore incomplete,
only a portion of the results obtained are reported here. In order
to simplify comparison of these data, total observed concentrations
for TCDDs, HCDDs, H7CDDs, and OCDD are given for all samples that
have been analyzed in Figures 3 through 6, respectively. Sample

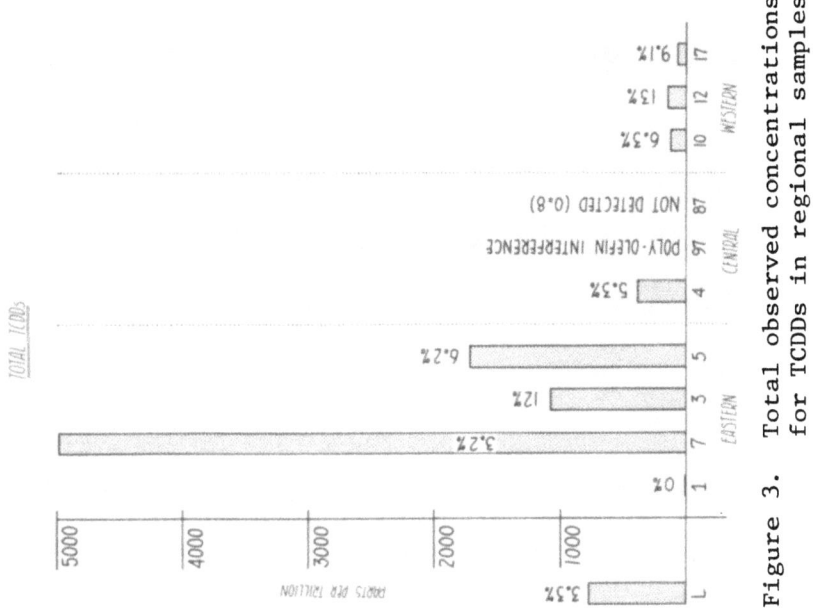

Figure 3.   Total observed concentrations
for TCDDs in regional samples

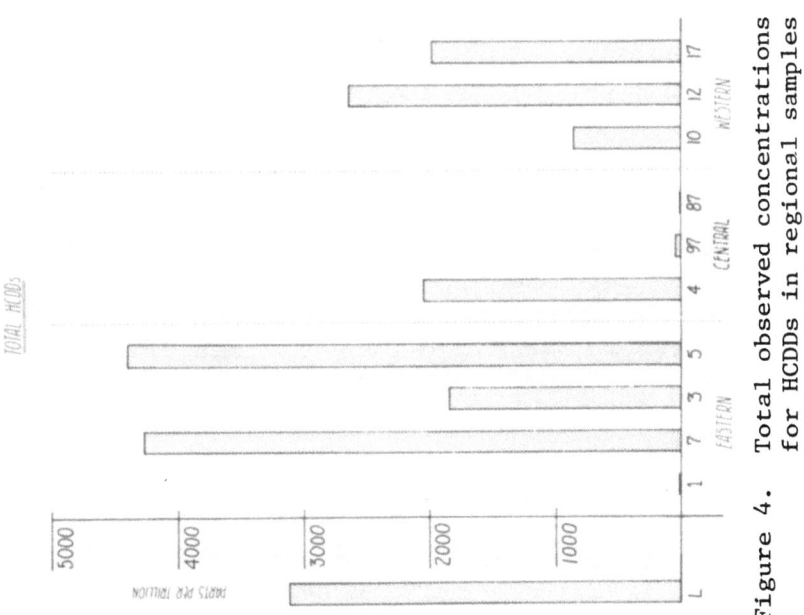

Figure 4.   Total observed concentrations
for HCDDs in regional samples

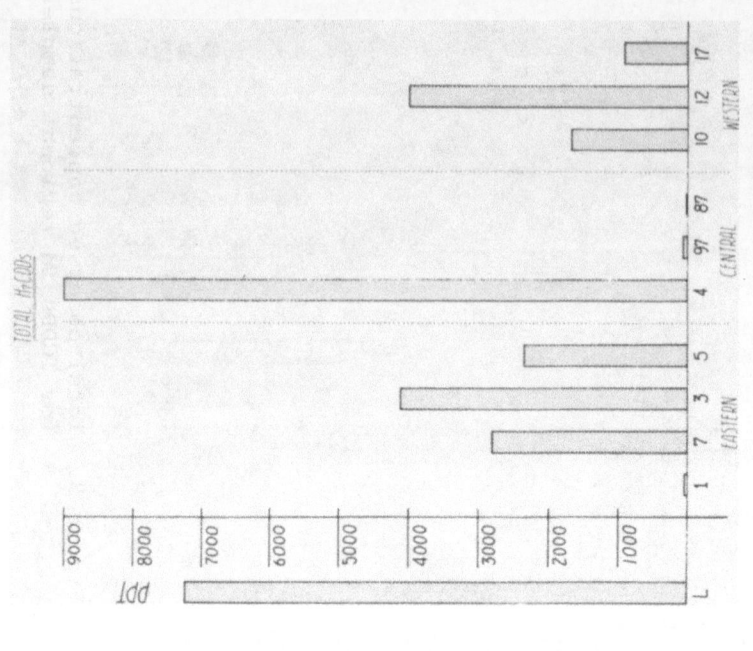

Figure 5.   Total observed concentrations
for H7CDDs in regional samples

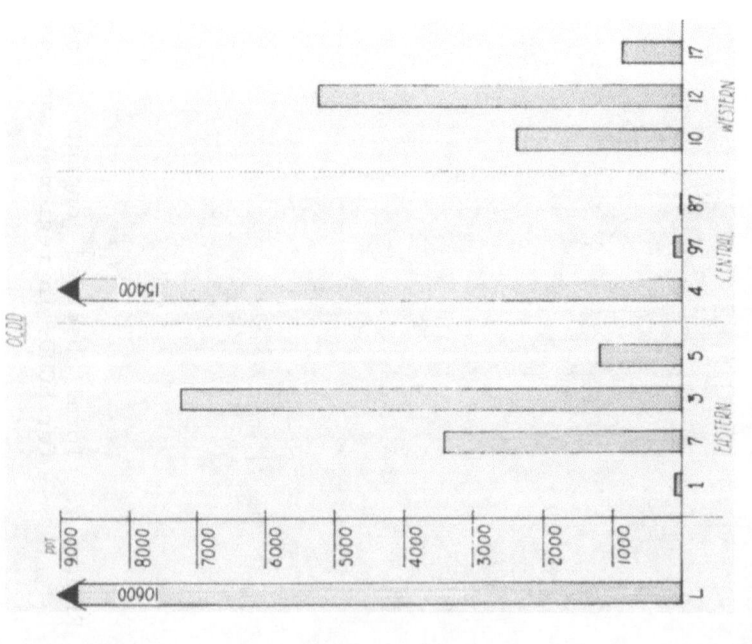

Figure 6.   Total observed concentrations for
OCDD in regional samples

identification numbers are listed on the x-axis and total observed
concentrations in ppt are annotated in bar-graph format on the y-asis.
The bar on the extreme left of each Figure (labeled "L") is the result obtained for chimney particulates collected from the furnace
near Midland, Michigan, which was examined during preliminary experimentation. The percentage figure associated with each bar for total
TCDDs (Figure 3) represents the fractional abundance of 2,3,7,8-TCDD.
The numerical values inside arrowed bars for total OCDD (Figure 6)
are observed concentrations for those samples exceeding 9000 ppt.

   In view of the generalized regional sample characteristics presented in Table 3, the CDD concentration data in Figures 3 through 6
do not present any easily discernible patterns or relationships other
than the anticipated higher CDD levels for those samples from masonry
chimneys. It is interesting to note that all of the Eastern samples
were collected from systems fueled with wood obtained from the same
woodlot throughout the entire heating season prior to sampling. There
were no obvious differences in equipment, operation parameters, or
residence location between Eastern #1 and the other samples from that
region. As indicated in Figure 3, the fractional abundance for
2,3,7,8-TCDD for those samples in which at least one TCDD isomer was
detected ranged from a low of 0 percent to a high of 13 percent. The
minimum total number of TCDD isomers observed in these samples are:
Eastern #1 = 1, Eastern #7 = 20, Eastern #3 = 20, Eastern #5 = 20,
Central #4 = 20, Central #87 = 0, Western #10 = 20, Western #12 = 19,
and Western #17 = 19. Representative isomer-specific TCDD results
for two samples from each region are given in Table 4.

   Although the TCOF is commonly associated with the formation of
CDDs, other species may also arise from combustion reactions. As an
integral part of the method development for this work, many residue
fractions were examined by HRGC-LRMS to determine the nature of major
components present. Hence several crude CDD fractions from Eastern
samples were examined prior to HPLC fractionation. The finding that
CDFs were major components present in these residues is illustrated
by the HRGC-LRMS SIM mass chromatogram in Figure 7. Additional mass
spectral analyses (LRMS) confirmed that these compounds were not
chlorinated diphenyl ethers. Though no CDF recoveries have been determined for this procedure, the observed concentration for total
TCDFs in Eastern #3 is greater than 200 ppb using 2,3,7,8-TCDF as a
calibration standard. Note that the total TCDDs in this sample are
1 ppb. HRGC-HRMS at a mass resolution of ∼11,000 confirmed the presence of T3CDFs and TCDFs in this sample as shown in Figure 8. In
spite of the concentration disparity between TCDFs and TCDDs in
Eastern #3, the apparent isomer distribution for TCDFs is similar
to that for TCDDs, in that a large number of isomers are present
(38 possible TCDFs) at reasonably similar levels. This observation
also indicates that the origin of these TCDFs is unlikely to be related to the presence of specific man-made precursor chemicals,

Table 4. Representative CDD Results from Regional Chimney
          Particulates. East = Eastern. Cent = Central.
          West = Western.

CDD Isomers	Concentration in ppt (pg/g)			
	East #3	East #7	Cent #87	Cent #4
^{13}C-2378 Recovery	69%	63%	68%	65%
2378-TCDD	90	100	ND (1.0)	13
1469-TCDD	8.2 (2.0)	28	ND (0.8)	2.2 (1.4)
1269-TCDD	36	140	ND (0.8)	6.4 (1.4)
1267-TCDD	24	92	ND (0.8)	4.0 (1.4)
1289-TCDD	25	110	ND (0.8)	4.3 (1.4)
1369-TCDD	39 (7.7)	210	ND (0.6)	13 (1.5)
1247+1248-TCDDs	77	480	ND (0.6)	28
1278-TCDD	58 (7.7)	290	ND (0.6)	15
1268-TCDD	59 (7.7)	320	ND (0.6)	15
1237-1238-TCDDs	82	540	ND (0.6)	34
1279-TCDD	78	240	ND (0.6)	12 (1.5)
1246-TCDD	34 (4.4)	130	ND (0.5)	5.8 (1.1)
1478-TCDD	12 (4.4)	65	ND (0.5)	6.8 (1.1)
1236-TCDD	36 (4.4)	160	ND (0.5)	7.2 (1.1)
1239-TCDD	34 (4.4)	150	ND (0.5)	7.3 (1.1)
1249-TCDD	30 (4.4)	150	ND (0.5)	7.6 (1.1)
1368-TCDD	59 (6.7)	530	ND (0.7)	72
1379-TCDD	74	620	ND (0.7)	63
1378-TCDD	140	390	ND (0.7)	37
1234-TCDD	46 (6.7)	190	ND (0.7)	20
Total TCDDs	1041	4925	ND	374
^{13}C-HCDD Recovery	77%	76%	83%	76%
Total HCDDs	1800	4300	2.7 (1.1)	2000
Total H7CDDs	4100	2800	8.9 (1.0)	9000
^{13}C-OCDD Recovery	74%	60%	59%	80%
OCDD	5300	2000	14 (1.5)	12400

Table 4.   Representative CDD Results from Regional Chinmey
(Cont.)   Particulates. East = Eastern.   Cent = Central.
            West = Western.

CDD Isomers	Concentration in ppt (pg/g)	
	Western #10	Western #12
^{13}C-2378 Recovery	74%	62%
2378-TCDD	6.0	11
1469-TCDD	1.8 (0.7)	ND (1.6)
1269-TCDD	4.9 (0.7)	5.6 (1.6)
1267-TCDD	2.2 (0.7)	3.4 (1.6)
1289-TCDD	3.2 (0.7)	4.4 (1.6)
1369-TCDD	7.0 (1.2)	5.0 (1.3)
1247+1248-TCDDs	9.7 (1.2)	9.9 (1.3)
1278-TCDD	7.6 (1.2)	7.1 (1.3)
1268-TCDD	8.0 (1.2)	6.9 (1.3)
1237+1238-TCDDs	13	11 (1.3)
1279-TCDD	11 (1.2)	13
1246-TCDD	3.2 (0.7)	3.5 (1.1)
1478-TCDD	4.0 (0.7)	5.1 (1.1)
1236-TCDD	2.9 (0.7)	5.0 (1.1)
1239-TCDD	4.0 (0.7)	5.2 (1.1)
1249-TCDD	2.6 (1.0)	4.8 (2.0)
1368-TCDD	13	9.1 (1.3)
1379-TCDD	10	7.7 (1.3)
1378-TCDD	8.3	15
1234-TCDD	2.0 (0.6)	4.4 (1.3)
Total TCDDs	124	137
^{13}C-HCDD Recovery	68%	61%
Total HCDDs	850	2600
Total H7CDDs	1700	4000
^{13}C-OCDD Recovery	69%	77%
OCDD	1700	4000

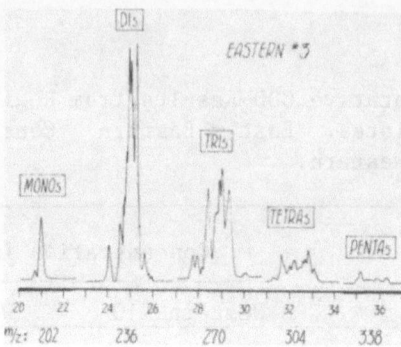

Figure 7.   HRGC-LRMS SIM Mass Chromatogram of CDFs Present in
            Crude CDDs Fraction Prior to RP-HPLC Fractionation

especially in view of the fact that ∿10 ppm (parts per million) of
unchlorinated dibenzofuran was identified in this sample by HRGC-
LRMS during subsequent analyses. The fact that no unchlorinated
dibenzo-p-dioxin could be found (at a similar sensitivity to that
for dibenzofuran) may indicate that CDDs and CDFs are produced by
significantly different reaction mechanisms during the combustion
of natural wood. The observed concentration profiles with respect
to degree of chlorination for CDDs and CDFs in Eastern #3 also support this conclusion. CDD concentrations increase with increasing
degree of chlorination, while CDF concentrations appear to increase
with decreasing degree of chlorination.

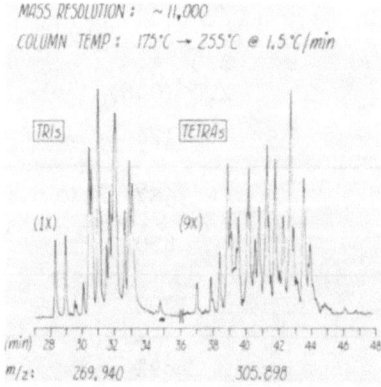

Figure 8.   HRGC-HRMS MPM Mass Chromatogram of T3CDFs
            and TCDFs, Sample Same as Figure 7.

CONCLUSIONS

    Although this survey is not complete, the analytical data ob-
tained to date, without exception, support the conclusion that the
common combustion of naturally occurring wood in domestic furnaces
results in the emission of CDDs into the environment. Concentrations
of tetra- through octachlorinated dioxins in chimney particulates
have been observed in the ppb range. The relatively constant distribution pattern observed for TCDD isomers, including 2,3,7,8-TCDD,
suggests that the production of all 22 isomers occurs as the result
of a random statistical process. The observation that no specific
TCDD isomers routinely exist at concentrations significantly greater
than other strongly supports the conclusion that their production is
not related to fuel contamination by distinct chemical precursors of
man-made origin. Variations of CDD concentrations observed thus far
are anticipated to be largely the result of our attempt to collect
particulates within the source from a point where only a small fraction of the total emissions are retained, rather than differences in
wood-type or combustion conditions. In view of these preliminary
findings, we expect that reasonably similar CDD emissions can also
result from either controlled, or uncontrolled, open burning of wood,
which increases the environmental significance of this source. Obviously the degree of environmental impact associated with residential
wood combustion CDD emissions cannot be ascertained from this survey
study. However, Cooper (1980) has reported:
        "Currently available information suggests a substantial environ-
        mental impact from residential wood combustion emissions. Air
        pollution from this source is widespread and increasing. Cur-
        rent ambient measurements, surveys, and model predictions in-
        dicate winter respirable (<2µm) emissions from residential wood
        combustion can easily exceed all other sources."

REFERENCES

Chlorinated Dioxin Task Force. The trace chemistries of fire - a
    source of and routes for the entry of chlorinated dioxins into
    the environment. Midland, Michigan: The Dow Chemical Company,
    1978.
Cooper, J.A. Environmental impact of residential wood combustion
    emissions and its implications. JAPCA, 30(8):855-861, 1980.
Lamparski, L.L., and Nestrick, T.J. Determination of tetra-, hexa-,
    hepta-, and octachlorodibenzo-p-dioxin isomers in particulate
    samples at parts per trillion levels. Anal. Chem., 52:2045-
    2054, 1980.
Lustenhouwer, J.W.A., Olie, K., and Hutzinger. O. Chlorinated
    dibenzo-p-dioxins and related compounds in incinerator effluents.
    Chemosphere, 9:501-522, 1980.

Rappe, C., Buser, H.R., and Bosshardt, H.P. Polychlorinated dibenzo-
    p-dioxins (PCDDs) and dibenzofurans (PCDFs): occurrence forma-
    tion and analysis of environmentally hazardous compounds.
    Presented at CIPAC - Symposium in Baltimore, Maryland, June
    5/6, 1979.

SAMPLING AND ANALYTICAL METHODOLOGIES FOR PCDDs AND

PCDFs IN INCINERATORS AND WOOD BURNING FACILITIES

D.K.W. Wang, D.H. Chiu, P.K. Leung, R.S. Thomas
and R.C. Lao
Environment Canada
Air Pollution Control Directorate
Ottawa, Ontario, Canada

ABSTRACT

Representative stack samples were taken from typical combustion
sources and wood burning stoves using isokinetic sampling techniques.
Each segment of the sampling trains and coincident collected flyash
samples were analyzed for polychlorinated dibenzo-p-dioxins (PCDDs)
and polychlorinated dibenzofurans (PCDFs). Analyses indicated that
the bulk of the PCDDs/PCDFs were directly associated with particulate
matter in the exhausted effluent and that woodstove effluents contained only components with high chlorine numbers. Samples were
submitted to cytosol receptor bioassay and produced responses far
higher than the expected values based on the 2,3,7,8-tetrachloro-
dibenzo-p-dioxin (TCDD) concentrations measured. Results from solvent extraction and cleanup column recovery studies are reported.
Data for PCDDs/PCDFs concentrations are reported and discussed.

INTRODUCTION

PCDDs and PCDFs have been found in flyash and flue gas samples
from muncipal incinerators and industrial heating facilities throughout the world. Their mechanisms of formation and measurements were
summarized in the review by Kooke et al, (1981). Since then, more
data have been published in the literature (Cavallaro et al, 1980
and Liberti et al, 1981). Progress has also been made in cleanup
procedures (Kooke et al, 1981) and chromatographic separations
(Buser and Rappe, 1980, Lamparski and Nestrick, 1980). In Canada,
PCDDs and PCDFs have been measured in flyash collected from municipal
incinerators (Eiceman et al, 1979). This has prompted the initiation
of testing programs to establish PCDD/PCDF stack sampling and analyt-

ical methodologies to evaluate emission rates for these pollutants
arising from combustion of coal, wood and municipal refuse. Samples
of the gas phase and particulate matter in the stack effluent were
drawn using methods designed to quantitatively collect representative
integrated samples of both phases. Concurrent precipitator flyash
samples were taken to provide a total combustion byproduct profile
for each class of combustion process.

The plants selected for this study are typical of four classes
of stationary combustion sources:

    1. A large modern coal-fired thermal generating station,
    2. A 30-year old coal boiler at a central heating plant,
    3. A large modern heat recovery incinerator, and
    4. A small modular controlled-air incinerator.

At each site a minimum of six individual stack tests were run.
During the period of the tests, plant operation was closely monitored
to ensure that all equipment was functioning in a steady, representative manner.

In addition to the large sources listed above, a selected line
of domestic wood burning stoves of various configurations were also
sampled. As the cost of operating conventional heating systems based
on petrochemical fuels continues to rise, the use of a renewable resource fuel, notably wood, has become an alternative for domestic
heating. To assess the potential environmental impact of a proliferation of wood combustion sources in areas of high population,
samples of woodstove particulate emissions and residual ash have
been analyzed for PCDDs and PCDFs in addition to other organic toxic
substances.

PROCESS DESCRIPTIONS OF SAMPLED SITES

Coal-Fired Thermal Station

The example studies has two 150,000 units and produces on average about 2×10^9 kilowatt-hours annually while consuming 726,000
metric tons of coal. Electrostatic precipitators are located in the
flue gas passage to the stack and the cleaned flue gases are exhausted to the atmosphere through a 152 m stack.

Central Heating Plant

This plant has three 70×10^6 BTU/hr boilers which annually consume 19,000 metric tons of coal. At the time of the survey and sampling this plant had no flue gas-cleaning system. Measurements of
particulate matter in the stack effluent show that the major protion
of the ash produced falls by gravity to the boiler ash pits and is
drawn off periodically for use as over-burden at a sanitary landfill.

Heating Recovery Incinerator

The incinerator has four furnace and boiler systems, each rated at 250 metric tons/day capacity. Each furnace is equipped with an electrostatic precipitator to remove flyash from the effluent gas stream. The treatment time for the gases entering the system is approximately five seconds with a designed exit particulate concentration of 0.2 Kg/1000 Kg gas at 50% excess air. The cleaned gases leaving the precipitators are exhausted to the atmosphere through a 32-m refractory-lined metallic stack. After burning, the residue remaining and flyash collected from various places are moved to a storage silo.

Controlled-Air Municipal Incinerator

The unit was designed for a capacity of 12.5 metric tons waste refuse per 10-hour work-day. This type of incinerator is made up of three main segments: a feed system, a primary combustion zone and a secondary combustion chamber. Waste material is fed to the horizontal primary combustion chamber using a ram injection technique and the chamber is isolated from the outside atmosphere. Volatilized gases and particulate matter produced in the primary process pass into the secondary chamber which contains a modulated fuel burner and air fan. Once secondary combustion is complete, the effluent gases are fed to the stack where they are axially cooled and mixed with ambient air and exhausted. Residual solid waste is removed to an adjacent landfill site.

SAMPLING

Procedures

For each process site representative samples of stack effluent were drawn using a multipoint, isokinetic stack sampling technique and a modifided EPA sampling train #5. The train was modified to incorporate improvements developed by Dow Chemical Co. (1978) and during previous inhouse evaluation testing (1974, 1980), and allowed the collection of gaseous and particulate PCDDs and PCDFs present in the process streams sampled.

The particulate matter in the sample is collected in the front half of the train. Some particles adhere to the walls of the nozzle and probe. A 0.3-μm pore size glass fibre filter is used to entrain the smaller particles. Any PCDDs or PCDFs that are in the gaseous state and escape filtration are collected in the back half of the sampling train in a series of six impingers maintained at 0°C in an ice bath. The first of these impingers contains 250 ml of distilled water which quenches the hot stack gases. This impinger is followed by three similar impingers, each containing 350 ml of 2,2,4 trimethylpentane (isooctane), that serve to trap the gaseous organic components in the sample. Backing this set, two final impingers, the first empty to prevent liquid carry-over and the final one containing silica gel, complete the train. The silica gel is intended to knock out any residual water from the sample gas. Following the impinger set, a cartridge containing granular ion-exchange resin (Amberlite XAD-2) is placed in series to adsorb any unretained organic vapours and the train is connected through a vacuum line to various instruments for measuring and controlling of sampling conditions. All glass components for sampling were wrapped with aluminum foil during and after sampling to minimize any opportunity for photodegradation

During sampling, the probe temperature was maintained above 106°C and the filter box temperature was regulated at about 114°C. The impingers were connected in series and placed in an ice/salt water bath. The temperature at the outlet of the last impinger was maintained at 5°C in an attempt to prevent freezing of the water in the impingers while, at the same time, reducing volatilization of the isooctane solutions as much as possible. If volatilization was significant, additional isooctane was added to the impingers during the test. For each test, a total volume of about 3 to 4 m^3 of stack gases was sampled. Other stack gas parameters determined for each test included: velocity, temperature, moisture content, volumetric flowrate and molecular weight.

Sample Recovery

At the completion of sampling, the probe was disconnected from the train and all openings were sealed with aluminum foil. The equipment was then carefully removed to an on-site laboratory. The apportionment of the individual samples and the recovery procedures used are summarized in Table 1. All liquid samples were stored in sealed amber glass bottles which were wrapped in aluminum foil. Filter samples were stored in sealed petri dishes covered with foil.

Sampling Locations

At all sources the attempt was made to obtain samples from points in the stack which conformed to the reference criteria for site selection (EPS, 1974), that is, sampling ports should be located eight stack diameters downstream and two diameters upstream from any flow disturbance such as a bend, expansion, contraction, visible flame or stack exit. Because of limited access, however, it was not possible to meet these criteria at all sources.

Sampling Site 1. Sampling ports were located at an elevation of 76 m above the ground level. At this point the stack diameter is 4.6 m and the site is well within the selection criteria. Each test involved four horizontal traverses spaced 90° apart, each traverse consisting of eight points giving a total of 32 sampling points per test.

Table 1.   Apportionment of Samples and
Recovery Procedures

Sample No.	Recovery Procedures
1	Nozzle, probe and front-half of filter holder brushed and rinsed with isooctane
2	Filter
3	Contents of impinger No. 1 plus isooctane rinse of impinger and connecting glassware
4	Contents of impinger No. 2 plus isooctane rinse of impinger and connecting glassware
5	Contents of impinger No. 3 plus isooctane rinse of impinger and connecting glassware
6	Contents of impingers Nos. 4 and 5 and isooctane rinse of impingers and connecting glassware
7	Contents of impinger No. 6 - silica gel and Amberlite
8	Rinse of front-half components with acetone, methanol and methylene chloride
9	Rinse of back-half glassware with acetone, methanol and methylene chloride.

Sampling Site 2.   Since no suitable locations on the stack were available, sampling was conducted at a location in one of the boiler exhaust ducts (rectangular 0.9 m x 0.7 m) using five ports spaced evenly along the long dimersion of the duct.   Each traverse consisted of four points providing a total of 20 sampling points per test.

Sampling Site 3.   The downstream criterion of eight diameters could not be met for sampling ports at this site.   However the up-stream criterion of two diameters was met.   Each test consisted of four horizontal traverses at 90° spacing and eight test points per traverse to provide 32 sampling points per test.

Sampling Site 4.   At this site two ports spaced 90° apart were located in the stack some 9 m above the ground level.   Since the inside diameter of the stack at this point is 1.2 m, these apertures would not meet the normally accepted site selection criteria. Further, since the physical limitations of the sampling platform would permit sampling through one port only, the number of traverse points per test were increased by 50%, and in one case by 100%, to give 64 points/test, to ameliorate the poor site availability.

  Woodstoves. These units presented a different set of problems from the previous four sites. To assess the actual effluent to the environment, samples were drawn at the top of the stack using an EPA #5 sampling train with a filter preceeding three impingers in series filled with distilled water, 2,2,4 trimethylpentane and the last one filled with cyclohexane. Each stove configuration was evaluated under several load conditions with wood feed stock of defined type and moisture content. In addition to measured consumption rates and stove exit temperatures, excess air precentages, total carbon monoxide and total hydrocarbons were monitored over the test period.

ANALYSIS

  Many analytical schemes have been developed to extract, separate and quantify PCDDs, PCDFs and other related compounds from particulate samples collected from incinerators (Kooke et al, 1981). In this study the original methodology developed by Dow Chemical scientists (Lamparski and Nestrick, 1980) and modified by Kooke et al, (1981) was used for sample preparation for PCDDs and PCDFs at ppt concentration in particulate matter. The method has been further altered in this laboratory to optimize the exclusion of such interferants as PCB, chlorinated biphenyl ethers, DDE and other chlorinated pesticides.

  All solvents were purchased from Caledon Ltd, distilled-in-glass quality. Laboratory chemicals were ACS reagent grade from Fisher Scientific Ltd. n-Hexadecane was spectrophotometric grade from Aldrich Chemical Co. The procedures for preparation of reagents and clean-up of column packings were the same as those used by Lamparski and Nestrick (1980) and Nestrick et al, (1979). Dioxin standards including isotope-enriched ^{13}C and ^{37}Cl, were obtained from KOR Inc. and their purities were checked by GC and GC/MS.
  A Hewlett-Packard Model 5830 GC was used for sample screening purposes. The instrument is equipped with an ECD, Silar 10C soda lime glass capillary column 60 m x 0.25 mm.

  A Finnigan Model 4023 HRGC/LRMS system coupled with an Incos data system, operated in total scan and select ion mode at unit resolution, was used for the measurement of PCDDs and PCDFs.

  The analytical procedure used by Kooke et al (1981) and Nestrick (1980) involves three main steps: extraction, clean-up, and identification/measurement. A schematic representation of this analytical technique is presented in Figure 1.

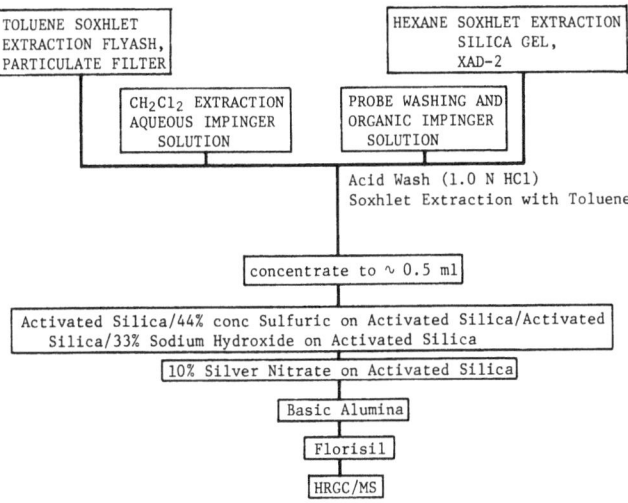

Figure 1. Analytical Scheme for the
Determination of PCDD and PCDF

Extraction

Both particulate and liquid samples were obtained from each of
the sources sampled. The following preparation and extraction procedures were used for the different sample types.

Flyash: Aliquots of between 30 and 100 g are acid-extracted with
1 N HCl (200-500 ml), centrifuged, washed and dried (Hutzinger, 1980).
The flyash samples are extracted in a soxhlet apparatus for 48 hours
using 250-500 ml of distilled-in-glass toluene. The resulting extract
is carefully reduced to a small volume in a rotary evaporator.

Particulate filter/probe wash. The combined sample is separated
by filtration. The filter with entrained particulate is soxhlet extracted as from the flyash. The extract is combined with probe wash
filtrate and carefully reduced to a small volume.

Silica gel and Amberlite (XAD-2) samples: These samples are
soxhlet extracted for 24 hours with 500 ml distilled-in-glass hexane
and the resultant extract is reduced carefully to a small volume.

Aqueous impinger samples: Solutions from the water impingers
are serially extracted with three portions of 150 ml of dichloro-
methane. The organic layers are combined, dried with sodium sulfate
and carefully reduced to a small volume.

Organic impinger samples: The isooctane from the impingers is
combined with the glassware washings, dried with sodium sulfate and
then reduced to a small volume.

Woodstoves: The four sample segments of this sampling train are
handled in a like manner to the corresponding fractions from stack
sampling.

Sample Cleanup

Each of the sample extracts is chromatographed using silica gel,
acid and base, silver nitrate, modified silica gel and basic alumina
as previously reported (Lamparski and Nestrick, 1979, 1980). The same
sizes of Super-Macro and High Aspect liquid columns were used in the
present study. The eluent obtained above may be further cleaned by
a final column which is packed with 7.5 g of Florisil prewashed with
100 ml hexane. First chlorinated interferants are eluted with 40 ml
of 10% methylene chloride (V/V), then PCDDs and PCDFs are eluted with
100 ml of methylene chloride. The solution is carefully reduced to
near dryness by a high purity filtered nitrogen stream and the final
sample volume is made up to between 50 and 200 µl with isooctane.

Extracts are sealed and stored in the dark at $0^{o}C$ for subsequent
GC screening tests or GC/MS quantitative analyses.

GC/MS Quantitation of PCDDs and PCDFs

The prepared samples are spiked with D_{10}-anthracene and/or
$2.3,7,8$-TCDD-$^{37}Cl_4$ for retention and quantitation studies and are
analyzed using either a J&W 30 m x 0.25 mm SE-54 or DB-5 fused silica
column and Finnigan 4023 GC/MS/DS system. The glass-lined tubing in
the Finnigan MS interface oven was removed from the triaxial inlet
system and the fused silica capillary was fed directly into the ion
block to a point approximately 2.0 mm from the inside edge of the ion
volume. Initially, samples were introduced to the column using Grob
splitless injection techniques; however, low yields for 7Cl- and
8Cl-dioxin standards in comparison with on-column injection methods
indicated discrimination against 7Cl- and 8Cl-dioxin species. The
column conditions were $40^{o}C$ for 2.0 minutes; 40 to $210^{o}C$ at $10^{o}C/min$,
210 to 260 at $6^{o}C/min$ and hold $260^{o}C$; helium carrier linear velocity
adjusted to 45cm/sec and $260^{o}C$. All temperature zones, injector,
interface oven, transfer line and ion sources were held at $300^{o}C$.

MS data was collected for both full scan and MID runs on each
sample. For full scan the quadrupole was scanned from 10 to 500 amu
at 1 sec/scan. For MID runs a series of eight MID descriptors containing a 200 ms dwell/ion for each of four ions were daisychained to
monitor two ions each for dibenzodioxins and dibenzofurans from 2Cl
to 8Cl, and the internal standards. The MS data were processed using
the INCOS data system to produce peak ion and area integrals for ion

pairs; for the PCDDs 2Cl (252, 254), 3Cl (286, 288), 4Cl (320,322),
5Cl (356,358(, 6Cl (388,390), 7Cl (424,426), 8Cl (458,460), and for
PCDFs: 2Cl (236,238), 3Cl (270,272), 4Cl (306,308), 5Cl (338,340),
6Cl (374,376), 7Cl (408,410), 8Cl (444,446). Furthermore, cluster
isotope ratios were screened to ensure that the identified components
were within experimental tolerances before quantitation against the
internal reference standards. Where concentrations of the 6Cl-,
7Cl- and 8Cl-PCDDs and -PCDFs were at the detection limit for EI/MS,
negative CI/methane was used to enhance the detection limits of the
quadrupole mass spectrometer.

BIOLOGICAL ANALYSIS

The toxicity and biological activity of 2,3,7,8-TCDD is well
established. However, many other PCDD and PCDF congeners contribute
to the overall biological activity of environmental samples. In this
study biological testing was done on five flyash samples from different combustion sources using the cytosol receptor binding method
(Safe et al, 1981) as a short-term screening assay for assessing
activity.

DISCUSSION

PCDD and PDCF Concentration Levels

Preliminary results indicated that PCDDs and PDCFs were present
in the flyash of the municipal incinerators and one of the two power
plants surveyed, with concentrations of the tetrachloro isomers ranging from 2 to 85 ppb and total concentration for all PCDDs ranging
from 30 to 900 ppb. These values may represent only minimum levels
due to uncertainties about analytical methodologies. The PCDFs were
present in all samples containing PCDDs. Concentrations for PCDF
isomers were typically 2-5 times higher than for the corresponding
PCDD isomeric groupings. These data are far higher than the published
literature values, which are mostly European, and no doubt reflect
the high chlorine content of "throw-away" plastics in typical North
American municipal waste. Besides, it is reported that the incinerator often accepts inductrial waste for burning. The data are also
higher than other Canadian data (Eiceman et al, 1979, 1981); this
underlines the importance of optimizing extraction efficiencies and
the dangers of calibrations based on "representative" responses where
standards are unavailable. Aliquot portions of these flyash samples
were analyzed by two contract laboratories in Europe and validated by
high resolution MS. The resultant data agree (within experimental
error) with the data reported here.

No PCDDs or PCDFs were found in sample 2, taken from the 30-year old coal fired boiler, which is in agreement with the negative bioassay results for this sample. Sample 3 did, however, contain a broad cross-section of chlorobenzenes and polycyclic aromatic hydrocarbons, a number of which are known carcinogens, but do not interfere with the dose/response specific for PCDDs/PCDFs, of the bioassay procedure.

The concentrations of PCDDs/PCDFs entrained on the sampling train filter and those from probe wash solutions, that is the particulates which had escaped precipitation and would be exhausted to the atmosphere, are in proportion to concentrations measured in the corresponding source flyash. There was little carry-over to the impinger train except in the case of the incinerator samples which had the highest concentrations of all sources studied. For the woodstove samples, only octachlorodibenzo-p-dioxin was present and only on the particulate entrained on the primary filter. Detailed studies are under way for these samples.

GC/MS Operations

Preliminary investigations using standard solutions of PCDDs with splitless injection techniques and combined GC/MS/DS indicated some apparent discrimination from 1Cl to 8Cl substitution and a rolling off of the EI/MS detection limits as the Cl number increased. In spite of thorough system deactivation and use of elevated temperatures throughout the chromatographic pathway to the ion volume, response by probe analysis of the individual components at equivalent concentrations and matched detection parameters were far higher than TIC responses generated by capillary chromatographed compounds. Injection of the samples directly onto the fused silica column resulted in marked improvement in the detection limits for hepta- and octachloro- PCDDs and PCDFs.

Recoveries from the Clean-Up Procedure

Several blends of PCDDs and PCDFs were used to evaluate potential losses of 1Cl to 8Cl compounds passed through the multicolumn clean-up scheme. Standard amounts of each component (including isotope-labled compounds) were added to blank solvent and submitted to the complete sample preparation procedure. The mean percentage recoveries for 15 PCDDs and PCDFs are shown in Tables 2 and 3. Unfortunately, the complexity of the flyash extracts appears to alter the characteristics of the clean-up columns and these percentage recoveries, although indicative, could not be used as recovery factors in the final concentration calculations. Using isotope-spiked extracts, studies directed toward developing a recovery factor schedule are continuing.

Table 2.   Recovery of PCDDs and PCDFs
from Super Macro Columns

Compounds	% Recovery	
	I	II
1-M_1CDD	31	17
3,6 & 2,8-D_2CDF	57	47
2,7-D_2CDD	61	51
2,3-D_2CDD	57	45
1,2,4-T_3CDD	54	43
1,3,7,8-T_4CDD	81	75
2,3,7,8-T_4CDF	79	70
1,2,3,4-T_4CDD	76	67
1,2,3,7 & 1,2,3,8-T_4CDD	81	75
2,3,7,8-T_4CDD	85	77
1,2,7,8-T_4CDD	81	72
1,2,3,7,8-P_5CDD	92	85
1,2,3,4,7,8-H_6CDD	97	90
1,2,3,4,6,7,8-H_7CDD	101	102
OCDD	105	125
OCDF	104	124

Table 3.   Recovery of PCDDs and PCDFs from
Super Macro and Florisil Columns

Compounds	% Recovery	
	I	II
1-M_1CDD	20	--
3,6, & 2,8-D_2CDF	45	38
2,7-D_2CDD	62	48
2,3-D_2CDD	56	43
1,2,4-T_3CDD	3	24
1,3,7,8-T_4CDD	91	79
2,3,7,8-T_4CDF	87	78
1,2,3,4-T_4CDD	81	74
1,2,3,7 & 1,2,3,8-T_4CDD	94	81
2,3,7,8-T_4CDD	91	82
1,2,7,8-T_4CDD	89	79
1,2,3,7,8-P_5CDD	98	85
1,2,3,4,7,8-H_6CDD	83	73
1,2,3,4,6,7,8-H_7CDD	32	24
OCDD	--	--
OCDF	--	--

Although the Florisil column does not affect the recoveries of
tetra-, and penta-PCDDs and PCDFs, it is effective in removing chlorinated hydrocarbons which interfere with their analyses. Recoveries
of hepta- and octa-isomers proved to be very low with the Florisil
column, thus the column is used only when a large number of interfering compounds are present in a sample and the additional column clean-
up operation is necessary.

Efficiencies of Extraction of PCDDs and PCDFs from Flyash

Several schemes have been developed to produce data on the degree
to which PCDDs and PCDFs are removed from the flyash matrix prior to
clean-up. Kooke et al, (1981) have reported that pretreatment of the
flyash with acid prior to extraction resulted in improvements in extraction efficiencies ranging as high as 10%. He also evaluated a
number of extraction solvents and concluded that benzene or toluene
provided marked improvements in extraction efficiency over dichloro-
methane. Two areas appear to have been neglected: the effects of
acid-type and of reduction of particle size by grinding (Reese et al,
1981). Preliminary data on the effects of acid treatment appear in
Table 4. These data are based on determinations of octachlorodibenzo-
p-dioxin, which represents the most difficult of the components to
extract efficiently. These preliminary measurements indicate marked
improvements are possible and work is proceeding to provide the corresponding data for 1Cl to 7Cl PCDD and PCDF isomers in flyash
matrices from various sources.

Table 4. Acid Extraction of OCDD from Flyash

Pretreatment of Flyash	Relative Ratio of OCDD
No Acid Treatment	1.0
1 N HCl	1.1, 2.0
12 N HCl	2.4
12 N H_2SO_4	2.0

Bioassay

The work was performed by Safe et al, (1981) and the results
are given in Table 5.

Table 5.   PCDDs and PCDFs Cytosol Receptor
Binding Bioassay Results

Fly Sample	2,3,7,8-TCDD[a] Bioequivalent (ng/g)	2,3,7,8-TCDD[b] (estimated, ng/g)	Enhancement Factor (estimated)[c]
A	32.1	0.25	128
B	--	--	--
C	3.53	0.08	44
D	65.5	0.39	167
E	11.1	0.07	159

[a] Determined from the dose-response competition between flyash extract and (^{3}H)-2,3,7,8-TCDD.

[b] Determined by multiplying (Cl_4DD) value measured times 3%, the average concentration of 2,3,7,8-TCDD in the total (Cl_4DD) isomers.

[c] Defined as $EF = \dfrac{2,3,7,8\text{-}Cl_4DD \text{ bioequivalents (ng/g)}}{2,3,7,8\text{-}Cl_4DD \text{ (estimated)(ng/g)}}$

These results underline the apparent contributions of many other
PCDD and PCDF isomers to the overall biological activity of flyash
and emphasize the necessity of performing broad-based bioassays
since specific measurement focused on 2,3,7,8-TCDD alone may seriously
underestimate the potential biological and environmental impact of the
collective PCDD/PCDF compounds found in flyash samples.   Further tests
on air emission samples are now underway to assess their biological
activities.

CONCLUSION

The current results and data from other incinerators must be
considered of a semi-quantitative nature.   The sampling methodology
is still in a developmental stage and systematic studies are needed
for possible interferences, validated clean-up procedures and high
resolution mass spectrometric techniques for confirmation.   However,
the present data do demonstrate that PCDDs and PCDFs are present, and
they indicate minimum levels for the various congeners, and, in some
cases, reasonable levels of specific isomers in a sample.   A program
of interlaboratory studies is needed to validate the results for the
measurement of PCDDs/PCDFs in incinerators.

REFERENCES

Air Pollution Control Directorate, Environment Canada, Standard
    Reference Methods for Source Testing: Measurement of Emissions
    of Particulates from Stationary Sources, EPS 1-AP-74-1, Feb.
    1974.
Air Pollution Control Directorate, Environment Canada, Collection
    of Organic Particulate Samples from a Municipal Incinerator
    and a Coal-Fired Boiler, Internal Report, File No. 4172-2,
    Dec., 1980.
Buser, H.R., and Rappe, C. Anal. Chem., 52 (14):2257-2262, 1980.
Cavallaro, A., Bandi. G., Invernizzi, G., Luciani, L., Mongini, E.,
    and Gorni, A. Chemosphere, 9:611-621, 1980.
Eiceman, G.A., Clement, R.E., and Karasek, F.W. Anal. Chem., 15 (14):
    2343-2350, 1979.
Eiceman, G.A., Clement, R.E., and Karasek, F.W. Anal. Chem., 53 (7):
    955-959, 1981.
Kooke, R.M.M., Lustenhouwer, J.W.A., Olie, K., and Hutzinger, O.
    Anal. Chem., 53 (3):461-463, 1981.
Lamparski. L.L., and Nestrick, T.J. Anal. Chem., 52 (13):2045-2054,
    1980.
Liberti, A., Brocco, D., Cecinato, A., and Passanzini, M.
    Mikrochimica Acta., 271-280, 1981.
Nestrick, T.J., Lamparski, L.L. and Stehl, R.H. Anal. Chem., 51 (13):
    2273-2278, 1979.
Reese, G.A.V., Smillie, D., Tosine, H. and Osborne, J. Abstract,
    28th Canadian Spectroscopy Symposium, Ottawa, September 28-30,
    1981.
Safe, S., Bandiera, S., Hutzinger, O., Olie, K., Lustenhouwer, J.W.A.
    and Okey, A.B. Chemosphere, 10:19-25, 1981.
The Dow Chemical Co., Midland, Michigan. Stack Gas Sampling Proce-
    dure for the Determination of Chlorinated Dioxin Emissions from
    Stationary Source, ML-AM 78-64, Nov., 1978.

COMPARATIVE MONITORING AND ANALYTICAL METHODOLOGY

FOR 2,3,7,8-TCDD IN FISH

H. Tosine, D. Smillie and G.A.V. Rees

Ministry of the Environment
Laboratory Services Branch
Rexdale, Ontario, Canada

INTRODUCTION

Most agencies involved in monitoring 2,3,7,8-tetrachlorodibenzo-p-dioxin (TCDD) residues in fish would now appear to have a fairly thorough understanding of the problems and pitfalls involved with this very demanding type of analysis. Unofficial results from exchange samples would indicate that laboratories monitoring Great Lakes fish can produce compatible data from split samples. There still exists, however, one major stumbling block to direct comparison of TCDD data. Due to their varying mandates and historical development, different agencies sample fish for contaminant analysis in different ways. It is the intention of this paper to examine 2,3,7,8-TCDD residue data produced in one laboratory using two of the more common sampling techniques for the same fish. The 2,3,7,8-TCDD data are evaluated for length, weight and lipid content of samples and for evaluation of data variability. A comparison of data from the two sampling methods is made.

Table 1 illustrates the agencies in the U.S. and Canada that currently monitor TCDD in fish, their mode of sampling, and the use of the data.

Canadian Agencies

The Canadian Dept. of Fisheries and Ocean is responsible for testing all fish taken by commercial fishermen for sale to the public or for export. Representative samples of fish from a catch are filleted, the dorsal fillets are pooled and homogenized prior to analysis.

127

Table 1. Dioxin Monitoring Agencies

Agency	Fish Sample	Use of Data
Canadian		
Fisheries & Oceans Canada	whole edible fillet	commercial regulations
Environment Canada (a) Inland Waters Directorate	whole fish	modelling studies
(b) Freshwater Institute	dorsal fillet	modelling studies
Ontario Ministry of the Environment	dorsal fillet	consumption guidelines, sports fish
American		
FDA	whole fillet skin on	interstate commerce
N.Y. State Dept. of Health		
N.Y. State Dept. of Environmental Conservation	whole fillet skin on	consumption guidelines, sports fish

Environment Canada (a) Inland Waters Directorate, Burlington.
This group has responsibility for freshwater research, especially
in the Great Lakes. Their interest in fish is with the general
health and dynamics of the fish population. Since total body burden
of pollutants is of major concern in this context, they analyse individual whole fish homogenates.

Environment Canada (b) Freshwater Institute, Winnipeg. Their
interests are similar to that of the Inland Water Directorate, but
concentrate on western Canadian water bodies other than the Great
Lakes, and analyze the dorsal fillet from individual fish.

Ontario Ministry of the Environment (MOE) provides guidelines for the public consumption of sports fish from all parts of the province of Ontario. Fish are collected primarily by the Ontario Ministry of Natural Resources, analysed by MOE, and the health implications of residue levels evaluated jointly by the Ontario Ministries of the Environment, Natural Resources, Health and Labour. MOE samples individual fish (or composites in the case of forage fish). The dorsal fillet above the lateral line, skin and fat removed, is used for analysis. Previous work (70,000 samples over 8 years) has shown that results for other contaminants (PCB, Hg, etc.) using this fillet are more reliable that were results obtained for whole fish or other portions. Composition of the dorsal fillet appears to change less than that of other fillets or steaks throughout the life of the fish. It has proven easier to make predictions of possible human consumption of contaminants from various sizes of fish of a given species using a limited number of measurements (10-20 individuals) than from data on more variable fish portions.

Because of its policy of using this relatively lean, fat removed fillet, MOE guidelines for fish consumption caution against consumption of fatty tissue due to its potentially higher pollutant levels. MOE currently has a contaminants data base from dorsal fillet samples of:

60,000	Analyses for Hg
6,000	Analyses for PCB
4,500	Analyses of organochlorines
150	Analyses for 2,3,7,8-TCDD

U.S. Agencies

U.S. agencies whose mandates and techniques we considered were limited to those supplying dioxin data on the Great Lakes, namely the U.S. FDA and New York State Department of Environmental Conservation.

U.S. FDA, Their mandate includes the regulation of contaminants in fish for interstate commerce. U.S. FDA "edible fillet" comprises individual fish, sampled as a fillet from dorsal to pectoral fin, gill to tail, skin on but bones removed.

New York State, New York State officials have a similar mandate to that of MOE: advice and regulation to the public on consumption of sports fish. They sample individual fish. The whole side of the fish is taken, less head, bones, viscera and pectoral fins. Skin and belly fat is left on the fillet.

From these large differences in sampling techniques, it is understandable why analyses for the concentration of TCDD in fish would

show great variations between agencies. When a round robin was conducted among agencies surrounding Lake Ontario, it became evident
that, as expected, a large difference existed in the concentration of
TCDD in a whole fish as compared to a MOE dorsal fillet.

In order to determine the extent of the variations resulting from
analysing different fillets, an investigation of the distribution of
2,3,7,8-TCDD in various portions of sports fish - Lake Trout and Rainbow Trout from Lake Ontario was undertaken. The MOE fillet, the N.Y.
State fillet, the livers, where possible the fat, and, due to the
small size of fish brains, a 9 brain composite sample were all analyzed for 2,3,7,8-TCDD.

ANALYTICAL METHODOLOGY

In establishing an analytical methodology for the routine monitoring of TCDD in fish tissue, the following points had to be
considered:

Due to the existing instrumentation in the laboratory, a low
resolution GC/MS (Hewlett Packard 5992 benchtop model, methodology had to be tailored for low resolution GC/MS, which meant
using high efficiency cleanup.

The method had to lend itself to batch operation to allow for
a "high" throughput, and also had to allow for future automation.

Although the method was initially designed soley for 2,3,7,8-
TCDD, it has to allow for future isomer specificity as other
TCDD isomers became available.

Therefore, the method documented by Lamparski et al (Anal. Chem.
51, 1453, (1979)) was considered.

Some of the modifications that were made to this method are outlined below:

Digestion/Extraction

Carbon 13 - labelled 2,3,7,8-TCDD was added at a level of 50
ppt, as an internal standard for individual sample recovery determination. The fish homogenated was allowed to digest at room
temperature for 16 hours with concentrated HCL. This was followed
by extraction with 2x50-mL portions of hexane. The hexane was
separated and evaporated to a small volume (approx. 1-2 mL) before
open column chromatographic cleanup. Formation of emulsions was a
problem occasionally, if samples with a particularly high lipid

content were digested. The most satisfactory solution for this
problem was allowing the sample to stand in warm water, until the
emulsion cleared.

Open Column Chromatography

The H_2SO_4/silica columns (Figure 1) were designed to remove
lipids and oxidizable components from the hexane fraction. The wider
column packed with 22% H_2SO_4/silica allows for a faster elution; the
narrower 44% H_2SO_4/silica column removes the final lipids. Since
the dimensions of these columns were optimized for extraction of the
fattiest of the MOE type fillets, no problems were encountered with
overloading. For this particular study, modifications had to be
made to increase the amount of 44% H_2SO_4 for cleanup of the liver
and fat samples.

The $AgNO_3$/silica column (Figure 2) was reputed to remove un-
saturated compounds, sulphur compounds and DDE. The DDE is only
removed by the $AgNO_3$ column in combination with the alumina column.
The alumina columns allow for a complete fractionation of PCB and
pesticides. Recovery of 99.9% of the PCB was obtained through this
type of chromatographic procedure.

The fact that this procedure is very manpower intensive is a
definite drawback, but the resulting sample extract was very clean
of interfering compounds. The columns also show promise for future
automation.

HPLC

The HPLC conditions used were: Water Model 6000 instrument;
254 nm (fixed) wavelength detector; 2 - 4.6 x 250 mm Spherisorb 5 u
ODS (Brownlee C_{18} Reverse Phase) columns; 100% HPLC grade methanol
solvent; 1 mL/min flow rate.

Liquid chromatographic polishing was done using two Brownlee
(5 u x 25 cm) columns with methanol used isocratically as the mobile
phase. The elution of 2,3,7,8-TCDD was first established using 1,2,
3,4-TCDD as a marker; high concentrations of 2,3,7,8-TCDD standard
need not be used daily to determine retention times.

Only the reverse phase columns were used, as the method is
2,3,7,8-TCDD specific. As our bank of tetra-isomers increases,
we may also use normal phase columns for isomer separation. At
present, modifications are being made to install an autoinjector and
autocollector for the HPLC cleanup. As is apparent from Figure 3,
the elution area around the TCDD was free from other compounds.
The peaks at shorter retention times represented residual PCB and
some uncharacterized compounds. We have had no problems with the
HPLC cleanup step.

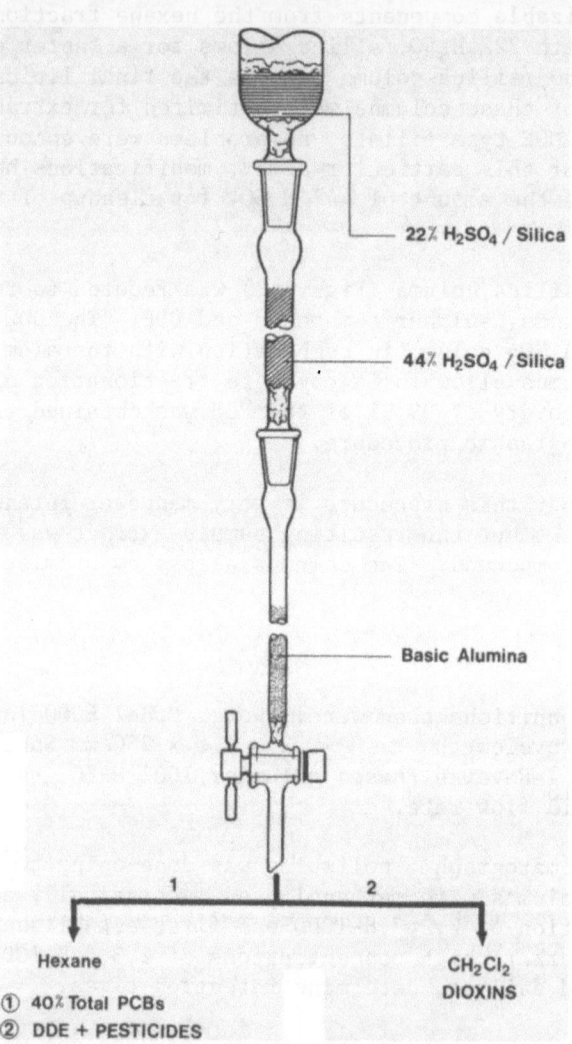

Figure 1.   H₂SO₄/Silica Open Column Chromatographic Clean-Up

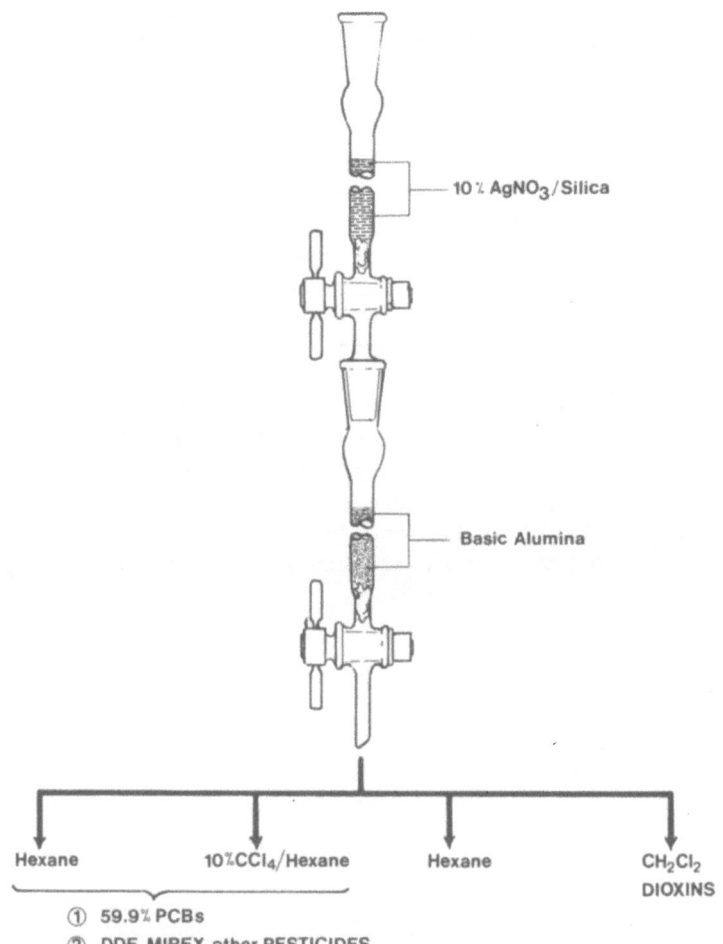

Figure 2. A$_g$NO$_3$/Silica Open Column Chromatographic Clean-Up

GC/MS

Low resolution mass spectrometry coupled with fused silica capillary columns was used for confirmation of the presence of 2,3,7,8-TCDD in sample extracts. Four different degrees of specificity must be obtained in order for a signal to be definitely assigned to TCDD; these are:

The retention time for 2,3,7,8-TCDD must be correct to $\pm$ 0.05%.

The response at m/z 320, m/z 322 and m/z 324 must be in the correct isotopic ratio, i.e. 78:100:46 $\pm$15%.

Signal to noise must be greater than 3:1.

The ^{13}C-2,3,7,8-TCDD internal standard must coelute at the correct retention time.

Initially, all confirmations of samples were conducted on the Hewlett Packard 5992. The sensitivity of the Hewlett Packard 5992 was 10 ppt for these samples. This was inadequate as a large portion of our samples from last year's monitoring program contained less than 10 ppt.

To improve sensitivity and accuracy of measurements in the 1-20 ppt range, a Finnigan 4023 GC/MS was recently obtained. A detection limit of 1-2 ppt for fish using the conditions listed in Table 2 is now possible. The Finnigan has the fused silica capillary column going directly into the source, and gives excellent peak shapes (Figure 4).

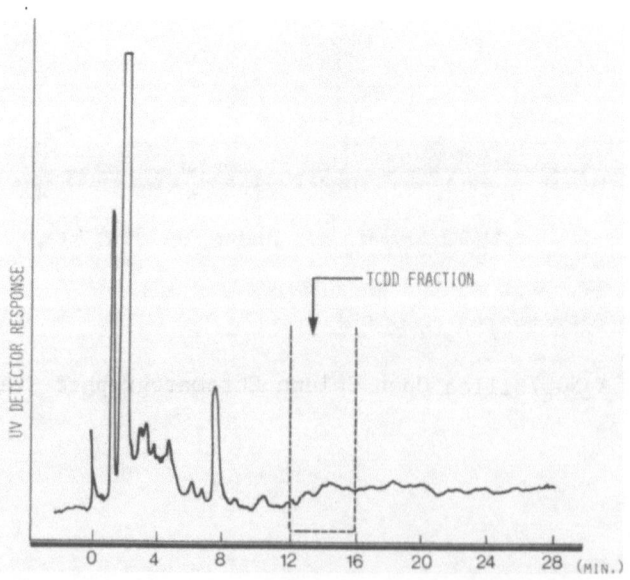

Figure 3.   HPLC Chromatogram of a "Cleaned" Fish Extract

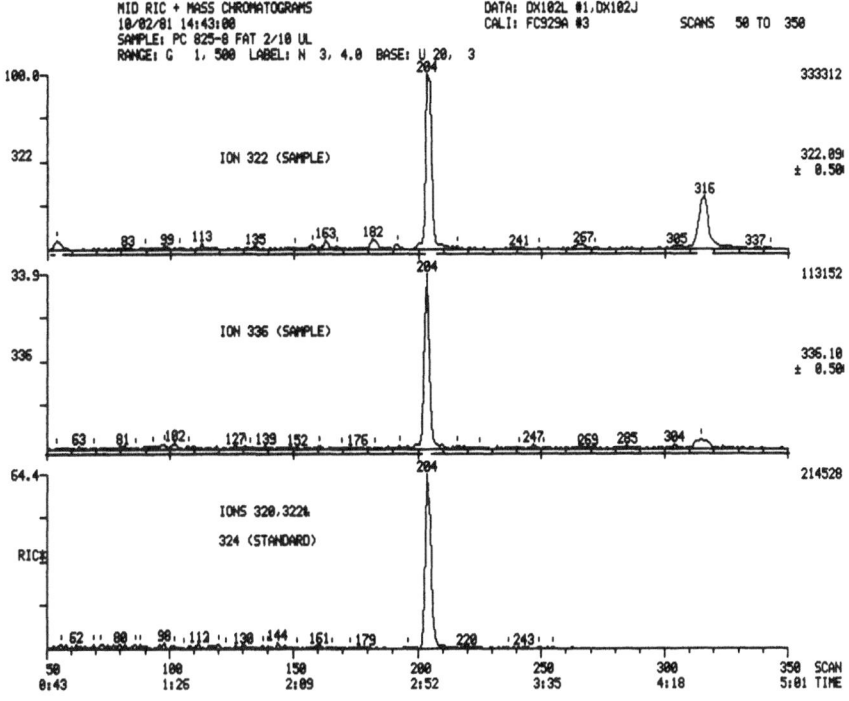

Figure 4. GC/MS Chromatogram of 2,3,7,8-TCDD in Sample and Standard

Table 2. GC/MS Conditions

<u>Finnigan 4023</u>

Capillary Conditions:

Injection Port:	260°
Column:	DB 1, 30m, fused silica
Program:	75° 225° @15°/min
	225° 290° @5°/min

MS Conditions:

EM:	1350V
Scan Time/ion:	0.210 msec
Ions Monitored:	319.9
	321.9
	323.9
	335.9

RESULTS AND DISCUSSION

Table 3 lists 2,3,7,8-TCDD concentrations for the various portions of 9 Lake Trout. These trout were all taken off Port Credit, Ontario, and were obtained from sports fishermen during the Lake Ontario Salmon Derby in late August of 1981. Data shown in this table is for 2,3,7,8-TCDD levels in wet tissue for MOE fillet, N.Y.S. fillet, liver and fat. A composite of brain tissue was also analyzed, yielding a level of 24 ppt wet weight for 2,3,7,8-TCDD. The recovery data for these analyses were quite consistent, as presented in Table 4.

The concentration of 2,3,7,8-TCDD was plotted against length and weight for both the MOE and NYS fillets (Figure 5). For the MOE fillet, the concentration of dioxin appears to be independent of the weight and length (over the range of 1-3 kg and 20-60 cm). In fact, it could be stated that the dioxin concentration, for this limited size range, is constant at 19 ppt +12%. On the other hand, similar plots for N.Y.S. data indicate a first order correlation coefficients are only 0.5 - 0.6.

Figure 6 is a representation of the same data recalculated to give dioxin concentration relative to the extractable lipid content of the various fillets. The MOE dioxin concentration variability is now reduced in both dioxin/length and dioxin/weight plots. The variation in dioxin concentration of the New York fillet data does not appear to improve significantly. When the dioxin concentration

Table 3.   2,3,7,8-TCDD Concentrations in Lake Ontario Lake Trout

	Length (cm)	Weight (kg)	MOE fillet (ppt)	NY State fillet (ppt)	Liver (ppt)	Fat (ppt)	Brain (ppt)
1.	62.5	2.9	15	28	73	discarded	
2.	46.0	1.9	18	28	38	24	
3.	43.0	1.85	19	18	24	23	
4.	37.5	0.92	18	14	–	8	
5.	44.0	1.7	26	4	21	53	
6.	57.5	2.75	26	32	–	52	
7.	46.2	1.85	16	17	14	68	
8.	51.0	2.0	23	38	25	61	
9.	59.0	2.3	16	20	21	56	24 ppt (9 brain comp)

Table 4. Quality Assurance Data

Percent C^{13} Recovery at 95% Confidence Limits of C^{13} 2,3,7,8-TCDD

MOE Fillet	NYS Fillet	Liver	Fat
77+10%	70+11%	77+11%	89+10%

Reproducibility of Sample Analysis

Sample No.	ppt 2,3,7,8-TCDD in MOE fillet	
2	16 18 20	$\bar{x}$ = 18 ppt $\pm$ 10%
5	16 20	$\bar{x}$ = 18 ppt $\pm$ 12%

---

obtained from both methods are compared statistically, there is a 99% probability that the data sets from each method are statistically different.

There are limited data to show that this phenomenon is not limited to lake trout. Concentrations of 2,3,7,8-TCDD in MOE fillets from Rainbow Trout from Lake Ontario (Ganaraska River) also show no correlation with weight and length, but a constant level of 8 ppt $\pm$15%, as shown in Figure 7.

Although only a limited range in length (40-60 cm) of Lake and Rainbow Trout have been sampled, finding a constant level of 2,3,7,8-TCDD rhroughout this range is somewhat at variance with MOE normal findings for other lipophilic pollutants in the Great Lakes. Usually, a first order relationship between length, weight and pollutant concentration is seen. This suggests a "threshold" or equilibrium concentration for 2,3,7,8-TCDD has been developed in these species .in Lake Ontario. The explanation for the 2,3,7,8-TCDD concentrations we have observed in selected Lake Ontario fish may be the following. There has been information published which indicates that for large, high lipid, long-lived fish like Lake Trout, a point of growth is reached after which total body burden of pollutants increases proportionally to weight increase, and that for tissues with a relatively

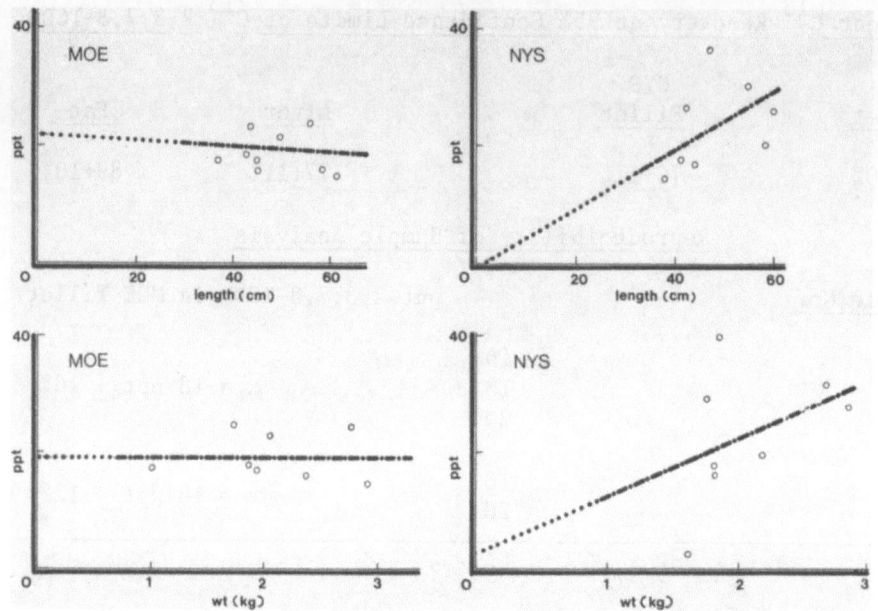

Figure 5.   Dioxin Concentration in Wet Weight Fillet (Lake Trout)

constant fat level (dorsal muscle) the pollutant concentration may
also remain constant.

CONCLUSIONS

    For the MOE fillet of the population of Lake and Rainbow Trout
sampled, the concentration of 2,3,7,8-TCDD appears to be independent
of weight and length.

    This lower variance obtained from the lean dorsal fillet may
indicate that fish consumption guidelines based on a lean fillet,
may be more valid than those based on a fatty fillet. The data also
suggest it is indeed prudent to advise the public to trim fat and
avoid eating fatty tissue.

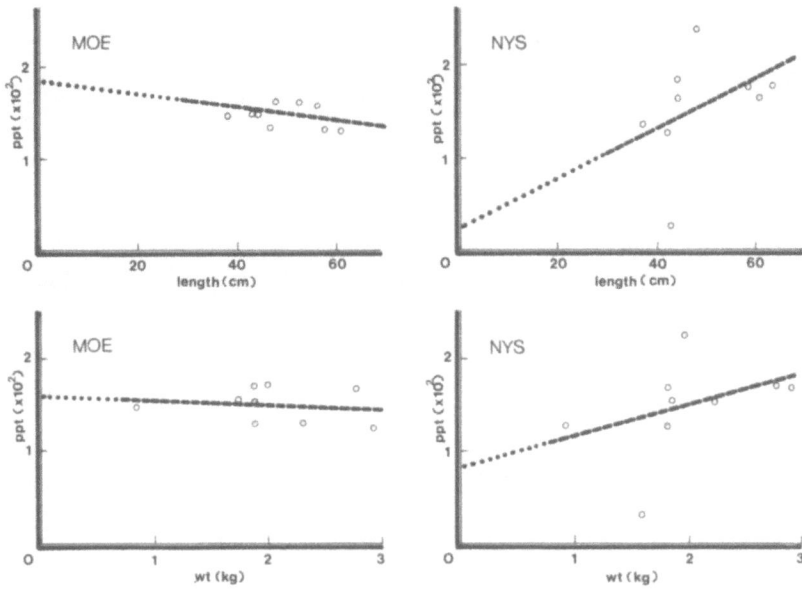

Figure 6. Dioxin Concentration in Extractable Lipid (Lake Trout)

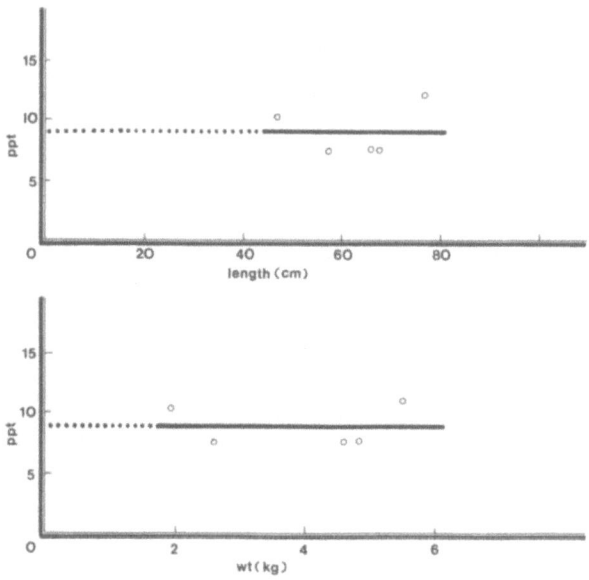

Figure 7. Dioxin Concentration in Rainbow Trout

ENVIRONMENTAL CHEMISTRY

Chairman: Phillip C. Kearney
           U.S. Department of Agriculture
           Beltsville, Maryland USA

EMISSIONS OF PCDDs and PCDFs

FROM COMBUSTION SOURCES

David P. Redford,* Clarence L. Haile** and
Robert M. Lucas***

*United States Environmental Protection Agency,
  Washington, DC   USA
 **Midwest Research Institute, Kansas City, Missouri, USA
***Research Triangle Institute, Research Triangle Park,
  North Carolina, USA

ABSTRACT

Gaseous, aqueous, and solid samples were collected for
analysis of polycyclic organic matter (POM), including
polychlorinated dibenzodioxins (PCDDs) and polychlorinated
dibenzofurans (PCDFs) from two facilities which burn municipal
waste. PCDDs and PCDFs were only observed in the stack gas and
associated particulates from the facility burning 100 percent raw
refuse. At the times sampled in this facility, 2,3,7,8-TCDD
averaged 0.41 ng/dscm (dry standard cubic meter) in the stack
effluent and other dioxins and furans ranged from 2.5 ng/dscm for
the otachlorodibenzodioxins to 300 ng/dscm for the trichlorodibenzofurans. No PCDDs or PCDFs were detected in any of the
samples from the second facility, which burned coal and refuse
derived fuel, where the detection limit in the stack gas samples
was less than 0.25 ng/dscm for TCDDs and TCDFs.

INTRODUCTION

In 1979, the Environmental Protection Agency (EPA) began a
program to determine what polycyclic organic matter (POM) was
being emitted from stationary conventional combustion sources and
to estimate these emissions on a national scale. As a first step,
various combustion sources were categorized and ranked according
to the relative likelihood of emitting POMs, based on
characteristics such as combustion temperature, fuel feed size,
residence time and mixing efficiency (Shih, 1980). These
categories include refuse, coal and wood combustion.

In 1980, as part of the above program, EPA's Office of Toxic
Substances and Office of Research and Development jointly
initiated a pilot study to determine the optimum sampling and
analysis techniques for use in quantifying POM emissions from
combustion facilities. This study included the development of a
statistical sampling strategy which could be used to estimate
national POM emission rates using data from a limited number of
sites. The categories of concern in this pilot study were
coal/refuse combustion and municipal waste combustion.

Two facilities were chosen for the pilot study. One burned
raw municipal waste and produced only enough electricity to run
the plant itself, and the other burned 85 percent coal and 15
percent refuse derived fuel (RDF) for large scale conversion of
heat to electricity. This RDF was prepared by removing bottles
and cans etc. from the raw refuse and milling the remainder.
These facilities were chosen because it was anticipated that the
variability between these two particular facilities would be
greater than between most facilities in just the incineration or
coal combustion categories because of the vast difference in
operating conditions. Thus, the pilot results can be used to
describe the greatest variability one might expect in either
category.

The coal/RDF boiler burned approximately 2,300 kg of RDF per
hour (14 percent of fuel by weight) and 14,000 kg of coal per hour
at a temperature near $1200°C$. The RDF was fed into the boiler and
burned in suspension in the coal fire ball. It then fell onto a
grate where it continued to burn until the ash was removed. Fly
ash was removed from the combustion gases in an electrostatic
precipitator at approximately $190°C$ (Table 1).

Approximately 17,000 kg of refuse was burned per hour in the
municipal waste boiler at a temperature near $650°C$. The refuse
burned on a reverse reciprocating grate with a residence time of
about 20 minutes. Fly ash was removed from the combustion gases
in an electrostatic precipitator at approximately $230°C$ (Table 1).

Since the overall objective of the pilot study was to
describe the variability in the emissions at each site and between
the two sites, a single compound, or class of compounds had to be
used as a tool to describe variability. Total organic chlorine
(TOCL) was used as the variability indicator. More specifically,
TOCL represents the extractable and chromatographable organic
halogens present in the sample. This was used because a majority
of the POM compounds of interest contained chlorine and because
TOCL analyses could be performed on all the samples collected
under a limited budget. It was also chosen because the
concentration of any one specific compound might yield a high
level of variation, and since many compounds may only be present

near the minimum detection limit, measurement error could be very
significant. The TOCL analysis is very sensitive and the effect
of adding together the concentrations of many compounds would
decrease variability as well as decrease the influence of
measurement error.

It is very important to note the fact that this pilot program
was designed to be the basis for a national monitoring program to
be used to estimate POM emissions and not just the PCDD and PCDF
emissions, described in this report. For instance, the sample
train used to collect the substances present in the flue gas and
particulates has been widely used by EPA and others in the
detection and quantification of PCBs, PAHs and other compounds.
The collection efficiency of the resin trap has been documented
for a wide range of compounds under EPA's Source Assessment
Sampling System (SASS) (Adams et al, 1977).

No experiments have been performed to ascertain the specific
collection efficiency of PCDDs or PCDFs in this train to date.
Based on the efficiencies described in the literature for similar
compounds and our data describing the absence of chlorinated
compounds in the water impingers at the back of the train, it is
estimated that the collection efficiency is comparable to that of
similar compounds. Further work in this area is necessary.

The Quality Assurance/Quality Control (QA/QC) techniques used
in this study also had to be modified to describe the efficiencies
of POM extraction and detection, and not just PCDDs and PCDFs.
None of the samples could be spiked with a PCDD or PCDF compound
before extraction or the TOCL analysis would contain chlorine from
the spike compound. Therefore, two compounds were used as
surrogate compounds in order to estimate the extraction
efficiency. The compounds were d_8-naphthalene and d_{12}-chrysene,
neither of which are truly similar to PCDDs or PCDFs. However,
the results of recovery tests where fly ash was spiked with
several compounds and extracted using various methods, showed that
by using the extraction method utilized in this study, the
recovery efficiency for 1,2,3,4-TCDD (81 percent) was slightly
better than that for chrysene (73 percent). Based on these tests
and the chemical properties of the compounds involved, the
extraction efficiency for the PCDDs and PCDFs is likely to be as
good or better than that for either surrogate. Average extraction
efficiencies were 47% for d_8-naphthalene and 85% for d_{12}-chrysene
from the sample train components in the coal/refuse facility, and
27% and 62%, respectively, in the municipal waste combustor. The
difference between these recoveries is understandable since the
volatility of naphthalene is greater than that of chrysene and it
is more susceptible to loss during extraction.

The use of surrogate compounds such as TOCL, naphthalene, and chrysene, to decrease the resources necessary for large scale monitoring programs is still in its beginning stages. EPA is currently evaluating their use through programs such as this one.

SAMPLING METHODS

The coal/refuse and municipal refuse burning facilities were sampled for 10 and 9 days, respectively, by TRW Inc. under contract to EPA's Industrial Environmental Research Laboratory in Research Triangle Park, North Carolina. Samples of fly ash, bottom ash, aqueous influents and effluents, and fuel feeds were taken on a statistically determined periodic basis. Ambient air was sampled using high volume air samplers placed on the roof of each facility. Flue gas and particulates were sampled using two modified EPA Method 5 sample trains which drew gas and particulate from the stack. Two trains were used in order to sample a volume of 20 dscm over an 8-hour period.

Each sample train consisted of: a heated probe which could traverse the stack isokinetically and not permit water to condense; a heated cyclone to collect particulates; a heated filter to collect small particulates; a condensor; a cooled XAD-2 resin trap to collect organic compounds; and several impingers to protect the pump and other equipment from substances (Figure 1).

Upon completion of a day's run, each train was disassembled, and the filters and resin traps were stored on ice (4°C). The probe and glassware were rinsed with water, acetone and cyclohexane. The extracts were also stored on ice.

The aqueous contents of the first impinger were stored on ice for subsequent analysis to determine if any organic compounds had passed through the resin trap.

Field blanks were run on the sample trains, solvents, laboratory air, sample containers and filters.

ANALYTICAL METHODS

Coolers containing the filters, resin traps, aqueous extracts, ashes and water samples were delivered to two sub-contract laboratories. Southwest Research Institute, San Antonio, Texas, received the samples from the coal/refuse plant

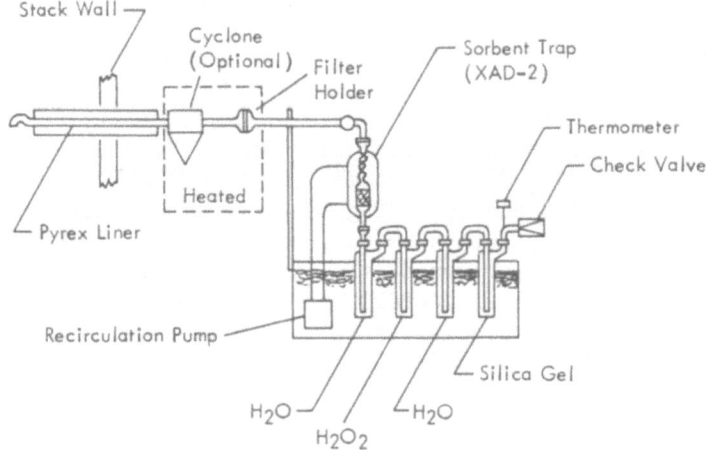

Figure 1. Modified Method 5 Train for Sampling Organics

and Gulf South Research Institute, New Orleans, Louisiana,
received the samples from the municipal waste combustor.

Each laboratory spiked all the samples with d_8-naphthalene d_2-
and d_1 chrysene. Solid samples (filters, resins and ashes) were
Soxhlet extracted with benzene for 8 to 24 hours depending on
sample type. Aqueous samples were extracted in separatory funnels
using cyclohexane. Portions of all extracts were then analyzed
for total organic chlorine (TOCL), or, more descriptively, total
organic halogens, by short packed column gas chromatography using
a Hall (halide mode) electrolytic conductivity detector.

The remaining portions of the extracts were then combined
into daily composites for each media (e.g., sample trains, fly
ash, bottom ash, combustion air, etc.), and analyzed in a phased
analytical scheme.

It should be noted that the particulate and vapor phase
extracts from the sample trains were combined, thus no data
describing phase partitioning are available.

These extracts were dried by passage through short columns of
anhydrous sodium sulfate, then concentrated to 5 mL in a kuderna -
Danish evaporator and further concentrated to 1 mL under purified
nitrogen.

The extracts of these composits were then analyzed by High
Resolution Gas Chromatography-Mass Spectrometry (HRGC-MS) using a
Hewlett Packard quardrupole GC-MS system at unit resolution and a
fused silica capillary column coated with SE-54. All extracts
showing possible PCDD or PCDF responses (0.25 ng/dscm detection
limit in flue gas) were cleaned using base and acid treatment
followed by chromatography on alumina (Harless et al, 1980).
Before this cleanup was performed, the extract was split, and one
portion was spiked with 1,2,3,4-TCDD and Octa-CDD for recovery
estimation. The average recoveries in this step were 60% for
1,2,3,4-TCDD and 25% for Octa-CDD.

The cleaned extracts were analyzed at Midwest Research
Institute by HRGC-HRMS on a Finnigan/MAT 311A using fused silica
WCOT columns with the spectrometer operated at 10,000 mass
resolution (10 percent valley) in the selected ion monitoring
mode. The two most abundant ions in each molecular cluster were
monitored. Prior to analysis, each extract (1.0 mL volume) was
spiked with 10 ng of $^{37}Cl_4$-2,3,7,8-tetrachlorodibenzo-p-dioxin
to serve as an internal standard. All dioxins and furans were
analyzed using a 12-m column coated with SE-54. TCDDs were also
analyzed using a 60-m column coated Carbowax 20 M in order to
better separate the 2,3,7,8-isomer. Responses for the internal
standard were the same in the standards samples.

Quantitation was based on responses of authentic standards or
the most similar compound available.

RESULTS AND DISCUSSION

Coal/RDF Burning Facility

No PCDDs or PCDFs were detected in any of the samples from
this facility in any medium at an instrumental detection limit of
less than 10 pg. In order to lower the detection limit in the
flue gas analyses, and to reduce the number of expensive analyses,
the ten daily extracts were combined into 5 extracts, by
compositing two days' samples (representing 40 m^3 of stack
effluent). The detection limit for PCDDs and PCDFs in the stack
emissions corresponded to a stack concentration of less than 0.25

ng/dscm, or a mass emission rate of less than 151 µg/hr, and the
detection limit in the fly ash was less than 0.5 ng/g.

Municipal Waste Combustor

No PCDDs or PCDFs were detected in the extracts of fly or
bottom ashes or aqueous effluents from this facility, but they
were detected in the flue gas sample extracts. No PCDD or PCDF
analyses were performed on the raw refuse due the high TOCL
variability found in the refuse. Detection limits in fly ash were
less than 0.5 ng/g.

In order to lower the detection limit in flue gas analyses
for PCDDs and PCDFs on HRGC/HRMS, the daily extracts were combined
into three extracts, by compositing three days' samples
(representing 60 m^3 of stack effluent). Tri-, tetra-, hexa-,
hepta-, and octa-chlorodioxins and furans were detected in all
three extracts (Table 2).

CONCLUSIONS

PCDDs and PCDFs were detected and quantified in a municipal
waste combustor, but were not detected in a coal/RDF burning
facility. The PCDDs and PCDFs observed in the former were only
present in the stack gas and associated particulates and not in
the fly ashes as described in studies conducted at other waste
combustion facilities (Bumb et al, 1980; Lustenhouwer et al, 1980).

The absence of detectable levels of PCDDs or PCDFs in all the
samples from the coal/RDF facility are in agreement with the
results of Junk and Richard (1981) at the same facility. Other
studies performed to date in facilities burning strictly coal
(Harless and Lewis, 1980; Kimble and Gross, 1980), have shown that
these facilities also do not contain detectable levels of PCDDs or
PCDFs in their fly ash.

Based on the procedures used in this pilot study and the
resulting detection limits, it appears that the coal/RDF burning
facility is emitting far less (if any) PCDDs or PCDFs per kg of
refuse burned than the municipal waste facility.

Table 1. Characteristics of the Two Facilities Selected by
EPA for a Pilot Study Quantifying Polycyclic Organic
Matter Emissions from Combustion Facilities.

| Facility | Fuel | | Combustion | Combustion | Electrostatic Precipitator |
	type	feed rate kg/hr)	Mechanism	Temperature°C	Temperature°C
Coal/Refuse Derived Fuel	Coal	14,000 (86%)	Suspension	1200	190
	RDF	2,300 (14%)			
Municipal Waste	raw refuse	17,000	Reverse Reciprocating Grate (20 minute Residence time)	650	230

Table 2. PCDD and PCDF Concentrations Observed in Flue Gas in a Municipal Waste Combustion Facility

Compound	Mean[1,2] ng/dscm	Std. Dev.	% of Total PCDD[3]	% of Total PCDF[3]	% of Total PCDD + PCDF[3]	Mean μg/hr. Emitted	Minumum[4] number of isomers
Tri-CDD	13	2.1	29		2.6	1100	3
Tri-CDF	300	44		65	59	26,000	7
Tetra-CDD	6.3	0.90	14		1.3	540	9
2,3,7,8-TCDD	0.41	0.10	0.9 (6.5 % of TCDD)		0.08	34	
Tetra-CDF	90	6.0		20	18	7600	5
Hexa-CDD	16	4.0	35		3.2	1400	5
Hexa-CDF	62	21		14	12	5200	5
Hepta-CDD	7.6	0.32	17		1.5	640	2
Hepta-CDF	7.5	0.46		1.6	1.5	640	1
Octa-CDD	2.5	0.39	5.5		0.49	220	1
Octa-CDF	0.60	0.13		0.13	0.12	52	1
Total PCDD	45				9	3,900	
Total PCDF	460				91	39,492	
Total PCDD +PCDF	505					43,392	

1 Mean of 3 data points
2 Not corrected for recovery
3 Sum does not equal 100 due to rounding
4 Minimum number of major isomers observed using SE-54 column

REFERENCES

Adams, J., Menzie, K., and Levins, P. Selection and evaluation of
    sorbent resins for the collection of organic compounds. EPA
    600/7-77-044, prepared by A.D. Little Inc., for
    U.S.E.P.A.-IERL, April 1977.

Bumb, R., Crummet, W., Cutie, S., Gledhill, J., Hummel, R., Kagel,
    R., Lamparski. L., Luoma, E., Miller, D., Nestrick, T.,
    Shadoff, L., Stehl, R., and Woods, J. Trace Chemistries of
    fire: A source of chlorinated dioxins. Science, 210:
    385-390, 1980.

Cavallaro, A., Bandi, G., Invernizzi, C., Luciani, L., Mongini,
    E., and Gorni, A., Sampling, occurrence and evaluation of
    PCDDs and PCDFs from incinerated solid waste. Chemosphere, 9
    (10):611, 1980.

Harless, R., and Lewis, R.G. Quantative Determination of
    2,3,7,8-TCDD Residues by Gas Chromatograpy/Mass Spectrometry
    In: Chlorinated Dioxins and Related Compounds - Impact on
    the Environment. O. Hutzinger, R.W. Frei, E. Merian and
    F. Pocchiar (eds.). Pergamon Series on Environmental Science,
    5:25-35, 1981.

Harless, R., Oswald, E., Wilkinson, M., Dupuy Jr., A., McDaniel,
    D., and Han Tai. Sample preparation and gas chromatography -
    Mass spectrometry determination of 2,3,7,8-TCDD. Anal. Chem.,
    52:1239-1245, 1980.

Junk, J.A., and Richard, J.J. Dioxins not in Effluents from Coal
    Refuse Combustion. Chemosphere, in press, 1981.

Lustenhouwer, I., Ilie, K., and Hutzinger, O. Chlorinated dibenzo
    p-dioxins and related compounds in incinerator effluents: A
    review of measurements and mechanisms of formation.
    Chemosphere, 9:501-522, 1980.

Shih, C., Ackerman, D., Scinto, L., Moon, E., and Fishman, E. POM
    emissions from stationary conventional combustion processes
    with emphasis on polychorinated compounds of dibenzo-p-dioxin
    (PCDDs), biphenyl (PCBs), and dibenzofuran (PCDFs). CCEA
    Issue Paper. Prepared by TRW Inc., for U.S.E.P.A.-IERL,
    January 1980.

CHANGES IN DIOXIN ISOMER GROUP RATIOS IN THE ENVIRONMENT:

AN UPDATE AND EXTENSION OF THE PRESENT THEORY

D. I. Townsend

Process Development
Dow Chemical Company
Midland, Michigan, USA

ABSTRACT

In 1979 a theory was developed to explain differences in
dioxin isomer group concentrations observed in samples taken from
combustion sources and the environment. From that work it was
hypothesized that simultaneous reduction-oxidation chemistry
tailored the composition of dioxin residue to conform with the
reductive conditions of the local environment. As a consequence
of this it was postulated that samples taken from reductive
sources should contain higher concentrations of the lower
chlorinated dioxin species than environmental samples exposed to
the oxidative conditions of the atmosphere. A recent study of
dioxins produced by wood burning stoves now makes it possible to
further test this theory as well as to compare the dioxin isomer
group ratios observed in different types of combustion units.

INTRODUCTION

Chlorinated dioxins (CDD's) are a group of 75 compounds which
have been found in a wide range of samples taken from both combustion sources and the environment. Because of the toxicity of
some of the CDD's, major scientific efforts have been underway
for several years to understand the conditions and chemistry
which led to the formation of the molecules from simpler chemical
species. Incineration has been identified as a source of CCD's
and high temperature reactions are generally accepted to be one
set of conditions under which CCD's are formed. The mechanism(s)
for formation of CDD's in the combustion process is not completely

153

understood. However, the observation that CDD's are widely
spread throughout the environment and come from several sources
is well documented.

   In 1979 the author developed a theory to tie together CDD's
found in the environment with those observed in combustion
sources (Townsend, 1979). The chemical equilibration concepts
advanced in the original hypothesis agree with results of a 1980
study of source and environmental samples (Crummett, 1980).
Further testing and extension of theory is now possible using
recently reported data for source samples taken from wood burning
equipment.

   The 75 CDD isomers can be divided into groups containing the
same number of chlorine atoms in the molecule. This defines eight
"isomer groups" upon which quantitative measurements are based."

   As discussed by Townsend (1980), the theory starts from the
observation that the mean relative concentration of the CCD
isomer groups found in environmental samples (taken at moderate
distances from combustion sources) have a geographically
independent average composition that can be estimated from the
observed ratios:

$$\frac{H_7CDD}{OCDD} = .152 \qquad \frac{HCDD}{OCDD} = .014 \qquad \frac{TCDD}{OCDD} = .002$$

   Where TCDD is the sum of the tetrachlorodibenzo-p-dioxin
isomers, HCDD is the sum of the hexachlorodibenzo-p-dioxin
isomers, H_7CDD is the sum of the heptachlorodibenzo-p-dioxin
isomers, and OCDD is the octachlorodibenzo-p-dioxin isomer.

   It is important to note that the proportionalities do not
apply to either source samples or biological samples. On the
contrary, source samples are observed to vary in composition and
typically contain higher amounts of the lower chlorinated CDD
isomers. It is this difference in the relative composition
between source and environmental samples that suggests an
equilibration process may be taking place in the atmosphere. As
shown in Table 1, samples taken at progressive distances from
combustion sources support the equilibration hypothesis and show
a shift in relative composition that approaches the ratios
projected for the environmental background.

Table 1. The Shift of Isomer Group Ratios with Distance
          from Source

INCINERATOR

Sample Identification	$H_7CDD/OCDD$
Major Metro #1, 30m NE of Incinerator	.340
Major Metro #2, 60m NE of Incinerator	.240
Major Metro #3, 120m NE of Incinerator	.150
Major Metro #4, 300m NE of Incinerator	.165

POWERHOUSE

Urban #1, Downtown 180m ENE of Powerhouse	.800
Urban #2, Downtown 270m ENE of Powerhouse	.240
Urban #3, Downtown 450m ENE of Powerhouse	.150
Environmental Mean (Table 4)	.152

     Such behavior could be accounted for by simultaneous
reduction-oxidation reactions where trace quantities of CDD's are
reduced to lower chlorinated species while the major portion of
CDD's are oxidized to non-dioxin by-products. Stated
differently, the oxidation process lowers the concentration of
all species while the reduction process thermochemically
redistributes the residual CDD isomers in a parallel process
whose rate depends on the reductive conditions of the local
environment. Results of a sample laboratory experiment that
demonstrates such chemical behavior is possible are presented in
Table 2.

Table 2. Products of Combustion of OCDD With N-Propanol

	Wt. Species/Starting Wt. OCDD Secondary Temperature			
Dioxin Isomer Group	$600^{\circ}C$	$700^{\circ}C$	$800^{\circ}C$	$900^{\circ}C$
OCDD	2×10^{-2}	3×10^{-5}	1×10^{-5}	ND (2×10^{-6})
H_7CDD	2×10^{-2}	1×10^{-4}	5×10^{-6}	ND (6×10^{-7})
HCDD	7×10^{-3}	9×10^{-5}	3×10^{-6}	ND (3×10^{-7})
TCDD	2×10^{-4}	2×10^{-4}	2×10^{-7}	ND (2×10^{-7})

These data show that co-combusion of pure octachlorodibenzo-*p*-dioxin with alcohol fuel not only oxidizes CDD's, but simultaneously redistributes CDD isomers not consumed by the destructive process. The experiment also implies that time, temperature, and other factors may all play an important role in the complex chemistry and account for variability of samples taken from "reductive sources." Conversely, environmental mean ratios should be independent of both source and geography if atmospheric interaction is responsible for the destruction of dioxins on airborne particulates as they travel into the environment. This hypothesis has been at least partially tested using data presented in a recent study by Crummett (1980). A sample of "Urban Particulates" collected by the National Bureau of Standards and one of Milwaukee Milorganite (dried municipal sludge) fertilizer both contained dioxins in the projected isomer group concentrations. It is assumed that these samples contain random particulates, a mixture of which creates the environmental background. Unlike environmental samples, particulates taken from reductive sources presumably reflect reductive conditions within those sources.

ISOMER GROUP RATIOS OBSERVED IN SOURCE SAMPLES

A summary of isomer group ratios observed in source samples (Crummett et al., 1978) is presented in Table 3. When compared to the background ratios, these data exhibit relatively high concentrations of lower chlorinated species compared to background levels as anticipated for samples taken from a reductive environment. Although the ratio values vary, the results are remarkably similar considering the probable differences in feed

Table 3. Source Sample Isomer Group Ratios from 1978 and 1980 Dow Studies

	$\dfrac{H_7CDD}{OCDD}$	$\dfrac{HCDD}{OCDD}$	$\dfrac{TCDD}{OCDD}$
Filtered Scrubber Water Table XII	.97	.19	.11
Rotary Kiln Scrubber Water w/o Supplemental Fuel Table XI	.62	.08	.06
Rotary Kiln Scrubber Water with Supplemental Fuel	.81	.17	.04
Muffler #5 (rust) Table XV	.11	.03	.16
Muffler #6 (Diesel) Table XV	.53	.10	.12
Fireplace A Table XVI	.64	.14	.01
Powerhouse	.16	.08	1.58
Nashville Incinerator	.93	.47	.26

and operating conditions of the various units. This reflects the importance of the role that redistribution chemistry may have in establishing the composition of dioxins entering the environment from combustion sources.

These data provide an overview on the similarities in performance of unrelated combustion equipment. Of equal interest is the difference in behavior of similar units operating under different conditions, which will now be reviewed for wood burning systems.

Isomer Group Ratios of Samples Taken From Wood Burning Units

Samples of particulate emissions from wood burning units have been collected and analyzed by Nestrick and Lamparski (these Proceedings). None of the wood has been treated with preservatives, and none of the wood was obtained from chemically sprayed areas to the knowledge of the owners. A summary of the isomer group ratios for the samples is presented in Table 4.

From the data of Table 4, it can be noted that the H_7CDD/OCDD (hepta) ratio for wood burning units varies from .37 to 1.95 with a mean value of .77.

Table 4. Dioxin Isomer Group Ratios in Samples Taken from
          Wood Burning Units

	$\dfrac{H_7CDD}{OCDD}$	$\dfrac{HCDD}{OCDD}$	$\dfrac{TCDD}{OCDD}$
Lamparski chimney	.68	.30	.07
Eastern #1	.44	.17	.01
Western #10	.70	.36	.05
Central #4	.58	.13	.02
Eastern #7*	.82	1.25	1.47
Western #12	.77	.51	.03
Central #97	.53	.34	--
Eastern #3	.57	.26	.15
Eastern # 5	1.95	3.66	1.42
Western #17*	1.06	2.78	.08
Central #87	.37	.11	--

*Copper based chimney cleaner used.

These results are very similar to the data of Table 3 for
miscellaneous sources where the ratios vary from .16 to .93 with
a mean of .60. Although the values range over a factor of 12,
this variance is relatively small compared to the concentration
range over which the ratios are observed. For example the OCDD
level in the sample "scrubber water Table XI" is 42,000,000 parts
per trillion (ppt) while the OCDD level in Central #87 is 24 ppt
which represents a million fold range in concentration. Apparently, Central #87 (hepta ratio of .37) can be interpreted as a
"dilute" version of the scrubber water sample (hepta ratio of
.61) despite the enormous difference in absolute concentration.
On the other hand, the mean ratio of all combustion samples is
.70 which is significantly different than .52 (range .10-.35)
reported for enviornmental samples (Townsend, 1980). This

difference in mean compositions is expected from the reduction-oxidation theory. A similar observation can be made with the HCDD/OCDD ratio.

In the source samples HCDD/OCDD ratios vary from .03 to 3.66, while corresponding range in environmental samples is .0034-.0414. The means of source and environmental samples are .59 and .014, respectively. A comparison of both the mean and the range of the data show that significantly more of the lower chlorinated species are present in source samples than environmental samples. It is interesting to note that three source samples from wood burning equipment have exceptionally high HCDD/OCDD ratios. Although two of these samples were obtained from systems where a copper based chimney cleaner was used after the burn, the relationship does not hold for the TCDD/OCDD ratios and the cause of the variations are not yet understood.

Perhaps the most interesting feature of the ratio variations is obtained from the TCDD/OCDD data. The Eastern #7, Eastern #5, and Powerhouse data all have tetra ratios greater than 1.0 and are "separated" from the rest of the data. The data in Tables 3 and 4 have no ratios between .2 to 1.0. The reason for this is not known, and may be coincidental.

Mean Ratios for Samples

The $H_7CDD/OCDD$ ratio data for source samples have a mean value of .77, but unlike the corresponding environmental mean, it is not to be thought of as being constant except perhaps in a statistical sense.

If one accepts the reduction-oxidation hypothesis, variations of the hepta ratio in environmental samples represent dioxins on particles trapped in various stages of oxidative decay and a mixture of these particles creates the environmental background. Accordingly, the average of many environmental samples should be related to the mean composition of the background. This value should be constant regardless of location. On the other hand, source ratios are hypothesized to represent differences in operating conditions within units and, therefore, vary with both the equipment and its operation. This is a random variation and the average is a statistical value which is interpreted as representing a typical combustion unit.

CONCLUSIONS

All of the data presented herein follow the trends projected from the reduction-oxidation theory and contain comparatively high concentrations of the lower chlorinated CDD isomers.

In the author's opinion, the source samples show a remarkable similarity in bulk composition considering the differences in the combustion processes under study. There are exceptions to the "typical" distribution found in reductive sources, but even the exceptions have common characteristics which may suggest a common cause. These observations suggest that the redistribution chemistry may be very important in tailoring the bulk CDD concentrations found in residues from combustion sources.

REFERENCES

Crummett, W. and Co-workers, 1978, "The Trace Chemistries of Fire a Source of and Routes for the Entry of Chlorinated Dioxins Into the Environment," Dow Chemical Company, Midland, Michigan.

Crummett, W., 1980, Environmental chlorinated dioxins - the trace chemistries of fire hypothesis, in: "Chlorinated Dioxins and Related Compounds - Impact on the Environment," O. Hutzinger, R. W. Frei, E. Merian and F. Pocchiari, eds., Pergamon Series on Environmental Science, 5:253-263, Pergamon Press, Oxford.

Townsend, D. I., 1980, The use of dioxin isomer group ratios to identify sources and define background levels of dioxins in the environment, J. Environ. Sci. Health, B15 (15): 571-609.

ANALYSIS FOR 2,3,7,8-TETRACHLORODIBENZO-P-DIOXIN

RESIDUES IN ENVIRONMENTAL SAMPLES

R. L. Harless and R. G. Lewis*; A. E. Dupuy and
D. D. McDaniel**

*U.S. Environmental Protection Agency
Health Effects Research Laboratory (MD-69)
Research Triangle Park, North Carolina, USA

**U.S. Environmental Protection Agency
Toxicant Analysis Center
Bay St. Louis, Mississippi, USA

ABSTRACT

    Small scale environmental monitoring studies have been
conducted to quantitatively determine the presence or absence of
2,3,7,8-tetrachlorodibenzo-p-dioxin (2,3,7,8-TCDD) in fish and
deer. 2,3,7,8-TCDD was detected in a concentration range of 3 to
42 parts-per-trillion (ppt) in 19 of 20 fish samples collected in
the State of Michigan. Two of 27 ppt levels of 2,3,7,8-TCDD were
detected in various parts of the bodies of deer which were used
in a study conducted in California to determine the effects of an
aerial application of 2,4,5-T. Average minimum limits of
detection for these analyses were: fish, 3 ppt; deer, 2 ppt. The
analytical methodology, criteria for confirmation of 2,3,7,8-TCDD
and the results of each study are discussed.

INTRODUCTION

    The analytical methodology and the results generated for
parts-per-trillion (ppt) levels of 2,3,7,8-tetrachlorodibenzo-
p-dioxin (2,3,7,8-TCDD) have been the subject of intense scrutiny
by the scientific community during the past decade. The
isolation and identification of ppt levels of 2,3,7,8-TCDD in
sample matrices that contain million-fold higher levels of other
naturally occurring compounds and/or other chlorinated industrial

pollutants have presented a challenging task for analytical
chemists. The analytical methods employed to date have utilized
efficient sample preparation procedures coupled with highly
sensitive and specific direct probe high resolution mass
spectrometry and gas chromatography-mass spectrometry detection
techniques. Analytical methodologies have evolved rapidly since
the early 1970's. Specificity for 2,3,7,8-TCDD determinations
may now be incorporated into the sample preparation procedure or
into the detection technique. The Environmental Protection
Agency (EPA) methodology (Harless et al, 1980) utilizes an
efficient but general type sample preparation procedure and
relies on high resolution gas chromatography (HRGC) coupled to
high resolution mass spectrometry (HRMS) for conclusive
identification and quantification of ppt levels of 2,3,7,8-TCDD
alone or in the presence of other isomers. Validation studies,
analytical criteria for confirmation of 2,3,7,8-TCDD, multiple
laboratory participation, and incorporated quality assurance
programs have been used to provide credibility and to validate
the analytical methodology and results.

    This report provides brief background information and
describes the analytical results that have been generated in
studies involving freshwater fish and deer. The states from
which samples were collected are Michigan and California. The
analytical results are confined to those generated in the U.S.
EPA laboratories and do not include those from samples analyzed
by contract laboratories (University of Nebraska and Wright State
University). It should be made clear that this is not an
official U.S. EPA report, but instead, it is a brief description
and summary of work performed in specific monitoring studies by
EPA scientists.

EXPERIMENTAL METHODS

Sample Collection

    The samples in each study were collected and documented
according to chain-of-custody protocols. The samples were then
shipped in a frozen state to the EPA Toxicant Analysis Center in
Bay St. Louis, Mississippi, for sample preparation and for
incorporation of appropriate quality assurance (QA) programs.
The extracts of test samples and QA samples were then shipped to
the EPA Health Effects Research Laboratory, Research Triangle
Park, North Carolina, for HRGC-HRMS determination of 2,3,7,8-TCDD
residues.

The Quality Assurance (QA) Program

The QA programs were instituted prior to sample preparation
and involved: fortification of actual test samples and QA
samples with 2.5 to 10 ng of $^{37}Cl_4$-TCDD[1]; fortification of
QA samples with 0 to 400 pg of 2,3,7,8-TCDD; and submission of
test samples and QA sample extracts (60 $\mu\ell$) to the HRGC-HRMS
laboratory in a blind fashion (i.e., there was no way for the
mass spectroscopist to distinguish between WA and actual test
samples). The validity of 2,3,7,8-TCDD analyses was enhanced by
the incorporated QA programs.

Sample Preparation and Determination of 2,3,7,8-TCDD

The comprehensive protocols and methods of analysis employed
in these studies are described elsewhere (Harless et al, 1980).
Briefly, the sample preparation involved: (1) fortification of a
2.5 to 10-g sample with 2.5 to 10 ng of $^{37}Cl_4$-TCDD for
determination of extraction and cleanup efficiency; (2)
saponification with hot alkali, followed by extraction with
hexane; (3) treatment with concentrated sulfuric acid; (4) chromatographic cleanup on alumina; and (5) concentration of the
alumina column extract to 60 μ for HRGC-HRMS analysis. In
addition, a neutral acetonitrile hexane partitioning extraction
and cleanup procedure was utilized on fish samples to provide
additional confirmation of TCDD residues. The sample preparation
procedures were developed for specific use with HRMS detection
techniques.

A Varian MAT 311A MS, directly coupled to a Varian Model 2700
GC, was used for these analyses. The GC was equipped with 30-m x
0.25-mm i.d., SE-30, OV-101 or OV-17 WCOT glass columns. The MS
was operated in the electron impact (EI) mode and utilized an
eight-channel hardware multiple ion selection (MIS) device. The
HRGC-HRMS parameters were: injection port temperature, 265°C;
GC oven program 6 min at 80°C, then programmed to 270°C at
34°C/min or alternatively programmed to 150°C at 34°C/min
and then programmed to 270°C at 4°C/min; GC transfer line to
MS ion source, 255°C; ion source temperature, 240°C; variable
acceleration voltage, 3 kV maximum; electron energy, 70 eV;
filament emission, 1 mA; mass resolution, 8000-10,000; multiplier
gain greater than 10^6.

The HRGC-HRMS MIS operating parameters were established
utilizing a $^{37}Cl_4$-TCDD and 2,3,7,8-TCDD quantification standard

---

[1] $^{37}Cl_4$-TCDD = labeled 2,3,7,8-$^{37}Cl_4$-TCDD, isotopic purity
greater than 98 percent

(i.e., 2 μℓ of 125 pg/μℓ $^{37}Cl_4$-TCDD and 5 pg/μℓ 2,3,7,8-TCDD)
at 8000 to 10,000 mass resolution with perflourokerosene (PFK) m/z
318.9793, as the reference mass. The masses used in these
measurements were: m/z 319.8965 and 321.8935 corresponding to
TCDDs; and m/z 327.8847 corresponding to full labeled $^{37}Cl_4$-
TCDD, the internal standard. A wide time window (i.e. 2 to 4
minutes before and after the elution of 2,3,7,8-TCDD) was
utilized in the analysis for detection of other TCDD isomers, if
present.

The exact masses, m/z 256.9327, 258.9298, 319.8965 and
321.8935, along with specific HRGC retention time, were used to
determine the COCl loss, which is indicative of 2,3,7,8-TCDD
structure in some fish samples. The HRGC-HRMS peak matching
technique was used in real time to confirm the presence of exact
masses, m/z 319.8965, and 321.8935, corresponding to the
elemental composition of 2,3,7,8-TCDD at the correct HRGC
retention time in a portion of the positive samples. TCDD
isomers, if present, could also be detected in COCl loss and
peak matching analyses.

Quantification of $^{37}Cl_4$-TCDD was based on both external
and internal standards (standard addition technique). The
concentration of 2,3,7,8-TCDD and the minimum limit of detection
were corrected for any losses in sample preparation efficiency.
The minimum limit of detection was defined as that amount of
2,3,7,8-TCDD that would provide clearly defined peak shapes for
the masses m/z 320 and 322, in the proper isotopic ratio and with
a signal to noise ratio greater than 2.5:1.

RESULTS AND DISCUSSION

Efficient sample preparation procedures incorporating the
$^{37}Cl_4$-TCDD internal standard and the direct-coupled HRCG-HRMS
detection technique provide a sensitive and specific analytical
method for unambiguous confirmation and quantification of ppt to
ppm levels of 2,3,7,8-TCDD in complex sample matrices. The
analytical criteria that were used to confirm the presence of
2,3,7,8-TCDD in these sample extracts are defined and shown in
Table 1. Each HRGC-HRMS analysis had to satisfy criterion I
through V to be considered a confirmed positive sample. Criteria
A and B provide the maximum degree of confirmation and were
applied to specific samples.

Fish

Twenty fish were collected from the Tittabawassee and Saginaw
Rivers and the Saginaw Bay in the State of Michigan in 1981. The
edible portions of fish samples, (catfish, carp, northern pike and

Table 1. Criteria for Confirmation of 2,3,7,8-TCDD Residues*

I   Correct HRGC-HRMS retention time for 2,3,7,8-TCDD.

II   Correct HRGC-HRMS multiple ion monitoring response for
      $^{37}Cl_4$-TCDD and 2,3,7,8-TCDD masses (simultaneous
      response for elemental composition of m/z 320, m/z 322,
      and m/z 328).

III   Correct chlorine isotope ratio for the molecular ions (m/z
      320 and m/z 322).

IV   Correct responses for the coinjection of sample fortified
      with $^{37}Cl_4$-TCDD and 2,3,7,8-TCDD standard.

V   Response of the m/z 320 and m/z 322 must be greater than 2.5
      times the noise level.

---

*Supplemental criteria which were applied to specific sample
 extracts:
 (a) COCl loss indicative of TCDD structure, and
 (b) HRGC-HRMS peak matching analysis of m/z 320 and m/z 322 in
     real time to confirm the exact masses corresponding to
     elemental compositions of TCDD.

sucker) were subjected to analysis for 2,3,7,8-TCDD. The
analytical results summarized in Table 2 show that 19 of 20
samples contained detectable quantities of 2,3,7,8-TCDD. The
concentrations ranged from 3 to 142 ppt. Seven samples contained
concentrations greater than 50 ppt.

    The QA samples incorporated in this study consisted of ocean
perch fortified with 0 to 30 ppt levels of 2,3,7,8-TCDD and
method blanks. The analytical results, also shown in Table 2,
show that reasonably accurate values were obtained for the
various fortification levels of 2,3,7,8-TCDD. Also, it was not
detected in "blank" ocean perch and method blanks.

    The analytical results shown in Table 2 are somewhat similar
to those previously reported (EPA News Release, 1978; Harless and
Lewis, 1980). In both EPA Studies, 1978 and the present one,
2,3,7,8-TCDD was also detected in fish from the Saginaw River and
the Saginaw Bay, commerical fishing waters. The 1978 EPA study
had been conducted in response to announcements from the Dow
Chemical Company reporting the presence of TCDD in Tittabawassee
River fish (Dow Chemical Company, 1978). The following year
(Kuehl et al, 1979) reported the presence of tetra- through

Table 2. Analytical Results for 2,3,7,8-TCDD Residues in Fish

Location and Species	No. of Fish	No. of Positive Samples	Sample Preparation Efficiency[a]	TCDD Detected (ppt)[b] Low	High	Mean	TCDD Minimum Limit of Detection (ppt)[b]
Tittabawassee River							
Carp	8	8	88	7	142	71	4
Sucker	3	3	79	3	10	7	2
Saginaw River							
Carp	5	5	91	18	63	31	3
Sucker	1	0	98	ND	--	--	2
Northern Pike	1	1	100	4	--	--	2
Saginaw Bay							
Carp	1	1	100	50	--	--	2
Catfish	1	1	98	61	--	--	5
Quality Assurance Program							
North Atlantic Ocean Perch	2	0	100	ND	--	--	2
Ocean Perch Fortified with 30 ppt 2,3,7,8-TCDD	1	1	68	28	--	--	2
Method Blank	2	0	72	ND	--	--	3
Standard Solution Containing 5 ppt 2,3,7,8-TCDD	1	-	88	5	--	--	2

[a] Mean percent recovery for 5 ng $^{37}Cl_4$-TCDD added to 10 g samples prior to sample preparation.
[b] Corrected for losses in sample preparation efficiency. ND = not detected. ppt = parts per trillion.

octa-CDDs in fish from the Tittabawassee River. Recently (Buser, 1980) also reported the presence of 2,3,7,8-TCDD in a carp fish collected from the Tittabawassee River.

Blodgett Forest Study

Bioaccumultion of 2,4,5-T and its contaminant, 2,3,7,8-TCDD, in the food chain has been a major concern during the past decade. This project was designed to provide in-depth and documented knowledge on the behavior of the herbicide and its contaminant.

An eleven acre plot in the University of California's Bldogett Experimental Forest, located in El Dorado County, was used to study the effects of an aerial application of 2,4,5,-T in 1978. An intensive monitoring project was conducted to document the occurrence of 2,4,5-T in air, water, soil and vegetation before an after the aerial application. 2,4,5-T containing less than 0.1 ppm 2,3,7,8-TCDD was applied at a rate of 3.4 kilograms per hectare to approximately three hectares of the enclosed plot.

Twelve deer were placed in the enclosure prior to the application of 2,4,5-T. One deer died 2 days later of unknown causes. Deer were sacrificed prior to and at specific intervals during the course of the 30 day study. Muscle, adipose tissue, liver and bone marrow of respective deer were subjected to analysis for 2,3,7,8-TCDD.

The analytical results generated by this laboratory are summarized in Table 3. Extremely low concentrations of 2,3,7,8-TCDD were detected in these deer: The frequency of detection was highest in adipose tissue and lowest in muscle and liver. It was not detected in bone marrow of five deer that were analyzed. Although not shown in this summary, some of the highest concentrations were detected in deer that were sacrificed at the conclusion of the 30 day study. These results are not surprising. They show that 2,3,7,8-TCDD residues can accumulate in animals that live and graze in areas exposed to phenoxy-herbicides contaminated with 2,3,7,8-TCDD. Investigations reported by (Ryan et al, 1974; Bartleson et al, 1975) and other unpublished reports also who that 2,3,7,8-TCDD can accumulate in tissue of wildlife and domestic animals exposed to phenoxyherbicides and 2,3,7,8-TCDD. The QA program incorporated in this study is shown in Table 4.

SUMMARY

The results showing ppt levels of 2,3,7,8-TCDD in fish from the Tittabawassee River are in agreement with those reported by

Table 3.   Analytical Results for 2,3,7,8-TCDD Residues in Blodgett Forest Deer Study

Sections of Eleven Deer in Study	No. of Deer Samples Analyzed	No. of Positive Samples	Concentration Range of 2,3,7,8,-TCDD Detected (ppt)[a]	Limit of Detection Range (ppt)[a]
Muscle	11	3	12 to 27	0.5 to 5
Adipose Tissue	10	8	3 to 12	1 to 3
Liver	11	4	2 to 5	0.4 to 4
Bone Marrow	5	0	ND	1 to 3

[a]Results corrected for sample preparation efficiency.
ND = not detected.
ppt = parts per trillion.

Table 4. Quality Assurance Program for Blodgett Forest Deer Study

	Experimental Results		
Sample Preparation Efficiency[a]	Minimum Limit of Detection for 2,3,7,8-TCDD (ppt)[b]	2,3,7,8-TCDD Detected (ppt)[b]	2,3,7,8-TCDD Fortification Level (ppt)[c]
92	2	ND	0
81	3	35	30
91	3	15	15
112	3	19	16
96	3	44	43
70	2	ND	0
67	4	18	28
105	1	20	14
62	3	60	48
108	2	12	7
87	3	16	10

[a]Percent recovery of 2.5 to 10 ng $^{37}Cl_4$-TCDD from 5 or 10 g deer adipose samples.
[b]Results corrected for sample preparation efficiency.
[c]Fortification levels of 2,3,7,8-TCDD.
ND = not detected.
ppt = parts per trillion.

other independent laboratories. In both EPA studies, 1978 and
the present one, ppt levels of 2,3,7,8-TCDD were also detected in
fish from the Saginaw River and the Saginaw Bay, commerical
fishing waters. Other isomers of TCDD were not detected in this
study. Investigations to determine any preferential retention of
specific TCDD isomers by fish have not been performed and/or
reported. Therefore, the source or sources of the 2,3,7,8-TCDD
contamination in fish remains unknown.

The findings for the Blodgett Forest Study show that low ppt
levels of 2,3,7,8-TCDD can accumulate rapidly in tissue of deer
intentionally exposed to one aerial application of 2,4,5-T.
These findings provide a preliminary data base for future
monitoring activities. The analytical methodology was adequate.
However, the detection limits should be 0.5 ppt or lower in
future studies and the QA program should be designed for residue
levels of 0.5 to 20 ppt.

Complex problems can be encountered in ppt level analysis and
this is especially true in monitoring projects involving large
numbers of samples. Multiple and independent laboratory
participation and confirmation of results enhance the credibility
and validity of ppt level analysis.

REFERENCES

1. F. D. Bartleson, D. D. Harrison, and J. D. Morgan, Field
    Studies of Wildlife Exposed to TCDD Contaminated Soil,
    Technical Report, AFATL-TR-75-49. Air Force Armament
    Laboratory, Armament Development and Test Center, Eglin
    Air Force Base, Florida 32542, 1975.
2. H. R. Buser, High-Resolution gas chromatography of the 22
    tetrachlorodibenzo-p-dioxin (TCDD) isomers. In:
    "Chlorinated Dioxins and Related Compounds - Impact on the
    Environment", O. Hutzinger, R.W. Frei, E. Merian, and F.
    Pocchiari, eds., Pergamon Series on Environmental Science
    5:15-24, Pergamon Press, Oxford, 1981.
3. Dow Chemical Company, The Trace Chemistries of Fire -- A
    Source of and Routes for the Entry of Chlorinated Dioxins
    into the Enviornment, The Chlorinated Dioxin Task Force,
    Michigan Division Dow Chemical Company, Midland, Michigan, 1978.
4. R. L. Harless, E. O. Oswald, M K. Wilkerson, A. E. Dupuy, D. D.
    McDaniel, and Han Tai. Anal. Chem. 52:1239-1245, 1980.
5. R. L. Harless, and R. G. Lewis, Quantitative determination of
    2,3,7,8-tetrachlorodibenzo-p-dioxin residues by gas chromato-
    graphy/mass spectrometry. In: "Chlorinated Dioxins and Related
    Compounds, O. Hutzinger, R. W. Frei, E. Merian, and F.
    Pocchiari, eds. Pergamon Series on Environmental Science 5:25-
    35, Pergamon Press, Oxford, 1981.

6. D. W. Kuehl, R. C. Dougherty, Y. Tondeur, D. L. Stallings, L. M. Smith, and C. Rappe. Negative Chemical Ionization Studies of Polychlorinated dibenzo-p-dioxins, dibenzofurans and naphthalenes in environment samples. Unpublished report. D. W. Kuehl, U.S. EPA Environmental Research Laboratory, 6201 Congdon Blvd., Duluth, Minnesota 55804, 1979.

7. J. F. Ryan, F. J. Biros, and R. L. Harless, Proceedings 22nd Annual Conference on Mass Spectrometry and Allied Topics, Philadelphia, Pennsylvania, May 19-24, 1974.

LONG-TERM STUDIES ON THE PERSISTENCE AND

MOVEMENT OF TCDD IN A NATURAL ECOSYSTEM

A. L. Young

Office of Environmental Medicine
Veterans Administration
Washington, D.C., USA

ABSTRACT

Field studies of the persistence and movement of 2,3,7,8-
tetrachlorodibenzo-p-dioxin (TCDD) were conducted during 1973-
1979 on a unique 3.0 km^2 military test area (Test Area C-52A,
Eglin Air Force Base, Florida) that was aerially sprayed with
73,000 kg 2,4,5-trichlorophenoxyacetic acid (2,4,5-T) during the
period 1962-1970. Analysis of archived samples of the formulations
indicated that approximately 2.8 kg TCDD were applied as a contaminant of the herbicide. However, 2.6 kg of this TCDD were
applied to a 37 ha test grid (Grid I) from June 1962 through July
1964. Levels of 10 to 1,500 parts-per-trillion (ppt) could be
found in the top 15 cm of soil 14 years after the last application
of herbicide on this site. Nevertheless, analysis of 61 soil
samples suggested that less than one percent of the TCDD remained
on the test area. Photodegradation at the time of and immediately
after aerial application probably accounted for much of the disappearance of TCDD, although volatilization, wind and water erosion,
and biological removal may have also contributed to its
disappearance.

INTRODUCTION

The toxin 2,3,7,8-tetrachlorodibenzo-p-dioxin (TCDD) is known
to occur as both a contaminant of products made from trichlorophenol (Young et al, 1978) and as a by-product from low
temperature incineration of wastes containing chlorinated precursors (Esposito et al, 1980). The magnitude of environmental

173

contamination by the 2,3,7,8-TCDD isomer is currently the subject
of intense debate (Young, 1980). Although a number of TCDD sources
have been identified, environmental monitoring programs for TCDD have
generally been unsuccessful in documenting contamination (Esposito
et al, 1980). However, with continued development of sophisticated
instrumentation for detecting TCDD in picogram (1 x 10^{-12} g)
quantities and with renewed interest in monitoring improper
disposal of hazardous wastes, additional data on the distribution
of TCDD in the environment may soon be forthcoming. In the
interim, two long-term studies have been initiated on the fate and
persistence of TCDD in natural ecosystems. Fanelli et al, 1980,
and di Domenico et al, 1981, have continued to document the fate
of TCDD near Seveso, Italy, following an industrial accident in
1976 that resulted in the contamination of over one square kilometer
(km^2) of land in an industrial-agrarian community. The second
long-term study is of a unique military test site in Northwest
Florida, that received massive quantitites of the herbicides
2,4,5-trichlorophenoxyacetic acid (2,4,5-T) and 2,4-dichloro-
phenoxyacetic acid (2,4-D) in the course of developing defoliation
spray equipment for use in the Vietnam Conflict, 1962-1970. Data
from this study were initially released in 1974 (Young), 1975
(Young et al) and updated in 1979 (Young et al). The present
report is a final report of studies originally initiated in 1973 on
the fate of TCDD in the ecosystem of this unusual test site.

INVESTIGATIONS

    The Eglin Reservation in Northwest Florida has served various
military uses, one of them having been the development and
testing of aerial dissemination equipment in support of military
defoliation operations in Southeast Asia. It was necessary for
this equipment to be tested under controlled situations that would
simulate actual use conditions as near as possible. For this
purpose an elaborate testing installation, designed to measure
deposition parameters, was established on the Eglin Reservation
with the place of direct aerial application restricted to an area
of approximately 3 km^2 within Test Area C-52A in the southeastern
part of the reservation. Massive quantities of herbicide, used in
the testing of aerial defoliation spray equipment from 1962 through
1970, were released and fell within the instrumented test area. The
uniqueness of the area prompted the United States Air Force to set
aside the area in 1970 for research investigations. Numerous
ecological surveys have been conducted since 1970. As a result, the
ecosystem of this unique site has been well studied and documented.

Description of the Field Site

Test Area C-52A (TA C-52A) covers an area of approximately 8 km^2 and is a grassy plain surrounded by a forest stand that is dominated by longleaf pine (Pinus palustris), sand pine (Pinus clausa), and turkey oak (Quercus laevis). The actual area for test operations occupied an area of approximately 3 km^2 and was a cleared area, mechanically leveled and subsequently occupied mainly by broomsedge (Andropogon virginicus), switchgrass (Panicum virgatum), woolly panicum (Panicum lanuginosum) and low growing grasses and herbs. Today, most of the area used for test operations has been overgrown by a diverse vegetative community of grass, broadleaves, and shrubs. Much of the center of the range was established prior to 1960, but the sites used for the establishment of spray equipment testing grids were developed in 1961 and 1962. The test area is approximately 28 m above sea level with a water table at 1.5 to 3 m. The major portion of this test area is drained by five small creeks whose flow rates are influenced by an average rainfall of 150 cm. The mean annual temperature for the test area is 19.7^{o}C while the mean annual relative humidity is 70.8 percent. For the most part, the soil of the test area is a fine white sand on the surface, changing to yellow beneath. The soils of the range are predominantly well drained, acid sands of the Lakeland Association with a 0 to 3 percent slope. A typical one-meter soil core contained approximately 92 percent sand, 3.8 percent silt, and 4.2 percent clay with an organic matter content of 0.17 percent, an average pH of 5.6, and a cation exchange capacity of 0.8.

Herbicide Application

Although the total area for testing aerial dissemination equipment was approximately 3 km^2, the area actually consisted of four separate testing grids. The primary area was located in the southern portion of the testing area and consisted of a 37 ha instrumented grid. This was the first sampling grid and was in operation in June 1962. It consisted of four intersecting straight lines (flight paths) arranged in a circular pattern, each path being at a 45° angle from those adjacent to it. Although this grid was discontinued after 2 years, it received the most intense testing program. From 1962 to 1964, this grid (called Grid I) received 39,550 kg 2,4-D and 39,550 kg of 2,4,5-T as the Herbicide Purple formulation (50 percent n-butyl 2,4-D, 30 percent n-butyl 2,4,5-T and 20 percent iso-butyl 2,4,5-T). Two other testing grids were sprayed with Herbicide Orange (50 percent n-butyl 2,4-D and 50 percent n-butyl 2,4,5-T). Grid II was an area of 37 ha and located immediately north of Grid I. Grid II received 15,890 kg 2,4-D and 15,890 kg 2,4,5-T from 1964 through 1966. Grid IV was the largest and final Grid established on Test Area C-52A. It was approximately 97 ha and received 20,000 kg 2,4-D and 17,570 kg 2,4,5-T from 1968 through

1970. Grid III was an experimental circular grid that received 1,300 kg 2,4-D from 1966 through 1970. Thus, for the four spray equipment calibration grids, a total of approximately 73,000 kg 2,4,5-T and 77,000 kg 2,4-D were aerially disseminated during the period 1962-1970. These data are summarized in Table 1.

TCDD in Military Herbicide Formulations

Only a few archived samples of Herbicides Orange or Purple were available from the Eglin spray equipment test program when the project terminated in 1970. The single archived sample of Purple contained 45 ppm TCDD. However, as noted by Young et al in 1978, a more likely mean value of TCDD in Purple was 32.8 ppm. Four archived samples of Orange Herbicide remained from the Eglin program. The mean of these four samples (0.04, 0.04, 3.2 and 6.4 ppm) was 2.4 ppm TCDD. However, the analysis of TCDD of 490

Table 1. Approximate Amount of 2,4,5-T and 2,4-D and Estimated Amount of TCDD Applied to Test Area C-52A, Eglin AFB Reservation, Florida, 1962-1970

Test Grid	Grid Area (ha)	2,4,5-T[a] (kg)	2,4-D[a] (kg)	TCDD[b] (kg)
I	37	39,550 (1962-1964)[c]	39,550 (1962-1964)	2.613
II	37	15,890 (1964-1966)	15,890 (1964-1966)	0.078
III	37	- - - -	1,300 (1966-1970)	- - -
IV	97	17,570 (1968-1970)	20,000 (1968-1970)	0.087
TOTAL	208	73,010	76,740	2.778

[a]Amount of 2,4,5-T and 2,4-D calculated on weight of active ingredient in the military Herbicides Orange or Purple.
[b]Amount of TCDD calculated from data on mean concentration of TCDD in the formulation of Herbicides Purple or Orange, i.e., 32.8 ppm TCDD in Purple and 1.98 ppm TCDD in Orange.
[c]Years when the specific grid received the herbicide contaminated with TCDD.

archived Herbicide Orange samples from other Air Force programs
placed the weighted mean concentration of TCDD in Orange as
1.98 ppm (Young et al, 1978). Therefore, the calculation of
the amount of TCDD applied to Test Area C-52A is based upon the
mean values of 32.8 ppm and 1.98 ppm, for Herbicides Purple and
Orange, respectively. The TCDD data for the individual test
grids are reported in Table 1. The total amount of TCDD (2.8
kg) applied to the Eglin test sites necessitates studies of its
environmental fate.

Analysis of Soil Samples

Despite excellent records as to the number of missions and
quantity of herbicide per mission, there was no way to determine
the exact quantity of herbicide per mission deposited at any point
on the instrumented grids. The first residue studies of Test
Area C-52A involved analyses of soils for phenoxy herbicides by
both chemical and bioassay techniques. The problems encountered
in the residue studies centered on the heterogeneity of the test
grids. Not only were there small geologic differences (soil
types, contours, organic matter, and pH), and differences in
vegetation density and locations of water, but most important the
herbicides had been sprayed on specific test arrays (i.e., along
dictated flight paths) over a span of 8 years. By considering
the flight paths, the water sources and the terracing effects, it
was possible to divide the test area into 15 vegetation areas.
These areas formed the basis for the random selection of 48 soil
cores, plus three additional cores collected from a control site.
Sampling at each site consisted of removing 15 cm increments of
soil to a total depth of 90 cm. Each increment from each of
the 51 cores was bioassayed using soybeans and cucumbers. Many
were analyzed by gas chromatography. The initial experiment was
conducted in April 1970. Selected sites were re-sampled in
December 1970. Both the bioassay and chemical method indicated
that 2,4-D and 2,4,5-T were present throughout the soil core in
the parts-per-million level in April 1970. However, by December
1970, herbicide residue levels at all depths were in the low
parts-per-billion. The last detectable levels of 2,4-D or
2,4,5-T were recorded in 1971. The rapid invasion of herbicide-
sensitive plant species that occurred from 1971 to 1973 confirmed the disappearance of phytotoxic residues (Young et al, 1975)
For example, in 1971, 74 dicotyledonous species were collected on
the 208 ha area; in 1973, 107 dicotyledonous species were
collected.

In the spring of 1971, soil cores were selected that
previously had high levels of herbicide residues in them as
indicated by the bioassay studies. These cores were collected

from Grid IV, an area sprayed with Herbicide Orange from 1968 to
early 1970. The samples were analyzed by the United States
Department of Agriculture Pesticide Degradation Laboratory,
Beltsville, Maryland, and found to be negative at a detection limit
of 0.001 ppm TCDD. No additional samples were collected until
1973. Since data from studies conducted by other researchers
suggested that TCDD may persist in the soil, a critical review of
the history of the test range, combined with data suggesting that
Herbicides produced in the late 1950's and early 1960's were more
contaminated with TCDD than material produced later in the 1960's,
indicated the site to search would be Grid I. Simultaneously the
detection limit for TCDD was now approaching the level of parts-
per-trillion (ppt). The analysis of Grid I soils in 1973 confirmed
the presence of TCDD. Consequently, a variety of sampling and
residue monitoring studies for TCDD have been conducted on the
test area (Young, 1974; Bartleson et al, 1975; Young et al, 1975;
and Young et al, 1979). Because of the long-term nature of these
studies, it has been necessary for more than one laboratory to
provide analytical services. The TCDD analyses reported were
obtained through Air Force contracts for analytical services from
the following laboratories: The Interpretive Analytical Services,
Dow Chemical Company, Midland, Michigan (1972-1975); The Aerospace
Research Laboratories, Wright-Patterson AFB, Ohio (1972-1974); the
Brehm Laboratory, Wright State University, Dayton, Ohio (1975-1977);
Midwest Center for Mass Spectrometry, Department of Chemistry,
University of Nebraska, Lincoln, Nebraska (1977-1980); and The
Flammability Research Center, University of Utah, Salt Lake City,
Utah (1977-1979). The methods for analyses have been previously
described by Shadoff and Hummel, 1978.

RESULTS AND DISCUSSIONS

The analytical results of a 1974 sample from Grid I suggested
that most of the TCDD would be found in the top 15 cm of soil
profile (Table 2).

From June 1974 through April 1978 soil samples were collected
in an attempt to define the magnitude of TCDD concentration
remaining on the three grids that received 2,4,5-T herbicide. These
data are shown in Table 3 and validate that Herbicide Purple must
have contained concentrations of TCDD much greater than Herbicide
Orange (compare Grids I and IV in Tables 1 and 3).

Is it valid to "average" TCDD data from many dates (years) to
obtain the data in Table 3? The answer would depend upon the
amount of variation in TCDD concentration that occurs within a

Table 2.   Concentration of TCDD in a Soil Profile from
           Grid I, Test Area C-52A, Eglin AFB, Florida[a].

Depth (cm)	TCDD (Parts-per-Trillion)
0 - 2.5	150
2.5 - 5.0	160
5.0 - 10	700
10 - 15	44
15 - 90	ND[b]

[a]Grid I received 1,069 kg/ha of 2,4,5-T Herbicide during 1962-1964.
The soil samples were collected and analyzed in 1974.
[b]None detected, minimum detection limit   10 ppt.

Table 3.   Concentration of TCDD (parts-per-trillion) in Test
           Grid Soils, Test Area C-52A, Eglin AFB, Florida[a].

Grid	Number of Samples[b]	Range	Median	Mean
I	22	10 - 1,500	110	325
II	6	10 - 470	30	115
IV	26	10 - 150	20	30

[a]Source: Young et al, 1979.
[b]0 - 15 cm increment, collected during the period June 1974 through
April 1978.

grid, over time, and in the analysis of the soil samples. As previously noted, the aerial determination of herbicides on a test grid was neither uniform nor random but rather along discrete sampling arrays arranged to measure particle size and deposition. Moreover, since the flights also occurred either in-wind or cross-wind, and the testing of the aerial dissemination equipment for Grid I extended from June 1962 through July 1964, tremendous variation in residue levels would be predicted (note the range of soil TCDD levels for Grid I in Table 3). Indeed, sampling along an imaginary line interesecting where the flight paths were located on Grid I confirmed the variations in TCDD residue (Table 4).

The variations that occurred in the soil concentration of TCDD within a 1 ha area randomly sampled during a 5-year period are shown in Table 5. Even when samples are collected over time from small discrete locations, in this case near the center of Grid I, significant variation in TCDD concentration occurred. Data from 1 m^2 plots sampled in August 1974 and again in January 1978 are presented in Table 6. Although the data from analysis of the samples are consistent in showing a downward trend, the magnitude of decrease between samples is inconsistent. This suggests that

Table 4. Concentration of TCDD (parts-per-trillion) in Soil Samples Collected on an East-West Line that Would Have Intersected Flight Paths on Grid I.[a]

Depth cm	Location[b] 0-4S	0-7S	0-8S	0-9S	Mean[c]
0 - 5	930	90	130	180	335
5 - 10	930	110	220	100	340
10 - 15	630	250	210	80	295
Mean	830	150	185	120	320 ± 340

[a]Samples collected in December 1975.

[b]Samples collected 20 m south of permanent impinger posts on "0" row (northern boundary of Grid I).

[c]Means rounded to 5 ppt increment.

Table 5. TCDD Concentrations (parts-per-trillion) in Single
          Samples of the Top 15 cm of Soil Randomly Collected
          Within a 1 ha Area (Location 0 - 7S, Grid I) over a
          5-year Period.

Date (Month/Year)	TCDD (ppt)
October 1973	710
June 1974	300
December 1975	150
October 1977	150
April 1978	490
January 1979	410
Mean	$\overline{370} \pm 215$

Table 6. Disappearance of TCDD from Soils of Grid I
          (parts-per-trillion)[a]

PLOT NUMBER[b]	AUGUST 1974	JANUARY 1978
1	1,500	420
2	610	300
3	1,200	580
4	270	100
5	440	400
Mean	$\overline{805} \pm 525$	$\overline{360} \pm 175$

[a]Source: Young et al, 1979.
[b]Five subsamples from each 1-m^2 plot composited (1 - 10 cm depth).

factors other than actual disappearance of TCDD maybe involved.
Indeed, samples collected in August 1974 were analyzed by one
laboratory (in 1974) while the samples collected in January 1978
were analyzed by a different laboratory (in 1978). Although
presumably the method was similar, the laboratory that analyzed
soil samples in 1978 provided data variability encountered in the
analysis of a single soil sample. These data are provided in
Table 7. The rather large variability in this soil sample suggests
that the extraction of TCDD from soil (soil contaminated at least
14 years prior to extraction and analysis) and its subsequent
analysis (via high resolution mass spectrometry) are no easy tasks.

In summary, from data provided in Tables 4 through 7, and
re-emphasizing that the TCDD in the soils of the test grids was
deposited over a period of years, the data in Table 3 are probably
the most representative data for assessing the relative contamination of the grids of Test Area C-52A. Moreover, using the
data in the previous tables may provide information on the fate
of TCDD in the ecosystem associated with Test Area C-52A.

ROUTES OF TCDD DISAPPEARANCE FROM SOIL

Data in Table 1 provided an estimate of the amount of TCDD
likely to have been contained in the herbicide disseminated on the
Test Area. Data in Young et al, 1978, from a review of operational
records from evaluations of the spray equipment tested on Test

Table 7. Variation in the Analysis of a Soil Sample
Collected from Grid I in 1978.

Subsample[a]	TCDD (parts-per-trillion)
1	340
2	500
3	420
4	590
Mean	465 $\pm$ 105

[a]200 g soil was dried, sieved, and thoroughly mixed prior to sub-
sampling and analysis.

Area C-52A, indicated that approximately 87 percent of the
herbicide fell within the instrumented area. Therefore, if 87
percent of the TCDD in the herbicide applied to Grid I impacted the
test grid, then approximately 2.3 kg of TCDD needs to be
accounted (0.87 X 2.613 kg = 2.27 kg TCDD). If a mean value of
325 ppt TCDD (Table 3) is used as the concentration remaining in
the test grid 12 to 14 years after the last application of herbicide,
and if most TCDD is in the top 15 cm of soil, then less than 27 gms
of TCDD remained on Grid I (37 ha X 2.24 X 10^6 kg soil/ha X 325
X 10^{-9} g TCDD/kg soil = 26.9 g TCDD). Thus, approximately 1 percent of the amount of TCDD contained in the herbicide was still
present on the test grid 12-14 years after it was disseminated, or
more appropriately 99 percent of the TCDD had disappeared.

Numerous routes have been proposed for the disappearance of
TCDD from soil (Esposito et al, 1980; Ward and Matsumura, 1978; and
Young, 1980). The methods most likely responsible for TCDD disappearance include photodegradation, wind and water movement of
contaminated particles, volatilization, microbial degradation, and
biomass removal.

Photodegradation

Crosby and Wong (1977) found that Herbicide Orange containing
known amounts of TCDD and exposed to natural sunlight on leaves,
soil, or grass, lost most or all of the TCDD in a single day, due
prinicpally to photochemical dechlorination. Furthermore, they
have established three requirements for significant TCDD breakdown
in the environment; namely, dissolution in a light-transmitting
film, the presence of an organic hydrogen doner such as a solvent
or pesticide, and ultraviolet light. They noted that all three
conditions are normally met during the practical application of
2,4,5-T.

Young (1974) noted that during the major years of testing
spray equipment, 1963 through 1969, micrometeorological conditions
existing below 90 m over the test area were continuously monitored
by an Automated Meteorological Data Acquistion and Processing
System. This system automatically measured, processed, and stored
data from meteorological senosrs on a series of towers (2 to 90 m
in height). Thus a mechanism for monitoring temperature, wind,
rainfall, dew point and periods of sunlight was available in
scheduling test operations. Missions were generally scheduled and
conducted when environmental conditions were optimal. This suggests
that conditions favorable for dissemination of herbicide were
probably the same conditions favorable for photodegradation of TCDD.

From the above information, it is indeed likely that photodegradation may have been responsible for the majority of TCDD
disappearance. The failure of test operations personnel to observe

or record animal deaths or to have experienced readily detected
health problems, e.g. chloracne, suggest that significant TCDD
accumulation on the test grids did not occur (see Thalken and
Young, 1982).

Wind and Water Movement of Contaminated Particles

In studies to determine leaching and degradation rates of
Herbicide Orange and TCDD, Young et al, 1976, established field
plots in 1972 immediately adjacent to Grid II. These studies
confirmed that despite 150 cm annual rainfall (falling onto highly
porous sandy soil), no appreciable movement of TCDD was detected.
Although data in Tables 2 and 4 show TCDD dispersed within the
top 15 cm of soil, it is unlikely that these data on TCDD represent
leaching. Rather, a more likely explanation is that the TCDD
was deposited in layers, during and in subsequent years after
herbicide application, as a consequence of wind and water movement
of contaminated soil particles. Examination of the soil horizons
in excavated profiles of Grid I clearly show within the top 15 cm
discrete layers that differ from the parent soil. In reviewing
the climatic data from the Test Area, Young et al, 1975, plotted
the surface wind quadrants and concluded that direction and speed
of the prevailing winds probably resulted in soil moving not only
back and forth across Grid I, but because of a slight difference
in topography between Grid I and II, contaminated soils from Grid
II could have been deposited in low areas of Grid I. Certainly,
water moved contaminated particles into low lying areas of Grid
I. Similar observations of the soils and TCDD contamination of
Zone A, Seveso, Italy have been made by di Domenico et al, 1981.

Although the vegetation on Grid I has become rather dense (60
percent ground cover) soil removed from below the surface by the
burrowing activity of rodents, can be seen to disperse as a consequence of wind and water. In the digging of a typical burrow by
the dominant rodent, the beachmouse, Persomyscus polionotus, approximately 2,000 g of soil are removed and deposited on the surface.
Population studies of the beachmouse were conducted in 1973 through
1975 (Young et al, 1975). From these studies it is estimated that at
least 300 new burrows are dug annually on Grid I. Therefore
burrowing activity may result in at least 600 kg of soil being
annually exposed to the action of the wind, rain, and sun. If the
mound soil (soil removed in digging the burrow) contains 180 ppt
TCDD (see Thalken and Young, 1982) then 1.1×10^{-4} g TCDD are
lost annually from Grid I through this route of TCDD disappearance.

Young et al, 1975, have also documented TCDD movement through
water erosion into the aquatic system adjacent to Grid I. The
magnitude of this loss cannot be measured but is likely low
because monitoring studies have consistently detected levels from
non-detected levels (less than 10 ppt) to 35 ppt in the eroded soil.

Volatilization

As previously noted, 87 percent of the aerially-disseminated herbicide impacted the sampling arrays; presumably the unaccounted 13 prevent could have drifted or volatilized. Moreover, the "active" grids were generally denuded of vegetation due to mechanical clearing and herbicide residue, thus exposing the white sand to intense sunlight and subsequent heat. Undoubtedly this "hot" bare surface promoted volatilization. Nash and Beall (1978) in studies involving microagroecosystems concluded that approximately 10 percent of the applied TCDD volatilized. Once the TCDD volatilized it dechlorinated in the direct sun and apparently even in shade outdoors. Thus TCDD is sensitive to photodechlorination in the vapor phase even without the presence of ultraviolet light. Volatilization, like photodegradation was probably a major route in the disappearance of TCDD during and immediately after herbicide applications on the test grids.

Volatilization of TCDD may still be occurring on the test grids. In laboratory studies with ^{14}C-TCDD added to lake sediments, Ward and Matsumura (1978) detected a direct relationship between the total loss of radioactivity and the loss of water from various samples. The authors suggested that the observed relationship indicated that the loss of radioactivity may have been related to water-mediated evaporation of TCDD. As previously noted, the Eglin Test Area receives an average of rainfall of 150 cm. The intensity of sunlight and high temperatures on the partially denuded test grids certainly promotes conditions favorable to evaporations of soil water. During the process of evapotrans-portation does TCDD evaporate or codistill with water vapor? As noted in Figure 1, the water table is within 1 to 2 m of the soil surface. For water to be "pulled" through the soil profile during evaporative processes, it must penetrate the zone of TCDD. Hence, the opportunity for water mediated evaporation to take place.

Microbial Degradation

The role of microorganisms in enhancing the degradation of TCDD remains to be documented. The general consensus from laboratory studies (Young et al, 1978) is that TCDD is not readily metabolized by soil microorganisms. However, recently Matsumura et al (1982) have isolated two organisms capable of degrading the TCDD molecule. Young et al (1976) documented significant increases in microbial populations in soils adjacent to the Eglin AFB test grids that received 4,480 kg Herbicide Orange/ha in a simulated subsurface injection treatment. The limited data collected over a 4-year period (1972-1976) suggested that the half-life of TCDD in these soils was less than 200 days. These data, however, applied to a situation where the herbicide concentration was very

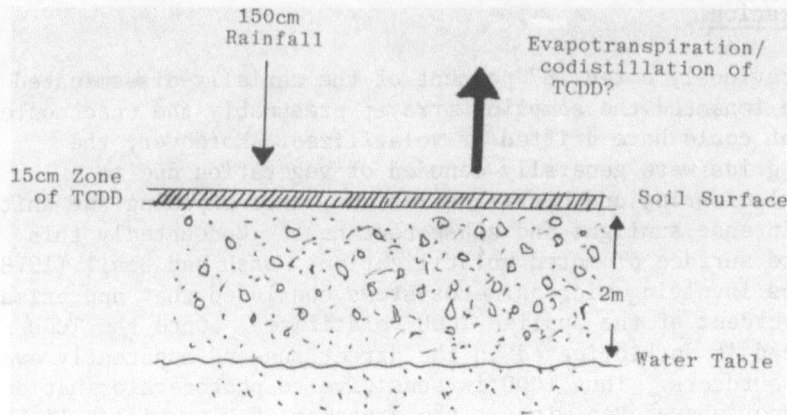

Fig. 1.   Water mediated evaporations is one route proposed
          for the disappearance of TCDD from the Eglin Test Grids

high, providing the microorganisms with a readily available carbon
source. Whether co-metabolism of TCDD occurred in this situation
remains to be determined. Certainly the present situation on
Grid I, where herbicide residues are negligible, TCDD residues are
in the parts-per-trillion, and the overall organic matter content
of the soil is approximately one percent, is not conducive to
microbial degradation of TCDD. Moreover, Poiger and Schlatter
(1980) have noted that there is an extremely strong adsorption of
TCDD onto soil particles that reduces its bioavailability.

Biomass Removal

    Biomass removal of TCDD has been documented on the test grids.
Two major biologic systems function in the removal of TCDD from
the soil. The first system involves the plant uptake of TCDD. In
November 1978 and again in January 1979, sites were selected on
Grid I for plant TCDD uptake studies. One-half meter square plots
were isolated by trenching around them to a depth of 25 cm.
Systematically the aerial vegetation by species and plant parts was
removed from the "island plots". This was followed by carefully
removing the 0-5, 5-10, and 10-15 cm increments of soil. As each
increment of soil was removed, roots were excavated and pooled by
species. TCDD analysis of the soil and plant parts was contracted
to the University of Nebraska's Midwest Center for Mass Spectrometry. Typical data for "grasses", small "broadleaf plants", and
soil increments are shown in Figure 2. The levels of TCDD in
roots and soils are similar, suggesting that a "passive" process
of uptake is responsible, e.g., the incorporation of contaminated

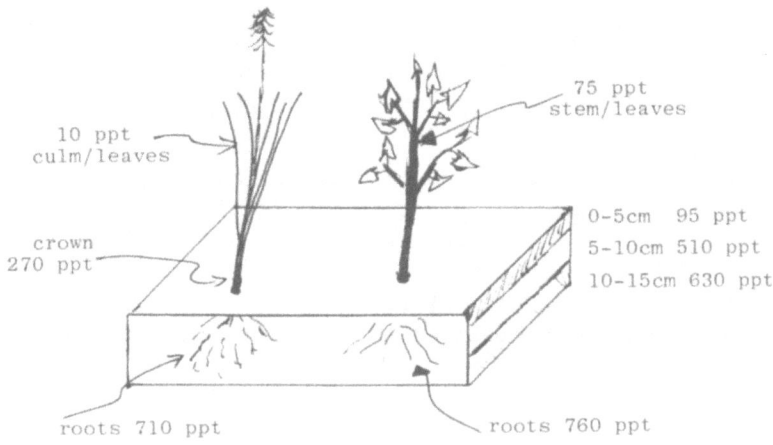

Fig. 2.   Typical plant TCDD update data for perennial grasses
          and broadleaves from studies conducted on selected
          sites of Grid I.

soil particles into the epidermis of the root tissue. Although
the above ground portions of vegetation were contaminated, these
species were perennial species and the levels of TCDD may reflect
soil particle contamination. Nevertheless, an estimate of the
magnitude of this TCDD removal mechanism can be calculated.
Vegetative studies of Grid I found that on the average, one square
meter yielded 360 g dry matter per year. Therefore for Grid I,
assuming a 40 ppt TCDD vegetative uptake, approximately 5.3 X
10^{-4} g TCDD are removed annually by the vegetation (37 ha X 1000
m^2/ha X 360 g/m^2 X 40 X 10^{-12} g TCDD/g = 5.3 X 10^{-4} g TCDD). How
much of this TCDD is returned to the soil of the test grid or
removed by animals cannot be estimated.

     Thalken and Young (1982) have studied the role of animals
especially, the beachmouse, Peromyscus polionotus, in the uptake
and removal of TCDD from soils of Grid I. Beachmice are continually
digging burrows that penetrate the zone of TCDD. In the process of
removing the soil from the burrow, the contaminated soil particles
adhere to the beachmouse pilage (fur) and are ingested by the
beachmouse in the process of grooming. A single female beachmouse
is capable of producing litters (3 to 4 pups per litter) every 26
days. Because the beachmouse constitutes a key position in the food
chain of many local predators (bird, fox, skunk, and snake), the
stress on population turnover is intense. Using data from Thalken
and Young (1982) and Young et al, (1975), assume 240 adult beachmice are removed annually from Grid I (a conservative estimate).
An average liver weight (the site of TCDD accumulation) is 1 g and
the average TCDD concentration within the liver is 1 ppb. The

pilage weighs approximately 3 g and has a mean concentration of 100 ppt TCDD. Therefore beachmice conservatively remove 3.12 X 10^{-7} g TCDD per year from Grid I (240 mice X 1 g liver/mouse X 1 X 10^{-9} g TCDD/g liver tissue = 2.4 X 10^{-7} TCDD + 240 mice X 3 g pilage/mouse X 1 X 10^{-10} g TCDD/g pilage = 2.4 X 10^{-7} g + 0.72 X 10^{-7} TCDD = 3.12 X 10^{-7} g TCDD). Similar calculations can be done for other dominant animals that remove TCDD from the soil of Grid I (e.g., the six-lined racerunner, (Cnemidophorus sexlineatus; and the hispid cotton rat, Sigmodon hispidus). These other animals conservatively remove 7 X 10^{-7} g TCDD. Therefore animal biomass probably results in the loss of 1.1 X 10^{-6} g TCDD.

The limitations in estimating loss of TCDD from soil via biomass removal is obvious. Nevertheless, such calculations based on the best available data do permit "crude" assessments of the role that living systems play in the removal of toxic chemicals from the soil.

CONCLUSION

Studies of TCDD in the ecosystem of Test Area C-52A confirm that significant levels of TCDD contaminate much of the biota of this unique site. Although actual data are not available on the amount of TCDD originally applied (1962-1970) as a contaminant of 2,4,5-T herbicide, the best estimates are that at least 2.8 kg TCDD may have been contained in the total product. Most of the TCDD in the formulation was apparently photodegraded at the time of herbicide dissemination. Today less than 1 percent of the TCDD remains in the ecosystem of the test site with the bulk of that 1 percent retained in the soil.

Clearly, mechanisms are operative for the removal of TCDD from the soils. The "crude" data available on the processes of water and wind transport of contaminated particles and biomass removal suggest that far less than 1 percent of the TCDD will disappear annually through these mechanisms. The role that volatilization and microbial degradation play in the removal of TCDD from soils is not clear. If the information from Nash and Beall (1978) and Ward and Matsumura (1978) are applicable, perhaps 5 percent of the TCDD is lost annually through volatilization. Such estimates would put the half-life for TCDD on the test grid at 10-12 years. Data from the analysis of six composited soil samples collected from Grid I in 1979 and compared to the previous mean data for Grid I, also suggest a half-life for TCDD of 10-12 years. It is now apparent that once TCDD becomes bound to soil particles, its persistence in the environment is significantly increased.

REFERENCES

Bartleson, F.D., D.D. Harrison and J.D. Morgan. 1975. Field
     studies of wildlife exposed to TCDD contaminated soils.
     Air Force Technical Report AFATL-TR-75-49. Air Force
     Armament Laboratory, Eglin AFB, Florida. 53p. Available
     from the National Technical Information Service, Springfield,
     Virginia.
Crosby, D.G. and A.S. Wong. 1977. Environmental degradation
     of 2,3,7,8-tetrachlorodibenzo-p-dioxin (TCDD). Science
     195:1337-1338.
di Domenico, A., G. Viviano, and G. Zapponi. 1981. Methodological
     problems in assessing 2,3,7,8-TCDD environmental contamination
     at Seveso. In: O. Hutzinger, R.W. Frei, E. Merian, and F.
     Pocchiari (eds.). Chlorinated Dioxins and Related Compounds-
     Impact on the Environment. Pergamon Series on Environmental
     Science. Pergamon Press, Oxford. 5:47-54.
di Domenico, A., G. Viviano, and G. Zapponi. 1981. Envrionmental
     persistence of 2,3,7,8-TCDD at Seveso. In: O Hutzinger,
     R.W. Frei, E. Merian, and F. Pocchiari (eds.). Chlorinated
     Dioxins and Related Compounds - Impact on the Environment.
     Pergamon Press, Oxford. 5:105-114.
Esposito, M.P., T.O. Tiernan, and F.E. Dryden. 1980. Dioxins.
     Environmental Protection Technology Series, Document
     EPA/600/2-80-197. Available from the National Technical
     Information Service, Springfield, Virginia.
Fanelli, R., M.G. Castelli, G.P. Martelli, A. Noseda, and S.
     Garattini. 1980. Presence of 2,3,7,8-tetrachlorodibenzo-
     p-dioxin in wildlife living near Seveso, Italy: a preliminary
     study. Bull. Environ. Contam. Toxicol. 24:460-462.
Matsumura, F., J. Quensen, and G. Tushimoto. 1982. Microbial
     degradation of TCDD in a model ecosystem. These Proceedings.
Nash, R.G. and M. L. Beall, Jr. 1978. Environmental distribution of
     2,3,7,8-tetrachlorodibenzo-p-dioxin (TCDD) applied with silvex
     to turf in microagrosystem. Final Report EPA-1AG-DG-0054.
     Agricultural Envirnomental Quality Institute, U.S. Department
     of Agriculture, Beltsville, Maryland.
Poiger, H. and Ch. Schlatter. 1980. Influence of solvents and
     adsorbents on dermal and intestinal absorption of TCDD.
     Fd. Cosmet. Toxicol. 18:477-481.
Shadoff, L.A. and R.A. Hummel. 1978. The determination of
     2,3,7,8-tetrachlorodibenzo-p-dioxin in biological extracts
     by gas chromatography mass spectometry. Biomed. Mass
     Spectrom. 5(1):7-13.
Thalken, C.E. and A.L. Young, 1982. Long-term field studies of a
     rodent population continuously exposed to TCDD. These
     Proceedings.

Ward, C.T. and F. Matsumura. 1978. Fate of 2,3,7,8-tetrachloro-
dibenzo-p-dioxin (TCDD) in a model aquatic enviornment.
<u>Arch. Environm. Contam. Toxicol.</u>, 7:349-357.

Young, A.L. 1974. Ecological studies on a herbicide-equipment
test area (TA C-52A), Eglin AFB Reservation, Florida. Air
Force Technical Report AFATL-TR-74-12, Air Force Armament
Laboratory, Eglin AFB, Florida. 146 p. Available from the
National Technical Information Service, Springfield, Virginia.

Young, A.L. 1980. The chlorinated dibenzo-p-dioxins. In:
R.W. Bovery and A.L. Young. The Science of 2,4,5-T and
Associated Phenoxy Herbicides. Wiley-Interscience, New York,
pp. 133-205.

Young, A.L., J.A. Calcagni, C.E. Thalken and J.W. Tremblay. 1978.
The toxicology, environmental fate, and human risk of
Herbicide Orange and its associated dioxin. Air Force
Technical Report OEHL-TR-78-92, USAF Occupational and
Environmental Health Laboratory, Brooks AFB, Texas. 247 p.
Available from the National Technical Information Service,
Springfield, Virginia.

Young, A.L., C.E. Thalken, E.L. Arnold, J.M. Cupello and L.G.
Cockerham. 1976. Fate of 2,3,7,8-tetrachlorodibenzo-p-
dioxin (TCDD) in the environment: Summary and decontamination
recommendations. Air Force Technical Report USAFA-TR-76-18,
United States Air Force Academy Colorado. 41 p. Available
from the National Technical Information Service, Springfield,
Virginia.

Young, A.L., C.E. Thalken and D.D. Harrison. 1979. Persistence,
bioaccumulation and toxicology of TCDD in an ecosystem treated
with massive quantities of 2,4,5-T herbicide. Paper
presented to the Symposium on the Chemistry of Chlorinated
Dibenzodioxins and Dibenzofurans. American Chemical Society,
178th National Meeting, 14 September 1979, Washington, DC
24 p.

Young, A.L., C.E. Thalken and W.E. Ward. 1975 Studies of the
ecological impact of repetitive aerial applications of
herbicides on the ecosystem of Test Area C-52A, Eglin AFB,
Florida. Air Force Technical Report AFATL-TR-75-142, Air
Force Armament Laboratory, Eglin AFB, Florida. 127 p.
Available from the National Technical Information Service,
Springfield, Virginia.

MICROBIAL DEGRADATION OF TCDD IN A MODEL ECOSYSTEM

F. Matsumura, John Quensen and G. Tsushimoto

Pesticide Research Center
Michigan State University
East Lansing, Michigan, USA

ABSTRACT

Microbial degradation of 2,3,7,8-tetrachlorodibenzo-p-dioxin
(TCDD) was studied by using pure culture isolates of microorganisms, terrestrial and aquatic model systems, and an outdoor
pond. In each case metabolic activities were recognized by the
appearance of metabolic products from ^{14}C-TCDD. In the outdoor
pond the apparent half-life of TCDD was in the order of one year,
recoveries of TCDD after 12 and 25 months being 49.7 and 29.4
percent, respectively. In model systems metabolic activities on
TCDD were stimulated by the addition of general nutrients such as
glucose, bactopeptone and mannitol yeast. The two microbial
isolates, Bacillus megaterium and Nocardiopsis sp., were found
to degrade TCDD. The most important factor found to promote
their metabolic activities was the nature of carrier solvent for
TCDD. In this regard ethyl acetate gave the best results under
the experimental conditions.

INTRODUCTION

When 2,3,7,8-tetrachlorodibenzo-p-dioxin (TCDD) was found to
be a contaminant of the widely used herbicide 2,4,5-trichloro-
phenoxyacetic acid (2,4,5-T), scientists became concerned whether
this chemical persisted for a long period in the environment.
There are several reports predicting the overall persistence of
TCDD using model materials and ecosystems. Kearney et al, (1973),
for instance, reported that TCDD was persistent in soil under
certain conditions with its half-life being in the order of one
year. Crosby and Wong (1977) showed that TCDD could photochemically

degrade at a relatively fast rate. According to the report by
Matsumura and Benezet (1973), TCDD was almost immobile in the
soil and the remains on the top-most layer. They concluded that
TCDD, because of its poor lipid solubility, is not likely to
bioaccumulate in a biological system as much as DDT. Isensee and
Jones (1975), on the other hand, concluded that TCDD accumulation
in fish is controlled by the amount of TCDD available in water.
In their model ecosystem, degradation of TCDD did not occur up to
one month. Using a model aquatic system, Ward and Matsumura
(1978) showed that most of TCDD deposited on the sediment and a
small amount of metabolites of TCDD in the sediment, were
released into water.

The TCDD that does not immediately photodegrade in the
environment is expected to accumulate in soil and aquatic
sediment in a manner similar to DDT, PCB and other organic
pollutants. Therefore, a question must be raised whether TCDD is
degraded by soil microorganisms, and if so whether there are ways
to stimulate such degradation activities to lower residue levels.
To answer this question microbial degradation in the environment
must be confirmed. The characteristics of microbial degradation
of TCDD in model ecosystem and by defined microbial cultures are
documented in this report.

MATERIALS AND METHODS

The fate of TCDD in an aquatic ecosystem was studied with two
sets of experimental conditions. The first one was an outdoor
pond experiment. The second one was with indoor model
ecosystems. The outdoor pond consisted of an artificial circular
pool 6.85 m in radius, lined with 30 cm-thick concrete wall. The
depth of the pond water was 1.1 m (at the beginning) and 1.4 m
(one year later). Another pond identical in size and structure
nearby (approximately 20 m west) served as a control pond. The
sediment consisted of a 5-cm clay, organic, loam top layer, and a
50-cm layer of sand on the bottom. The top layer's texture was:
sand, 79.3 percent; silt, 0 percent; clay, 20.7 percent; and 4.1
percent organic matter and 0.75 ppm nitrate at pH 7.6. The
predominant pondweed species were Elodea, Elodea nuttali, and
Hornwort, Ceratophyllum demersum. The biomass of pondweeds was
estimated to be about 8.3 g/m^3 (fresh weight) on October 18,
1978. The fish used for this experiment was the fathead minnow,
Pimephales promelas. For accumulation studies 50 fish were
placed in two separate cages and dipped in the upper water layer
about 40 cm below the surface. On October 18, 1978, 3.4 mCi
^{14}C-TCDD (8.7 mg) in 6.8 ml of anisole and 1 mL of Triton X-100
were added first to 20 L of pondwater, thoroughly mixed by
shaking, and spread over the pond surface homogeneously. The
specific activity of TCDD was 126 mCi/m mol (uniformly ring

labeled). The initial concentration of TCDD in water was 53.7 parts-per-trillion (ppt). On each sampling day, water (0.2-0.3 mn below the surface) and lower water (1.0-1.1 m below the surface), sediment (from the surface to about 5 cm depth), pondweeds, and fish were collected. For routine assay, the radioactivity in the 0.5 mL water was directly measured by liquid scintillation spectroscopy. To estimate TCDD concentration below 10 ppt level, 1 to 12 L of water samples were extracted with 500 mL of chloroform, dried over sodium sulfate, solvent evaporated, the residues picked up in liquid scintillation counting solution and the radioactivity was assayed. TCDD and its metabolites were extracted from 10 g of sediment by the series of solvents described by Ward and Matsumura (1978). A preliminary test established that the recovery of the added radioactivity was complete by this method; the recovery values being 95.1 percent in one experiment and 105 percent in another. For fish and pond-weeds experiments, two fish were collected and weighed, and 2 g of pondweeds were collected. TCDD and its metabolites in fish and pondweeds were extracted with 20 to 25 mL of chloroform by homo-genization. In all cases the data were calculated in concentration on fresh weight basis (e.g., ppt) based upon radioactivities and and expressed as the amount equivalent of the original TCDD.

For the metabolic study on TCDD, 5 L of upper water, 50 g sediment and 40 g of pondweeds were collected after one year. TCDD and its metabolites in the upper water were extracted with 500 mL of chloroform. In the cases of the sediment and pondweed samples, 150 mL of acetone and 400 mL of chloroform were used respectively. The solvent extracts were evaporated to dryness, and the residues were picked up in 2 mL aliquots of chloroform (in the case of water and weed), or acetone (in the case of sediment) for thin-layer chromatographic (TLC) analysis.

Polar metabolites were separated from TCDD on the TCL plate (LK5F-TLC, Whatman) by using carbon tetrachloride as a mobile phase (Matsumura and Benezet, 1973). TCDD on the plate was scraped from the area of R_f 0.36-1.0 and polar metabolites were similarly collected from the area of R_f 0-0.1. Radioactivity in the silica gel was measured by combustion and liquid scintillation counting.

The model ecosystem experiments were carried out in a glass bottle (radius: 16 cm, height: 23.6 cm). Each glass bottle contained 1100 g of sediment, 2.3 L of water, 10 g of pondweed and a fish (0.5-1.0 g fish). These materials were taken from the same outdoor pond before the experiment started. The model ecosystems were kept on the laboratory bench with a glass cover and a 100 w florescent light source about 40 cm above the top of the water surface. On the starting day of the experiment, ^{14}C-TCDD of which specific activity was 126 mCi/m mol (uniformly

ring labeled) was added to make final concentration of 61.36 ppt.
On each sampling day 10 g of sediment, 10 mL of water 1 g of
pondweed (fresh weight) and one fish were collected. Their
radioactivities were measured by the identical analysis methods
as the ones described for the outdoor pond experiment. The model
ecosystem experiment was continued up to 45 days.

The details of the design of the aquatic microbial ecosystem
have been described by Ward and Matsumura (1978). The system
consists of the sediment and the alke water kept in a loosely
capped 20 mL culture tube in darkness at room temperature.

The terrestrial microbial ecosystem consisted of 2 cm of
glass wool, 25 g of washed, ignited sea sand (Mallinckrodt, Inc.)
and 20 g of sieved soil, added in this order to a 50 mL Pyrex
beaker. The soil was collected from one of the areas on or near
the Michigan State University campus and represents farm, garden,
and oak woods types. Five mg of finely ground nephthalene was
gently stirred into the soil in half of the beakers. This gave a
concentration of 250 ppm of naphthalene in the soil. The soil
"plugs" were then covered with aluminum foil to exclude light but
not air and placed in a growth chamber at 30°C and 95 percent
relative humidity. Soil moisture was checked visually each week,
and distilled water was added to each beaker of soil. Extractions
were made two and four months after the addition of TCDD.

Soil, sand, and glass wool were extracted one with 100 mL
chloroform:methanol (2:1), again with 50 mL chloroform:methanol,
and a third time with 50 mL chloroform. All three extracts were
combined, evaporated, and the residue redissolved in 0.4 mL ethyl
acetate.

Separation of TCDD and metabolites was by TLC (LK5D-TLC,
Whatman) with carbon tetrachloride as the mobile phase. Radioactivity on the plates was visualized by scanning and/or autoradiography, and the appropriate areas scraped and counted by
liquid scintillation.

RESULTS

Fate of TCDD in the Outdoor Pond and in the Model Ecosystem

The fate of TCDD in the aquatic system was studied using an
outdoor pond and the result was compared with that of model
ecosystem studies. As shown in Figures 1 and 2, TCDD was
distributed among the sediment, water, pondweed and fish in the
outdoor ecosystem. The initial buildup of TCDD in the sediment
reached 2,700 ppt. The radioactivity in the sediment gradually

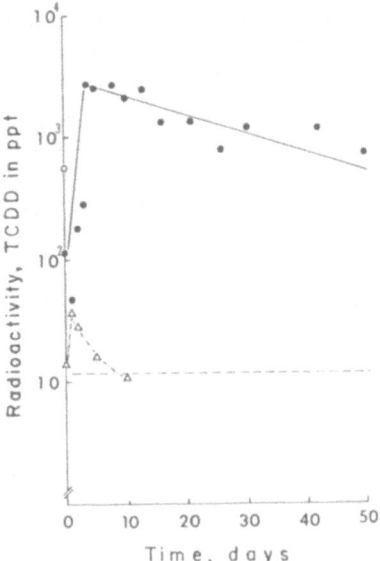

Fig. 1.   Short-term distribution of TCDD-derived radioactivity
among the sediment (closed circles), upper layer (open
circles) and lower layer of water (open triangles) in
the outdoor pond experiment. The data are expressed
in the equivalent of the originally-added TCDD as
assessed by the radioactivity. Dotted line shows a
detection limit.

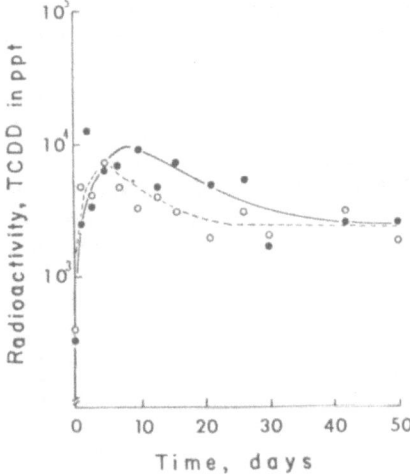

Fig. 2.   Short-term accumulation of TCDD-derived radioactivity by
fish (closed circles) and pondweed (open circles) in the
outdoor pond experiment.

from 2,700 to 500 ppt within 50 Days (Figure 1). The bioaccumulation of TCDD in the fish and pondweed proceeded rapidly as shown
in Figure 2. Accumulation of radioactivity by the fathead minnow
increased for about 10 days, and decreased thereafter to reach a
constant level after 40 days (2,500 ppt). The maximum accumulation
level in the minnow was 8,500 ppt after 10 days. The bioaccumulation of the radioactivity by the pondweeds reached a maximum
level (7,000 ppt) after five days. At the time of the initial
equilibrium (one month), the level of accumulation in the
pondweed was in the order of 2,500 ppt based upon an assumption
that the entire radioactivity was due to TCDD.

A similar tendency was observed in the model ecosystem study
with regard to fate of TCDD in the sediment and fish (Figures 3
and 4). However, the level of bioaccumulation of TCDD by the
fish in the bottle was somewhat higher than that found in the
outdoor pond experiment. The days needed for maximum bioaccumulation of TCDD in fish in our experimental conditions were five to
ten days.

In the case of pondweed in the model ecosystem experiment,
the level of the radioactivity was the highest at the starting
time. Thereafter, it decreased and reached a constant level (500
ppt) in about one month. This mode of accumulation was different
from the data obtained in the outdoor pond experiment. Also the
level of TCDD accumulation was much lower in the model ecosystem
than that obtained in the outdoor pond.

The results of the sixty-five days to one year monitoring
study on accumulation of TCDD in the sediment, pondweed, and fish
indicate that the level of accumulation of TCDD into each
component had reached a constant level under this experimental
condition (Figures 5 and 6). The levels in the biological
systems were as follows: sediment 70 dpm/g (80 ppt if all
radioactivity was derived from entire TCDD), pondweed 1,740 dpm/g
(2000 ppt) and fish 2,170 dpm/g (2500 ppt), respectively. The
level of radioactivity in the sediment reached a constant level in
about six months under the experimental condition.

The results of a general survey on the distribution of TCDD
has been summarized in Table 1. It can be seen that at the forty
to fifty days stage, the data obtained by using the model ecosystem
were roughly comparable to the ones from the outdoor test. One
exception was the data for the pondweed.

Metabolic changes of ^{14}C-TCDD were also studied. The ratio
of TCDD and its metabolites after one year were obtained and the
results are shown in Table 2. At this time, water contained 60
percent of the radioactivity as TCDD and the rest as polar
metabolites (40 percent). On the other hand, almost all of the

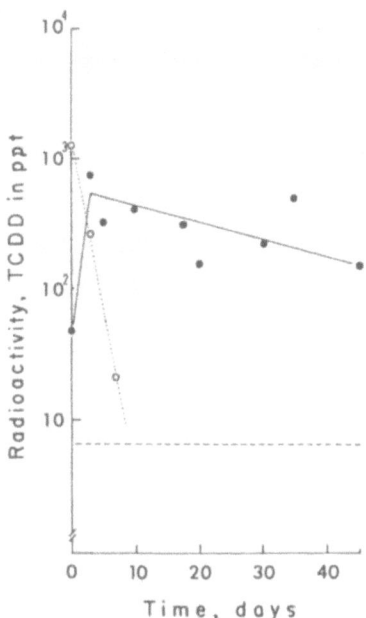

Fig. 3. Distribution of TCDD-derived radioactivity between water
(open circles) and sediment (closed circles) in the model
ecosystem experiment. Dotted Line shows a detection limit.

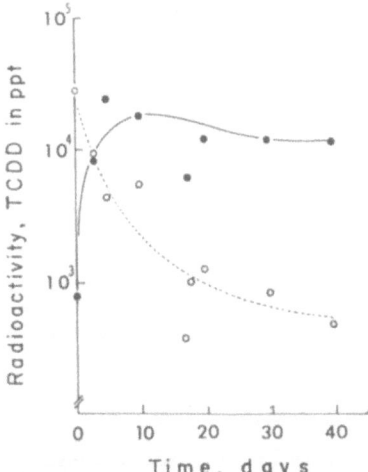

Fig. 4. Bioaccumulation of TCDD-derived radioactively in fish
(closed circles) and pondweed (open circles), in the
model ecosystem experiment.

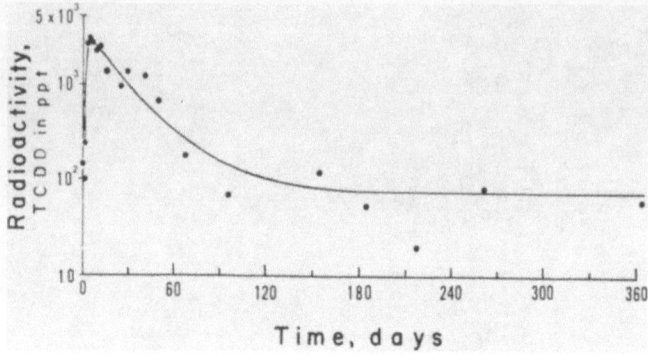

Fig. 5.   Long-term distribution of TCDD-derived radioactivity
among sediment in the outdoor pond experiment.

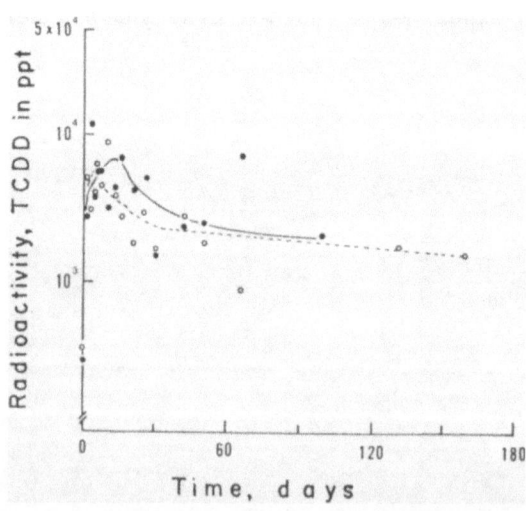

Fig. 6.   Long-term bioaccumulation of TCDD-derived radioactivity
in fish (closed circles) and pondweed (open circles)
in the outdoor pond experiment.

Table 1. Levels of Total Radioactivity Expressed in TCDD Equivalent

Source	Incubation Time (days)	TCDD Equivalent (ppt)			
		Water	Sediment	Fish	Pondweed
Outdoor	50	< 12	500	2500	2500
Outdoor	365	$0.066[0.026]^a$	$97^b[3.5]$	—	2000[794]
Outdoor	760	< 0.0043	$321^c[9.6]$	—	8.8[3.6]
Indoor	40	< 12	200	2000	500

[a]Parentheses show metabolites.
[b]This value was obtained in the top organic layer.
[c]The increase is due to death of pondweed and resulting accumulation of decayed plant-matter on the surface of the top layer.

Table 2. Ratio Between TCDD and its Metabolites after One Year Pond Experiment

	Upper Water	Sediment	Pondweed[b]
TCDD[a]	60.2%	96.4%	60.3%
Metabolites[a]	39.8%	3.6%	39.7%

[a]TCDD and its metabolites were isolated with chloroform in the case of water and pondweed, with aceton in the case of sediment. These were analyzed on the TLC plate.
[b]Pondweed was collected after 14 months.

radioactivity in sediment was due to unmetabolized TCDD itself.
This result confirms the observation made by Ward and Matsumura
(1978) that TCDD is metabolized in the sediment and the
metabolites are released into water. It is interesting to note
that the level of metabolites present in pondweed was almost
identical to that found in water. It is likely that both TCDD
and its metabolites are distributed by similar partition
coefficients between water and pondweed.

At the end of one year an effort was made to survey the
amount of TCDD left in the system. To carry out the task the
total weight of each component was estimated. In the case of the
sediment, TCDD was found to be exclusively present in the topmost
organic soil layer. The average depth of the layer was obtained
by sampling several stations. The sediment samples were obtained
by using a clear acrylamide tube (2 cm inner diameter). The top
layer of the sediment had a distinct dark color. This portion
was carefully collected apart from other soil materials. Within
the top layer the sample was mixed to obtain the average TCDD
concentration. The samples were spread over several layers of
paper towels and air dried for one day. The weights of the air
dried samples were taken for the calculation of TCDD levels. The
moisture content of the air dried sample was 13.5 ± 4.3 percent
as judged by oven drying at $200^{\circ}C$ for overnight.

The results (Table 3) clearly show that the bulk of TCDD is
tied up in pondweed. The total recovery of unmetabolized TCDD
was 49.7 percent of initially added quantity. At the end of a
25-month period, the same survey effort was made to assess the
amount of TCDD remaining. The most noticeable change in the pond
from the one-year post-treatment observation time was that a
massive death of the pondweed resulted in the accumulation of
decayed plant matter at the bottom of the pond. This extra layer
was clearly distinguishable from the top organic soil layer. Both
layers were found to contain radioactivity. They were radioassayed
separately, but the results are expressed as a combined figure
(Table 4). As a result of sedimentation of the dead plant
matter, the bulk of radioactivity was found to shift from the
pondweed to the sediment component. An intriguing observation is
that the newly formed sediment did not contain a high proportion
of polar metabolites. A calculation would show that if the
entire unmetabolized TCDD in the pondweed from the previous year
sedimented and added to the already existing TCDD in the top
organic soil layer it would give 2.8 mg of total TCDD. The value
is not too far above the actual figure (2.56 mg) found at the end
of the second year.

The total recovery of TCDD for the second year was 29.4
percent. The recovery figures are generally lower than the ones
produced by Ward and Matsumura (1978) by using a model system.

Table 3. Estimation of Total TCDD Remaining in Various Components After One Year

	Volume and Weight	Radioactivity	% Unmetabolized TCDD	Amount of TCDD (ppt)[a]	Total Amount of TCDD Remaining
Water	206 m^3	0.057 dpm/ml	60.2	0.04	8.2 μg
Pondweed	2.72 t	2000 dpm/g[c]	60.3	1390	3.78 mg
Sediment	5.75 t[b]	84 dpm/g	96.9	94	0.54 mg
					4.32 mg[d]

[a] 1 dpm/g TCDD was equivalent to 1.15 ppt. To obtain the value for water it was necessary to extract 5 ℓ of water with chloroform.
[b] This value was estimated by the weight of the top organic soil layer.
[c] This value was estimated by using Fig. 6.
[d] This value was 49.7% of initial TCDD added.

Table 4. Estimation of Total TCDD Remaining in Various Components After 25 Months

	Volume and Weight	Radioactivity	% Unmetabolized TCDD	Amount of TCDD (ppt)[a]	Total Amount of TCDD Remaining
Water	249 m^3	0.0037 dpm/ml	—	4.3×10^{-3}	1.07 μg[c]
Pondweed	0.1 t	7.7 dpm/g	59.5	5.2	0.52 μg
Sediment[b]	8.47 t	279 dpm/g	97.0	311	2.56 mg[d]
					2.56 mg

[a] 1 dpm/g TCDD was equivalent to 1.15 ppt.
[b] This sediment contained decayed plant matter and the top organic soil layer. The former layer formed as a result of a massive death of pondweed which took place sometime between 12th to 25th month posttreatment.
[c] Because of the low level of radioactivity the ratio of metabolites to TCDD could not be determined. The values shown assume that the entire radioactivity was due to TCDD and hence are overestimated.
[d] This value was 29.4% of initial TCDD added.

However, at this stage we cannot assure the dissipation of TCDD
according to the first order kinetics.

In summary, TCDD given at 53.7 ppt to water slowly dissipated
from the outdoor pond. The amount of TCDD found after one year
and two years were 49.7 and 29.4 percent, respectively. The
model laboratory system using the identical components from the
outdoor pond produced roughly comparable data for short-term
distribution behavior. The largest source of variable was
pondweed. This material produced the largest biomass changes due
to rapid growth and subsequent massive death, causing a large
shift in TCDD distribution and probably an anaerobic environment.
While the source of metabolic activity has not been determined,
the bulk of TCDD metabolites were detected in water and algae.
Metabolites appear to disappear from the system faster than TCDD
itself as attested to the decrease in the overall level of
metabolites in the system in the second year. In this case,
evaporation is the likely cause.

As for the difference in the TCDD persistence figures between
the laboratory produced data (Ward and Matsumura, 1978) and
outdoor ones, there is no solid evidence to single out the cause
since many factors could interact in the outdoor environment. If
one is allowed to speculate the most likely major cause for the
difference, one could select the effect of sunlight combined with
the influence of algae. Ward and Matsumura (1978) placed the
system under ordinary laboratory light in a thick Pyrex glass
container. In the outdoor environment sunlight intensity is, of
course, much stronger. Combined with the knowledge that TCDD is
relatively labile to photochemical attack and algae are known to
produce photosensitizers (Matsumura and Esaac, 1979), it would
not be too surprising to find that sunlight contributes
significantly to dissipation of TCDD in the outdoor environment,
making the half-life figure less for the outdoor experiment as
compared to the ones derived in the laboratory without
photochemical simulation devices.

Aquatic and Terrestrial Microbial Model Ecosystems

The above study has established that TCDD, though it is a
very stable chemical, is still slowly degraded in the
environment. In order to demonstrate that microbial activities
are important in this regard, two sets of microbial model
ecosystems were designed. The first system is an aquatic one.
For this system, the sediment samples were obtained from
University Bay, Lake Mendota and Lake Wingra, Madison Wisconsin
in large quantities and stored in a cold room at $4^{\circ}C$. For
incubation tests, 5 grams (wet weight) of the sediment and 18 mL
of lake water were added to each 20 mL culture tube, shaken and
equilibrated. ^{14}C-TCDD was added and the tubes were kept at

room temperature for a specified time period with screw caps. In
the results summarized in Figure 7, one can clearly see that
there is an overall decline in the level of TCDD recovered from
the ecosystem. It must be pointed out that the extraction
efficiency of this test series was rather good, the range of
recovery being 92 to 105 percent (Ward and Matsumura, 1978).

To ascertain that the microbial activities do contribute to
the overall disappearance of TCDD from the system, two types of
nutrients, glucose and bactopeptone, were added to the system,
and their influence on TCDD metabolism was studied (Table 5).
Despite the variation which is expected for this type of study,
the results indicate that the level of total metabolites formed
in the aqueous phase of the systems which have been amended with
these nutrients are higher than those found in the unamended
systems. It must be pointed out that in the above study only

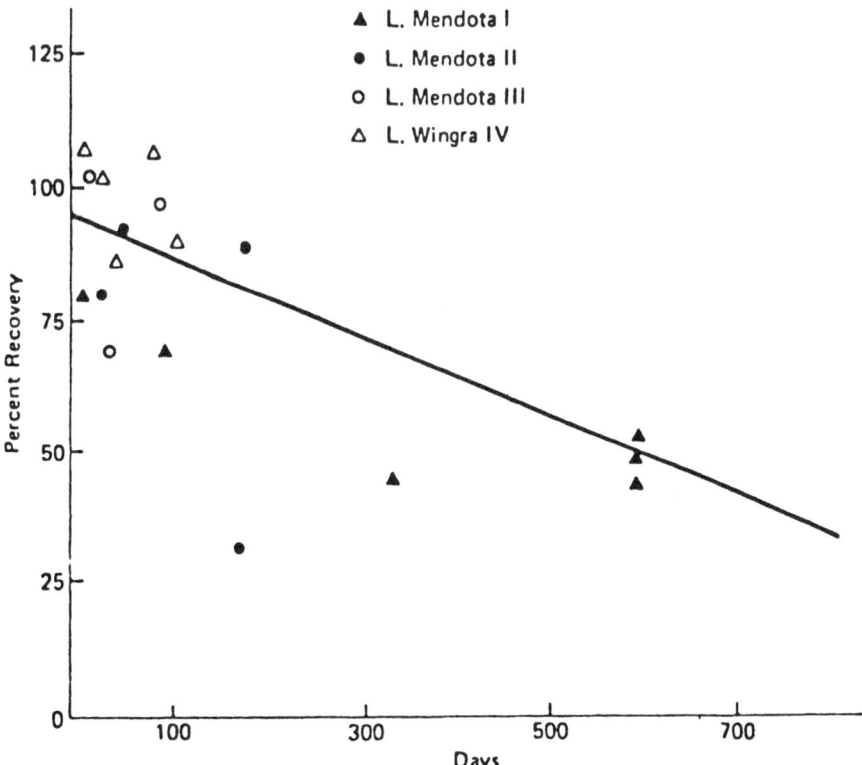

Fig. 7.   Percent recovery of TCDD from lake sediments in relation
          to time of incubation: the correlation coefficient of
          the line is 0.731.

Table 5. Metabolite Formation in the Aqueous Phase of Samples Incubated with Sediment and Water

Sample	Days Incubated	Radioactivity Recovered[a] (%)	TLC Distribution of Radioactivity TCDD (%)	Total Metabolites (%)
Glucose				
II40[b]	0	5.49	90.00	10.00
II26[b]	19	0.54	51.70	48.29
II32[b]	39	2.40	39.36	60.64
II26	163	1.34	1.73	98.27
II32	167	1.38	47.40	52.60
Bactopeptone				
II41[b]	0	1.20	94.14	6.14
II27[b]	19	3.22	22.37	77.63
II33[b]	39	0.54	13.85	86.15
II27	163	4.84	63.87	36.13
II33	167	1.41	11.08	88.92
Glucose + bactopeptone				
II28[b]	19	9.38	30.18	69.82
II34[b]	39	1.20	14.28	85.71
II34	167	1.78	43.09	56.90
No nutrients				
II43[b]	0	3.31	89.63	10.36
II29[b]	19	0.94	54.63	45.37
II35[b]	39	4.07	66.67	33.33
II35	167	0.32	54.25	45.74

[a]Percent administered radioactivity that was recovered in the aqueous phase.
[b]Water removed via freeze drying. Water in the other samples was removed by evaporation.

the radiocarbons found in the aqueous phase were analyzed (i.e., those in the sediment were not studied). Earlier it was noted that the ratio of metabolites to TCDD is much higher in the water phase than that found in the sediment. Yet, when TCDD was incubated with filtered lake water above, in the absence of sediment, metabolic products did not form (Ward and Matsumura, 1978), indicating the source of microbial activities was in sediment. This phenomenon has been interpreted to mean that the metabolic products, having higher water solubilities, are released to water from the sediment. This interpretation agrees well with the observation made earlier in the outdoor pond experiment. For comparison the overall metabolic patterns involved in the entire model ecosystem have been studied and the results are shown in Table 6. Samples II34 and II32 in this study are identical to the ones listed in Table 5. Here the percentage of the metabolic products found is much lower than that found in the aqueous phase because of the inclusion of a higher level of unmetabolized TCDD in the sediment.

Another factor which might be contributing to the overall loss of radiocarbon is evaporation. In this series of experiments some water losses were noted, particularly in those systems that have been kept over one year. The reason for this loss of water is that the screw caps of the tubes were not completely tightened in this model ecosystem to avoid explosion due to formation of gas during the incubation period. When the loss of water through evaporation was plotted against the percent recovery of TCDD (Figure 8), a general correlation between these two parameters was observed.

A model terrestrial, microbial ecosystem consisted of a soil plug (20 g) placed in a 50 mL beaker over a 2 cm layer of glass wool and a 25 g equivalent layer of washed ignited sand. The system was kept under a moist condition as outlined in the method section. Three different types of soil samples were used in the presence and absence of 250 ppm of naphthalene which was added to promote microbial activities to metabolize aromatic compounds. After 2 and 4 months of incubation the soil samples were extracted and ^{14}C-TCDD remaining unmetabolized was assessed by using thin-layer chromatography (Table 7). The result indicates that after four months approximately 66 percent of originally-added TCDD remained in soil. Naphthalene did not have a significant influence on the rate of soil degradation of TCDD. As a result of thin-layer chromatographic analysis of these samples some polar metabolic products were recognized. These two sets of experiment have established that TCDD does indeed degrade in aquatic sedimen and in soil. However, to prove that microorganisms are capable of degrading TCDD, the following tests using microbial cultures were conducted. In the experiment

Table 6. Metabolite Formation in the Sediment

Sediment Sample	Days Incubated	TLC Distribution of Radioactivity[a]		Total Recovery (%)	Nutrients[b]
		% TCDD	Total Metabolite (5)		
III60	0	92.52	7.48	101.78	—
IV66	0	93.67	6.33	102.14	—
I28	19	96.19	3.81	53.28	G,B
IV64	22	98.35	1.65	103.28	—
I34	39	91.71	8.29	91.36	G,B
IV64	67	97.47	2.53	107.96	—
IV64	98	89.04	10.96	90.03	—
I32	167	97.48	2.52	89.03	G
I1	333	76.78	23.22	43.09	—
I22	586	98.96	1.04	50.41	—
I11	588	98.74	1.26	44.47	—
I17	588	95.59	4.41	52.15	B

[a]Expressed as percent of radioactivity found in TCDD position and lower Rf positions on TLC plate, respectively.
[b]G=glucose, B=bactopeptone

Table 7. Results after Two and Four Months Incubation of Soil Samples, With and Without Addition of 250 ppm Naphthalene

Soil	Months	Aqueous[a] Phase	Solvent phase	
			Metabolite	TCDD
Woods	2	—[b]	0.6	54.8
	4	0.3	1.1	72.0
Woods–naphthalene	2	—	0.9	68.9
	4	2.1	0.9	56.6
Farm	2	—	1.2	72.2
	4	28.3	0.1	0.4
Farm–naphthalene	2	—	0.9	78.9
	4	0.9	3.0	64.8
Garden	2	—	0.8	68.1
	4	5.8	0.7	61.6
Garden–naphthalene	2	—	0.9	91.6
	4	1.8	1.1	75.8

[a]Moisture originally in soil, left after extraction and evaporation of solvent.
[b]The samples were not analyzed.

summarized in Table 8, 1 g of soil sample from each type of soil was added to the nutrient medium and incubated for 7 days. The samples were extracted and analyzed as before. As judged by the level of solvent-soluble metabolites formed, the addition of 40 percent # 2 medium had a stimulatory effect on TCDD metabolism. The microbial isolates in this medium also gave a higher metabolite production level than in the basal salt medium.

Metabolism of TCDD by Defined Microbial Isolates

A preliminary screening test established that among several isolates tested two, <u>Nocardiopsis</u> sp. (Beeman and Matsumura, 1981) and <u>Bacillus magaterium</u> (ATCC No. 13368) showed some TCDD-degrading activities. In the first experiment an effort was made to assess the effect of naphthalene, which was added to the medium as an alternative carbon source, and mannitol. The results shown in Table 9 indicate that there is a slight increase in the level of radiocarbon in the aqueous phase and a slight decrease in the level of TCDD remaining unmetabolized. The

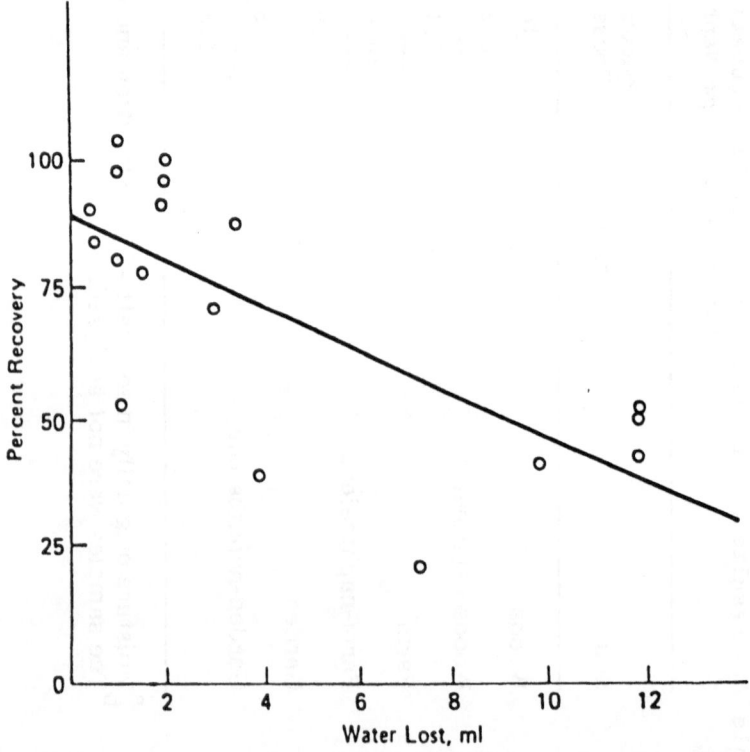

Fig. 8. Percent recovery of TCDD in relation to water loss from the incubation container: the correlation coefficient of the line is 0.716.

Table 8. Effects of Nutrient Media Added to Soil Samples on the Metabolic Activities of TCDD. As Controls Micobial Isolates were Obtained by Using Respective Media Transferred to Fresh Media after One Day and Incubated with TCDD for Seven Days.

Treatments (number of experiment)	Naphthalene ppm	Aqueous Phase	Solvent Phase Metabolites	TCDD
Soil[a] plus #2[b] medium (10)	100	4.0 ± 1.9	1.5 ± 1.1	58.1 ± 5.9
Soil plus basal salt (4)	1000	4.7 ± 4.1	0.2 ± 0.1	39.3 ± 9.9
Microbial isolates in #2 medium (10)	100	9.7 ± 5.4	0.9 ± 0.3	62.1 ± 5.8
Microbial isolates in basal salt (8)	1000	3.5 ± 4.2	0.3 ± 0.1	74.5 ± 14.9

[a] Soil samples used were four samples from mixed woods land, two dried river bottom sample, and four farm soil samples. Incubation period: 14 days.
[b] 40% #2 medium (yeast-mannitol) plus naphthalene as indicated.

Table 9. Effect of Varying Naphthalene Concentration on Metabolism of TCDD by Nocardiopsis. Twenty Percent #2 Mannitol-Yeast Medium was used in all Treatments. Results are Expressed as Means and Standard Deviations of Three Replicates.

	% of ^{14}C Activity Recovered		
	Aqueous Phase	Solvent Phase	
Treatment		Metabolites	TCDD
0 ppm naphthalene	1.9 + 0.6	5.2 + 2.3	88.8 + 9.1
10 ppm naphthalene	1.7 + 1.0	3.5 + 0.8	82.9 + 5.7
50 ppmm naphthalene	2.1 + 0.6	3.5 + 1.7	84.8 + 5.7
100 ppm naphthalene	2.4 + 1.2	3.3 + 2.0	80.1 + 6.4
500 ppm naphthalene	2.4 + 0.1	6.3 + 2.7	79.0 + 4.1
Standard	0	0.6 + 0.3	99.3 + 0.3

differences, however, were not statistically significant. In the next experiment the effect of decreasing the level of nutrient and replacing with the basal medium (i.e., salts without any carbon source) containing 10 ppm of naphthalene was assessed. The results (Table 10) indicate that the overall metabolic activity as judged by the level of solvent soluble metabolites found through thin-layer chromatographic analyses was higher, when there was some amount of mannitol-yeast medium present than in the case where naphthalene was the sole carbon source. At this stage, the effect of carrier solvent which was used to dissolve TCDD and used as a vehicle for administration was reexamined. The results (Table 11) indicate that both ethyl acetate and DMSO are better carriers than ethanol or corn oil as judged by the solvent-soluble metabolites and the amount of radiocarbons in the aqueous phases found. By using the same criteria ethyl acetate was judged as the better one of the two. A confirmatory experiment to this carrier effect was conducted with varying levels of soybean extract in the medium (Table 12). The result clearly shows that ethyl acetate is a superior carrier for TCDD, inducing a higher level of TCDD degradation than DMSO. The level of soybean extract did affect the metabolism particularly in the case of the ethyl-acetate treated sample, where the level of solvent-soluble metabolites increased when the content of soybean extract was reduced from 100% to below 50%. An autoradiogram of the thin-layer chromatogram of the solvent phase from one replication of this experiment is shown in Figure 9. It shows that metabolism of TCDD is extensive in the ethyl acetate containing samples and that several metabolic products are separated by this chromatography.

DISCUSSION

There is little doubt that TCDD is a relatively stable chemical against metabolism, and, as a result, it stays in the environment for a relatively long time period. This is in stark contrast to its susceptibility to photochemical reactions, making TCDD a unique evnironmental contaminant. However, even this microbially recalcitrant pollutant does slowly disappear from soil and aquatic environments as shown by our work here and by others (e.g., Kearney et al, 1973). Three factors considered to be important in this regard are evaporation, photochemical degradation and microbial degradation. Evaporation is the least studied phenomenon for TCDD, and yet it should be considered as an important process since it makes TCDD available to photochemical degradative forces in the atmosphere. Our preliminary data show that there is a water-mediated evaporation process for TCDD in the aquatic environment. Much more work would be needed, however, to assess the significance fo this process particularly from soil surfaces.

Table 10. Effects of Culture Medium on Metabolism of TCDD by Nocardiopsis. Mannitol-Yeast Medium (#2) was Diluted to Various Concentrations with Basal Medium. Results are Expressed as the Means and Standard Deviations of Three Replicates. Ten ppm Naphthalene were Present in all Cultures and DMSO was Used as the Solvent.

| | % of ^{14}C Activity Recovered | | |
| | | Solvent Phase | |
Treatment	Aqueous Phase	Metabolites	TCDD
0% #2	1.5 ± 0.6	4.2 ± 2.0	71.6 ± 6.3
10% #2	2.0 ± 0.5	7.1 ± 4.1	76.1 ± 6.5
20% #2	3.0 ± 0.4	8.2 ± 2.7	72.4 ± 5.4
30% #2	2.8 ± 0.5	7.5 ± 4.1	78.2 ± 2.8
40% #2	3.1 ± 2.5	7.1 ± 1.4	73.9 ± 9.6
Standard	0	2.6 ± 2.2	97.4 ± 2.2

Table 11. Effect of Solvent on Metabolism of TCDD by Bacillus megaterium in Yeast–Soybean Medium. The Amount of Solvent Used was 1 mL.

	Aqueous phase A[a]	Aqueous Phase B[b]	% of ^{14}C Activity Recovered Solvent Phase Metabolites	TCDD
Ethyl acetate				
1	24.0	3.3	6.4	32.3
2	0.7	5.2	11.3	71.2
DMSO				
1	1.4	2.9	9.5	72.6
2	0.7	0.9	8.2	91.1
Ethanol				
1	0.3	4.2	1.7	94.7
2	2.5	7.5	1.4	85.3
Corn oil				
1	1.7	0.3	—[c]	92.8[c]
2	0.3	0	—	91.2

[a] Water left after extraction and evaporation of solvents.
[b] Aqueous phase after partitioning against the solvent.
[c] Thin–layer chromatographic analysis was not possible because of the interference with corn oil.

Table 12. Effect of Solvent on Metabolism of TCDD by _Bacillus megaterium_. Various Percentages of the Standard Amount of Soybean Extract were Used in the Yeast-Soybean Extract Media. Results are Expressed as the Means and Standard Deviations of Three Replicates, Except two Replicates for 12% Soybean.

Treatment	% of ^{14}C Activity Recovered		
	Aqueous Phase	Solvent Phase	
		Metabolites	TCDD
Ethyl acetate			
100% soybean	10.5 ± 9.4	12.4 ± 7.9	58.7 ± 28.2
50% soybean	16.2 ± 6.7	41.6 ± 10.9	29.6 ± 20.8
25% soybean	18.0 ± 6.8	40.3 ± 13.7	28.2 ± 10.6
12% soybean	16.9 ± 0.6	35.9 ± 2.7	33.3 ± 7.1
(All ethyl acetate samples)	15.3 ± 6.8	32.2 ± 15.5	37.8 ± 22.1)
DMSO			
100% soybean	3.3 ± 1.6	13.3 ± 6.4	69.0 ± 4.8
50% soybean	2.1 ± 2.2	24.4 ± 22.7	63.1 ± 23.7
25% soybean	1.8 ± 1.7	8.4 ± 13.3	76.0 ± 7.0
12% soybean	1.0 ± 0.8	0.7 ± 0.5	85.4 ± 7.6
(All DMSO samples)	2.1 ± 1.7	12.7 ± 14.9	72.3 ± 14.1)

[a]Water left after extraction and evaporation of solvents.

As for the role of microbial activities in degrading TCDD we have shown both in aquatic sediment and terrestrial soil samples that TCDD does indeed degrade by the action of microbes. The addition of general nutrients such as glucose, bactopectone or mannitol-yeast increases the overall degradation activities. The amounts of metabolic products produced were higher when TCDD was directly given to microbial cultures than in the cases where it was added to soil or sediment. Such a result is expected as the total number of microorganisms in culture are much higher. Also important is the consideration that in soil and sediment some portion of TCDD present is bound to soil constituents and therefore is not immediately available to microorganisms.

As to the nature of TCDD metabolism by soil microorganisms, the addition of naphthalene either as an extra or a sole carbon source in a basal salt medium gave a variable result. Even at the level of pure isolates in culture, naphthalene was stimulatory in one species (Nocardiopsis sp.) or variable in another (B. megaterium). One possible interpretation of this event is that some strains of microorganisms which have been acclimated to grow on naphthalene as a carbon source could also metabolize TCDD, but others could recognize its difference from

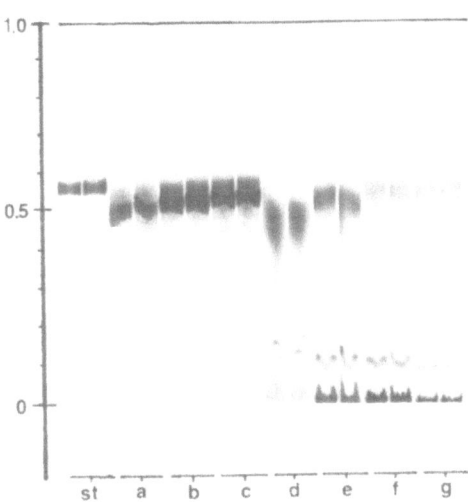

Fig. 9. Audtoradiogram of TLC plate of solvent phase for one replication of experiment summarized in Table 12. Band near the origin (d-g) represent polar metabolites. st: TCDD standard; a: DMSO, 50% soybean; b: DMSO, 25% soybean; c: DMSO 12% soybean; d: ethyl acetate, 100% soybean; e: ethyl acetate, 50% soybean; f: ethyl acetate, 25% soybean; g: ethyl acetate, 12% soybean.

naphthalene and as a consequence did not degrade TCDD. In this regard it is interesting to note that general nutrients such as glucose, mannitol and bactopeptone have shown stimulatory properties for microbial degradation of TCDD in all tests conducted so far. At this stage, therefore, one could tentatively conclude that processes involved in such cases are not likely to be very specialized ones. Much more work would be needed to confirm such a possibility.

The most conspicuous stimulatory effect on TCDD metabolism was observed when ethyl acetate was used as a carrier to TCDD. The nature of the carrier appears to be very important (see Table 11) as the use of a wrong carrier such as corn oil completely abolished TCDD-metabolizing activity by microorganisms. The most likely explanation of this pheomenon is that TCDD is not readily picked up by microorganisms or hardly penetrates through the microbial cell membrane. TCDD is a unique halogenated aromatic pollutant that has a relatively low lipid solubility (Matsumura and Benezet, 1973). The fact that its bioaccumulation potential is much lower than that of DDT supports the above hypothesis. Such a difficulty in microbial uptake could be the limiting factor in TCDD metabolism in the environment, and as such, in the future it requires much more attention than it has received in the past.

ACKNOWLEDGEMENT

This research was supported by the Michigan Agricultural Experiment Station under the Environmental Toxicology Program, and by research grants from Dow Chemical Company, Midland, Michigan, and from Syntex, Inc., Palo Alto, California.

REFERENCES

Beeman, R. W. and Matsumura, F. Metabolism of cis- and trans-chlordane by a soil microorganism. J. Agric. Food Chem. 29:84-89, 1981.
Crosby, D. G. and Wong, A. E. Environmental degradation of 2,3,7,8-tetrachlorodibenzo-p-dioxin (TCDD). Science 195:1337, 1977.
Isensee, A. R., Jones, G. H. Adsorption and translocation of 2,3,7,8-tetrachloro-p-dioxin in an aquatic ecosystem. Environ. Sci. Technol. 9:668, 1975.

Kearney, P. C., Woolson, E. A., Isensee, A. R., and Helling, C. S.
    Tetrachlorodibenzodioxin in the environment: sources, fate
    and decontamination. Environ. Health Perspect. 5:273, 1973.
Matsumura, F., and Benezet, J. J. Studies on the bioaccumulation
    and microbial degradation of 2,3,7,8-tetrachloro-p-dioxin.
    Environ. Health Perspect. 5:253, 1973.
Matsumura, F. and Esaac, E. G. Degradation of pesticides by
    algae and aquatic microorganisms, in: Khan, M A. Q., Lech,
    J. J., Menn, J. J., eds. Pesticide and Xenobiotic Metabolism
    in Aquatic Organisms. Am. Chem. Soc. Symp. Ser. 99:371, 1979.
Ward, C. T., and Matsumura, F. Fate of 2,3,7,8-tetrachlorodibenzo-
    p-dioxin (TCDD) in a model aquatic environment. Arch.
    Environ. Contam. Toxicol. 7:349-357, 1978.

RESIDUES OF POLYCHLORINATED DIBENZO-P-DIOXINS AND

DIBENZOFURANS IN LAURENTIAN GREAT LAKES FISH

D. L. Stalling, L. M. Smith, J. D. Petty, J. W. Hogan
J. L. Johnson, C. Rappe*, and H. R. Buser**

Columbia National Fisheries Research Laboratory
U.S. Fish and Wildlife Service
Columbia, Missouri, USA

*University of Umea, Umea, Sweden

**Swiss Federal Research Station, Wadenswil, Switzerland

ABSTRACT

Fish from the Laurentian Great Lakes, and related rivers, and
lakes were examined for residues of polychlorinated biphenyls
(PCBs), dibenzofurans (PCDFs), and dibenzo-p-dioxins (PCDDs).
Patterns of PCDF residues were more complex than those of PCDD
residues, and were dominated by isomers having 2,3,7,8-chlorine
substitution. PCDF residues were presented in samples from all
of the Great Lakes and exceeded 100 pg/g in fish from Lakes
Ontario and Huron, Michigan, and Saginaw Bay. A composite sample
of fish from the Tittabawassee River (Michigan) contained 290
pg/g of PCDFs and 223 pg/g of PCDDs. Ratios of PCDFs to PCBs
ranged from 1 to 20 X 10^6. Attempts to correlate PCDF residue
levels in biological samples with PCDF concentrations in PCB
preparations are complicated by an apparent preferential
retention or accumulation of 2,3,7,8-substituted congeners. The
PCDF concentrations in a fish from Lake Siskiwit (on Isle Royale,
Lake Superior) were the lowest detected in this study (15 pg/g).
Fish from Lakes Huron and Ontario, the Tittabawasee River and
Saginaw Bay contained the highest concentrations of 2,3,7,8-TCDF.
In another river system, the Housatonic River, Massachusetts, the
composition of PCDD and PCDF residues in sediments differed
radically from residues in fish from this river.

INTRODUCTION

Polychlorinated dibenzofurans (PCDFs) and polychlorinated dibenzo-p-dioxins (PCDDs) are two groups of toxic aromatic compounds composed of 135 and 75 individual isomers, respectively. Selected isomers are extremely toxic, particularly those with chlorine substituents in the 2,3,7,8-positions of the aromatic ring (Moore et al, 1979). PCDFs occur as trace contaminants in polychlorinated biphenyls (PCBs) (Bowes et al, 1975; Albro and Parker, 1979) and can be formed at significant levels from pyrolysis or incomplete combustion of PCBs (Rappe and Buser, 1980). PCDFs and PCDDs also occur as contaminants in the manufacture of chlorinated phenols (Buser and Bosshardt, 1976) and combustion or pyrolysis of chlorinated phenols (Rappe and Buser, 1980). During combustion of these formulations, PCDDs are primarily formed from thermal dimerization and conversion of chlorinated phenoxyphenols, whereas PCDFs are formed from dechlorinated diphenyl ethers. Diphenyl ethers are present also as impurities (Rappe and Buser, 1980). Pyrolysis or burning of purified chlorophenols generally yields higher quantities of PCDDs than of PCDFs. In one study, pyrolysis or burning of carefully purified chlorophenols yielded only PCDDs; no PCDFs could be identified (Rappe et al, 1978).

Recent studies by Helder (1980) have demonstrated that 2,3,7,8-TCDD is extremely toxic to fertilized fish eggs. Exposure of eyed eggs to concentrations of TCDD of 0.1 to 11 ng/L for 4 days resulted in decreased growth and survival, all fish exposed to the highest TCDD concentration died within 21 days. Impaired liver and tail fin development were also observed. No residue data were obtained from these studies to establish that TCDD had been bioaccumulated. Earlier studies in microcosms had established the potential for 2,3,7,8-TCDD to bioaccumulate in fish. Bioaccumulation factors of 2 to 2.6×10^4 were observed for snails, fish, and daphnids exposed for 3 to 6 days in model ecosystems (Isensee and Jones, 1975). Isensee (1978) also reviewed bioaccumulation of TCDD in relation to chlorinated phenoxy acids and their dioxins.

Fewer toxicological studies have been conducted with 2,3,7,8-TCDF than with TCDD, and no aquatic toxicity data exist to establish the toxicity of TCDF to fish from water exposures (McConnell, 1980). However, the similarities of these two compounds give reason for concern about the impact of PCDFs on aquatic resources. Also, no studies of PCDF bioaccumulation in aquatic organisms from water exposure have been reported. Both TCDF and TCDD cause similar symptoms of toxicosis--liver dysfunction and chloracne in mammals (Huff et al, 1980). Additionally, atrophy of the thymus and induction of liver enzyme systems, e.g., aryl hydrocarbon hydroxylase, result from exposure to these compounds (McConnell, 1980; Huff et al, 1980). These classes of compounds have, in general, a profoundly depressive effect on reproductive capacity in mammals, especially females, and on the viability of offspring (McConnell, 1980).

The impetus for an initial survey for PCDD and PCDF residues in fish (Dougherty et al, 1980) was the discovery that pyrolysis of Aroclors 1254 and 1260 produces many PCDF congeners of four of more chlorines that have 2,3,7,8-chlorine substitution (Rappe and Buser, 1980). Heating of PCBs can yield partial conversion to PCDFs (Buser et al, 1978). The potential magnitude of the impact of PCB pyrolysis to form PCDFs is indicated by the quantity of PCBs produced domestically. Total production of PCBs (Aroclors) by Monsanto Chemical Company is estimated at 1,400 million pounds during the period 1930-1975 of which 1,253 million pounds were marketed domestically (Brinkman and DeKok, 1980).

The recent detection of PCDFs and PCDDs, as well as chlorinated biphenylenes in soot formed from PCBs in an electrical fire demonstrates that environmental contamination by PCDFs occur when PCBs are pyrolyzed (Stalling, 1981a). This fire originated in a primary electrical distribution panel located in the basement of the New York Office Building in Binghamton, New York, and thermal pyrolysis to form PCDFs occurred upon leakage of PCBs from a transformer. The concentration of PCDFs was increased from about 5 µg/g-PCB to 7,500 µg/g-PCB as a consequence of the fire.

Because no data existed on the isomer distribution of PCDFs or total concentration of PCDD or PCDF residues in fish, we under-took an assessment of the environmental occurrence of these chemicals. This effort was a preliminary step toward determining the impact of these toxic chemicals on Great Lakes fisheries. The primary objectives of this study were to determine the distribution and composition of residues of PCDDs and PCDFs in the aquatic biota of the Great Lakes and other freshwater lakes and rivers. We also sought to correlate the concentration of PCDFs to the levels of PCBs known to be present in Great Lakes fish (Schmidt et al, 1981). Residues of TCDD and other PCDDs were also measured, to assess the extent of dioxin contamination.

Prior to the present study, our knowledge of PCDF isomers present in aquatic ecosystems was derived from analyses for PCDFs in the fat from a snapping turtle which revealed the presence of PCDFs having congeners with 2,3,7,8-chlorine substituents. The turtle fat contained 3 µg/kg of total PCDFs (Rappe et al, 1981). Residues of TCDD and other PCDDs were also measured, to assess the extent of dioxin contamination.

EXPERIMENTAL METHODS

Sample Sources

Tissue of herring gulls (Larus argentatus) collected from the
vicinity of Saginaw Bay, (Lake Huron) Michigan was provided by
Dr. Douglas Hallet, National Wildlife Research Center, Ottawa,
Ontario. Samples of common carp (Cyprinus carpio) from Saginaw
Bay and lake trout (Salvelinus namaycush) Lake Siskiwit, (Isle
Royale, Lake Superior), were provided by Dr. Waylon Swain,
Environmental Protection Agency, Large Lakes Laboratory, Grosse
Isle, Michigan. We were especially fortunate to obtain samples
of lake trout from Lake Siskiwit, an isolated lake, on an island
in northern Lake Superior. This island constitutes Isle Royale
National Park and is remote from direct sources of pollution
(Swain, 1980).

Samples of lake trout from the Great Lakes and Lake St. Clair
were provided by Mr. Wayne Wilford, U.S. Fish and Wildlife
Service, Great Lakes Fishery Laboratory, Ann Arbor, Michigan.
Lake trout were also collected from the following locations: Lake
Superior, Apostle Islands; Lake Michigan, Saugatuck, Michigan;
Lake Huron, Rock Port, Michigan; Lake Erie, Cedar Point,
Michigan; Lake Ontario, Oswego, New York; and Lake St. Clair,
Anchor Bay, Michigan. Other samples analyzed from the Great
Lakes were obtained from current National Pesticide Monitoring
Program sample archives and were provided by Dr. Larry Ludke and
Mr. Buford Seabolt, Columbia National Fisheries Research
Laboratory, Columbia, Missouri (Schmidt et al, 1981). A sediment
sample taken from Woods Pond, Massachusetts was provided by Dr.
Charles Frink, Connecticut Agriculture Experiment Station, New
Haven, Connecticut. Analysis of this sample provided a
preliminary comparison of PCDD and PCDF residues in fish and
sediment. Previously, information had been obtained on PCBs,
PCDFs, and PCDDs from analyses of fish samples from Woods Pond
(Stalling, 1980b). We thank those who provided samples for the
present study.

Sample Preparation

The fish extracts obtained from composites of three to five
whole fish for each sample were enriched by the method of
Stalling et al, (1981b). Residues of PCDFs and PCDDs in 1981
samples were extracted from fish and separated from interfering
substances in two series of sequential chromatographic processes.
The first process involved sequential passage of the extracted
oil through a column containing a segment of potassium silicate
prior to another segment of silica gel; the eluate then passes
directly through a column containing carbon dispersed on glass
fibers. The carbon adsorbs most fused-ring aromatic compounds

from the eluate of the first column whereas most biological
coextractives are not retained. The PCDFs and PCDDs, along with
similar chemical classes, are subsequently recovered from the
carbon by reverse elution with toluene. In the second chromatographic process, the sample is redissolved in hexane and applied
to two columns in tandem; the first column contains sulfuric acid
dispersed on silica gel and cesium silicate, and the second
contains alumina (Stalling et al, 1981). This second column
provides a means of separating the PCDDs and PCDFs from other
contaminants such as chlorinated naphthalenes. This sample
enrichment procedure was one of several procedures that were
compared for the analysis of TCDD by Brumley et al, (1981).

Sample Spiking and Procedural Blanks

Samples processed for quality control and recovery studies
were prepared by grinding frozen whole fish µbrook trout
(Salvelinus tontinalis), and grass carp (Ctenopharyngodon idella)ß
that were reared at the Columbia National Fisheries Research
Laboratory. After preparation of the mixture (4:1 w/w) of
anhydrous sodium sulfate and ground fish (Stalling et al, 1981)
the samples were spiked with a solution of PCDF and PCDD
congeners: 2,3,7,8-tetchloro-, 1,2,4,7,8-pentachloro-, 1,2,4,6,7,9-
hexachloro-, 1,2,3,4,6,7,9-heptachloro-, and octachlorodibenzo-
furans and 2,3,7,8-tetrachloro-(TCDD), 1,2,3,4,7,8-hexachloro-,
1,2,3,4,6,7-hexachloro-, and octachloro-dibenzo-p-dioxin (OCDD).
Other samples were spiked only with the ^{13}C-2,3,7,8-TCDD.
Toluene was the spiking solvent and the solution of standards was
applied to the sample mixture with a syringe after the sodium
sulfate-sample mixture was placed in the extraction column. In
addition to the standards listed above, the spike solution
contained either 20 or 100 pg/g ^{13}C-labeled 2,3,7,8-TCDD. The
^{13}C-labeled TCDD was provided by Dr. Ron Mitchum, National
Center for Toxicological Research, U.S. Food and Drug
Administration, Jefferson, Arkansas.

The results of analyses of quality control samples associated
with the Great Lakes samples are presented in Table 1. All
quality control samples were spiked with ^{13}C-TCDD as an
internal standard.

Recovery of the ^{13}C-TCDD internal standard ranged from 60
to 90 percent at the 20 parts-per-trillion fortification level
and was 80 percent at the 100 parts-per-trillion level. The
recovery of the ^{13}C-TCDD in procedural blanks was 44 percent.
The procedural blank and the grass carp tissue blank contained
extremely low levels of PCDDs and PCDFs. The contamination of
the brook trout control sample was tentatively traced to a PCB
contaminated feed which contained PCDF levels comparable to those
found in the trout. Recovery of PCDFs was lower than the

fortification level (50 to 75 percent). In general, the results obtained with the samples in the present study should be considered minimal values. Further improvements in the precision of the PCDF determinations should result from use of a ^{13}C-labeled dibenzofuran internal standard. However, at the time this study was conducted the ^{13}C-standard was not available.

Analysis

Detection of PCDDs and PCDFs was accomplished with a Finnigan 4023 gas chromatography-mass spectrometer (GC-MS) system equipped with negative and positive ion chemical ionization options. An INCOS Software System running on a Data General Nova 3/12 mini-computer was used to obtain spectra and to quantitate PCDF and PCDD congeners. Another Finnigan 4023 GC-MS system equipped with a PROMIM multiple ion monitoring unit was employed for analysis of specific isomers. In these isomer analyses, methane or isobutane was used as the reagent gas.

GC conditions were selected to minimize the amount of disk storage used for each GC-MS analysis. Approximately 30 minutes were required for analysis of a sample. After a general perspective of the PCDD and PCDF residues was obtained, aliquots of the prepared extracts were analyzed using a capillary gas chromatographic procedure capable of separating all 22 TCDD isomers (Buser, 1981). This procedure was also suitable for analyses (Rappe and Buser, 1980) of the tetrachloro- through octa-CDD and -CDF congeners isolated by the enrichment procedure described above.

The chromatographic and mass spectrometry conditions for the survey GC-MS analyses follow:

Gas chromatography parameters --

   Gas chromatography column - 30 m x 0.25 mm i.d.
   fused silica capillary column coated with DB-5
   liquid phase (J and W Scientific)

   Carrier gas - Helium at 6 psig, 22 cm/sec linear
   velocity.

   Temperature program - the initial temperature was
   held at 180°C for 5 min., then programmed to 225°C
   at 3°C/min. The rate was then changed to 12°C/min.
   until the oven reached 291°C and the final temperature
   was held for 10 min.

Table 1. Quality Control Data for Analysis of Great Lakes Fish Samples. Recovery and Level of Laboratory Contamination (pg/g) in Blanks.

Samples	PCDD Samples, pg/g						$\%$ ^{13}C						
	#Cl 4	5	6	7	8	Total	Recovered	#Cl 4	5	6	7	8	Total
Brook trout, spiked at 20 pg/g ^{13}C-TCDD	nd[a]	nd	nd	nd	nd	nd	90	nd	nd	5	21	4	30
Procedural blank spiked with 100 pg/g ^{13}C-TCDD	1.6	nd	nd	nd	10[b]	12	45	nd	nd	nd	nd	1.4	1.4
Lab-reared brook trout spiked at 20 pg/g ^{13}C-TCDD and 20 pg/g standard mixture	26	nd ns	31	nd ns	33	--	60	8.6	8.8	nd ns	10.2	8.3	--
Lab-reared brook trout (replicate)	20	nd ns	35	nd ns	67	--	60	8	9	nd ns	13	8.9	--
Lab-reared grass carp spiked at 20 pg/g ^{13}C-TCDD	nd	nd	nd	nd	5	5	80	0.55	0.4	0.2	1.3	1.4	3.8
Procedural blank (repeat) spiked with 100 pg/g ^{13}C-TCDD	1.6	nd	nd	nd	11	13	43	nd	nd	nd	nd	1.4	1.4

[a] nd = not detected (s/n = 3); ns = not spiked in congener group; -- = not totaled
[b] signal to noise = 2

Mass spectrometer parameters:

Electron multiplier voltage = 2200 volts
Electron energy = 55.1 volts
Source temperature = 250°C for electron impact and
200°C for negative chemical ionization mode (NCI)
Manifold temperature - 100°C
Methane gas pressure = 5.5 X 10^5 torr (NCI).

The multiple ion detection parameters for analysis of PCDDs
by electron impact (EI)-MS permitted the detection of tetra- to
octa-CDD congeners. Eight ions were monitored in various time
windows corresponding to elution times for the PCDDs. Ions for
the chlorine clusters of tetra-, penta-, and hexa-chlorobiphenyls
(the non-ortho ortho'-chlorine substituted isomers) and penta- and
hexa-chloro-dibenzofurans were included to keep total ion dwell
times in each MS scan equal to that used for the NCI analyses for
PCDFs. Full scan spectra were acquired by scanning mass to
charge ratios (m/z) 240 to 520 in 1.95 sec, allowing a 0.05-sec
interscan interval. Four chlorine isotope ions were monitored in
the multiple ion detection mode (MID) of analysis for quantitating
low concentrations of PCDF and PCDD residues. In all cases the
first ion selected in the MID analyses corresponded to the
35Cl-isotopic molecular weight of the compound. Evaluation of
the resolving power of the DB-5 column using all 22 TCDD isomers
and (UL) ^{13}C-2,3,7,8-TCDD demonstrated that the analytical
procedure employing this column provided a specific analysis for
2,3,7,8-TCDD.

RESULTS AND DISCUSSION

Detection of PCDFs and higher chlorinated PCDDs was greatly
enhanced by the use of NCI with methane or isobutane as the
reagent gas (Rappe and Buser, 1980). This technique afforded an
increase in sensitivity of approximately two order of magnitude
over detection by EI-MS and the accompanying selectivity for PCDFs
significantly decreased interferences from sample background.
The limit of detection (3 to 1 signal to noise ratio) was
approximately 0.5 pg/g for PCDFs, based on 2 µL into the GC-MS.
EI ionization-MS was required to measure tetra-CDDs because they
were insensitive to NCI-MS detection. The limit of detection for
TCDD under EI-MS ranged from 1 to 3 pg/g, but was approximately 5
to 10 pg/g for OCDD.

Residues of PCDFs were measured in composite whole fish
samples from each of the Great Lakes, and a limited number of
rivers in their watersheds. Concentrations of PCDFs were lowest
in a sample of walleye (Stizostedion v. vitreum) from Lake St.
Clair and lake trout from Lake Siskiwit (Tables 2-3). Levels of

Table 2.   Dibenzofuran Residues (pg/g) in Composite Samples from
           Lakes Superior, Ontario, and Erie

Sample and Site	#Cl	Dibenzofurans, pg/g					Total	Ratio 4Cl/Total
		4	5	6	7	8		
Bloater Keweenaw Bay, Lake Superior	26	9	2	1	2		40	26/40=0.65
Lake trout Isle Royale, Lake Siskiwit, Lake Superior	10	5	nd	nd	trace		15	10/15=0.67
Apostle Islands, Lake Superior	19	5	5	4	3		40	19/40=0.48
Oswego, New York Lake Ontario	34	48	29	6	2		119	34/119=0.28
Brook trout Roosevelt Beach, Lake Ontario	19	4	3	3	3		32	19/32=0.59
Walleye Cedar Point, Ohio Lake Erie	18	9	6	5	2		40	18/40=0.45
Common Carp Port Clinton, Lake Erie	5	5	2	4	2		18	5/18=0.28
Lake trout Saugatuck, Lake Michigan								
Sample #1	35	41	8	1	1		86	35/86=0.41
Sample #2	33	61	10	4	2		110	33/110=0.30
Walleye Cedar Point, Ohio Lake Erie	18	9	6	5	2		40	18/40=0.45
Common Carp Port Clinton Lake Erie	5	5	2	4	2		8	5/18=0.28

Table 3.   Dibenzofuran Residues (pg/g) in Composite Samples from
           Lakes Michigan and Huron

Sample and Site	#Cl	Dibenzofurans, pg/g					Total	Ratio 4Cl/Total
		4	5	6	7	8		
Lake trout Saugatuck, Lake Michigan								
Sample #1		35	41	8	1	1	86	35/86=0.41
Sample #2		33	61	10	4	2	110	33/110=0.30
Common carp Saginaw Bay		27	44	34	44	4	153	27/153=0.18
Bay Port		5	12	5	6	1	29	5/29=0.17
Tittabawassee River, Midland, Michigan		37	73	145	31	4	290	37/290=0.13
Herring gull Saginaw Bay								
Sample #1		16	28	57	17	3	121	16/121=0.13
Sample #2		15	50	40	7	5	117	15/117=0.13
Lake trout Rockport, Michigan		32	28	16	9	6	91	32/91=0.35
(duplicate analyses)		19	29	16	4	2	70	18/70=0.27
Walleye Anchor Bay, Michigan Lake St. Clair		4	4	0.7	0.7	3	12	4/12=0.33

PCDFs were greater in fish from Lakes Erie, Ontario, Michigan, and
Superior (concentration range 18-119 pg/g). The greatest PCDF
concentrations were observed in fish from Saginaw Bay in Lake Huron
(153 pg/g) and the Titabawassee River (290 pg/g).

The composition of the PCDF residues was similar in most of
the samples examined: however, the concentration and relative
proportions of hexa- and hepta-CDF congeners were greatest in
fish from Saginaw Bay (Table 3). This difference was reflected
by a lower ratio of tetra-CDF to total PCDF concentration. In
the Saginaw Bay and Tittabawassee River samples, this ratio was
0.18 and 0.13 and the total PCDF concentrations were 153 pg/g and
290 pg/g, respectively. The ratio of tetra-CDF to total furan
ranged from 0.17 (in Bay Port, Saginaw Bay, Lake Huron) to 0.65
in other samples from the Great Lakes.

Table 4. PCDF/PCB Ratios in Composite Samples from the Great Lakes

Sample and Site	$(PCDFs/PCBs) \times 10^6$
Lake Huron (Saginaw Bay)	
Common carp	$(153/37.1) \times 10^6 = 4.1$
Herring gull	
Sample #1	$(121/45.3) \times 10^6 = 2.6$
Sample #2	$(117/51.8) \times 10^6 = 2.2$
Tittabawasee River (Midland, Michigan)	
Common carp	$(290/103.7) \times 10^6 = 2.8$
Lake Superior (Isle Royale, Lake Siskiwit)	
Lake Trout	$(15/4.3) \times 10^6 = 3.5$
Lake Ontario (Roosevelt Beach)	
Brook trout	$(32/5.2 \times 10^6 = 6.2$
Lake Michigan (Saugatuck)	
Lake Trout	$(86/7.1) \times 10^6 = 12.1$

Because of the presence of PCDFs in PCBs as manufactured (Rappe and Buser, 1980) we believed it was important to determine the ratio of PCDFs to PCBs in these samples. An increase in this ratio may reflect the conversion of PCBs to PCDFs during use, incineration, or aging prior to entry of the PCBs into the environment.

The ratio of PCDFs to PCBs was determined to selected samples (Table 4). This ratio, determined in a limited number of samples, strongly suggests that PCDF residues correlate with PCB content. PCDF concentrations ranged 1 to 20 parts-per-trillion relative to the PCB content of the sample. However, caution should be exercised in comparing the concentration of PCDFs in Aroclors to environmental samples. Aroclors contain more than 30 PCDF isomers (Rappe and Buser, 1980), which are predominantly those having 2,3,7,8-chlorine substitution patterns.

In this investigation, confirmatory analyses were made by high resolution glass capillary GC-low resolution EI-MS. These

Table 5. Dibenzofuran Isomer Residues (pg/g) in Composite Samples
          Analyzed

| | Site and Sample | | | | |
Isomer[a]	Herring gull #1 Lake Huron	Herring gull #2 Lake Huron	Common Carp Lake Huron	Common Carp Lake Huron	Grass Carp Lab Control
Tetra-Cl					
2,3,7,8-	4(0.5)[b]	2(0.5)	11	11	1.5
2,3,6,7-			2(0.5)	3(0.5)	
other			(1)		0.5(0.4)
Penta-Cl					
1,2,4,7,8-/ 1,3,4,7,8-	2	2	2(2)	3	1
1,2,3,7,8-	2	2	5	1	2
2,3,4,7,8-	16(2)	20(2)	11	4(1)	1(1)
Hexa-Cl					
1,2,3,4,7,8/ 1,2,3,6,7,8-	7(3)	4(2)	5(2)	2(2)	2(1)
			2		
Hepta-Cl					
1,2,3,4,6,7,8-	4	4	3(3)	4	
1,2,3,4,6,8,9-					2(2)
Octa-Cl	8	8	4	8	3
Σ PCDF	27	26	40	24	10

[a] Measured by GC-NI-CI-MS
[b] Detection limit in ( )

analyses identified the major PCDD and PCDF isomers as those having
2,3,7,8-chlorine substitution (Tables 5 and 6). In all cases,
2,3,7,8-TCDD was the predominant or exclusive TCDD isomer observed.
The residues in many of the samples had isomer distributions
similar to PCDFs present in human liver samples from Yusho patients
(Rappe et al, 1979a) and seal and turtle fat (Rappe et al, 1981).

Table 6.   Dibenzo-p-dioxin Residues (pg/g) in Composite Samples
           Analyzed

Isomer[a]	Herring gull #1 Lake Huron	Herring gull #2 Lake Huron	Common Carp Lake Huron	Common Carp Lake Huron	Grass Carp Lab Control
Tetra-Cl					
2,3,7,8-	165(0.5)[b]	75(1)	28	3	0.4
1,3,6,8-			0.8(0.5)		
other				2(1)	
Penta-Cl					
1,2,3,7,8- other	20(2)	18(2)	11(1)	2(1)	2
Hexa-Cl					
1,2,3,4,7,8- other	11(2)	17(2)	5(2)	3(2)	3
Hepta-Cl					
1,2,3,4,6,7,8-	4	4	4(2)	3(2)	2
1,2,3,4,6,8,9-					2(2)
Octa-Cl	4	4	3(3)	5(5)	5(3)
Σ PCDF	196	110	52	17	5.4

[a]Measured by GC-NI-CI-MS
[b]Detection limit in ( )

    Fish from Lakes Superior and Siskiwit did not contain
measurable levels of TCDD or other PCDDs (Table 7). Samples from
Lake Michigan and Lake Ontario contained 2,3,7,8-TCDD at 5 and 33
pg/g. No other PCDD isomers were detected. Fish and herring gulls
from Saginaw Bay and fish from Lake Huron contained high
concentrations of 2,3,7,8-TCDD and more complex mixtures of other
PCDD congeners (Tables 6 and 7 and Figure 1). These isomer
specific analyses confirm that the tetrachlorodioxin residues are
composed predominantly -- or more often, exclusively -- of the
2,3,7,8-tetra CDD isomer.

Table 7. Dibenzo-p-dioxin Residues (pg/g) in Composite Samples
          from Lakes Michigan, Erie, Superior, Ontario and Huron

Sample and Site	#Cl	Dibenzo-p-dioxin, pg/g					Total	Ratio 4Cl/Total
		4	5	6	7	8		
Common carp Port Clinton, Lake Erie	nd[a]	9	nd	11	30	50		--
Lake trout Saugatuck, Lake Michigan	5	nd	nd	nd	nd	5		1
Bloater, Keweenaw, Lake Superior	nd	nd	nd	nd	nd	nd		--
Lake trout Lake Siskiwit Isle Royale, Lake Superior	nd	nd	nd	nd	trace	trace		--
Brown trout Roosevelt Beach, Lake Ontario	33	nd	nd	nd	nd	33		1
Common carp								
Saginaw Bay	94	157	122	12	nd	385	94/385=0.24
Bay Port	27	21	nd	31	32	111	27/111=0.24
Tittabawasee River	81	31	44	53	14	223	81/223=0.36
Herring gull Saginaw Bay								
Sample #1	160	nd	20	nd	19	199	160/100=0.8
Sample #2	70	nd	88	nd	28	186	70/186=0.38

[a]nd = not detected

The ratio of tetra-CDD to total PCDD residues ranged from 0.24
to 0.80 in samples from Lake Huron. Samples from Saginaw Bay and
the Tittabawassee River (Table 4) contained the most complex
mixture of PCDD and PCDF congeners. Total concentrations of PCDDs
in herring gull samples were 199 and 186 pg/g; the concentration of
PCDFs in these samples was somewhat lower -- 121 and 117 pg/g
(Tables 7 and 3). These residue data suggest a somewhat different

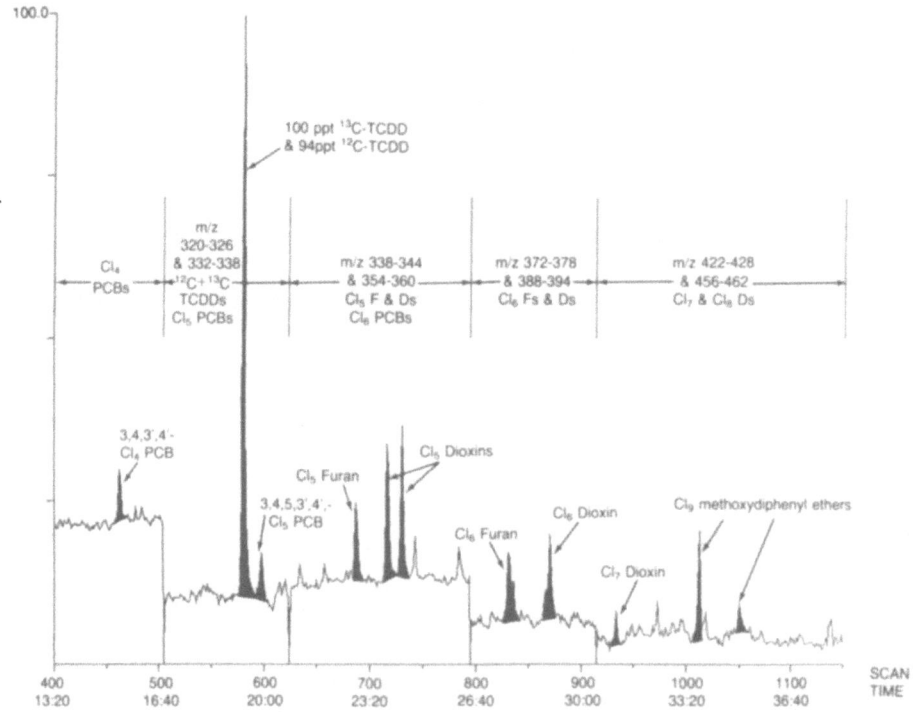

Fig. 1.   An Example of a GC-MS Scan Showing Dioxins and Furans in
          an Environmental Sample.

distribution of congeners in samples from Saginaw Bay (Lake Huron)
than in samples from the other Great Lakes. Pollution with PCDDs
is very low in the other Great Lakes, except for Lakes Ontario and
Huron.   However, PCDFs were detected in all of the Great Lakes
fish.   The occurrence of traces of methylated chlorophenoxyphenols
(Figure 1) suggests that commerical chlorophenols may contribute to
the environmental pollution by PCDDs and PCDFs in Lake Ontario and
Lake Superior.

    The occurrence of low levels of PCDFs in fish from Lake
Siskiwit suggests atmospheric transport and rainout of PCDFs along
with PCBs into the lake (Swain, 1980). The ratio of tetra-CDFs to
total PCDF residues is similar for the samples from Lakes Superior
and Siskiwit -- 0.65 and 0.48 vs 0.67, respectively (Table 2).
In view of the occurrence of PCDDs in material from various
combustion sources (Bumb et al, 1980; Rappe et al, 1979b), the
absence of PCDDs in fish samples in Lakes Siskiwit and Superior
raises questions about our understanding of sources of these
contaminants. These residue data suggest that the presence of
PCDDs in fish from other areas is strongly influenced by point
source discharges.

Table 8. Disporportionated PCDF Residues in Sediment and Fish
          from Woods Pond, Maine

PCDF Isomers	Sediment	PCDF Concentrations Yellow Perch	Aroclor 1260[a]
Tetra-CDFs	0.0001	1.06	290
Penta-CDFs	0.0009	0.64	1,330
Hexa-CDFs	0.15	0.44	1,810
Hepta-CDFs	0.92	0.0004	780
Octa-CDF	0.27	0.0001	29
TOTAL	1.35	2.15	4,240
PCB (μg/g)	60	170	(neat)
PCDF/PCB Ratio	22.5×10^{-6}	12.6×10^{-6}	4.24×10^{-6}

[a]Sample provided by Dr. David Alexrod, New York Department of Health

     We included data from another geographical area to illustrate
the complexity of the PCDD and PCDF distribution in the aquatic
environment. Major differences between PCDF and PCDD residues in
fish and sediment samples were detected in samples from Woods Pond,
Massachusetts, a part of the Housatonic River. The extent of PCB
pollution at this location is similar to that in the Hudson River.
PCB pollution resulted from electrical manufacturing operations
near Albany, New York (Horn et al, 1979).

     Measurement of the residues in fish and sediment samples by
electron capture GC revealed the presence of Aroclor 1260 in both
sample types. A fish sample from Woods Pond contained about 170
μg/g of Aroclor 1260 (Table 8). A similarly disproportioned
residue pattern between fish and sediment was also observed for the
PCDD congeners in these samples: i.e., the dominant isomer was
2,3,7,8-tetra-CDD (10 pg/g) in fish and octa-CDD in the sediment.

     A limited but preliminary assessment of the composition of
PCDFs in samples in relation to the PCDFs present in Aroclor 1260
was possible. We previously had obtained and analyzed the PCDFs in
a sample from one of the lots of this Aroclor, which the General

Electric Company had used in their electrical manufacturing operation. However, the composition of PCDFs in other various Aroclors used during the period of PCB use in the manufacturing operations at this site is unknown.

CONCLUSIONS

The presence of the lower chlorinated PCDF isomers in fish samples, particularly those having a 2,3,7,8-chlorine substitution pattern suggests greater biological retention of these isomers. This situation could result from decreased excretion of these isomers or a diminished ability to metabolize PCDFs with this pattern of chlorine substitution. A similar PCDF residue pattern was present in samples of liver from Yusho patients (Rappe et al, 1979b). Alternatively, the lower concentrations of the more highly chlorinated congeners in the fish could result from their lower water solubility and subsequent affinity for organic surfaces in aquatic systems which would reduce the relative bioavailability. However, such large differences among isomers of similar chlorine content seem unlikely. The observed residues in sediment might also reflect accumulation from fly ash particles contaminated with PCDFs and PCDDs.

These compounds would be tightly bound to carbonaceous moeities in the ash.

Additional study of the environmental significance of PCDFs is warranted because the observed components of PCDF residues in fish are primarily tetrachloro- and pentachloro-dibenzofuran congeners substituted in the 2,3,7,8-positions. Considering the mammalian toxicity of these chlorinated isomers, it seems prudent to determine the factors responsible for their occurrence and retention in fish and to assess the hazzard these residues may pose to organisms higher in the food chain. In addition, the potential adverse effects on reproduction and survival of Great Lakes fish and other aquatic organisms remains unknown.

REFERENCES

Albro, P. W., and Parker, C. E. Comparison of the compositions of Aroclor 1242 and Aroclor 1016. *J. Chromatogr.*, 169:161-166, 1979.

Bowes, G. W., Mulvihill, M. J. Simoneit, B. R. T., Burlingame, A. L., and Risebrough, R. W. Identification of chlorinated dibenzo-furans in American polychlorinated biphenyls. Nature, 256(5515): 305-307, 1975.

Brinkman, U. S. Th., and DeKok, A. Production, properties and usage. In: Kimbrough, R. D. ed. Halogenated Biphenyls, Terphenyls, Naphthalenes, Dibenzodioxins, and Related Products. Amsterdam, New York: Elsevier/North-Holland Biomedical Press, 1980. pp. 1-40.

Brumley, W. C., Roach, J. A. G., Sphon, J. A., Dreifuss, P. A., Andrzejewski, D., Neiman, R. A., and Firestone, D. J. Low resolution multiple ion detection gas chromatographic-mass spectrometric comparison of six extraction-cleanup methods for determining 2,3,7,8-tetrachlorodibenzo-p-dioxin in fish. Agric. Food Chem., 29(5):1040-1046, 1981.

Bumb, R. R., Crummett, W. B., Cutie, S. S., Gledhill, J. R., Hummel, R. H., Kagel, R. O., Lamparski, L. L., Luoma, E. V., Miller, D. L., Nestrick, T. J., Shadoff, L. A., Stehl, R. H., and Woods, J. S. Trace chemistries of fire: a source of chlorinated dioxins. Science, 210(4468):385-390, 1980.

Buser, H.R., and Bosshardt, H. P. Determination of polychlorinated dibenzo-p-dioxins and dibenzofurans in commerical pentachloro-phenols by combined gas chromatography-mass spectrometry. J. Assoc. Offic. Anal. Chem., 59(3):562-569, 1976.

Buser, H. R., Bosshardt, H. P., and Rappe, C. Formation of polychlorinated dibenzofurans (PCDFs) from the pyrolysis of PCBs. Chemosphere, 7:109-119, 1978.

Buser, H. R., and Rappe, C. High resolution gas chromatography of the tetrachlorodibenzo-p-dioxin isomers. Anal. Chem., 52:2257-2262, 1980

Dougherty, R. C., Whitaker, M. J., Smith, L. M., Stalling, D. L., and Kuehl, D. W. Negative chemical ionization studies of human and food chain contamination with xenobiotic chemicals. Environ. Health Persp., 36:103-117, 1980.

Helder, T. Effects of 2,3,7,8-tetrachlorodibenzo-p-dioxin (TCDD) on early life stages of the pike (Esox lucius L.). Sci. Total Environ., 14:255-264, 1980.

Horn, E. G., Hetling, L. J., and Tofflemire, T. J. The problem of PCBs in the Hudson River system. Ann. NY Acad. Sci., 320: 591-629, 1979.

Huff, J. E., Moore, J. A., Saracci, R., and Thomatis, L. Long-term hazards of polychlorinated dibenzodioxins and polychlorinated dibenzofurans. Environ. Health Perspec. 36:221-240, 1980.

Isensee, A. R. and Jones, G. E. Distribution of 2,3,7,8-tetrachloro-p-dioxin (TCDD) in aquatic model ecosystems. Environ. Sci. Technol., 9(7):668-672, 1975.

Isensee, A. R. Bioaccumulation of 2,3,7,8-tetrachlorodibenzo-p-dioxin. In: Ramel, C. ed. Chlorinated Phenoxy Acids and Their Dioxins. Ecol. Bull. No 27, Stockholm: Swedish Natural Science Research Council, 1978. pp. 255-262.

McConnell, E. E. Acute and chronic toxicity, carcinogenesis, repro-
     duction, teratogenesis and mutagenesis in animals. In:
     Kimbrough, R. D. ed. Halogenated Biphenyls, Terphenyls,
     Naphthalenes, Dibenzodioxins, and Related Products. Amsterdam,
     New York: Elsevier/North-Holland Biomedical Press, 1980. pp.
     109-150.
Moore, J. A., McConnell, E. E., Dalgard, D. W., and Harris, M. W.
     Comparative toxicity of three halogenated dibenzofurans in
     guinea pigs, mice and rhesus monkeys. Ann. NY Acad. Sci., 320:
     151-163, 1979.
Rappe, C., and Buser, H. R. Chemical properties and analytical
     methods. In: Kimbrough, R. D. ed. Halogenated Biphenyls,
     Terphenyls, Naphthalenes, Dibenzodioxins, and Related Products.
     Amsterdam, New York: Elsevier/North-Holland Biomedical Press,
     1980, pp. 47-76.
Rappe, C., Buser, H. R., Kuroki, H., and Masuuda, Y. Identification
     of polychlorinated dibenzofurans (PCDFs) retained in patients
     with Yusho. Chemosphere, 8:259-266, 1979a.
Rappe, C., Buser, H. R., and Bosshardt, H. P. Dioxins, dibenzofurans
     and other polyhalogenated aromatics: production, use, formation,
     and destruction. Ann. NY Acad. Sci., 320:1-18, 1979b.
Rappe, C., Buser, H. R., Stalling, D. L., Smith, L. M., and
     Dougherty, R. C. Identification of polychlorinated
     dibenzofurans in environmental samples. Nature, 292:524-526,
     1981.
Rappe, C., Gara, A., and Buser, H. R. Identification of poly-
     chlorinated dibenzofurans (PCDFs) in commerical chlorophenol
     formulations. Chemosphere, 7(12): 981-991, 1978.
Rappe, C., Marklund, S., Buser, H. R., and Bosshardt, H. P.
     Formation of polychlorinated dibenzo-p-dioxins (PCDDs) and
     dibenzofurans (PCDFs) by burning chlorophenolics. Chemosphere,
     1(3):269-281, 1978.
Schmidt, C., Ludke, L. J., and Walsh, D. F. Organochlorine
     residues in fish. National Pesticide Monitoring Program,
     1970-74. Pest monitoring J., 14(4):136-206, 1981.
Stalling, D. L. March 31, 1981, "Preliminary Report, March 31,
     Chlorinated Dibenzofurans and Related Compounds in Soot Formed
     in a Transformer Fire in Binghamton, New York," to State of New
     York, Department of Health, Office of Public Health, Albany,
     New York. Columbia, Missouri: Columbia National Fisheries
     Research Laboratory, 1981a.
Stalling, D. L. Personal communication: Report to Director, Fish
     and Wildlife Service on the occurrence of PCDFs in fish and
     sediments from Woods Pond, Massachusetts. Columbia, Missouri:
     Columbia National Fisheries Research Laboratory, 1981b.
Stalling, D. L., Petty, J. D., Smith, L. M. and Dubay, G. R.
     Contaminant enrichment modules, approaches to automation of
     sample extract cleanup. In: McKinney, J. D. ed. Environmental
     Health Chemistry, Ann Arbor: Ann Arbor Science Publishers,
     1981, pp. 177-193.

Swain, W. R. Chlorinated organic residues in fish, water, and
    precipitation from the vicinity of Isle Royale, Lake Superior,
    J. Great Lakes Res., 4(3-4):398-407, 1980.

IDENTIFICATION OF POLYCHLORINATED DIOXINS (PCDDs) AND
DIBENZOFURANS (PCDFs) IN HUMAN SAMPLES, OCCUPATIONAL EXPOSURE
AND YUSHO PATIENTS

C. Rappe*, M. Nygren*, H. Buser**, Y. Masuda****
H. Kuroki*** and P.H. Chen****

*Department of Organic Chemistry
 University Umea
 Umea, Sweden

**Swiss Federal Research Station
 Wadenswil, Switzerland

***Daiichi College of Pharmaceutical Sciences
 Fukuoka, Japan

****National Yang-Ming Medical College
 Taipei, Taiwan

ABSTRACT

A number of PCDDs and PCDFs have been identified in human
samples. The analytical technique used allows the identification
and quantification of individual isomers. The detection levels
were in the pg/g region (ppt). The samples analyzed include blood
plasma from occupationally exposed workers and also various samples
from the Yusho episodes in Japan in 1968 and in Taiwan in 1979.

INTRODUCTION

The polychlorinated dioxins (PCDDs) and dibenzofurans
(PCDFs) are two series of tricyclic aromatic compounds with similar
chemical, physical, biological, and toxicological properties. The
basic structures and numbering are given in Figure 1.

The number of chlorine atoms can vary between one and eight,
and the number of positional isomers is quite large. In all there

241

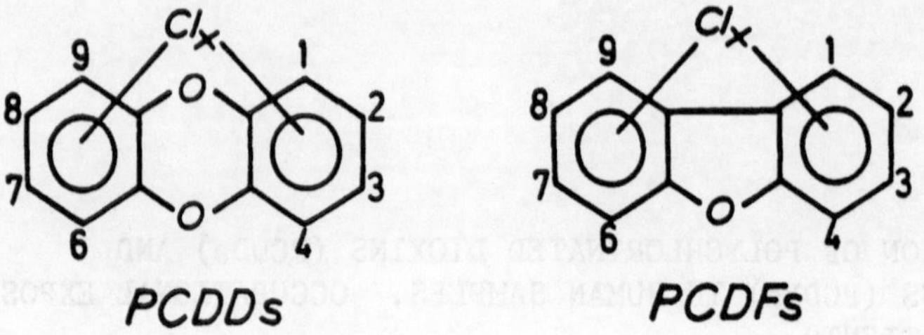

PCDDs   PCDFs

Fig. 1.   Basic Structures of the Polychlorinated Dioxins
          (PCDDs) and Polychlorinated Dibenzofurans (PCDFs).

Table 1.   Number of PCDD and PCDF Isomers for a Certain
           Number of Chlorine Atoms

	PCDDs	PCDFs
Mono-	2	4
Di-	10	16
Tri-	14	28
Tetra-	22	38
Penta-	14	28
Hexa-	10	16
Hepta-	2	4
Octa-	1	1
Total	75	135

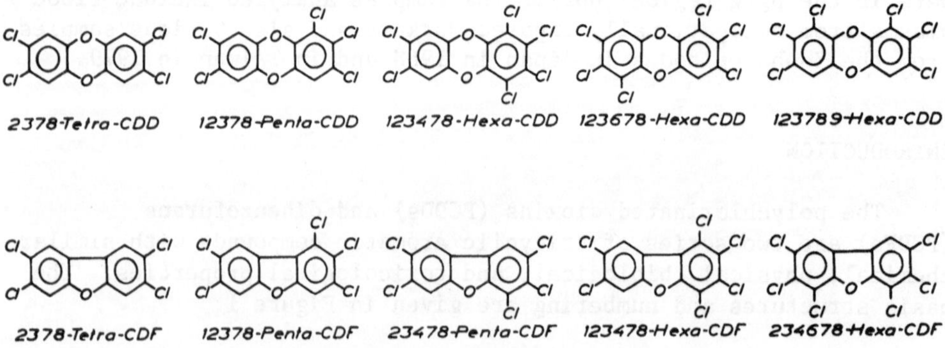

2378-Tetra-CDD   12378-Penta-CDD   123478-Hexa-CDD   123678-Hexa-CDD   123789-Hexa-CDD

2378-Tetra-CDF   12378-Penta-CDF   23478-Penta-CDF   123478-Hexa-CDF   234678-Hexa-CDF

Fig. 2.   The most toxic PCDD and PCDF isomers.

are 75 PCDDs and 135 PCDFs and the number of isomers for a given
number of chlorine atoms are given in Table 1.

Some of these compounds have extraordinarily toxic properties
and have been the subject of much concern. They are known as
highly stable contaminants in chlorinated phenols, phenoxy acids,
and in the polychlorinated biphenyls (PCBs). They have also been
identified in fly ash and other products from municipal incinerators
and accidental fires.

There is a pronounced difference in biological effects between
different PCDD and PCDF isomers. A factor of 10^3 – 10^4 can be found
for so closely related isomers as 2,3,7,8- and 1,2,3,8-tetra-CDD.
The isomers with the highest acute toxicity are 2,3,7,8-tetra-CDD,
1,2,3,7,8-penta-CDD, 1,2,3,4,7,8-, 1,2,3,6,7,8- and 1,2,3,7,8,9-
hexa-CDD, 2,3,7,8-tetra-CDF, 1,2,3,7,8- and 2,3,4,7,8-penta-CDF
and 1,2,3,4,7,8- and 2,3,4,6,7,8-hexa-CDF, see Figure 2. All these
isomers have their four lateral positions substituted for chlorine,
and they all have LD_{50}-values in the range 1-100 µg/kg for the most
sensitive animal species.

Human exposure to PCDDs or PCDFs may be due to either specific
exposure, mainly of occupational origin, accidental exposure or due
to general exposure of the public. No information is available
on the general background of these compounds in the general
population.

A general exposure to the public by 2,3,7,8-tetra-CDD has
previously been discussed in relation to the herbicide spraying
program in Vietnam in the 1960s and the explosion in a chemical
plant near Seveso in Northern Italy, July 1976. The Yusho oil
accidents in Japan in 1968 and in Taiwan in 1979 resulted in an
exposure to the public to various PCDFs.

Due to the extreme toxicity of some of the PCDD and PCDF
isomers, very sensitive and highly specific analytical techniques
are required. The different isomers of PCDDs and PCDFs are known
to vary greatly in their biological and toxicological properties;
consequently the separation, identification, and quantification of
the most toxic isomers become important.

In recent years a number of analytical methods have been
developed for trace analyses of PCDDs and PCDFs in a variety of
matrices. The most specific, sensitive, and selective of these
methods are utilizing high resolution gas chromotography – mass
spectrometry[1,4].

In the present study we report on PCDD and PCDF analyses of
human blood samples and tissue samples.

EXPERIMENTAL

Cleanup Procedure

Blood Plasma. The cleanup of the blood plasma samples is based on partitioning between n-hexane and acetonitrile.

n-Hexane (10 mL) is added to 10 mL of blood plasma. The sample is spiked with 0.1 ng of $^{13}C_{12}$-2,3,7,8-tetra-CDD. Acetonitrile saturated with n-hexane (20 mL) was added (three times) the mixture shaken and centrifugated. The combined acetonitrile phases were treated with 10 mL of aqueous sodium sulfate (1 percent) and extracted (three times) with 20 mL of n-hexane. The combined n-hexane layers were dried (sodium sulfate) and concentrated until 2 mL and added to a column (1 g of alumina). The column was eluted first with 10 mL of 20 percent methylene chloride in n-hexane and thereafter with 10 mL of 50 percent methylene chloride in n-hexane. The first eluate was discarded, the second concentrated and analyzed by GC/MS.

Tissue Samples. The cleanup of the tissue samples have been performed using methods described elsewhere[2,4].

GC/MS Analyses

The purified extracts were used directly for the final analyses using a Finnigan 4021 gas chromatograph-mass spectrometer (GC/MS) system equipped with a PROMIM multiple ion monitoring unit and negative chemical ionization options. Aliquots (1-2 µg) corresponding to up to 1 g of blood plasma or tissue sample were injected splitlessly into a glass capillary column (OV-17, Silar 10 c) or a fused silica column SP 2100, SE 54) leading directly into the ion source. Mass specific detection (mass fragmentography) was used for selective monitoring of M, M+2 and/or M+4 ions. The recovery of the cleanup procedure was studied from the internal standard.

The quantification was based on peak area measurements using external standards making the assumption that all isomers of a particular PCDD or PCDF (e.g., the tetrachloro isomers) have the same response. Isomer identification is based on retention time studies using polar and unpolar capillary columns and the qualitative standards available, about 40 PCDDs and 70 PCDFs. The 2,3,7,8-tetra-CDD isomer was separated from the other tetra-CDD isomers using a 55 m narrow bored Silar 10 c glass capillary column[5].

RESULTS

Occupational Exposure

Blood Plasma.   In earlier studies we have investigated the
levels of PCDDs and PCDFs in samples of blood plasma of workers
occupationally exposed to chlorinated phenols in the sawmill,
leather, and textile industry[6,7]. The levels were normally below
100 ppt (100 pg/g blood plasma) but a few workers directly exposed
to liquid formulations or solutions showed higher values, the
maximum was about 400 ppt (total level of all PCDDs and PCDFs).
This value was found for a worker in the textile industry.

We have also analyzed the chlorophenol formulation (pentachlorophenol laurate) used in this particular textile industry. Two
different qualities were used, the latter product has been purified,
the levels of impurities were reduced by approximately 99%, see
Table 2.

Samples of blood plasma were taken at two occasions, the first
set of samples was taken 5 months after the introduction of the
purified formulation, the second set was taken after an additional
11 months. In the meantime the purified product was the only product
used. The results of the blood plasma analyses are given in Table 3.

An examination of Tables 2 and 3 shows that the same isomers
are present both in the chlorophenol formulation and in the samples
of blood plasma. It is also found that the reduction of the blood
levels during the 9 months between the two sampling occasions were

Table 2.   Levels of PCDDs and PCDFs (in µg/g) in Chlorophenol
Formulations

Isomer	Formula- tion 1	Formulation 2 Batch 546	Batch 607
Octa-CDD	500	6	4.3
1,2,3,4,6,7,8-Hepta-CDD	150	0.7	0.5
1,2,3,4,6,7,9 -"-	25	0.1	0.05
1,2,3,6,8,9-Hexa-CDD	1	< 0.1	< 0.1
1,2,3,6,7,8- -"-	0.5	< 0.1	< 0.1
1,2,3,7,8,9- -"-	0.5	< 0.1	< 0.1
1,2,4,6,7,9- -"-	0.2	< 0.1	< 0.1
Octa-CDF	25	1.3	1.1
1,2,3,4,6,8,9-Hepta-CDF	15	0.8	0.5
1,2,3,4,6,7,8- -"-	4	< 0.05	< 0.05
1,2,4,6,8,9-Hexa-CDF	< 0.3	0.03	0.03

Table 3. Levels of PCDDs and PCDFs (in pg/g) in Blood Plasma From Workers in the Textile Industry

Isomer	Worker 1 a	Worker 1 b	Worker 2 a	Worker 2 b	Worker 3 a	Worker 3 b	Worker 4 a	Worker 4 b	Worker 5 a	Worker 5 b	Worker 6 a	Worker 6 b	Worker 7 a	Worker 7 b
Octa-CDD	14	6	6	2	304	35	3	3	10	1	105	25	30	5
1,2,3,4,6,7,8-Hepta-CDD	3	1	2	ND	59	7	ND	1	1	ND	15	3	6	1
1,2,3,4,6,8,9- -"-	1	ND	1	ND	ND	ND	ND	ND	ND	ND	ND	ND	ND	ND
Hexa-CDDs	1	ND	ND	ND	ND	ND	ND	ND	ND	ND	ND	ND	ND	ND
Octa-CDF	1	ND	1	ND	10	ND	ND	ND	ND	ND	ND	ND	ND	ND
1,2,3,4,6,7,8-Hepta-CDF	1	1	1	ND	25	5	ND	ND	ND	ND	1	1	3	1
1,2,3,4,6,7,9- -"-	ND	ND	ND	ND	ND	ND	ND	ND	ND	ND	ND	ND	ND	ND
1,2,3,4,6,8,9- -"-	ND	ND	ND	ND	8	ND	ND	ND	ND	ND	ND	ND	ND	ND
1,2,3,4,7,8,9- -"-	4	ND	3	ND	ND	ND	ND	ND	ND	ND	ND	ND	ND	ND
Hexa-CDFs	ND	ND	ND	ND	ND	ND	ND	ND	ND	ND	ND	ND	ND	ND

a = 5 months after exposure to heavy contaminated product

b = 16 -"- -"- -"- -"-

ND = Not Detected

60-90 percent, the values for the most exposed workers were 88 percent and 76 percent, respectively. The same reduction was found for all major isomers.

Tissue Samples. We have analyzed samples of liver and kidney tissue from a man occupied in the production of phenoxy acid herbicides like 2,4-D and MCPA and 2,4,6-trichlorophenol. The exposure ceased 3-4 years before the death of the man (the man died of a pancreatic tumor). The results of our analyses are collected in Table 4.

Of special interest is the observation of low levels of 2,3,7,8-tetra-CDD in these samples. This dioxin isomer has previously not been associated with the exposure to 2,4-D, MCPA and 2,4,6-trichlorophenol. However, the 2,4,6-trichlorophenol produced in this factory was found to be contaminated by 2,4,5-trichlorophenol (0.7-1.6 percent) and 2,3,4,6-tetrachlorophenol. We also analyzed a sample of the deposits in the ventilation system, which was found to be contaminated by four tetra-CDD isomers, one of these was the 2,3,7,8-isomer, see Table 5. These four isomers are the expected pyrolytic dimerization products from 2,4,6- and 2,4,5-trichlorophenol.

Table 5 also shows a comparison between the 2,4,6-trichlorophenol formulation produced, the deposit in the ventilation system and the liver sample.

Yusho Episodes

In 1968 more than 1500 persons in Southwest Japan were intoxicated by consuming a commerical rice oil accidentally contaminated by PCBs, PCDFs and polychlorinated quaterphenyls PCQs. In 1979 a similar episode wasreported from Taiwan, the number of persons involved approaching 2000.

Earlier analyses have proven that the Japanese rice oil contained more than 40 PCDF isomers, tri- to hexa-CDFs[8]. Analysis of liver samples taken from the Japanese patients about 18 months after the exposure showed a dramatic decrease in the number of PCDF isomers, see Figure 3[9]. Apparently most of the PCDF isomers have been metabolized or excreted during the period between exposure and sampling.

A comparison between the PCDF isomers found in the Yusho oil and in the liver samples revealed an interesting relationship between the isomers retained in the liver and those isomers excreted, see Figure 4. All the PCDF isomers excreted had two vicinal unchlorinated C-atoms in at least one of the two aromatic rings of

Table 4. Levels of PCDDs and PCDFs (in pg/g) in Tissue
          Samples (Values in Parenthesis are the Detection
          Limits)

Isomer	Liver	Kidney
2,3,7,8-Tetra-CDD	2(1)	1(1)
1,3,6,8- -"-	1(1)	1(1)
Penta-CDDs	< 2	< 2
1,2,3,6,7,8-Hexa-CDD	4(4)	< 2
1,2,3,4,6,7,8-Hepta-CDD	72(5)	8(4)
Octa-CDD	350(8)	15(4)
Tetra-CDFs	< 1	< 1
2,3,4,7,8-Penta-CDF	10(2)	7(2)
1,2,3,4,7,8- + 1,2,3,6,7,8-		
Hexa-CDFs	55(2)	8(2)
1,2,3,4,6,7,8-Hepta-CDF	100(4)	6(4)
Octa-CDF	< 3	< 3

Table 5. Comparison of PCDD and PCDF Isomers Found
          in Various Products

Isomer	2,4,6-Tri-chloro-phenol	Deposit Ventilation	Liver Sample
1,3,6,8-Tetra-CDD	-	+	-
1,3,7,9- -"-	-	+	-
2,3,7,8- -"-	-	+	+
1,3,7,8- -"-	-	+	-
Penta-CDDs	-	+	-
1,2,3,6,7,8-Hexa-CDD	-	+	+
1,2,3,4,6,7,8-Hepta-CDD	-	+	+
Octa-CDD	-	+	+
2,4,6,8-Tetra-CDF	+	+	-
2,3,6,8- -"-	+	+	-
2,3,7,8- -"-	+	+	-
1,2,4,6,8-Penta-CDF	+	+	-
2,3,4,6,8- -"-	+	+	-
2,3,4,7,8- -"-	+	+	+
1,2,3,4,6,8-Hexa-CDF	+	+	-
1,2,3,4,7,8- -"-	+	+	+
1,2,3,4,6,7,8-Hepta-CDF	+	+	+
1,2,3,4,6,8,9- -"-	+	+	-
Octa-CDF	-	+	-

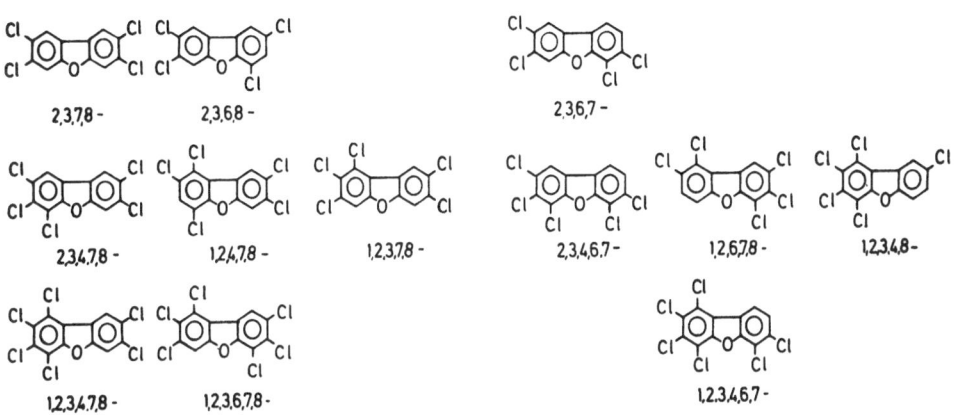

Fig. 4.   PCDF isomers retained (left) excreted (right)
          in Yusho oil.

the PCDF system.   On the contrary none of the isomers retained had
two <u>vicinal</u> unchlorinated C-atoms in any of the two aromatic rings.
Most of the latter isomers had all lateral (2-, 3-, 7-, and 8-)
positions substituted for chlorine.

We have also analyzed various organs of a "Yusho Baby" from
Taiwan.   In Table 6 we have collected the results of these PCDF
analyses together with PCB data.

It is interesting to notice that the PCB levels parallels
the fat content of the tissue.   However, PCDFs could only be detected
in the adipose tissue and in the liver.   The highest PCDF values
were found in the liver, but here we found the lowest PCB values.

Discussing the individual PCDF isomers, a comparison between
Figure 3 and Table 6 shows that we have the same isomers in all
samples with the exception of 2,3,6,8-tetra-CDF which was only
found in the Japanese liver samples.   However, it should also be
pointed out that in the samples of the Yusho Baby the ratio between
the penta-CDF isomers is not the same in the adipose tissue and the
liver sample, see Table 6.   In the liver sample the dominating
isomer is the 1,2,3,7,8-penta-CDF, while in the adipose tissue the
highest value was found for the 2,3,4,7,8-penta-CDF.

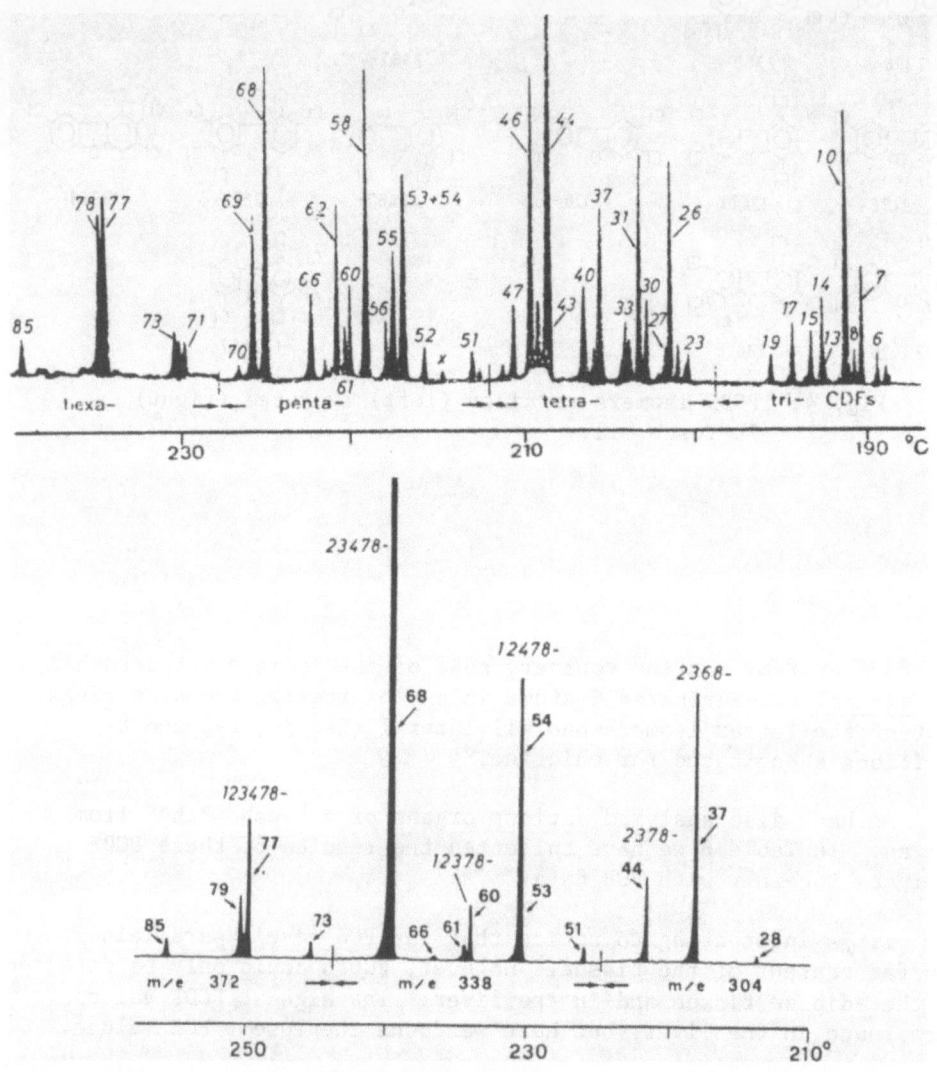

Fig. 3.   Mass fragmentogram of Yusho oil (top) and a
          liver sample (bottom).

Table 6. Levels of PCDFs (pg/g) and PCBs (ng/g) in Tissue Samples
          of Yusho Baby from Taiwan

Isomer	Adipose	Liver	Muscle	Omentum	Diaphragm
2,3,7,8-Tetra-CDF	17	60	ND*	ND	ND
1,2,4,7,8-Penta-CDF	14	42	ND	ND	ND
1,2,3,7,8-Penta-CDF	44	194	ND	ND	ND
2,3,4,7,8-Penta-CDF	68	91	ND	ND	ND
1,2,3,4,7,8-Hexa-CDF	88	193	ND	ND	ND
PCBs	316	27	38	64	46

*ND = Not Detected

Table 7. Levels of PCDFs (pg/g) in Blood Plasma from Yusho
          Patients (Values in Parenthesis are the Detection Limits)

| Isomer | Japan | | Taiwan | |
	A	B	C	D
Tetra-CDFs	3	3	30[a]	30[a]
1,2,4,7,8-Penta-CDF	ND*	ND	60	40
1,2,3,7,8-Penta-CDF	ND	ND	30	20
2,3,4,7,8-Penta-CDF	3(3)	3(3)	120	80
1,2,3,4,7,8-Hexa-CDF	6	6	150	60

*ND = Not Detected
[a] = high detection limits due to large amounts of overlapping PCBs
      and other chlorinated compounds

The rate of excretion of some of the PCDF isomers is very slow. We have analyzed a set of four blood samples, two collected in Japan in 1979 which means 11 years after the exposure, and two were collected in Taiwan, in 1980, 1 year after the exposure, see Table 7.

Low levels of 2,3,4,7,8-penta-CDF could be found in all samples, the levels in the Japanese samples were just above the detection limit. The penta- and hexa-CDF isomers identified were the same as in the other human Yusho samples discussed.

REFERENCES

1. Rappe, C. and Buser, H.R. Chemical Properties and Analytical Methods. In: Kimbrough, R., ed. Halogenated Biphenyls, Terphenyls, Naphthalenes, Dibenzodioxins and Related Products. Amsterdam, Elsevier/North-Holland, 1980, pp. 41-76.
2. Stalling, D.L., Petty, J.D., Smith, L.M., Rappe, C. and Buser, H.R. Isolation and Analysis of Polychlorinated Dibenzo-furans in Aquatic Samples. In: Hutzinger, O., Frei, R.W., Merian, E. and Pocchiari, F., eds. Chlorinated Dioxins and Related Compounds - Impact on the Environment. Pergamon, Press, Oxford 1982, pp. 77-85.
3. Rappe, C., Buser, H.R., Stalling, D., Smith, L.M. and Dougherty, R.C. Identification of Polychlorinated Dibenzo-furans in Environmental Samples. Nature, 292:524-526, 1981.
4. Stalling, D.L., Smith, L.M., Petty, J.D., Hogan, J.W., Johnson, J.L., Rappe, C. and Buser, H.R. Residues of Poly-chlorinated-p-dioxins and Dibenzofurans in Laurentian Great Lakes Fish. These Proceedings.
5. Buser, H.R. and Rappe, C. High-Resolution Gas Chromatography of the 22 Tetrachloro-dibenzo-p-dioxin Isomers. Anal. Chem. 52:2257-2262, 1980.
6. Rappe, C. and Buser, H.R. Occupational Exposure to Polychlori-nated Dioxins and Dibenzofurans. In: Choudhary, G. ed. Chemical Hazards in the Workplace, Measurements and Control. ACS Symposium Series Vol. 149. Washington, ACS, 1981, pp. 319-342.
7. Rappe, C., Nygren, M., Buser, H.R. and Kauppinen, T. Occupa-tional Exposure to Polychlorinated Dioxins and Dibenzofurans. In: Hutzinger, O., Frei, R.W., Merian, E. and Pocchiari, F. eds. Chlorinated Dioxins and Related Compounds - Impact on the Environment. Pergamon Press, Oxford. 1982, pp.495-513.
8. Buser, H.R., Rappe, C. and Gara, A. Polychlorinated Dibenzo-furans (PCDFs) in Yusho Oil and in Used Japanese PCB. Chemosphere, 7:439-449, 1978.

9. Rappe, C., Buser, H.R., Kuroki, H. and Masuda, Y. Identification
    of Polychlorinated Dibenzofurans (PCDFs) Retained in Patients
    With Yusho. Chemosphere 8:259-266, 1979.

SEVESO – AN ENVIRONMENTAL ASSESSMENT

H. K. Wipf and J. Schmid

Givaudan Research Company Ltd.
Dubendorf, Switzerland

ABSTRACT

An unforeseeable accident on July 10, 1976, in Seveso, Italy, led to an environmental contamination with caustic reaction products and 2,3,7,8-tetrachlorodibenzo-p-dioxin (TCDD). Original contamination of vegetation was in the order of 0.5 ppm TCDD.

The main effects caused by the contamination were the death of herbivores, due to oral uptake of contaminated vegetation and chloracne in humans, due to dermal contact. The cases considered severe in 1976 recovered within 5 years.

In Zone A (87.3 ha), 736 inhabitants were evacuated, two-thirds of which returned to their homes after decontamination in 1977. Agricultural and horticultural activities were suspended in Zones B (269.4 ha) and safety Zone R (1,430 ha).

By the end of 1976, the contamination had been transferred to the soil, where TCDD was strongly adsorbed, which reduced bio-availability. Plant uptake has not been a significant factor in the movement of TCDD. Moreover, spreading of the contamination by leaching and wind (dust) has been minimal.

Decontamination consists of agronomic measures in areas with low contamination and removal of topsoil where concentrations are high. In 1980-1981 levels of TCDD were practically non-detectable in soil from Zone R or in vegetation from Zones B and R (less than 1 part-per-trillion).

The Seveso accident was neither the most severe TCDD
accident nor has it produced the highest environmental contaminations. Zone R is contaminated with other PCDD's, which are unrelated
to the accident, at levels much higher than those of 2,3,7,8-TCDD.
Even so, the outlook is very optimistic.

INTRODUCTION

The Seveso, Italy, incident is probably the most widely
publicized industrial accident in history. Now, after more than 5
years, it seems worthwhile to re-examine the events of 1976 and
re-assess the situation as it presents itself today. We will limit
the assessment to the environmental aspects of the accident and
address in particular the following questions:

- What happened on July 10, 1976?

- What were the immediate effects?

- Why was the Seveso accident different from other industrial
  incidents?

- How does the TCDD contamination compare with other
  environmental episodes?

- What has been learned?

This assessment will have to stay preliminary by necessity: the
increase in basic knowledge is slow, technical progress has been
sluggish, and some of the most important documentation projects are
still underway.

THE ACCIDENT

The events leading to the accident during the production of
trichlorophenol were very complex[1,2]. The reaction had been
completed and part of the solvent removed. Because it was a
Saturday morning, removal of the solvent was discontinued. Heating
was stopped, atmospheric pressure reestablished, and stirring continued for 15 minutes. The reaction vessel was left with its contents at 158°C, a temperature considered absolutely safe in 1976.
In fact, the events leading to the accident more than 6 hours later
remain hypothetical even today. The proposed physicochemical
mechanism involves the combination of exothermic reactions (1)
starting well below the previously known 230°C threshold and a
thermal conduction-convection process (2) channelling the globally

negligible upper reactor wall heat into a small portion of the
reactor contents, yielding the temperature range conducive to these
exothermic reactions.

There was no explosion. After several hours, the increase in
temperature was sufficient to trigger further exothermic reactions.
The actual decomposition started at 280 - 290°C, causing a rapid
increase in pressure. At 3.8 bar the safety disk ruptured, leading
to the discharge of reaction products into the open.

Except for the aerosol cloud, which slowly drifted South, there
were no noticeable effects during the first hours. Many people
complained about the phenolic stench and closed the windows.
Few persons came in direct contact with the aerosol, and probably
none of them for long. ICMESA workers engaged in maintenance work
near the site of the accident took the first emergency measures.
None of them developed any health problems, not even chloracne.
There was no physical damage to speak of.

The immediate consequence of the accident was the contamination
of all surfaces which had been in contact with the drifting aerosol
cloud. The toxic deposit consisted of a rather caustic mixture of
trichlorophenol and reaction products, among them a small proportion
of highly toxic TCDD (2,3,7,8-tetrachlorodibenzo-p-dioxin). Contaminated was mainly the vegetation, but also roads and the outer
surfaces of buildings.

All the ensuing problems can be deduced logically from the
original distribution of the deposit, its toxicity and its physico-
chemical properties which result in an extremely high persistence of
TCDD in the environment.

ACUTAL RISKS IN JULY 1976

Immediately after the accident, the chemical deposited in the
environment represented a risk in two ways: by its caustic properties and by the particular toxicological properties of the TCDD
component. However, effects from either of these threats were slow
in developing.

There was only minor damage to plants, and only in the immediate
vicinity of the ICMESA factory. The phytotoxic effect of the
reaction mixture is clearly unimportant and can be neglected.

First effects to be noticed were skin and eye irritations
related to trichlorophenol. Next was the death of herbivores,
probably from the combined action of the mixture of components.
Slowest to develop were the typical TCDD effects such as chloracne
in humans and delayed deaths of animals.

From what is known today, it seems that during this early phase the dermal toxicity of the reaction mixture constituted the higher risk to humans than the oral uptake of TCDD. Children, who had direct skin contact with the contaminated vegetation during their play in the open, were at much higher risk than the rest of the population. Correspondingly, most of the dermatitis and chloracne cases were children. Although some (particularly mild) cases of chloracne were not discovered until several months after the accident, they were probably due to exposure during the first few days. No new cases developed after 1978.

The acute oral toxicity risk to humans was directly associated with the consumption of contaminated food, primarily from vegetables. Although some ingestion must have occurred during the few days before evacuation there were no acute health problems as the doses obviously had remained low, possibly due to the fact that vegetables were washed before consumption. Subsequent exposure by the oral route was blocked by the evacuation and restriction measurements taken by the local authorities.

The risk for herbivores was of an entirely different magnitude. By feeding directly and in many cases exclusively on the contaminated vegetation, these animals ingested sizeable amounts of toxic compounds. Even for them TCDD was not the only danger; in many cases chemical burns in the throat and inflamed sites on their extremities have been observed which must have been due to the caustic main components of the aerosol.

It is therefore not surprising that the first and most numerous casualties were rabbits. There was also noticeable damage to herbivorous wildlife. On the other hand, only few of the larger domestic animals died (Table 1).

The danger emanating from the directly contaminated vegetation persisted for several weeks. When the first rains set in after a few days the water-soluble fraction of the reaction mixture was washed off but most of the very lipophilic TCDD remained on the plants.

FIRST COUNTER-MEASURES

The first counter-measures were designed to minimize contact with the toxic deposit and to keep contaminated food out of the food chain.

Evacuations were carried out between July 26 and August 2, 1976, on the basis of our analytical results[3]. The evacuated zone was declared off-limits and was to become "Zone A" of Seveso (Table 2).

Table 1.   Fate of Domestic Animals from Zones A, B,
and R (as of June 30, 1978)

ANIMALS	ORIGINAL NUMBER (10 JULY 1976)	DIED AFTER THE ACCIDENT*	STILL IN THE ZONE	EVACUATED FOR STUDY PURPOSES	SLAUGHTERED PROPHYLACTICALLY BEFORE 30 JUNE 1978
SMALL ANIMALS	80'430	3'281 (4.1 %)	-	71	77'078 (95.8 %)
CATTLE	349	6 (1.7 %)	36	9	298 (85.4 %)
HORSES / DONKEYS	49	2 (4.1 %)	-	-	47 (95.9 %)
SWINE	233	3 (1.3 %)	-	3	227 (97.4 %)
SHEEP / GOATS	70	1 (1.4 %)	-	3	66 (94.3 %)
TOTAL	81'131	3'293 (4.1 %)	36	86	77'716 (95.8 %)

* SPONTANEOUS DEATHS OR EMERGENCY SLAUGHTER

Table 2.   Contaminated Zones in the Seveso Area

Zone	TCDD Contamination	Approximate Soil Concentrations (1977)	Area Affected	Inhabitants (Evacuated)
A	$> 15 \ \mu g/m^2$	> 150 ppt	87.3 ha	736 (736)
B	$5 - 15 \ \mu g/m^2$	$50 - 150$ ppt	269.4 ha	4'699 (-)
R	$< 5 \ \mu g/m^2$	< 50 ppt	1'430 ha	31'800 (-)

In "Zone B", which is adjacent to Zone A, the TCDD contamination
was lower and people were not evacuated. Certain restrictions were
imposed in order to limit exposure of the most vulnerable part of
the population (small children and pregnant women).

Even in "Zone R", the major part of which had never been
measurably contaminated with 2,3,7,8-TCDD, all agricultural and
horticultural activities were suspended for safety reasons.

Similarly a ban on the consumption of meat and animal products from Zones A, B and R was issued. This restriction was enforced by extermination of all animals from the three zones within 2 years of the accident (see Table 1).

The restrictions on consumption of vegetable materials from the areas of highest risk were clearly needed in 1976. These measures were apparently successful as no significant health effects have been observed in the population of the restricted zones. It could be argued, however, that the restrictions could have been relaxed in the subsequent years when the TCDD content in vegetable material was lower by several orders of magnitude, because the vegetation was no longer directly contaminated by the aerosol cloud[4].

EXTENT OF CONTAMINATION

Extent and degree of contamination can be seen from Table 2 and the corresponding map in Figure 1.

Results were originally expressed on a weight per weight basis of TCDD in plant material (ppm or ppb). This was the natural basis for the first 2 months or so because almost the entire fallout of the aerosol cloud had been deposited on the vegetation. At the same time these values were a direct measure of toxicological risk.

The TCDD contamination of vegetation, as measured by our laboratories during the first month after the accident, are given in Table 3. The highest single value detected was 15.8 ppm in a grass sample close to the ICMESA plant. The median of all values from the area to be declared "Zone A" was in the order of 0.5 ppm (500 ppb) in plant material. The detection limit at that time was 1 ppb.

Table 3. Vegetation Samples Analyzed within the First Month After the Accident (July 1976)

TCDD Contamination in Vegetation	Number of Analytical Results
≥ 1 ppm	11
0.1 - 0.99 ppm	15
0.01 - 0.099 ppm	18
0.001 - 0.0099 ppm	39
< 0.001 ppm	30

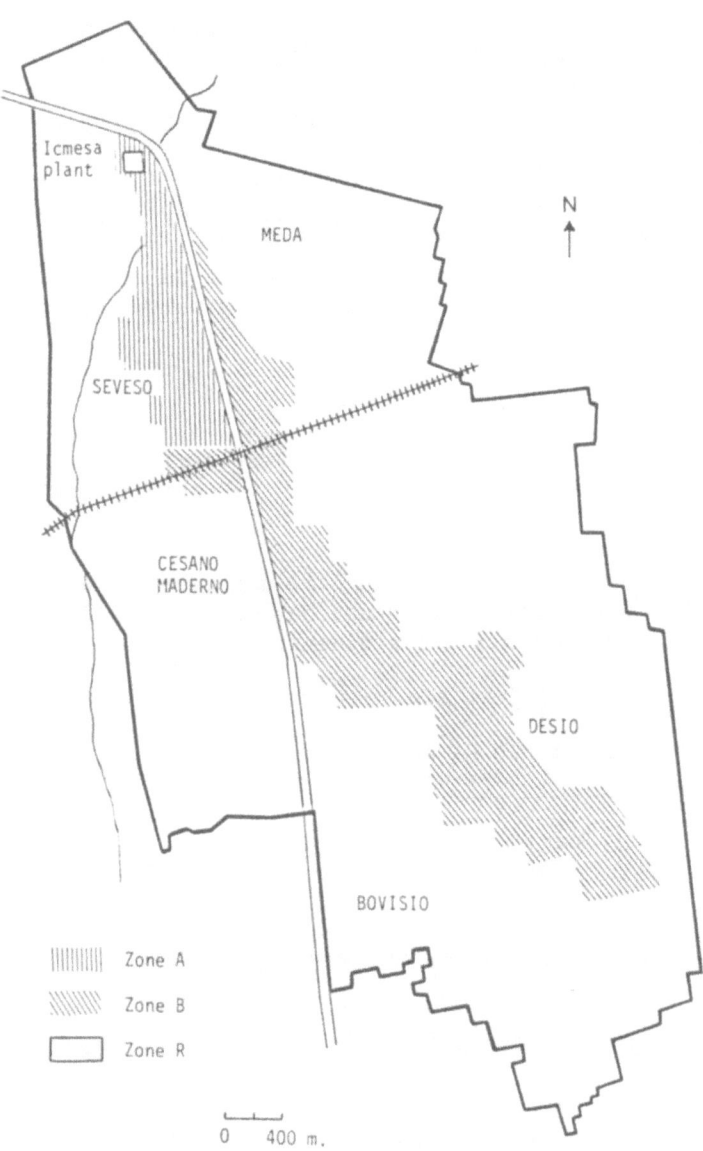

Fig. 1. Map of the Seveso Area (Zones A, B and R) following
Contamination with TCDD in July 1976.

Official mapping was carried out on a weight per area basis ($\mu g/m^2$) as a more constant measure of general contamination. The final definition of the restriction Zones A, B and R was based on these values.

In the subsequent years, when the relative risk in the zones was directly related to the TCDD concentration in soil, the measure of contamination should probably have been changed again, to a weight per weight basis in soil, e.g., parts-per-trillion (ppt).

The interconversion of the three measures of contamination is not straightforward. However, as a general rule, 10 ppt in the soil corresponds under standard sampling conditions to about 1 $\mu g/m^2$ on a weight per area basis, which in turn corresponds to about 1 ppb in the directly contaminated original vegetation.

COMPARISON WITH OTHER TCDD EPISODES

It is of interest to compare the Seveso accident and its consequence with other known industrial incidents and environmental TCDD episodes.

Table 4 summarizes all the tangible damages of the accident as they are known today, and in fact as they were basically known within about 3 months after the accident.

Table 4. Measurable Damages Due to the Seveso Accident

Human Deaths	0
Serious Injuries	0
"Severe" Chloracne (all recovered, some scarring)	15
Total Chloracne, Confirmed (mostly mild, all recovered)	183
Domestic Animals Killed	ca. 3'000
Phytotoxic Damages	1 - 2 ha
Physical Damages	~ 0

Nobody died and nobody was seriously injured; therefore, the actual number and seriousness of human casualties cannot be the reason why the Seveso accident was considered special. There are literally thousands of industrial accidents which are more severe.

Not even the death of 3,000 animals, as unfortunate as the fact is, made the event outstanding. For instance tens of thousands of birds have been killed in oil spills. However, the deaths of domestic animals were a clear sign of a toxin, which could not be noticed otherwise. This naturally created an atmosphere of fear and anxiety for the inhabitants of the Seveso area.

It seems therefore that the particular toxic properties of TCDD made "Seveso" different. The first and most typical effect of this compound in humans is chloracne. Table 5 lists the confirmed cases in the Seveso region[5].

Although the number of cases (ca. 180) seems large, most of them were mild or very mild. In fact, considering the background occurrence in this region of about 0.5% in children[6], the only clear effect was observed in Zone A. Even the 15 cases considered severe have healed completely within the 5 years since the accident, in contrast to other TCDD accidents where chloracne in some cases has lasted for at least 15 years[7,8]. So, if we compare the Seveso accident with the more than 20 other TCDD accidents[9] we are faced with the fact that there have been several more serious TCDD accidents.

Table 5. Occurrence of Chloracne in the Seveso Region

Zone	Population		Confirmed Chloracne	
	Total	Children	Total	Children
A	736	214	61 (8,3 %)	42 (19,6 %)
B	4'699	1'468	9 (0,2 %)	8 (0,5 %)
R	31'800	8'680	64 (0,2 %)	63 (0,7 %)
-	182'843	48'263	47 (<0,1 %)	46 (0,1 %)
Milano	(controls)	1'054		5 (0,5 %)
Como	(controls)	1'964		2 (0,1 %)
Lecco	(controls)	1'719		8 (0,5 %)

There remains one major difference. In all those other accidents the contamination has remained contained within the factory where the event had happened. At Seveso the contamination had spread over a large area in the environment. Fortunately, by this very same process the TCDD contamination had been diluted so that the actual effects remained smaller than originally feared.

Here then lies the real "Seveso" problem; what was known about TCDD in 1976 was either from animal studies or from effects on humans in accidents with high concentrations of TCDD in an enclosed environment. Modern toxicology, however, has not resolved the problem of inter-species differences nor mastered the extrapolation from small populations exposed to high concentrations to large populations exposed to small concentrations. Thus, immediately after the accident, experts did not know what long-term consequences to expect, and the mass media took up the case, thereby perpetuating the atmosphere of fear and uncertainty.

There are a few other environmental TCDD episodes with which "Seveso" can be compared. They were not due to industrial accidents. In spite of the small number of well-documented cases, it turns out that the Seveso accident has not even produced the highest environmental contaminations as Table 6 shows[10,11,12,13].

Table 6. Degree of TCDD Contamination in Various Places

LOCATION	REASON FOR CONTAMINATION	CONTAMINATION WEIGHT PER WEIGHT	SAMPLING DEPTH (SOIL DENSITY = 1.4)	CONTAMINATION PER UNIT AREA
HORSE ARENA C (MISSOURI)	WASTE OIL	540 PPB	30 CM →	220 MG/M²
EGLIN AFB (FLORIDA) HARDSTAND 7 (LOCALIZED)	AGENT ORANGE SPILLS	170 PPB	10 CM →	24 MG/M²
SEVESO, ZONE A (ITALY) HIGHEST VALUE, CLOSE TO FACTORY	INDUSTRIAL ACCIDENT	55 PPB	7 CM ←	5.5 MG/M²
SEVESO, ZONE A₂/A₃ (ITALY) HIGHEST VALUE IN FORMERLY INHABITED AREA	INDUSTRIAL ACCIDENT	20 PPB	7 CM ←	2.0 MG/M²
MIDLAND (MICHIGAN) SOIL OUTSIDE PLANT, HIGHEST VALUE	INCINERATORS ?	100 PPB	(1 CM) →	MIN 1 MG/M²
EGLIN AFB (FLORIDA) GRID I, HIGHEST VALUE	PRACTISE AREA FOR HERBICIDE APPLICATION	1.5 PPB	10 CM →	0.21 MG/M²
SEVESO, BORDERLINE ZONE A/B LIMIT FOR EVACUATION	INDUSTRIAL ACCIDENT	0.15 PPB	7 CM ←	0.015 MG/M²

If we go to areas with lower contamination (Zone R) we find ourselves in a region which seems to fall within the limits of ordinary contamination from municipal incinerators and other combustion sources (Table 7).

In fact, Zone R, closed to agricultural and horticultural activities for 5 years due to the accident, is contaminated with other chlorinated dioxins which have nothing to do with the accident at levels which are much higher than those of 2,3,7,8-TCDD from the accident (Table 8).

Although these isomers and homologues are probably somewhat less toxic, their combined effect can probably be compared to that of the component from the industrial accident[14]; in particular, the toxic 1,2,3,6,7,8- and 1,2,3,7,8,9-hexachlorodibenzo-p-dioxin (HCDD) isomers[15,16] are among the prominent components identified. It may be significant that a municipal incinerator is actually located within Zone R; however, it is by no means certain that this is the only source of chlorinated dioxins as small furniture factories in the region may well burn wood shavings treated with chlorinated phenols.

ENVIRONMENTAL BEHAVIOR OF TCDD

Given the situation of extended contamination as it presented itself after the accident, and given the successful immediate

Table 7. Degree of TCDD Contamination in Various Places

LOCATION	REASON FOR CONTAMINATION	CONTAMINATION WEIGHT PER WEIGHT	SAMPLING DEPTH (SOIL DENSITY = 1.4)	CONTAMINATION PER UNIT AREA
EGLIN AFB (FLORIDA) LARGE GRID, HIGHEST VALUE	PRACTISE AREA FOR HERBICIDE APPLICATION	0.47 PPB	10 CM	0.066 MG / M^2
SEVESO, BORDERLINE ZONE A / B LIMIT FOR EVACUATION	INDUSTRIAL ACCIDENT	0.15 PPB	7 CM	0.015 MG / M^2
SEVESO, BORDERLINE ZONE B / R	INDUSTRIAL ACCIDENT	0.05 PPB	7 CM	0.005 MG / M^2
MAJOR METROPOLITAN AREA (MICHIGAN); HIGHEST VALUE	INCINERATOR	0.03 PPB	(1 CM)	MIN 0.003 MG / M^2
APPROXIMATE DETECTION LIMIT FOR SOIL	——	0.003 PPB	7 CM	0.0003 MG / M^2
NORMAL APPLICATION OF HERBICIDES (2,4,5-T; 1 KG / HA; < 0.1 PPM TCDD	IMPURITY	0.0001 PPB	7 CM	0.00001 MG / M^2
SEVESO; LIMIT INSIDE HOUSES	INDUSTRIAL ACCIDENT	——	——	0.00001 MG / M^2

Table 8.   PCDD Contamination in Soil from Zone R (1981)
(Values in ppt)

Sample	2,3,7,8 TCDD[⊛] [*]		PCDD	HxCDD	HpCDD	OCDD	Total
S 1	0.8	0.3	0.4	6.0	1.4	1.7	10.6
S 2	3.4	1.0	0.7	9.5	2.1	2.2	18.9
S 3	4.0	1.9	1.2	8.2	8.6	27.0	50.9
S 4	2.3	0.8	0.6	10.2	2.1	2.0	18.0
S 5	≤0.1	≤0.3	0.5	9.5	1.9	1.4	13.7
S 6	6.3	1.5	1.1	10.4	2.6	1.3	24.8
S 7	1.7	1.0	0.8	12.4	1.9	1.8	19.6
S 8	2.2	0.4	0.8	8.8	1.8	0.8	14.8
S 9	1.0	2.8	2.3	21.2	9.6	13.5	50.4

[⊛] 2,3,7,8-TCDD, probably related to accident

[*] Isomers other than 2,3,7,8-TCDD, not related to accident

Table 9.   Properties of TCDD Which Determine its
Environmental Behavior

Factors Determining Stability

UV-Degradation   : Moderate

Microbiological Degradation : Slow

Hydrolysis   : Very Slow

Other Chemical Degradation : Improbable

Properties Determining Mobility

Water Solubility   : Low

Lipophilicity   : High

Adsorption on Soil   : Strong

Volatility   : Low

actions to limit exposure to and consumption of contaminated
material, it became most important to assess the longer-term risks
and to minimize them. On the toxicological side of the issue experts
did not agree on what long-term effects of TCDD were to be expected.

On the environmental side things were clearer. Although
detailed knowledge of chemical, physico-chemical, and biochemical
properties of 2,3,7,8-TCDD may have been somewhat more limited in
1976 than now, the most relevant facts were clear[17-21]. The properties dominating the behavior of TCDD in the environment are
summarized in Table 9.

From these properties two things became evident: if nothing
was done, the contamination would stay there for a relatively
long time, but it would probably not spread.

All processes leading to natural decontamination were
estimated to be slow. However, based on the limited data available
[18,22] it was expected that microbial degradation might still be the
fastest process in operation. Even though this may still be true,
the half-life times must definitely be longer than the original
estimate of 200 - 300 days[23].

MEASURES TO REDUCE CONTAMINATION

Immediate Possibilities

Since the aerosol had contaminated the vegetation, the main
approach for decontamination has been to collect and confine as
much of the plant material as possible. The limiting factor here
was the enormous size of the task. Time was also an important
factor, because plant material that could not be collected before
the start of the winter season would be incorporated into the soil.

Beside this purely mechanical approach, there had been one
serious attempt to reduce the contamination by a photo-chemical
process. From the literature it was known that TCDD could be degraded by UV-light[19], particularly in the presence of a hydrogen
donor. The idea was to spray olive oil as a natural hydrogen donor
and thereby to accelerate the degradation by sunlight. The process
worked in the laboratory. But although even a pilot trial had
shown promising results[24], the local authorities blocked a large-
scale spraying operation even after the demonstration that leaching
of TCDD would not be increased after the treatment with olive oil.
The idea finally had to be abandoned in the fall, when the sunlight
grew dimmer and the decaying plant material had reached the ground.

<u>After Winter 1976-1977</u>

During the winter of 1976-1977 the situation had stabilized inasmuch as all the contamination had now been transferred to the ground and TCDD was strongly adsorbed to soil particles. TCDD could only be leached or transported with the soil particles it was adsorbed to. Any remaining risk would emanate from the soil rather than from plants. Contamination of vegetable material was lower by many orders of magnitude as compared to July 1976. TCDD was measurable only on the vegetation from the most heavily contaminated Zone A (Table 10). Remaining small contamination of fruit and vegetables was probably due to local dust[4,25,26] as TCDD was detected only on the skin of the fruit analyzed.

Reduction of contamination meant reduction of TCDD in the soil, which proved to be a much more difficult task than the mechanical decontamination of houses[27], an operation which was carried out in 1977.

The methods finally chosen did reduce the danger by decreasing the soil concentration of TCDD in the surface layers. In the highly contaminated areas this was accomplished by physically removing the topsoil. In areas of low contamination agronomic methods were used. Repeated plowing and subsequent agricultural use of the land not only decreased the TCDD concentrations by dilution but also favored any microbiological degradation.

SITUATION AFTER FIVE YEARS

In 1980-1981 the TCDD contamination in safety Zone R was no longer measurable, except in a few localized spots (Table 11). Vegetables and agricultural crops grown in Zone R during the past 4 years for experimental purposes have invariably been found to be free of 2,3,7,8-TCDD at the 1 ppt level (Table 12). About one-third of Zone R has been released for agricultural purposes and official liberalization of the rest of Zone R is expected soon.

In Zone B the ban on the use of the land continues. Measurable amounts of TCDD are still found in the soil, typically in the order of 10 - 50 ppt. Measurements of plant samples[4,25] showed in 1977 and 1978 that TCDD was not present at the 1 ppt level. Considering the no-effect level of 70 ng per man/day as adopted by the FDA[28] there can be no doubt that Zone B could also be liberated immediately. Even Commoner[29] in 1977 suggested that a limit of 500 ppt in the soil would be an acceptable risk level.

Part of Zone A was decontaminated in 1977 and two-thirds of the evacuated houses were reinhabited. However, an area of about 50 ha is still fenced-in and decontamination operations are underway.

Table 10. TCDD Distribution in Fruit From Zone A_2/A_3

FRUIT	SOIL CONCENTRATION (APPROXIMATE)	CONCENTRATION IN			% IN PEELS
		PEELED FRUIT	PEELS	WHOLE FRUIT	
APPLES	10'000 PPT	< 2.5 PPT	137 PPT	37 PPT	> 95%
PEARS I	10'000 PPT	< 1.5 PPT	95 PPT	25 PPT	> 98%
PEARS II	10'000 PPT	< 0.5 PPT	79 PPT	21 PPT	> 96%
PEACHES	10'000 PPT	< 0.8 PPT	102 PPT	30 PPT	> 98%

Table 11. 2,3,7,8-TCDD in Soil from Zone R (1980)
           (Official Results, Confirmed by Capillary
           GC/MS)

Degree of Contamination	Frequency
> 10 ppt	-
5 - 10 ppt	4 (1 %)
1 - 5 ppt	8 (2 %)
< 1 ppt	348 (97 %)

Table 12. TCDD Analysis of Plant Samples from Zones
           B and R (1977-1980)

Year	Zone	Type of Vegetation	Number of Samples			Detection Limit
			measured	pos. for TCDD	neg. for TCDD	
1977	B + R	Wild Growing Vegetables	10	0	10	0,4 ... 5 ppt
1978	B + R	Oats, Peas, Vetch	37	0	37	0,2 ... 1,1 ppt
1979	R	Carrots, Radish, Sugar Beet	10	0	10	0,6 ... 1,8 ppt
1980	R	Rye, Wheat	41	0	41	0,3 ... 1,8 ppt

Plans are to remove the topsoil in the most contaminated parts of Zone A and bury it locally in plastic-lined basins. The less contaminated parts will be treated by agronomic methods. Finally, the entire zone will be covered with new topsoil and a green area created, that will be donated to a foundation.

LESSONS LEARNED SINCE 1976

Although few entirely new facts about TCDD have emerged since the Seveso accident in 1976, a lot has been learned to confirm, adjust, or accentuate earlier assessments and interpretations.

The original assessment of the situation with respect to persistence and lack of mobility of TCDD in the environment[17] proved to be essentially correct. It has been confirmed that the amount of leaching is very small[30]. The slow vertical movement of TCDD is probably due to mechanical transport of soil particles on which the TCDD is adsorbed.

The strong adsorption of TCDD to the soil has also been demonstrated in toxicological experiments[31]. Bioavailability of TCDD is reduced both for oral and dermal uptake. Increased contact time of TCDD with soil was demonstrated to decrease bioavailability.

The actual decrease in the concentration of of TCDD in the soil is slow. While early experiments[3,23] were consistent with a half-life time of 200 - 300 days as seemed probably in 1976[17,18,20,22], newer observations[23] show that the environmental half-life must be longer. A similar trend has been observed in the studies conducted in Florida by Young[32].

Studies documenting microbiological degradation have not been very successful. Although a metabolite has finally been demonstrated[33] it has become clear that the role of microbiological degradation under natural conditions has been overestimated. The possibility of a breakthrough in this area is still open but does not seem imminent[34].

Experiments on plant-uptake have remained somewhat controversial. The low residue levels of TCDD found in plants[3,4,25] suggest that plant-uptake cannot be a significant factor, confirming earlier experiments with radio-labelled TCDD[22,35]. Our results with carrots (Table 13) demonstrate that the edible part of root vegetables contains much less TCDD than the surrounding soil[3,25]. Only Cocucci maintains that plant-uptake is significant[36].

Monitoring TCDD levels in the atmosphere[24] has confirmed that dust esentially reflects the local soil concentrations and is no

Table 13. TCDD Distribution in Carrots Grown in Heavily
Contaminated Soil of Zone A (Estimated Contamination:
1,000-5,000 ppt in Soil)

TYPE OF MATERIAL	AMOUNT OF MATERIAL	TCDD FOUND	TCDD (IN PPT)	CUMULATIVE AMOUNT IN PPT	PERCENT OF TOTAL
CENTRAL CYLINDER	191 G	1.9 NG	10 PPT		1 %
OUTER EDIBLE PORTION	591 G	3.8 NG	6.5 PPT		2 %
TOTAL OF EDIBLE PARTS (PEELED)	(782 G)	(5.7 NG)		(7.5 PPT)	(3 %)
CARROT PEELS	218 G	48.8 NG	224 PPT		29 %
TOTAL OF SCRUBBED CARROTS	(1'000 G)	(54.5 NG)		(54.5 PPT)	(32 %)
WASH WATER		112.5 NG			68 %
TOTAL OF CARROTS, UNWASHED	(1'000 G)	(167 NG)		(167 PPT)	(100 %)

significant factor in spreading the contamination. Only one station
in the center of Zone A has consistently shown measurable amounts
of TCDD; even there human exposure by inhalation is estimated to
be not larger than 1.4 pg TCDD/day.

Advances have been made in the area of toxicology. Long-term
studies are now available[37]. Extrapolation from animal experiments
to the human remains a controversial issue.   However, much has been
learned from the health-surveillance program in the Seveso area
itself[12,38]. The absence of measurable health effects in the
exposed population shows that the doomsday prognoses were fortunately
in error. Additional information has been gained from comparison
with other episodes where much higher concentrations of TCDD have
been involved[11,39,40]. Extrapolations from such studies show that
long-term effects are not to be expected.

So, although the results of some long-term studies will not be
known for several years the outlook is much brighter than it was in
1976. Some problems have actually been solved, some are currently
being tackled, and a few have lost their importance because of the
availability of new, more detailed knowledge.

REFERENCES

1. Salomon, Ch.M. "Bericht uber den Storfall in Seveso",
    Sonderkilloquium (Dechema-Haus, Frankfurt am Mai, Nov. 4, 1980)
    and Meeting of the Industrial Health and Safety Group of
    Soc. Chem. Ind. London, 1981.
2. Theofanous, T.G. Nature, 291:640-642, 1981.
3. Homberger, E., Reggiani, G., Sambeth, J., and Wipf, H.-K.
    Ann. Occup. Hyg., 22:327-370, 1979.
4. Wipf, H.-K., Homberger, E., Neuner, N., Ranalder, U.B., Vetter,
    W., and Vuilleumier, J.P. "TCDD-Analysis in Vegetation Samples
    from the Seveso Area", CIPAC Proceedings Symposium Series 1
    (W.R. Bontoyan, Ed.), CIPAC Publications, 1979.
5. Fara, G.M. Int. J. of Epidemiology, 10:135-142, 1981.
6. Pocchiari, F. Int. Conf. on Health Effects of Halogenated
    Aromatic Hydrocarbons, June 24-27, 1978. New York Academy of
    Sciences, New York, New York.
7. Goldman, P.J. Arbeitsmed. Sozialmed. Arbeitshyg., 7:12-18,
    1972.
8. Pazderova-Vejlupkova, J., Nemcova, M., Pickova, J., Jirasek,
    L., and Lukas, E. Arch. Environ. Health, 36:5-11, 1981.
9. Young, A.L., Calcagni, J.A., Thalken, C.E., and Tremblay, J.W.
    "The Toxicology, Environmental Fate, and Human Risk of Herbicide
    Orange and its Associated Dioxin", Technical Report OEHL-TR-78-
    92. Occupational and Environmental Health Laboratory,
    Brooks Air Force Base, Texas 78235. 247p.
10. Bartleson, F.D. Jr., Harrison, D.D., and Morgan, J.D. "Field
    Studies of Wildlife Exposed to TCDD Contaminated Soils",
    Technical Report AFATL-TR-75-49, Air Force Armament Laboratory,
    Eglin AF Base, Florida 32542. 53p.
11. Kimbrough, R.D., Carter, D.C., Liddle, J.A., Cline, R.E., and
    Phillips, P.E. Arch. Environ. Health, 32:77-85, 1977.
12. Angeli, F. (Editor) "Disastro Icmesa", Franco Angeli Editore,
    Milano, Italy, 1979.
13. Bumb, R.R., Crummett, W.B., Cutie, S.S., Gledhill, J.R., Hummel,
    R.H., Kagel, R.O., Lamparsky, L.L., Lumoma, E.V., Miller, D.L.,
    Nestrick, T.J., Shadoff, L.A., Stehl, R.H., and Woods, J.S.
    Science, 210:385-390, 1980.
14. Olie, K., Lustenhouwer, J.W.A.,"Polychlorinated dibenzo-p-
    dioxins and Related Compunds in Incinerator Effluents. In:
    O. Hutzinger, R.W. Frei, E. Merian, and F. Pocchiari (eds.)
    Chlorinated Dioxins and Related Compounds - Impact on the
    Environment. Pergamon Series on Environmental Science.
    Pergamon Press, Oxford. Volume 5: 227-244. 1981.
15. Leng, M.L. "Comparative Toxicology of Various Chlorinated
    Dioxins as Related to Chemical Structure", CIPAC Proceedings
    Symposium Series 1 (W.R. Bontoyan, Ed.), CIPAC Publications
    1979.

16. Kearney, P.C. Symposium on Human Health Aspects of Accidental Chemical Exposure to Dioxins, Oct. 4-7, 1981, Bethesda, Maryland.

17. Kearney, P.C., Isensee, A.R., Helling, C.S., Woolson, E.A., and Plimmer, J.R. Adv. Chem. Ser., 120:105-111, 1973.

18. Kearney, P.C., Woolson, E.A., Isensee, A.R., and Helling, C.S., Environ. Health Perspect., 5:273-277, 1973.

19. Crosby, D.G., Moilanen, K.W., and Wong, A.S. Environ. Health Perspect., 5:259-266, 1973.

20. Helling, C.S., Isensee, A.R., Woolson, E.A., Ensor, P.D.J., Jones, G.E., Plimmer, J.R., and Kearney, P.C., J. Environ. Quality, 2:171-178, 1973.

21. Plimmer, J.R., Klingebiel, U.I., Crosby, D.G., and Wong, A.S. Adv. Chem. Ser., 120:44-54, 1973.

22. Young, A.L., Thalken, C.E., Arnold, E.L., Cupello, J.M., and Cockerham, L.G. "Fate of 2,3,7,8-Tetrachlorodibenzo-p-dioxin (TCDD) in the Environment: Summary and Decontamination Recommendations", Technical Report USAFA-TR-76-18, United States Air Force Academy, Colorado 80840. 41p.

23. di Domenico, A., Silano, V., Viviano, G., and Zapponi, G. Ecotoxicol. Environ. Safety, 4:339-345, 1980.

24. Wipf, H.K., Homberger, E., Neuner, N., and Schenker, F. "Field Trials on Photodegradation of TCDD on Vegetation After Spraying with Vegetable Oil" in "Dioxin: Toxicological and Chemical Aspects" (F. Cattabeni, Ed.), Spectrum Publications Inc., New York, London, 1978.

25. Wipf, H.K., Homberger, E., Neuner, N., Ranalder, U.B., Vetter, W., and Vuilleumier, P.P. TCDD-Levels in Soil and Plant Samples from the Seveso Area. In: O. Hutzinger, R.W. Frei, E. Merian, and F. Pocchiari (eds.) Chlorinated Dioxins and Related Compounds - Impact on the Environment. Pergamon Series on Environmental Science. Pergamon Press, Oxford. Volume 5: 105-114. 1981.

26. di Domenico, A., Silano, V., Viviano, G., and Zapponi, G. Ecotoxicol. Environ Safety, 4:346-356, 1980.

27. di Domenico, A., Silano, V., Viviano, G., and Zapponi, G. Ecotoxicol. Environ. Safety, 4:321-326, 1980.

28. Colby, A. Symposium on Human Health Aspects of Accidental Chemical Exposure to Dioxins, October 4-7, 1981, Bethesda Maryland.

29. Commoner, B., and Scott, R.E. "US Air Force Studies on the Stability and Ecological Effects of TCDD (Dioxin): an Evaluation Relative to the Accidental Dissemination of TCDD at Seveso, Italy", November 13, 1976. Dioxin Information Project, Scientists' Institute for Public Information, New York, New York.

30. di Domenico, A., Silano, V., Viviano, G., and Zapponi, G. Ecotoxicol. Environ. Safety, 4:327-338, 1980.

31. Poiger, H., and Schlatter, C., Ed. Cosmet. Toxicol., 18: 477–481, 1980.

32. Young, A.L. Long–Term Studies on the Persistence and Movement of TCDD in a Natural Ecosystem. These Proceedings, 1982.

33. Philippi, M., Schmid, J., Wipf, H.K., and Hutter, R. Submitted to Experientia for publication, 1981.

34. Hutter, R., and Philippi, M. Studies on Microbial Metabolism of TCDD under Laboratory Conditions. In: O. Hutzinger, R.W. Frei, E. Merian, and F. Pocchiari (eds.) Chlorinated Dioxins and Related Compounds – Impact on the Environment. Pergamon Series on Environmental Science. Pergamon Press, Oxford. Volume 5:87–94.

35. Isenesee, A.R., and Jones, G.E. J. Agr. Food Chem., 19: 1210–1214, 1971.

36. Cocucci, S., di Gerolamo, F., Verderio, A., Cavallaro, A., Colli, G., Gorni, A., Invernizzi, G., and Luciani, L. Experientia, 35:482–484, 1979.

37. Kociba, R.J., Keyes, D.G., Beyer, J.E., Carreon, R.M., Wade, C.E., Dittenber, D.A., Kalnins, R.P., Frauson, L.E., Park, C.N., Barnard, S.D., Hummel, R.A., and Humiston, C.G. Toxicol. Appl. Pharmacol., 46:279–303, 1978.

38. Donatelli, L., Lampa, E., and Creso, E. "Considerations on the Epidemiology of the Seveso Accident", Toxicol. Europ. Res. III (1), 1981.

39. Zack, J.A., and Suskind, R.R. J. Occup. Med., 22:11–14, 1980.

40. Cook, R., Townsend, J.C., Ott, M.G., and Silverstein, L.G. "Mortality Experience of Employees Exposed to 2,3,7,8–Tetrachlorodibenzo-p-dioxin (TCDD)", J. Occup. Med., 22: 530–532, 1980.

ENVIRONMENTAL TOXICOLOGY

Chairman: Eugene E. Kenaga
          Dow Chemical Company
          Midland, Michigan USA

ENVIRONMENTAL TOXICITY OF TCDD

E. E. Kenaga* and L. A. Norris**

*Health and Environmental Sciences
The Dow Chemical Company
Midland, Michigan, USA

**U.S. Department of Agriculture
Forestry Sciences Laboratory
Corvallis, Oregon, USA

ABSTRACT

Data on toxicity of TCDD to organisms, concentrations in the organisms, and concentrations in the media in and on which the organisms live can help determine the importance of measured levels of TCDD in the environment as part of the hazard assessment process. These toxicity and environmental fate data relationships vary depending on the media (soil, food, water, air) in which TCDD occurs. These variations are indicated by differences among bioconcentration factors, and the no-effect concentrations of TCDD for organisms in different media. Such data are summarized in this report.

INTRODUCTION

There is a need for more accurate techniques for hazard evaluation of 2,3,7,8-TCDD (2,3,7,8-tetrachlorodibenzo-p-dioxin) and other chemicals in the environment. Important data, often missing, are the toxicity to the organism, the concurrent related concentration in the organism, or specific tissue, and the concentration in the media in which the organism lives or through which it is exposed (food, soil, water, air). LC_{50} data are frequently the only data

available. These concentrations in the media which result in
50 percent lethality to the organisms do not reveal the
concentrations in the organisms which are lethal. Equally,
the lethal concentration measured in the organism or tissue
does not necessarily reveal the concentration at the specific
site in the organism responsible for lethality. However,
such concentrations as the latter are rarely known because of
lack of our knowledge of the mode of toxic action of TCDD and
most other chemicals and the sequences leading to effects
such as death.

There appears to be more data available in the aquatic
environment than in the terrestrial or aerial environment.
Due to the toxicity of TCDD and extreme care needed to be
used by humans handling it, toxicity tests for other
organisms are limited. The test methods have often been
different than standard tests, especially for aquatic
organisms. Because of these difficulties the test results
often have no uniform basis for comparative toxicity with
other chemicals. Nevertheless, we believe that the
information in this paper is summarized in a way to help use
bioconcentration factors (BCF), environmental concentrations
in different media, and concentrations toxic to aquatic and
terrestrial organisms in their respective media in the hazard
assessment process. Important data from field and laboratory
toxicity and environmental fate tests (including bioconcentration factors) are correlated for use in predicting results
based on one or the other piece of information. While acute
toxicity may be important, chronic toxicity is even more
important with persistent chemicals because of delayed
toxicity. Likewise, the no-observable-effect concentration
(NOEC) is more important than the LC_{50} or LD_{50}. Consequently,
in this paper more attention is given to the chronic
no-observable-effect concentrations. Due to lack of
information on environmental residues and toxicity of other
isomers of the variously chlorinated dibenzo-p-dioxins and
dibenzofurans only 2,3,7,8-TCDD is considered in this
evaluation. Norris (1981) has provided a comprehensive review
of the movement, persistence and fate of TCDD in the forest.
Bovey and Young (1980) reviewed the presence of TCDD in other
environments.

LABORATORY TOXICITY TESTS ON ANIMAL AND PLANT ORGANISMS

Toxicity to Mammals

Kociba et al, (1978) fed rats a diet containing 22 parts-
per-trillion (ppt) TCDD for two years and produced no
irreversible toxic effects. Rats fed 210 ppt showed several

symptoms of toxicity, including liver and lung lesions. Rats
fed 2200 ppt TCDD suffered mortality, carcinomas and acute
effects. These diets were equivalent to 0.001, 0.01, and 0.1
μg TCDD/kg/day, respectively. Using a maximum food bioconcentration factor (BCF) of about 25 (based on measured fat or
liver residues) the body burden of TCDD would be 0.55 to a
maximum of 5.2 μg/kg at steady state.

Murray et al, (1979) studied the three-generation
reproduction of rats receiving 0.001, 0.01, 0.1 μg TCDD/kg body
weight/day. The 0.1 dosage produced no effect on the adults,
but neonatal survival was poor. The 0.01 dosage significantly
decreased fertility in the second and third generation, but not
the first. The 0.001 dosage caused no-effect on any of the
three generations. The total dosage per generation was 0.1
μg/kg/day x 130 days = 13 μg/kg, 0.01 μg/kg/day x 130 days =
1.3 μg/kg and 0.001 μg/kg/day x 130 days = 0.13 μg/kg TCDD,
respectively.

Beach mice, Peromyscus polionotus having alumina dust
containing 2.5 ppb TCDD applied to their pellage 10 times in 28
days in laboratory studies suffered no mortality, nor cellular
changes, but there were statistically significant differences
in liver to body weight ratios, (Cockerham et al, 1980).

The toxicity of TCDD to organisms is highly influenced by
its solvent or adsorbent carriers, either dermally or by
intestinal absorption. When TCDD was administered orally to
rats in ethanol the amount adsorbed was twice that when the
TCDD was administered in soil particles. Adsorption onto
activated charcoal almost completely prevented uptake of the
compound. Similar results were obtained dermally. Threshold
levels for the induction of chloracne lesions on the skin of
rats were 160 times greater when TCDD was adsorbed on charcoal
than when administered as pure TCDD (Poiger and Schlatter,
1980). Such results indicated that TCDD is strongly adsorbed
on soil and organic matter and is not easily removed by contact
or through ingestion.

Toxicity to Birds

Acute and subacute oral toxicity. The acute oral toxicity
of 2,4,5-trichlorophenoxyacetic acid (2,4,5-T) to mallards was
determined by Tucker and Hudson (1970). No mortality occurred
at dosages of 2000 mg/kg 17 to 19 days after treatment.
Assuming a TCDD impurity of <0.1 ppm in the 2,4,5-T, the TCDD
dosage would be <0.2 μg/kg causing essentially no mortality in
this test. Seventeen to nineteen days seems a sufficient

period of time to observe delayed symptoms or mortality typical
of TCDD. Other tests with 2,4,5-T on chickens are shown by
Kenaga (1975) with similar low levels of toxicity.

TCDD fed as one dose orally each day for 21 days to 3-day
old Leghorn chicks produced no-observable-effect at 0.1
μg/kg/day (2.1 μg/kg total), 80 percent mortality and chick
edema at 1 μg/kg/day for 21, days and 100 percent mortality at
10 μg/kg/day (in 15 days) (Schwetz et al, 1973).

These two pieces of information suggest that one would not
expect mortality to occur with a total dose of less than 2.1
μg/kg, whether from a single or multiple dose. This is not
meant to imply a "cumulative effect" from multiple doses, only
to recognize that TCDD excretion is slower than the excretion
of 2,4,5-T.

The results of tests on the chick edema effects of TCDD
have been summarized by the National Research Council of Canada
(1981) but are not reported here.

Bird/egg contact tests. A review of the effects of
2,4,5-T, (and TCDD impurities) by injection of pheasant and
grouse eggs indicates that this method is a poor and unreliable
index of toxicity relative to the practical use of the
chemical. Spraying with heavy concentrations of 2,4,5-T on
eggs of chickens and pheasants does not cause greater mortality
or malformed embryos greater than occurs in control eggs.
Concentrations of TCDD in the eggs, if present, have not been
identified, (Kenaga, 1975).

Dietary feeding tests. The main oral intake of birds is
via their diet, and thus dietary tests represent a more
realistic route of exposure for birds than acute oral, or egg
injection tests. The following tests (See Table 1) indicate a
rather low order of toxicity from dietary feeding of 2,4,5-T to
mallards and gallinaceous birds. 2,4,5-T assumed to contain
0.1 or less ppm TCDD caused no observable effect on chickens or
turkeys at calculated concentrations of about 200-300 ppt TCDD
in the diet in 11 to 21-day feeding studies (Roberts and
Rogers, 1957; Whitehead and Pettigrew, 1972).

Reproduction tests using bobwhite in an 18-week dietary
study using 50 ppm 2,4,5-T containing 0.06 ppm TCDD (equals 3
ppt TCDD) caused no observable effects at the highest
concentration tested (Kenaga, 1975).

Table 1. Effects of TCDD Administered in 2,4,5-T on Various Species of Birds in Dietary Feeding Studies

Species	Concentration in the diet 2,4,5-T (ppm)	TCDD (ppt)	Duration of Test[a]	Effect[b]	References
Bobwhite Colinus virginianus	2776[c]	167	5 day/8 day	LC50	Kenaga, 1975
Bobwhite	50[c] 5[c]	3 0.3	18 wk/18 wk	No effect on reproduction	Kenaga, 1975
Mallard Anas platyrhynchus	4640[c]	278	5 day/8 day	LC10	Kenaga, 1975
Turkey Malleagris gallopavo	2500[d]	>259	11 day/11 day	EC0	Roberts & Rogers, 1957
Chicken Gallus domesticus	5000[d] 2000[d] 1000[d]	>500 >200 >100	21 day/21 day	LC90[e] LC0[e] LC0	Whitehead & Pettigrew, 1972

a/ Duration of exposure/Duration of observation.

b/ LC50 = Concentration causing 50% lethality, EC0 = no effect.

c/ 2,4,5-T containing 0.06 ppm TCDD.

d/ 2,4,5-T samples of this vintage usually contained >0.1 to 40 ppm TCDD.

e/ Also reduced feeding and growth.

Toxicity to Fish

The toxic response of fish to TCDD is characteristically slow with a long latent period for lethality after a short exposure. Death may be delayed from a few days to a few weeks or months. Other characteristic responses are edema, and reduced growth. These responses occur in water containing TCDD in the ppt to ppb range (Table 2).

Fish Dietary Treatment Tests

Rainbow trout consuming a diet containing 2.3 ppt or 2.3 ppb TCDD for 105 days showed no effect on food consumption, growth or survival, however, 50 to 88 percent mortality occurred in 61 and 77, days respectively, in fish exposed to 2.3 ppm TCDD (Hawkes and Norris, 1978). These authors expressed fish exposure data on a dryweight basis. We calculate these food concentration levels are equivalent to 0.0000064, 0.0072 and 4.2 µg TCDD/kg/day, respectively, on a fresh weight basis, assuming live weight is 5 times dry weight. The approximate no-effect oral intake level then would be 0.0072 µg/kg/day x 105 days of feeding or a total of 0.76 µg TCDD/kg live weight.

Aquatic Exposures

Hiltibran (1967) exposed small bluegills for 12 days to 50 ppm of the sodium salt of 2,4,5-T (45.9 ppm acid equivalent) causing no mortality. Assuming the 2,4,5-T contained 0.1 ppm TCDD, the concentration in the water would be 4.6 ppt TCDD. Miller et al, (1979) reported young coho salmon were more sensitive, with approximately 50 percent mortality 60 days after 96 hours exposure to 5.4 ppt TCDD in water, but 0.54 ppt for 96 hours had no effect in 114 days. A fish bioconcentration factor of 6400 for TCDD (Isensee, 1978) can be used to convert TCDD concentrations in water to concentrations in fish.

All of the water treatment tests in Table 2 on TCDD were conducted under static water conditions. Concentrations as low as 0.001 - 0.056 ppt caused some mortality (Miller et al, 1973; Helder, 1981). Yockim et al, (1978) measured tissues of dead mosquito fish and catfish which contained 7.2 (ppb) µg/kg and 4.4 (ppb) µg/kg, respectively. Thus, it appears that effect levels between 0 and 100 percent mortality for fish based on the body burden is approximately between 0.006 - 29.4 µg/kg.

Toxicity to Amphibians

Intraperitoneal injections of 1 mg TCDD/kg of body weight in bullfrog tadpoles and adults showed no effect indicating

Table 2. Effects of TCDD on Aquatic Organisms

Species	Test Duration[a]	Effect	Concentration in Water (ppb) (µg/l)	Reference
Snail, Physa sp.	48 d/48 d	EC35, reduced reproduction	0.200	Miller et al, 1973
Oligochaete worm, Paranais sp.	49 d/55 d	EC35, reduced reproduction	0.200	" "
Mosquito (larvae) Aedes aegypti	17 d/39 d	No effect on pupation	0.200	" "
Guppy, Poecilia reticulata	5 d/37 d	Feeding decline, skin discoloration, fin necrosis, mortality	10	" "
Coho Salmon Oncorhynchus kisutch	24-96 hr/40 d	LC100	0.056	" "
Coho Salmon	24-96 hr/60 d	LC55	0.0056	Norris and Miller, 1974
Guppy	5 d/5 d	LC8	0.100	" "
Guppy	5 d/21.7 d	LC50	0.100	" "
Guppy	5 d/37 d	LC100	0.100	" "
Guppy	5 d/11.6 d	LC50	1.0	" "
Guppy	5 d/18.2 d	LC50	10.0	" "
Alga Oedogonium cardiacum	32 d/32 d	None	1.33	Isensee, 1978

(Continued)

Table 2. Effects of TCDD on Aquatic Organisms (Continued)

Species	Test Duration[a]	Effect	Concentration in Water (ppb) (µg/l)	Reference
Pond weeds, Elodea nuttali and Certaophyllum emersum	Months	none	0.0537	Tsushimoto et al., 1981
Snail	32 d/32 d	none	1.33	Isensee, 1978
Daphnia magna	32 d/32 d	none	1.33	" "
Mosquito fish	14 d/14 d	LC100	0.0025-0.0042	" "
Catfish	6 d/6 d	none	0.24	" "
Channel catfish Ictalurus punctatus	15 d/15 d	LC100	0.0042	Yockim et al., 1978
Mosquito fish Gambusia affinis	15 d/15 d	LC100	0.0028	" "
Rainbow Trout	61-71d/61-71 d	50-88% mortality feeding decline, weight loss, fin erosion, fungal growth liver degeneration	2,300[b] (in food)	Hawkes & Norris, 1977
Rainbow Trout	15 wk/15 wk	No effect	2.3[b] (in food)	" "
Rainbow Trout	15 wk/15 wk	No effect	0.0023[b] (in food)	" "

Species	Test Duration[a]	Effect	Concentration in Water(ppb) (μg/l)	Reference
Coho salmon	96 hr/114 d	LC50	.0056 (5.4[c])	Miller et al., 1979
Coho salmon	96 hr/114 d	No effect (Feeding growth, survival)	.00056 (0.54[c])	" "
Northern pike (eggs) Esox lucius	96 hr/23 d	LC11	0.0001	Helder, 1980
Northern pike (eggs)	96 hr/23 d	LC42	0.001	" "
Northern pike (eggs)	96 hr/23 d	LC98	0.01	" "
Rainbow trout (juvenile)	96 hr/72 d	LC12	0.01	Helder, 1981
Rainbow trout (juvenile)	96 hr/21 d	LC50	0.100	" "
Rainbow trout (juvenile)	96 hr/27 d	LC100	0.100	" "
Rainbow trout (juvenile)	96 hr/73 d	EC54 (growth)	0.01	" "
Rainbow trout (eggs)	96 hr/24 wk	EC45 (growth)	0.01	" "
	96 hr/24 wk	EC8 (growth)	0.001	" "
	96 hr/24 wk	No effect (growth)	0.0001	" "
American bullfrog Rana catesbeiana	35 d	No mortality for larva & adults	1,000[b] (IP)	Beatty et al., 1976

[a]Test durations are expressed as follows: duration of exposure/duration of test; e.g., 36 day/48 day = 36 days exposure, 48 day test duration (12 day post-exposure observation). LC = lethal concentration, EC = some type of effect concentration other than lethal. IP = Interperitoneal injection.

[b]Dosage expressed in μg/kg of organism or food (i.e., not concentration in water).

[c]ng TCDD in water per gram of wet body weight of fish (i.e., not concentration in water).

that bullfrogs are relatively insensitive to TCDD compared to
fish (Beatty et al, 1976) (Table 2.)

Toxicity to Aquatic Invertebrates

The species of snails, worms, cladocerans, and insects
tested seem to be considerably less sensitive to TCDD than
fish, although some effect was seen on reproduction on snails
and worms at high levels of exposure (Table 2). No observable
effect was seen on mosquito larvae at 0.2 ppb in water, the
highest concentration tested (Table 2).

Toxicity to Aquatic Plants

Plants are relatively insensitive to TCDD compared to
animals. No attempt has been made to determine the maximum no-
effect levels. Plants tested at low concentrations of TCDD
were not affected (Table 2).

Toxicity to Algae

The algae tested were not affected by the low
concentrations (ppt) of TCDD (Table 2).

Toxicity to Soil Microorganisms

Bollen and Norris (1979) tested the respiratory effect of
2,4,5-T containing < 0.1 ppm TCDD on the microbial population of
a forest floor and soil. No effect on CO_2 evolution was
noted at concentrations of TCDD of 131 ppt in the forest floor
or 52 ppt in the forest soil, the highest concentrations
tested.

NO-EFFECT DOSAGE FROM DIETARY FEEDING IN MAMMALS, BIRDS, AND
FISH

The total body intake of TCDD in daily dietary food giving
a semi-chronic no-effect level for TCDD appears to be 0.55 -
5.2 µg/kg in mammals; 2.1 µg/kg in birds; and 0.76 µg/kg in
fish. These values are remarkably similar.

AQUATIC AND TERRESTRIAL FIELD TOXICITY AND BIOCONCENTRATION
TESTS ON ANIMAL AND PLANT ORGANISMS

Young et al, (1975) conducted field investigations on
rodents, insects, aquatic organisms, and plant species
associated with a unique 3 km^2 military test site (Test

Area C-52A, Eglin Air Force Base, Florida) that was sprayed
with 73,000 kg 2,4,5-T and 77,000 kg 2,4-D between 1962 - 1970.
Although neither 2,4-D nor 2,4,5-T residues could be detected
in the soils in 1973 or 1974, significant levels (10-710 parts-
per-trillion - ppt) of TCDD were found within the top 15 cm of
test site soils although in some instances 10 years had elapsed
since the last aerial application of 2,4,5-T.

A detailed study of the field effects of the herbicide and
TCDD was conducted on populations of beach mice, Peromyscus
polionotus and hispid cotton rats, Sigmodon hispidus. Liver
tissue from rodents inhabiting the test site contained 210-1,300
ppt TCDD. However, no gross or histological evidence of
teratogenesis or toxicity was found in 122 adults and 87
fetuses. An analysis of variance of liver and spleen weights
for the beach mouse indicated significant differences between
control and TCDD-exposed animals. Analysis of plant seeds
revealed no detectable levels of TCDD (minimum detection limit
of 1 ppt TCDD). TCDD accumulation in liver tissue was thought
to be associated with pelt contamination from burrowing and
subsequent ingestion of soil particles via grooming. Species
diversities and food chain studies were conducted in two
aquatic ecosystems draining the test area.

Deposition of soil form runoff occurred into a pond on the
test area and into a stream immediately adjacent to the area.
TCDD levels of 10 to 35 ppt were found in silt of the aquatic
systems but only at the point where eroded soil entered in
the water. Species diversity studies of the stream were
conducted in 1969, 1970, 1973, and 1974. Insect larvae,
snails, diving beetles, crayfish, tadpoles, and major fish
species (by body parts) from both aquatic systems were analyzed
for TCDD.

Species diversity studies indicated no significant change
in the composition of ichthyofauna between these dates relative
to a control stream. Concentrations of TCDD (12 ppt) were
found in only two species of fish from the stream, Notropis
hypselopterus (sailfin shiner) and Gambusia affinis (mosquito
fish). Samples of skin, muscle, gonads, and gut were obtained
from Lepomis punctatus (spotted sunfish) from the test grid
pond. Levels of TCDD in those body parts were 4, 4, 18, and 85
ppt, respectively. Gross pathological observations of the
sunfish revealed no significant lesions or abnormalities.

BIOCONCENTRATION

Bioconcentration factors are important because they
provide the link between residues of the chemical in organisms

compared to concentrations in other media such as water, soil,
plants etc. These concentrations are linked to toxicity.

Norris (1981) reviewed environmental residues of TCDD in
the field and concluded that bioaccumulation of TCDD in excess
of 10 ppt in the majority of the animal population is not
occurring in or near forest areas treated with 2,4,5-T or
silvex.

Bioconcentration in Terrestrial Mammals from Food

Kociba et al, (1978) in two-year feeding studies with rats
consuming dietary concentrations of TCDD at 2,200, 210, and 22
ppt found liver bioconcentration factors (BCF's) of 10.9 -
24.5, and fat BCF's of 3.7 - 24.5.

Kenaga (1980) reported a BCF of 3.5 based on a fat
accumulation of 84 ppt of TCDD in cows after feeding a diet
containing 24 ppt for 28 days. Bioconcentration factors in
animals, based on dietary intake are of several different
orders of magnitude less than those from water. Kenaga (1980)
provides an equation for converting between terrestrial and
aquatic bioconcentration factors.

Bioconcentration in Aquatic Animals from Water

Matsumura and Benezet (1973) calculated BCF's in
organisms from simulated aquatic-terrestrial ecosystems exposed
in different ways. The main source of uptake appears to be
from water. Only these data are given in Table 3. The highest
BCF figure for fish was 2850.

Isensee and Jones (1975), Isensee (1978), and Yockim et al,
(1978) conducted static water tests in which soil was treated
and organisms accumulated TCDD. Wet weight BCF's in several
organisms (Table 3) ranged form 2200 - 6400, the latter for
mosquito fish.

Isensee (1978) reported 6-14% degradation of ^{14}C-TCDD
in his BCF tests. This infers that most but not all of the
[14]C count found in animals used in calculating BCF data is
based on TCDD.

Bioconcentration in Aquatic Plants from Water

It is unlikely that the TCDD accumulation for algae and
duckweed shown in Isensee and Jones (1975) and Yockim et al,
(1978) represent much more than adsorption of TCDD onto the
plant. Since the surface area ratio of foliage and algae are

Table 3. TCDD Bioconcentration Factors for Water for Aquatic Organisms.

Species	Concentration in Water	Bioconcentration Factor[a]	Exposure Duration (days)	Reference
Algae, Oedogonium cardiacum	2.42 ppt	2,075	7	Isensee, 1978[d] (Table 2)
Algae	0.05 ppt - 239 ppt	3,268[b] (aver.)	31	Isensee, 1978 (Table 1)
Pondweeds, Elodea nuttali & Ceratophyllum emersum	53.7 ppt	130 (max. 5 d)	5 - >60	Tsushimoto et al., 1981
Snail, Physa sp.	2.42 ppt	2,095	7	Isensee, 1978 (Table 2)
Snail	0.05 ppt - 239 ppt	6,106[b] (aver.)	31	Isensee, 1978 (Table 1)
Daphnia magna	2.42 ppt	7,070[b]	7	Isensee, 1978 (Table 2)
	0.05 ppt - 239 ppt	4,438[b] (aver.)	31	Isensee, 1978 (Table 1)
Channel catfish, Ictalurus punctatus	4.2 ppt	1,048	32	Yockim et al., 1978
	2.6 ppt	2,269	15	" "
Mosquito fish, Gambusia affinis	2.4 ppt	4,875[b]	7	" "
	0.05 ppt - 239 ppt	6,970[b] (aver.)	3	Isensee, 1978 (Table 1)
	2.42 ppt	4,850	7	Isensee, 1978 (Table 2)
Silversides Labidesthes sicculus	1300 ppt	545[e]	4 - 7	Matsumura & Benezet, 1973
Mosquito larvae Aedes aegypti	1300 ppt	2,846[e]	4 - 7	Matsumura & Benezet, 1973

a/ Based on ^{14}C count as ^{14}C TCDD, whole body, average values, wet weight.
b/ Dry wt/water of Isensee & Jones (1975), converted to wet weight.
c/ Calculated.
d/ Soil treated with TCDD and put in aquatic ecosystem.
e/ Values above 0.2 ppb exceed water solubility of TCDD and are probably low and not included in summary table.

great compared to their weight, the residue accumulated from
adsorption from water is several thousand fold. The
bioconcentration data concerning organisms in water are in
Table 3.

Bioconcentration in Aquatic Animals from Sediment

Most tests of this type of exposure have not involved
organisms that "live" in or on sediment. Fish or other free
swimming types of organisms have been most commonly used. In
laboratory tests TCDD has usually been adsorbed on sand or soil
which was then placed in an aquarium and water and organisms
added. Exposure probably involved desorption of TCDD from the
sand or soil into the water and subsequent uptake from the
water by the organism. In tests like this the BCF has usually
been based on the level of TCDD in the water. Based on the
level of TCDD in the soil or the sand, BCF values of less than
1 to 10 for snails, Daphnia, mosquito fish and catfish were
found by Isensee and Jones (1975) and for brine shrimp,
mosquito larvae, and fish by Matsumura and Benezet (1973). In
these same tests the BCF values for transfer from water to
organisms were much higher as discussed in a previous section.
In a field test, TCDD residues of 10 to 35 ppt in sediment were
associated with residues of 4 to 85 ppt in fish from the same
site (BCF of less than 3).

Bioconcentration in/on Terrestrial Plants from Soil

TCDD apparently is not translocated in appreciable
quantities in terrestrial plants. The amount found could be
accounted for by volatility from the soil, or by rain splashed
soil particles. This view is bolstered by the work of Kearney
et al, (1973) who found that plants in soil fortified with high
levels of TCDD do not readily take up TCDD and translocate it
to aerial plant parts. Young et al, (1975) and Homberger et
al, (1979) also found negligible BCF from plants on heavily
treated areas.

Bioconcentration in Mammals from Soil and/or Food

Young et al, (1975) reported little bioconcentration by
mammals from treated soil containing TCDD at the Eglin Air
Force test site. Fanelli et al, (1980) reported 4.5 ppb TCDD
in mice collected in an area where the concentration in soil
was 3.5 ppb near Seveso (BCF about 1).

Conclusions from Bioconcentration Data

Bioconcentration of chemicals in organisms is highly variable and dependent upon the metabolism of the organism, molecular stability of the chemical, surface area to weight ratio of the organism exposed, food intake rate and the medium in which the organism lives, etc. The accuracy of BCF's are also dependent on the reliability of the analytical information. For organisms in water the BCF of TCDD is moderately high because of TCDD's hydrophobic nature (Tables 3 and 4). TCDD is also highly adsorbed on soil and particles.

BCF's of TCDD from soil to mammals, sediment to aquatic organisms, food to mammals and soil to plants are very low. Such low BCF's in organisms exposed to TCDD from soil is related to high soil adsorption and consequently to the low availability of TCDD and consequent low mammalian exposure to TCDD.

CORRELATION OF THE BCF AND THE NO-EFFECT CONCENTRATION OF TCDD IN ORGANISMS AND IN THE MEDIA CONTACTED BY THE ORGANISMS

The purpose of this correlation is to relate the no-observable-effect concentrations in water, soil, and food to the approximate no-effect concentrations occurring in organisms and to BCF's. Too often one measurement is available and the others are not. Where all three kinds of concentration data are available they have been tabulated (Table 4). Since these values vary among different species and test methods the values are useful as estimation, not as exact figures.

From the data in the tables and text, as summarized in Table 5, it is shown that no-observable-effect concentrations of TCDD in aquatic organisms are several hundred or thousand times greater than those in water. These differences correspond to the BCF values in whole body, liver or fat tissues as designated. It is also found that the TCDD no-observable-effect concentrations in organisms are no higher than 25 times that in their food from dietary uptake, whether aquatic or terrestrial organism, or form soil contacted by the soil exposed organism. BCF values for a given chemical vary greatly depending on the media to which the organism is exposed. Thus, by selection of the BCF for a specific organism (o) category (aquatic or terrestrial) and its specific media (m) category (water, food, or soil), the following equations are applicable:

Table 4. Media – Organism – Concentration – Toxicity Relationships with TCDD.

Media	TCDD Concentration in Media	Organism	TCDD Concentration in Organism[a]	BCF[b]	Duration[c]	Type of Test	Toxic Effect	Reference
1) Aquatic Habitat								
Water	2.42 ppt	Daphnia	17,100 ppt	7,070(7d)	32d/32d	Static model ecosystem	LC0	Isensee, 1978
Water	2.42 ppt	Snail, Physa	5,100 ppt	2,095 (7d)	32d/32d	Static model ecosystem	LC0	"
Water	2.42 ppt	Algae	5,000 ppt	2,075 (7d)	32d/32d	Static model ecosystem	LC0	"
Water	53.7 ppt	Pondweeds	7,000 ppt	130	>60d	Outdoor pond	LC0	Tsushimoto et al., 1981
Food	2.3 ppb	Rainbow trout	1.57 ppb	≤ 1	105d	Dietary Feeding	None	Hawkes et al, 1977
Food	2.3 ppt	Rainbow trout	63 ppt	27(?)	105d	Dietary feeding	None	"
Sediment	10-35 ppt	Mosquito fish	12-35	1-3	life	Field	None?	Young et al., 1975
2) Terrestrial Habitat								
Food	22 ppt	Rat	540[e,f]	24.5[e,f]	2 yr	Chronic & oncogenic	None	Kociba et al., 1978
Food	210 ppt	Rat	5,100[e] / 1,700[f]	24.3[e] / 8.1[f]	2 yr	Chronic & oncogenic	Decreased neonatal size, etc.	"
Food	24 ppt	Cattle	84 ppt[f]	3.5	4 wk	Residue	None	Kenaga, 1980
Food & Soil	0.22-2.3ppt[d,g] plants(initially)	Mountain beaver	<3-17 ppt[e]	<1-14	45-59d	Field	None	Newton et al., 1979
Food & Soil	10-710 ppt(soil) <1 ppt(food)	Beachmouse & Cotton rat	210-1300 ppt[e]	2-21	multigeneration, teratogenesis	Field reproduction	None	Young et al., 1975
Soil	2-200 ppt	Various plants	<2.5 ppt	<0.01	several months	Field	None	Homberger et al., 1979
Soil	10-710 ppt	Plants (many species)	<1 ppt	<0.1	multigeneration	Field reproduction	None	Young et al., 1975
Explosion from factory into air	Variable (2-200 ppb)	Plants	up to 15 ppm	?	several weeks	Monitoring	Practically none	Homberger et al., 1979

a/ wet weight assumed. b/ Average figures. c/ Duration of exposure/duration of test (see Table 2 footnote a/.) d/ in 2,4,5-T containing 0.1 ppm or less TCDD. e/ liver. f/ fat. g/ calculated.

Table 5. Summary of No-Effect Concentration in Media and
          Organisms, and Bioconcentration Factors of TCDD

Organism Tested	Approximate No Observable Effect Concentrations (From Tables and Text)		BCF[c]
Aquatic Species	In Organism	In Medium	
		Water	
Fish	.1 ppb - 9.3 ppm[a,b]	0.1 ppt - 1.33 ppb	1048 - 6970
Daphnia	5.9 - 9.4 ppm[b]	1.33 ppb	4438 - 7070
Mosquito larvae	>569 ppb[b]	>200 ppt	>2850
Snail	>0.88 ppb <1.22 ppm[b]	>4.2, <200 ppt	2095 - 6106
Algae	>4.2 ppm	>1.3 ppb	2075 - 3268
Pondweed	>7 ppb[b]	>53.7 ppt	130
		Sediment	
Fish	12-85 ppt	10-35 ppt	1 - 3
		Food	
Fish	63 ppt - 1.6 ppb	0.002-2.3 ppb	<1 - 27
Terrestrial Species			
	Organism	Food	
Mammals	>7.7-<5250 ppb[b,d]	>22 - <210 ppt	3.5 - 25[d]
	Organism	Soil	
Mammals	>1300 ppt[d]	>710 ppt	2 - 21[d]
Plants	>1 ppt	>710 ppt	<0.1

[a] ppb = μg/kg body weight.

[b] Calculated from BCF and amount in medium.

[c] BCF = $\dfrac{\text{concentration in organism}}{\text{concentration in media}}$ (See data in Tables 3 and 4)

[d] BCF based on fat and/or liver concentrations.

$$BCF = \frac{NOEC_o \text{ (no-observable-effect concentration in organism)}}{NOEC_m \text{ (no observable-effect-concentration in the media)}}$$

or $NOEC_o = BCF \times NOEC_m$

or $NOEC_m = \dfrac{NOEC_o}{BCF}$

The predictive value of these and other recently discovered relationships are considerable. Once the BCF has been determined or estimated only one no-observable-effect concentration for similar habitat and organism value is needed to estimate other unknown values. Thus, the no-effect concentration for the medium and for the organism are both easily estimated to within a reasonable and practical limit.

If the BCF has not been established it can now be reasonably estimated for persistent organic chemicals by regression equations from water solubility, octanol-water partition coefficient (K_{ow}), soil-water adsorption coefficient (K_{oc}), or high pressure liquid chromatography (Kenaga and Goring, 1980; and Veith and Morris, 1979). Values for K_{ow} can be predicted from organic structure alone or from linear free energy (Leo et al, 1971). There may be limits to these correlations on polychloroaromatic compounds, such as biphenyls, dibenzofurans, and dibenzodioxins above the molecular weight of 450-500 and with limited water solubility resulting in decreased penetration and distribution of such chemicals in animal tissues, thus lower BCF's.

Values for water solubility not ascertained experimentally can be accurately predicted for organic chemicals from melting point and molecular surface area (Yalkowsky and Valvani, 1979).

If a BCF is known for aquatic species, BCF's can be calculated by regression equations for terrestrial species from their food (Kenaga, 1980), which are reasonably similar to those from soil.

COMPARISON OF EXCESSIVE CONCENTRATIONS OF TCDD IN VARIOUS MEDIA AND TOXIC EFFECTS WITH TCDD CONCENTRATIONS FROM RECOMMENDED USES OF 2,4,5-T

A comparison of different TCDD dosages per acre; residue concentrations in soil, vegetation, and animals; and toxicological effects are useful for practical consideration in

hazard evaluation. Such measurements were made for the massive
contamination by TCDD from the "Seveso Accident" (Homberger et
al, 1979) and the Eglin Air Force Base test site from Agent
Orange (Young et al, 1975) in contrast to maximum
concentrations expected from normal EPA registered uses for
control of vegetation by aerial application of 2,4,5-T to
range (Table 6). It should be emphasized that no animal or
plant mortality ascribed to TCDD alone has occurred at these
sites. No long-term toxic effects such as teratogenesis or
reproductive effects are known from animals living their entire
life cycle on areas treated with among the highest
concentrations of TCDD known to occur. The lack of effects at
these sites illustrate the potential for substantially
overestimating risk by the undue extrapolation of acute oral
toxicity of TCDD to toxicity which might result from daily
contact with TCDD in feed and soil in the natural environment
of the animal.

     The dosage difference between high concentrations of TCDD
in the Seveso and Eglin Air Force Base sites, and the maximum
concentrations expected from the EPA registered dosages is over
1000 fold. It appears that concentrations from 2,4,5-T
applications registered by EPA result in TCDD concentrations
which are far below those from which adverse effects on animals
or plants could be expected from exposures to water, soil or
food in the natural environment.

CONCLUSIONS

     The data on TCDD summarized here establishes approximate
no-observable-effect concentrations for the food and media of
organisms and the bioconcentration factors to be expected
between various media and organisms. With the use of the
predictive relationships described, given one or two of these
factors or no-effect levels the other no-effect level or BCF
can be determined for practical purposes. None of the
registered uses of 2,4,5-T will result in TCDD concentrations
in the environment similar to those resulting from the Seveso,
Italy explosion or the testing at the Eglin Air Force Base
Agent Orange test site. Despite more than a 1000-fold greater
concentration of TCDD at these sites, compared to those
expected from the registered uses of 2,4,5-T, there appears to
be little evidence of reduction of animal or plant organism
populations due to TCDD. These findings are consistent with
those predicted by the relationship between bioconcentration
factor and no-observable-effect concentration of TCDD in the
organism and in the medium through which it is exposed.

Table 6. Concentrations in Various Media and Toxic Effects of TCDD.

Location	lb. TCDD/A (average)	ppt TCDD Residues			Toxicological Effects - Terrestrial Animals
		Soil	Vegetation[a]	Animals	
Seveso, Italy	0.000026[b] 0.00056[b,c]	23[b] 10,000[b,c]	1 - 137[b] 15.8 (ppm)[b,c]	58,000 - 215,000[b]	Initial mortality of small animals, none after 2 mo.[a]
Eglin Air Force Base, Florida, U.S.A.	0.00025[d]	10 - 710[d]	<1[d]	210 - 1300[d]	No mortality or teratogenesis; livers enlarged[d]
EPA approved aerial application of 2,4,5-T to rangeland.	0.0000002[c]	0.18[c,e,f]	23[c,e,g]	4[e]	None known from many years of use

a/ No toxic effects seen at any concentrations due to TCDD. Acute effects ascribed to sodium trichlorophenate and alkalinity from explosion.

b/ Measured shortly after explosion in Zone A, area of highest contamination.

c/ Maximum values.

d/ Measured 1973, 4 up to 10 years after last application of 2,4,5-T totaling 160,948 lb/A; 2,4,5-T assumed to contain 1 ppm TCDD.

e/ Assume 2 lb. 2,4,5-T applied/A, containing 0.1 ppm TCDD.

f/ In top 20 cm of soil.

g/ See Hoerger & Kenaga (1972), 1 lb/A = 110 ppm[c] on grass, immediately after spraying.

REFERENCES

Beatty, P.W.; Holscher, M.A.; Neal, R.A.   Toxicity of 2,3,7,8-
    Tetrachlorodibenzo-p-dioxin in Larval and Adult Forms of
    Rana catesbeiana.   Bull. Environ. Contam. Toxicol. 16:
    578-82, 1976.
Bollen, W.B.; Norris, L.A.   Influence of 2,3,7,8-Tetrachloro-
    dibenzo-p-dioxin on Respiration in a Forest Floor and Soil.
    Bull. Environ. Contam. Toxicol. 22:648-652, 1979.
Bovey, R.W.; Young, A.L.   The Science of 2,4,5-T and Associated
    Phenoxy Herbicides.   Environmental Sciences and
    Technology Monograph Series.   John Wiles & Sons.   New York
    462 p.   1980.
Cockerham, L.G.; Young, A.L.; Thalken, C.E.   Histopathological
    and Ultrastructural Studies of Liver Tissue from TCDD-
    Exposed Beach Mice Peromyscus polionotus.   Frank J. Seiler
    Research Laboratory, United States Air Force Academy,
    Colorado.   Technical Report FJSRL-TR-80-0008.   61 p. 1980.
The Dow Chemical Company.   U.S. Environmental Protection Agency
    Approved Label for ESTERON* 2,4,5 Herbicide for the Control
    of Trees, Brush, and Broadleaf Weeds.   (Contains propylene
    glycol butyl ether esters of 2,4,5-T).   Registration 86-
    1064.   January 1980.
Fanelli, R.; Castelli, M.G.; Martelli, G.P; Noseda, A.; Garattini,
    S.   Presence of 2,3,7,8-Tetrachlorodibenzo-p-dioxin in
    wildlife Living Near Seveso, Italy: A Preliminary Study.
    Bull. Environ. Contam. Toxicol. 24:460-462, 1980.
Hawkes, C.L.; Norris, L.A.   Chronic Oral Toxicity of 2,3,7,8-
    Tetrachlorodibenzo-p-dioxin (TCDD) to Rainbow Trout.   Trans.
    Am. Fish Soc. 106(6): 641-645, 1977.
Helder, T.   Effects of 2,3,7,8-Tetrachlorodibenzo-p-dioxon (TCDD)
    on Early Life Stages of the Pike (Esox lucius L.).   The
    Science of the Total Environment 14:255-264, 1980.
Helder, T.   Effects of 2,3,7,8-Tetrachlorodibenzo-p-dioxin (TCDD)
    on Early Life Stages of Rainbow Trout (Salmo gairdneri,
    Richardson).   Toxicology 19:101-112, 1981.
Hiltibran, R.C.   Effects of Some Herbicides on Fertilized Fish
    Eggs and Fry.   Trans. Am. Fish Soc. 96(4):414-416, 1967.
Hoerger, F.D; Kenaga, E.E.   Pesticide Residues on Plants:
    Correlation of Representative Data as a Basis for
    Estimation of their Magnitude in the Environment.
    Environ. Qual. Safety 1:9-28, 1972.
Homberger, E.; Reggiani, G.; Sambeth, J.; Wipf, H.K.   The Seveso
    Accident: Its Nature, Extent and Consequences.   Ann.
    Occup. Hyg. 22:327-370, 1979.

---

*Trademark of the Dow Chemical Company

Isensee, A.R. Bioaccumulation of 2,3,7,8-Tetrachlorodibenzo-
    para-dioxin. In: Ramel, C. (ed.). Chlorinated Phenoxy
    Acids and Their Dioxins. Ecological Bulletin
    (Stockholm) 27:255-262, 1978.

Isensee, A.R.; Jones, G.E. Distribution of 2,3,7,8-Tetrachloro-
    dibenzo-p-dioxin (TCDD) in Aquatic Model Ecosystem. Environ.
    Sci. Technol. 9(7):668-672, 1975.

Kearney, P.C., Woolson, E.A., Isensee, A.R., Helling, C.S.
    Tetrachlorodibenzodioxin in the Environment: Source,
    Fate, and Decontamination. Environ. Health Perspect. 5:
    273-277, 1973.

Kenaga, E.E. The Evaluation of the Safety of 2,4,5-T to Birds
    in Areas Treated for Vegetation Control. Residue Rev. 59:
    119, 1975.

Kenaga, E.E. Correlation of Bioconcentration Factors of
    Chemicals in Aquatic and terrestrial Organisms With Their
    Physical and Chemical Properties. Environ. Sci. Technol,
    14:553-556, 1980.

Kenaga, E.E.; Goring, C.A.I. Relationship Between Water Solu-
    bility, Soil Sorption, Octanol-Water Partitioning, and
    Concentration of Chemicals in Biota. American Society
    for Testing and Materials, Special Technical Publication
    707. pp. 78-115, 1980.

Kociba, R.J.; Keyes, D.G,; Beyer, J.E.; Carreon, R.M.; Wade,
    C.E.; Dittenber, D.A.; Kalnins, R.P.; Frauson, L.E.; Park,
    C.N.; Barnard, S.D.; Hummel, R.A.; Humiston, C.G. Results
    of a Two-year Chronic Toxicity and Oncogenicity Study of
    2,3,7,8-Tetrachlorodibenzo-p-dioxin in Rats. Toxicol.
    Appl. Pharmacol. 46:279-303, 1978.

Leo, A.; Hansch, C.; Elkins, D. Partition Coefficients and
    Their Uses. Chem. Rev. 71(6):525-616, 1971.

Matsumura, F.; Benezet, H.J. Studies on the Bioaccumulation
    and Microbial Degradation of 2,3,7,8-Tetrachlorodibenzo-
    p-dioxin. Environ. Health Perspect. 5:253-258, 1973.

Miller, R.A.; Norris, L.A.; Hawkes, C.L. Toxicity of 2,3,7,8-
    Tetrachlorodibenzo-p-dioxin (TCDD) in Aquatic Organisms.
    Environ. Health Perspect. 5:177-186, 1973.

Miller, R.A.; Norris, L.A.; Loper, B.R. The Response of Coho
    Salmon and Guppies to 2,3,7,8-Tetrachlorodibenzo-p-dioxin
    (TCDD) in Water. Trans. Am. Fish Soc. 108:401-407, 1979.

Murray, F.J.; Smith, F.A.; Nitschke, K.D.; Humiston, C.G.;
    Kociba, R.J.; Schwetz, B.A.. Three-Generation
    Reproduction Study of Rats Given 2,3,7,8-Tetrachlorodibenzo-
    p-dioxin (TCDD) in the Diet. Toxicol. Appl. Pharmacol.
    50:241-252, 1979.

National Research Council Canada. Polychlorinated Dibenzo-p-
    Dioxins: Criteria for Their Effects on Man and His
    Environment. NRCC No. 18574, Ottawa, Canada. 251 p.
    1981.

Newton, M.; Snyder, S.P. Exposure of Forest Herbivores to 2,3,7,8-Tetrachlorodibenzo-p-dioxin (TCDD) in Areas Sprayed with 2,4,5-T. Bull. Environ. Contam. Toxicol. 20: 743-750, 1978.

Norris, L.A. The Movement, Persistence, and Fate of the Phenoxy Herbicides and TCDD in the Forest. Residue Rev. 80:65-135, 1981.

Norris, L.A.; Miller, R.A. The Toxicity of 2,3,7,8-Tetrachlorodebenzo-p-dioxin (TCDD) in Guppies (Peocilia reticulatus Peters). Bull. Environ. Contam. Toxicol. 12(1):76-80, 1974.

Poiger, H.; Schlatter, C. Influence of Solvents and Adsorbents on Dermal and Intestinal Absorption of TCDD. Food Cosmet. Toxicol. 18:477-481, 1980.

Roberts, R.E.; Rogers, B.J. The Effects of 2,4,5-T Brush Spray on Turkeys. Poult. Sci. 36:703. 1957.

Schwetz, B.A.; Norris, J.M.; Sparschu, G.L.; Rowe, V.K.; Gehring, P.J.; Emerson, J.L.; Gerbig, C.G. Toxicology of Chlorinated Dibenzo-p-dioxins. Environ. Health Perspect. 5:87-99, 1973.

Tsushimoto, G.; Matsumura, F.; Sago, R. Fate of 2,3,7,8-Tetrachlorodibenzo-p-dioxin (TCDD) in the Aquatic Environment. (Submitted to Env. Tox. Chem.). 1981.

Tucker, R.K.; Hudson, R.H. 2,4,5-T Acute and Oral Toxicity to Mallard Ducks. Private Communication. November 15. Denver Wildlife Research Center, Colorado. U.S. Department of Interior. 1970.

Veith, G.D.; Morris, R.T. A Rapid Method for Estimating Log P for Organic Chemicals. Water Res. 13:43-47, 1979.

Whitehead, C.C.; Pettigrew, R.J. The Subacute Toxicity of 2,4-Dichlorophenoxyacetic Acid and 2,4,5-Trichlorophenoxyacetic Acid to Chicks. Toxicol. Appl. Pharmacol. 21:348. 1972.

Yalkowsky, S.H.; Valvani, S.C. Solubilities and Partitioning 2. Relationships Between Aqueous Solubilities, Partition Coefficients, and Molecular Surface Areas of Rigid Aromatic Hydrocarbons. J. Chem. Eng. Data. 24(2): 127-129, 1979.

Yockim, R.S., Isensee, A.R.; Jones, G.E. Distribution and Toxicity of TCDD and 2,4,5-T in an Aquatic Model Ecosystem. Chemosphere, 7(3):215-220, 1978.

Young, A.L.; Thalken, C.E.; Ward, W.E. Studies of the Ecological Impact of Repetitive Aerial Applications of Herbicides on the Ecosystem of Test Area C-52A, Eglin AFB, Florida. Technical Report AFATL-TR-75-142. Available from the National Technical Information Service, Springfield, Virginia, U.S.A. Publication AD-A032-773. 142. pp. 1975.

THE THEORY OF ACCUMULATION AND ITS RELATIONSHIP TO THE
CHOICE OF MONITORING MATRICES FOR DIOXINS

J.R. Roberts and M.J. Boddington

Environmental Secretariat, National Research
Council Canada Ottawa, Canada

ABSTRACT

There now exists a sufficient body of theoretical knowledge
to translate an environmental input into a matrix concentration.
The chemical dynamics of an environmental contaminant will depend
on first, the physico-chemical properties of the chemical and
secondly, the properties of the different ecosystem compartments.
In the case of dioxins in aquatic systems, the combination of a
few fundamental measurements such as molecular weight, vapor
pressure, solubility, quantum yield, and octanol-water partition
coefficient can be used to predict a chemical's fate and
persistence patterns in water, sediments, and biota. While their
low water solubility combined with their high octanol-water
partition coefficient indicate a high affinity for sediments and
biota, theory predicts that the pattern should be homologue
specific and a wide range of accumulation patterns should be
observed. Even though one would predict that sediments would
have a much higher concentration than the biota at equilibrium,
the theory of sediment sorption versus that for bioaccumulation
suggests that the equilibrium would be reached only after a long
period of time. Consequently, in the short-term, biota could be
a more appropriate monitoring matrix. Additionally, the
bioaccumulation potential of various types of organisms can be
modelled on the basis of their metabolic requirements. Thus,
because fish depend on water to satisfy their respiratory
requirements, they appear more likely to be useful indicators of
aquatic contamination than organisms higher on the food chain.

INTRODUCTION

        Chlorinated dioxins have been detected in food products,
biota from numerous aquatic ecosystems, and some highly
contaminated terrestrial systems in North America (Esposito et
al, 1980; and NRCC, 1981a). Detection does not imply a risk per
se for it is the duration of the exposure combined with the toxic
potency that defines the toxic response and the degree of hazard.
However, given that dioxins have been detected, there must be
some accompanying interpretation. In this interpretation the
problem is the reliability of assumed similarities of 2,3,7,8-TCDD
to a homologue as a whole. There is a high degree of structural
specificity displayed with respect to the toxic potency of
dioxins and perhaps 30-40 percent of all chlorinated dioxins may
be potentially hazardous (NRCC, 1981a). There is no reason to
believe that such differentials will not be paralleled in their
overall behavior. Therefore, rather than dealing with one isomer
or eight homologues, we may have to interpret residue data for up
to 30 or more specific dioxins.

        Given that we are sure that a specific congener has been
detected (a subject covered by Crummett in the analytical section
and exhaustively in NRCC, 1981b), we must also be aware that:
(1) differential sequestering patterns may exist for the
different congeners; and (2) that the residue patterns can change
with time.

        In this paper, we examine our ability to predict the
temporal accumulation patterns of chlorinated dioxins in aquatic
ecosystems; the data required to make such predictions; and how
such predictions allow for monitoring indicators to be
identified.

DEGRADATION PROCESSES, TRANSFER PROCESSES AND PARTITIONING

        There are a variety of ways to conceptualize the
interactions of degradative and removal processes. Their
derivation depends upon the end-use of the interpretation. For
instance, if thinking of a specific matrix, then all the
sequential steps operative prior to the entrance of the pollutant
into the matrix of interest could be envisaged. The cascade in
Figure 1 envisages some of the potential transfer and degradative
processes prior to accumulation in a number of matrixes
potentially useful in monitoring. The processes can be divided
into two broad categories: abiotic and biotic. As microbial
degradation is not thought to play a significant role, biotic
processes do not appear to contribute greatly to the gross

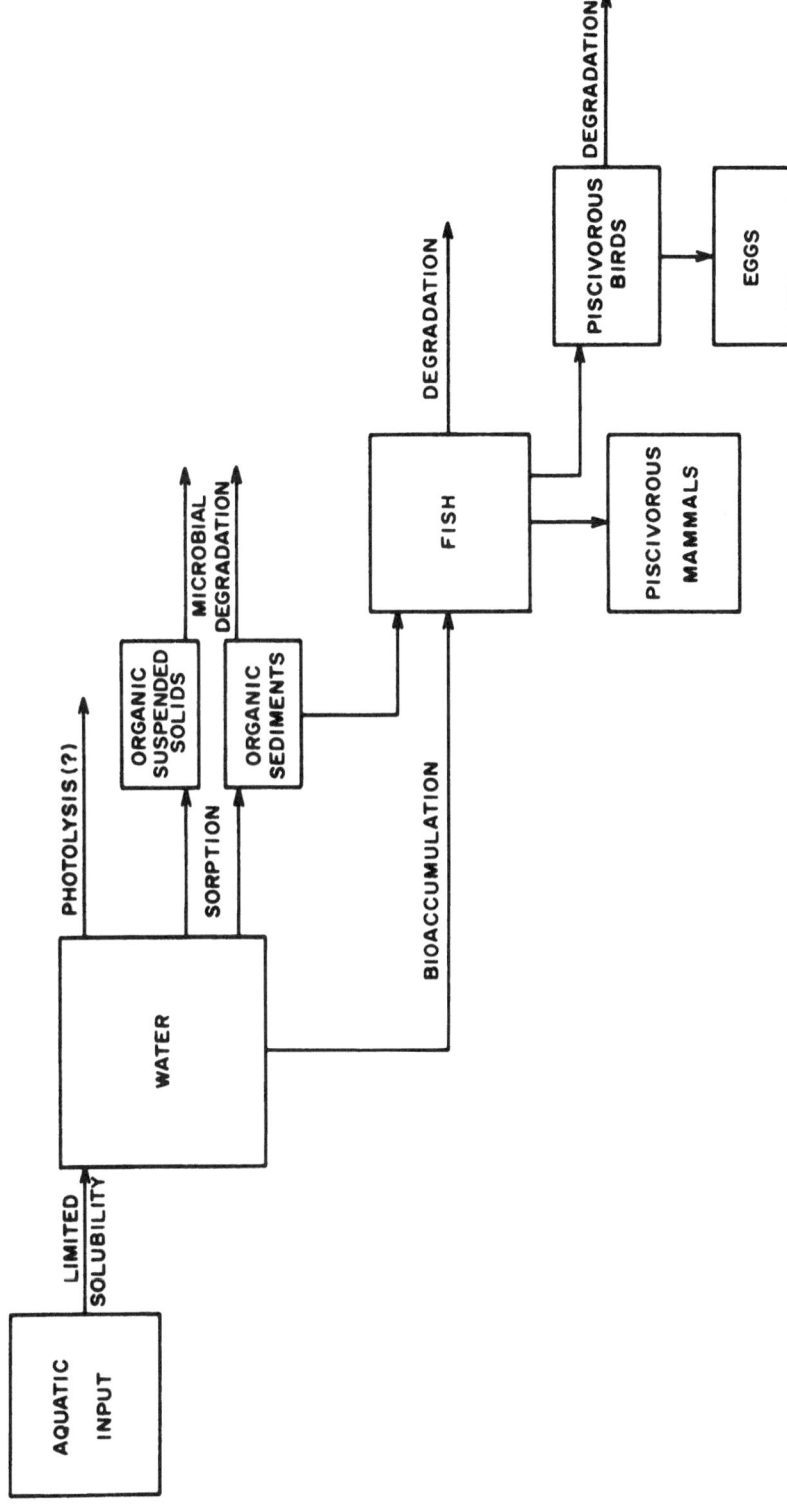

Fig. 1. Cascade of potential removal and degradation processes preferential to bioaccumulation of a pollutant.

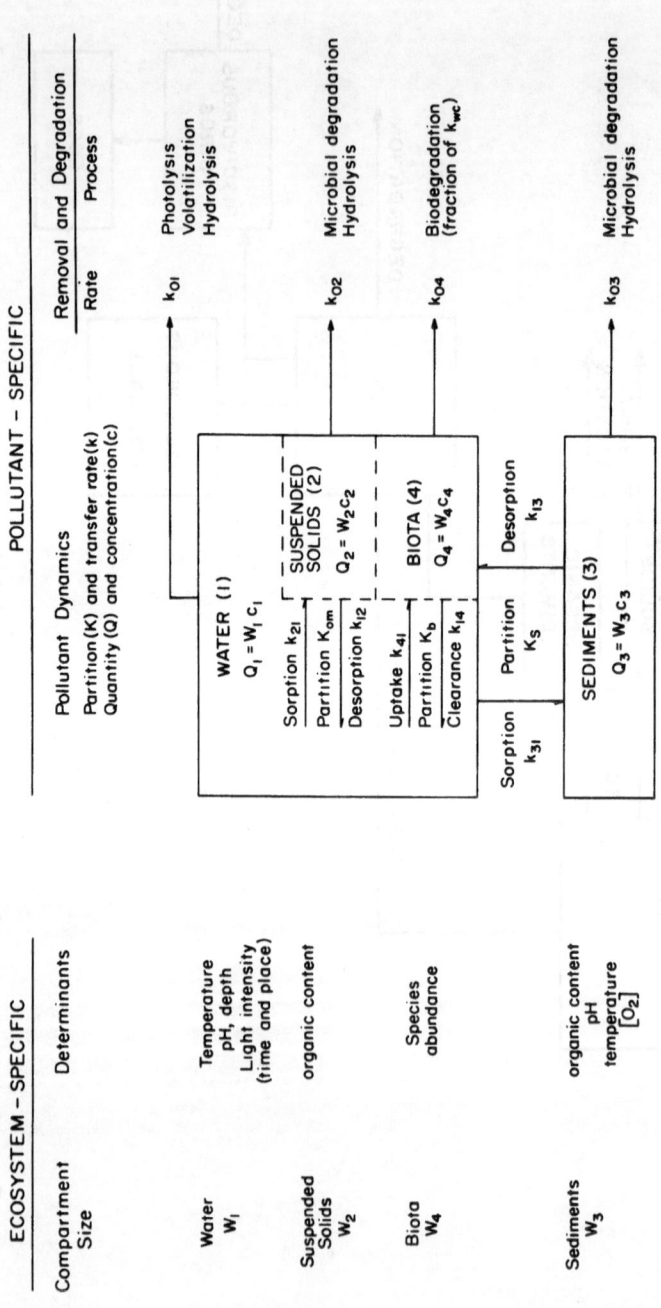

Fig. 2 Schematic of model determinants for pollutant dynamics.

Table 1. Minimum Input Requirements for Selected Processes that
          Remove Pollutants from the Environment

Process	Predictor	Prediction
Photolysis	- Extinction coefficient for 290 - 400 nm	- 1st order rate constant
	- Quantum yield	
Hydrolysis	- 1st order rate constant	- 1st order rate constant
Volatilization	- Molecular weight	- 1st order rate constant
	- Octanol-water partition Coefficient or solubility	
	- Vapor pressure	
Biotic processes	- Octanol-water partition Coefficient or solubility	- Bioconcentration factor - 1st order clearance
	- Fraction of pollutant Biodegradable	- Transfer rate constants
Sediment processes	- Octanol-water partition Coefficient or solubility	- Sediment-water partition Coefficient
		- Removal rate constants
Other	- "Catchall" definition	- "Catchall" partition Coefficients
		- Transfer rate constant
		- Removal rate constant
	- Melting point	

Table 2. Physical and Chemical Parameters for 2,3,7,8-TCDD

Parameter		
Molecular weight		321.9
Melting point (°C)		360
Vapor pressure (torr)		1.7×10^{-6}
Quantum yield		0.0 - 0.1
Molar extinction		
coefficient at	290	3476
	300	5330
	310	5562
	320	2085
	330	230
	340	-
	350	-
	360	-
	370	-
	380	-
	390	-
	400	-
Octanol-water partition coefficient		1.4×10^{6}
Sorption rate constant (g water g sediment $^{-1}d^{-1}$)		1×10^{1}

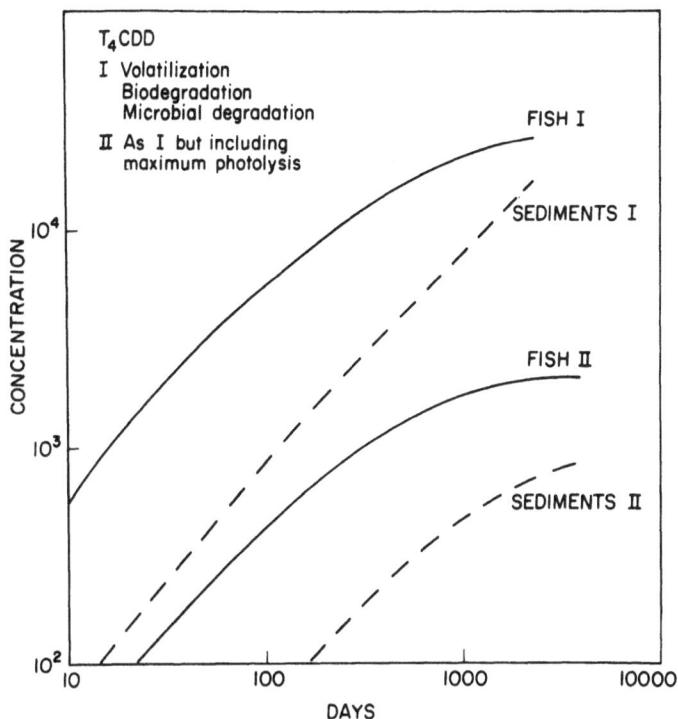

Fig. 3.   Hypothetical effectiveness of maximum photolysis on
          accumulation of 2,3,7,8-TCDD in sediments and fish
          under constant loading as compared to the effect of
          volatilization (see Table 3 for conditions).

their own predictive relation (Yalkowsky et al, 1979). Numerous
relations interrelate these parameters to volatilization,
sorption coefficients, and the bioaccumulation potential of
various organisms (Kenaga and Goring,. 1980; Roberts et al,
1981b).

    By way of example, using the equations of Veith et al,
(1979) the octanol-water coefficient of 2,3,7,8-TCDD can be used
to predict a bioaccumulation potential in fish of 33,000. The
clearance rate constant can be similarly predicted using the
equation of Roberts et al, (1981a), to give a value of 5.9 x
$10^{-3}g^{0.2}d^{-1}$. Together, these values can be used to
predict an uptake rate constant of 1.9 x $10^{2}g^{0.2}d^{-1}$. The
bioaccumulation potential of fish is considered to be somewhat
lower with a maximum observed value of 15,000 (Isensee and Jones

1975), but an experimental uptake rate constant (Neely, 1979) gives almost exactly the same value as predicted: 2.2 x $10^2 g^{0.2} d^{-1}$. The experimental clearance rate constant is four times faster than predicted, about 25 days (Neely, 1979). If the experimental uptake and clearance rate constants are used to determine the bioaccumulation potential, a value of 10,000 results. This indicates that the predictive equation using the octanol-water partition coefficient results in rather high values.

In fact, none of the predictive relations (Table 3) are validated for "super lipophilic compounds" such as the highly chlorinated dioxins. Specifically, valid data on compounds with octanol-water partition coefficients in excess of about 10^6 are generally not sufficiently developed to establish these correlations for the "super lipophilic compounds".

When degradation and dissipation processes (dilution, etc.) are slower than the rates of the major transfer process occurring between compartment, i.e. when a dynamic equilibrium exists, these relations could, in theory, form a basis for the preliminary prediction of accumulation patterns of chlorinated dioxins in the sediment, water, and biotic compartments. Given the gross uncertainties inherent in the determination of the water solubility of hydrophobic chemicals such as chlorinated dioxins (Schoor, 1975), the octanol-water partition coefficient is potentially the better indicator of relative accumulation patterns. We have used it throughout these analyses to determine what information can be gleaned from today's data base.

The work of Hansch and Leo (1979) suggests that the partition coefficients might increase about five-fold with each additional chlorine atom. This ignores isomeric differences which the work of Bruggeman et al, (1981) suggests needs inclusion, at least, in the case of bioaccumulation. Specific structural relationships are unavailable and it is impossible to refine the analysis to isomeric levels at this time. On this basis (Table 3), the relative partitioning patterns of the various homologues in various compartments at equilibrium can be analyzed (Table 4).

The octanol-water partition coefficient can also be used in conjunction with data on vapor pressures (Firestone, 1977) to develop a first estimate of volatilization rates as a function of depth and degree of turbulence (Roberts et al, 1981b). The vapor pressures of dioxins have only been measured indirectly, without validation, on the basis of gas chromatographic retention times and some caution is necessary in interpreting the results.

Table 3. Predictors for the Accumulation of a Pollutant

Relation	Range (Indicator)	Correlation coefficient	Reference
Log BCF fish = 0.85 Log Kow - 0.7	10^0–10^7	0.95	Veith et al, 1979
Log Koc = 1.00 Log Kow - 0.21	10^2–10^6	1.00	Karickhoff et al, 1979
Log S = 0.0095 (MP) - 0.9874 Log Kow + 0.7178	10^2–10^6	1.00	Yalkowsky et al, 1979
Log BCF algae = 4.94 - 0.33 Log S	10^0–10^6	0.68	Korte et al, 1978
Log BCF daphnia = 0.987 + 0.679 Log BCF fish	10^2–10^5	0.83	Kenaga and Goring, 1980

Table 4. Relative Partitioning Potential of Chlorinated Dioxins

Homologue	Fish	Organic Sediments	Water
O_8CDD	300	800	0.001
H_6CDD	20	30	0.04
T_4CDD	1	1	1
D_2CDD	0.06	0.04	30

In an examination of today's data base, we have used relations descriptive of aquatic systems with air-water, water-air exchange coefficients 10 to 30 times those expected for quiescent systems but considerably lower than those expected for very turbulent systems (Southworth, 1979), i.e., we have used the mean values suggested by Liss and Slater (1974) from field observations. While the use of any value is arbitrary, these values do provide a consistent base with which to examine relative distribution patterns (Roberts et al, 1981b). As noted, the role of degradative processes has only been studied in a qualitative manner and their influence on the dynamics cannot be assessed quantitatively with any certainty within the context of such screens of persistence, albeit some information on the maximum potential influence of photolysis can be developed (see above).

Clearly, the absence of validated information on the chemical properties of dioxins severely limits the use of any modelling approach to the delineation of distribution patterns. As a first step, the octanol-water correlating relations (Table 3) would need to be established for compounds with properties similar to those of the higher homologues. The octanol-water partition coefficient of the homologues should be determined empirically and the information on the chemical's vapor pressures validated because volatilization is demonstrated to be a key removal process operative within aquatic systems.

PREDICTION OF ACCUMULATION PATTERNS OF 2,3,7,8-TCDD

Even in the overwhelming absence of any validated input parameters, some useful information can still be obtained from models about accumulation patterns and the time scale at which equilibration may be established.

This is particularly true for 2,3,7,8-TCDD (Figures 3, 4 and Table 5) and the model suggests that, in the absence of other exchange processes (e.g., sedimentation), sediments equilibrate very slowly when a continuous input of 2,3,7,8-TCDD occurs in the water phase. Fish reach levels approaching equilibrium much more quickly and are more likely to reflect the 2,3,7,8-TCDD concentrations in the water than those in the sediment shortly

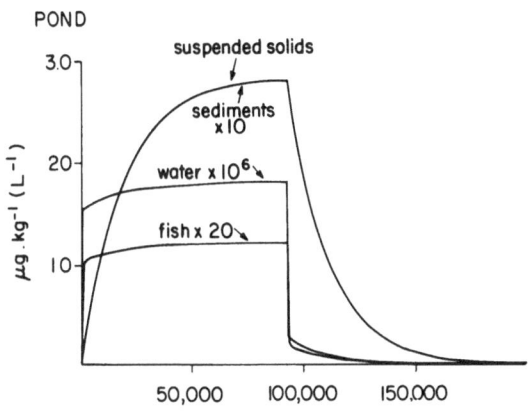

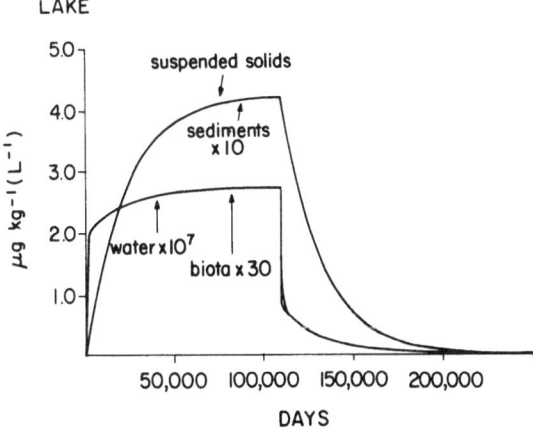

Fig. 4. Comparative accumulation to equilibrium and duration of 2,3,7,8-TCDD in the relevant compartments of a standard lake and pond as predicted by a computer model (following NRCC 1981a,c), (see Table 3 for conditions).

Table 5. Comparative Conditions at Equilibrium When There is a
          Constant Input of TCDD

	Pond [a,b]	Lake [a,c]
Input concentration $\mu g \cdot L^{-1} \cdot d^{-1}$	0.74	2×10^2
Input volume (L)	10	10
Load $(\mu g \cdot d^{-1})$	7.4	2×10^3
Loading concentration to water $(\mu g \cdot L^{-1} \cdot d^{-1})$	2×10^{-6}	2×10^{-7}
Approximate time to equilibrium of slowest compartment	(Sediment)	(Sediment)
days	90110	100400
years	247	275
Concentration in:		
Water $(ng \cdot L^{-1})$	1.8	0.21
Suspended solids $(\mu g \cdot kg^{-1})$	2.8	4.2
Sediments $(\mu g \cdot kg^{-1})$	0.28	0.42
Fish $(ng \cdot kg^{-1})$	61	91
Fraction (%) of total load in:		
Water	0.03	0.21
Suspended solids	2.47	1.69
Sediments	97.49	98.06
Fish	0.01	0.04
Fraction degraded or lost by:		
Volatilization	100	100
Photolysis	0	0
Biogradation fish	0	0
Approximate residence tiem (days)	2868	6287
Half-life of system clearance (days)	1987	4357

[a]Latitude, 45 N; Temperature, 20 C.
[b]Mean depth, 1 m; volume, 3.7×10^6 L; suspended organics, 50 mg
L^{-1}; light attenuation, high; sediment, depth 1 cm, 10%
organics; fish 18.5 kg.
[c]Mean depth, 15 m; volume 10×10^{10} L; suspended organics,
5.0 mg L^{-1}; light attenuation, low; sediment, depth 1 cm, 10%
organics; 5×10^4 kg.

after loading has begun. In fact there is a cross-over point,
only after which, sediments can serve as reliable indicators of
gross 2,3,7,8-TCDD contamination. Before this point, the levels
in the sediments are actually lower than those in fish and in
this instance sediment samples would clearly be of limited use in
defining overall contamination patterns.

On the other hand, if the loading is to the sediments, many
fish are not likely to be the best early indicators because of
the potentially slow release of dioxins from the sediments. In
this case, organisms which rapidly equilibrate with the benthos
could be better indicators of 2,3,7,8-TCDD contamination in the
short term. A similar conclusion is implied by the microcosm
study of Isensee and Jones, 1975. Obviously, the choice of an
indicator matrix on the basis of gross accumulation potential,
estimated solely on the basis of correlating relations and steady
state assumptions, could be misleading in the case of 2,3,7,8-TCDD.
They may not reflect the level of loading. The period elapsed
after the beginning of the loading and the mixing rates within
the system become key considerations in identifying the most
relevant indicators for 2,3,7,8-TCDD and their sampling
locations.

ACCUMULATION PATTERNS OF OTHER DIOXINS

The model also allows some initial qualitative comparisons
to be developed about the possible relative distribution patterns
of the other dioxins. As can be seen from Figure 5, the time
required for equilibration to occur and the point of cross-over
will be a property of the PCDD in question. Explicitly, besides
the complications that arise due to the range of partitioning
properties possessed by the various chlorinated dioxins, it has
been suggested that photolysis rates are highly dependent on the
dioxin's structure (Dobbs and Grant, 1979). Additionally,
dioxins with fewer chlorines may be more susceptible to
biodegradation (NRCC, 1981a). Thus, at least in shallow systems,
the cross-over points and the time to equilibrium could be quite
variable due to differences in the rates of degradation of the
different dioxins.

Until the structural relations governing their
susceptibility to photolysis and microbial degradation are
clearly delineated, the interpretation of the patterns of
residues obtained in the specific matrix, in terms of general
patterns within the various other compartments of the system,
cannot be made with certainty. It is clear that when a mixture
of PCDD homologues is released into a ecosystem, there should be

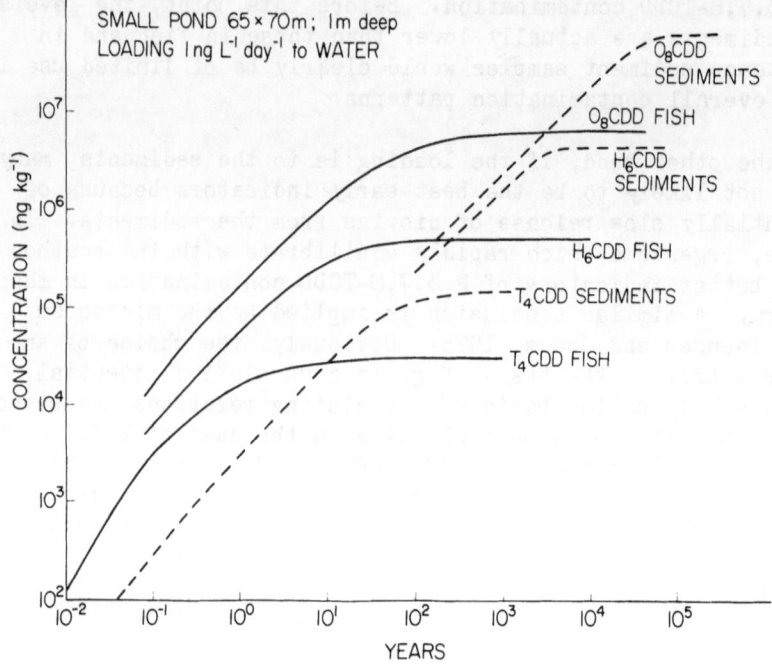

Fig. 5.   Accumulation and distribution patterns of TCDD, HCDD
          and OCDD to equilibrium in sediments and fish under
          constant loading where volatilization is the only
          removal process

extensive shifts in the homologue patterns on the basis of
variations in sorptive properties and the ease with which they
are removed. Until the equilibrium is achieved, these shifts
will also vary as a function of time. Obviously, particular care
must be exercised in interpreting data from any monitoring study.
Unless the data is normalized for differences in the rate at
which the specific chlorinated dioxins accumulate, little useful
information is obtained about either the patterns of ambient
levels of chlorinated dioxins within the various other
compartments or the loading rates. In fact, given the
multiplicity of complications, results showing the presence of a
particular chlorinated dioxin in a matrix and indicating the
absence of other chlorinated dioxins do not necessarily suggest a
higher loading rate for the identified PCDD. For example,
situations where sorptive processes dominate and the loading

rates are similar, higher homologues will be much more likely to be detected than the less hydrophobic chemicals.

In the case of fish and other organisms, another complicating factor may be the saturation of the uptake mechanisms with the higher chlorinated dioxins. For example, the uptake of some halogenated hydrocarbons by fish is reported to be inefficient (e.g., Zitko and Hutzinger, 1976), as is the uptake of OCDD by rats (see Esposito et al, 1980). Although studies have not been carried out with fish and "super lipophilic chemicals" like OCDD, such complications are certainly a possibility and the precise nature of shifts in congener patterns that may occur in fish cannot be predicted.

Fish and other organisms will be useful indicators of the general contamination in the system only if saturation does not occur and then only when the results are adjusted for the various accumulation potentials of the various PCDDs and the nature of the exposure.

As noted, an added consideration to the interpretation of residue patterns in biotic matrices is the possibility that specific isomers or homologues will degrade more readily than others with grossly similar structures. Relevant studies do not exist in the case of PCDDs. It is known that the susceptibility of chlorinated hydrocarbons in mammals and birds is related to subtle changes in structure (e.g., PCBs) (Ballschmiter et al, 1978; Nisbet, 1976; and Roberts et al, 1978). Patterns reflect not only the chlorine content of the molecule, but also the position of substitution. If analogies can be drawn from work with the chlorinated furans (e.g., Nagayama et al, 1980), greater than 1,000-fold shifts might exist in patterns between compartments.

There is no universal matrix useful for all monitoring needs. Ultimately, the data's end-use governs the choice of matrix. The requirements of a screening program used to identify sources of pollution are drastically different from those encountered in a program designed to delineate man's dietary exposure to chlorinated dioxins. Uthe et al, (1981) have presented a detailed overview of the considerations determining the choice of an indicator species in aquatic systems. The need to detect the presence of specific toxic congeners at levels approaching or exceeding our analytical capabilities (see NRCC, 1981a) is an additional constraint. If monitoring programs are to provide the maximum degree of safety, the chosen matrices must include good accumulators. The choice must be made with an understanding of the dynamics that will occur within the system.

The use of poor accumulators provides only cursory information on the level of contamination in a system. In fact, given today's analytical capabilities (see NRCC, 1981a), the detection of PCDDs in a water sample means that other components are severely contaminated and that a hazard potentially exists. For example, if the water contained 2,3,7,8-TCDD at 100 $ng \cdot L^{-1}$, the fish would be expected to contain in excess of 1 $mg \cdot kg^{-1}$. On the other hand, if fish contained 5 $ng \cdot kg^{-1}$ of 2,3,7,8-TCDD (approximately the detection limit), the water could contain less than 0.0005 ng of 2,3,7,8-TCDD$\cdot L^{-1}$, considerably below our analytical capabilities. Obviously, fish are a better indicator of the problem than water.

Feeding patterns are additional considerations to be reflected in the choice of matrix and the interpretation of the results. For example, experience with mirex contamination in the Great Lakes (Norstrom et al, 1978) suggested that birds (e.g. herring gulls, Larus argentatus) and fish (e.g., coho salmon, Oncorhynchus kisutch; alewives, Alosa pseudoharengus; and rainbow smelt, Osmerus mordax) tend to be better indicators of system-wide contamination than of site-specific or local sources of contamination. In other words, movements of these organisms make them integrators reflecting system-wide trends in contamination. For example, while the diet of herring gulls reportedly contains a preponderance of fish, depending on location, conditions and season, it can also include significant quantities of terrestrial items such as small mammals, birds, earthworms and even "garbage" (Allen, 1977, 1978).

Thus, unless thorough documentation of the specific seasonal feeding habits on the colonies are available, and levels of contamination of the food are also documented, it is conceivable that the bird residues could reflect terrestrial sources of contamination. This problem is common to the interpretation of data using any animal with opportunistic feeding habits. There is no simple solution, except the careful documentation of feeding and movement patterns and the use of other indicators as supportive evidence.

Results of such programs are most useful in identifying system-to-system patterns of contamination. In the studies of mirex contamination of Lake Ontario, the organic component of the sediment was considered to be the best indicator of local site-specific contamination patterns. Parenthetically, sediment translocation is a continual process and the interpretation of results from such monitoring must eventually reflect the hydrodynamics within the system (e.g., Horn et al. 1979).

Today, there is insufficient information on the structural dependence of the attenuating factors to permit confident extrapolations to be made about the relative contribution of the various congeners to the overall PCDD loading from the monitoring of organisms associated with aquatic systems. The data can demonstrate general trends in the levels of specific isomers, but, without more information, cannot be used with confidence in an examination of the relative exposure encountered by the indicator organism and hence the relative loading.

FOOD CHAINS IN THE SELECTION OF INDICATORS

Indictors have at times been selected solely upon the basis of food chains in the mistaken belief that biomagnification is the dominant process. This is now argued to be a potentially misleading simplification by a number of authors who have examined the fundamental processes which govern accumulation (Roberts et al, 1981a; Bruggeman et al, 1981; Macek et al, 1981; Roberts et al, 1979; and NRCC 1981a). They conclude that in the case of fish, the food vector is unlikely to be more important than the water vector unless the octanol-water partition coefficient is in excess of 10^5 x $10^{\pm 1}$. In fact, the dietary biomagnification factor for fish is not predicted to exceed one unless the octanol-water partition coefficient exceeds about 10^5 (Figure 6). This situation arises because of the relatively low energy content of a gram of water as compared to that required to catabolize a gram of food (about 1 to 60,000). In other words, the need for the energy obtained by the animal from the two vectors to be balanced required that a fish must clear about 60,000 times as much water for oxygen as it consumes food to maintain an energy balance.

The regression relation obtained in Figure 6 for the dietary vector arises directly from reducing the intercept of the relation for the water vector by the log of the relative volume of the two vectors - log 60,000 (4.8) - i.e.,

$$\log BCF_{fish/food} = 0.85 \log P - 0.7 - 4.8$$

Experimentally, in young trout (Salmo gairdneri), the final body burden of 1.38 µg 2,3,7,8-TCDD.g^{-1} food concentration of 2.3 µg 2,3,7,8-TCDD.g^{-1} food (Hawkes and Norris, 1977), shows bioaccumulation from the dietary vector to be low, only about 0.6 which is in agreement with the predictions of this relation.

As a consequence, the levels of pollution in the food of a fish must exceed the level in the water by about 10^5 before

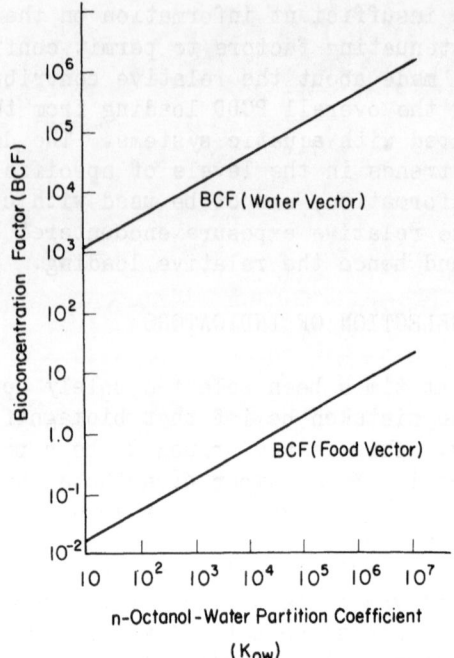

Fig. 6.   Steady-state bioconcentration factors (BCF) associated
          with the water vector (2.8×10^3 day^{-1}) and the
          the food vector (4.5×10^{-2} day^{-1}) as functions
          of the octanol-water partition coefficient, K_{ow}.

the two vectors will even compete. Since such large
bioconcentration factors are generally only reported for the
"super lipophilic compounds", the water vector may often be more
important than food chains in defining residue patterns in fish
and hence useful indicators.

CONCLUSION

     The theoretical considerations discussed in this paper, and
the analogies that can be drawn between dioxins and other
chlorinated hydrocarbons, argue for the occurrence of large
structurally related shifts in the residue patterns between the
various compartments of aquatic systems. Unfortunately, the
validated data required even for preliminary assessments of
shifts in dioxin patterns are absent. As a consequence, whole
ecosystem contamination patterns cannot be determined with
certainty from the monitoring of residues in a single matrix or
closely related matrices, e.g., fish and piscivorous birds.
Additionally, the results of such monitoring cannot be used to
assess, with traditional scientific rigour, the relative loading
rates of the various dioxins to a system.

REFERENCES

Allen, L.J. The comparative feeding ecology relative to population dynamics of the herring and ring-billed gulls (Larus argentatus and L. delawarensis) and some comparisons with the Caspian Tern. MSc Thesis: Queen's Univ. (Kingston, Canada), 1977.

Allen, L.J. Food of the ring-billed and herring gulls nesting on Chantry Island, Lake Huron, 1978. MS to contract #KL229-8-7448, Canadian Wildlife Service, Environment Canada, 1978.

Ballschmiter, K., Zell, K.M., and Neu, H.J. Presistence of PCBs in the ecosphere: Will some PCB components "never" degrade? Chemosphere, 7(2): 173-176, 1978.

Bruggeman, W.A., Martron, L.B.J.M., Kooiman, D., and Hutzinger, O. Accumulation and elimination kinetics of di-, tri- and tetrachlorobiphenyls by goldfish after dietary and aqueous exposure. Chemosphere, 10(8):811-832, 1981.

Dobbs, A.J and Grant, C. Photolysis of highly chlorinated dibenzo-p-dioxins by sunlight. Nature, 278:163-165, 1979.

Esposito, M.P., Tiernan, T.D., and Dryden, F.E. Dioxins. U.S. Environmental Protection Agency 600/2-80-197, 1980. p. 351.

Firestone, D. Chemistry and analysis of pentachlorophenol and its contaminants. F.D.A. By-lines, 2:57-89, 1977.

Hansch, C., and Leo, A. Substitution constants for correlation analysis in chemistry and biology. Toronto: J. Wiley and Sons, 1979. pp. 339.

Hawkes, C.L., and Norris, L.A. Chronic oral toxicity of 2,3,7,8-tetrachlorodibenzo-p-dioxin (TCDD) to rainbow trout. Trans. Am. Fish Soc., 106:641-645, 1977.

Horn, E.G., Hetling, L.J. and Tofflemire, T.J. The problem of PCBs in the Hudson River system, Ann. NY Acad. Sci., 320:591-609, 1979.

Isensee, A.R., and Jones, G.E. Absorption and translocation of root and foliage applied 2,4-dichlorophenol, 2,7-dichloro-dibenzo-para-dioxin and 2,3,7,8-tetrachlorodibenzo-para-dioxin. J. Agric. Food Chem., 19:1210-1214, 1971.

Isensee, A.R., and Jones, G.E. Distribution of 2,3,7,8-tetra-chlorodibenzo-para-dioxin (TCDD) in an aquatic model ecosystem. Environ. Sci. Technol., 9:668-672, 1975.

Karickhoff, S.W., Brown, D.S. and Scott, T.A. Sorption of hydro-phobic pollutants on natural sediments. Water Res., 13:241-248. 1979.

Kenaga, E.E., and Goring, C.A.I. Relationship between water solubility, soil sorption, octanol-water partitioning and bioconcentration of chemicals in biota. Aquatic Toxicology ASTM STP, 707:78-115, 1980.

Korte, F., Freitag, D., Geyer, H., Klein, W., Kraus, A.G. and Lahaniatis, E. Ecotoxicologic profile analysis. Chemosphere, 1:79-102, 1978.

Leighton, P.A. Photochemistry of Air Pollution. New York:
    Academic Press, 1961. pp. 300.
Liss, P.S., and Slater, P.G. Flux of gases across the air-sea
    interface. Nature, 247:181, 1974.
Macek, K.J., Petrocelli, S.R., and Sleigth, B.H.III. Considerations
    in assessing the potential for and significance of
    biomagnification of chemical residues in aquatic food
    chains. ASTM STP, 667:251-268, 1979.
Mackay, D. Finding fugacity feasible. Environ. Sci. Technol.,
    13:1218-1223, 1979.
Matsumura, F., and Benezet, H.J. Studies on the bioaccumulation
    and microbial degradation of 2,3,7,8-tetrachlorodibenzo-para-
    dioxin. Environ. Health Persp., 5:253-258, 1973.
Nagayama, J., Todudome, S., and Kuratsune, M. Transfer of polychlor-
    inated dibenzofurans to the fetuses and offspring of mice.
    Food Cosmet. Toxicol., 18:153-157, 1980.
Neely, W.B. Estimating rate constants for the uptake and clearance
    of chemicals by fish. Environ. Sci. Technol., 13:1506-1510,
    1979.
Nisbet, I.C.T. Criteria document for PCBs. United States
    Environmental Protection Agency 440/9-76-021, 1976. pp. 583.
Norstrom, R.J., Hallett, D.J., and Sonstegard, R.A. Coho salmon,
    (Oncorhynchus kisutch), and herring gulls, (Larus argentatus)
    as indicators of organochlorine contamination in Lake Ontario.
    J. Fish Res. Board. Can., 35:1401-1409, 1978.
NRCC. Polychlorinated dibenzo-para-dioxins: Criteria for their
    effects on man and his environment. Ottawa: Associate
    Committee on Scientific Criteria for Environmental Quality,
    National Research Council of Canada, #18574, 1981a. pp. 248.
NRCC. Polychlorinated dibenzo-para-dioxins: Limitations to the
    current analytical techniques. Ottawa: Associate Committee
    on Scientific Criteria for Environmental Quality, National
    Research Council of Canada, #18576, 1981b. pp. 248.
NRCC. A screen for the relative persistence of organic chemicals
    in aquatic ecosystems: An analysis of the role of simple
    computer models. Ottawa: Associate Committee on
    Scientific Criteria for Environmental Quality, National
    Research Council of Canada, #18570, 1981c. pp. 302.
Roberts, J.R., and Marshall, W.K. Retentive capacity: An index
    of chemical persistence expressed in terms of chemical-
    specific and ecosystem-specific parameters. Ecotoxicol.
    Environ. Safety, 4:158-171, 1980.
Roberts, J.R., Rodgers, D.W., Bailey, J.R. and Rorke, M.A. Poly-
    chlorinated biphenyls: Biological criteria for an assessment
    of their effects on environmental quality. Ottawa: Associate
    Committee on Scientific Criteria for Environmental Quality,
    National Research Council of Canada, #16077, 1978. pp. 172.

Roberts, J.R., deFreitas, A.S.W. and Gidney, M.A.J. Control factors on uptake and clearance of xenobiotic chemicals by fish. In: Animals as Monitors of Environmental Pollutants. Washington: National Academy of Sciences, 1974. pp. 3-14.

Roberts, J.R., McGarrity, J.T., and Marshall, W.K. An introduction to process analyses and their use in preliminary screens of chemical persistence. In: NRCC 1981c.

Roberts, J.R., Mitchell, M.S., Boddington, M.J., Ridgeway, J.M., and Miller, D.R. A simple computer model as a screen for persistence. In: NRCC, 1981c.

Schoor, W.P. Problems associated with low-solubility compounds in aquatic toxicity tests: Theoretical model and solubility characteristics of Aroclor 1254 in water. Water Res. 9:937-944, 1975.

Southworth, G.R. The role of volatilization in removing polycyclic aromatic hydrocarbons from aquatic environments. Bull. Environ. Contam. Toxicol., 21:507-514, 1979.

Uthe, J.F., Chou, C.L., and Scott, D.O. The state of art of environmental pollution level monitoring using resident biota populations. MS. Paper presented at Statistical Aspects of the Use of Biological Indicators in Pollution Monitoring. ICES, Nantes, France, 1981.

Veith, G.D., De Foe, D.L., and Bergstedt, B.V. Measuring and estimating the bioconcentration factor of chemicals in fish. J. Fish Res. Board Can., 36:1040-1048, 1979.

Ward, C.T., and Matsumura, F. Fate of 2,3,7,8-tetrachlorodibenzo-para-dioxin (TCDD) in a model aquatic environment. Arch. Environ. Contam. Toxicol., 7:349-357, 1978.

Yalkowsky, S.H., Orr, R.J., and Valvani, S.C. Solubility and partitioning. 3. The solubility of halobenzenes in water. Am. Chem. Soc., 18:351-353, 1979.

Zepp, R.G., and Cline, D.M. Rates of direct photolysis in the aquatic environment. Environ. Sci. Technol., 11:359-366, 1977.

Zitko, V., and Hutzinger, O. Uptake of chloro- and bromobiphenyls, hexachloro- and hexabromobenzene by fish. Bull. Environ. Contam. Toxicol., 16:665-673, 1976.

A FIELD STUDY OF SOIL AND BIOLOGICAL SPECIMENS FROM A
HERBICIDE STORAGE AND AERIAL-TEST STAGING SITE FOLLOWING
LONG-TERM CONTAMINATION WITH TCDD

D.D. Harrison and R.C. Crews

United States Air Force Armament Laboratory
Eglin Air Force Base, Florida, USA

ABSTRACT

A study was made of the residual levels of 2,3,7,8-tetrachloro-
dibenzo-p-dioxin (TCDD) in soil and biological samples collected
from an area on Eglin AFB, Florida, that was used for storing and
loading herbicides from 1962-1970. The study deals only with the
immediate vicinity of the storage and loading area and its
associated drainage system. Soil and sediment samples were
collected and analyzed to determine residual TCDD levels.
Various tissues from specimens representing 12 species of the
native fauna were also analyzed for TCDD content.

Soil concentrations above 1 parts-per-billion (ppb) TCDD were
largely confined to a small area immediately surrounding the
concrete loading pad. Nevertheless this area where storage and
loading operations had occurred, soil concentrations of TCDD as
high as 275 ppb were found with concentrations of one-third that
amount present to a depth of one meter at two of the nine
sampling sites.

TCDD was present in many of the organisms collected from the
upper portions of the drainage system of the contaminated area,
but no organisms collected from Tom's Bayou, 3000 meters
downstream, contained detectable levels of TCDD. Visual
observations revealed no gross abnormalities on any of the
specimens. Histopathological studies of tissues from five
contaminated specimens, representing five species, showed no
microscopic anomolies.

Isomeric analyses of soil, sediment, biological samples, and Herbicide Orange held in storage for at least 10 years verified that the predominant isomer was 2,3,7,8-TCDD.

INTRODUCTION

Eglin Air Force Base in Northwest Florida was used for extensive testing of military herbicide spray equipment from 1962 to 1970. One site on Eglin known as Hardstand 7, was used for storing drums of herbicide and transferring the contents to the aircraft spray tanks. Preliminary checks of pumps and spray nozzles occurred prior to flying each mission. As a result of the storage, handling, and transfer of more than 130,000 liters (35,000 gal.) of herbicides consisting of 50 percent 2,4,5-trichloro-phenoxyacetic acid (2,4,5-T), the area surrounding this hardstand was contaminated with TCDD.

Nearby vegetation showed herbicide damage during the 1962-1970 period, but little or no visible effects are present today. The decision to conduct this study was based on a scientific interest in contamination levels of TCDD rather than a response to documented observable effects from the contaminant. The purpose of the study was to determine the extent of the contamination, both in the soil surrounding the storage and loading site, and in the drainage system and its associated aquatic fauna. After this determination, the primary objective was to find out if the contamination had affected the organisms in the aquatic system.

Because of the wide variation of TCDD concentrations in the soil around the hardstand, the lack of information on the time and quantity of herbicide spillage, and the relatively short span of time in which the samples were collected, no attempts were made to calculate degradation rates. As more samples are analyzed in the future, using data from this study as a baseline, valuable conclusions on degradation rates maybe established.

This study began with soil and sediment samples collected in 1974. Since that time the program has continued and has revealed much about the movement of TCDD in an aquatic environment and body burdens of TCDD in certain organisms under field conditions.

DESCRIPTION OF AREA

Hardstand 7 is an asphalt and concrete aircraft parking area located west of the north-south runway on the main Eglin airdrome, approximately twenty meters above sea level (Figure 1). This hardstand was the most extensively used site for herbicide

Fig. 1. Aerial View of hardstand 7 and Hardstand Pond,
         Eglin Air Force Base, Florida

storage and loading operations during the 1962-1970 spray
equipment testing program at Eglin. Several hundred 55-gallon
drums of various types of herbicide were stored around Hardstand
7 for later transfer of their contents to tanks aboard spray
aircraft. Much of the area immediately surrounding this
hardstand was contaminated with herbicide due to accidental
spills during loading operations and transfer procedures, leaking
drums, and purging of spray systems before and after missions. A
pit was dug in 1969 to the south-west of the hardstand as a
temporary means of preventing the excess herbicides from entering
a samll the stream immediately behind the hardstand (Figure 2).
After several months of use, the pit was filled with soil.

    The soil surrounding the area is sandy with excellent
drainage properties. Directly behind the hardstand is a ravine
that drops off approximately fiteen meters to a small pond.
Because of the packing caused by vehicular traffic and the
water-repellent nature of the oil-based herbicide contamination,
runoff of excess water caused an erosion problem in some spots
which led to the frequent use of fill dirt. Eventually, a dike
covered with asphalt was constructed on the rim of the ravine for
soil stabilization. A storm drain was installed to help control
erosion. The Hardstand Pond directly behind Hardstand 7 drains
into a small stream which flows north until it enters Beaver
Pond, a man-made reservoir. It then flows into Tom's Bayou and
Choctawhatchee Bay (Figure 3).

METHODS

Sampling

    Samples for this study were collected from 1974-1980. Soil,
sediment, and biological samples were collected by Eglin
personnel. Florida Game and Freshwater Fish Commission personnel
assisted in the collection of some biological specimens.

    Soil samples were collected using a metal spatula or a 10 cm-
diameter core sampler. All soil samples were taken over a 10 cm
range in depth unless otherwise specified.

    Sediment samples from deep water were collected with a 15.6
kg Ponar grab sampler, whereas those from shallow water were
taken with a metal spatula.

    Several types of biological specimens were collected by
whatever means proved to be most suitable; e.g. a shotgun was
used to get an adult beaver. All specimens were fresh when
collected and immediately put in an ice chest and kept cool until

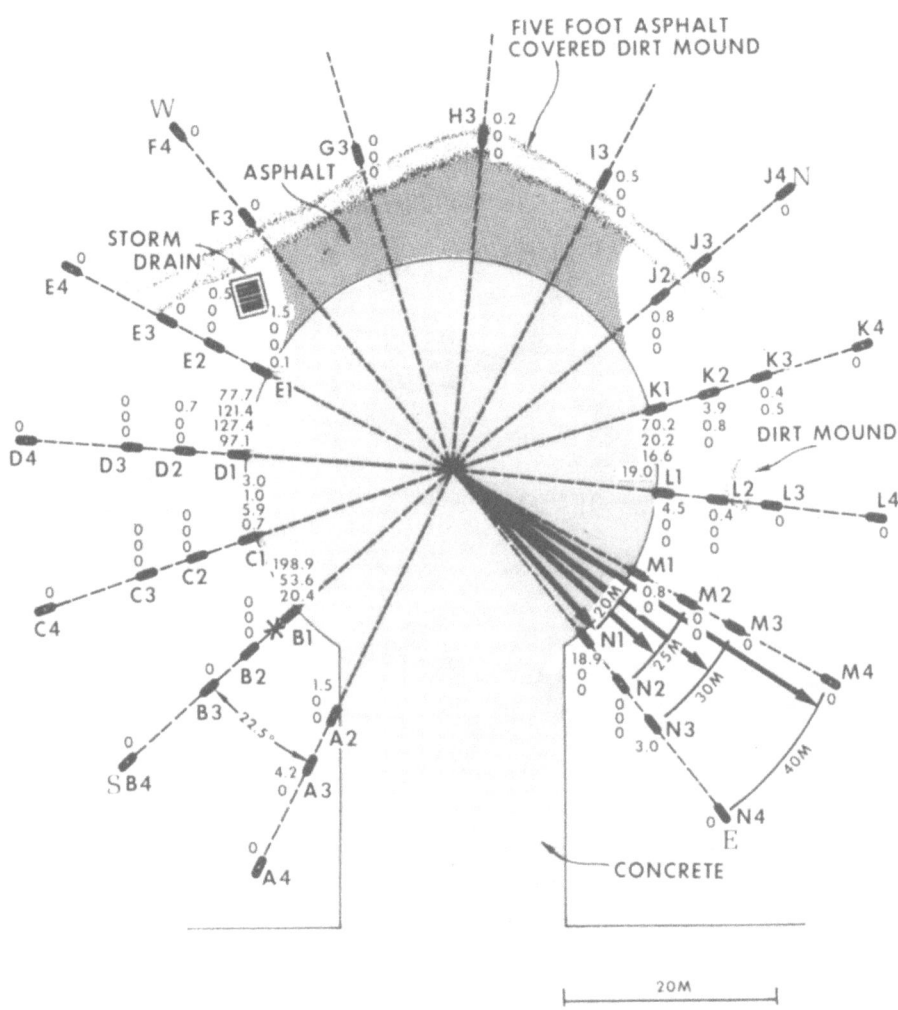

Figure 2.   1978 Sampling Sites and TCDD Concentrations at
             Hardstand 7, Eglin Air Force Base, Florida.

             *275 ppb (taken in 1977)
             Note: 1.   Disposal pit was located near D1.
                     2.   Soil concentrations of TCDD in ppb are
                          given at each site by depth from top to
                          bottom. Depths sampled were 0-10 cm,
                          20-30 cm, 55-70 cm, and 95-110 cm. All
                          depths were not sampled at each site,
                          however.

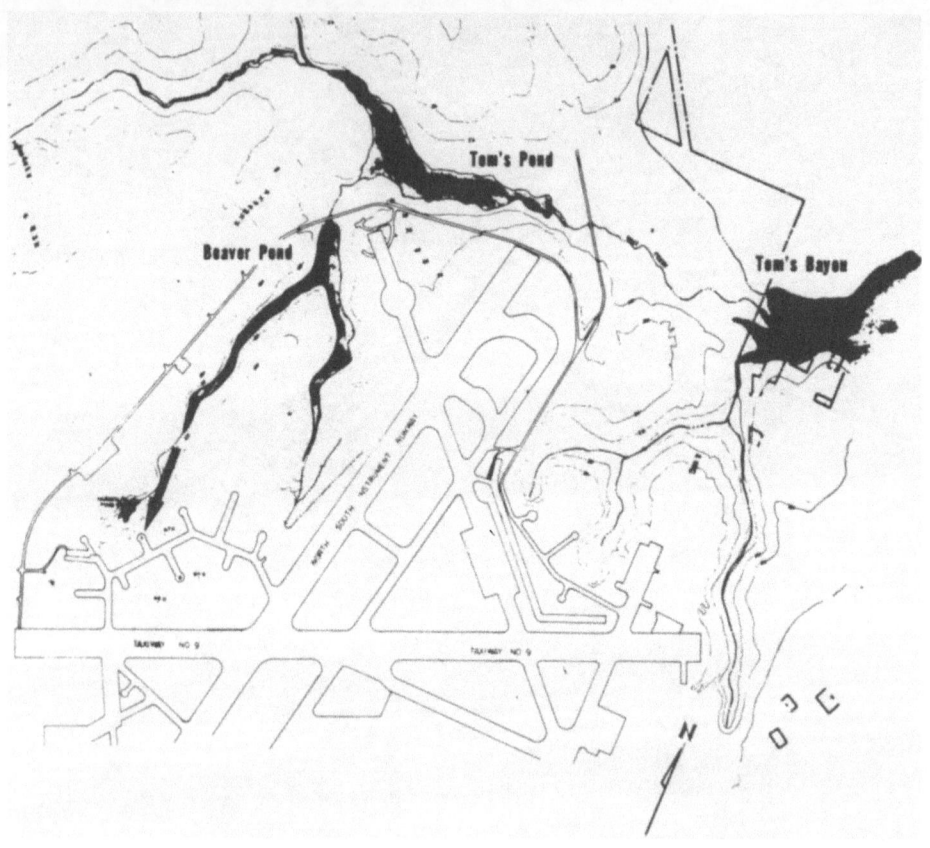

Fig. 3.   Map of hardstand 7 and its Associated Drainage System on
          Eglin Air Force Base, Florida.   The Arrow Locates
          Hardstand 7

dissection. Care was taken during dissection to avoid cross-contaminating the tissue with external portions of the body.

Samples for TCDD analysis were frozen and those for histopathological studies were fixed in 10 percent buffered formalin. The first sampling was done in 1974 to get an idea of the contamination level around Hardstand 7 and to see if fish from the drainage area were contaminated. The soil samples were taken from locations where herbicide spills had been greatest and sediment samples were collected to determine if TCDD had gotten into the aquatic system. Water samples were not taken because of the relative insolubility of TCDD in water. Specimens of two fish species were collected from the Hardstand Pond and immediately downstream. Soil samples were collected from additional sites in 1977 to see if TCDD were present at depths of 40-50 cm.

Based on data obtained from these samples it was decided to conduct a more thorough mapping effort. A grid system was established around Hardstand 7 (Figure 2) to facilitate the monitoring effort. Beginning at the center of the hardstand pad, 14 radians were established at 22.5 degree intervals, designated A through N, with A being the south southeast radian and proceeding clockwise. Sampling points on each radian were established at 20, 25, 30, and 40 meters from the center point. These sampling points were designated by the numbers 1, 2, 3, or 4, respectively, following the letter designating the radian. The third designation for each sample identification indicates the depth at which the sample was taken: top number = 0-10 cm, second number = 20-30 cm, third number = 55-70 cm, and fourth number = 95-110 cm. Soil samples were collected and identified using this grid system during January 1978. Some of the designated sampling sites were not sampled since some were covered with asphalt and others had been covered with fill dirt.

Biological specimens were collected from Tom's Bayou and Tom's Pond in 1977-1978 to determine the extent of biological contamination. As a result of data from the Tom's Pond specimens, other similar samples were collected in 1979 from Hardstand Pond and Beaver Pond where specimens were suspected of having higher contamination levels. Samples from these specimens were analyzed for TCDD concentrations as the previous samples had been, but tissue samples were also examined histopathologically.

In the final phase of this study the specific isomers of TCDD were determined. It had been assumed that all TCDD previously reported was of the 2,3,7,8-TCDD isomer. However, to confirm this assumption, samples consisting of Herbicide Orange, soil, sediment, and biological tissue were collected for specific isomer analyses. The samples were collected in March 1980, with

with the exception of Herbicide Orange sample which was a sample
remaining from the 1962–1970 spray missions and that had been
maintained in a chemical storage room.

Sample Analysis

Except for the samples collected in 1974, which had been
analyzed by Dow Chemical Company Midland, Michigan, all samples
were analyzed for TCDD by Brehm Laboratory, Wright State
University, Dayton, Ohio. Brehm Laboratory also performed the
specific isomer analyses with the following calibration standards
available: 1,3,6,8; 1,3,7,9; 1,4,6,9; 1,3,7,8; 1,2,6,8; 1,2,7,9;
1,2,6,9; 1,2,3,4; 2,3,7,8; 1,2,6,7; and 1, 2,8,9–TCDD. A Varian
3740 Gas Chromatograph–High Resolution Mass Spectrometer–Nicolet
1074 System was employed for specific isomer analyses. Histo-.
pathological analyses of biological tissues were performed by the
Veterinary Pathology Department, Armed Forces Institute of
Pathology, Washington, D.C.

RESULTS AND DISCUSSION

At the beginning of this study it was known that herbicides,
including Herbicide Orange, had been stored, loaded, and spilled
in the Hardstand 7 area and TCDD contamination was probable.
Results from preliminary soil analysis in 1974 verified that
Hardstand 7 was contaminated and indicated that the distribution
was spotty and concentrations were highly variable (Figure 4).
The highest concentration found in these first samples was 170
ppb in a low spot where the asphalt had deteriorated. As expected,
TCDD was not detected on the sandy slope from the hardstand to
the pond because this soil had been hauled in to repair erosion
damage after the stored herbicide had been removed. In 1977 soil
samples from near the surface around Hardstand and had
concentrations of TCDD as high as 275 ppb, while extensive
sampling in 1978 (Table 1) showed up to 199 ppb.

The highest concentration of TCDD found in the soil
surrounding Hardstand 7 was 275 ppb in the surface sample at site
B1. Several other sites near the concrete pad were also heavily
contaminated. This was expected since drainage of spills on the
concrete pad would flow to those areas and also because of
leakage from drums that had been stored directly in those
locations. The presence of TCDD at depths to 100 cm at site D1
was also expected because of a pit which had been dug at that
location to restrain herbicide runoff.

The high concentrations in the deeper samples at sites B1 and
K1 are not as easily explained. Probably several factors are

involved. The most obvious explanation is the probable
removal of, at least, the more highly chlorinated dioxins from
aquatic systems (Matsumura and Benezet, 1973; Ward and Matsumura
1978; Esposito et al, 1980; and NRCC, 1981a). Although the
absolute quantities in any specific matrix may not be severely
affected by the influence of biota, especially in light of the
importance of volatilization and perhaps photolysis (Roberts et
al, 1981a), the concentrations, and hence accumulation in
specific organisms or food chains, can still be significant and,
hence, potentially important as exposure vectors.

An extension of this analysis of degradation and transfer
processes, which can provide a quantitative definition of the
distribution patterns, is gained through the use of simple models
(e.g., Figure 2). These can be used to screen for chemical
persistence in aquatic ecosystem (Roberts and Marshall, 1980;
Mackay, 1979; and Roberts et al, 1981a). Alternatively, such
models can be used not only to examine the relative distribution
patterns between different compounds, such as the different
dioxin congeners or homologues, but also examine the usefulness
of a data base; its extent and validity.

This examination can be carried out at two levels. First,
one can examine specific rate constants for processes such as
photolysis. In this respect, it is pertinent to say that even
for the most well characterized dioxin, 2,3,7,8-TCDD, validated,
environmentally relevant, rate constants do not exist. For
example, photolysis is known to occur under certain laboratory
conditions, but the role of hydrogen donors in the environment is
unknown and makes any firm quantitative assessment impossible.
However, a number of predictive equations exist that allow the
estimation of rate constants for certain processes. Thus,
secondly, one can examine the data base for process predictors
and derive hypothetical rate constants that can later be
teted and validated. For example, in the case of photolysis, an
ultraviolet extinction curve and quantum yield are required for
predictive purposes (Leighton, 1961; Zepp and Cline, 1977; and
Roberts et al, 1981b). However, even in the absence of a quantum
yield or any knowledge of the role of hydrogen donors, one can
study the hypothetical maximum effectiveness of photolysis as a
removal process by arbitrarily setting the quantum yield at one
(Figure 3).

Photolysis represents a special case as it is one of the
most complex predictive relations used in such models (see
Roberts et al, 1981b). In general, very few parameters are
required for any single compound (Tables 1 and 2). Perhaps the
most important are the octanol-water partition coefficient and
water solubility, which are themselves interchangeable through

Table 1.   Results of TCDD Determinations in Soil Samples Collected January 1978 from Hardstand 7, Eglin Air Force Base, Florida

Sample	TCDD Concentration		Sample	TCDD Concentration	
	TCDD (ppb)	Detection Limit (ppb)		TCDD (ppb)	Detection Limit (ppb)
A21	1.5	-[b]	F33	ND[a]	1.0
A22	ND[a]	1.0	F41	ND[a]	1.0
A23	ND[a]	1.0	G31	ND[a]	1.0
A31	4.2	-[b]	G32	ND[a]	1.0
A32	ND[a]	1.0	G33	ND[a]	1.0
A41	ND[a]	1.0	H31	0.2	-[b]
B11	198.9*	-[b]	H32	ND[a]	1.0
B12	53.6	-[b]	H33	ND[a]	1.0
B13	20.4	-[b]	I31	0.5	-[b]
B21	ND[a]	1.0	I32	ND[a]	1.0
·B22	ND[a]	1.0	I33	ND[a]	1.0
B23	ND[a]	1.0	J21	0.8	-[b]
B31	ND[a]	1.0	J22	ND[a]	1.0
B41	ND[a]	1.0	J23	ND[a]	1.0
C11	3.0	-[b]	J31	0.5	-[b]
C12	1.0*	-[b]	J32	ND[a]	1.0
C13	5.9*	-[b]	J41	ND[a]	1.0
C14	0.7*	-[b]	K11	70.2*	-[b]
C21	ND[a]	1.0	K12	20.2*	-[b]
C22	ND[a]	1.0	K13	16.6*	-[b]
C23	ND[a]	1.0	K14	19.0	-[b]
C31	ND[a]	1.0	K21	3.9	-[b]
C32	ND[a]	1.0	K22	0.8	-[b]
C33	ND[a]	1.0	K23	0.030**	-[b]
C41	ND[a]	1.0	K31	0.4	-[b]
D11	77.7*	-[b]	K32	0.5*	-[b]
D12	121.4	-[b]	K41	ND[a]	1.0
D13	127.4	-[b]	L11	4.5	-[b]
D14	97.1*	-[b]	L12	0.133**	-[b]
D21	0.7	-[b]	L13	ND[a]	1.0
D22	0.079**	-[b]	L21	0.4	-[b]
D23	ND[a]	1.0	L22	ND[a]	1.0
D31	ND[a]	1.0	L23	ND[a]	1.0
D32	ND[a]	1.0	L31	ND[a]	1.0
D33	ND[a]	1.0	L41	ND[a]	1.0
D41	ND[a]	1.0	M11	0.8	-[b]
E11	1.5*	-[b]	M12	ND[a]	1.0
E12	ND[a]	1.0	M21	ND[a]	1.0
E13	ND[a]	1.0	M22	ND[a]	1.0
E14	0.1	-[b]	M31	ND[a]	1.0
E21	0.5	-[b]	M41	ND[a]	1.0
E22	ND[a]	1.0	N11	18.9	-[b]
E23	ND[a]	1.0	N12	ND[a]	1.0
E31	ND[a]	1.0	N13	ND[a]	1.0
E32	ND[a]	1.0	N21	ND[a]	1.0
E33	ND[a]	1.0	N22	ND[a]	1.0
E41	ND[a]	1.0	N23	ND[a]	1.0
F31	ND[a]	1.0	N31	3.0	-[b]
F32	ND[a]	1.0	N41	ND[a]	1.0

[a]Not Detected

[b]Not Applicable

*Samples contained varying quantities of red dye. Red dye was used as an indicator for test purposes during the 1962 through 1970 spray missions. A slightly different extraction technique was used for these samples.

**Samples were subjected to high resolution mass spectrometry (GC-HRMS).

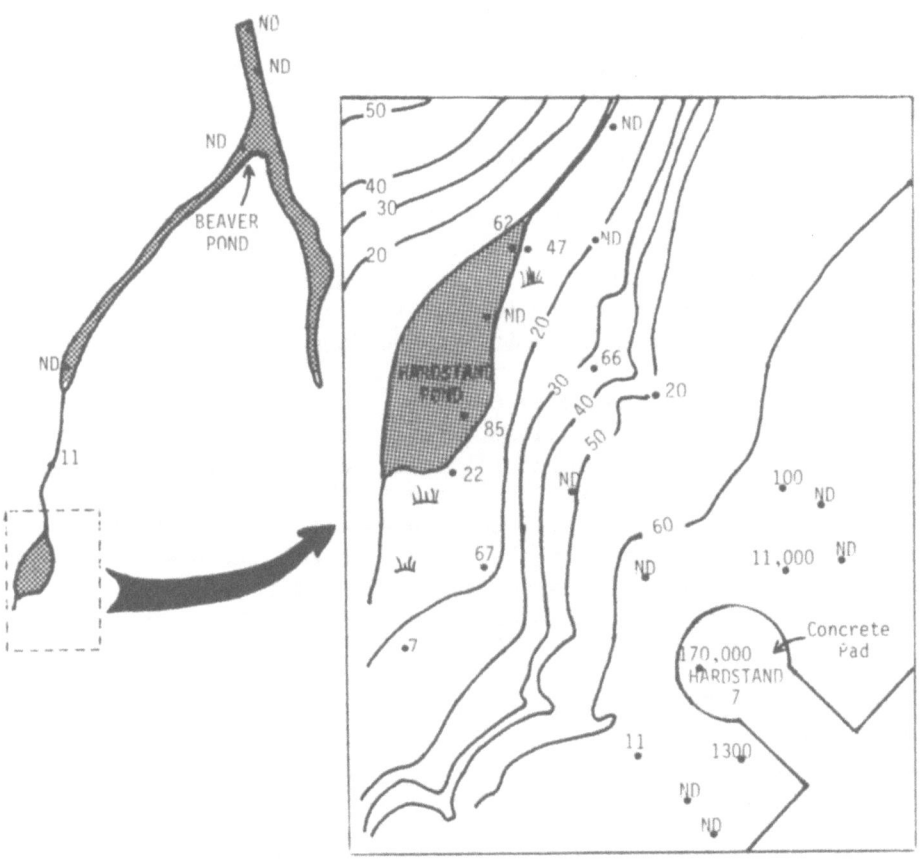

Fig. 4. Maps Showing Hardstand 7, Hardstand Pond, and Beaver Pond.
         Numbers along the Irregular Lines Show Approximate
         Elevation in Feet Above Hardstand Pond Water Level.
         Numbers by the Dots Represent TCDD Contamination Levels
         (parts-per-trillion) in Soil or Sediment. ND Means TCDD
         was Not Detected at Those Locations.

saturation of the soil with herbicide to a considerable depth
when the hardstand was being used for loading operations. TCDD
could have been transported to those depths via the herbicide
solvent. Following degradation of the herbicide, the TCDD would
have remained. Because of its low solubility in water,
subsequent transport of TCDD would be minimal.

Detection limit was 1 ppb for the soil samples collected from
Hardstand 7 in 1978 (Table 1). Therefore, TCDD in concentrations
less than 1 ppb could have been present in samples reported as
not detected. TCDD contamination above 1 ppb was found
predominantly in an area 3 to 4 meters wide surrounding the
perimeter of the concrete pad. Soils outside this area were
largely below the detection limit of 1 ppb.

Sediment samples indicated that TCDD had migrated into the
drainage system of Hardstand 7 (Table 2). The difference in
sedimentation rates of sampling locations, as influenced by a
series of beaver dams, and the small number of samples taken
probably account for the disparity in the data.

No biological samples collected from Tom's Bayou or Tom's
Pond contained TCDD in detectable quantities except for a
cottonmouth (Akistrodon piscivorus) collected from Tom's Pond in
1979 that gave a liver sample showing 0.151 ppb. The natural
range of this species suggests that the source of contamination
could have been upstream from the site where the sample was
collected. Other biological specimens collected from Hardstand
Pond or Beaver Pond contained concentrations of TCDD up to 1.5
ppb in some tissue (Table 3).

Snapping turtles are near the top of the food chain in the
aquatic system draining the contaminated hardstand. Therefore,
the potential for TCDD accumulation from both the environment and
contaminated food existed for this species. This may explain the
high TCDD concentrations found in this species in comparison to
other species sampled. The specimens collected were old enough
to have been living during the 1962 through 1970 spray program,
although it cannot be determined if they actually lived in the
Hardstand Pond/Beaver Pond area all that time. Even though there
seems to be agreement that certain organs tend to accumulate TCDD,
data from this study were not sufficient to draw any conclusions
on preferential accumulation in tissues.

Although TCDD was not present in any of the biological
samples collected from Tom's Bayou or Tom's Pond in 1977 (Table
3), it was felt that a more thorough study of aquatic organisms
closer to the source of TCDD contamination should be conducted.
Therefore, in 1979, biological tissue samples of representative
organisms collected from Hardstand Pond, Beaver Pond, and Tom's

Table 2.   TCDD Analyses of Sediment Samples Collected from
            Hardstand 7 Drainage Area

Sample Location	Date Collected	TCDD Concentration	
		TCDD (ppt)	Detection Limit (ppt)
Hardstand Pond	1974	85	--
Hardstand Pond (Immediately upstream from beaver dam)	1974	62	--
Stream (200 meters NE of Hardstand Pond)	1974	11	--
Hardstand Pond (NE edge)	1974	ND	50
Stream (50 meters NE of Hardstand Pond)	1974	ND	3
Beaver Pond	1974	ND	7
Beaver Pond	1974	ND	10
Beaver Pond	1974	ND	10
Beaver Pond	1974	ND	10
Hardstand Pond (Detritus)	1977	ND	370
Hardstand Pond (Sand)	1977	ND	37
Tom's Pond	1977	650	--
Tom's Bayou	1977	ND	31
Hardstand Pond	1980	ND	70

Table 3. TCDD Analysis of Biological Organisms Collected from Hardstand 7 Drainage Area

SPECIES	SAMPLES	LOCATION	DATE COLLECTED	TCDD (ppt)	D.L (ppt)
Mosquito fish (Gambusia affinis)	Whole body 20 fish	Hardstand Pond and 200 Meters downstream	1974	150	-
Sunfish (Lepomis sp.)	Whole body 4 fish	Beaver Pond	1974	14	-
	fillets 8 fish		1974	ND*	5
	Liver 25 fish		1974	ND	5
Sunfish (Lepomis sp.)	6 whole bodies + 18 Livers	Hardstand Pond	1974	150	-
	Liver + fat 12 fish		1974	740	-
Snails (Neritina reclavata)	Whole bodies (except shell)	Tom's Bayou	1977	ND	11
Fish (Alewife) (Pomolobus pseudoharengus)	Liver	Tom's Bayou	1977	ND	15
	Muscle	Tom's Bayou	1977	ND	109
Clams (Polymesoda caroliniana and Rangia cuneata)	Whole body (except shells)	Tom's Bayou	1977	ND	15
Crab (Callinectus sapidus)	Muscle	Tom's Bayou	1977	ND	26
	Viscera		1977	ND	30
Fish (Bass) (Micropterus salmoides)	Liver	Tom's Pond	1977	ND	17
	Muscle		1977	ND	10
Turtle (Chelydra serpentina)	Fat	Hardstand Pond	1978	1500	-
	Liver	Hardstand Pond	1978	ND	260
	Muscle	Hardstand Pond	1978	ND	9
	Testes	Hardstand Pond	1978	ND	14
Beach Mice (Peromyscus poliontus)	Liver	Hardstand 7	1978	550	-
	Skin		1978	53	-
Sunfish (Lepomis sp.)	Muscle	Beaver Pond	1978	ND	8
	Viscera		1978	ND	7
Fish (Bass) (Micropterus salmoides)	Liver	Beaver Pond	1979	ND	19
	Muscle		1979	18	-
Catfish (Ictalurus natalis)	Liver	Beaver Pond	1979	47	-
	Muscle		1979	104	-
Turtle (Chelydra serpentina)	Muscle	Beaver Pond	1979	ND	7
	Fat**		1979	271	-
Beaver (Castor canadensis)	Liver	Hardstand Pond	1979	32	-
	Fat**		1979	259	-
Sunfish (Lepomis punctatus)	Liver	Tom's Pond	1979	ND	7
	Muscle		1979	ND	7
Catfish (Ictalurus natalis)	Liver	Tom's Pond	1979	ND	31
	Muscle		1979	ND	25
Cottonmouth (Agkistrodon piscivorus)	Liver**	Tom's Pond	1979	151	-
	Visceral fat			ND	7
Turtle (Chelydra serpentina)	Liver **	Hardstand Pond	1980	420	-
	Muscle			ND	20
Catfish (Ictalurus natalis)	Muscle	Hardstand Pond	1980	60	-
	Liver			340	

* Not Detected.
** Isomeric analysis performed on these samples

Table 4. TCDD and Histological Analyses of Biological Samples Collected in 1979 from Hardstand 7 Drainage Area, Eglin Air Force Base, Florida

SAMPLES	LOCATION	TCDD(ppt)	HISTOPATHOLOGICAL RESULTS
Bass (Micropterus salmoides) liver, heart	Beaver Pond	ND (liver) 18 (muscle)	Mild, encapsulated granuloma in liver, parasitic; probably nematode larvae
Catfish (Ictalurus natalis) liver, intestine	Beaver Pond	47 (liver) 104 (muscle)	Granulomas, multiple, mild, parasitic, liver, intestinal wall; probably due to nematodes
Turtle (Chelydra serpentina) liver, eggs, intestine, lung, fat, heart, spleen, ovary,	Beaver Pond	0 (muscle) 271 (fat)	Essentially normal tissue
Beaver (Castor canadensis) liver, spleen, intestine, heart, lung, kidney, oviduct	Hardstand Pond	32 (liver) 259 (fat)	1. Pnuemonitis, multifocal, mild, subacute, lung 2. Pigmentation, hemosiderin, mild to moderate, hepatic and Kupfer cells, liver 3. Fibrosis, multifocal, mild, heart non-specific as to etiology, probably not of clinical significance
Catfish (Ictalurus natalis) liver, intestine	Tom's Pond	0 (liver) 0 (muscle)	Essentially normal tissue
Cottonmouth (Agkistrodon piscivorus) liver, heart	Tom's Pond	151 (liver) 0 (fat)	Essentially normal tissue

Table 5. TCDD Isomers Present in Selected Samples from Eglin
         Air Force Base, Florida

SAMPLE	DATE COLLECTED	TOTAL TCDD (ppb)	TCDD ISOMERS DETECTED
Hardstand 7 Soil Site D-1 (0-5 cm)	1980	120	2,3,7,8
Hardstand 7 Soil Site D-1 (10-15 cm)	1980	85	2,3,7,8
Contaminated Sand Buried For 8-10 Years on Hardstand 7	1980	1410	2,3,7,8
Herbicide Orange (from sample held in storage 10 years)	1980	6400	2,3,7,8
Turtle Liver (Hardstand Pond)	1980	0.42	2,3,7,8
Turtle Fat	1979	0.271	2,3,7,8
Cottonmouth Liver (Tom's Pond)	1979	0.151	2,3,7,8 Major 1,2,3,4 1,3,6,8 Minor
Beaver Fat (Hardstand Pond)	1979	0.259	2,3,7,8

Pond were subjected to TCDD and histopathological analyses.
Organisms and tissues examined are listed in Table 4.
No gross abnormal pathological findings were noted by Eglin and
Air Force Academy personnel in visual observations during
specimen dissection prior to histopathological analysis by the
Armed Forces Institute of Pathology (AFIP). AFIP analysis
indicated no histopathologic anomalies attributable to TCDD in a
any of the tissues even though TCDD analysis revealed that all
but one of the specimens contained TCDD. Other significant facts
about the specimens collected for histopathologic analysis were
noted during dissection. Eggs were present in the oviduct of the
turtle and the cottonmouth contained fetuses, all of which
appeared to be normal.

Isomeric analyses of soil, sediment, biological samples, and
Herbicide Orange held in storage for at least 10 years (Table 5)
verified that the predominant TCDD present was the 2,3,7,8-TCDD.
This was the only isomer detected in any of the samples except
for trace quantities of 1,2,3,4,- and 1,3,6,8-TCDD in the
cottonmouth liver.

The most significant findings from this study are:

1. More than 10 years after herbicide had been stored at the
    hardstand, TCDD still remained.

2. TCDD was not found more than 3000 meters downstream from
    the original source of contamination.

3. For animals, no harmful effects were traced to TCDD.

4. The 2,3,7,8-TCDD isomer was the predominant isomer found
    in soil or biologic specimens.

CHICK EDEMA DISEASE AND HEPATIC PORPHYRIA IN

LAKE ONTARIO HERRING GULL EMBRYOS IN THE EARLY 1970s

M. Gilbertson* and G. A. Fox**

*Toxic Chemicals Management Centre
Department of Environment
Ottawa, Canada

**Wildlife Toxicology Division, Canadian Wildlife Service
Department of Environment
Ottawa, Canada

ABSTRACT

The severity, characteristics, and possible causes of reproductive failure among fish-eating birds on the lower Great Lakes were
investigated in the early 1970s. Examination of embryos and chicks
from Herring Gull (Larus argentatus) colonies in Lake Ontario and
Port Colborne in eastern Lake Erie suggested that a highly embryotoxic, porphyrinogenic, chick edema-active agent or agents was
present in this food chain. Retrospective analysis of stored eggs
has revealed the presence of significant quantities of 2,3,7,8-TCDD,
as well as DDE, PCBs, mirex, hexachlorobenzene and a number of as
yet unidentified contaminants. The relative importance of these
contaminants in the observed reproductive failure is discussed.

INTRODUCTION

Historical Background

During the past decade there have been a series of investigations to document the causes of severe reproductive failures among
fish-eating birds on the Great Lakes, particularly on Lake Ontario.

The initial discovery of the reproductive failure was in 1970
in Common Terns nesting on islands in Hamilton Harbour, Lake Ontario

(Gilbertson and Reynolds, 1972). Subsequent detailed investigations in 1971 revealed the geographic extent and severity of the failure and the possible interrelationship with organochlorine compounds (Gilbertson 1974a, and unpublished) including DDT and its metabolites DDE and DDD, heptachlor epoxide, hexachlorobenzene, dieldrin, and PCBs. The breeding failure was characterized (Gilbertson, 1974b) by disappearance of eggs, early embryonic mortality, mortality at hatching associated with growth retardation and congenital abnormalities (Gilbertson et al, 1976).

A monitoring program based on the Herring Gull, Larus argentatus, as an indicator species (Gilbertson, 1974b) has become an integral part of the Canadian Wildlife Service investigations on the Great Lakes. These investigations include long-term monitoring of biological effects, trends in organochlorine residue levels, and retrospective analysis for previously unidentified or undetected compounds. In addition, intensive investigations were undertaken on the Herring Gull population on Scotch Bonnet Island in eastern Lake Ontario where the breeding success in 1972 was about 0.05 young per pair (Gilbertson, 1974b). Studies undertaken on Scotch Bonnet Island in 1973 revealed (Gilbertson and Hale, 1974a and b) that (1) 20 percent of embryos died during the first week of incubation, (2) that all eggs contained embryos at some stage of development, (3) egg disappearance was high, (4) failure was characterized by the loss of the whole clutch, (5) in nests where at least one egg hatched, 22 percent died prior to pipping and 16 percent died after pipping, and (6) survival of hatched young was very low. The overall reproductive effort resulted in 0.06 chicks fledged per pair in marked contrast to that of 1.0 to 1.5 observed in other studies of this species in North America and Europe. Chemical analysis of the eggs revealed the presence of high levels of PCBs (over 500 ppm on a dry matter basis) as well as other organochlorine compounds (Gilbertson, 1974).

Rationale

Chlorinated dibenzo-p-dioxins and chlorinated dibenzofurans have been isolated from toxic fats and/or experimentally shown to cause the symptoms of chick edema disease (Cantrell et al, 1969; Verrett, 1970; Vos et al, 1970; and Flick et al, 1972). Chlorinated dibenzofurans or chick edema producing fractions have been isolated from five commercial PCB mixtures (McCune et al, 1962; Flick et al, 1965; Kohanawa et al, 1969; Vos et al, 1970; and Bowes et al, 1975).

Chick edema disease is a syndrome of poultry, caused by the consumption of feed containing fats or oils contaminated with chlorinated dibenzofurans and chlorinated dibenzo-p-dioxins. The disease is characterized by a high rate of mortality and necropsy

Findings of subcutaneous edema, hydropericardium (pericardial edema), ascites (peritoneal edema), and liver necrosis with fatty degeneration. Hepatic porphyria may also be present. Injection of a chick edema factor into fertile chicken eggs reduces hatchability, and those embryos which fail to hatch show edema and deformities of the beak, legs, and eyes; those that hatch have retarded growth (Firestone, 1973).

The high embryonic mortality of Lake Ontario Herring Gull embryos and the presence of congenital anomalies in chicks of other terminal predators in the same food chain led us in 1973 to ask if the high levels of PCBs found in Lake Ontario Herring Gull eggs were accompanied by highly embryotoxic and teratogenic trace contaminants such as the chlorinated dibenzofurans, and whether developing embryos might exhibit signs of chick edema disease, have hepatic porphyria, and show congenital anomalies.

METHODS

In 1974, eggs were sampled from two colonies on Lake Ontario (Scotch Bonnet Island and West Brothers Island), two colonies in eastern Lake Erie near Port Colborne, and two relatively uncontaminated "control" colonies: one at Bathurst, New Brunswick, the other in Namur Lake in northern Alberta (Gilbertson and Fox, 1977). Eggs were incubated artificially at $37.5^{\circ}C$ and 50 percent relative humidity. Newly hatched chicks ("hatched") were sacrificed within minutes of hatching. Embryos which had pipped the shell but become moribund were sacrificed when it became apparent that they would not hatch ("moribund"). Eggs which failed to pip were opened and the presence and condition of the embryo noted. Those embryos which had not undergone excessive decomposition were necropsied ("dead"). The methodology for the necropsy of chicks employed in the standard bioassay for detecting the presence of a chick edema factor in feeds (Horowitz, 1965) was modified for Herring Gull chicks and embryos. The skin over the thoracic and abdominal cavities was cut and peeled back to permit detection of sub-cutaneous edema. The chest cavity was exposed by cutting each side of the rib cage. The pericardial fluid volume was measured by aspiration through an 18 guage blunt-ended needle into a 1.0 mL tuberculin syringe. The peritoneal cavity was examined for the presence of ascitic fluid.

Edema was scored on an arbitrary basis. Absence of any sub-cutaneous fluid accumulation in the vicinity of the pectoral muscles and abdomen was denoted by a score of 0. Marked sub-cutaneous fluid accumulation which lifted the integument from the underlying muscles was assigned a score of 1. Large subcutaneous accumulations of fluid extending into the region of the neck or legs, or which was mucoid or serious in nature was denoted by a score of 2. Absence of ascites was denoted by a score of 0. A noticeable accumulation of ascitic fluid was denoted by a score of 1. Marked ascites, producing distention of the abdomen, or of a cloudy nature was assigned a score of 2. Pericardial fluid volumes of 50 mL or less were considered normal and denoted by a score of 0. Volumes of 51 to 70 mL, 70 to 100 mL, and 100 mL or more, were assigned pericardial edema scores of 1, 2, and 3 respectively.

Tarsus length, body weight, and liver weight were measured. Sections of liver were preserved in formalin and sectioned for light microscopy. One lobe of liver was retained for total porphyrin determination (Abbritti and de Matteis, 1971-1972), the other lobe and brain tissue were used for residue analysis.

In 1975 eggs were collected from Scotch Bonnet Island and eggs from a Prince Edward Island colony served as a control. Artifical incubation conditions were those used in 1974. The number of deaths prior to, and during pipping were recorded. Newly hatched chicks were sacrificed and necropsied as in 1974, but pericardial fluid volumes, hepatic porphyrin levels, and pesticide residues were not determined. Only a small proportion of moribund or dead embryos were examined.

RESULTS

Hatchability of Eggs

As in previous years, the percentage of embryos from Scotch Bonnet Island which died prior to pipping was very high (Table 1). Of embryos which reached the pipping stage, those from both Great Lakes locations suffered twice the mortality observed in the controls. This resulted in a markedly depressed hatching success for eggs from Lake Ontario.

One of the 51 eggs artifically incubated in 1974 from Scotch Bonnet Island produced a dead embryo with a crossed beak. Two pipping, moribund embryos with congenital anomalies were found in the 90 eggs examined from Scotch Bonnet Island in 1975.

Pathology

In both years, newly hatched chicks from Lake Ontario were significantly smaller, as measured by tarsus length, than the controls (Table 1). In addition, chicks from both the Great Lakes locations had hepatomegally. Significant hepatic prophyria was detected in newly hatched chicks from Lake Ontario in 1974.

Table 1. Two-Year Study of Variation in Mortality and Pathology of Herring Gull Embryos Collected from the Great Lakes.

	CONTROLS 1974	CONTROLS 1975	LAKE ONTARIO 1974	LAKE ONTARIO 1975	LAKE ERIE 1974
Eggs					
Number of eggs incubated	14	109	47	149	25
% of eggs that pipped	85	70	39*	30*	83
% of pipped eggs that hatched	85	90	67	87	64
% of incubated eggs that hatched	69	61	26*	26*	53
Newly Hatched Chicks					
Tarsus length as % of control	100	100	93.3*	93.7*	99
Hepatomegally Index** as % of control	100	100	119*	119*	124*
Mean total liver porphyrins n moles/gm	0.13	nd***	0.68*	nd***	0.15

* $P < 0.05$, control vs contaminated colony

** Liver weight/tarsus length

*** nd = not determined

In 1974 subcutaneous edema in the region of the pectoral muscle, abdomen, and neck was observed in 36 percent of the embryos from Lake Ontario and Port Colborne combined in contrast to 8 percent of those from the control areas. In addition, newly hatched chicks from the Great Lakes colonies exhibited fragility of the skin. Subcutaneous edema was not observed in 1975. Peritoneal edema (ascites) was observed in 47 percent of the embryos from Lake Ontario but was absent from Port Colborne and the control areas in 1974. It was not observed in 1975. Pericardial edema was observed in some Great Lakes embryos in 1974 but was not measured in 1975.

Lake Ontario embryos in 1974 had significant porphyrin accumulations in their livers, regardless of hatching status; those that died prior to pipping had the greatest accumulations. Some embryos from Port Colborne had slightly elevated porphyrin accumulations, and the single moribund control embryo had a significant accumulation. Because porphyrin levels were not measured in 1975 embryo livers, it is not known to what extent porphyria was present after 1974.

A diffuse hepatic lipidosis (fatty degeneration) was present in 19 Of 20 individuals from artifically incubated 1974 Lake Ontario hatchlings. It was present in 6 of 15 Port Colborne hatchlings. The remaining specimens showed a degree of lipidosis within the range of the four control birds. In several of the livers with diffuse hepatic lipidosis there was evidence of mild hyperplasia of bile duct epithelium. A second lesion consisted of extreme vacuolation of the cytoplasm among Lake Ontario birds, 8 of 20 individuals showed the lesion. A single small focus of these cells was present in one of the four sections from New Brunswick birds.

In 1974 newly hatched chicks and embryos from eggs naturally incubated in Scotch Bonnet Island were also examined and exhibited the same pathological findings, suggesting that our experimental results were not an artifact of the artificial incubation conditions.

The median chlorinated residue content of the brains and livers of embryos analyzed are summarized in Table 2. Residues of PCBs, DDE, and HCB but not dieldrin were significantly lower in Port Colborne embryos than those from Lake Ontario. PCBs were the most abundant contaminant in all groups. Comparisons of residues between those embryos which hatched and those which died, by location, gave no indication that any residue measured was responsible for death.

A comparison of pathology findings of all embryos, regardless of source, by hatching status (Table 3) suggests that death prior to pipping was associated with severe hepatic porphyria and

Table 2. Selected Chlorinated Hydrocarbon Residues Found in Herring Gull Embryo Tissues Collected in 1974

Tissue	Location	Median residue content, PPM, on a dry-matter basis				
		PCB	DDE	HCB	Dieldrin	
Brain	Scotch Bonnet Island (n=18)	160.5*	31.80*	0.36*	0.36	
	Port Colborne (n=11)	66.0*	11.54*	0.08*	0.43	
	Control (n=10)	7.0	3.81	0.04	0.19	
Liver	Scotch Bonnet Island (n=18)	969.5*	211.0*	4.31*	1.87*	
	Port Colborne (n=11)	578.0*	87.5*	1.37*	1.76*	
	Control (n=10)	35.5	18.3	0.22	0.12	

*Significantly higher than Control group $P < 0.05$

Table 3. Variation in Pathology of Herring Gull Embryos with Hatching Status, All Locations Combined, 1974

Sign or Parameter	Pipped		Failed to Pip
	Hatched	Moribund	Dead
Pericardial Edema : incidence	32.5% (13/40)	86.4% (19/22)*	not measurable
mean severity score (0-3)[a]	0.57	2.04*	
Ascites : incidence[b]	7.5% (3/40)	15.4% (4/26)	68.4% (13/19)*
mean severity score (0-2)	0.15	0.19	0.74*
Subcutaneous Edema : incidence	35% (14/40)	77% (20/26)*	74% (14/19)*
mean severity score (0-2)	0.45	1.27*	1.16*
Total liver porphyrins n moles/gm (n)	0.31(19)	0.38(15)	8.81*(6)
Tarsus length as % of those which hatched (n)	100(40)	92*(26)	not comparable

*Statistically significantly different from those that hatched (P > 0.05)

peritoneal and subcutaeous edema. Death after pipping was
associated with retarded growth and pericardial and subcutaneous
edema. We concluded that the signs of edema, growth retardation,
liver necrosis and fatty degeneration, porphyria, and the
existence of congenital anomalies in embryos from Lake Ontario and
Port Colborne in 1974 were consistent with chick edema disease and
thus with the presence of a chick edema factor in the food chain of
the gulls nesting in these colonies.

DISCUSSION

     This is the only documented observation of chick edema disease
and hepatic porphyria in wild birds. Documentation of chick edema
disease requires a specific study protocol, which to our knowledge,
has been undertaken by few other environmental toxicologists.
However, fish-eating birds suspected of dying of polyhalogenated
hydrocarbon poisoning have been routinely examined for pericardial
edema and hepatic porphyria by Dutch investigators but no positives
have been found (Koeman et al, 1969; and Strik, personal communication.

     The reduced hatching success of eggs from both Lake Ontario
and Port Colborne under controlled incubation conditions was
indicative of the presence of an intrinsic embryotoxic agent. The
concentration or toxicity of this agent or agents was higher in
Lake Ontario eggs than those from Port Colborne since most of the
mortality occurred early in development in the Lake Ontario eggs.

     The hepatomegally observed in newly hatched chicks from Great
Lakes locations was consistent with the high levels of pollutants
found in their livers. Moribund embryos from both lakes were
significantly smaller than those which hatched, suggesting that their
reduced size was related to toxicity; the embryos from Scotch
Bonnet Island being the most affected.

     Subcutaneous edema was observed significantly more often in the
Great Lakes embryos than from the control areas in 1974, and was
significantly more severe in moribund and dead embryos. Ascites
was only observed in embryos from Lake Ontario and was most severe
and most frequently observed in dead embryos. Pericardial edema
was observed in embryos from both Great Lakes locations in 1974 and
was associated with death after pipping. The possibility that the
subcutaneous edema and ascites were postmortem artifacts seems
highly improbable in that ascites also occurred in hatched and
moribund embryos, and the incidence and severity of subcutaneous
edema did not differ between moribund and dead embryos. Accumulation
of peritoneal fluid and excessive muco-serous exudate were not
observed in the small sample of embryos examined from the
artificially incubated 1975 Lake Ontario eggs.

Hepatic porphyria has never been observed in dead wild birds, including Cormorants (Phalacrocorax carbo) whose livers contained 2.5-28 ppm of HBC and 50-470 ppm PCB, and Herons (Ardea cinera) whose livers contained 0.4 to 24 ppm HCB and 0.8 to 500 ppm PCB (Koeman et al, 1972). It was not induced by feeding Cormorants 100-500 ppm PCB (Phenoclor) for 55-124 days (accumulating 210-290 ppm PCB in the liver), Herons 1500 ppm Phenoclor for 152-164 days (accumulating 1140-1430 ppm PCB in the liver), or in a Black-headed Gull (Larus ridibundus) dosed with 100 ppm HCB for 7 days (Strik, personal communication). This suggests that these compounds do not produce hepatic porphyria in adult fish-eating birds even at high concentrations. However, no work has been done with embryos of fish-eating birds. The levels of hepatic porphyria observed in Lake Ontario embryos (200 to 25,000 ng/g) are greater than those observed in most experimental studies with avian embryos, regardless of the toxicant used, and imply sustained induction of aminolevulinic acid synthetase. Levels observed in most embryos from Port Colborne were in the 100-200 ng/g range while those in 50 percent of the control embryos were less than 60 ng/g. The latter value is comparable with those observed in control embryos in other avian studies. These observations suggest that an extremely porphyrinogenic agent capable of sustained induction was responsible for the hepatic porphyria observed in the Great Lakes gull embryos.

Possible Etiology

Chick edema-active compounds are aromatic organohalogens that induce aryl hydrocarbon hydroxylase and cytochrome P-448 (McKinney et al, 1976; and Goldstein, 1980). It appears that aromatic organo-halogen compounds that induce aryl hydrocarbon hydroxylase have specific conformational requirements for binding to the receptor. These include coplanarity, rectangular dimensions of about 10x3 Angstroms, and halogen substitution at the lateral positions on the dibenzofuran and dibenzo-p-dioxin molecules and at the meta and para positions on the biphenyl and azobenzene molecules (McKinney and Singh, 1981).

Hepatic porphyria induced by aromatic polyhalogenated hydro-carbons is characterized by a delayed onset and an accumulation of porphyrins with 7 and 8 carboxyl groups. The step-wise mechanism is thought to be as follows: (1) induction of aminolevulinic acid synthetase, (2) induction of the mixed function oxidase system, particularly aryl hydrocarbon hydroxylase and cytochrome P-448, (3) formation of phenolic, sulfur-containing, or otherwise reactive intermediates or metabolites which inactivate or inhibit uroporphyrinogen decarboxylase, leading to (4) hepatic porphyrin accumulation. The inhibition of uroporphyrinogen decarboxylase represents the first biochemical indication of liver mitochondrial damage (Strik et al, 1980).

Compounds that are chick edema active are also porphyrinogenic.
In addition, hexachlorobenzene, which is not chick edema active,
is porphyrinogenic. Hexachlorobenzene levels are higher in Lake
Ontario than Lake Erie but are not such that severe porphyria is
likely to be induced. Other macrocontaminants are not porphyrinogenic but could contribute to hepatic porphyria through the production of phenolic, sulfur-containing, or otherwise reactive
intermediates which may inactivate or inhibit uroporphyrinogen
decarboxylase if sufficient induction of the aminolevulinic
acid precursor was provided by a porphyrinogenic agent.

Due to the apparent absence of edema in 1975 artifically
incubated embryos, two reasonable hypotheses can be put forward:
either chick edema factor(s) had decreased below a threshold level
in the eggs for appearance of gross edema in hatched chicks, or the
level of compound(s) potentiating the appearance of symptoms had
decreased significantly between 1974 and 1975. Macrocontaminant
residues (PCBs, DDE, mirex, photomirex, HCB, dieldrin) in Scotch
Bonnet Island eggs have declined between the period 1974-1976
(Weseloh et al, 1979). Of the major identified contaminants, only
mirex is notably higher in Lake Ontario eggs than eggs from the
other Great Lakes (Norstrom et al, 1980), where reproductive success
was much higher. Mirex, which is not a chick edema-active agent,
is known to interfere with the excretion of PCB metabolites in
isolated, perfused rat livers (Mehendale, 1976), but recent experiments feeding a mirex/PCB mixture to Ring Doves did not produce a
significant reduction in hatching success (McArthur et al, in
preparation).

Of much greater significance was the finding that 2,3,7,8-TCDD
was present at levels in the order of 60 ng/g (ppt) in Lake
Ontario Herring Gull eggs collected in 1980 (Norstrom et al, in
press). Levels in eggs from the other Great Lakes were in the order
of 10 ppt, with the exception of Saginaw Bay, Lake Huron, where
eggs had levels as high as 86 ppt. Since 2,3,7,8-TCDD is the most
embryotoxic, chick edema-active, teratogenic, hepatotoxic,
porphyrinogenic substance known in avian embryos, it is the prime
suspect chemical for the pathology observed and for the poor
reproductive success. Analyses of Scotch Bonnet Island eggs stored
in the Canadian Wildlife Service Tissue Bank revealed that the
levels of 2,3,7,8-TCDD were following a temporal trend nearly
identical to that of the major organochlorine contaminants (Hallett
and Norstrom, in preparation). The levels of 2,3,7,8-TCDD were in
the order of 200-1200 ppt in a period when reproductive success
was low, declining thereafter, with a half-life of 2.5 years. The
relatively high incidence and concentration of mirex suggest that
the gulls nesting in this colony feed in Lake Ontario or the
Niagara River for some period of the year and would thus be
exposed to the same contaminants as Lake Ontario gulls, but to
a lesser extent.

Reproductive success in the Scotch Bonnet and other Lake
Ontario colonies was greatly improved in 1976, and continued to
improve in successive years. In addition, the incidence of
congenital abnormalities in other species feeding in this food web
was markedly different between the periods 1971-1973 and 1976-
1980 (Table 4), suggesting that the concentration of teratogenic
substances had dropped below a threshold level by 1976. The
combined incidence rates of congenital abnormalities for Common
Tern, Caspian Tern, and Black-crowned Night Heron (Table 4) in
1972-1973 and 1976-1980 were 12.1 and less than 0.5 per 1000
chicks, respectively. One abnormality was noted in 418 Herring
Gull chicks from Scotch Bonnet Island in 1979. The recent
incidence rates can be compared to the incidence (0/720) for Common
Terns banded in Alberta (Fox, unpublished), 0.08 (1/12785) for
Herring Gulls banded in Newfoundland, 1966-1972 (Threlfall,
personal communication) and less than 0.13 (0/7463) for California
Gulls banded in Saskatchewan, 1955-1973 (Houston, personal
communication), suggesting that the incidence rate of congential
abnormalities in Lake Ontario Herring Gulls after 1974 was
approaching background levels.

Egg exchanges between Scotch Bonnet Island and less
contaminated, successful colonies suggested that factors intrinsic
and extrinsic to the egg were involved in the reduced hatching
success of Scotch Bonnet Island eggs (Peakall et al, 1980).
Intrinsic factors were of no importance after 1976. The extrinsic
factor was shown to be parental behavior, and the degree of parental
neglect was correlated with macrocontaminant burdens (Fox et al,
1978). The ability of the combination of the major organochlorine
contaminants found in Lake Ontario Herring Gulls to induce endocrine
dysfunction and similar, related parental behavioral anomalies has
been confirmed for Ring Doves in the laboratory (McArthur et al,
in preparation).

Taking the biological and chemical evidence to date, we are
left with more questions than answers. If we hypothesize that
2,3,7,8-TCDD was the main causative chemical for the chick edema
disease observed in 1974, a corollary must be that this disease,
as indicated by its gross pathology, was not the sole cause of
reproductive failure, since the symptoms were absent in 1975,
while reproduction was equally poor. Another difficulty with this
hypothesis is that, although there is evidence for a dose-response
relationship between dioxins and severity of chick edema disease
in domestic foul, there appears to be a sharp disappearance
of symptoms at a time when 2,3,7,8-TCDD levels were declining
slowly and regularly. Conversely, the chick edema symptoms
observed in 1974 Herring Gull chicks, although themselves probably
not the proximate cause of embryonic death, strongly indicate
that this disease and/or the accompanying hepatic porphyria was the
main cause of a high mortality observed in this year. It is possible

Table 4. Variation in the Incidence of Congenital Anomalies in Lake Ontario Fish-eating Birds in the 1970s.

Species	1971-1973		1975-1980		
	Incidence	Rate per 1000 chicks	Incidence	Rate per 1000 chicks	
Common Tern	11/900	12.2	0/629	1.6	P< 0.01
Caspian Tern	1/100	10.0	0/766	1.3	P<0.01
Black-crowned Night Heron	1/72	13.9	0/574	1.7	P<0.005
Herring Gull	?	*	1/1608	0.62	

*If we assume an incidence rate of 15.0, then a minimum of 60 chicks would have to be examined to detect any. Since hatching success was 15-20% and 40% of eggs layed disappeared or were damaged, approximately 830 eggs would be required to produce 60 chicks or roughly 300 nests. This exceeds the total number of pairs nesting in the colonies examined in any year.

that there are more chemicals than just 2,3,7,8-TCDD and other
pathologies than chick edema disease and hepatic porphyria
involved. However the presence of this highly embryotoxic,
porphyrinogenic, and teratogenic compound at concentrations up to
1200 ppt indicates that it was probably significantly involved in
the observed pathology associated with reproductive dysfunction of
Lake Ontario Herring Gulls in the early 1970's.

ACKNOWLEDGEMENTS

We thank P.A. Pearce, and R.D. Morris for collecting eggs,
L.M. Reynolds, Ontario Research Foundation, for organochlorine
residue analyses; D. Grant, Health and Welfare Canada for porphyrin
determinations; J.P. Couillard for preparation of the histological material; and F. Leighton, Department of Pathology, New
York State College of Veterinary Medicine for histopathological
consultation. David Peakall provided encouragement and support
during the preparation of this manuscript. He and R.J. Norstom,
A.P. Gilman, K. Marshall, D.J. Hallett and J. Ellenton provided
useful comments for which we are grateful.

REFERENCES

Abbritti, G. and DeMatteis, F. Decreased levels of cytochrome
    P-450 and catalase in hepatic porphyria caused by acetamides
    and barbiturate. Chem. Biol. Interact., 4:281-286, 1971-1972.
Bowes, G. W., Mulvihill, J. J., Simoneit, B. R. T., Burlingame,
    A. L. and Risebrough, R. W. Identification of chlorinated
    dibenzofurans in American polychlorinated biphenyls. Nature
    256:305-307, 1975.
Cantrell, J. S., Webb, N. C. and Mabis, A. J. Identification and
    crystal structure of a hydropericardium-producing factors:
    1,2,3,7,8,9-hexachlorodibenzo-p-dioxin. Acta Crystallogr.,
    Section B 25(1):150-156, 1969.
Firestone, D. Etiology of chick edema disease. Environ. Health
    Perspect., 5:59-66, 1973.
Flick, D. F., Firestone, D. and Higgingbotham, G. R. Studies of
    the chick edema disease: 9. Response of chicks fed or singly
    administered synthetic edema-producing compounds. Poult. Sci.,
    51:2026-2034, 1972.
Flick, D. F., O'Dell, R. G. and Childs, V. A. Studies of the chick
    edema disease: 3. Similarity of symptoms produced by feeding
    chlorinated biphenyl. Poult. Sci., 44:1460-1465, 1965.
Fox, G. A., Gilman, A. P., Peakall, D. B. and Anderka, F. W.
    Behavioral adnormalities of nesting Lake Ontario Herring Gulls.
    J. Wildl. Manage., 42:477-483, 1978.

Gilbertson, M. Seasonal changes in organochlorine compounds and
     mercury in Common Terns of Hamilton Harbour, Ontario. Bull.
     Environ. Contam. Toxicol., 12:726-732, 1974a.

Gilbertson, M. Pollutants in breeding Herring Gulls in the
     lower Great Lakes. Can. Field Nat., 88:273-280, 1974b.

Gilbertson, M. and Fox, G. A. Pollutant-associated embryonic
     mortality of Great Lakes Herring Gulls. Environ. Pollut.,
     12:211-216, 1977.

Gilbertson, M. and Hale, R. Early embryonic mortality in a
     Herring Gull colony in Lake Ontario. Can. Field Nat., 88:
     354-356, 1974a.

Gilbertson, M. and Hale, R. Characteristics of the breeding
     failure of a colony of Herring Gulls on Lake Ontario.
     Can. Field Nat., 88:356-358, 1974b.

Gilbertson, M., Morris, R. D. and Hunter, R. A. Abnormal chicks
     and PCB residue levels in eggs of colonial birds on the lower
     Great Lakes (1971-1973). Auk, 93:434-442, 1976.

Gilbertson, M. and Reynolds, L. M. Hexachlorobenzene (HCB) in the
     eggs of Common Terns in Hamilton Harbour, Ontario. Bull.
     Environ. Contam. Toxicol., 7:371-373, 1972.

Goldstein, J. A. Structure-activity relationships for the bio-
     chemical effects and the relationship to toxicity. In:
     Kimbrough, R. J. (ed). Halogenated biphenyls, terphenyls,
     naphthalenes, dibenzodioxins and related products. Topics in
     Environmental Health. 4. Elsevier/North-Holland Biomedical
     Press. Amsterdam, New York, Oxford, 1980. pp.151-190.

Horowitz, W.(ed). Official Methods of Analysis of the Association
     of Official Agricultural Chemists. Washington, D.C., Sections
     26. 087-26.091, 1965.

Koeman, J. H., Bothof, T., DeVries, R., Van Velzen-Blad, H. and
     Vos, J. G. The impact of persistent pollutants on piscivorous
     and molluscivorous birds. TNO Nieuws, 27:561-569, 1972.

Koeman, J. H., ten Noever der Brauw, M. C. and de Vos, R. H.
     Chlorinated biphenyls in fish, mussels and birds from the
     River Rhine and the Netherlands coastal area. Nature, 221:
     1126-1128, 1969.

Kohanawa, M., Shoya, S., Yonemura, T., Nishimura, K. and Tsushio,
     Y. Poisoning due to an oily by-product of rice bran. II
     Tetrachlorodiphenyl as toxic substance. Nat. Inst. Anim.
     Health Q, 9:220-228, 1969.

McCune, E. L., Savage, L. E. and O'Dell, B. L. Hydropericardium
     and ascites in chicks fed a chlorinated hydrocarbon. Poult.
     Sci., 41:295-299, 1962.

McKinney, J. D., Chae, K., Gupta, B. N., Moore, J. A. and
     Goldstein, J. A. Toxicological assessment of hexachlorobiphenyl
     isomers and 2,3,7,8-tetrachlorodibenzofuran in chicks. 1.
     Relationship of chemical parameters. Toxicol. Appl. Pharmacol.,
     36:65-80, 1976.

McKinney, J.D. and Singh, P. Structure-activity relationships in halogenated biphenyls: unifying hypothesis for structural specificity. Chem. Biol. Interact., 33:271-283, 1981

Mehendale, H. M. Mirex-induced suppression of biliary excretion of polychlorinated biphenyl compounds. Toxicol. Appl. Pharmacol., 36:369-381, 1976.

Norstrom, R. J., Hallett, D. J., Onuska, F. I. and Comba, M. C. Mirex and its degradation products in Great Lakes Herring Gulls. Environ. Sci. Technol., 14:860-866, 1980.

Norstrom, R. J., Hallett, D. J., Simon, M. and Mulvihill, M. J. Analysis of Great Lakes Herring Gull eggs for tetrachloro-dibenzo-p-dioxins. In: Hutzinger, O., R.W. Frei, E. Merian and F. Pocchiari (eds.). Chlorinated Dioxins and Related Compounds - Impact on the Environment. Pergamon Series on Environmental Science Volume 5, Pergamon Press, New York and Oxford, 1981. pp. 173-181.

Peakall, D. B., Fox, G. A., Gilman, A. P., Hallett, D. J. and Norstrom, R. J. Reproductive success of Herring Gulls as an indicator of Great Lakes Water Quality. In: B. K. Afghan and D. Mackay (eds.) Hydrocarbons and Halogenated Hydrocarbons in the Aquatic Environment. Plenum Press. New York and London, 1980. pp.337-344.

Strik, J. J. T. W. A., Debets, F. M. H. and Koss, G. Chemical porphyria. In: Kimbrough, R. J. (ed). Halogenated biphenyls, terphenyls, naphthalenes, dibenzodioxins and related products. Topics in Environmental Health 4. Elsevier/North-Holland Biomedical Press. Amsterdam, New York, Oxford, 1980. pp. 191-239.

Verrett, M. J. Hearings before the Subcommittee on Energy, Natural Resources and the Environment of the Committee on Commerce, U.S. Senate, Serial 91-60, Government Printing Office, Washington, D.C., 1970.

Vos, J. G., Koeman, J. H., van der Maas, H. L., ten Noever de Brauw, M. C. and de Vos, R. H. Identification and toxicological evaluation of chlorinated dibenzofuran and chlorinated naphthalene in two commerical polychlorinated biphenyls. Food Cosmet. Toxicol., 8:625-633, 1970.

Weseloh, D. V., Mineau, P. and Hallett, D. J. Organochlorine contaminants and trends in reproduction in Great Lakes Herring Gulls, 1974-1978. Trans. North Am. Wildl. Nat. Resour. Conf., 44:543-557, 1979.

LONG-TERM FIELD STUDIES OF A RODENT POPULATION

CONTINUOUSLY EXPOSED TO TCDD

C. E. Thalken* and A. L. Young**

*USAF Occupational and Environmental Health Laboratory
Brooks Air Force Base, San Antonio, Texas, USA

**Office of Environmental Medicine
Veterans Administration Washington, DC USA

ABSTRACT

Field investigations were conducted during 1973-1978 on
populations of the beachmouse, Peromyscus polionotus, from a
unique 3.0 km² military test area (Test Area C-52A, Eglin AFB,
Florida) that was sprayed with 73,000 kg 2,4,5-trichloro-
phenoxyacetic acid (2,4,5-T) herbicide during the period
1962-1970. No residues of 2,4,5-T were detected at a lower
detection limit of 10 parts per billion in any soil sample
collected during 1971-1972. Residues of 2,3,7,8-tetrachloro-
dibenzo-p-dioxin (TCDD) were still present in 1978. During
1974-1978, 54 soil samples were collected to a depth of 15 cm on
the test area. TCDD levels ranged from ˜10 to 1,500 parts per
trillion (ppt). Liver tissue from 36 individual beachmice
inhabiting the test site contained 300 to 2,900 ppt TCDD. A close
relationship between soil and liver levels of TCDD was observed,
i.e., high liver levels of TCDD were consistent with high soil
levels of TCDD, bioconcentration factors (mean liver concentrations divided by mean soil concentrations) ranged from 6 for
females to 18 for males. Whole body analysis of fetuses from test
area females indicated apparent placental transport of TCDD.
Histopathological examinations were performed on 255 adult or
fetal beachmice from the test area and a control area. Examinations were performed on the heart, lungs, trachea, salivary
glands, thymus, liver, kidneys, stomach, pancreas, adrenals, large
and small intestine, spleen, genital organs, bone, bone marrow,
skin and brain. Initially the tissues were examined on a blind

study basis. All microscopic changes were recorded including
those interpreted as minor or insignificant. The tissues were
then reexamined on a control versus test basis, which demonstrated
that the test and control mice could not be distinguished histopathologically. The mean number of fetuses per observed pregnancy
was 3.1 and 3.4 for the test area and a control area, respectively.
A single female beachmouse is capable of producing a litter every
26 days. At this frequency, the animals collected in 1978 could
have been 50 generations removed from the population studied in
1973. A two-factor (treatment and year) dispropor- tional
analysis of covariance of organ weights revealed that liver
weights for pregnant beachmice from the test area were
significantly heavier (P<.01) than liver weights of pregnant
females from the control area, and these differences were
consistent over the five years of observation. These studies
suggest that long-term, low level exposure to TCDD under field
conditions has had minimal effect upon the health and reproduction
of the beachmouse.

INTRODUCTION

     Since 1970 hundreds of laboratory studies have been conducted
on the toxic contaminant 2,3,7,8-tetrachlorodibenzo-p-dioxin
(TCDD) found in trichlorophenol. Although numerous commercial
products are made from trichlorophenol, including the herbicide
2,4,5-trichlorophenoxyacetic acid (2,4,5-T), controversy remains
as to what extent the TCDD found in these products has impacted
humans or their environment.

     Laboratory data for rodents strongly suggest a correlation
between histological lesions in the liver and lymphatic system and
the amount of TCDD ingested. Unfortunately, data relating to any
actual effects on wild populations in their natural habitat are
lacking. The problem of finding a field site where a wild
population of rodents has been exposed to significant quantities
of TCDD is improbable because of (1) low levels of TCDD (<0.1 ppm)
found in currently produced phenoxy herbicide, and (2) low rates
of 2,4,5-T applied for brush control on rangelands or for
reforestation (1.1 to 2.2 kilogram (kg)/hectare (ha). This
report, however, documents the effects of residual TCDD on a
rodent population inhabiting a unique test site: a site
previously treated with massive quantities of 2,4,5-T herbicide
and located on the Eglin Air Force Base Reservation, Florida.

     The Eglin Reservation has served various military uses, one
of them having been the development and testing of aerial
dissemination equipment in support of military defoliation
operations in Southeast Asia. It was necessary for this equipment
to be tested under controlled situations that would simulate

actual use conditions as near as possible. For this purpose an
elaborate testing installation, designed to measure deposition
parameters, was established on the Eglin Reservation with the
place of direct aerial application restricted to an area of
approximately 3 square kilometers (km^2) within Test Area C-52A
in the southeastern part of the reservation. Massive quantities
of herbicide, used in the testing of aerial defoliation spray
equipment from 1962 through 1970, were released and fell within
the instrumented test area. The uniqueness of the area has
prompted continued ecological surveys since 1967. As a result,
few ecosystems have been so well studied and documented.

INVESTIGATION

Description of Field

     Test Area C-52A (TA C-52A) covers an area of approximately 8
km^2 and is a grassy plain surrounded by a forest stand that is
dominated by longleaf pine (Pinus palustris), sand pine (Pinus
clausa), and turkey oak (Quercus laevis). The actual area for
test operations occupies an area of approximately 3 km^2 and is a
cleared area occupied mainly by broomsedge (Andropogon
virginicus), switchgrass (Panicum virgatum), woolly panicum
(Panicum lanuginosum) and low growing grasses and herbs. Much of
the center of the range was established prior to 1960, but the
open range as it presently exists was developed in 1961 and 1962.
The test grid is approximately 28 m above sea level with a water
table at 1.5 to 3 m. The major portion of this test area is
drained by five small creeks whose flow rates are influenced by an
average rainfall of 150 cm. The mean annual temperature for the
test area is 19.7C while the mean annual relative humidity is 70.8
percent. For the most part, the soil of the test grid is a fine
white sand on the surface, changing to yellow beneath. The soils
of the range are predominantly well drained, acid sands of the
Lakeland Association with a 0 to 3 percent slope. A typical
one-meter soil core contained approximately 92 percent sand, 3.8
percent silt, and 4.2 percent clay with an organic matter content
of 0.17 percent, an average pH of 5.6, and a cation exchange
capacity of 0.8.

     Although the total area for testing aerial dissemination
equipment was approximately 3.0 km^2, the area actually consisted
of four separate testing grids. The primary area was located in
the southern portion of the testing area and consisted of a 37 ha
instrumented grid. This was the first sampling grid and was in
operation in June 1962. It consisted of four intersecting
straight lines in a circular pattern, each being at a 45° angle
from those adjacent to it. Although this grid was discontinued
after two years, it received the most intense testing program.

From 1962 to 1964, this grid (called Grid I) received 39,550 kg
2,4-D and 39,550 kg of 2,4,5-T as the Herbicide Purple formulation
(50 percent n-butyl 2,4-D, 30 percent n-butyl 2,4,5-T, and 20
percent iso-butyl 2,4,5-T). Two other testing grids were sprayed
with Herbicide Orange (50 percent n-butyl 2,4-D and 50 percent
n-butyl 2,4,5-T). Grid II was an area of 37 ha and located
immediately north of Grid I. Grid II received 15,890 kg 2,4,5-T
from 1964 through 1966. Grid IV was the latest and final Grid
established on Test Area C-52A. It was approximately 97 ha and
received 17,570 kg 2,4,5-T from 1968 and 1970. These data are
presented in Table 1.

Dioxin Residues

As previously noted, the testing of aerial spray equipment
terminated in early 1970. Bioassay and chemical studies of the
soils of the test area during 1970 indicated that although the
soil concentrations of phenoxy herbicides were initially high
(parts per million) in April 1970, these residues rapidly
decreased so that in December 1970 only parts per billion of
phenoxy herbicide (primarily 2,4,5-T) could be detected.

What about the dioxin residues? In the spring of 1971, we
selected soil cores that previously had high levels of herbicide
residues in them as indicated by the bioassay studies. These
cores were collected from Grid IV, an area sprayed with Herbicide
Orange from 1968 to early 1970. The samples were analyzed by the
USDA Pesticide Degradation Laboratory, Beltsville, Maryland, and
found to be negative at a detection limit of 0.0005 ppm TCDD. No
additional samples were collected until 1973. Data from studies
conducted by other researchers suggested that TCDD does persist in
the soil. We critically reviewed all our data on the history of
the test range. It became obvious that if the herbicides produced
in the late 1950's and early 1960's were more contaminated with
TCDD than materials produced later in 1960's, the site we should
search would be Grid 1. Simultaneously the detection limit for
TCDD was now approaching the level of parts per trillion (ppt).

Table 2 shows the first positive analysis of TCDD on the test
area. The data in the table suggested soil penetration of TCDD.
Hence, the following summer (1974), this site on Grid I was
re-sampled. The data in Table 3 show that TCDD was confined to
the top 15 cm and that sampling procedures accounted for the
previous results showing contamination below 15 cm. Over the next
few years, numerous soil samples were collected from each of the
test grids that had received 2,4,5-T Herbicide. A summary of
these data is shown in Table 4. Although more 2,4,5-T was applied
to Grid I than to the other grids (see Table 1), it was applied
years earlier. Note that the mean value between Grids I and IV is
approximately 1 magnitude in difference as is the maximum value

Table 1.   Approximate Amount of 2,4,5-T and Estimated Amount of
           TCDD Applied to Test Area C-52A, Eglin AFB Reservation,
           Florida, 1962-1970

Test Grid	Grid Area (ha)	2,4,5-T[a] (kg)	TCDD[b] (kg)
I	37.25	39,550 (1962-1964)[c]	2.613
II	37.25	15,890 (1964-1966)	0.078
IV	97.0	17,570 (1968-1970)	0.087
	----	------------	-----
Total	171.5	73,010	2.778

[a] Amount of 2,4,5-T calculated on weight of active ingredient
2,4,5-T in the military herbicides Orange or Purple.
[b] Amount of TCDD calculated from data on mean concentration of
TCDD in the formulation of herbicides Purple or Orange,
i.e., 32.8 ppm TCDD in Purple and 1.98 ppm TCDD in Orange.
[c] Years when the specific grid received the herbicide
contaminated with TCDD.

between the two grids. Considering the time differences (periods)
of application) the data suggest that indeed levels of TCDD in
Herbicide Orange were significantly less than the levels of TCDD
in Herbicide Purple.

Animal Studies

     Studies of the animals of Test Area C-52A began in 1970.
However, detailed investigations of key species (e.g., the
beachmouse, Peromyscus polionotus, and the six-line racerunner,
Cnemidophorus sexlineatus) did not begin until 1973. Key species
have been repeatedly studied in subsequent years (1974, 1975 and
1978). Birds of the test area were studied in 1974 and 1975.
Insect studies were conducted in 1971 and 1973, while aquatic
communities were initially examined in 1970 and again in 1973 and
1974. List of species, description of habitats and TCDD residue
analysis were conducted throughout all years of study.

Table 2. Levels (parts per trillion) of TCDD
          in Soil from TA C-52A Collected in either June or
          October 1973

Depth cm	Location				
	C-9[a]	F-7[a]	0-7[b]	Grid-I	Control
0-15	<10[c]	11	30	710	20[d]
15-30	ND[e]	ND	<10	140[f]	<10
30-45	ND	ND	<10	72[f]	<10
45-60	ND	ND	<10	<10	<10
60-75	ND	ND	<10	<10	<10
75-90	ND	ND	<10	<10	<10

[a] Located on Grid IV.
[b] Located on Grid II.
[c] Lower limit of detection in parts per trillion TCDD.
[d] Probable interference from excessive organic matter.
[e] ND = Not determined.
[f] Probable contamination due to sampling method.

        Probably, the most startling observation about Test Area
C-52A, is that biological organisms are abundant. The composition
of species is diverse and the distribution extensive. In February
1969, we initiated a "list of species" for the test grids.
Whenever a species was observed on or associated with the grids,
that species was recorded. Over the years of observation,
approximately 341 species or organisms have been observed and
identified as associated with the test area. The sheer number of
species testifies to the extensiveness of the ecological studies
that have been conducted on this unique area. To date 290
biological samples (plants and animals) have been analyzed for
TCDD. TCDD residues have now been found in a wide spectrum of
animals collected from the test area. Approximately one-third
(21) of the different species examined for TCDD residue have been
positive. In general, the levels of TCDD in the organisms
appeared to be close to the mean levels of TCDD found in the
soils.

Table 3. Concentration of TCDD in a Soil Profile from
Grid I, Test Area C-52A, Eglin AFB, Florida

Depth (cm)	Parts per Trillion (ppt) TCDD
0 - 2.5	150
2.5 - 5.0	160
5.0 - 10	700
10 - 15	44
15 - 90	ND[b]

[a]Grid I received 1,062 kg/ha of Herbicide Purple per acre
during 1962-1964. The soil samples were collected and
analyzed in 1974.
[b]None detected, minimum detection limit <10 ppt.

Table 4. Concentration of TCDD (ppt) in Test Grid Soils[a]

Grid	Number of Samples	Range	Median	Mean
I	22	10 - 1,500	110	326
II	6	10 - 470	30	117
IV	26	10 - 150	19	27

[a]Samples collected during the period 1974-1978 with each
sample composed of soil from the top 15 cm of profile.

Studies of the Beachmouse

In the laboratory, teratogenesis has been documented only in the mouse. For other species, fetotoxicity occurs rather than teratogenesis. Thus, we reasoned that a study of the beachmouse at our Eglin test site might provide a key to the impact of TCDD in the environment. Certainly significant concentrations of TCDD have been present in the soils for at least 14 years. Animal populations are diverse and many have been shown to be heavily (by normal field standards) contaminated (up to 2900 ppt TCDD). Many generations of these animals have existed on the test range. When the suspension of Herbicide Orange occurred, by the Department of Defense in April 1970, it was over the concern of toxic effects alleged to be associated with spraying 2,4,5-T contaminated with TCDD. Discussions were held with personnel who had been associated with the spray-equipment testing program on Test Area C-52A. These individuals were asked if they could remember any unusual animal deaths associated with the test grids. They could not. Nevertheless, numerous field trips were conducted on the test grids with the sole purpose of observing the wildlife and search for dead or dying animals. These studies were negative. In all subsequent trapping programs for animals, gross observations for defects, illnesses, and overall health status were made. Nothing out of the "ordinary" (e.g., parasites) has been observed. In 1973, when the data on TCDD in soils first became available, an extensive research effort was initiated on the beachmouse. Subsequent studies were conducted in 1974, 1975, and 1978.

It is clear from our studies that the liver is the site of TCDD accumulation in the beachmouse. The magnitude of those levels apparently depends upon the levels of TCDD in the soil. Our present analytical capability permits an analysis of a single liver sample. The significance of this capability can be seen in Tables 5 and 6. The mound soil is that soil removed during the course of digging the burrow. If it represents mean exposure, then the concentration factor for animals from site 0-4 is between 6 and 7 (500 divided by 75 = 6.67) for females and 18 to 19 for males (1400 divided by 75 = 18.67). The concentration factors for site 0-7 is between 6 and 7 (1900 divided by 285 = 6.67) for females and approximately 9 for males (2600 divided by 285 = 9.12). It is assumed in these studies that body burden levels of TCDD are actually liver levels of TCDD. Beachmice obtained from the field were not found to contain significant levels of body fat, which is consistent with other studies of wild mouse populations.

How long does it take beachmice to accumulate these levels of TCDD, assuming no previous exposure? In 1974 we obtained beachmice from our control site at Eglin and raised an animal

Table 5.   TCDD (ppt) in Soil and Beachmice, Site 0-4,
Table   Grid I, Test Area C-52A, April 1978

SOIL			BEACHMICE				
0-5 cm	=	150	Burrow 1.	Female:	Liver	=	500
					Pelt	=	110
5-10 cm	=	155					
				Pups:	Liver	=	500
10-15 cm	=	70			Pelt	=	150
Mound Soil =		75		Fetuses:	Whole		
					Body	=	40
			Burrow 2.	Female:	Liver	=	490
					Pelt	=	40
				Fetuses:	Whole		
					Body	=	90
			Composite Males:		Liver	=	1400
					Pelt	=	160

Table 6.   TCDD (ppt) in Soil and Beachmice, Site 0-7, Grid I,
            Test Area C-52A, April 1978

SOIL			BEACHMICE			
0-5 cm	=	510	Female:	Liver	=	1900
				Pelt	=	160
5-10 cm	=	520				
			Fetuses:	Whole		
10-15 cm	=	440		Body	=	150
Mound Soil =		285	Male:	Liver	=	2600
				Pelt	=	150

colony in our laboratory. In October 1975, we transported
"tagged" animals to Grid I and released them. Three months later,
we recaptured a small number of these animals. At the time of
recapture we also captured animals apparently endigenous (native)
to the site. The results are shown in Table 7. The data suggest
that body burden levels are obtained at this site within 3 months.
Note that the body burden levels are between 5 and 8 for tagged
and natives, respectively. However, the tagged consists of both
sexes.

     Do these levels of residue impact the health and reproduction
of the beachmouse? The approach we took to answer this question
was to collect beachmice from the test grid and a control site and
compare them for as many parameters as possible. In those females
that were pregnant, the fetuses were also critically examined.
All animals were prepared for examination using a cervical
dislocation procedure to accomplish humane euthanasia.
Euthanatized animals were photographed, weighed, measured, and
systematically examined for developmental defects such as cleft
palate, cleft lip, polydactyly, and microphthalmia. All internal
organs were examined for gross lesions and individually weighed.
Representative sections of each tissue were placed in neutral 10
percent buffered formalin and processed for microscopic study by
the Veterinary Pathology Division, Armed Forces Institute of
Pathology, Washington, D.C. 20305. Remaining liver tissues from

Table 7. A TCDD Exposure Study of Beachmice Released and
          Recaptured on Grid I, Test Area C-25A, Eglin AFB,
          Florida
_____

          MEAN SOIL CONCENTRATION (0-15 cm) = 326 ppt

               BEACHMICE RELEASED – 16 Sept 75

               BEACHMICE CAPTURED – 22 Dec 75

                           RESULTS

          TAGGED: Liver = 1700 ppt (Male/Female)
                   Pelt = 200 ppt

          NATIVE: Liver = 2600 ppt (Male)
                   Pelt = 190 ppt

_____

mice captured in the test and control areas were placed in glass vials, frozen, and retained for TCDD analysis.

Histopathological examinations were performed on 255 adult, immature, or fetal beachmice from the test area and a control area. The animals examined are those listed in Tables 8 and 9 for control and test, respectively. Examinations were performed on the heart, lungs, trachea, salivary glands, thymus, liver, kidneys, stomach, pancreas, adrenals, large and small intestine, spleen, genital organs, bone, bone marrow, skin and brain (Table 10). Initially, the tissues were examined on a blind study basis. All microscopic changes were recorded including those interpreted as minor or insignificant. The tissues were then reexamined on a control versus test basis, which demonstrated that the test and control mice could not be distinguished histopathologically.

Tables 11 and 12 provide body and organ weights for the pregnant female beachmice and the mature male beachmice, respectively. These were the two largest segments of the population that were captured. A two-factor (treatment and year) disproportional analysis of covariance of organ weights revealed that liver weights for pregnant females were significantly heavier ($P<.01$) between the control and test area beachmice, and these differences were consistent over the years of observation. These data are displayed in Table 13. The increase in liver weight may reflect an increase in enzymatic activity associated with low level exposure to TCDD. An ultrastructural study of liver tissue from the test and control site females found no morphologic differences.

The mean number of fetuses per observed pregnancy was 3.4 and 3.1 for the control area and the test area, respectively. A single female beachmouse is capable of producing litters every 26 days. At this frequency, the animals collected in 1978 may have been at least 50 generations removed from the population studies in 1973.

CONCLUSIONS

We believe that our data from studies of TCDD in the natural environment permit the following conclusion:

a. When massive quantities of 2,4,5-T were aerially applied to a spray-equipment testing grid at Eglin AFB, Florida, detectable levels of TCDD could be found in some soils 14 years after the last application of herbicide.

b. Organisms that come into direct and intimate contact with TCDD-contaminated soil generally become contaminated themselves.

Table 8.   Number of Beachmice, <u>Peromyscus polionotus</u>, Collected
           from Control Area, Test Area C-52A, Eglin AFB, Florida

| | | YEAR | | | |
MATURITY, SEX	1973	1974	1975	1978	Total
Mature					
Male	4	11	3	2	20
Female	3(3)[a]	8(3)	3(1)	2(2)	16(9)
Immature					
Male	1	1	0	0	2
Female	0	2	0	0	2
Fetuses	12	11	3	5	31

Grand Total   71

[a]Number of Pregnant Females
Fetuses/Pregnancy = 3.4

Table 9.   Number of Beachmice, <u>Peromyscus polionotus</u>, Collected
           from Test Grid I, Test Area C-52A, Eglin AFB, Florida

| | | YEAR | | | |
MATURITY, SEX	1973	1974	1975	1978	Total
Mature					
Male	18	14	7	7	46
Female	15(6)[a]	9(6)	6(4)	6(6)	36(22)
Immature					
Male	8	3	7	6	24
Female	1	4	3	3	11
Fetuses	25	25	12	21	67

Grand Total 184

[a]Number of Pregnant Females
Fetuses/Pregnancy = 3.1

Table 10.   Histological Parameters Examined in the Beachmouse,
            Peromyscus polionotus, from Test Area C-52A,
            Eglin AFB, Florida[a]

Heart	Pancreas
Lungs	Adrenals
Trachea	Large/Small Intestine
Salivary Glands	Spleen
Thymus	Genital Organs
Liver	Bone
Kidneys	Bone Marrow
Stomach	Skin
	Brain

[a]All miscroscopic examinations were conducted by the Armed
Forces Institute of Pathology, Washington, D.C.

c.   For the beachmouse bioconcentration factors (mean liver
TCDD concentrations divided by mean soil TCDD concentrations)
ranged from 6 for females to 18 for males. In these cases,
it appears that soil is directly ingested so that
accumulation occurred in the absence of food chains.

d.   From studies of organisms that ingested TCDD-contaminated
organisms, the data suggest a simple concentration mechanism
consisting of a single stage, e.g., birds eat insects
contaminated with TCDD-contaminated soil particles.
Biomagnification, i.e., orders-of-magnitude increases of
residue through trophic levels, does not occur.

e.   The ecological studies conducted on Test Area C-52A, have
found no significant adverse chronic toxic effects of TCDD in
animal populations exposed to soil concentrations of TCDD in
the range of 0.1 to 1.5 parts per billion.

f.   The ecological studies conducted on Test Area C-52A,
suggest that long-term, low level exposure to TCDD under
field conditions has had minimal effect upon the health and
reproduction of the beachmouse.

Table 11. Mean Body Weights and Organ Weights of Mature Male Beachmice, Peromyscus polionotus, From Control and TCDD-exposed Field Sites, Test Area C-52A, Eglin AFB, Florida

| Field Site | Year | Animals | Weight (g) | ORGAN WEIGHTS | | | | |
				Liver (mg)	Heart (mg)	Lung (mg)	Spleen (mg)	Kidney (mg)
Control	1973	4	11.88+1.03	708+114	78+15	150+14	33+10	120+27
	1974	11	12.02+1.21	611+111	101+13	99+14	17+6	191+20
	1975	3	13.25+0.24	837+103	133+6	135+1	12+3	268+6
	1978	2	11.91+0.32	667+5	135+8	163+74	21+1	225+10
Test Grids	1973	18	11.84+1.12	819+223	93+23	154+43	35+10	140+60
	1974	14	11.49+0.93	664+150	94+22	100+21	23+15	193+20
	1975	7	11.67+0.83	774+112	134+14	106+25	17+7	225+17
	1978	7	12.25+1.35	756+130	144+37	130+48	22+9	227+39

Table 12. Mean Body Weights and Organ Weights of Mature Male Beachmice, Peromycus polionotus, From Control and TCDD-exposed Field Sites, Test Area C-52A, Eglin AFB, Florida

Field Site	Year	Number of Animals	Body Weight (g)	Liver (mg)	Heart (mg)	Lung (mg)	Spleen (mg)	Kidney (mg)
						ORGAN WEIGHTS		
Control	1973	3	16.29+2.66	955+308	110+35	156+12	33+15	136+40
	1974	3	11.98+0.85	640+220	95+1	103+19	21+14	186+18
	1975	1	15.84	955	149	210	56	311
	1978	2	13.94+3.48	868+140	139+13	144+48	13+3	243+23
Test Grids	1973	6	15.45+2.00	1253+175	113+12	188+35	63+29	248+96
	1974	6	15.75+1.33	1035+98	114+10	108+20	30+16	235+37
	1975	4	16.05+1.74	1138+58	158+4	138+27	26+11	308+15
	1978	6	15.67+2.20	1115+171	173+35	132+45	26+12	294+39

Table 13. Mean Liver Weights (mg) from Pregnant Beachmice, Test Area C-52A, Eglin AFB, Florida

Location	Year	Liver Weight (mg)[a]
Control	1973	955
	1974	640
	1975	955
	1978	868
Grid I	1973	1253
	1974	1053
	1975	1138
	1978	1115

[a]Differences in liver weights between Control and Grid I beachmice were statistically significant at the .01 level of probability.

ULTRASTRUCTURAL COMPARISON OF LIVER TISSUES FROM

FIELD AND LABORATORY TCDD-EXPOSED BEACH MICE

L. G. Cockerham* and A. L. Young**

*Armed Forces Radiobiology Research Institute
 Bethesda, Maryland, USA

**Office of Environmental Medicine
 Veterans Administration
 Washington, DC, USA

ABSTRACT

Quantitative studies were conducted on tissue and organs for beach mice, Peromyscus polionotus, exposed to the toxin 2,3,7,8-tetrachlorodibenzo-p-dioxin (TCDD) in field and laboratory environments. Hepatic tissue from 52 animals was examined using an ultrastructural stereological technique to determine differences in smooth endoplasmic reticulum (SER), rough endoplasmic reticulum (RER) and mitochondria. Organs were examined for differences in weights. Histopathological examinations were also performed on various tissues. To support the studies, chemical analysis of the soil, liver and pelt samples for TCDD content as well as the determination of the TCDD concentration in the alumina gel used in the dusting study was conducted. Fifteen animals were collected from a unique military test site in Northwest Florida where they had been continuously exposed to soil levels of 10 to 710 parts-per-trillion (ppt) TCDD. Twelve other animals from a field control site were exposed 10 times in 28 days to 2.5 parts-per-billion (ppb) TCDD applied as an alumina gel dust to their pellage. All remaining animals were from the same field control site and were not exposed to TCDD in the field or in the laboratory and therefore served as controls. The levels of TCDD in composite liver samples from mice collected in the field varied from 960 ppt for females to 1300 ppt for males, while a composite liver level of TCDD for the laboratory animals was 125 ppt. No significant histopathological lesions were observed. When field animals were compared with laboratory

animals, significant differences were noted in certain organ
weights and cellular structural components. Some of these
differences could possibly be attributed to dissimilarities in
field and laboratory environments.

INTRODUCTION

It is well know that animal species vary in their sensitivity
to toxic substances and that it is difficult to extrapolate data
from one species to another. Likewise, increasing evidence
points to the importance of such factors as handling, caging,
noise, light, etc. as factors in determining and predicting
toxicity (Doull, 1980). Even in a rigidly controlled laboratory
environment the reaction of any species may vary considerably
from that found in the species when in its natural habitat.
Accordingly, this study was designed to assess the validity of
extrapolating data from animals exposed to TCDD in the laboratory
to predict reactions of animals exposed to TCDD in their natural
field environment.

A suitable field site for this study must necessarily (1) be
contaminated with readily detectable quantities of TCDD, (2) have
an endemic animal population present, and (3) be isolated from
human activity, yet available for investigation. A unique site
in Northwest Florida possessing these criteria has been reported
by Young (1974). In support of programs testing aerial spray
systems for use in the Vietnam conflict, Test Area (TA) C-52A,
Eglin Air Force Base, Florida, received massive quantities of
military herbicides. This test area of approximately 2.6 km^2
received approximately 73,000 kg 2,4,5-trichlorophenoxyacetic
acid (2,4,5-T) containing TCDD as a contaminant. Of major
interest was a 0.4 km^2 plot, known as Grid I, located in the
southern portion of the testing area. Significant levels (10-710
parts-per-trillion, ppt) of TCDD in the top 15 cm of the soil
from Grid I were reported by Young (1974). Grid I, then, was
selected as the field test site.

Four designated areas approximately 800 to 1600 meters east
of TA C-52A had not received applications of 2,4,5-T and
therefore were considered as control areas for obtaining animals.
A more detailed description of TA C-52A, its history and prsent
status, may be found in reports by Young (1974) and Young et al,
(1975, 1978, 1979).

The most common mammalian species reported by Pate et al,
(1972) and Young (1974) on TA C-52A is the beach mouse,
Peromyscus polionotus. This was the animal of choice for
investigating effects of TCDD in the field and in the laboratory
because mice have been used extensively in toxicological studies

of TCDD by such investigators as Greig and DeMatteis (1973), Gupta et al, (1973); Schwetz et al, (1973); and Vos et al, (1974) and thus provide known indicators of toxicity.

MATERIALS AND METHODS

Soil and Seed Analysis

To determine the actual levels of TCDD in the soil in June 1974, samples of the top 0-15 cm of soil were taken from six sites on Grid I. One of these sites was also subsampled at increments of 0-2.5, 2.5-5.0, 5.0-10.0, and 10.0-15.0 cm. These soil samples, along with soil samples from the four designated control areas approximately 800 to 1600 meters east of Grid I, were later analyzed for TCDD concentrations.

Animal Description

The beach mouse is a small rodent weighing about 13 grams (g) approximately 120 mm in length, with brown (adult) or dark gray (juvenile) fur on the back, and pale gray to white fur on the ventral region and legs. It may be found in old field habitats and in areas of 5 percent to 60 percent vegetative cover, preferring sandy areas.

Field work for this study was conducted in June and July 1974. Havahart traps (Havahart Traps, Dept. 1, P.O. Box 551, Ossing, New York 10562), sizes 0 and 1, for small animals, were used to trap the rodents. The traps were baited with a mixture of peanut butter and oatmeal and then randomly placed on areas of the test grid where 20 percent to 80 percent vegetative coverage was present, or near openings to mouse burrows. The four designated control areas approximately 800 to 1600 meters east of Grid I were trapped in the same manner as was Grid I.

Traps were checked daily and were moved to other locations within the test and control areas after 4 days failure to catch an animal. Fifty-three live mice were captured and taken to the laboratory for histopathologic examinations, hepatic ultra-structural study, and chemical analysis of the tissue. Fifteen of the mice captured from Grid I were designated as treated field animals and the first 15 mice captured from the control area were designated as control field animals. The remaining 23 mice from the control area were selected to be used as subjects in a laboratory dusting study.

Laboratory Study

Twenty-three of the beach mice captured from the designated

control areas were brought into the laboratory and individually
placed in separate Iso-cages (Carworth, Division of Becton,
Dickinson and Co., New York) and maintained on laboratory chow
(Ralston Purina Company, General Offices, Checkerboard Square,
St. Louis, Missouri). The 23 animals were weighed, sexed, and
randomly divided (using a random numbers table) into a "control"
group of 11 animals (four female and seven male) and a "test"
group of 12 animals (five female and seven male). These animals
were observed for 2 or 3 weeks (depending on date captured) in
the laboratory to determine grooming habits and to allow for
metabolic stabilization after change in diet before dusting was
initiated.

The fur on the ventral thoracic and abdominal regions, sides,
back and tail on each test animal was dusted with 100 mg of
alumina gel containing 2.5 parts-per-billion (ppb) TCDD by
analysis. Control animals were dusted in the same areas but with
alumina gel alone. All dusting was accomplished using a camel
hair artist's brush. The 100 mg application per animal resulted
in an approximate exposure of 60 mg of gel at each application
per animal (based on average weight of recovered residue
following dusting).

The dusting procedure was repeated every third day for a
total of 10 applications during a 28-day period. On the 29th day
the 22 mice (one control animal died apparently as a result of
handling) were sacrificed and prepared for examination.

Animal Preparation and Examination

The 30 mice selected for the field study and the 22 mice from
the laboratory study were prepared for examination using a
cervical dislocation procedure to accomplish humane euthanasia.
All animals were then weighed, skinned and systematically
examined for gross developmental defects such as cleft palate,
cleft lip and polydactyly. Body and organ weights were recorded,
internal organs were examined for gross lesions and representative
sections of each tissue were placed in neutral 10 percent
buffered formalin and processed for ultrastructural studies. All
remaining liver tissues and pelts were pooled according to the
study, sex and treatment, placed in glass jars, frozen and
submitted for TCDD analysis.

Hepatic Ultrastructural Study

After the liver was removed from the 52 beach mice and
weighed, a section approximately 1 mm thick was taken from across
the central lobe (Lobus centralis). This section, to be used for
the ultrastructural study, was prepared for photographic analysis
using procedures outlined by Cockerham (1979).

Data for analysis were obtained from the electron micrographs using morphometric procedures described by Cockerham (1979). This modified technique employed a method of extrapolating from areas to volumes using a system of "point counting." The volume fraction of each cell structure is considered by Meek (1976) to be the ratio between the point count of that structure and the total point count of the cytoplasm.

After the volume fraction was determined for each structure of each cell photographed, the means of the volume fractions or ratios were then computed for each animal. In this manner, the ratio of mitochondrial volume (MITO) to cytoplasmic volume (TOT) of the hepatic parenchymal cell was determined for each animal as was the ratio of damaged mitochondrial volume (d.MITO) to total mitochondrial volume, rough endoplasmic reticulum (RER) to cytoplasm, smooth endoplasmic reticulum (SER) to cytoplasm and RER to SER. These volume fractions or ratios were used as quantitative measurements of the structures to compare the hepatic parenchymal cells from treated animals with those from control animals.

TCDD and Histopathological Analyses

To support the ultrastructural studies, analysis of the soil, seed, liver, and pelt samples for TCDD content, as well as the determination of the TCDD concentration in the alumnina gel used in the dusting study, was conducted by Interpretive Analytical Services, Dow Chemical, Midland, Michigan, USA. Histopathological examination of internal organs was accomplished by the Veterinary Pathology Division, Armed Forces Institute of Pathology, Washington, DC, USA.

Statistical Analysis

The Wilcoxon Rank Sum Test was used to analyze statistically the hepatic morphometric data. The body weight and organ weight data were analyzed statistically by Analysis of Covariance performed using the body weight as a covariate thereby eliminating the variations in organ weight caused by variations in body weight (Shirley, 1977).

RESULTS

Since this study is concerned with a comparison between field and laboratory animals, only the pertinent data having a bearing on that comparison will be presented here. For further comparison and discussion refer to Cockerham et al, (1980) and Cockerham and Young (1982).

Soil Analysis

There were wide fluctuations in TCDD concentrations in the mixed soil from Grid I, with TCDD concentrations of 10, 25, 70, 70, 110 and 710 ppt. The unmixed 15 cm core, obtained from the site having 110 ppt TCDD, showed 150, 160, 700, and 44 ppt TCDD detected at depths of 0-2.5, 2.5-5.0, 5.0-10.0, and 10.0-15.0 cm, respectively. TCDD was not detected in soil samples taken from the designated control areas.

Analysis of Livers and Pelts

Livers, as well as the pelts of beach mice captured from Grid I, where significantly high soil levels of TCDD were found, displayed evidence of accumulation of TCDD. The male beach mice from Grid I displayed a hepatic TCDD level of 1300 ppt while the level for the females was 960 ppt. The pelt levels were 130 ppt and 140 ppt for the male and female mice, respectively.

The livers of both male and female mice from the control area also contained TCDD, but at a much lower level than those from Grid I, with the males having a TCDD level of 51 ppt and the females 83 ppt. For the males this was only 3.9 percent of the level found in the test animals and for the females only 8.6 percent. With the minimum level of detection at 40 ppt, TCDD was not detected on the pelts of either the control males or the control females.

No TCDD was found in the livers and pelts from beach mice dusted 10 times in a period of 28 days with alumina gel containing no TCDD. The animals dusted with alumina gel containing 2.5 ppb TCDD had detectable levels on their pelts of 45 ppt for males and 89 ppt for females. The pooled sample of liver tissue contained 125 ppt TCDD. (Due to the small amounts of liver tissue available, analysis by sex for TCDD in the liver was not possible.)

Histopathology

The supporting histopathological studies were performed by the Veterinary Pathology Division, Armed Forces Institute of Pathology on both test and control mice with no distinction being made between the animals from the field study and the animals from the laboratory (dusting) study. A series of histological examinations were performed on the heart, lungs, trachea, salivary glands, thymus, liver, kidneys, stomach, pancreas, adrenals, large and small intestines, spleen, genital organs, bone, bone marrow, skin, and brain.

Initially, the tissues were examined on a random basis without the knowledge of whether the mouse was a control or test animal.

All microscopic changes, including those interpreted as minor or insignificant, were recorded. Following the recording of all microscopic findings, the tissues were reexamined on a control and test basis. Results of both studies determined that the test and control mice could not be distinguished on a microscopic basis.

Significant lesions were found in only one mouse, a test mouse from the field study. The liver displayed moderately severe, multifocal, necrotizing hepatitis. Sections from the liver of this animal were stained from a variety of stains in attempts to identify an etiologic agent. Neither bacterial or fungal organisms were identified and the lesions were considered viral induced as they resembled the lesions seen in viral hepatitis of laboratory mice.

The gross lesions observed in the kidney of one test mouse from the field study proved to be severe ectasia of renal veins. Microscopically, the vascular dilatation was interpreted as being of little functional significance. All other lesions observed in both control and test mice were minor and insignificant and of the type normally observed when a large group of animals is examined at the microscopic level.

Organ Weight Analysis

The basic body weight and organ weight data for the field and laboratory studies are shown in Table 1. An analysis of body weights per se was not attempted since the ages of the beach mice were not known and the animals could only be classified by sex and treatment. The means of the four groups are compared graphically, expressed as percent of body weight, in Figures 1 and 2.

The organ weight data were examined with an analysis of covariance using the body weight as the covariate. At the 95 percent level of confidence, using this procedure of analysis, the only difference between the field control and laboratory control groups was in adrenal weight. The laboratory control group had a significantly greater adrenal weight than did the field control group.

When it was noted that the data from one animal of the field-treated group deviated from the mean by 2.5 or more standard deviations, the organ weight data from that animal was omitted and the data from the two treated groups were examined using an analysis of covariance. Again, as in the control groups, the labortory group had a significantly greater adrenal weight (p = 0.05). Additionally, with a 95 percent level of confidence, a difference was noted between liver weights and heart weights. The field-treated animals displayed a significantly greater liver weight and smaller heart weight than did the laboratory treated group.

Table 1. Mean Body Weights and Organ Weights of Peromyscus polionotus from Control or TCDD-Treated Laboratory or Field Studies

Treatment	Body wt (gm)	Liver wt (gm)	Spleen wt (mg)	Adrenal wt (mg)	Kidney wt (mg)	Heart wt (mg)	Lung wt (mg)
Field Control							
Male (9)*	12.29 ±1.17	0.614 ±0.122	16.78 ±3.03	20.44 ±6.58	193.33 ±19.22	102.11 ±13.87	100.11 ±15.05
Female(6)	11.48 ±2.97	0.675 ±0.201	23.83 ±5.31	19.83 ±6.34	180.33 ±42.20	79.0 ±18.35	95.0 ±16.61
Laboratory Control**							
Male (6)	13.17 ±0.64	0.656 ±0.055	14.33 ±4.32	29.00 ±7.32	182.83 ±32.59	109.83 ±31.29	102.17 ±16.81
Female(4)	15.00 ±2.77	0.840 ±0.170	21.00 ±5.29	41.00 ±9.27	236.00 ±26.50	123.00 ±27.39	98.25 ±14.06
Field TCDD-Exposed							
Male (10)	11.05 ±1.39	0.710 ±0.160	22.5 ±12.84	22.00 ±6.83	196.2 ±18.52	93.5 ±23.27	96.6 ±18.08
Female(5)	12.69 ±3.50	0.843 ±0.210	30.8 ±21.78	17.00 ±5.10	203.2 ±46.00	101.0 ±14.30	93.4 ±20.73
Laboratory TCDD-Exposed***							
Male (7)	14.72 ±1.84	0.578 ±0.124	25.71 ±6.78	36.86 ±11.48	218.29 ±18.59	127.00 ±17.44	99.86 ±13.33
Female(5)	13.04 ±1.33	0.760 ±0.121	14.20 ±3.27	28.80 ±5.54	226.40 ±35.38	113.00 ±12.79	95.40 ±15.87

*   Number of animals in treatment in parenthesis
** Dusted with alumina gel
*** Dusted with alumina gel plus 2.5 ppb TCDD

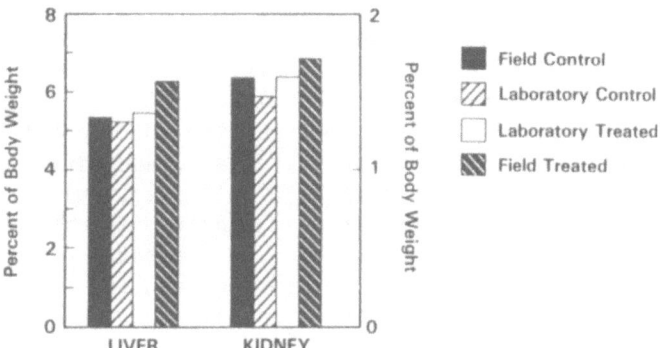

Figure 1.   Comparison of Organ Weights, Expresed as Percent of
            Body Weight, for Control and TCDD-Exposed Beach Mice

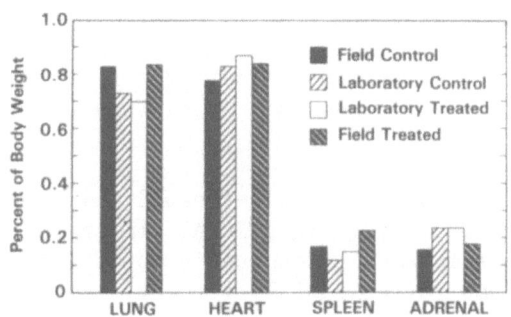

Figure 2.   Comparison of Organ Weights, Expressed as Percent of
            Body Weight, for Control and TCDD-Exposed Beach Mice

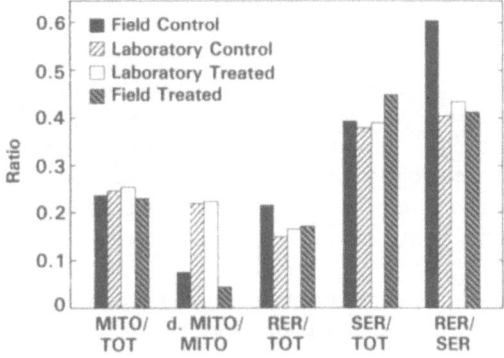

Figure 3.   Hepatic Morphometric Data, Comparison of Means, for
            Control and TCDD-Exposed Beach Mice

Hepatic Morphometric Analysis

The hepatic morphometric data for the field and laboratory
studies are presented as mean values in Table 2. Since morphometric analysis is concerned with the volume fraction of each
structure in question or the ratio between the point count of that
structure and the total count of the cytoplasm, and since the
count for each structure could vary with cell size, only the total
cytoplasmic count was statistically analyzed for differences.
Using the Wilcoxon Rank Sum Test to examine the total counts ($p =$
0.05), no significant difference was seen between the field and
laboratory animals, both control and treated.

After the volume fraction was determined for each required
structure of the photographed cells, the means were computed for
each group and presented in Table 3. Comparisons of the means for
the four groups are presented graphically Figure 3.

Using the Wilcoxon Rank Sum Test to examine the two control
groups, differences in data were noted for RER/cell volume and
RER/SER ratios. For both ratios the data for field control
animals was significantly larger ($p = 0.05$) than that for
laboratory control animals.

When morphometric data from field-treated and laboratory-
treated animals were examined, significant differences were noted
between the groups in damaged mitochondria to total mitochondria
and SER to cell volume ratios. At 95 percent level of confidence
the field-treated animals displayed a smaller damaged mitochondria
to total mitochondria ratio and a larger SER/cell volume ratio.

DISCUSSIONS AND CONCLUSIONS

A factor of concern in interpreting the data was the sample
size for both the field study and the laboratory study. The
number of beach mice in each group, when separated by sex and
treatment, ranged from 5 to 10 in the field study and from 4 to 7
in the laboratory study. In such small samples the deviation of
one individual will strongly influence the data for the entire
group. For this reason, caution must be used in the
interpretation of the results.

The soil samples from the test area displayed wide fluctuations
in TCDD concentrations, probably as the result of unequal
distribution of the herbicide during aerial dissemination. Three
major flight paths intersected at Grid I and the soil samples were
taken from areas thought to be on the flight paths. However, if
the samples were obtained from an area outside the flight paths or

Table 2. Mean Hepatic Morphometic Data of <u>Peromyscus polionotus</u>
          from Control or TCDD-Treated Laboratory or Field Studies

Treatment	Total Count	MITO Count	Damaged MITO Count	RER Count	SER Count
Field Control*					
Male (9)	372.1 +96.02	83.76 +29.85	5.29 +3.98	72.44 +17.63	152.69 +62.33
Female(6)	338.5 +120.47	79.12 +25.84	5.9 +5.87	81.65 +42.70	129.45 +48.60
Laboratory Control**					
Male (6)	426.10 +60.87	108.43 +18.76	26.73 +17.50	63.90 +12.43	165.37 +39.86
Female(4)	383.80 +81.16	90.95 +25.02	21.42 +8.50	54.38 +15.28	142.22 +23.43
Field TCDD-Exposed					
Male (10)	339.1 +81.74	77.12 +19.39	4.06 +5.45	52.22 +18.88	150.3 +48.40
Female (5)	349.2 +97.80	78.48 +15.89	3.06 +2.43	66.48 +17.75	167.06 +65.43
Laboratory TCDD-Exposed***					
Male (7)	380.37 +41.77	101.80 +18.35	24.94 +6.02	65.77 +10.70	148.14 +23.42
Female(5)	402.88 +80.05	95.18 +25.28	22.30 +13.44	59.92 +9.22	159.60 +37.30

*   Number of animals in treatment in parenthesis
**   Dusted with alumina gel
*** Dusted with alumina gel plus 2.5 ppb TCDD

Table 3. Mean Hepatic Morphometric Data, Expressed in Ratios, of
Peromyscus polionotus from Control or TCDD-Treated
Laboratory or Field Studies

Treatment	MITO/ TOT	d.MITO/ MITO	RER/ TOT	SER/ TOT	RER/ SER
Field Control*					
Male (9)	0.228 +0.052	0.066 +0.055	0.207 +0.064	0.399 +0.076	0.584 +0.314
Female(6)	0.251 +0.098	0.083 +0.084	0.231 +0.080	0.383 +0.058	0.640 +0.275
Laboratory Control**					
Male (6)	0.253 +0.033	0.213 +0.105	0.152 +0.022	0.383 +0.038	0.408 +0.084
Female(4)	0.239 +0.038	0.232 +0.046	0.144 +0.015	0.373 +0.034	0.397 +0.062
Field TCDD-Exposed					
Male (10)	0.230 +0.057	0.046 +0.062	0.157 +0.046	0.446 +0.081	0.381 +0.153
Female (5)	0.231 +0.037	0.045 +0.042	0.202 +0.033	0.455 +0.075	0.477 +0.138
Laboratory TCDD-Exposed*					
Male (7)	0.269 +0.047	0.230 +0.046	0.174 +0.027	0.387 +0.033	0.460 +0.098
Female(5)	0.234 +0.023	0.212 +0.092	0.156 +0.037	0.396 +0.028	0.400 +0.117

*   Number of animals in treatment in parenthesis
**   Dusted with alumina gel
*** Dusted with alumina gel plus 2.5 ppb TCDD

from the intersection of all three flights paths, the TCDD levels
would be expected to vary considerably. Nevertheless, analysis of
the soil samples did show that the beach mice from Grid I were
exposed to concentrations of TCDD up to 710 parts-per-trillion
(ppt) in the soil. In contrast, the soil from the control areas
did not contain TCDD at a minimum detection level of six ppt and
therefore did not provide a source of exposure for the control
animals.

The mice from Grid I continually contaminated themselves with
soil containing TCDD through recurrent burrowing in and out of
their nests. The levels of TCDD in the pelt samples from mice
trapped on Grid I confirm this method of contact. In contrast,
TCDD was not detected in pelt samples from control animals.

Although the TCDD levels in the pelt samples from the treated
animals in the laboratory study were not as high as from mice
collected from Grid I, TCDD was not detected in samples from the
laboratory control animals. The relatively short exposure time
(28 days) was probably responsible for the laboratory-treated
animals having lower TCDD levels than the field treated animals.

Kociba et al, (1978), in a chronic 2-year study showed that
rats given 0.01 µg TCDD/kg/day had an average TCDD content of 5100
ppt in the liver. Rats given 0.001 µg TCDD/kg/day had an average
of 540 ppt in the liver. The livers from beach mice collected
from Grid I in this study had a TCDD content of 1300 ppt for males
and 960 ppt for females. Extrapolation of the data would then
give the beach mice a daily TCDD intake dose of approximately
0.0012 µg/kg. Although extrapolation between species is not
always advisable, Fries and Marrow (1975) did state that total
retention of TCDD was closely related to total intake and Leng
(1979) showed that TCDD accumulation in the liver was approximately 2.5 percent of the total intake over a 2-year period.

The 125 ppt TCDD found in the livers of the treated animals of
the laboratory study fall far short of the 540 ppt TCDD in the
livers of rats given 0.001 µg TCDD/kg/day by Kociba et al, (1978),
a dose level that caused no cellular effects considered to be of
any toxicologic significance and within the limits of variation
seen in the controls. Although the actual oral dose in this
dusting study could not be determined, it was probably well below
the 0.001 µg/TCDD/kg level.

TCDD was also found in the livers of the beach mice collected
from the field control area, although at a much lower level. The
presence of TCDD in these pooled samples may have been due to high
levels in one or more mice that could have migrated from the test
area to the control area. A previous trapping study in this area

(Young, 1974) reported the longest random travel distance observed
to be slightly over 900 meters. A travel distance of this
magnitude was considered rare but could account for the presence
of TCDD in the control animals. Nevertheless, even though the
levels in the control mice were low compared to the levels in the
test animals, the use of these mice as true controls must be viewed
with caution.

Histopathological examination confirmed the absence of
significant differences between the four groups of beach mice.
Except for one report of viral hepatitis and one of renal vein
ectasia, all lesions were of the minor or insignificant type
normally observed in microscopic surveys of large numbers of field
animals. Neither of the more serious lesions observed was
considered to result from exposure to TCDD. This is in agreement
with investigators using comparable exposure levels (Kociba et
al, 1978).

Statistically significant differences in organ weight to body
weight ratios were noted between field and laboratory beach mice.
The laboratory mice, both control and treated groups, had a
significantly greater adrenal weight than did the field animals.
Almost any kind of stress could have resulted in this apparent
hypertrophy of the adrenals. The change of environment from field
to laboratory, daily handling and an apparent increase in
population are some of the many factors that could have been
responsible for the increase in adrenal weight.

The liver weight of the field-treated mice was significantly
larger than for the laboratory-treated mice. Since the field-
treated mice showed a liver content of TCDD approximately 10 times
greater than did the laboratory-treated mice and were exposed for
a considerably longer period of time, it is highly probable that
the increased liver weight could be attributed to differences in
dose levels. The differences in laboratory and field environments
probably were not contributing factors since a significant
difference could not be demonstrated between the two control
groups.

The laboratory-treated animals had a significantly higher
heart weight than did the field-treated animals. Although
statistical analysis did not show a difference between the two
groups of control animals, the means displayed in Figure 2 support
the same relationship seen between the two groups of treated
animals. The small size of the laboratory control group may be
responsible for the lack of statistically significant differences.
However, the difference does exist and may be attributed to the
same stress factors responsible for the increased adrenal weights.

The increased number of damaged mitochondria seen in the laboratory-treated animals compared to field-treated animals was not seen in the comparison of the two control groups. However, when the means are examined in Figure 3, it becomes evident that there is a great difference between the two control groups. Hutterer et al, (1969) advanced the concept that morphologic alterations of mitochondria such as bizarre shapes and abnormalities of the inner and outer membranes, indicators of mitochondrial damage, were associated with a hypoactive, hypertrophic smoother endoplasmic reticulum. It is probably, therefore, that some treatment or exposure to the laboratory is responsible for the increase in damaged mitochondria since the same laboratory animals actually showed a decrease in SER rather than an increase.

Van Logten et al, (1981) showed that a difference in diet could influence detoxification of TCDD by the liver. It stands to reason then, that if detoxification is influenced by diet, then the morphologic responses of mitochondria, smooth endoplasmic reticulum and rough endoplasmic reticulum would also be influenced by diet. Since the laboratory diet and field diet differed, this could be one of the factors producing a difference in cellular response.

The apparent increase in smooth endoplasmic reticulum seen in the field-treated mice may be attributed to the increased dose and duration of exposure. This explanation agrees with Hutterer et al, (1969) who state that a proliferation of smooth endoplasmic reticulum occurs with an increase in dose and duration of exposure up to the point of transition from physiological adaptation to toxicity.

Stenger (1970) asserted that although disorganization of the rough endoplasmic reticulum is a common reaction of liver parenchymal cells to injury it is not a specific reaction to chemical agents. He further states that reduced amounts of rough endoplasmic reticulum are not indicative necessarily of hepatocellular injury. If this is true, then, the apparent increase in RER of the field control group may actually be a decrease in RER in the other three groups.

Differences between the two treated groups that do not also appear between the two control groups may be explained by their positive relationship with dose level of TCDD and exposure time. However, since the laboratory control animals were initially taken from the same population as the field control animals, any differences in the two control groups can be attributed to the effects of their different environments. Therefore, this study affirms that animals of the same strain exhibit different cellular responses with changing environments.

If an animal in its natural environment is considered as a
true control, then the laboratory must be considered as a test
environment, for, even as some toxicity-influencing factors have
been eliminated, others have been introduced. Rarely is the
opportunity presented for a toxicity study of an animal in its
natural environment; consequently, the laboratory must be employed
to evaluate toxicity. Likewise, it is not always permissible or
possible to examine animals of the same taxonomic level, therefore,
extrapolation of data must occur, not only from laboratory to
natural environment but from species to species. This becomes
increasingly difficult as the difference in taxonomic levels
increases and all known toxicity-inducing factors must be
considered before a prediction of toxicity can be made.

REFERENCES

Cockerham, L. G., 1979, Ultrastructural study of selected tissues
    mice exposed to 2,3,7,8-tetrachlorodibenzo-p-dioxin (TCDD),
    Doctoral Dissertation, Colorado State University, Fort Collins,
    Colorado, December.
Cockerham, L. G., Young, A. L., and Thalken, C. E., 1980, Histo-
    pathological and ultrastructural studies of liver tissue from
    TCDD-exposed beach mice (Peromyscus polionotus). FJSRL-TR-80-
    0008, Frank J. Seiler Research Laboratory, USAF Academy,
    Colorado, March.
Cockerham, L. G., and Young, A. L., 1982, The absence of hepatic
    cellular anomalies in TCDD-exposed beach mice -- a field
    study. (Submitted for publication).
Doull, J., 1980, Factors influencing toxicology, in: "Casarett and
    Doull's Toxicology, The Basic Science of Poisons," Doull, J.,
    Klassen, C. D., and Amdur, M. O., eds., 2nd ed. New York:
    Macmillan Publishing Co., Inc., 70-83.
Fries, G. F., and Marros, G. S., 1975, Retention and excretion of
    2,3,7,8-tetrachlorodibenzo-p-dioxin by rats, J. Agr. Food Chem.,
    23(2):265-267.
Greig, J. B., and DeMatteis, F., 1973, Effects of 2,3,7,8-tetra-
    chlorodibenzo-p-dioxin on drug metabolism and hepatic micro-
    somes of rats and mice, Environ. Health Perspect., 5:211-219.
Gupta, B. N., Vos, J. G., Moore, J. A., Zinki, J. G., and Bullock,
    G. C., 1973, Pathologic effects of 2,3,7,8-tetrachlorodibenzo-
    p-dioxin in laboratory animals, Environ. Health Perspect.,
    5:125-140.
Hutterer, F., Klion, F. M., Wengraf, A., Schaffner, F., and
    Popper, H., 1969, Hepatocellular adaptation and injury:
    structural and biochemical changes following dieldrin and
    methyl butter yellow, Lab.Invest., 20:455-464.

Kociba, R. J., Keyes, D. G., Beyer, J. E., Carreon, R. M., Wade,
    C. E., Dittenber, D. A., Kalnins, R. P., Frauson, L. E., Park,
    C. N., Barnard, S. D., Hummel, R. A., and Humiston, C. G.,
    1978, Results of a two year chronic toxicity and oncogenicity
    study of 2,3,7,8-tetrachlorodibenzo-p-dioxin in rats, Toxicol.
    App. Pharmacol., 46:279-303.
Leng, M. L., 1979, Comparative toxicology of various chlorinated
    dioxins as related to chemical structure. Paper presented at
    symposium of Collabortive International Pesticide Advisory
    Council (GIPAC), Baltimore, Maryland, June 7.
Meek, G. A., 1976, Stereology, in: "Practical Electron Microscopy
    for Biologists," 2nd ed. New York: John Wiley and Sons, 409-410.
Pate, B. D., Voight, R. C., Lehn, P. J., and Hunter, J. H., 1972,
    Animal Survey of Test Area C-52A, Eglin AFB Reservation, Florida,
    AFATL-TR-72-72, Air Force Armament Laboratory, Eglin AFB,
    Florida, April.
Schwetz, B. A., Norris, J. M., Sparschu, G. L., Rowe, V. K.,
    Gehring, P. J., Emerson, J. L., and Gerbig, C. G., 1973,
    Toxicology of chlorinated dibenzo-p-dioxins. Environ. Health
    Perspect., 5:87-99.
Shirley, E., 1977, The Analysis of orgen weight data, Toxicology,
    8:13-22.
Stenger, R. J., 1970, Organelle pathology of the liver: the endo-
    plasmic reticulum, Gastroenterology, 58:554-574.
Van Logten, M. J., Guypta, B. N., McConnell, E. E., and Moore,
    J. A., 1981, The influence of malnutrition on the toxicity
    of 2,3,7,8-tetrachlorodibenzo-p-dioxin (TCDD) in rats,
    Toxicology, 21(1):77-88.
Vos, J. G., Moore, J. A., and Zinkl, J. G., 1974, Toxicity of
    2,3,7,8-tetrachlorodibenzo-p-dioxin (TCDD) in C57Bl/6 mice,
    Toxicol Appl. Pharmacol., 29(1):229-241.
Young, A. L., 1974, Ecological studies on a herbicide equipment
    test area (TA C52-A), Eglin AFB Reservation Florida. AFATL-
    TR-74-12, Air Force Armament Laboratory, Eglin AFB, Florida,
    January.
Young, A. L., Thalken, C. E., Ward, W. E., 1975, Studies of the
    ecological impact of herbicides on the ecosystem of test area
    C-52A, Eglin AFB, Florida. AFATL-TR-75-142, Air Force
    Armament Laboratory Eglin AFB, Florida, October.
Young, A. L., Calcagni, J. A., Thalken, C. E., and Tremblay, J. W.,
    1978, The toxicology, environmental fate and human risk of
    Herbicide Orange and its associated dioxin, OEHL-TR-78-92,
    USAF Occupational and Environmental Health Laboratory, Brooks
    AFB, Texas, October.
Young, A. L., Thalken, C. E., and Harrison, D. D., 1979,
    Persistence, bioaccumulation and toxicology to TCDD in an eco-
    system treated with massive quantities of 2,4,5-T herbicide.
    Paper presented at the 178th American Chemical Society
    National Meeting, Washington, D.C., September 13-14.

BIOCHEMISTRY AND METABOLISM

Chairman: Stephen Safe
            Texas A & M University
            College Station, Texas   USA

PCDDs AND RELATED COMPOUNDS: METABOLISM AND BIOCHEMISTRY

*S. Safe, *L. Robertson, *T. Sawyer, **A. Parkinson,
*S. Bandiera, *L. Safe and **M. Campbell

*College of Veterinary Medicine
*Texas A&M University
 College Station, Texas   USA

**Hoffman laRoche Corporation
  Nutley, New Jersey USA

***Guelph-Waterloo Centre for Graduate Work
  In Chemistry
  University of Guelph
  Guelph, Ontario, Canada

ABSTRACT

    Polychlorinated dibenzo-p-dioxins and related halogenated aromatic compounds are a group of chemicals which exhibit similar physical and environmental properties and elicit a number of common toxic and biologic responses. Moreover within each group of halogenated aromatics, there are a multiplicity of isomers and congeners and their metabolism and biochemical effects are remarkable dependent on structure. The metabolism of individual halogenated aromatics thus depends on their degree of halogenation and the specific halogen substitution pattern. Complementary results obtained for the polychlorinated dibenzo-p-dioxins and biphenyls show that the toxic congeners are approximate isostereomers of 2,3,7,8-tetrachlorodibenzo-p-dioxin; furthermore, there exists a correlation between the toxicity of halogenated aromatics and their binding activities to a cytosolic receptor protein and potencies as inducers of microsomal aryl hydrocarbon hydroxylase.

INTRODUCTION

The halogenated aromatic pollutants include the polychlorinated biphenyls (PCBs), naphthalenes (PCNs), terphenyls (PCTs) and benzenes (CB) and the polybrominated biphenyls (PBBs) which have been used widely in industry (Parkinson and Safe, 1981). This group also includes the polychlorinated azo- and azoxybenzenes (PABs), dibenzo-furans (PCDFs) and dibenzo-p-dioxins (PCDDs). The PABs are formed as by-products in the synthesis of chlorinated aniline-derived pesticides, the PCDDs (and PCDFs) are formed as by-products in the synthesis of chlorinated phenol-derived products, PCDFs, PCBs and PCDDS have all been detected as residues on municipal incinerator flyash. One of the major problems associated with the analytical chemistry, biochemistry, metabolism and toxicology of halogenated aromatics is the multiplicity of possible isomers and congeners (Table 1). Thus an understanding of this important and controversial group of chemicals requires a thorough investigation of the effects of structure on their activities.

Table 1. Halogenated Aromatics: Multiplicity of Isomers and Congeners

Halogenated Aromatic	PCBs	PBBs	PABs	PCNs	PCDDs	PCDFs
No. of Congeners	209	209	209	75	75	135

METABOLISM

The metabolism of PCDDs and PCDFs have not been investigated in detail (Tulp and Hutzinger, 1978; Poiger and Schlatter, 1979; Veerkamp et al, 1981); however, in vivo studies have shown: 1) the metabolism of lower chlorinated PCDDs and PCDFs yields hydroxylated metabolites; 2) the limited data suggest that hydroxylation is favored at the 2,3,7 and 8 positions; 3) the rate of hydroxylation of specific substrates tends to decrease with increasing ring chlorination although this is also dependent on the chlorine substitution pattern; 4) 2,3,7,8-tetrachlorodibenzo-p-dioxin (TCDD) is metabolized in vitro (Poland and Glover, 1979) by microsomal proteins to yield polar products and incorporation of [^{3}H]-2,3,7,8-TCDD into cellular macromolecules. The results strongly suggest that PCDDs and PCDFs, in common with many drugs and related xenobiotics, are metabolized by the microsomal mixed function oxidase (MFO) enzymes (Figure 1). However, it is not clear if this metabolic process generates potentially toxic electrophilic intermediates (e.g., arene oxides) or toxic metabolites.

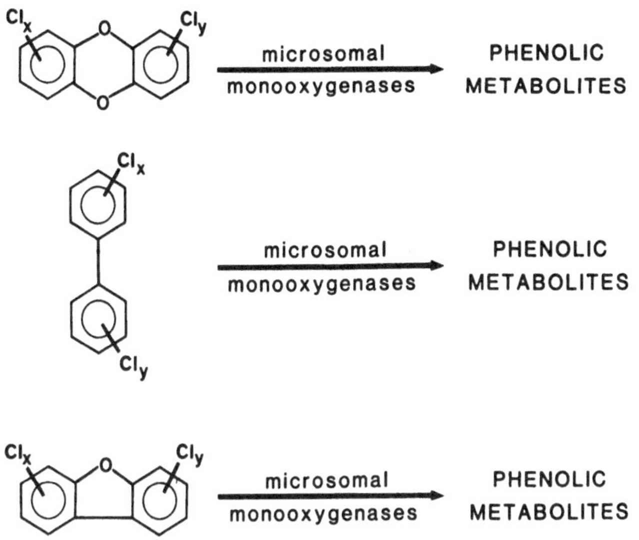

Figure 1.   Metabolism of PCDDs and Related Compounds

        PCB metabolism has been more intensively studied than that of
any of the other halogenated aromatics (Safe, 1980; Sundstrom et al,
1976).   Hydroxylated PCB metabolites have been detected in wildlife
excreta and tissues (Jensen and Hansson, 1976; Jansson et al, 1975;
Jensen et al, 1979), methyl sulfone metabolites were also present in
wildlife excreta and in human milk (Yoshida and Nakamura, 1979).
These metabolites are degradation products of the parent hydrocarbon
mixtures which are routinely found as contaminants in biota.   Administration of commercial PCB preparations to laboratory animals results
in the preferential degradation of the lower chlorinated biphenyls to
give phenolic urinary and fecal metabolites (Burse et al, 1974; Safe
et al, 1975).   The in vivo and in vitro metabolism of radiolabeled
PCB mixtures not only yields polar hydroxylated products but also
results in the formation of covalent PCB-protein adducts (Shimada
and Sato, 1978).   The in vivo and in vitro metabolism of individual
PCB isomers and congeners has been reviewed (Safe, 1980; Sundstrom
et al, 1976).   PCBs are metabolized to a variety of products including
phenols, catechols, dihydrodiols, hydroxylated methyl ethers and
sulfur-containing sompounds.   A number of studies (Safe, 1980) have
confirmed that: 1) PCBs are metabolized by the microsomal cytochrome
P-450-dependent monooxygenases to arene oxide intermediates which
subsequently react to yield the observed metabolites; 2) hydroxylation tends to be favoured at the para position of the less substituted ring unless this site is sterically hindered; 3) metabolic
rates are enhanced if there are two adjacent unsubstituted carbon
atoms available for oxidation; 4) the rate of PCB hydroxylation tends
to decrease with increasing chlorination; however, this is dependent
on the factors noted in 2) and 3).

The metabolism of PBB mixtures has not been thoroughly investigated; however, the metabolism of the major congeneric components of the mixtures has been reported (Moore and Aust, 1978). The results suggest that the effects of bromine substitution were comparable to those of chlorine substitution. One of the more rapidly metabolized PBBs, 2,2',4,5,5'-pentabromobiphenyl, contains two vicinal unsubstituted carbon atoms and this structural feature clearly facilitates the metabolism of the compound and its rapid removal from the adipose tissue of exposed animals. The metabolism of PBBs by microsomal monooxygenases via arene oxide intermediates is also supported by in vivo and in vitro data (Safe et al, 1976, 1978; Kohli et al, 1978).

The metabolism of PCN mixtures also yields phenolic metabolites (Chu et al, 1976, 1977, 1977a; Ruzo et al, 1975, 1976, 1976a) and the results with individual congeners are consistent with the pathways noted for the other halogenated aromatics. The metabolism of 1,4-dichloronaphthalene, 1,4-dibromonaphthalene and 1-chloro-4-[^{2}H]-naphthalene all yield hydroxylated products in which the halogen or ^{2}H have undergone 1,2-migrations from the site of hydroxylation to the adjacent carbon atom (Ruzo et al, 1976). The NIH-shift of these substituents was also observed for other halogenated aromatics and is consistent with their common metabolic pathways as summarized in Figure 1.

Although it is apparent that halogenated aromatics are degraded to yield phenolic products as major metabolites, the effects of structure on the metabolism of PBBs, PCDFs, PCDDs and PCNs has not been determined. Moreover, the role of metabolism in the toxicity of these compounds and the toxicity of the metabolites also requires further investigation.

BIOCHEMISTRY

Administration of halogenated aromatic pollutants to animal species elicits numerous biologic effects (Parkinson and Safe, 1981). Particularly noteworthy is the induction of several drug-metabolizing enzymes including the cytochrome P-450-dependent monooxygenases, glucuronyl transferases, epoxide hydrolases, glutathione S-transferases and NADPH-cytochrome c reductase (Parkinson and Safe, 1981). In a series of excellent papers, Poland and his coworkers investigated the toxic and biologic effects of several PCDD isomers and congeners and observed a remarkable dependence of structure on activity (Poland and Glover, 1973, 1980; Poland et al, 1974, 1976, 1979). Moreover, it was also noted that there was an excellent correspondence between the relative toxicity of an individual PCDD congener and its activity as an inducer of hepatic benzo[a]pyrene hydroxylase (or aryl hydrocarbon hydroxylase, AHH) and its binding affinity to a cytosolic receptor protein. Table 2 illustrates this

comparative data. It is apparent that structure plays an important
role in the activity of PCDDs, i.e., the most active PCDD is substituted in the 2,3,7 and 8 lateral positions; the introduction of
additional chloro substituents or the removal of one or more of the
lateral substituents results in diminished activity. It has also
been proposed that receptor binding plays a critical role in initiating the toxic and biologic effects of 2,3,7,8-TCDD and related
toxic halogenated aromatics (Poland et al, 1979). Using the genetically inbred C57BL/6J and DBA/2J mice strains, it was shown that the
former responsive strain was highly susceptible to the effects of
2,3,7,8-TCDD whereas the latter nonresponsive strain was much less
susceptible. The C57BL/6J mice contain relatively high levels of the
hepatic cytosolic receptor protein compared to the DBA/2J mice (Okey
et al, 1980; Greenlee and Poland, 1979; Hannah et al, 1981; Poland et
al, 1976; Carlstedt-Duke et al, 1981). Thus it was proposed that the
activity of 2,3,7,8-TCDD and related toxic halogenated aromatics
which are approximate isostereomers of 2,3,7,8-TCDD tend to segregate
with the Ah locus which codes for the Ah receptor protein. The proposed mechanism of action of 2,3,7,8-TCDD requires an initial binding of the organic ligand to the receptor, the translocation of the
ligand-receptor protein into the nucleus followed by a coordinate
gene expression or pleiotypic response.

Table 2. PCDDs: Effects of Structure on Activity

PCDD Congener	Relative Binding[1] Affinity	Relative Biological[2] Potency
2,3,7,8-tetrachloro-dibenzo-p-dioxin	100	100
1,2,3,7,8,9-hexachloro-dibenzo-p-dioxin	20	22
octachlorodibenzo-p-dioxin	∿ 0	∿ 0
2,3,7-trichlorodibenzo-p-dioxin	14	0.06
2,7-dichlorodibenzo-p-dioxin	∿ 0	∿ 0

[1]Measured by competitive displacement of [^{3}H]-TCDD from B6
 hepatic cytosol (Poland et al, 1979).
[2]Chick embryo AHH induction (Poland et al, 1979).

Several studies have confirmed that the biologic and toxic effects of PCBs are also dependent on their structure (Poland and
Glover, 1977; Goldstein et al, 1977; Goldstein, 1979; Yoshimura
et al, 1979; Parkinson et al, 1980, a,b,c, 1981). Figure 2 illustrates the four PCB congeners which induce AHH activity, namely
3,3',4,4'-tetra-(TCBP-1), 3,3',4,4',5-penta-(PCBP-1), 3,3',4,4',
5,5'-hexa-(HCBP-1) and 3,4,4',5-tetrachlorobiphenyl (TCBP-2). Enzyme
induction studies in the immature male Wistar rat and with rat
hepatoma H-4-II-E cells in culture showed that all four of these
PCBs induced AHH activity although the 3,4,4',5-tetrachlorobiphenyl
congener was the least active inducer (Sawyer and Safe, in press).
The three most active PCB congeners are approximate isostereomers
of 2,3,7,8-TCDD in their planar conformation and their potencies as
AHH inducers are paralleled by their binding to the cytosolic receptor protein (Bandiera et al, in press) and their reported toxicities (Yoshimura et al, 1978, 1979; Goldstein, 1979; Biocca et al,
1981). Although it was suggested that ortho-chloro substituted
analogs of the four AHH inducers would be inactive due to lack of
coplanarity, subsequent studies in our laboratory have shown this
not to be the case. All the mono-ortho-chloro substituted PCBs
(Figure 2) were AHH inducers (in vivo and in vitro) and bound to the
receptor protein (Parkinson et al, 1980, a,b,c; Bandiera et al, in
press). The toxicities of these compounds have not been systematically studied, but 2,3',4,4'-penta-, 2,3,3',4,4',5-hexa- and
2,3,4,4',5-penta-chlorobiphenyl are toxic to rats or mice (Ax and
Hansen, 1975; Yoshimura et al, 1978, 1979; Yoshibara et al, 1979

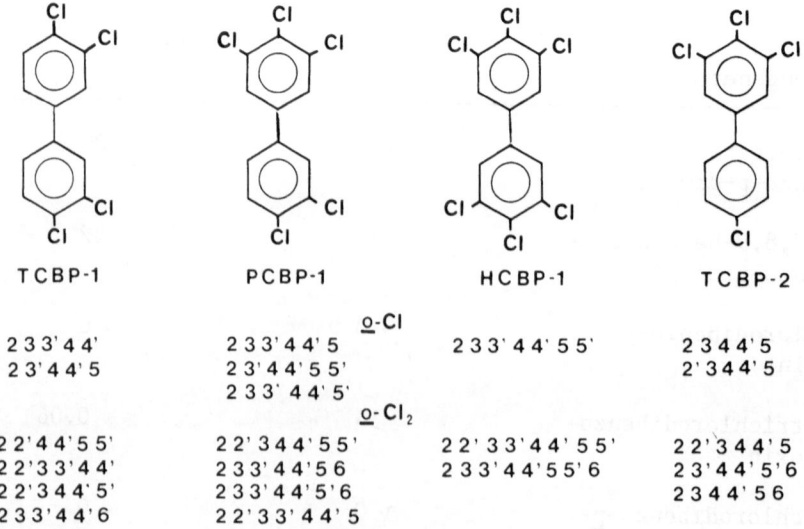

TCBP-1	PCBP-1	HCBP-1	TCBP-2

2 33'4 4'	2 33'4 4' 5	2 33' 4 4' 5 5'	2 3 4 4' 5
2 3' 4 4' 5	2 3'4 4' 5 5'		2' 3 4 4' 5
	2 3 3' 4 4' 5'		

o-Cl (above center column, over PCBP-1)

o-Cl₂ (above center column, over PCBP-1)

2 2'4 4' 5 5'	2 2'3 4 4' 5 5'	2 2' 3 3'4 4' 5 5'	2 2'3 4 4' 5
2 2'3 3'4 4'	2 3 3'4 4' 5 6	2 3 3'4 4'5 5' 6	2 3' 4 4' 5'6
2 2'3 4 4' 5'	2 3 3'4 4'5'6		2 3 4 4' 5 6
2 3 3'4 4'6	2 2'3 3' 4 4' 5		

Figure 2. PCBs as AHH Inducers:
Effects of Structure on Activity

Yamamoto et al, 1976).   We have also shown that some diortho-chloro
substituted PCBs, namely the 2,2',3,3',4,4'-, 2,2',3,4,4',5'-,
2,3,3,3,4,4,6-, 2,3,4,4',5,6-hexachlorobiphenyls and 2,2',3,3',4,4',5-
heptachlorobiphenyl were also relatively "weak" inducers of AHH
activity in the immature male Wistar rat (Parkinson et al, 1981).

Current studies in our laboratory indicate that PCBs which induce
AHH activity in the rat and the rat hepatoma cell cultures elicit comparable effects in C57BL/6J mice but not in DBA/2J mice.   This data
and the results summarized in Table 3 clearly supports the proposed
common mechanism of action for the halogenated aromatics (Poland
et al, 1979).   Despite the complementary data obtained for the PCBs
and PCDDs there are still a number of critical issues which must be
resolved, e.g.: 1) the effects of structure on the activity of PCDFs,
PCNs, PCTs and PBBs; 2) the structural factors which facilitate the
binding of the inducer to the receptor protein; 3) the interaction
between the receptor protein-inducer complex and nuclear material;
and 4) the mechanism of the pleiotypic response to these toxins.

Table 3.   PCB Isomers and Congeners: Effects of Structure
           on Activity as AHH Inducers and Receptor Binding
           Avidity

PCB Congeners	Relative Percent Activity		
	AHH	Induction	Receptor[a] Binding
MC-Type AHH Inducers (3)	$100 - 1^b$	$+++^c$	100 - 35
TCBP-2	2.2×10^{-5}	++	0.9
Mixed-Type AHH Inducers (8)	$0.3-2.4 \times 10^{-5}$	++	6 - 1.5
PB-Type Inducer[d]	Inactive	0	≤ 0.3
Non-Inducers[e]	Inactive	0	≤ 0.3
2,3,7,8-TCDD	400	+++++	2500

[a] To rat hepatic cytosolic protein
[b] Rat hepatoma H-4-11-E cells
[c] Immature male Wistar rats
[d] 2,2',4,4'-tetra and 2,2',4,4',5,5'-hexachlorobiphenyl
[e] 2,2',5,5'-tetra- and 2,3',4,4',5',6-hexachlorobiphenyl

REFERENCES

Ax, R.L. and Hansen, L.G. Effects of purified PCB analogs on chicken reproduction, Poultry Sci., 54:895-900, 1975.

Bandiera, S., Safe, S. and Okey, A.B. Binding of polychlorinated biphenyls classified as either PB-, MC- or Chem-Biol. Inter. (In press).

Biocca, M., Gupta, B.N., Chae, K., McKinney, J.D. and Moore, J.A. Toxicity of selected hexachlorobiphenyl isomers in the mouse, Toxicol. Appl. Pharmacol., 58:461-474, 1981.

Burse, V.W., Kimbrough, R.D., Villanueva, E.C., Jennings, R.W., Linder, R.E. and Sovocol, G.W. Polychlorinated biphenyls, Arch. Environ. Health, 29:301-307, 1974.

Carlstedt-Duke, J.M.B., Harnemo, U.B., Högberg, B. and Gustafsson, J.A. Interaction of the hepatic receptor protein for 2,3,7,8-tetrachlorodibenzo-p-dioxin with DNA, Biochem. Biophys. Acta., 672:131-141, 1981.

Chu, I., Secours, V. and Viau, A. Metabolites of chloronaphthalene, Chemosphere, 5:439-444, 1976.

Chu, I., Secours, V., Villeneuve, D.C. and Viau, A. Metabolism and tissue distribution of [1,4,5,8-^{14}C]-1,2-dichloronaphthalene in rats, Bull. Environ. Contam. Toxicol., 18:177-182, 1977.

Chu, I., Villeneuve, D.C., Secours, V. and Viau, A. Metabolism of chloronaphthalenes, J. Agr. Food Chem.,25:881-883, 1977a.

Goldstein, J.A. The structure-activity relationship of halogenated biphenyls as enzyme inducers, Ann. NY Acad. Sci., 320:164-178, 1979.

Goldstein, J.A., Hickman, P., Berman, H., McKinney, J.D. and Walker, M.P. Separation of pure polychlorinated biphenyl isomers into two types of inducers on the basis of induction of cytochrome P-450 or P-448, Chem-Biol. Interact., 17:69-87, 1977.

Greenlee, W.F. and Poland, A. Nuclear uptake of 2,3,7,8-tetrachloro-dibenzo-p-dioxin in C57BL/6J and DBA/2J mice, J. Biol. Chem., 254:9814-9821, 1979.

Hannah, R.R., Nebert, D.W. and Eisen, H.J. Regulatory gene product of the Ah locus. Comparison of 2,3,7,8-tetrachlorodibenzo-p-dioxin and s-methylcholanthrene binding to several moieties in mouse liver cytosol, J. Biol. Chem., 250:4584-4590, 1981.

Kohli, J., Wyndham, C., Smylie, M. and Safe, S. Metabolism of bro-mobiphenyls, Biochem. Pharmacol., 27:1245-1249, 1978.

Jansson, B., Jensen, S., Olsson, M., Renberg, L., Sundstrom, G. and Vaz, R. Identification by GC-MS of phenolic metabolites of PCB and p.p'-DDE isolated from Baltic guillemot and seal, Ambio., 4:93-97, 1975.

Jensen, S., Jansson, B. and Olsson, M. Methyl sulfone metabolites of PCB and DDE, Ambio, 5:257-260, 1976.

Jensen, S., Jansson, B. and Olsson, M. Number and identity of anthro-pogenic substances known to be present in Baltic seals and their possible effects on reproduction, Ann. NY Acad. Sci., 320:436-448, 1979.

Moore, R.W. and Aust, S.D. Purification and structural character-
    ization of polybrominated biphenyl congeners, Biochem. Biophys.
    Res. Commun., 84:936-942, 1978.
Okey, A.B., Bondy, G.P., Manson, M.E., Nebert, D.W., Forster-Gibson,
    C.J., Muncan, J. and Dufresne, M.J. Temperature-dependent
    cytosol-to-nucleus translocation of the Ah receptor for 2,3,7,8-
    tetrachlorodibenzo-p-dioxin in continuous cell culture lines.
    J. Biol. Chem., 255:11415-11422, 1980.
Parkinson, A. and Safe, S. Aryl hydrocarbon hydroxylase induction
    and its relationship to the toxicity of halogenated aryl hydro-
    carbons, Toxicol. Environ. Chem., 4:1-46, 1981.
Parkinson, A., Robertson, L., Safe, L. and Safe, S. Polychlorinated
    biphenyls as inducers of hepatic microsomal enzymes: Effects
    of diortho substitution, Chem. Biol. Interact., 35:1-12, 1981.
Parkinson, A., Cockerline, R. and Safe, S. Induction of both 3-
    methylcholanthrene- and phenobarbitone-type microsomal enzyme
    activity by a single polychlorinated biphenyl isomer, Biochem.
    Pharmacol., 29:259-262, 1980.
Parkinson, A., Cockerline, R. and Safe, S. Polychlorinated biphenyl
    isomers and congeners as inducers of both 3-methylcholanthrene-
    and phenobarbitone-type microsomal enzyme activity, Chem. Biol.
    Interact., 29:277-289, 1980a.
Parkinson, A., Robertson, L. and Safe, S. Hepatic microsomal enzyme
    induction by 2,2',3,3',4,4'- and 2,2',3',4,4',5-hexachlorobi-
    phenyl, Life Sciences, 27:2333-2337, 1980b.
Parkinson, A., Robertson, L., Safe, L. and Safe, S. Polychlorinated
    biphenyls as inducers of hepatic microsomal enzymes: Structure-
    activity rules, Chem. Biol. Interact., 30:271-285, 1980c.
Poiger, H. and Schlatter, Ch. Biological degradation of TCDD in rats,
    Nature, (London) 281:706-707, 1979.
Poland, A. and Glover, E. An estimate of the in vivo covalent bind-
    ing of 2,3,7,8-tetrachlorodibenzo-p-dioxin to rat liver protein,
    ribosomal RNA and DNA, Cancer Res., 39:3341-3344, 1979.
Poland, A. and Glover, E. Chlorinated dibenzo-p-dioxins: Potent in-
    ducers of δ-aminolevulinic acid synthetase and aryl hydrocarbon
    hydroxylase II. A study of the structure-activity relationship,
    Mol. Pharmacol., 9:736-747, 1973.
Poland, A. and Glover, E. Chlorinated biphenyl induction of aryl
    hydrocarbon hydroxylase activity: A study of the structure acti-
    vity relationship, Mol. Pharmacol., 13:924-938, 1977.
Poland, A. and Glover, E. 2,3,7,8-tetrachlorodibenzo-p-dioxin:
    Segregation of toxicity with the Ah locus, Mol. Pharmacol.,
    17:86-94, 1980.
Poland, A., Greenlee, W.F. and Kende, A.S. Studies on the mechanism
    of toxicity of the chlorinated dibenzo-p-dioxins and related
    compounds, Ann. NY Acad. Sci., 320:214-230, 1979.
Poland, A., Glover, E. and Kende, A.S. Stereospecific, high affinity
    binding of 2,3,7,8-tetrachlorodibenzo-p-dioxin by hepatic

cytosol. Evidence that the binding species is receptor for induction of aryl hydrocarbon hydroxylase, J. Biol. Chem., 251:
4936-4946, 1976.

Poland, A., Glover, E., Robinson, J.R. and Nebert, D.W. Genetic
expression of aryl hydroxylase activity. Induction of monooxygenase activities and cytochrome P-450 formation by 2,3,7,8-
tetrachlorodibenzo-p-dioxin in mice genetically nonresponsive
to other aryl hydrocarbons, J. Biol. Chem., 249:5599-5606, 1974.

Ruzo, L.O., Safe, S., Hutzinger. O., Platonow. N. and Jones, D.
Hydroxylated metabolites of chloronaphthalenes (Halowax 1031)
in pig urine, Chemosphere, 4:121-123, 1975.

Ruzo, L.O., Jones, D., Safe, S. and Hutzinger, O. The metabolism
of chlorinated naphthalenes, J. Agr. Food Chem., 24:581-583,
1976.

Ruzo, L.O., Safe, S., Jones, D. and Platonow, N.S. Uptake and distribution of chloronaphthalenes and their metabolites in pigs,
Bull. Environ. Contam. Toxicol., 16:233-239, 1976a.

Safe, S. Metabolism, uptake, storage and bioaccumulation. In:
Kimbrough, R., ed. Halogenated Biphenyls, Naphthalenes, Dibenzodioxins and Related Products. North Holland: Elsevier, 1980.
pp. 77-107.

Safe, S., Jones, D. and Hutzinger, O. Metabolism of 4,4'-dihalo-
biphenyls, J. Chem. Soc. Perkin. Trans., 1:357-359, 1976.

Safe, S., Kohli, J and Crawford, A. FireMaster BP-6: fractionation,
metabolic and enzyme induction studies, Environ. Health
Perspect., 23:147-152, 1978.

Safe, S., Platonow, N., Hutzinger, O. and Jamieson, W.D. Analysis
of organochlorine metabolites in crude extracts by high resolution photoplate mass spectrometry, Biomed. Mass Spectrom.,
2:201-203, 1975.

Sawyer, T. and Safe, S. PCB isomers and congeners induction of aryl
hydrocarbon hydroxylase and ethoxyresorufin O-deethylase activities in rat hepatoma cells, Toxicol. Letters (Submitted).

Shimada, T. and Sato, R. Covalent binding in vitro of polychlorinated biphenyls to microsomal macromolecules, Biochem.
Pharmacol, 27:585-593, 1978.

Shimada, T. and Sato, R. Inhibition of monooxygenase activities by
1,1,1-trichloropropene-2,3-oxide, an inhibitor of epoxide
hydrase in rat liver microsomes, Biochem. Pharmacol., 28:1777-
1781, 1979.

Sundstrom, G., Hutzinger, O. and Safe, S. The metabolism of chloro-
byphenyls review, Chemosphere, 5:267-298, 1976.

Tulp, M.Th.M. and Hutzinger, O. Identification of hydroxylated
chlorodibenzo-p-dioxins, chlorodibenzofurans, chlorodiphenyl
ethers and chloronaphthalenes as the methyl ethers by gas chromatography mass spectrometry, Biomed. Mass Spectrom., 5:224-
231, 1978.

Veerkamp, W., Wever, J. and Hutzinger, O. The metabolism of some
chlorinated dibenzofurans by rats, Chemosphere, 10:397-403, 1981.

Yamamoto, H., Yoshimura, H., Fujita, M. and Yamamoto, T. Metabolic
    and toxicologic evaluation of 2,3,4,3',4'-pentachlorobiphenyl
    in rats and mice, Chem. Pharm. Bull,24:2168-2174, 1976.
Yoshida, S. and Nakamura, A. Residual status after parturition of
    methylsulfone metabolites of polychlorinated biphenyls in the
    breast milk of a former employee in a capacitor factory, Bull
    Environ. Contam. Toxicol., 21:111-115, 1979.
Yoshihara, S., Kawano, K., Yoshimura, H., Kuroki, H. and Masuda, Y.
    Toxicological assessment of highly chlorinated biphenyl congeners
    retained in the Ysho patients, Chemosphere, 8:531-538, 1979.
Yoshimura, H., Ozawa, N. and Saeki, S. Inductive effect of poly-
    chlorinated biphenyls mixture and individual isomers on the
    hepatic microsomal enzymes, Chem. Pharm. Bull., 26:1215-1221,
    1978.
Yoshimura, H., Yoshihara, S., Ozawa, N. and Miki, M. Possible cor-
    relation between induction modes of hepatic enzymes by PCBs
    and their toxicity in rats, Ann. NY Acad. Sci., 320:179-192,
    1979.

2,3,7,8-TETRACHLORODIBENZO-P-DIOXIN-INDUCED WEIGHT LOSS:

A PROPOSED MECHANISM

Mark D. Seefeld and Richard E. Peterson

School of Pharmacy and Environmental Center
University of Wisconsin
Madison, Wisconsin USA

ABSTRACT

TCDD caused a dose-related decrease in body weight, food intake, motor activity and total and resting oxygen consumption in rats. In animals treated with 15 µg/kg, the depression in food intake and oxygen consumption lasted 2 weeks. Thereafter, the rats consummed the appropriate amounts of food and oxygen for their reduced weights. However, in rats treated with 50 µg/kg, food intake was insufficient to maintain weight. The results suggest that TCDD causes weight loss by a decrease in energy intake (as opposed to an increase in energy expenditure). The decrease in intake of energy could be due to mal-absorption of nutrients from the gastrointestinal tract and/or reduced food intake. Of these, reduction in food consumption is considered to be the main factor involved. This is because weight in pair-fed control rats stabilizes at a level only slightly higher than in TCDD-treated animals. To determine if TCDD is altering a regulatory system for body weight (as opposed to a system controlling food intake), rats had their body weights reduced by food restriction prior to TCDD (25 µg/kg). Immediately after TCDD they were given food ad lib and they exhibited hyperphagia and weight gain. Thus, the ability to feed is not directly inhibited by TCDD. We propose that TCDD lowers a regulated level or "set-point" for body weight and that the change in food intake reflects the rat's effort to reach the new weight level.

INTRODUCTION

One of the most sensitive signs of TCDD toxicity in laboratory animals is a progressive loss of body weight (Greig et al, 1973;

405

Gupta et al, 1973; Harris et al, 1973; McConnell et al, 1978a,b; Seefeld et al, 1979, 1980). The mechanism by which this occurs is unknown. Some investigators suggest that TCDD-treated animals continue to eat and drink until the terminal stages of toxicity, but food and water intake data were not given in their studies (Allen et al, 1975; McConnell et al, 1978a; McConnell and Shoaf, 1981). Others showed that food consumption was reduced in rats treated with single lethal doses of TCDD (Harris et al, 1973; Courtney et al, 1978 and in rats treated chronically with TCDD (Kociba et al, 1976). Food consumption in the latter studies was reported as cumulative food intake or an intake on only certain days after TCDD treatment.

The question raised by the above results is: Why has a decrease in food intake not been given greater consideration as the main cause of weight loss in TCDD-treated animals? One reason is that the reduction in intake was not considered to be great enough to account for the loss of weight (Harris et al, 1973). Furthermore, the following results suggested that depressed absorption of nutrients from the gastrointestinal tract may be involved: (1) TCDD-treated rats lose weight more rapidly and to a greater extent than pair-fed control rats (Van Logten, Gupta and Moore, unpublished results cited by McConnell et al., 1978a; Gasiewicz et al., 1980); (2) forced oral administration of a liquid diet to TCDD-treated rats does not prevent weight loss (Courtney et al, 1978); (3) when gastrointestinal absorption of nutrients is bypassed by feeding rats intravenously with a total parenteral nutrition solution, TCDD-treated rats gain weight at the same rate as control rats (Gasiewicz et al, 1980); (4) in TCDD-treated rats the assimilation of ingested dietary lipid by the intestine is depressed (McConnell and Shoaf 1981).

The overall objective of the present study was to gain greater insight into the mechanism of body weight loss in TCDD-treated rats. The results will show that TCDD decreases body weight, food intake and whole body energy expenditure in a dose-ralated fashion. Based on these findings a hypothesis is proposed whereby TCDD lowers a regulated level or "set-point" for body weight. The change in food intake is viewed as a secondary response to the lowered "set-point" and represents the animal's effort to reduce its weight to the lower level determined by TCDD.

MEASUREMENT OF FOOD INTAKE AND EFFECTS ON FOOD SPILLAGE

Male Sprague Dawley rats (250-275 g) were housed, one rat per cage, in a windowless room kept at 70°F and lighted from 0700-9000 hr. Upon arrival they were provided with ground Purina Rat Chow and water ad lib. After 5-7 days of acclimation to the ground food, each rat received 40 g of food/day in a 100 ml glass food cup that was wired to the cage. Food intake and body weight were determined daily between 0900-1100 hr for 7-10 days before treatment with a single, oral dose of TCDD or vehicle. Food intake was corrected for spilled food which was caught in a plastic pan (14 cm x 14 cm) placed beneath each food cup.

To determine the effect of TCDD on food spillage, three groups of rats were used. They were treated with a single, oral dose of vehicle (control) or with 25 or 50 µg/kg of TCDD. Relative to the control group, it was observed that the TCDD-treated rats spilled more food per day. This increase in spillage was first detected 6 days after treatment and steadily increased until the rats were spilling 2-3 times as much food as the controls. This was not an insignificant effect inasmuch as the amount of spilled food on day 15 for the 25 and 50 µg/kg TCDD groups was approximately 12 and 18 g/day, respectively, as compared to 5 g/day in the control group. If these differences in spillage were not accounted for, food intake in TCDD-treated rats would have been overestimated.

EFFECTS ON FOOD INTAKE AND BODY WEIGHT

Rats were treated with a single. oral dose of vehicle or TCDD (5, 15, 25 and 50 µg/kg). TCDD produced a dose-related depression in food intake and body weight. The depression in intake was detected on day 1 and became progressively greater until day 10. There-after, food intake in the groups treated with 5 and 15 µg/kg stablize at a lower daily intake level. On the other hand, in the 50 µg/kg group, food intake continued to increase until death. The depression in body weight had a slower onset and persisted over the 35-day post-treatment period. Lethality was not observed in rats treated with 5 and 15 µg/kg of TCDD. However, in animals treated with 25 and 50 µg/kg, cumulative lethalities at day 35 were 25 percent and 75 percent, respectively. The median times to death for rats that died in the 25 and 50 µg/kg groups were 28 and 20 days, respectively. These were adult animals and their body weights at the time of death were approximately 50 percent of that of the controls.

To determine if the persistent depressant effect of TCDD on food intake was due to the lower level of maintained weight, intakes were expressed relative to the rat's metabolic body size (MBS). MBS is defined as body weight in kilograms raised to the 0.75 power (Kleiber 1975). Our rationale for expressing food intake as g food/MBS/day was that resting metabolic rate has been shown to be directly proportional to MBS (Kleiber 1975), and it follows that this should also hold true for the amount of food required for weight maintenance.

When food intake was expressed relative to MBS, there was a dose dependent depression for up to 20 days after TCDD treatment. Howeve from 20-35 days, food intake was similar in control rats and rats

treated with 5 and 15 µg/kg of TCDD. This indicates that these TCDD
groups were consuming the appropriate amount of food for their reduced
body weights. In other words, these rats were regulating food intake
as if they were at their proper levels of maintained weight. The
intakes of rats treated with 25 and 50 µg/kg never returned to levels
sufficient to maintain body weight.

COMPARISON OF BODY WEIGHT LOSS IN CONTROL RATS PAIR-FED TO TCDD-TREATED RATS

Pair-feeding experiments were conducted in adult, male rats
(300 g). Pair-fed controls lost weight at the same rate and to the
same extent as their weight matched TCDD-treated partners (25 or
50 µg/kg) until day 10 after treatment. Thereafter, body weights of
the two groups began to diverge. This divergence of weights resulted
in the pair-fed control group having body weights that were 20-30 g
higher than the corresponding TCDD group 20-35 days after treatment.
Cumulative lethality on day 35 was higher in the TCDD groups (25 µg/kg
33 percent; 50 µg/kg - 75 percent) than in the respective pair-fed
control groups (0 and 15 percent).

A pair-feeding experiment was also done in young rats (100 g).
Here the pair-fed control group was matched to rats treated with
25 µg/kg of TCDD. In the younger animals, TCDD caused a depression
in food intake and weight gain that was immediate in onset and persistent. Weight gain in the pair-fed control and TCDD groups were
comparable over the 35-day post-treatment period. Cumulative lethality at the end of the study was 13 and 65 percent, respectively, for
the pair-fed control and TCDD treatment group.

The above results suggest that the reduction in food intake
caused by TCDD is primarily responsible for the loss of body weight
or depressed growth rate of rats. These results differ from those of
other pair-feeding studies (Gasiewicz et al, 1980) where it was found
that rats treated with TCDD lost weight more rapidly and to a much
greater extent than pair-fed control animals. A possible explanation
for this difference is that the doses of TCDD used in the present
study (25 and 50 µg/kg) were less than the dose (100 µg/kg) used by
Gasiewicz et al, (1980). Alternatively, food intake may not have
been as rigorously corrected for spillage in the study by Gasiewicz
et al, (1980). If the latter is true, the pair-fed control group may
have been overfed relative to the TCDD group.

EFFECTS ON WHOLE BODY OXYGEN CONSUMPTION

Rats used in the energy expenditure (oxygen consumption) study
were housed in a room lighted from 0800-2000 hr. The room temperature

was kept constant at 81°F. At this thermoneutral temperature the
rat's thermoregulatory mechanisms are inactive so that essentially
no energy is required to maintain body temperature. After daily food
intake and body weight had been collected for each animal for one
week, the rats were divided into 3 groups matched for body weight
(275-300 g) and were treated with a single, oral dose of vehicle or
TCDD (15 or 50 µg/kg). Food intake and body weight were recorded
daily and motor activity, total oxygen consumption and resting oxygen
consumption were determined continuously for 24 hr. on designated days
before and after treatment. To measure the mean oxygen consumption of
an individual rat, the animals was placed in an air tight, plexiglass
chamber for 24 hr. with ground food (40 g/day) and water available
ad lib. The chamber was approximately 4 liters in volume and was
equipped with a urine trap, a screen to collect spilled food and feces,
and a food cup. Compressed air was passed through the chamber at a
constant rate of 1500 ml/min. After exiting the chamber, the air was
dried in a $CaSO_4$ column and passed through a paramagnetic oxygen
analyzer. The volume of oxygen consumed by the animal was calculated
by determining the rate of air flow through the chamber and the change
in oxygen concentration in air entering and exiting the chamber.
Oxygen consumption was sampled continuously for 10 min. once every
30 min. and the data was stored on magnetic tape by a microprocessor
for later analysis. Oxygen consumption values preceded by at least
5 min. of motor inactivity were used to estimate resting oxygen consumption. Estimates of total oxygen consumption were based on data
collected during periods of motor activity as well as inactivity.
The total and resting oxygen consumptions calculated for a single rat
on a designated day after TCDD treatment were the mean values for the
24 hr. sampling period and were expressed as ml O_2/min/rat. Oxygen
consumption was also expressed relative to metabolic body size (MBS)
as ml O_2/MBS/min. This transformation of the data was used to directly compare oxygen consumption in rats of different body weights.
Motor activity was also measured over the same 24 hr. period as oxygen
consumption. This was done by having the chamber containing the rat
rest on compression springs with a velocity transducer at the top.
This permitted the measurement of all chamber movements caused by
physical activity of the animal. The electrical output of the velocity transducer was recorded on magnetic tape by the microprocessor
as counts of motor activity.

    When rats were maintained at a thermoneutral temperature (81°F),
TCDD treatment produced the expected decrease in food intake and body
weight. A new finding was that motor activity in the TCDD group was
depressed relative to the control group. The magnitude of the depression appeared to be dose-related and was most pronounced in the
50 µg/kg group during the one-week period before death. Another new
finding was that TCDD produced a dose-related decrease in total and
resting oxygen consumption. When oxygen consumption was expressed
as ml O_2/mim/rat it was depressed in both dosage groups and at all
times after treatment. However, when it was expressed as ml

O_2/MBS/min it was similar in the control and 15 µg/kg group from 17–27 days after treatment. This implies that the TCDD-treated animals were regulating energy expenditure as if they were at their proper level of weight maintenance. When O_2 consumption was expressed in the 50 µg/kg group as ml O_2/MBS/min it remained lower than the control group at all times after treatment, until the animals died. Thus, these results show that weight loss in TCDD-treated rats is not due to increased energy expenditure because motor activity and oxygen consumption are depressed.

REDUCTION IN THE REGULATED LEVEL OR "SET-POINT" FOR BODY WEIGHT: A POSSIBLE MECHANISM FOR TCDD-INDUCED WEIGHT LOSS

As an alternative to the nutrient malabsorption hypothesis for weight loss (see Introduction) we suggest that TCDD lowers the regulation level or "set-point" for body weight in the rat in a dose-dependent fashion. The "set-point" monitors existing body weight and defends against differences between existing weight and the "set-point" by adjusting food intake to raise or lower body weight. Our hypothesis is that by lowering the body weight "set-point", TCDD creates a mismatch between the "set-point" and existing body weight. This, in turn, leads to decreased food intake and weight loss until the mismatch is corrected. Once corrected, TCDD-treated rats are postulated to maintain their weight at the reduced level determined by their lower body weight "set-point". Thus, TCDD and control rats may differ, not in how they regulate body weight, but rather in terms of the level of maintained weight each is prepared to defend.

To test this hypothesis, the experiment shown in diagrammatic form in Figure 1 was conducted. Control rats weighing 250 g were divided into 2 groups. One group was fed ground chow ad lib for one week and gained weight to 300 g. Food consumption during this time was about 25 g/day. The other group was food restricted in that they were given 10 g of food per day. After one week the restricted group weighed 210 g which was 70 percent of the weight of the ad lib group. At this time both groups were (1) treated with 25 µg/kg of TCDD and (2) provided with food ad lib for the next 25 days. The ad lib fed rats (300 g) became hypophagic and lost weight after TCDD treatment until their body weight leveled off at 260 g. The response of the food restricted rats (210 g) was different. Beginning immediately after TCDD, these animals consumed more food per day than the ad lib fed group and they also gained weight. Approximately 2 weeks after TCDD, both groups were consuming a comparable amount of food and their body weights were similar, about 260 g. These different responses to TCDD in the ad lib and restricted groups (hypophagia and weight loss versus relative hyperphagia and weight gain) are viewed as normal regulatory responses which enable the animal to reach the new level of weight maintenance determined by TCDD. According to our hypothesis, the decrease in the regulation

level or "set-point" for body weight is considered to be the primary
response of the rat to TCDD. The ensuing change in food intake is
thought of occur secondarily to the change in "set-point".

SUMMARY

The main findings of this study were: (1) TCDD increased food
spillage which, if not corrected for, would have led to an overesti-
mation of food intake. (2) TCDD caused a dose-dependent decrease in
food intake and body weight at lethal as well as sublethal doses.
(3) The decrease in food intake could be detected as early as one
day after TCDD treatment. In rats given 5 and 15 µg/kg, intake recovered to a level that was sufficient to maintain a reduced level
of body weight. At a lethal dose, 50 µg/kg, the depressant effect
on food intake was not reversible. (4) The primary cause of weight
loss in TCDD-treated rats was reduced food intake. This was suggested
by pair-fed control rats having body weights that approached those
of the TCDD treatment groups. (5) Weight loss was associated with
decreased motor activity and oxygen consumption which indicated that
it was not due to an increase in energy expenditure. (6) Rats that
had their body weights reduced by food restriction prior to TCDD
treatment exhibited hyperphagia and weight gain after being given
TCDD and provided with food ad lib. Thus, inhibition of a system
controlling food intake is not the primary effect of TCDD.

The hypothesis proposed to explain the above results is that TCDD
lowers a regulated level or "set-point" for body weight. The following lines of evidence support the hypothesis. (1) Although rats
given 5 and 15 µg/kg of TCDD displayed reduced food intake and weight
loss immediately after treatment, intake eventually recovered to near
the control level while body weight stablized at a level below that
of control. Maintenance of weight at this lower level was sustained
and parallel to the control group. (2) TCDD-treated rats made adjust-
ments in food intake that were appropriate in order to reach a new
level of maintained weight. Rats treated with TCDD at a normal body
weight became hypophagic and lost weight after treatment. On the
other hand, rats whose body weight had been reduced prior to treatment became hyperphagic and gained weight. Eventually food intake
and body weight in the two groups stabilized at the same level. (3)
Two to four weeks after rats were treated with 15 µg/kg of TCDD their
energy balance was appropriate to their reduced level of weight maintenance. In other words, when their food intake and oxygen consumption were referenced to their reduced metabolic body size it was the
same as that of control rats.

Finally, we wish to emphasize that the notion of TCDD decreasing
the level of regulated body weight in the rat is now a hypothesis.
Before it can be given serious consideration as the mechanism for
TCDD-induced weight loss, it has to be rigorously tested. It will

be necessary to show that the reduced level of weight in TCDD-treated rats is not only maintained but also defended with the same precision that control animals display in defending their normal weight level.

REFERENCES

Allen, J.R., Van Miller, J.P., and Norback, D.H. Tissue distribution, excretion and biological effects of [^{14}C] tetrachlorodibenzo-p-dioxin in rats. Food Cosmet. Toxicol., 13:501-505, 1975.

Courtney, K.D., Putnam, J.P., and Andrews, J.E. Metabolic studies with TCDD (dioxin) treated rats. Arch. Environ. Contam. Toxicol. 7:383-396, 1978.

Gasiewicz, T.A., Holscher, M.A., and Neal, R.A. The effect of total parenteral nutrition on the toxicity of 2,3,7,8-tetrachlorodi-benzo-p-dioxin in the rat. Toxicol. Appl. Pharmacol, 54:469-488, 1980.

Greig, J.B., Jones, G., Butler, W.H., and Barnes, J.M Toxic effects of 2,3,7,8-tetrachlorodibenzo-p-dioxin. Food Cosmet. Toxicol., 11:585-595, 1973.

Gupta, B.N., Vos, J.G., Moore, J.A., Zinkl, J.G., and Bullock, B.C. Pathological effects of 2,3,7,8-tetrachlorodibenzo-p-dioxin in laboratory animals. Environ. Health Perspect., 5:125-140, 1973.

Harris, M.E., Moore, J.A., Vos, J.G. and Gupta, B.N. General bio-logical effects of TCDD in laboratory animals. Environ. Health Perspect. 5:101-109, 1973.

Kleiber, M. The Fire of Life, New York: R.E. Krieger Co., 1975

Kociba, R.J., Keeler, P.A., Park, C.N., and Gehring, P.J. 2,3,7,8-tetrachlorodibenzo-p-dioxin (TCDD): Results of a 13-week oral toxicity study in rats. Toxicol. Appl. Pharmacol. 35:553-574, 1976.

McConnell, E.E., Moore, J.A., Haseman, J.K., and Harris, M.W. The comparative toxicity of chlorinated dibenzo-p-dioxins in mice and guinea pigs. Toxicol. Appl. Pharmacol., 44:335-356, 1978a.

McConnell, E.E., Moore, J.A., and Dalgard, D.W. Tocicity of 2,3,7,8-tetrachlorodibenzo-p-dioxin in rhesus monkeys (Macaca mulatta) following a single oral dose. Toxicol. Appl. Pharmacol., 43:175-187, 1978b.

McConnell, E.E., and Shoaf, C.S. Studies of the mechanism of 2,3,7,8-tetrachlorodibenzo-p-dioxin (TCDD) toxicity-lipid assimilation. I. Morphology. The Pharmacologist, 23(3):342, 1981.

Seefeld, M.D., Albrecht, R.M., and Peterson, R.E. Effects of 2,3,7,8-tetracholorodibenzo-p-dioxin on indocyanine green blood clearance in rhesus monkeys. Toxicol., 14:263-272, 1979.

Seefeld, M.D., Albrecht, R.M., Gilchrist, K.W., and Peterson, R.E. Blood clearance tests for detecting 2,3,7,8-tetrachlorodibenzo-p-dioxin hepatotoxicity in rats and rabbits. Arch. Environ. Contam. Toxicol., 9:317-327, 1980.

ACKNOWLEDGEMENTS

We express our gratitude to Dr. Richard E. Keesey and Mr. Stephen Corbett, Psychology Department, University of Wisconsin, for their expert advice on various aspects of this research and for permitting us to use their facilities for conducting the energy expenditure experiments. We are also grateful to Mrs. Jan Dickins and Mrs. Connie Wierman for their technical assistance. This research was supported by NIH grant ES01332. Salary support for M. Seefeld was provided by NIH training grant ES07071.

STUDIES OF THE MECHANISM OF ACTION OF HEPATOTOXICITY OF 2,3,7,8-

TETRACHLORODIBENZO-P-DIOXIN (TCDD) AND RELATED COMPOUNDS

G. D. Sweeney and K. G. Jones

McMaster University
Hamilton
Ontario, Canada

ABSTRACT

TCDD causes liver damage, porphyria, chloracne, thymic atrophy and damage to other organs in rats and mice. The porphyria due to TCDD reflects interference with an enzyme, uroporphyrinogen decarboxylase, converting uroporphyrinogen to coproporphyrinogen. We studied TCDD liver toxicity in an inbred strain of mice using histrological criteria and porphyria as indices of severity. Toxicity of TCDD was enhanced in C57Bl/6J (responsive) mice in comparison with DBA/2J (non-responsive) mice. This difference is attributed to the high affinity binding site in hepatic cytosol of C57 mice. Interaction with this site is required for induction of aryl hydrocarbon hydroxylase (AHH) and other coordinately expressed functions under control of a regulatory gene ('Ah') which differs in C57 and DBA mice. We compared the effects of TCDD in C57/Bl mice depleted of iron by venesection with control animals not deficient in iron. Low iron animals were protected against porphyria and liver damage but not against thymic atrophy or chloracne. Dependence upon the Ah gene and on non-heme tissue iron has suggested a free radical mechanism initiating lipid peroxidation. The finding that dietary supplements of butylated hyroxy anisole also protect against histological changes in the liver and porphyria is consistent with this hypothesis.

INTRODUCTION

The specific organ toxicity of 2,3,7,8-tetrachlorodibenzo-p-dioxin (TCDD) differs widely among species. In rats and mice, doses from 0.3 to 1.0 μmole/kg lead to severe liver damage and it is the mechanism of this effect of the chlorinated aromatic hydrocarbon

which is the main focus of this paper. The histological changes
described include deposition of fat, cell swelling, necrosis,
collapse of parenchymal tissue, bile duct proliferation and associated regenerative phenomena (Jones and Butler 1974; Jones and
Greig 1975). Abnormal liver function has been described in humans
contaminated with dioxin in industrial accidents (May 1973) but slow
recovery appears to have occured. Histological changes have not
been reported. A more specific biochemical manifestation of liver
toxicity has also been reported in exposed humans, namely porphyria
cutanea tarda (PCT) (Bleiberg et al 1964; Jirasek et al 1974);
Goldstein and her collegues demonstrated that this liver disorder
also occurs in mice and rats exposed to TCDD (Goldstein et al 1973;
Goldstein et al 1976).

During the past 4 years we have performed experiments in which
the toxic effects of TCDD on mouse liver have been significantly
modified by various strategies. This paper will summarized the
findings in these experiments and discuss possible mechanisms of
TCDD toxicity which can be inferred from these results. The extent
to which extrapolation to man may be valid, and the significance of
our findings for population surveillance, will also be considered.

Two quite separate criteria have been used to judge the extent
of an effect of TCDD on mouse liver. Tissues have been examined
histologically and, where an intervention has been a clear-cut
effect, this endpoint has been useful. However, the presence or
absence of abnormal heme biosynthesis has provided a simple and quan-
titive end point which has permitted much clearer segregation of
mice subjected to the maneuvers to be described. A brief description of the porphyria caused by TCDD may be useful.

PORPHYRIA CUTANEA TARDA

Porphyria cutanea tarda (PCT) is a skin disease affecting sun-
exposed areas of the body when increased amounts of polar
photosensitizing porphyrins circulate. PCT occurs sporadically in
human populations where it tends to be associated with alcohol abuse
or exposure to synthetic estrogens. In some patients, predisposi-
tion to PCT is familial (Elder 1977). The association between
PCT and halo-aromatic hydrocarbon toxicity was highlighted by a
major epidemic of the disease in Turkey where peasants were exposed
to wheat treated with a fungicide containing hexachlorobenzene (HCB).
When rats were fed an HCB-containing diet they developed disturbed
porphyrin metabolism similar to that seen in the Turkish peasants.

The porphyrias are a group of biochemical diseases which affect
heme synthesis only in the liver and is associated with reduced
activity of uroporphyrinogen decarboxylase (UD) (Elder 1977). This
cytosolic enzyme catalyses 4 steps in the conversion of uroporphyin-
ogen. When activity of UD becomes rate-limiting, partially

decarboxylated intermediates accumulate, oxidize to porphyrins and spill into urine or feces. Further, one of these intermediates which accumulates is a 5-COOH porphyrinogen which is metabolised to an abnormal 4-COOH porphyrin, isocoproporphyrin. In this porphyria, activity of the enzyme which regulates flux through the heme biosynthetic pathway, δ-aminolevulinate synthase (ALA-S) need not be significantly increased until interference with heme synthesis is severe.

Severity of the process is easily monitored by measuring urine porphyrin. Total liver porphyrin and the activity of UD can be determined when animals are killed; the latter assay is presently rather complex. It is important not to confuse this specific biochemical lesion with the porphyrin accumulation which occurs in chicken embryos exposed to many types of foreign chemicals. Similarities may exist but cannot be assumed.

The importance of PCT in relation to the toxicity of TCDD is: (a) that the process may be monitored non-invasively; (b) that the target enzyme (UD) is known and thus the mechanism of its inactivation is accessible to study at a molecular level; and (c) the process has occured in humans exposed to TCDD and may be of value for population surveillance.

When animals are exposed to TCDD a latent period precedes porphyrin accumulating in the liver and appearing in the urine. This latent period depends upon dose and other factors. The defect in UD activity is associated with a characteristic pattern of porphyrins excreted in the urine, which not only permits the process to be monitored by a simple spectrophotometric test (Jones and Sweeney 1979), but can be readily recognized when urine porphyrin is analyzed chromatographically.

In summarizing our experiments we shall attempt to establish 4 points: (1) that the toxicity of TCDD depends upon a product of the regulator gene referred to as "Ah"; (2) that this gene product requires the presence of tissue iron for toxicity to manifest; (3) that a free radical mechanism involving lipid peroxidation appears to be involved in toxicity of TCDD in the liver; and (4) our results suggest that TCDD may exert its toxic effects in liver by mechanisms different from those operating in other tissues.

EVIDENCE FOR INVOLVEMENT OF A PRODUCT OF THE "Ah" GENE

In his important early studies of the mechanism of TCDD toxicity, Poland demonstrated the exquisite sensitivity to induction by TCDD of ALA-S (the enzyme regulating heme biosynthesis) and aryl hydrocarbon hydroxylase (AHH) in embryonic avian tissue (Poland and Kende 1976). He argued that this response required a specific binding site with high affinity for TCDD and identified such a site in

cytosol of mouse liver. Also studying induction of AHH, Robinson and
others (1974) were able to categorize inbred strains of mice as
"responsive" or "nonresponsive" depending upon whether they responded
to 3-methylcholanthrene treatment with induction of AHH. Poland
showed that AHH could be induced in all mouse strains but that
"responsive" mice were about ten times more sensitive than "nonresponsive." He has postulated that the difference depends upon the
affinity for TCDD of a cytosol binding site coded for by the Ah
gene; this gene is thus regarded as regulating expression of AHH
activity (Poland et al 1976).

        We argued that is a product controlled by the Ah gene was in-
volved in liver toxicity of TCDD, then responsive strains of mice
would be more susceptible to toxicity than non-responsive strains.
This hypothesis was tested in C57Bl/6J (responsive) and DBA/2J
(non-responsive) strains treated with 77 nanomoles/kg of TCDD weekly
as well as in heterozygotes derived from appropriate interbreeding
and pheontyping. We showed that C57 and DBA mice segregated in their
sensitivity to develop porphyria and to manifest histological changes in the liver (Jones and Sweeney 1980). However, we were unable
to demonstrate a difference between homozygotes and heterozygotes
for responsiveness and the reason for this is not entirely clear. Although we have hypothesized that the monooxygenase utilizing P_1-450
as terminal oxygenase is concerned in the development of hepatotoxicity, this is speculative, as under the conditions of dosage used,
differences in either the level of P_1-450 or activities associated
with this hemoprotein could not be demonstrated.

TISSUE IRON IS INVOLVED IN THE HEPATOTOXICITY OF TCDD

        Clinical experience with the human disease PCT has demonstrated
that reduction of tissue iron stores reverses the biochemical abnormality (Sweeney and Jones 1979). This suggested that we should
compare the toxicity of TCDD for responsive mice with either low or
normal tissue iron stores. Such experiments clearly showed that
C57Bl/6J mice given .007 µmole TCDD weekly were protected from developing porphyria if they had previously been venesected over a
3-week period to reduce hemoglobin to roughly 7 g/dl. Somewhat more
surprising to us was the extend to which these animals were also
protected from histological damage due to the toxin (Sweeney et al
1979). These experiments have been repeated under a variety of conditions and, accepting the proviso already made with regard to
histological endpoints, the validity of the conclusion has been up-
held. The toxicity of TCDD is not reduced in animals which have
been venesected and supplemented prior to treatment with dioxin by
i.m. injections of iron-dextran complex equivalent in iron content
to the blood withdrawn. We conclude that the effect of TCDD on UD
is decreased or abolished in the low iron state and that tissue
damage is also decreased but not abolished (Jones et al in press).

We have tested the effect of iron deficiency on damage to other organ systems, particularly the skin, and the thymus. In these experiments the data have been quite clear; reduction of tissue iron by venesection has no effect on either thymic atrophy induced by TCDD or upon the development of chloracne in hairless mice (Jones et al in press).

EVIDENCE FOR A FREE RADICAL PROCESS IN DIOXIN TOXICITY

The mechanism for the foregoing processes is presently obscure. Nevertheless, the involvement of the gene which regulates the induction of P_1-450 together with other coordinately expressed enzymes, the requirement for a non-heme form of iron and the histogoical evidence consistent with lipid peroxidation (accumulation of droplets of neutral lipid) led us to speculate that a free radical mechanism is initiated. The question arose whether a tissue antioxidant might interfere with this process and we have performed experiments to investigate possible protection by the antioxidant, butylated hydroxyanisole (BHA). Initial experiments with this antioxidant administered as a 0.75 percent additive to the diet clearly showed protection against the biochemical disturbance porphyria, and this diet also was associated with decreased deposition of neutral fat droplets in tissues examined after 3 weeks treatment with TCDD. However, protection against TCDD toxicity by BHA is incomplete: in a subsequent experiment designed to investigate the dose-response relationship, we found that only 4 out of 6 mice were protected against porphyria despite 0.75 percent BHA in the diet whereas 0.25 percent BHA provided complete protection. It should be added that BHA fed at this level has major effects on mouse liver including a large weight increase. The weight gain due to BHA is additive with the weight gain due to BHA is additive with the weight gain due to BHA is additive with the weight gain due to TCDD but has not been found to affect hepatic monooxygenase systems. We have performed one experiment in which animals were supplemented with vitamin E (0.01 percent w/w added to standard laboratory diet) but this antioxidant regime did not prevent porphyria developing.

DISCUSSION

For reasons which are currently unclear, an enzyme (UD) in the cytosol of hepatic parenchymal cells in zone III of the acinus (pericentral cells) is particularly susceptible to damage when animals are treated with TCDD. This effect appears most likely to involve a free radical reaction and is potentiated by some product of the Ah regulatory gene operating in conjunction with tissue iron. We consider it unlikely that depletion of hemoproteins is part of this process. Damage to the liver appears to be quite distinct from damage to other target organs in terms of mechanism but definitive studies have been limited to skin and thymus.

At present, the case for the experiental findings presented above being relevant to the problem of human exposure to TCDD is not particularly strong. Certainly TCDD can cause PCT in humans but workers in whom this has been presented as an occupational disease were heavily exposed (see page 1). PCT has been sought but not yet reported in the Seveso population although this population was sufficiently heavily exposed for many individuals to develop chloracne.

Goldstein et al (1975) have shown that PCT is also caused by polychlorinated biphenyl (PCB) mixtures and by those specific PCB isomers which are potent inducers of AHH activity. As hexachlorobenzene also induces AHH activity and causes PCT, it has been tempting to draw the general conclusion that all haloaromatic hydrocarbons which induce AHH will demonstrate this effect. However, PCT has not been reported as a feature of the Yusho epidemic of PCB poisoning in Japan despite the prominent abnormalities of liver function manifested by some of these patients (Urabe et al 1979). Failure to recognise PCT may reflect inadequate methodology; in Michigan, where a comparison of the porphyrins in urine from PBB-exposed and control residents was made (Strik et al 1979), significant differences were not found when a simple quantitative test was used. However, the PBB-exposed group had a higher frequency of samples with $(COOH)_7$ porphyrin > $(COOH)_5$; this represents a reversal of the normal pattern and is consistent with decreased activity of UD.

It is possible that if more refined analytical methods are used when populations are screened for possible toxic effects of TCDD on the liver, consistent abnormalities may be found. An alternative is that PCT occurs in man only in response to massive exposure unless an underlying susceptibility to develop PCT co-exists or a higher burden of storage iron is carried.

ACKNOWLEDGEMENT

Supported by the Medical Research Council of Canada.

REFERENCES

Bleiberg, J., Wallen, M., Brodkin, R., and Applebaum, I.L. Industrially acquired porphyria. Arch Dermatol 89:793-797, 1964.
Elder, G.H. Porphyrin metabolism in porphyria cutanea tarda. Seminars in Hematol, 14:227-242, 1977.
Goldstein, J.A., Hickman, P., Bergman, H., and Vos, J.G. Hepatic porphyria induced by 2,3,7, 8-tetrachlorodibenzo-p-dioxin in the mouse. Res Commun in Chem Pathol & Pharmacol, 6:919-928, 1973.

Goldstein, J.A., Hickman, P. and Bergman, H. A comparative study of
    two polychlorinated biphenyl mixtures (Aroclors 1242 and 1016)
    containing 42% chlorine on induction of hepatic porphyria and
    drug metabolizing enzymes. Toxicol & Appl Pharmacol, 32:461-
    473, 1975.
Goldstein, J.A., Hickman, P. and Bergman, H. Induction of hepatic
    porphyria and drug-metabolizing enzymes by 2,3,7,8-tetrachlorodi-
    benzo-p-dioxin (TCDD). Fed Proc, 35:708, 1976.
Jirasek, L., Kalensky, J., Kubec, K., Pazderova, J., and Lukas, E.
    Acne chlorina porphyria cutanea tarda and other manifestations
    of general intoxication during the manufacture of herbicides.
    II. Cesk Dermatol, 49:145-147, 1974.
Jones, G., and Butler, W.H. A morphological study of the liver lesion
    induced by 2,3,7,8-tetrachlorodibenzo-p-dioxin in rats. J Pathol,
    112:93-97, 1974.
Jones, G., and Greig, J.B. Pathological changes in the liver of mice
    given 2,3,7,8-tetrachlorodibenzo-p-dioxin. Experientia, 31:1315-
    1317, 1975.
Jones, K.G., and Sweeney, G.D. Quantitation of urinary porphyrins by
    use of second-derivative spectroscopy. Clin Chem, 25:71-74,
    1979.
Jones, K.G., and Sweeney, G.D. Dependence of the porphyrinogenic
    effect of 2,3,7,8-tetrachlorodibenzo(p)dioxin upon inheritance
    of aryl hydrocarbon hydroxylase responsiveness. Toxicol & Appl
    Pharmacol, 53:42-49, 1980.
Jones, K.G., Cole, F.M., and Sweeney, G.D. The role of iron in the
    toxicity of 2,3,7,8-tetrachlorodibenzo-(p)-dioxin (TCDD).
    Toxicol & Appl Pharmacol, (in press).
May, G. Chloracne from the accidental production of tetrachlorodi-
    benzodioxin. Brit J Industr Med, 30:276-283, 1973.
Poland, A., and Kende, A. 2,3,7,8-tetrachlorodibenzo-p-dioxin:
    environmental contaminant and molecular probe. Fed Proc, 35:
    2404-2411, 1976.
Poland, A., Glover, E., and Kende, A.S. Stereospecific, high affini-
    ity binding of 2,3,7,8-tetrachlorodibenzo-p-dioxin by hepatic
    cytosol. Evidence that the binding species is receptor for
    induction of aryl hydrocarbon hydroxylase. J Biol Chem, 251:
    4936-4946, 1976.
Robinson, J.R., Considine, N., and Nebert, D.W. Genetic expression
    of aryl hydrocarbon hydroxylase induction. Evidence for the
    involvement of other genetic loci. J Bio Chem, 249:5851-5859,
    1974.
Strik, J.J.T.W.A., Doss, M., Schraa, G., Robertson, L.W., von
    Tiepermann, R., and Harmsen, E.G.M. Coproporphyrinuria and
    chronic hepatic porphyria type A found in farm families from
    Michigan (U.S.A.) exposed to polybrominated biphenyls (PBB).
    In: Strik, J.J.T.W.A. and Koeman, J.H., eds. Chemical Porphyria
    in Man. Elsevier/North Holland Biomedical Press, pp. 29-53,
    1979.

Sweeney, G.D. and Jones, K.G. Porphyria cutanea tarda: clinical and
    laboratory studies. Can Med Assoc J, 120:803-807, 1979.
Sweeney, G.D., Jones, K.G., Cole, F.M., Basford, D., and Krestynski,
    F. Iron deficiency prevents liver toxicity of 2,3,7,8-tetra-
    chlorodibenzo(p)-dioxin (TCDD). Science 204: 332-335, 1979.
Urabe, H., Koda, H., and Asahi, M. Present state of Yusho patients.
    Annals N Y Acad Sci, 320:273-276, 1979.

THE Ah RECEPTOR: A SPECIFIC SITE FOR ACTION

OF CHLORINATED DIOXINS?

Allan B. Okey

The Hospital for Sick Children
Department of Pediatrics
Division of Clinical Pharmacology
Research Institute
Toronto, Ontario, Canada

ABSTRACT

The Ah receptor is a major regulatory gene product of the Ah gene complex. Its best-known function is regulation of the induction of cytochrome P_1-450 (aryl hydrocarbon hydroxylase) and several associated drug-metabolizing enzyme activities. TCDD (2,3,7,8-tetrachlorodibenzo-p-dioxin) and related halogenated aromatic compounds are high affinity ligands for the Ah receptor. Generally, those tissues with high concentrations of Ah receptor are the most susceptible to TCDD toxicity, but presence of the receptor, in itself, does not ensure that toxicity from halogenated aromatics will occur. Toxicity of various ligands (TCDD, other halogenated dibenzo-p-dioxins, halogenated dibenzofurans, polychlorinated biphenyls, polycyclic aromatic hydrocarbons) also generally is correlated with the affinity with which the specific chemical binds to Ah receptor. However, certain compounds such as 3-methylcholanthrene bind to Ah receptor with an affinity sililar to that of TCDD, yet are far less toxic in vivo. Thus binding affinity alone is not the sole determinant to toxicity. Overall, binding to Ah receptor appears to be an essential early step in the mechanism of toxicity of chlorinated dioxins and related compounds. Some subsequent receptor-regulated event(s) must be required for full expression of toxicity, but the specific nature of such events is unknown.

INTRODUCTION

The mechanism by which 2,3,7,8-tetrachlorodibenzo-p-dioxin (TCDD) and related compounds exert tocicity is one of the most perplexing unresolved problems in toxicology. Risk assessment in humans exposed to chlorinated dioxins would be greatly facilitated if we understood the mechanism of toxicity in any model system.

Chlorinated dioxins produce a diverse spectrum of biological effects (as described throughout this volume), but the nature of the toxic response is very obscure. Are there specific sites at which chlorinated dioxins interact with biological systems? Are these specific sites mediators of the toxic response?

A POTENTIALLY SPECIFIC SITE OF ACTION: THE Ah RECEPTOR

In the liver and several other tissues TCDD binds with high affinity to a cytoplasmic protein termed the Ah receptor (Poland et al, 1976; Carlstedt-Duke et al, 1978; Okey et al, 1979). Less than 1 percent of TCDD in the liver is bound to Ah receptor sites (Okey et al, 1979); thus, it is necessary to employ an assay method which distinguishes TCDD bound to receptor from the high background of TCDD bound to other cellular macromolecules. We prefer sucrose density gradient centrifugation as a means of quantifying and characterizing the Ah receptor (Okey et al, 1979; Okey et al, 1980; Tsui & Okey 1981). As shown in Figure 1, 2-hour separations in a vertical tube rotor yield gradient profiles virtually identical to those originally obtained with 16-hour analyses in a swinging bucket rotor. Vertical tube rotor separations are the standard assay procedure presently used in our laboratory.

When analyses are done on sucrose density gradients prepared in low ionic strength buffer, the TCDD Ah-receptor complex sediments near 9S (Figure 1). TCDD binding in the 9S region is competitively inhibited by low concentrations of compounds which induce cytochrome P_1-450 (aryl hydrocarbon hydroxylase), but is affected little by chemicals that are not P_1-450 inducers (Okey et al, 1979).

Our major interest in the Ah receptor has been this receptor's role in regulating induction of cytochrome P_1-450 and other structural gene products of the Ah gene complex (reviewed in Nebert et al, 1981). TCDD is exceptionally potent as a P_1-450 inducer (Poland & Glover 1974) and the initial characterization of the Ah receptor was done using [3H]TCDD as the radioligand. In addition to the receptor's role in regulating P_1-450 induction, however, binding of TCDD (and other halogenated aromatic compounds) to the Ah receptor also has been postulated as an essential step in their toxic action (Poland et al, 1979).

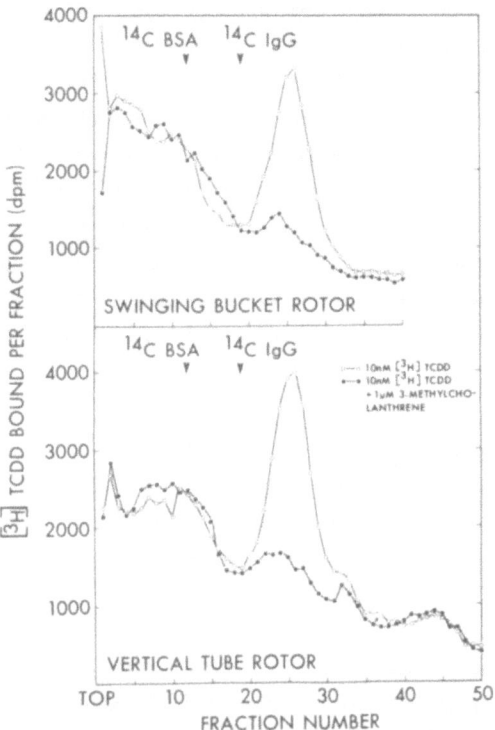

Figure 1.   Specific Binding of [³H]TCDD to the Ah Receptor as
             Detected by Sucrose Density Gradient Centrifugation.

Hepatic cytosol (5 mg protein/ml) from C57BL/6J mice was incubated in vitro with 10 nM [³H]TCDD in the presence or absence
of a 100-fold molar excess of nonradioactive MC. Samples then
were treated with charcoal-dextran before analysis with a swinging bucket rotor (16-hour centrifugation) or a vertical tube
rotor (2-hour centrifugation). The Ah receptor is represented
by the peak sedimenting near "9S" (fraction 25). Chemicals
which are inducers of cytochrome P_1-450 (e.g. TCDD, MC, etc.)
eliminate the "9S" peak by competing with [³H]TCDD for the
small number of Ah receptor sites. Chemicals which are not
P_1-450 inducers are poor competitors in this region of the gradient. Nonspecific, high capacity binding often occurs in the
"4-5S" region of the gradient (near the [¹⁴C]BSA marker) and
does not represent Ah receptor. (From Tsui & Okey, 1981)

If the Ah receptor plays a key role in toxicity, two expectations
are that: (1) the receptor should be present in reasonable concentration in those species and tissues which are most susceptible to
toxicity from halogenated aromatic compounds; (2) the receptor should
preferentially bind those chemicals which elicit the characteristic
toxic response.

TISSUE OCCURRENCE OF THE Ah RECEPTOR

C57BL/6 Versus DBA/2 Mice

The Ah receptor was first identified and characterized in liver from C57BL/6 mice (Poland et al, 1976; Okey et al, 1979). C57BL/6 mice are approximately 15 times more sensitive to P_1-450 induction by TCDD than are DBA/2 mice (Poland et al, 1974), and C57BL/6 mice also are more susceptible to TCDD-induced thymic involution (Poland & Glover 1980) and to TCDD-induced porphyria (Jones & Sweeney 1980). The Ah receptor is detectable in hepatic cytosol from C57BL/6 mice, but is not detectable in hepatic cytosol from DBA/2 mice (Figure 2).

The low sensitivity of DBA/2 mice to P_1-450 induction by TCDD appears to be due to a defect in the regulatory gene of the Ah gene comple, i.e., the gene which codes for Ah receptor. This defect does not cause total absence of Ah receptor in DBA/2 mice and other genetically "nonresponsive" strains. Rather, a form of receptor is present, but it is altered to such an extent that most P_1-450 inducers (e.g. 3-methylcholanthrene (MC), benzo(a)pyrene, β-naphthoflavone, etc.) are not bound with affinity sufficient to evoke a· response.

Although Ah receptor cannot be detected in DBA/2 cytosols by in vitro assays with [^{3}H]TCDD as the radioligand (Table 1), the nuclear form of Ah receptor is detectable in liver, lung and kidney of DBA/2 mice injected with high doses of [^{3}H]TCDD (Figure 3). It is known that Ah receptor appears in the nuclear compartment only after translocation from the cytoplasm (Okey et al, 1979; Okey et al, 1980). Thus, recovery of Ah receptor from nuclei of DBA/2 mouse tissues implies that cytosol binding of [^{3}H]TCDD must occur in vivo to a degree sufficient to generate a measureable quantity of translocated nuclear Ah receptor. This finding may explain the ability of DBA/2 mice to exhibit P_1-450 induction when treated with high doses of TCDD.

In C57BL/6 mice, cytosolic Ah receptor concentrations are highest in liver, lung and kidney (Table 1). Thymus and small intestine have cytosolic receptor concentrations about one-fourth that in liver. Cytosolic Ah receptor is not detectable by in vitro assays (sucrose density gradient analyses) on skin, skeletal muscle, heart, brain or testis of C57BL/6 mice, nor is cytosolic Ah receptor detectable in any tissue of DBA/2 mice (Table 1 from Mason & Okey, in press).

In summary, both C57BL/6 and DBA/2 mouse strains have forms of the Ah receptor. An alteration in the Ah receptor in DBA/2 mice makes this strain less sensitive to P_1-450 induction by TCDD and also may make them less susceptible to the toxic effects of TCDD.

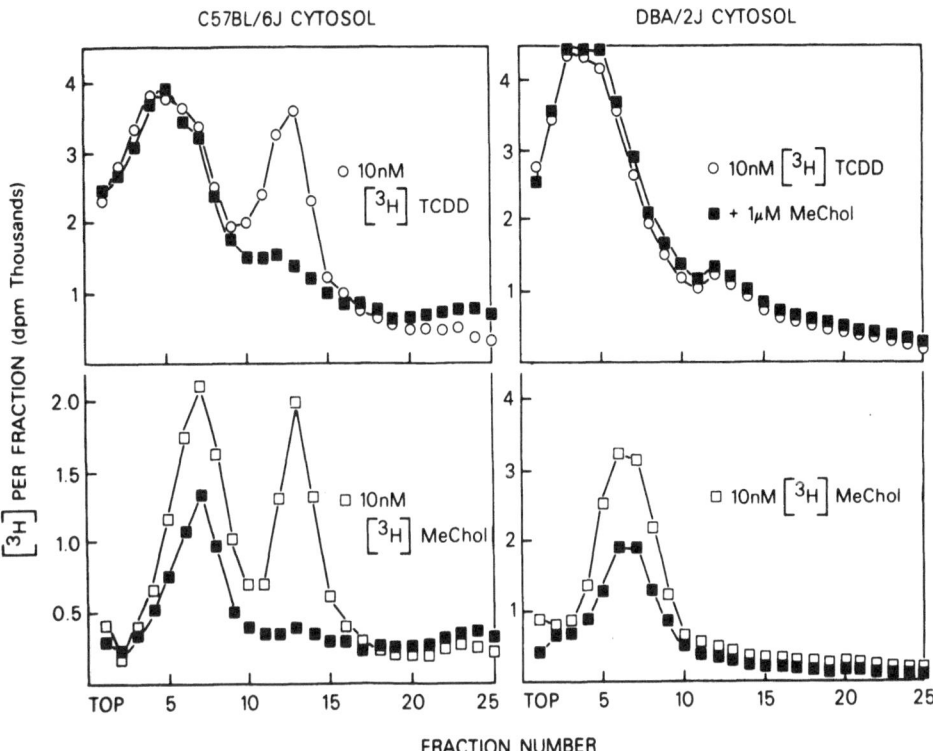

Figure 2.   Genetic Differences in the Binding of [^{3}H]TCDD
and [^{3}H]MC to Mouse Hepatic Cytosols.

Hepatic cytosols from genetically "responsive" C57BL/6J or
genetically "nonresponsive" DBA/2J mice were incubated with
10 nM [^{3}H]TCDD (50 Ci/mmol) or 10 nM [^{3}H]MC (37 Ci/mmol) in
the presence or absence of a 100-fold molar excess of nonradioactive MC.   Samples were analyzed by vertical tube rotor sucrose density gradient centrifugation.   The Ah receptor is
represented by the peak sedimenting near "9S" (fraction 13 in
these assays).   This peak is competitively eliminated by incubation in the presence of potent cytochrome P$_1$-450 inducers
such as MC (illustrated) and TCDD (not illustrated).   Binding
of [^{3}H]MC in the "4-5S" region of the gradient (near fraction
7) is partially eliminated by a 100-fold molar excess of nonradioactive TCDD (data not illustrated).   Neither MC nor TCDD
inhibit binding of [^{3}H]TCDD in the "4-5S" region of the gradient.   Note that nonspecific binding of [^{3}H]TCDD and [^{3}H]MC
in the "4-5S" region is present in both "responsive" C57BL/6J
mice and "nonresponsive" DBA/2J mice, whereas specific Ah receptor is detectable only in cytosol from C57BL/6J mice.
(From Okey et al, 1981.)

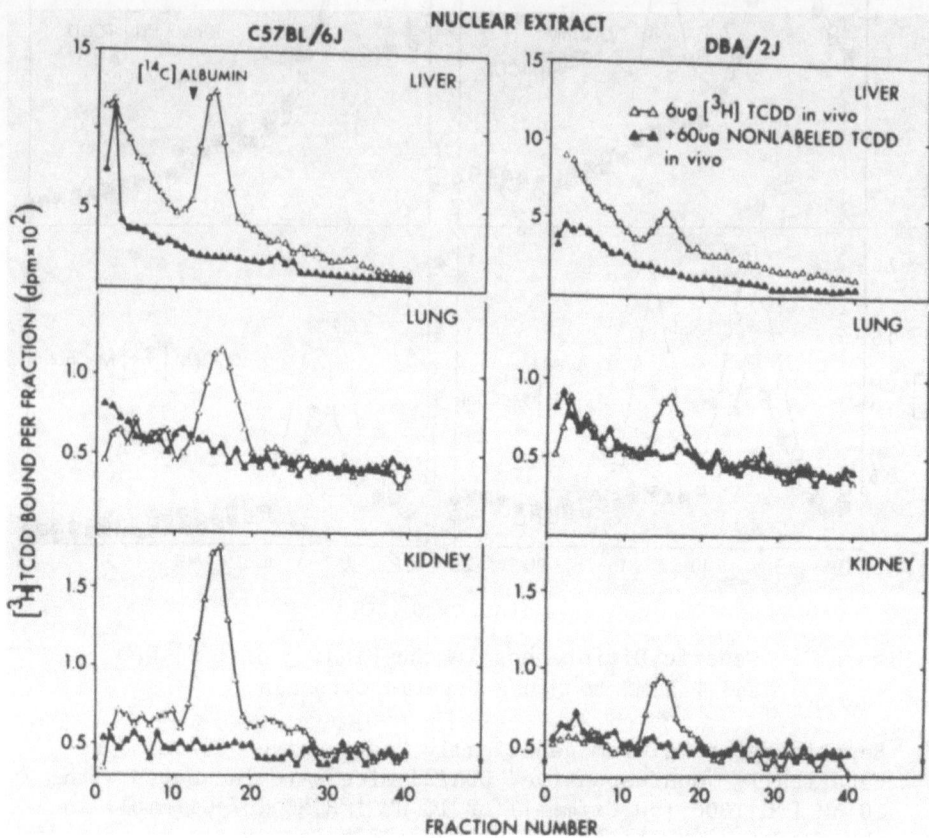

Figure 3.   Nuclear Ah Receptor in Tissues from C57BL/6J
             and DBA/2J Mice Injected with [³H]TCDD.

[³H]TCDD was injected intraperitoneally (6 μg/mouse = 0.75
μmol/kg body weight) 18 hours before tissues were removed for
analysis.   Nuclear Ah receptor is represented by the radioactive peak sedimenting near "5-6S" (fraction 15 in these assays).   To test saturability of binding, pair-matched mice of
the same age as those receiving [³H]TCDD only were injected
with a 10-fold molar excess of nonradioactive TCDD as a competitor six hours before injection of [³H]TCDD.   (From Mason
& Okey, in press.)

Table 1.   Ah Receptor Concentration in Cytosols
           Prepared from Various Rat and Mouse Tissues

Receptor concentrations are determined by in vitro incubation
of 10 nM [^{3}H]TCDD with cytosols prepared from different tissues.
Analysis was done by sucrose density gradient centrifugation as
described in Mason & Okey (in press).

	Ah Receptor Concentration (fmol/mg cytosol protein)		
Tissue	C57BL/6J Mice	DBA/2J Mice	Sprague-Dawley Rats
Liver	32	0	39
Lung	23	0	47
Kidney	10	0	1
Intestine	8	0	15
Thymus	8	0	54
Prostate	1	0	0
Adrenal	NT	NT	0
Heart, Brain Skeletal Muscle Testis	0	0	0

"0" indicates that no specific Ah receptor was detectable in
that tissue, although several tissues have components which
bind [^{3}H]TCDD nonspecifically.

NT = not tested

See Mason & Okey, European J. Biochem. (in press, 1982) for
further details.

Ah Receptor in Rat Tissues

Ah receptor is present in several tissues of Sprague-Dawley rats
(Carlstedt-Duke, 1979). The thymus is especially interesting since
cytosolic Ah receptor concentrations in rat thymus exceed that in
liver from the same animals (Carlstedt-Duke, 1979; Mason & Okey, in
press) (Table 1). As illustrated in Figure 4, the [^{3}H]TCDD Ah
receptor complex can translocate from cytoplasm into nucleus in the
thymus of rats injected with [^{3}H]TCDD in vivo. The translocation

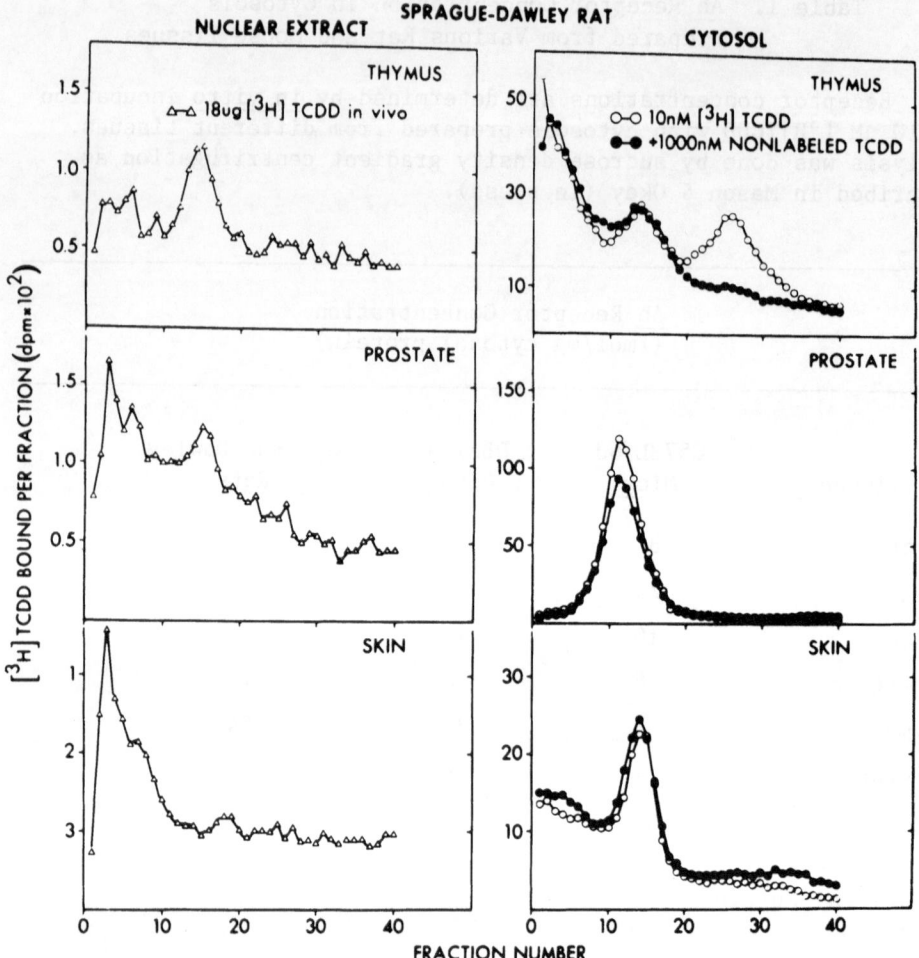

Figure 4.   [³H]TCDD Binding to Cytosols (in vitro) and Nuclear
            Extracts (in vivo) in Tissues from Sprague-Dawley Rats.

Cytosols were incubated in vitro with 10 nM [³H]TCDD in the
absence or presence of a 100-fold molar excess of nonradioactive TCDD. Nuclear extracts were prepared from tissues of
rats injected intraperitoneally with [³H]TCDD (18 μg/rat =
0.56 μmol/kg body weight). Specific Ah receptor binding was
detectable only in thymus cytosol and is represented by the
radioactive peak sedimenting near fraction 25. The large
peaks sedimenting near fractions 10-15 in cytosols from prostate and skin are not specific for chemicals which induce
cytochrome P_1-450 and do not represent Ah receptor. Nuclear
Ah receptor is represented by the peaks sedimenting near
fraction 15. (From Mason & Okey, in press.)

step, which appears to be essential for cytochrome P_1-450 induction, was first described in liver, but appears also to be a general mechanism in other responsive tissues (Mason & Okey, in press).

In the rat, thymic atrophy is one of the most sensitive indicators of TCDD exposure (Harris et al, 1973). Although Ah receptor levels are high in rat thymus and the inducer-receptor complex translocates into the nucleus, the thymus has very low cytochrome P_1-450 levels (Okey et al unpublished). Ah receptor levels in rat thymus obviously are greatly out of proportion to the induction of P_1-450-linked enzymes. It is not known what role, if any, the high levels of Ah receptor might play in rat thymic tissue.

Liver, lung and intestine of Sprague-Dawley rats also have substantial concentrations of Ah receptor (Table 1). No specific Ah receptor binding is detectable in rat prostate or skin cytosols, but both of these tissues have high levels of other components which bind [^{3}H]TCDD nonspecifically (Figure 4). The identity, function and significance of these other components is unknown.

Ah Receptor in Tissues of Other Strains and Species

In addition to C57BL/6N and C57BL/6J mice, cytosolic Ah receptor is detectable in several other genetically "responsive" inbred mouse strains. These include: CBA/J, A/J and C3H/HeJ (Okey et al, 1979). Cytosolic Ah receptor also is detectable in F_1 offspring from the C57BL/6N x DBA/2N cross and in heterozygous responsive Ah^b/Ah^d offspring from the B6D2F$_1$ x D2 backcross.

Cytosolic Ah receptor is not detectable in genetically "nonresponsive" DBA/2N, DBA/2J, AKR/J, SWR/J or RF/J inbred strains, nor is cytosolic receptor detectable in homozygous nonresponsive Ah^d/Ah^d offspring from the B6D2F$_1$ x D2 backcross (Okey et al, 1979). Susceptibility of mice to TCDD-induced thymic involution and cleft palate generally correlates well with the presence of cytosolic Ah receptor (Poland & Glover 1980). An exception is CBA/J mice which have Ah receptor, but are resistant to cleft palate induction.

As has been described earlier in this report, the cytosolic Ah receptor concentration in Sprague-Dawley rat liver is similar to that in C57BL/6J mice. Wistar rats also have receptor concentrations of 30-40 fmol/mg cytosol protein, similar to that in Sprague-Dawley rats (Bandiera et al, in press). Cytosolic Ah receptor also has been demonstrated in New Zealand white rabbits and in the "cotton rat" Sigmoden hispedis (Kahl et al, 1980).

Guinea pigs are the species most sensitive to the toxic effects of TCDD (Neal et al, 1979) and there is great interest in understanding the reasons for the high sensitivity of this species. Ah receptor is present in hepatic cytosols of adult male, adult female

and pregnant frmale guinea pigs in concentrations similar to that in
rat liver; Ah receptor also is found in high concentration in cytosols
from fetal guinea pig liver, lung and kidney (Okey et al, in preparation).   The high concentration of Ah receptor in guinea pig tissue
cytosols is somewhat surprising since guinea pigs generally show
little induction of cytochrome P_1-450 when treated with TCDD or MC
(Abe & Watanabe 1981).   As in the case of rat thymus, it is unknown
what role high concentrations of Ah receptor might play in tissues
which do not exhibit the P_1-450 induction response.

CHEMICALS WHICH INTERACT STRONGLY WITH THE Ah RECEPTOR

TCDD Compared with 3-Methylcholanthrene

As illustrated in Figures 2 & 5, the Ah receptor can be detected,
quantified and characterized by direct labeling either with [^{3}H]TCDD
or with [^{3}H]MC.   Specific binding peaks for [^{3}H]TCDD sediment at the
same position in sucrose density gradients as those for [^{3}H]MC and
the concentration of binding sites is the same using [^{3}H]MC as when
using [^{3}H]TCDD.   This would suggest that [^{3}H]TCDD and [^{3}H]MC are binding at the same molecular site.   It was considered possible, however,
that [^{3}H]TCDD and [^{3}H]MC might each have their own receptor, i.e.
receptors which had similar molecular properties but which were distinct and unique species for each radioligand.   Thus we tested the
ability of 10 nonradioactive compounds to compete against [^{3}H]TCDD
and compared this with their ability to compete against [^{3}H]MC.   As
shown in Figure 6, the rank-order competitive potency of the 10
compounds against [^{3}H]TCDD is the same as their rank-order against
[^{3}H]MC.   The fact that the 10 competitors rank the same against
[^{3}H]TCDD as against [^{3}H]MC is strong evidence that the two radioligands are indeed binding to the same (Ah receptor) site.

In rats the maximal level of induced cytochrome P_1-450 achieved
is the same when MC is injected as when TCDD is injected.   The dose
required to reach this maximum, however, is much higher for MC than
for TCDD.   On a molar basis TCDD is approximately 30,000 times more
potent than MC in vivo (Poland & Glover 1974).

This great difference in potency does not appear to be due to
differences in the affinity with which MC and TCDD bind to the Ah
receptor.   Scatchard plot analyses (Figure 7) reveal that [^{3}H]TCDD
and [^{3}H]MC bind with very similar affinities to the Ah receptor in
rat cytosol in vitro.

Interaction of Other Chemcials with the Ah Recptor

[^{3}H]TCDD and [^{3}H]MC are the only radioligands which have been
shown to interact with the Ah receptor in direct binding assays
(Hannah et al, 1981; Okey et al, 1981).   Binding of other compounds

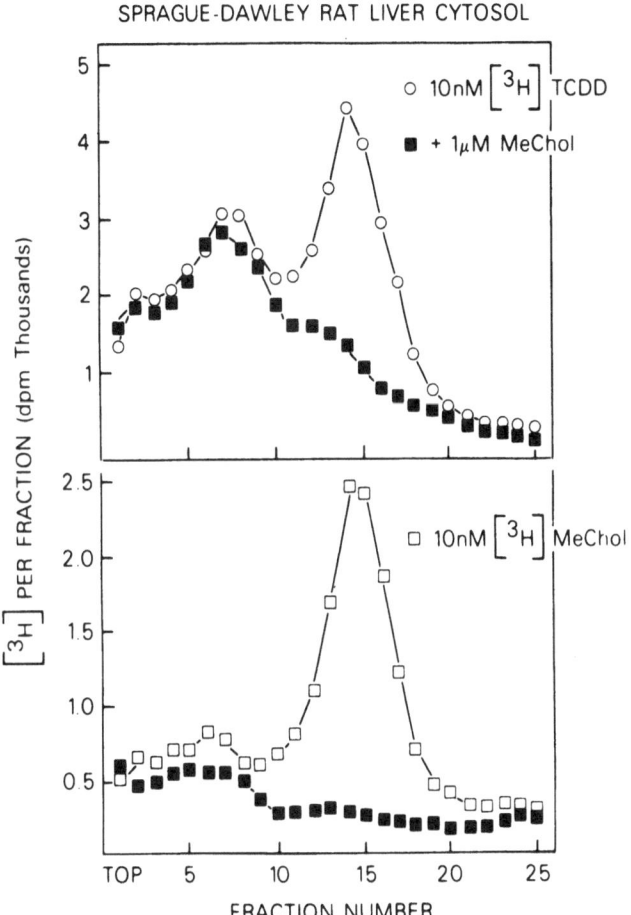

Figure 5.   Comparison of [³H]TCDD and [³H]MC as Radioligands
            for Ah Receptor in Sprague-Dawley Rat Liver Cytosol.

Hepatic cytosol from a male Sprague-Dawley rat was incubated
either with 10 nM [³H]TCDD (50 Ci/mmol) or with 10 nM [³H]MC
(37 Ci/mmol) in the absence or presence of a 100-fold molar
excess of nonradioactive MC. The radioactive peak sedimenting
near fraction 13 was completitively inhibited by nonradioactive MC (illustrated), nonradioactive TCDD (not illustrated)
and other chemicals which induce cytochrome P_1-450. Calculated specific binding in the peak-13 region is 32 fmol/mg
protein with [³H]TCDD as the radioligand and 31 fmol/mg
protein with [³H]MC as the radioligand. (From Okey et al,
1981).

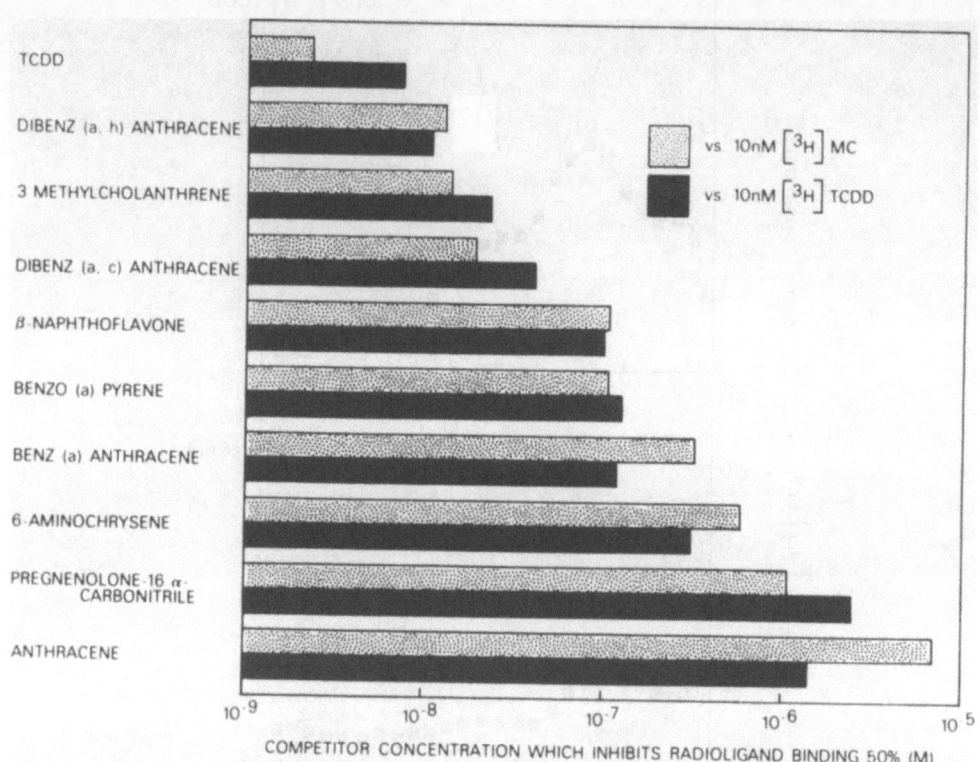

Figure 6.   Rank-Order Potency of Various Chemicals as Competitors
             with [³H]TCDD Or [³H]MC for <u>Ah</u> Receptor Sites.

A large pool of hepatic cytosol was prepared from C57BL/6J
mice. The ability of various chemicals to compete for <u>Ah</u>
receptor sites was assessed by determining the chemical's
ability to inhibit binding of [³H]TCDD or [³H]MC in the "9S"
region of sucrose density gradients. Full dose-response assays were performed for each competitor over the range of 1 nM
to 10 µM against both [³H]TCDD and [³H]MC. The EC_{50} for each
competitor was determined by interpolation from the dose-response curves. (From Okey et al, 1981.)

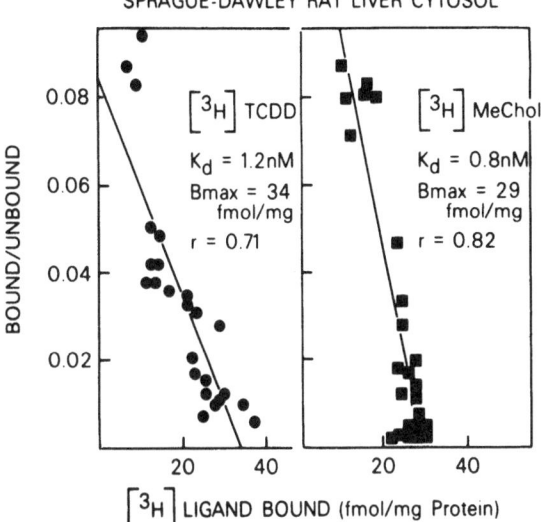

SPRAGUE-DAWLEY RAT LIVER CYTOSOL

Figure 7.   Comparative Scatchard Plot Analyses of [³H]TCDD Versus
            [³H]MC Binding to Ah Receptor in Sprague-Dawley Rat
            Liver Cytosol.

Aliquots from a pool of cytosol were incubated with [³H]TCDD
or [³H]MC over a concentration range of 0.5 to 15 nM. Specific
binding was determined by sucrose density gradient centrifugation analysis for each sample. Calculated binding parameters
are shown in the Figure. (From Okey et al, 1981.)

to the receptor is inferred from their ability to compete with
[³H]TCDD or [³H]MC for the small pool of specific receptor sites.
Competition studies have been very useful in determining the range
of chemcial structures which have significant interactions with
the Ah receptor.

By competition studies it has been shown that several chemical
classes contain agents capable of binding to the Ah receptor. Poland
et al, (1979) have shown that tetrahalogenated aromatic hydrocarbons
from a variety of classes are among the highest affinity ligands for
Ah receptor. These include halogenated dibenzo-p-dioxins (such as
TCDD) halogenated dibenzofurans, halogenated azo- and azoxybenzenes
and halogenated biphenyls.

In the halogenated biphenyl class, PCBs that induce an "MC-type"
enzyme induction pattern bind Ah receptor with highest affinity.
"Phenobarbital-type" PCBs have weak interactions with the Ah receptor,
whereas PCB congeners classed as "mixed-type" inducers (i.e. produce
a response like that of MC coadministered with phenobarbitol) exhibit
an intermediate affinity (Bandiera et al, in press).

Some unsubstituted polycyclic aromatic hydrocarbons such as dibenz(a,h)anthracene have affinities approaching that of TCDD (Figure 6). Benzo(a)pyrene and benz(a)anthracene, which are widely studied both as P_1-450 inducers and as chemical carcinogens, are approximately 10-fold weaker than MC or dibenz(a,h)anthracene as Ah receptor ligands (Figure 6).

Details of the structure-activity relationships for Ah receptor ligands are presented elsewhere in this volume. Generally speaking, the affinity with which a given chemical binds to the Ah receptor correlates well with that chemical's potency as a cytochrome P_1-450 inducer and with the chemical's toxicity. However, exceptions to this generalization may be important in interpreting the role of the Ah receptor in toxicity.

INTERPRETATIONS AND CONCLUSIONS: IS BINDING TO THE Ah RECEPTOR A KEY EVENT IN THE TOXICITY OF CHLORINATED DIOXINS AND RELATED COMPOUNDS?

This review has briefly described the nature of the Ah receptor, its occurrence in different tissues and the spectrum of chemicals which bind to the receptor. Receptor-bound TCDD constitutes less than one percent of all TCDD in the liver after in vivo injection. Is this small, receptor-bound fraction indeed responsible for TCDD's toxic activity?

Tissue Occurrence of Ah Receptor: Implications for the Mechanism of Toxicity

Ah receptor concentrations generally are high in those tissues that are most sensitive to TCDD toxicity, e.g. rat thymus, guinea pig liver, etc. Mouse strains that have a "normal" form of the Ah receptor (e.g. C57BL/6) are more susceptible to the toxic effects of TCDD than are strains with a "defective" receptor (Poland & Glover 1980; Jones & Sweeney 1980).

However, several cell lines in culture appear to have Ah receptor, yet are highly resistant to TCDD toxicity (Knutson & Poland 1980a). Thus presence per se of the Ah receptor does not necessarily ensure that a given cell will exhibit toxicity from TCDD.

Perhaps this is not surprising since it previously has been demonstrated that presence of Ah receptor does not guarantee that cytochrome P_1-450 induction will be expressed. In the adult rabbit, Ah receptor is present, but cytochrome P_1-450 induction no longer occurrs when adults are injected with β-naphthoflavone. Cytochrome P_1-450 induction is expressed in the newborn rabbit. This led to the proposal that a "temporal gene" exists in the Ah gene complex. The

function of the temporal gene is to dictate at what developmental
stage receptor-mediated P_1-450 induction can be expressed (Kahl et
al, 1980).

It is possible that an event (or events) leading to TCDD toxicity
is simply one of the battery of pleiotypic responses regulated by the
Ah receptor. Induction of cytochrome P_1-450 (the best known pleiotypic response regulated by Ah receptor) would not seem to be an
important event in TCDD toxicity since toxicity can be expressed in
the absence of P_1-450 induction. For example, guinea pigs are highly
susceptible to TCDD toxicity, but show little P_1-450 induction, whereas cells in culture may display highly inducible P_1-450 without being
susceptible to toxicity.

Knutson and Poland (1980b) have propsoed that hyperplasia and/or
differentiation of epithelial cells may be key receptor-mediated
events which lead to TCDD toxicity. Overall evidence supports their
view that binding of halogented aromatic compounds to cytosolic Ah
receptor is an essential step in toxicity but that receptor binding
in itself is not sufficient to explain the toxic response. Thus cells
must have Ah receptor in order to be susceptible to TCDD, but not all
cells with receptor will exhibit the toxic response.

Although the precise nature of receptor-regulated events leading
to toxicity is not certain, expression of the key events may be modified by developmental programs (i.e. temporal gene expression) or by
other factors such as sex of the animal.

Much of the preceding discussion may have implied that TCDD
causes toxicity by stimulating some pathway which leads to the toxic
response. It also is possible that TCDD is toxic, not because it
activates some system, but rather because it blocks some essential
physiologic function. By occupying Ah receptor sites, TCDD may prevent some endogenous (physiologic) ligand from binding, thereby interfering with a critical (but presently unknown) physiologic process
(Nebert et al, 1981).

Affinity of Binding to the Ah Receptor in Relation to Toxicity

With the exception of β-naphthoflavone, all chemicals which are
known to be high-affinity ligands for the Ah receptor are highly toxic
and/or carcinogenic. As a generalization, binding affinity correlates with toxicity. This has been established within the dibenzo-
p-dioxin series, dibenzofuran series, azo- and azoxybenzene series
(Poland et al, 1979) and for PCB congeners (Bandiera et al, in
press).

Although the generalization holds, it is important to realize
that the affinity of binding in itself is not the sole determinant
of a compound's toxicity. As described earlier in this review, TCDD

and MC appear to bind Ah receptor with very similar affinities in an
extra-cellular system in vitro. In vivo, however, TCDD is very much
more potent than MC, both in inducing cytochrome P_1-450 (Poland &
Glover 1974) and in inducing thymic involution (Poland & Glover 1980).

In whole animals or in intact cells in culture, factors other
than simple binding affinity may substantially alter the relative
potency of toxic xenobiotics. For example, Knutson and Poland (1980b)
have shown that MC can induce keratinization in XB/3T3 mixed-cell
cultures, but only if metabolism and cytotoxicity of MC are inhibited
by inclusion of 7,8-benzoflavone in the medium. TCDD induces keratinization in this system without the requirement for any additional
chemical treatments.

As suggested by Poland et al (1979), it is possible that the
potency of TCDD in vivo is due to the compound's long biological
half-life and sustained occupancy of Ah receptor sites.

Although binding affinities generally correlate with toxicity,
the affinity of binding alone does not explain toxicity.

POSSIBLE HUMAN RELEVANCE OF Ah RECEPTOR INVOLVEMENT IN TOXICITY

There is little doubt that the Ah receptor plays a critical role
in the toxicity of chlorinated dioxins and related compounds in experimental animals and in cells in culture. Binding to the receptor
can be viewed as an essential early step in the toxicity mechanism,
but binding in itself is not sufficient to account for the toxic response. Some event(s) subsequent to receptor binding are required
for expression of toxicity. Although these events appear to be
regulated by the Ah receptor, their nature still is unknown.

If the major events leading to toxicity do indeed require Ah
receptor, it would be fortunate if the human population were devoid
of receptor or had low receptor levels. Carlstedt-Duke et al (1980)
reported that some human lymphocytes had specific [^{3}H]TCDD binding at
levels similar to that in rat liver, but that binding frequently was
absent in human lymphocytes. We have assayed a small number of
samples from human thymus and none yet has had significant levels of
specific [^{3}H]TCDD binding.

There is ample evidence for genetic variation at the Ah locus in
humans (Nebert & Jensen 1979). The frequency of "high inducibility"
phenotypes in the human population is not known, but these would be
the individuals most likely to possess high levels of Ah receptor.
If the Ah receptor is required for toxicity of chlorinated dioxins,
we would hope that high concentrations of this receptor are a rare
occurrence in the human population.

REFERENCES

Abe, T., and Watanabe, M. Microsomal aryl hydrocarbon hydroxylase and cytochrome P-448 in guinea pig liver. Fifth Internat. Symp. on Microsomes and Drug Oxidations, Tokyo, Japan, July 1981 (abstract).

Bandiera, S., Safe, S., and Okey, A.B. Binding of polychlorinated biphenyls classified either as PB-, MC- or mixed-type inducers to cytosolic Ah receptor. Chem-Biol. Interact. (in press).

Carlstedt-Duke, J.M.B. Tissue distribution of the receptor for 2,3,7,8-tetrachlorodibenzo-p-dioxin in the rat. Cancer Res., 39:3172-3176, 1979.

Carlstedt-Duke, J., Elfstrom, G., Snochowski, M., Hogberg, B., and Gustafsson, J.-A. Detection of the 2,3,7,8-tetrachlorodibenzo-p-dioxin (TCDD) receptor in rat liver by isoelectric focusing in polyacrylamide gels. Toxicol. Lett., 2:365-373, 1978.

Carlstedt-Duke, J., Gillner, M., Hansson, L.-A., Toftgard, R., Gustafsson, S., Hogberg, B., and Gustafsson, J.-A. The molecular basis for the induction of aryl hydrocarbon hydroxylase: Characteristics of the receptor protein for 2,3,7,8-tetrachlorodibenzo-p-dioxin (TCDD). In: Biochemistry, Biophysics and Regulation of Cytochrome P-450. (J.-A. Gustafsson, J., Carlstedt-Duke, A. Mode, and J. Rafter, eds.). Amsterdam: Elsevier/North-Holland Biomedical Press, 1980. pp. 147-154.

Hannah, R.R., Nebert, D.W., and Eisen, H.J. Regulatory gene product of the Ah complex. Comparison of 2,3,7,8-tetrachlorodibenzo-p-dioxin and 3-methylcholanthrene binding to several moieties in mouse liver cytosol. J. Biol. Chem., 256:4584-4590, 1981.

Harris, M.E., Moore, J.A., Vos, J.G., and Gupta, B.N. General biological effects of TCDD in laboratory animals. Envir. Health Perspect., 5:101-109, 1973.

Jones, K.G., and Sweeney, G.D. Dependence of the porphyrogenic effect of 2,3,7,8-tetrachlorodibenzo(p)dioxin upon inheritance of aryl hydrocarbon hydroxylase responsiveness. Toxicol. Appl. Pharmacol., 53:42-49, 1980.

Kahl, G.F., Friederici, D.E., Bigelow, S.W., Okey, A.B., and Nebert, D.W. Ontogenetic expression of regulatory and structural gene products associated with the Ah locus. Comparison of rat, mouse, rabbit and Sigmoden hispedis. Develop. Pharmacol. Therap., 1:137-162, 1980.

Knutson, J.C., and Poland, A. 2,3,7,8-tetrachlorodibenzo-p-dioxin: Failure to demonstrate toxicity in twenty-three cultured cell types. Toxicol. Appl. Pharmacol., 54:377-383, 1980a.

Knutson, J.C., and Poland, A. Keratinization of mouse teratoma cell line XB produced by 2,3,7,8-tetrachlorodibenzo-p-dioxin: An in vitro model of toxicity. Cell, 22:27-36, 1980b.

Mason, M.E., and Okey, A.B. Cytosolic and nuclear binding of 2,3,7,8-tetrachlorodibenzo-p-dioxin to the Ah receptor in extrahepatic tissues of rats and mice. Eur. J. Biochem. (in press).

Neal, R.A., Beatty, P.W., and Gasiewicz, T.A.   Studies on the mech-
    anisms of toxicity of 2,3,7,8-tetrachlorodibenzo-p-dioxin (TCDD).
    Ann. NY Acad. Sci., 320:204-213, 1979.
Nebert, D.W., and Jensen, N.M.   The Ah locus: Genetic regulation of
    the metabolism of carcinogens, drugs, and other environmental
    chemicals by cytochrome P-450-mediated monooxygenases. In:
    G.D. Fasman, ed. CRC Critical Reviews in Biochemistry. Vol. 6.
    Cleveland, Ohio: CRC Press, 1979. pp. 401-437.
Nebert, D.W., Eisen, H.J., Negishi, M., Lanq, M.A., Hjelmeland, L.M.,
    and Okey, A.B.   Genetic mechanisms controlling the induction of
    polysubstrate monooxygenase (P-450) activities. Ann. Rev.
    Pharmacol. Toxicol., 21:431-462, 1981.
Okey, A.B., Bondy, G.P., Mason, M.E., Kahl, G.F., Eisen, H.J.,
    Guenthner, T.M., and Nebert, D.W.   Regulatory gene product of
    the Ah locus. Characterization of the cytosolic inducer-
    receptor complex and evidence for its nuclear translocation.
    J. Biol. Chem., 254:11636-11648, 1979.
Okey, A.B., Bondy, G.P., Mason, M.E., Nebert, D.W., Forster-Gibson,
    C.J., Muncan, J., and Dufresne, M.J.   Temperature-dependent
    cytosol-to-nucleus translocation of the Ah receptor for
    2,3,7,8-tetrachlorodibenzo-p-dioxin in continuous cell culture
    lines. J. Biol. Chem., 255:11415-11422, 1980.
Okey, A.B., Choi, C.K., and Vella, L.M.   [^{3}H]3-Methylcholanthrene
    binding to the Ah receptor in hepatic cytosol. Proc. Amer.
    Assoc. Cancer Res., 22:36, 1981 (abstract).
Poland, A., and Glover, E.   Comparison of 2,3,7,8-tetrachlorodibenzo-
    p-dioxin, a potent inducer of aryl hydrocarbon hydroxylase, with
    3-methylcholanthrene. Mol. Pharmacol., 10:349-359, 1974.
Poland, A., and Glover, E.   2,3,7,8-Tetrachlorodibenzo-p-dioxin:
    Segragation of toxicity with the Ah locus. Mol. Pharmacol.,
    17:86-94, 1980.
Poland, A., Glover, E., and Kende, A.S.   Stereospecific, high-affinity
    binding of 2,3,7,8-tetrachlorodibenzo-p-dioxin by hepatic
    cytosol. Evidence that the binding species is receptor for
    induction of aryl hydrocarbon hydroxylase. J. Biol. Cheml,
    251:4936-4946, 1976.
Poland, A., Greenlee, W.E., and Kende, A.S.   Studies on the mechanism
    of action of the chlorinated dibenzo-p-dioxins and related
    compounds. Ann. NY Acad. Sci., 320:214-230. 1979.
Tsui, H.W., and Okey, A.B.   Rapid vertical tube rotor gradient assay
    for binding of 2,3,7,8-tetrachlorodibenzo-p-dioxin to the Ah
    receptor. Can. J. Physiol. Pharmacol., 59:927-931, 1981.

ACKNOWLEDGEMENTS

Research in the author's laboratory is supported by grants from the
National Cancer Institute of Canada and the Medical Research Council
of Canada.

GENETIC DIFFERENCES IN ENZYMES WHICH METABOLIZE DRUGS,

CHEMICAL CARCINOGENS, AND OTHER ENVIRONMENTAL POLLUTANTS

Daniel W. Nebert, Masahiko Negishi, and Howard J. Eisen

National Institute of Child Health
  and Human Development
National Institutes of Health
Bethesda, Maryland USA

ABSTRACT

In the literature dozens of examples exist in which gene differences are reflected as variations in individual risk of cancer
or pharmacologic response. One of the most well-characterized is the
Ah locus, which controls the induction of a group of drug-metabolizing
enzymes by polycyclic aromatic compounds such as 2,3,7,8-tetrachloro-
dibenzo-p-dioxin (TCDD) and 3-methylcholanthrene. While studying
this genetic system, we have examined the effects of TCDD in cell
culture and the metabolism of TCDD by mouse liver microsomes in vitro.
The principal genetic defect among certain inbred mouse strains involves a cytosolic receptor for TCDD. The level of receptor has been
shown in mice and perhaps in man to reflect individual differences
in risk of certain chemically induced cancers, mutations, drug toxicities, and birth defects.

With one of the structural genes (cytochrome P_1-450) of the Ah
locus, recently cloned in this laboratory, we hope to understand the
genetic regulation of P-450 induction and the evolution of this
enzyme system which responds to various adverse environmental chemicals. We hope to develop sensitive clinical assays for predicting
individual genetic differences in risk of cancer, drug toxicity, and
birth defects caused by certain environmental pollutants.

INTRODUCTION

The study of the genetic control of drug metabolism is often
called pharmacogenetics. In a single sentence, pharmacogenetic

research involves the attempt to understand the hereditary basis for
two individuals responding differently to drugs or other foreign
chemicals. These responses inlcude therapeutic effects of a drug,
e.g. anticoagulation or control of seizures, but also unwanted deleterious effects such as increased risk of cancer or drug toxicity.

The experimental system to be examined in detail here involves
principally genetic differences in concentration and/or affinity of
a receptor to which 2,3,7,8-tetrachlorodibenzo-p-dioxin (TCDD) binds
avidly. Because of this, there are large genetic differences in the
biotransformation and pharmacokinetics of certain drugs and other
environmental pollutants, resulting in important differences in
tendency toward cancer, drug toxicity, mutation, and birth defects.

We first describe the general characteristics of the drug-metabolizing enzymes. Secondly, the genetic differences in this model
system in mice are examined. Third, the induction response of TCDD
in cell culture and TCDD metabolism in vitro are reviewed. Fourth,
detailed characteristics of the Ah receptor are described. Fifth,
numerous examples of environmentally caused cancer and toxicity associated with this mouse genetic system are listed. Sixth, current
evidence for this genetic difference in man is briefly assessed.
Lastly, the use of recombinant DNA technology is suggested as a new
means of studying pharmacogenetics.

P-450-Mediated Monooxygenases and Coordinated Enzymes

Many drugs and other foreign compounds are chemicals that are so
fat-soluble they would remain in the body indefinitely were it not
for metabolism resulting in more polar derivatives. The drug-metabolizing enzyme systems, which are localized principally in the liver,
are usually divided into two groups: Phase I and Phase II. During
Phase I metabolism, one or more polar groups (such as hydroxyl) are
introduced into the hydrophobic parent molecule, thus allowing a handle, or position, for the Phase II conjugating enzymes (such as UDP
glucuronosyltransferase) to attack. The conjugated products are
sufficiently water-soluble, so that these detoxified chemicals are
now excreted from the cell and from the body (Williams, 1959).

One of the most interesting of the Phase I enzyme systems is a
group of enzymes known collectively as the cytochrome P-450-mediated[1]
monooxygenases (cf. Lu and West, 1980; Mannering, 1981; Nebert et al,
1981; and Nebert et al, 1982b for recent reviews). These membrane-
bound enzyme systems are known to metabolize almost everything in the
Merck Index--from small molecules such as ethanol, acetone, and carbon disulfide to large molecules such as TCDD, benzo[a]pyrene, and
synthetic steroids. Virtually all drugs are substrates. Endogenous
substrates include steroids, fatty acids, biogenic amines, indoles,
and thyroxine.

Evidence is growing that metabolism to reactive intermediates by cytochrome P-450-mediated[1] monooxygenases is a prerequisite for mutagenesis, carcinogenesis, and toxicity caused by numerous drugs, polycyclic hydrocarbons, and other environmental pollutants. These reactive intermediates bind covalently to numerous cellular macromolecules. Most of this binding is probably random, but some may be nonrandom, i.e. specific binding dependent upon the chemical structures of the reactive intermediate and the cellular macromolecule.

The steady-state levels of these reactive intermediates, and, consequently, the rates at which they interact with the critical subcellular target(s) are dependent upon a delicate balance between their formation and removal (Figure 1). Changes in the balance between toxification and detoxication in any particular tissue of an individual may therefore affect his risk of tumorigenesis or toxicity.

The Ah Locus

The [Ah] system is an experimental model that has provided several good examples of the delicate balance between genetic and environmental factors in the etiology of cancer, drug toxicity, and birth defects (Nebert and Jensen, 1979). The Ah locus regulates the induction (by polycyclic aromatic compounds such as 3-methylcholanthrene, benzo[a]pyrene or TCDD) of numerous drug-metabolizing enzyme "activities" associated with two or more new induced forms of cytochrome P_1-450. The Ah system comprises regulatory, structural, and probably temporal genes that may or may not be linked.

Several studies indicate that the fundamental genetic difference lies in the Ah regulatory gene, which encodes a cytosolic receptor (Figure 2) capable of binding to inducers such as 3-methylcholanthrene, benzo[a]pyrene and TCDD (Poland et al, 1976). To our knowledge, only foreign chemicals bind to this receptor with high affinity (less than 1 nM). The Ah receptor appears to be defective in the DBA/2N mouse and other "Ah-nonresponsive" inbred strains that are relatively insensitive to these inducers. Translocation of the inducer-receptor complex into the nucleus has been demonstrated in the phenotypically responsive heterozygote and homozygote (Okey et al, 1979). A temperature-dependent nuclear translocation of the inducer-receptor complex has been demonstrated (Okey et al, 1980) in tissue culture studies.

---

[1]In this discussion cytochrome P-450 is defined as all forms of CO-binding hemoproteins associated with membrane-bound NADPH-dependent monooxygenase activities. We define cytochrome P_1-450 as all forms of CO-binding hemoprotein that increase in amount concomitantly with rises in induced aryl hydrocarbon hydroxylase activity following polycyclic aromatic inducer treatment. Since there appears to be more than one form of P_1-450 (Negishi and Nebert, 1979), it is emphasized that this definition of P_1-450 is simplistic.

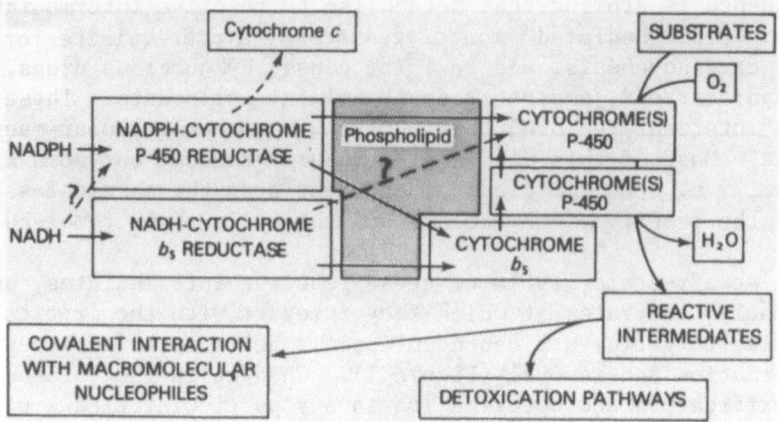

Figure 1. Scheme for the membrane-bound multicomponent monooxygenase
system(s) and the various possibly important pathways for
hydrophobic substrates (Nebert, 1981). For any given substrate, the relative balance between toxification and detoxication likely would vary among different tissues,
strains, and species. Age, genetic expression, nutrition,
hormome concentration, diurnal rhythm, pH, saturating
versus nonsaturating conditions of the substrate, K_m and
V_{max} for each enzyme, subcellular compartmentalization of
each enzyme, efficiency of DNA repair, and the immunological competence of the animal--may all be important factors affecting this balance [Reproduced with permission
from National Institute of Environmental Health Sciences].

What happens in the nucleus is not yet known, but somehow the "information" (that these inducers of P_1-450 exist in the cell's micro-
environment) is received; the response is transcription of specific
mRNA's, translation of these mRNA's into specific enzymes such as
P_1-450, and incorporation of P_1-450 into cellular membranes. These
induced enzymes may aid in detoxication or they may generate increased amounts of reactive intermediates.

     The induction of aryl hydrocarbon (benzo[a]pyrene) hydrozylase
(AHH) activity with its associated P_1-450 and at least two dozen
other monooxygenase activities occurs in 3-methylcholanthrene-treated
C57BL/6N and other genetically "responsive" inbred strains and is
always much lower in 3-methylcholanthrene-treated DBA/2N and other
genetically nonresponsive strains (at any given dose of inducer).
Besides the liver, this genetic expression is seen in such tissues
as lung, kidney, intestine, lymph nodes, skin, bone marrow, pigmented
epithelium of the retina, brain, mammary gland, uterus, ovary, and
testis. The genetic response is therefore called "systemic," or

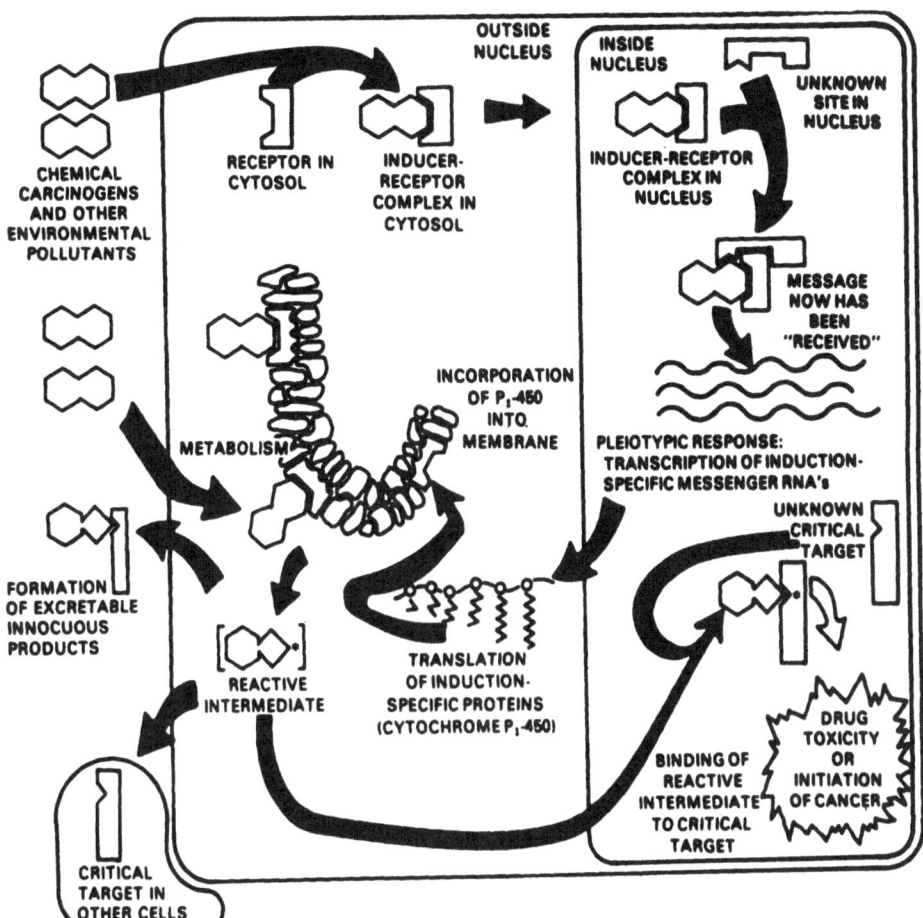

Figure 2.   Diagram of a cell and the hypothetical scheme by which
            the cytosolic Ah receptor binds to inducer (Nebert, 1979).
            Depending upon the half-life of the reactive intermediate,
            the rate of formation of the intermediate, and the rate of
            conjugation and other means to detoxify the intermediate--
            important covalent binding may occur in the same cell in
            which metabolism took place, or in some distant cell.
            Although the "unknown critical target" is illustrated
            here in the nucleus, there is presently no experimental
            evidence demonstrating unequivocally the subcellular lo-
            cation of a critical target(s) required for the initiation
            of drug toxicity or cancer [Reproduced with permission
            from Dr. W. Junk Publishers].

occurring throughout virtually all tissues of the animal. Responsiveness to aromatic hydrocarbons has been designated the Ah locus:
Ah^b is the dominant allele; Ah^d is the recessive allele; the Ah^b/Ah^d

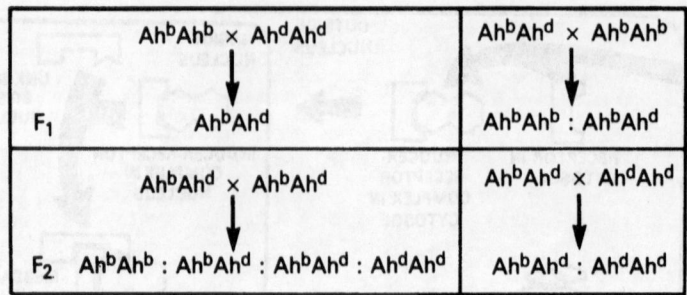

Figure 3. Simplified genetic scheme for "Ah-responsiveness" when
           C57BL/6N and DBA/2N inbred mouse strains are used. Cross
           between the C57BL/6N (Ah^b/Ah^b) and DBA/2N (Ah^d/Ah^d) at
           upper left yields the heterozygote (Ah^b/Ah^d), which has
           half as much receptor but the same amount of AHH induci-
           bility as the Ah^b/Ah^b mouse. Backcross between the F_1
           hybrid and the C57BL/6N parent (upper right) yields all
           Ah-responsive progeny. Backcross between the F_1 and the
           DBA/2N parent (lower right) yields children, half of which
           are responsive heterozygotes and the other half nonrespon-
           sive homozygotes. The F_1 x F_1 intercross (lower left)
           yields the F_2 population, three-fourths phenotypically
           responsive as measured by AHH inducibility and one-fourth
           nonresponsive homozygotes.

heterozygote is phenotypically similar to the Ah^b/Ah^b homozygote in
terms of degree of responsiveness (Figure 3). This type of inheritance is termed autosomal dominant.

AHH Induction by TCDD in Cell Culture

     The kinetics of AHH induction by TCDD are very similar to those
by 3-methylcholanthrene among 10 established cell lines, as well as
fetal primary cultures derived from hamster, rat, chick, rabbit, and
four inbred starins of mice, and among cultured human lymphocytes
(Niwa et al, 1975). The TCDD-inducible process is sensitive to
actinomycin D and cycloheximide at levels of inhibitor similar to
those previously reported with 3-methylcholanthrene in culture
(Nebert and Gielen, 1971). The induced AHH activity in cells treated
with TCDD plus 3-methylcholanthrene is not greater than that in cells
exposed to either inducer alone. There exists no relationship between the cytotoxicity of TCDD and the level of inducible AHH
activity in culture (Niwa et al, 1975). This finding was recently
confirmed in studies with 15 other cell culture lines (Knutson and
Poland, 1980).

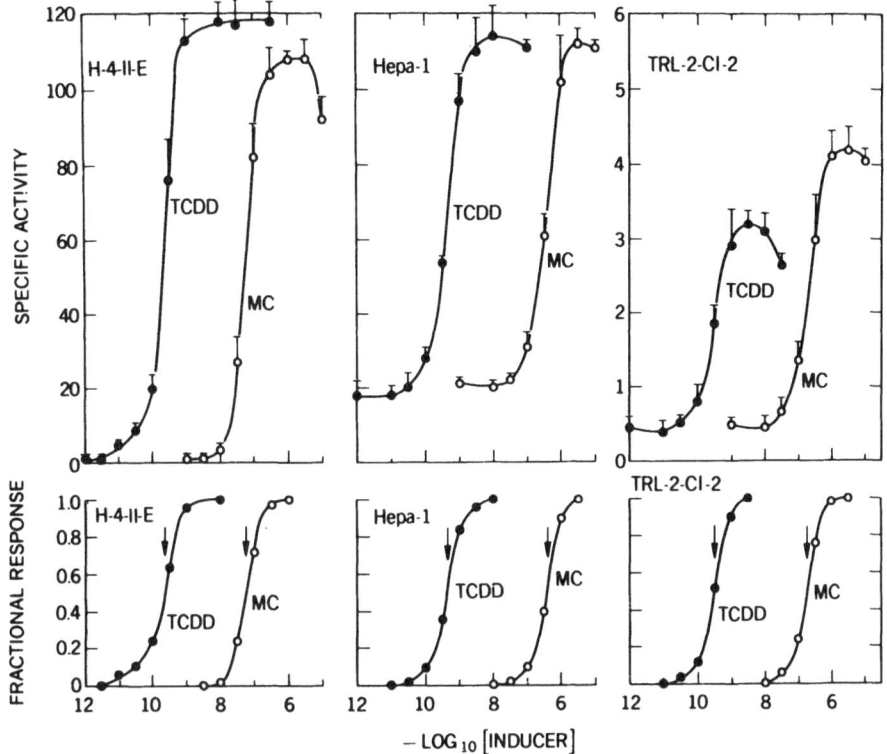

Figure 4.   Basal and inducible AHH activities and fractional response
             of induction, as a function of TCDD or 3-methylcholan-
             threne (MC) concentration (Niwa et al, 1975). H-4-II-E
             cells are derived from a rat Reuber hepatoma; Hepa-1 cells
             are derived from a mouse hepatoma; TRL-2-Cl-2 cells are
             derived from normal 10-day-old rat liver. The vertical
             arrows denote the estimated ED_{50} for each inducer. The
             enzyme activities in H-4-II-E cells treated with 3.0 pM
             TCDD or 3.0 nM MC, and in Hepa-1 or TRL-2-Cl-2 cells
             treated with 10pM TCDD or 10 nM MC, were not statistically
             different from the enzyme activities in these respective
             cell lines grown in control medium alone [Reproduced with
             permission from Academic Press].

        Estimated ED_{50} values for AHH induction by TCDD range from about
0.12 nM in C57BL/6N mouse cell cultures and 0.23 nM in the H-4-II-E
cell line to more than 100 nM in the VERO and HTC cell lines (Niwa
et al, 1975). In several cell lines (Figure 4) and in primary cultures, the responsiveness of AHH induction to TCDD is between 250
and 900 times greater than that to 3-methylcholanthrene. The responsiveness of AHH induction to TCDD in C57BL/6N mouse-derived
cultures is about 16 times greater than the responsiveness to TCDD

in DBA/2N mouse-derived cultures; this difference in responsiveness
to TCDD is very similar to that seen in these two mouse strains in
vivo (Poland et al, 1974). A bioassay with H-4-II-E cells was suggested (Niwa et al, 1975) for the detection of minute levels
(10^{-14} mol) of TCDD.

TCDD Metabolism by Mouse Liver Microsomes In Vitro

Using tritiated TCDD of very high specific radioactivity
(Guenthner et al, 1979), this laboratory reported that TCDD is
metabolized by the mouse liver P_1-450-mediated monooxygenase system
to reactive intermediates which bind covalently to cellular macromolecules. Although very difficult to quantitate (Table 1), the
presumably covalent binding to microsomal protein occurs between 120
and 2,640 times more readily than binding to deproteinized DNA in
the in vitro reaction. Because of the extremely high rate of binding
to protein rather than to DNA, it is visualized that TCDD metabolites
may be so reactive that they bind in or near the P-450-active site
where the TCDD is monooxygenated. This extreme reactivity may preclude the formation of detectable quantities of phenols, dihydrodiols,
or conjugated products.

The rate of TCDD metabolism is estimated to be between 9,000
and 36,000 times lower than the rate of P-450-mediated benzo[a]pyrene
metabolism. To our knowledge, this was the first demonstration
(Guenthner et al, 1979) that TCDD is metabolized in any organism.
There remains the possibility, however unlikely, that this covalently-
bound radioactivity represents metabolites of contaminants--present
in the radiolabeled TCDD sample in very minute amounts--rather than
metabolites of tritiated TCDD itself. A similar study was then carried out with rat liver in vivo (Poland and Glover, 1979); the conclusion was that the amount of TCDD bound covalently to cellular
macromolecules was so miniscule as to preclude any association between TCDD metabolism and either toxicity or cancer caused by TCDD.

THE Ah RECEPTOR

Velocity Sedimantation Analysis (Sucrose Density Gradient Centrifugation)

Sucrose density gradient analysis following dextran-charcoal
treatment was found (Okey et al, 1979) to be far more reliable in
characterizing the Ah receptor than DEAE-cellulose column chromatography or previously published procedures (Poland et al, 1976;
Guenthner and Nebert, 1977) involving dextran-charcoal adsorption.
Our laboratory had continued to study the physico-chemical characteristics of the Ah receptor, whereas Poland and coworkers have proceeded with attempts to understand the mechansim of TCDD toxicity
and the true function of the Ah receptor.

Table 1. Amount of TCDD Metabolites Covalently Bound to Deproteinized "DNA" or Microsomal Protein In Vitro Following Metabolism by Mouse Liver Microsomal P-450[a]

Source of Liver Microsomes from 3-Methylcholanthrene-Treated Mice	TCDD Equivalents Bound to "DNA"		TCDD Equivalents Bound to Protein		Detectable TCDD Metabolites, fmol formed/min/ mg microsomal protein (adjusted for blank)	Specific AHH Activity[b]
	dpm/mg	pmol/mg (adjusted for blank)	dpm/mg x 10^{-6}	pmol/mg (adjusted for blank)		
C57BL/6N	10,120	0.074	2.1	9.3	209	2,610
DBA/2N	2,860	0.017	1.8	7.6	169	420
C57BL/6N (proteinase K)[c]	1,600	0.016				
C57BL/6N (ANF)[d]	3,910	0.028	0.46	2.0	45	470
C57BL/6N (no NADPH)	660		0.88			32

[a] In vitro incubations included 5 mg of microsomal protein and 20 mg of deproteinized salmon sperm DNA (Guenthner et al, 1979) [Reproduced with permission from S. Karger Publishers].
[b] AHH specific activities are expressed as pmol of phenolic benzo[a]pyrene formed/min/mg microsomal protein.
[c] Further digestion of "pure DNA" by proteinase K.
[d] Incubation of 2.0 μM [^{3}H]TCDD with microsomes, deproteinized DNA, and 500 μM α-naphthoflavone, a relatively specific inhibitor of P_1-450-mediated metabolism (Nebert and Jensen, 1979).

Sucrose density gradients clearly separate a class of high affinity, low capacity binding sites from nonsaturable binding sites for [^{3}H]TCDD. With this assay it was found that the hepatic cytosolic receptor: (a) binds most specifically to foreign polycyclic aromatic compounds that are effective inducers of cytochrome P_1-450 (Figure 5); (b) is saturable; (c) possesses high affinity, low capacity binding sites; (d) appears to be composed principally of protein; (e) is highly thermolabile, more so in the absence of inducer; (f) does not appear to increase during in vivo treatment with inducers of P_1-450; (g) is present in five responsive inbred mouse strains examined, plus the Sprague-Dawley rat; (h) is not detectable in five nonresponsive inbred strains examined; and (i) is present in the Ah^b/Ah^d homozygote from the (C57BL/6N)(DBA/2N)F_1 x DBA/2N backcross (Okey et al, 1979).

Large differences in sedimentation properties and in the estimated number of binding sites were shown to depend upon (a) choice of the buffer system (stablized especially in the presence of glycerol), (b) ionic strength, and (c) protein concentration during the gradient analysis (Okey et al, 1979). Whereas hepatic cytosolic receptor levels range between 11 and 72 fmol/mg, the kidney and lung were found to have about 8 and 90 fmol/mg of cytosolic protein, respectively. The apparent mean number of sites detected by saturation analysis in C57BL/6N mice corresponds to about 5,500/hepatic cell (60 fmol/mg of cytosolic protein). The apparent K_d for TCDD binding is estimated to be approximately 0.7 nM (Okey et al, 1979).

[^{3}H]TCDD treatment in vivo of C57BL/6N or Ah^b/Ah^d mice for 2 to 18 h results in the appearance of a hepatic nuclear inducer-receptor complex that sediments at ∿6 S on sucrose density gradients of high ionic strength following dextran-charcoal treatment (Figure 6). A much smaller peak occurs in similarly treated DBA/2N or Ah^d/Ah^d mice. These data suggest that the cytosolic receptor, and its translocation with specifically bound inducer into the nucleus, are essential during the sequence of cytochrome P_1-450 induction and other events controlled by the Ah locus (Okey et al, 1979; Okey et al, 1980).

Gel Permeation Chromatography

[^{3}H]TCDD or [^{3}H]3-methylcholanthrene binding to the Ah receptor and other moieties in hepatic cytosol was examined by gel permeation chromatography, velocity sedimentation (sucrose density gradient centrifugation), dextran-charcoal adsorption, and anion-exchange chromatography (Hannah et al, 1981). In the liver of Ah-responsive C57BL/6N and the Ah^b/Ah^d heterozygote, both radioligands bind to three major components (Figures 7 and 8): peak I, a large aggregate which is eluted in the void volume of Sephacryl S-300 columns and which sediments as a residue to the bottom of sucrose density gradients; peak II, an asymmetric protein (M_r∿245,000) with a Strokes

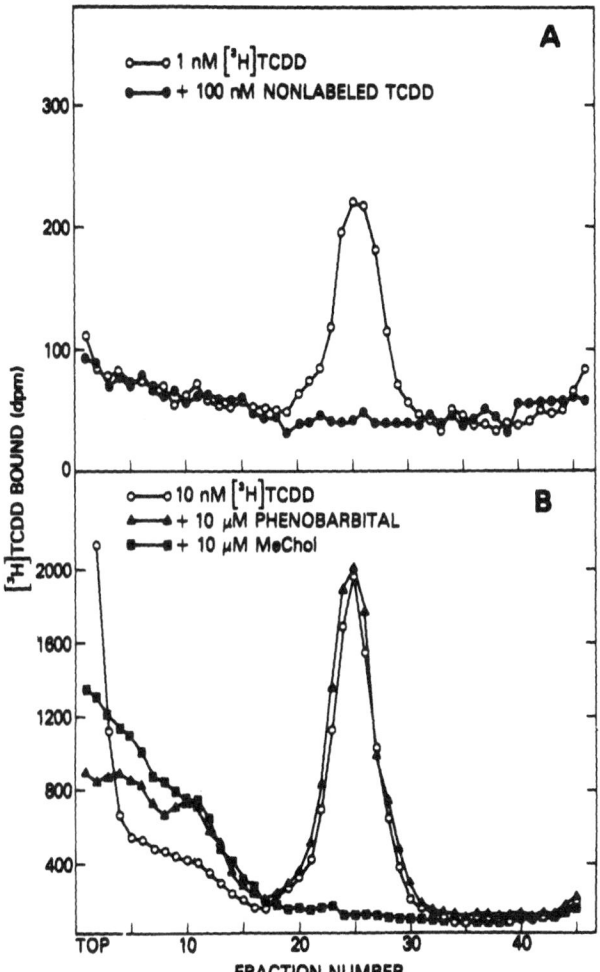

Figure 5. Detection of specific [^{3}H]TCDD binding of a component
from C57BL/6N hepatic cytosol (Okey et al, 1979). A,
cytosol (1 mg of protein/ml) from C57BL/6N was incubated
with 1 nM [^{3}H]TCDD in the absence of competitor (○—○)
or in the presence of 100 nM nonlabeled TCDD (●—●).
Following dextran-charcoal treatment, gradients were centrifuged and fractionated. B, elimination of specific
binding peak by 3-methylcholanthrene, but not by phenobarbital. Cytosol (5 mg of protein/ml) was incubated with
10 nM [^{3}H]TCDD in the absence of competitor (○—○), and in
the presence of 10 μM phenobarbital (▲—▲) or 10 μM 3-
methylcholanthrene (■—■) [Reproduced with permission
from American Society of Biological Chemists].

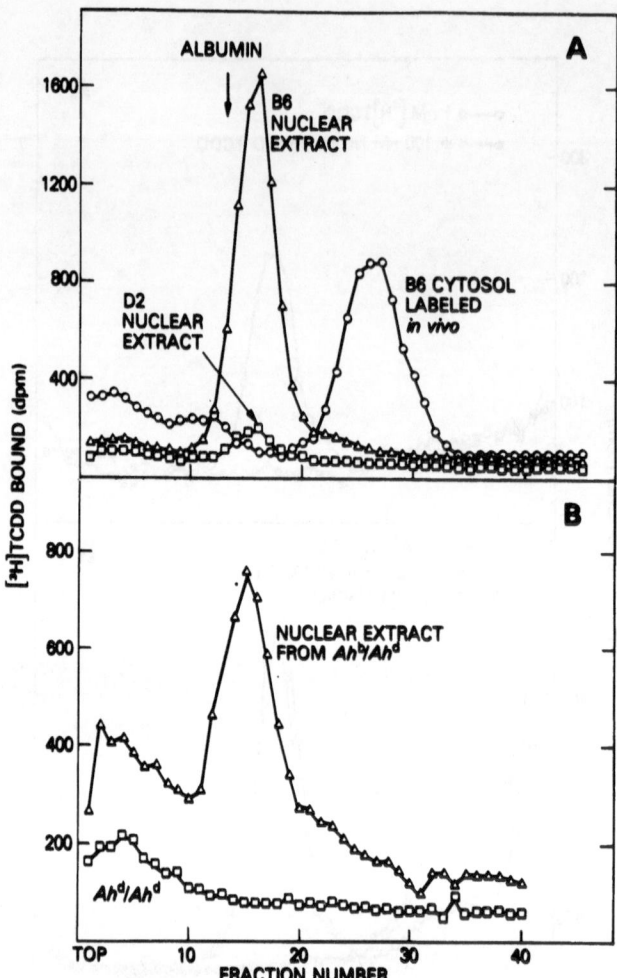

Fig. 6. Genetic differences in the nuclear binding of [^{3}H]TCDD in vivo
(Okey et al, 1979). $\underline{A}$, nuclear extracts (approximately 5 mg
of protein/ml) from responsive C57BL/6N (B6) and nonresponsive
DBA/2N (D2) liver were treated with dextran-charcoal and then
centrifuged on sucrose density gradients containing 0.4 M KCl.
B6 cytosol (labeled in vivo, 15 mg of protein/ml) was treated
with dextran-charcoal and centrifuged as usual on a gradient
prepared in buffer without KCl. $\underline{B}$, hepatic nuclear extracts
from a responsive $\underline{Ah}^b/\underline{Ah}^d$ and a nonresponsive $\underline{Ah}^d/\underline{Ah}^d$ individual from the B6D2F$_1$ x D2 backcross. The extracts (6 mg of
protein/ml) following dextran-charcoal treatment were centrifuged on gradients prepared in buffer containing 0.4 M KCl.
The B6 and D2 mouse had each received 2 μg of [^{3}H]TCDD (approximately 0.3 μmol/kg of body weight) and were killed 2 h
latter. The backcross animals had each received 5 μg of [^{3}H]
TCDD (about 0.75 μmol/kg of body weight) and were killed 3 h
later. These backcross mice had been phenotyped (Robinson and
Nebert, 1974) more than 1 week earlier [Reproduced with permission from American Society of Biological Chemists].

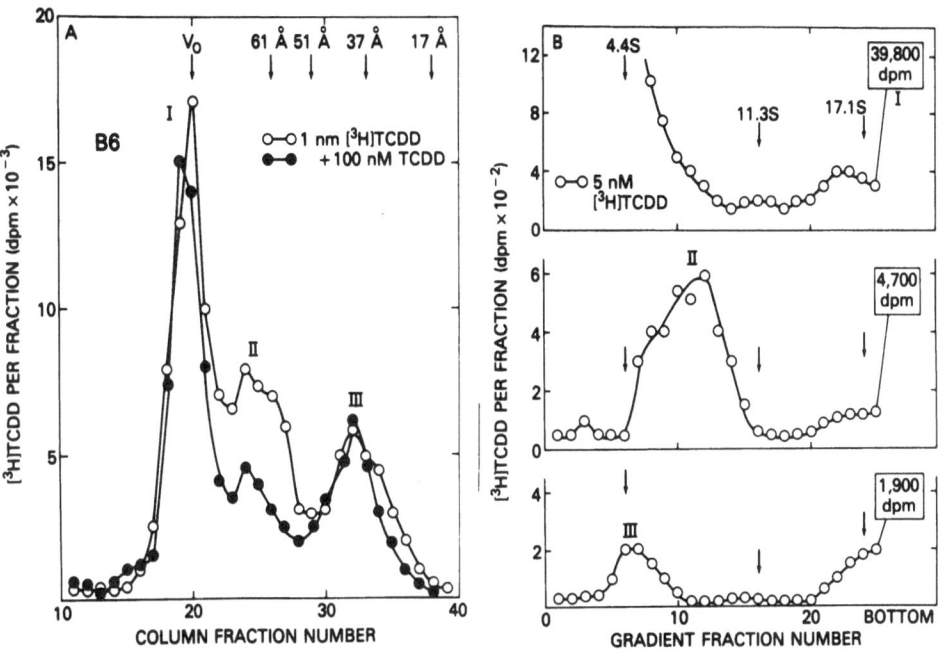

Figure 7.   [^{3}H]TCDD binding to C57BL/6N (B6) cytosol (Hannah et al, 1981). A, gel permeation chromatography of B6 hepatic cytosol treated with [^{3}H]TCDD. V_0, void volume determined with blue dextran, thyroglobulin, ferritin, bovine serum albumin, and cytochrome c, with each of their Stokes radii indicated in Å, were used to calibrate the columns. B, analysis of peaks I, II, and III by velocity sedimentation. Cytosol (15 mg protein/ml) was treated with 5 nM [^{3}H]TCDD for 1 h at 4 °C and then chromatographed on a Sephacryl S-300 column. Samples from the peak I (fractions 16 to 22), peak II (fractions 23 to 27), and peak III (fractions 28 to 35) regions were then added to sucrose density gradients (5% to 20%). In addition to the usual gradient fractions, the bottoms of the centrifuge tubes were cut off, soaked in Aquasol, and counted for radioactivity (dpm shown in boxes). Approximate sedimentation values are shown for bovine serum albumin (4.4 S), bovine liver catalase (11.3 S), and ferritin (17.1 S); these standards were centrifuged in a separate gradient [Reproduced with permission from American Society of Biological Chemists].

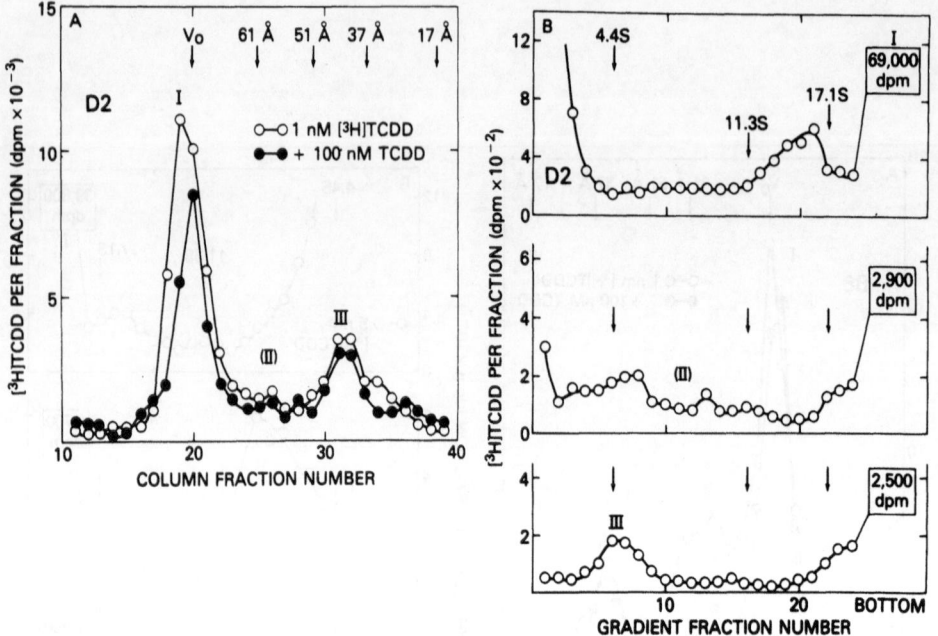

Figure 8. [³H]TCDD binding to DBA/2N (D2) cytosol (Hannah et al,
           1981). A, gel permeation chromatography of D2 hepatic
           cytosol treated with [³H]TCDD. B, analysis of peaks I,
           II, and III by velocity sedimentation. These experiments
           with D2 mice were identical to those with B6 mice illus-
           trated in Figure 7 [Reproduced with permission from
           American Society of Biological Chemists].

radius of about 75 Å; and peak III, a globular protein ($M_r \sim 87,000$)
with an estimated Stokes radius of 40 Å.

    The peak I aggregate is not adsorbed by dextran-coated charcoal
and therefore represents the large proportion of nonsaturable radioligand binding measured by dextran-charcoal adsorption (Hannah et al,
1981). The peak II protein has a size of about 9.0 S in low ionic
strength, high affinity for TCDD, and saturability at TCDD concentrations greater than about 1.0 nM. The peak II protein is not
detectable in the liver of Ah-nonresponsive DBA/2N and the Ah^d/Ah^d
homozygote (Figure 8) and therefore represents the Ah receptor. The
peak III protein has an estimated size of 5.0 S, is not saturable
with either TCDD or 3-methylcholanthrene under the conditions of
these experiments, and is not associated with the Ah^b allele. 3-
Methylcholanthrene binds to the peak III protein to a greater extent
than TCDD.

These data (Hannah et al, 1981) explain the discrepancies between the dextran-charcoal adsorption and sucrose density gradient
assays. We therefore believe that any further studies of the function of the Ah receptor and these other ligand-binding moieties
(e.g. nuclear translocation) should include gel permeation chromatography in order to distinguish among the various binding components.

IMPORTANCE OF THE ROUTE OF ADMINISTRATION

Numerous cancer and toxicity studies in the mouse are categorized in Table 2. When the carcinogen (or other toxic drug) is placed
in relatively direct contact with the tissue being studied, the
genetically responsive Ah^b/Ah^b or Ah^b/Ah^d mouse is at increased risk
for developing a tumor or toxicity in that tissue, compared with the
nonresponsive Ah^d/Ah^d receiving the same dose of xenobiotic. On the
other hand, if the malignancy or toxicity is found at a site distant
from the administered drug, the Ah^d/Ah^d mouse is at increased risk,
compared with the Ah^b/Ah^b or Ah^b/Ah^d individual receiving the same
dose of foreign chemical. In this latter case, we believe the data
are explainable by the "first-pass effect," also termed "presystemic
drug elimination" (reviewed in Routledge and Shand, 1979). Fundamentally, presystemic elimination reflects the metabolism and excretion of a drug before the drug reaches its site of action.

Proof of the first-pass effect came from studies with radiolabeled benzo[a]pyrene in the diet (Nebert et al, 1980). More than
10 times as much drug reaches the marrow (and spleen) in Ah^d/Ah^d
mice, as compared with Ah^b/Ah^d mice receiving the same oral dose.
Because much more drug is metabolized as a result of the (highly
induced) AHH activity in intestinal epithelium and liver of the Ah-
responsive mouse than the Ah-nonresponsive mouse, the relative dose
of benzo[a]pyrene reaching the marrow is very much less in the Ah-
responsive than the Ah-nonresponsive individual.

The data summarized in Table 2 demonstrate that P_1-450 induction
represents a "double-edged sword." In other words, tissue sites in
direct contact with a carcinogen develop cancer more readily in responsive animals because of induced P_1-450; tissues in distant sites
of the body may develop malignancy more readily in nonresponsive
animals because more carcinogen reaches that tissue due to decreased
P_1-450 induction in portal-of-entry tissues and therefore a decrease
in detoxication. Hence, not only the dose, but the route of admin-
istration, the timing of the dosage, and the site of the tumor or
toxicity--relative to the site of administered drug--are all very
important in the interpretation of data from carcinogenesis or toxicity experiments involving P_1-450 inducers such as polycyclic aromatic compounds. These data point out the complexity of drug-metabolizing enzyme systems: increased metabolism of a carcinogen may be

Table 2. Summary of Toxicity and Tumorigenesis in the Mouse Associated with the Ah Locus[a]

Individual at Increased Risk	Tumor or Toxicity	Route of Administration	
Ah^b/Ah^b and Ah^b/Ah^d	Skin inflammation	Topical	7,12-Dimethylbenzo[a]anthracene
	Fibrosarcomas	Subcutaneous	3-Methylcholanthrene or benzo[a]pyrene
	Pulmonary tumors	Intratracheal	3-Methylcholanthrene >> benzo[a]pyrene
	In utero fetal toxicity	Intraperitoneal	Benzo[a]pyrene, 3-methylcholanthrene, 7,12-dimethylbenzo[a]anthracene
	Primordial oocyte depletion	Intraperitoneal	7,12-Dimethylbenzo[a]anthracene, 3-methylcholanthrene, benzo[a]pyrene
	Epidermal carcinoma	Topical	Benzo[a]pyrene
	Cleft palate in fetus	Intraperitoneal	2,3,7,8-Tetrachlorodibenzo-p-dioxin
	Experimental porphyria	Intraperitoneal	Chlorinated aromatic compounds
Ah^d/Ah^d	Lymphoma, lymphosarcoma	Intraperitoneal	7,12-Dimethylbenzo[a]anthracene
	Bone marrow toxicity	Oral	Benzo[a]pyrene
	Leukemia	Subcutaneous	3-Methylcholanthrene
	Leukemia	Oral	Benzo[a]pyrene

[a] References for each of these studies can be found in Nebert, 1980; Nebert, 1981 [Reproduced with permission from Academic Press, Inc.].

good or bad. These data also emphasize the utility of intact animal studies, in which pharmacokinetic and enzyme induction differences may occur among various tissues; such subleties cannot be detected in short-term bacterial or tissue culture testing (Nebert, 1980; Nebert, 1981).

THE HUMAN Ah LOCUS

With the use of 20 to 40 cc of drawn blood, peripheral lympho-cytes have been cultured in the presence of mitogens and an inducer of AHH phenotype. In spite of the shortcomings of this assay method (reviewed in Atlas and Nebert, 1978), a growing list of clinical disorders (Table 3) appears to be associated with the human Ah locus.

There clearly exists sufficient evidence that heritable vari-ation of AHH inducibility occurs in man. Experimental difficulties, however, make it impossible at this time to be certain whether AHH induction is controlled by a single genetic locus or by two or more loci (i.e., polygenic). Until one can increase the range of induci-bility of AHH activity and/or decrease the magnitude of day-to-day variability of "control" AHH activity, however, AHH inducibility in cultured mitogen-activated lymphocytes or any other similar test system cannot be used as a promising biochemical marker for pre-dicting who is at risk for aplastic anemia, leukemia, bronchogenic carcinoma, or other types of environmentally-caused toxicity or malignancy.

A high ratio of P_1-450 to other forms of P--450 exists in many, if not all, extrahepatic tissues in vivo, just as appears to be the case in cultured lymphocytes, monocytes, pulmonary macrophages, and even skin fibroblasts. We believe that an alternative assay for assessing the human Ah phenotype (such as a receptor assay, a radio-immunoassay for induced P_1-450, detection of induced P_1-450 mRNA or analysis of the chromosomal P_1-450 gene itself by use of the cloned P_1-450 gene) might be more successful than the existing commonly performed AHH inducibility assay.

RECOMBINANT DNA TECHNOLOGY AND THE Ah SYSTEM

Using partially purified mouse liver 23 S mRNA known (Negishi and Nebert, 1981) to be associated with the Ah locus and 3-methyl-cholanthrene-induced cytochrome P_1-450, double-stranded cDNA was synthesized; the DNA was then inserted into pBR322 plasmid DNA and cloned in E. coli LE392 (Negishi et al, 1981). By both genetic and immunologic criteria, we have proved that clone 46 (Figure 9) con-tains part of the structural gene for 3-methycholanthrene-induced P_1-450. With the use of this cloned sequence (Tukey et al, 1981),

Table 3. Human Disorders That Appear To Be
Associated With The Ah Locus

Disorder	Association with High or Low Aryl Hydrocarbon Hydrozylase
Malignancy	
Bronchogenic carcinoma	High[a]
Laryngeal carcinoma	High[b]
Cancer of oral cavity	High[b]
Cancer of renal pelvis or ureter	No association found
Cancer of urinary bladder	No association found
Acute leukemia of childhood	Low[a]
Toxicity	
Zoxazolamine-induced fetal hepatic necrosis	Unknown[c]
Earlier onset of menopause among cigarette smokers	Unknown[d]
Infertility among cigarette smokers	Unknown[d]
Acetaminophen-induced diffuse bilateral cataracts	Unknown[d]

[a]Consistent with genetic data from inbred strains of mice.
[References given in Nebert, 1981).
[b]Studies of these disorders in mice have not been specifically
carried out, but the human data are consistent with what is
known (Nebert, 1980) about environmental carcinogens and their
effect on local and distant tissue sites in genetically "Ah-
responsive" and "Ah-nonresponsive" mice.
[c]Genetically responsive mice are at decreased risk for
zoxazolamine-induced muscle paralysis (Robinson and Nebert, 1974).
[d]Genetically responsive mice are at increased risk for these
disorders (Nebert and Jensen, 1979). In retrospect, it would have
been of interest to know the Ah phenotype of afflicted clinical
patients [Reproduced with permission from Environmental Health
Perspectives].

we recently have shown that P_1-450 induction is under transcriptional
control rather than a result of gene amplification or some form of
gross DNA rearrangement.

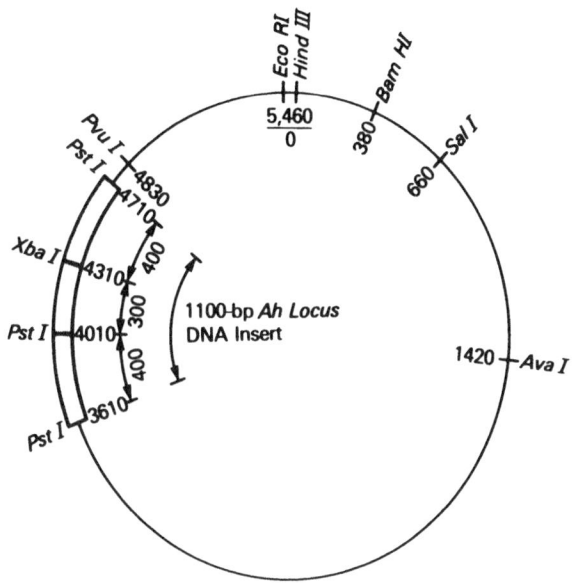

Figure 9.   Restriction map of pBR322–clone 46 (Negishi et al, 1981).
[Reproduced with permission from the Proceedings of the
National Academy of Sciences].

The size and many properties of the genomic structural gene for
cytochrome P_1-450 have been studied (Nebert et al, 1982a). Further
characterization of the mouse P_1-450 structural gene and the identification and characterization of other P-450 structural genes should
provide valuable insight and understanding of the genetic regulation
of P-450 induction, the ultimate number of possible (basal and induced) forms of P-450 and their evolution (Nebert, 1979), and potentially useful assays for predicting genetic differences in risk for
individual humans of experiencing various types of drug- or chemically-
induced cancer, drug toxicity, and birth defects.

REFERENCES

Atlas, S.A., and Nebert, D.W. Pharmacogenetics: A possible pragmatic
perspective in neoplasm predictability. Sem. Oncol., 5:89-106,
1978.
Guenthner, T.M., Fysh, J.M., and Nebert, D.W. 2,3,7,8-tetrachloro-
dibenzo-p-dioxin: Covalent binding of reactive metabolic inter-
medidates principally to protein in vitro. Pharmacology,
19:12-22, 1979.

Guenthner, T.M., and Nebert, D.W. The cytosolic receptor for aryl hydrocarbon hydroxylase induction by polycyclic aromatic compounds. Evidence for structural and regulatory variants among established cell culture lines. J. Biol. Chem., 252:8981-8989, 1977.

Hannah, R.R., Nebert, D.W., and Eisen, H.J. Regulatory gene product of the Ah complex. Comparison of 2,3,7,8-tetrachlorodibenzo-p-dioxin and 3-methylcholanthrene binding to several moieties in mouse liver cytosol. J. Biol. Chem., 256:4584-4590, 1981.

Knutson, J.C., and Poland, A. 2,3,7,8-tetrachlorodibenzo-p-dioxin: Failure to demonstrate toxicity in twenty-three cultured cell types. Toxicol. Appl. Pharmacol., 54:377-383, 1980.

Lu, A.Y.H., and West, S.B. Multiplicity of mammalian microsomal cytochromes P-450. Pharmacol. Rev., 31:277-295, 1980.

Mannering, G.J. Hepatic cytochrome P-450 linked drug-metabolizing systems. In: Jenner, P., and Testa, B., eds. Concepts in Drug Metabolism. Part B. New York: Marcel-Dekker, Inc., 1981.

Nebert, D.W. Genetic differences in susceptibility to chemically induced myelotoxicity and leukemia. Environ. Health Perspect, 39:11-22, 1981.

Nebert, D.W. Multiple forms of inducible drug-metabolizing enzymes. A reasonable mechanism by which any organism can cope with adversity. Mol. Cell. Biochem., 27:27-46, 1979.

Nebert, D.W. The Ah locus: Genetic differences in toxic and tumorigenic response to foreign compounds. In: Coon, M.J., and O'Brien, P.J., eds. Microsomes, Drug Oxidations, and Chemcial Carcinogenesis. Vol II. New York: Academic Press, 1980. pp. 801-812.

Nebert, D.W., Eisen, H.J., Negishi, M., Lang, M.A., Hjelmeland, L.M., and Okey, A.B. Genetic mechanisms controlling the induction of polysubstrate monooxygenase (P-450) activities. Annu. Rev. Pharmacol. Toxicol., 21:431-462, 1981.

Nebert, D.W., and Gielen, J.E. Aryl hydrocarbon hydroxylase induction in mammalian liver cell cutlure. II. Effects of actinomycin D and cycloheximide on induction processes by phenobarbital or polycyclic hydrocarbons. J. Biol. Chem., 246:5199-5206, 1971.

Nebert, D.W., and Jensen, N.M. The Ah locus: Genetic regulation of the metabolism of carcinogens, drugs, and other environmental chemicals by cytochrome P-450-mediated monooxygenases. In: Fasman, G.D., ed. CRC Critical Reviews in Biochemistry. Vol. 6. Cleveland, Ohio: CRC Press, Inc., 1979. pp. 401-437.

Nebert, D.W., Jensen, N.M., Levitt, R.C., and Felton, J.S. Toxic chemical depression of the bone marrow and possible aplastic anemia explainable on a genetic basis. Clin. Toxicol., 16:99-122, 1980.

Nebert, D.W., Nakamura, M., Altieri, M., Ikeda, T., Tukey, R.H., and Negishi, M. Characterization of the genomic cytochrome P_1-450 structural gene in the mouse. J. Supramol. Struct. Cell Biochem. in press, 1982a [Abstract].

Nebert, D.W., Negishi, M., Lang, M.A., Hjelmeland, L.M. and Eisen, H.J. The Ah locus, a multigene family necessary for survival in a chemically adverse environment: Comparison with the immune system. Advanc. Genet., in press, 1982b.

Negishi, M., and Nebert, D.W. Structural gene products of the Ah locus. Genetic and immunochemical evidence for two forms of mouse liver cytochrome P-450 induced by 3-methylcholanthrene. J. Biol. Chem., 254:11015-11023, 1979.

Negishi, M., and Nebert, D.W. Structural gene products of the Ah complex. Increases in large mRNA from mouse liver associated with cytochrome P_1-450 induction by 3-methylcholanthrene. J. Biol. Chem., 256:3085-3091, 1981.

Negishi, M., Swan, D.C., Enquist, L.W., and Nebert, D.W. Isolation and characterization of a cloned DNA sequence associated with the murine Ah locus and a 3-methylcholanthrene-induced form of cytochrome P-450. Proc. Nat. Acad. Sci. USA, 78:800-804, 1981.

Niwa, A., Kumaki, K., and Nebert, D.W. Induction of aryl hydrocarbon hydroxylase activity in various cell cultures by 2,3,7,8- tetrachlorodibenzo-p-dioxin. Mol. Pharmacol., 11:399-408, 1975.

Okey, A.B., Bondy, G.P., Mason, M.E., Kahl, G.F., Eisen, H.J., Guenthner, T.M., and Nebert, D.W. Regulatory gene product of the Ah locus. Characterization of the cytosolic inducer-receptor complex and evidence for its nuclear translocation. J. Biol. Chem., 254:11636-11648, 1979.

Okey, A.B., Bondy, G.P., Mason, M.E., Nebert, D.W., Forster-Gibson, C., Mucan, J., and Dufresne, M.J. Temperature-dependent cytosol-to-nucleus translocation of the Ah receptor for 2,3,7,8-tetrachlorodibenzo-p-dioxin in continuous cell culture lines. J. Biol. Chem., 255:11415-11422, 1980.

Poland A., and Glover, E. An estimate of the maximum in vivo covalent binding of 2,3,7,8-tetrachlorodibenzo-p-dioxin to rat liver protein, ribosomal RNA, and DNA. Cancer Res., 39:3341-3344, 1979.

Poland, A.P., Glover, E., and Kende, A.S. Sterospecific, high affinity binding of 2,3,7,8-tetrachlorodibenzo-p-dioxin by hepatic cytosol. Evidence that the binding species is the receptor for the induction of aryl hydrocarbon hydroxylase. J. Biol. Chem., 251:4936-4946, 1976.

Poland, A.P., Glover, E., Robinson,J.R., and Nebert, D.W. Genetic expression of aryl hydrocarbon hydroxylase activity. Induction of monooxygenase activities and cytochrome P_1-450 formation by 2,3,7,8-tetrachlorodibenzo-p-dioxin in mice genetically "nonresponsive" to other aromatic hydrocarbons. J. Biol. Chem., 249:5599-5606, 1974.

Robinson, J.R., and Nebert, D.W. Genetic expression of aryl hydrocarbon hydroxylase induction. Presence or absence of association with zoxazolamine, diphenylhydantoin, and hexobarbital metabolism. Mol. Pharmacol., 10:484-493, 1974.

Routledge, P.A., and Shand, D.G. Presystemic drug elimination.
     Annu. Rev. Pharmacol. Toxicol., 19:447-468, 1979.
Tukey, R.H., Nebert, D.W., and Hegishi, M. Structural gene product
     of the [Ah] complex. Evidence for transcriptional control of
     cytochrome P$_1$-450 induction by use of a cloned DNA sequence.
     J. Biol. Chem., 256:6969-6974, 1981.
Williams, R.T. Detoxication Mechanisms. 2nd Edition. New York:
     John Wiley & Sons, 1959. 796pp.

FACTORS AFFECTING THE DISPOSITION AND PERSISTENCE

OF HALOGENATED FURANS AND DIOXINS

H.B. Matthews and L.S. Birnbaum

National Institute of Environmental
  Health Sciences
Research Triangle Park
North Carolina   USA

ABSTRACT

A number of interrelated factors determine the disposition and persistence of halogenated dioxins and furans. Among the more important of these factors are solubility, lipid solubility and metabolism to more polar compounds which can be excreted. Halogenated dioxins and furans are very insoluble in all solvents and their solubility tends to decrease with increasing halogenation. Since compounds which are not in solution are not readily absorbed, solubility may selectively limit the absorption of some of the more highly halogenated dioxins and furans. However, that portion of the dose which is in solution is sufficiently lipid soluble to be passively absorbed from the gastrointestinal tract and too lipid soluble to be excreted prior to metabolism to more polar compounds. The rate of metabolism of dioxins and furans varies greatly with molecular structure and among species. The more readily metabolized dioxins and furans have adjacent unsubstituted carbon atoms; furans may be more readily metabolized than dioxins, and rats and hamsters metabolize and excrete these compounds more rapidly than do guinea pigs. The relation of toxicity to metabolism is complex, but it is apparent that metabolism constitutes a route of detoxication for dioxins and furans.

INTRODUCTION

Halogenated dibenzo-p-dioxins (dioxins) and dibenzofurans (furans) share many chemicals and physical properties with other members of their chemical class, the halogenated aromatic hydro-

463

carbons. These chemical and physical properties are major determinants of absorption, blood transport and partition from blood into tissues, i.e. tissue/blood ratios. Therefore, the disposition of dioxins and furans is determined by many of the same factors which have been considered in detail in more extensive studies of other members of this chemical class. The following is a review, discussion and analysis of those factors which determine the disposition of halogenated aromatics and how the disposition of dioxins and furans resembles or differs from that of other members of this chemical class.

SOLUBILITY

Most halogenated aromatics are sufficiently lipid-soluble to be absorbed from the gastrointestinal tract by passive diffusion across cell membranes. Once absorbed, lipid solubility is a primary determinant of tissue distribution (Matthews and Kato 1979). Halogenated aromatics without polar substituents tend to be sequestered in adipose tissue. At equilibrium the tissue/blood ratios for these compounds may favor adipose tissue by several hundred-fold (Table 1). This affinity for adipose tissue acts as a protective mechanism by effectively isolating a major portion of the body-burden from the site(s) of toxic action for these compounds. However, high lipid solubility also extends the biological half-life and increases the persistence of these compounds by reducing the concentration of substrate available to the enzymes responsible for their metabolism. These highly lipid-soluble compounds are not excreted prior to metabolism to more polar compounds because the major routes for their excretion, urine and bile, involve an aqueous medium. If these lipid-soluble compounds enter urine or bile during formation they partition out of this aqueous environment into the surrounding tissues, thus preventing excretion.

Halogenated aromatics with one or more polar moieties are facilitated in their excretion as the parent compounds or as conjugates of the parent compounds (Courtney 1970, Sauerhoff et al, 1976). Few if any of the polar halogenated aromatics have been implicated in major incidents of environmenatl contamination or human intoxication.

The dioxins and furans may be considered to belong to a third group of halogenated aromatics. These compounds are too lipid-soluble to be excreted prior to metabolism to more polar compounds, but their tissue distribution is not restricted primarily to adipose tissue. These compounds concentrate in liver and other possible sites of action as well as adipose tissues. This intermediate lipid solubility has the effect of extending the biological half-life for these compounds while also increasing the effective dose at the site of action.

Table 1. Tissue Distribution of Halogenated Aromatics

Chemical	Species	Liver/Blood[a]	Adipose/Blood[a]	Adipose/Liver[b]	Reference
2,4,5-Trichloro-phenoxyacetic Acid	Rat	0.24	0.11	0.46	Sauerhoff et al, 1976
Pentachlorophenol	Rat	1.6	0.18	0.11	Larsen et al, 1972
2,3,7,8-Tetrachloro-dibenzo-p-dioxin	Rat			1.0	Rose et al, 1976
2,3,7,8-Tetrachloro-dibenzo-p-dioxin	Hamster	34	24	0.71	Olson et al, 1980
2,3,7,8-Tetrachloro-dibenzofuran	Rat	74	21	0.28	Birnbaum et al, 1980
Hexachlorobenzene	Rat	19	163	8.6	Mehendale et al, 1975
2,4,5,2',4',5'-Hexachlorobiphenyl	Rat	12	400	33	Lutz et al, 1977

[a] Tissue/Blood ratio
[b] Adipose/Liver ratio

A cytosolic receptor which is concentrated in some tissues, particularly thymus, lung, liver and kidney, has a high affinity for dioxins and furans (Poland and Glover 1976, Carlstedt-Duke 1979, Poland and Glover 1980). However, this portion of the tissue burden is too small to exert a direct effect on the amount of substrate available for metabolism and excretion. A more significant limiting factor on the amount of substrate available for metabolism and excretion is most probably the amount of nonspecific "binding" to various proteins and membranes. Since highly lipid-soluble chemicals such as dioxins and furans are not soluble in the aqueous portions of biological media, they adsorb onto the surfaces or lipophilic portions of proteins or dissolve in various lipid fractions of cell membranes or contents. This is not true binding but it serves to reduce the amount of substrate available for metabolism, thereby reducing the amount of metabolism and excretion.

We conducted a brief experiment to determine if the nonspecific binding of 2,3,7,8-tetrachlorodibenzo-p-dioxin (TCDD) could be saturated thereby increasing the substrate available for metabolism. In this experiment adult male rats were anesthetized, the bile ducts were cannulated, the rats were administered a dose of 12 or 120 µg/kg of radiolabeled TCDD by intravenous injection, and the excretion of TCDD-derived radioactivity in bile was followed for three hours. After three hours the animals were sacrificed and the concentration of TCDD was determined in liver and blood.

The livers of animals that had been treated with TCDD at 12 µg/kg contained 45.3+2.8% of the total dose, blood contained 0.68+ 0.04%, and the liver/blood ratio was 82.8+4.8. On the other hand, livers of animals that had received TCDD at 120 µg/kg contained only 23.7+1.2% of the total dose, blood contained 1.35+0.02% of the total dose and the liver/blood ratio was 33.2+8.7. The lower percent of the total dose in liver and the higher percent in blood with the greater dose indicated either saturation of nonspecific binding in liver or less hepatic uptake of the higher dose. However, animals that had received the greater dose of TCDD excreted 3.2+0.4% of the dose in bile in 3 hr. whereas animals that had received the lower dose excreted only 2.3+.3%, Figure 1. In each case virtually all of the TCDD-derived radioactivity excreted in bile was in the form of TCDD metabolites.

Animals receiving a 10-fold greater dose of TCDD had approximately five times higher concentrations of TCDD in their livers. The higher concentrations of TCDD in liver resulted in the metabolism and excretion of approximately thirteen times as much TCDD in bile. This nonproportional increase in metabolism and excretion implies that nonspecific binding may limit TCDD metabolism when the concentration of TCDD in liver is low. We have not established an optimum concentration for TCDD metabolism because we could not put higher concentrations of TCDD in solution for intravenous administration,

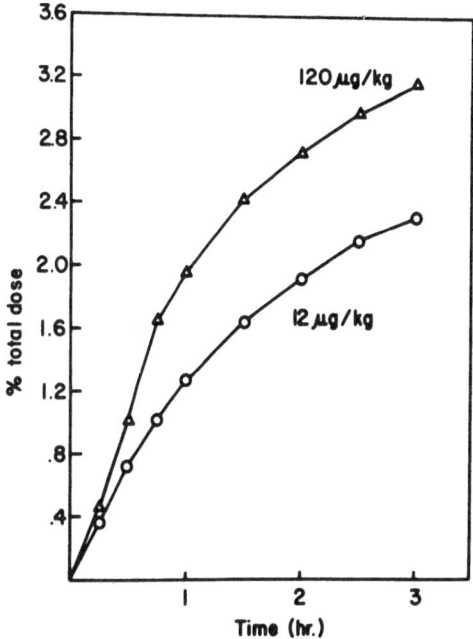

Figure 1. Effect of Dose on TCDD Metabolism and Excretion.
Adult male Sprague-Dawley rats (200-250 g) were
anesthetized with pentobarbital (50 mg/kg) and
the common bile ducts cannulated. TCDD, 12 or 120
µg/kg in Emulphor EL-620 (GAF): ethanol was administered 0.5 ml/kg by iv injection into an ex-
teriorized femoral vein. Total bile was collected
at each timepoint and the data represent average
values obtained from three animals treated at each
dose. TCDD-derived radioactivity excreted in bile
was confirmed as TCDD metabolites by extraction
and chromatography.

and lower concentrations were difficult to quantitate in tissues
and bile.

With the exception of those halogenated aromatics that contain
a polar moiety, all of these compounds have limited solubility in
water, but most are readily soluble in organic solvents. Exceptions
include the very highly halogenated biphenyls and naphthalenes and
certain halogenated dioxins and furans. Some of these compounds
have limited solubility in both lipid and aqueous mediums. Compounds
that are not in solution in the gastrointestinal tract are poorly
absorbed by adult animals. The solubility of dioxins and furans varies
with degree and position of halogenation. Therefore, some members of
the group are more readily absorbed than others and the body burden

may be determined by the degree of absorption as well as the level of exposure.

A number of workers have observed that the more highly halogenated dioxins and furans tend to be less soluble and may be poorly absorbed, but the lack of a variety of radiolabeled isomers has made this a difficult phenomenon to study and quantitate. However, one interesting example of the effect of absorption on the tissue concentration of furans is seen in the work of Morita and Oishi (1977). These investigators administered a mixture of chlorinated dibenzofurans to mice by ip injection. The components of this mixture were identified only by peak number. However, the results of this study show that the various peaks were absorbed and cleared at different rates. One of the major constituents of the mixture, peak 52, was rapidly concentrated in liver and then cleared from liver at an appreciable rate showing a steady decrease with time to an almost undetectable trace by 8 weeks after administration (Figure 2). On the other hand, the liver concentration of peak 62 increased for two weeks after administration and remained relatively constant thereafter (Figure 2).

This observation was most probably the product of variances in the rates of both metabolism and absorption. That is, peak 52 was readily absorbed, accumulated in liver and was metabolized and

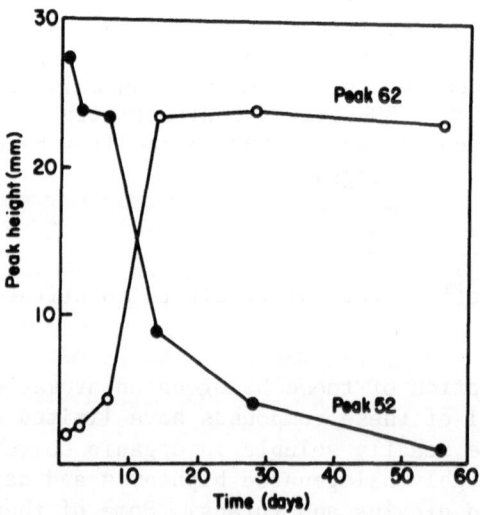

Figure 2.   Effect of Solubility on Furan Metabolism and Disposition.
            Relative concentrations in liver versus time for two
            chlorinated furans injected ip into mice. Peak heights
            taken from Fig. 2 of a manuscript by Morita and Oishi
            (1977).

excreted at an appreciable rate, whereas, peak 62 was both poorly absorbed and slowly metabolized. It is interesting to note that these results may not have been observed if the dose had not been administered by ip injection. If it had been administered orally, most of peak 62 might have passed through the gastrointestinal tract prior to a similar degree of absorption and thus, would not have accumulated to a comparable level in liver.

We do not intend to indicate that highly halogenated dioxins and furans are not absorbed from the gastrointestinal tract, only that absorption of some of these compounds may be less complete than the absorption of many other halogenated aromatics. The poor absorption of certain highly halogenated dioxins and furans is a fortunate circumstance because once absorbed these compounds may be cleared from tissues only at very slow rates. Therefore, these compounds may have a greater potential for bioaccumulation than other members of this group. This is an important consideration in evaluating chronic environmental exposures.

METABOLISM

Dioxins: Major factors controlling the rate of metabolism and excretion of lipid-soluble halogenated aromatics are the degree and position of halogenation. This fact was first demonstrated in the laboratory of R.T. Williams in studies of tri- and tetrachloro-benzenes (Jondorf et al, 1955, Jondorf et al, 1958). These studies demonstrated that whereas increasing chlorination tended to inhibit metabolism, the position of chlorination was of major importance. The critical factor for metabolism was the presence of two adjacent unsubstituted carbon atoms. It was subsequently demonstrated that an arene oxide intermediate is involved in the major route of metabolism for these compounds (Kohli et al, 1976, Ruzo et al, 1976a) and that similar requirements for adjacent unsubstituted carbon atoms are exhibited by polychlorinated biphenyls (PCBs) (Matthews and Anderson, 1975) and halogenated naphthalenes (Ruzo et al, 1976b).

Some of the first efforts to study the metabolism of dioxins indicated that unchlorinated or slightly chlorinated dioxins were metabolized but that the extremely toxic TCDD might not be metabolized (Vinopal and Casida 1973) or at best was metabolized very slowly (Piper et al, 1973). Tulp and Hutzinger (1978) made a systematic study of the metabolism of dibenzo-p-dioxin and a series of mono-, di-, tri and tetrachlorodibenzo-p-dioxins, all having adjacent unsubstituted carbons atoms, and octachlorodibenzo-p-dioxin. In this study they observed that the less chlorinated dioxins were metabolized to mono- and dihydroxy derivatives with the hydroxyl groups in the 2 and 3 positions and that there was no evidence for metabolism of octachlorodibenzo-p-dioxin or for fission of the carbon-oxygen bonds. They concluded from these studies that dioxin metabolism occurs

primarily via epoxide intermediates and that the metabolism of TCDD
may be hindered by the positions of the chlorine atoms. TCDD does
not have two adjacent unsubstituted carbon atoms, a major requisite
for ready metabolism of most halogenated aromatics.

Rose et al (1976) provided evidence for the importance of metabolism to TCDD excretion in a study comparing the disposition of an
acute oral dose of 30 μg/kg with multiple oral doses of 0.01, 0.1
and 1.0 μg/kg of radiolabeled TCDD administered daily 5 days a week
for 7 weeks. They found that in all cases virtually all of the radioactivity remaining in liver, the only tissue assayed, was unmetabolized TCDD whereas most of the radioactivity excreted was in the form
of one or more polar metabolites. More recent studies with radiolabeled TCDD have demonstrated that the TCDD-derived radioactivity
is excreted almost exclusively in bile and as compounds that are more
polar than TCDD whereas the radioactivity retained in the tissues is
almost exclusively TCDD (Poiger and Schlatter 1979, Ramsey et al
1979, Olson et al 1980). Olson et al (1980) speculated that the
enhanced rate of TCDD metabolism and excretion observed in the hamster may contribute in part to the greater resistance of this species
to TCDD toxicity. Olson et al (1981) also studied TCDD metabolism in
isolated hepatocytes from rats, detected the same 3 or 4 metabolites
that are extracted from rat bile and provided evidence that TCDD is
metabolized by a hepatic cytochrome P-450-dependent, mixed-function
oxidase system.

Furans

Furans have not been the subject of as much laboratory research
as have the dioxins. This is due at least in part to the unavailability of radiolabeled furans. However, because of the large-scale
human exposure to furan-contaminated PCBs in the Yusho incident,
there are some data available on the fate of halogenated furans in
humans (Nagayama et al, 1977, Kuroki and Masuda 1978). Chlorinated
dibenzofurans detected in the tissues of persons who were exposed to
PCB-contaminated oil in the Yusho incident and died 1 to 9 years
latter included tetra-, penta- and hexa- isomers (Kuroki and Masuda,
1978, Rappe et al, 1979). However, the original oil contained a
more extensive list of chlorinated furans, several of which were
present in higher relative concentrations than those observed in the
tissues of Yusho patients. Some tetra-, penta- and hexachlorodibenzo-
furans that were present in the oil were not present in tissues.
Rappe et al, (1979) pointed out that none of the furans detected in
human liver approximately 1 year after exposure had adjacent unsubstituted carbon atoms.

Studies with laboratory animals indicate that the toxicity of
2,3,7,8-tetrachlorodibenzofuran (TCDF) may vary more widely among
species than does that of TCDD. Guinea pigs are exquisitely sensitive to TCDF intoxication, whereas mice are relatively resistant
and monkeys have an intermediate resistance to TCDF (Moore et al,

1976, Moore et al, 1979). Morita and Oishi (1977) studied the fate
of a synthetic mixture of chlorinated furans in mice and observed
that mice clear some isomers more rapidly than others. Of particular
interest was the observation that a peak thought to be TCDF, one of
the most acutely toxic furans, was cleared with an estimated half-
life of 1 week. A study of TCDF in primates indicated that the
Rhesus monkey might also have some capacity to clear TCDF. In a study
of chronic TCDF administration, intoxicated monkeys recovered more
rapidly upon removal of TCDF from the diet than was observed in
similar studies of TCDD (McNulty et al, 1981).

     A series of studies investigating metabolism and disposition of
radiolabeled TCDF in rats, monkeys, guinea pigs and two strains of
mice were conducted in our laboratory. The persistence and toxicity
of TCDF were observed to be inversely proportional to the rate at
which the respective species metabolized and excreted TCDF. Of the
species we have studied, the rat and mouse metabolize and excrete
TCDF most readily, the monkey metabolized TCDF at an intermediate
rate and the guinea pig shows little or no capacity for TCDF metabolism (Birnbaum et al, .1980, Birnbaum, et al, 1981, Decad et al,
1981a, Decad et al, 1981b). This species-specific ability to metabolize and excrete TCDF inversely parallels the relative sensitivity
of these species to intoxication by TCDF as reported by Moore et al
(1979), with the guinea pig being most sensitive and rats and mice
being least sensitive. The half-lives of TCDF in these species vary
from less than 2 days in rats and mice to greater than three weeks
in guinea pigs with an intermediate half-life of approximately 8 days
in monkeys.

     An additional factor affecting the whole-body half-life was
observed in our study of TCDF in two strains of mice. Mice of the
DBA/2J strain have as much as 70% more adipose tissue than do mice
of the C57BL/6J strain, and significantly more of the TCDF was sequestered in adipose tissue of the DBA/2J strain. Thus, even though TCDF
was cleared from the livers of each strain at comparable rates, the
whole-body half-life of TCDF was significantly longer in the DBA/2J
strain (Decad et al, 1981b). These results indicate that adipose
tissue effectively competes with liver for TCDF. Since metabolism
of TCDF in adipose tissue is negligible, the greater the volume of
adipose tissue the longer the half-life of lipid soluble compounds
such as TCDF. The same explanation should hold true for the effect
of adipose tissue volume on acute intoxication by TCDF and most
probably also for TCDD since adipose tissue is a storage depot and
not a site of toxic action for either of these compounds. Thus the
results described above may offer an additional explanation of why
Poland and Glover (1980) observed the C57BL/6J strain to be more
sensitive than the DBA/2J strain to intoxication by TCDD. We point
out though that this is just an additional factor since sensitivity
to intoxication is also associated with the Ah locus (Poland and
Glover, 1980).

The half-life of TCDF is significantly shorter in rats, mice and monkeys than is the half-life of TCDD. On the other hand, the half-lives of TCDF and TCDD in the guinea pig are comparable (Gasiewicz and Neal 1979, Decad et al, 1981a). These differences are apparently the result of species-specific variations in rates of metabolism between TCDF and TCDD in the first three species contrasted with a lack of metabolism of either compound by the guinea pig. Since neither TCDD nor TCDF have adjacent unsubstituted carbon atoms, it is apparent that, for the species that can metabolize TCDF, there is a fundamental difference in the metabolism of these two compounds. It has yet to be established what structural feature of TCDF facilitates its metabolism. However, based on past experience with other halo-genated aromatics, it must be assumed that either an arene oxide is more easily formed between carbon one and the carbon-carbon bond joining the two benzene rings or that the carbon-oxygen bridge of the furan is strained sufficiently by the sterochemical configuration of TCDF to increase its susceptibility to enzymatic attack. There is no apparent reason why an arene oxide would be more easily formed between an unsubstituted carbon and a chlorinated carbon of TCDF than of TCDD, but the final facts will await the isolation and identification of the respective metabolites.

SUMMARY

It is apparent from the foregoing discussion that a number of interrelated factors determine the disposition of halogenated dioxins and furans. All of these compounds are sufficiently lipid-soluble to be absorbed from the gastrointestinal tract by passive diffusion across cell membranes and all are too lipid-soluble to be readily excreted prior to metabolism to more polar compounds. However, the dioxins and furans are quite insoluble in all mediums and this limited solubility tends to decrease further as the degree of halogenation increases. Since compounds which are not in solution are not readily absorbed, insolubility may act to limit selectively the rate and degree of absorption of the more highly halogenated dioxins and furans. On the other hand, the less halogenated dioxins and furans are usually metabolized and excreted more rapidly than the more hal-ogenated members of the group. Therefore, once absorbed, the highly halogenated members of the group may be more persistent and thus may have the greater potential for bioaccumulation.

Metabolism to more polar compounds that can be excreted is a critical determinant of the biological half-life of most lipid-soluble halogenated aromatics, those dioxins and furans that have adjacent unsubstituted carbon atoms are more rapidly metabolized and excreted than those that do not. Neither TCDD nor TCDF have adjacent unsub-stituted carbon atoms. However, the rates at which these compounds are metabolized, and thus their biological half-lives, vary greatly with species. The metabolism of TCDD varies from slow in the rat and hamster to negligible in the guinea pig, and TCDD has a half-life of several weeks in most species. On the other hand, the metabolism of TCDF varies from relatively rapid in rats to negligible in guinea pigs and its half-life varies from less than 2 days to several weeks in the respective species. Therefore, some structural feature must favor the metabolism of TCDF over that of TCDD. In either case, sensitivity to intoxication by these compounds is inversely proportional to the ability of the given species to metabolize and excrete them. Therefore, metabolism and excretion must be assumed to be an important detoxication mechanism for the halogenated dioxins and furans.

REFERENCES

Birnbaum, L.S., Decad, G.M., and Matthews, H.B. Disposition and excretion of 2,3,7,8-tetrachlorodibenzofuran in the rat. Toxicol. Appl. Pharmacol., 55:342-352, 1980.

Birnbaum, L.S., Decad, G.M., McConnell, E.E., and Matthews, H.B. Fate of 2,3,7,8-tetrachlorodibenzofuran in the monkey. Toxicol. Appl. Pharmacol., 57:189-196, 1981.

Carlstedt-Duke, J.M.B. Tissue distribution of the receptor for 2,3,7,8-tetrachlorodibenzo-p-dioxin in the rat. Can. Res., 39:3172-3176, 1979.

Courtney, D.K. 2,4,5-T in the rat: Excretion pattern, serum levels, placental transport and metabolism. In: Pesticides Symposia, pp. 277-283. Hales and Associates, Florida, 1970.

Decad, G.M., Birnbaum, L.S. and Matthews, H.B. 2,3,7,8-Tetrachlorodibenzofuran tissue distribution and excretion in guinea pigs. Toxicol. Appl. Pharmacol., 57:231-240, 1981a.

Decad, G.M., Birnbaum, L.S. and Matthews, H.B. Distribution and excretion of 2,3,7,8-tetrachlorodibenzofuran in C57BL/6J and DBA/2J mice. Toxicol. Appl. Pharmacol., 59:564-573, 1981b.

Gasiewicz, T.A., and Neal, R.A. 2,3,7,8-Tetrachlorodibenzo-p-dioxin tissue distribution, excretion, and effects on clinical parameters in guinea pigs. Toxicol. Appl. Pharmacol., 51:329-339, 1979.

Jondorf, W.R., Parke, D.V., and Williams, R.T. Studies in detoxication 66. The metabolism of halogenobenzenes. 1:2:3-, 1:2:4-, and 1:3:5-trichlorobenzenes. Biochem. J., 66:512-521, 1855.

Jondorf, W.R., Parke, D.V., and Williams, R.T. Studies in detoxication 76. The metabolism of halogenobenzenes. 1:2:3:4-, 1:2:3:5- and 1:2:4:5-tetrachlorobenzenes. Biochem. J., 69:181-189, 1958.

Kohli, J., Jones, D., and Safe, S. The metabolism of higher chlorinated benzene isomers. Can. J. Biochem., 54:203-208, 1976.

Kuroki, H. and Masuda, Y. Determination of polychlorinated dibenzofuran isomers retained in patients with Yusho. Chemosphere, 10:771-777, 1978.

Larsen, R.V., Kirsch, L.E., Shaw, S.M., Christian, J.E., and Born, G.S. Excretion and tissue distribution of uniformly labeled ^{14}C-pentachlorophenol in rats. J. Pharmaceut. Sci., 61:2004-2006, 1972.

Lutz, R.J., Dedrick, R.L., Matthews, H.B., Eling, T.E. and Anderson, M.W. A preliminary pharmacokinetic model for several chlorinated biphenyls in the rat. Drug Metab. Dispos., 5:386-396, 1977.

Matthews, H.B. and Anderson, M.W. Effect of chlorination on the distribution and excretion of polychlorinated biphenyls. Drug Metab. Dispos., 3:371-380, 1975.

Matthews, H.B., Kato, S. The metabolism and disposition of halogenated aromatics. Ann. N.Y. Acad. Sci., 320:131-137, 1979.

Matthews, H.B., Kato, S., Morales, M.M., and Tuey, D.B. Distribution and excretion of 2,4,5,2',4',5'-hexabromobiphenyl, the major component of Firemaster BP-6. J. Toxicol. Environ. Hlth., 3:599-605, 1977.

McNulty, W.P., Pomerantz, I., and Farrell, T. Chronic toxicity of 2,3,7,8-tetrachlorodibenzofuran for Rhesus macaques. Fd. Cosmet. Toxicol., 19:57-65, 1981.

Mehendale, H.M., Fields, M., and Matthews, H.B. Metabolism and effects of hexachlorobenzene on hepatic microsomal enzymes in the rat. Agr. Fd. Chem., 23:261-265, 1975.

Moore, J.A., Gupta, B.N., and Vos, J.G. Toxicity of 2,3,7,8-tetrachlorodibenzofuran - preliminary results. Proceedings of the National Conference on Polychlorinated Biphenyls, November 1976, 77-79.

Moore, J.A., McConnell, E.E., Dalgard, D.W., and Harris, M.W. Comparative toxicity of three halogenated dibenzofurans in guinea pigs, mice and rhesus monkeys. Ann. N.Y. Acad. Sci., 320:151-163, 1979.

Morita, M. and Oishi, S. Clearance and tissue distribution of polychlorinated dibenzofurans in mice. Bull. Environ. Contam. Toxicol., 18:61-66, 1977.

Nagayama, J., Kuratsune, M., and Masuda, Y. Determination of chlorinated dibenzofurans in tissues of patients with "Yusho." Fd. Cosmet. Toxicol., 15:195-198, 1977.

Olson, J.R., Gudzinowicz, M., and Neal, R.A. The in vitro and in vivo metabolism of 2,3,7,8-tetrachlorodibenzo-p-dioxin in the rat. The Toxicol., 1:69-70. 1981.

Olson, J.R., Gasiewicz, T.A., and Neal, R.A. Tissue distribution excretion, and metabolism of 2,3,7,8-tetrachlorodibenzo-p-dioxin (TCDD) in the Golden Syrian Hamster. Toxicol. Appl. Pharmacol., 56:78-85, 1980.

Piper, W.N., Rose, J.Q., and Gehring, P.J. The excretion and tissue distribution of 2,3,7,8-tetrachlorodibenzo-p-dioxin in the rat. In: Advan. Chem. Ser., Chlorodioxins - Origin and Fate (Blair, E.H., ed). Vol. 120, pp. 85-91. American Chemical Society, Washington, D.C., 1973.

Poiger, H. and Schlatter, C. Biological degradation of TCDD in rats. Nature, 281:706-707, 1979.

Poland, A. and Glover, E. Stereospecific, high affinity binding of 2,3,7,8-tetrachloro-p-dioxin by hepatic tyrosol. J. Biol. Chem., 251:4936-4946, 1976.

Poland, A. and Glover, E. 2,3,7,8-tetrachlorodibenzo-p-dioxin: Segregation of toxicity with Ah locus. Mol. Pharmacol. 17: 86-94, 1980.

Ramsey, J.C., Hefner, J.G., Karbowski, R.J., Braun, W.H., and Gehring, P.J. The in vivo biotransformation of 2,3,7,8-tetrachlorodibenzo-p-dioxin in the rat. Toxicol. Appl. Pharmacol 48:A162, 1979.

Rappe, C., Buser, H.R., Kuroki, H. and Masuda, Y. Identification of polychlorinated dibenzofurans (PCDFs) retained in patients with Yusho. Chemosphere, 4:259-266, 1979.

Rose, J.Q., Ramsey, J.C., Wentzler, T.H., Hummel, R.A., and Gehring, The fate of 2,3,7,8-tetrachlorodibenzo-p-dioxin following single and repeated oral doses to the rat. Toxicol. Appl. Pharmacol., 36:209-226, 1976.

Ruzo, L.O., Safe, S. and Hutzinger, O. Metabolism of bromobenzenes in the rabbit. J. Agric. Food Chem., 24:291-293, 1976a.

Ruzo, L.O., Jones, D., Safe, S., and Hutzinger, O. Metabolism of chlorinated naphthalenes, J. Agric. Food Chem., 24:581-583, 1976b.

Sauerhoff, M.W., Braun, W.H., and Lebeau, J.E. Dose-dependent pharmacokinetics profile of silvex following intravenous administration in rats. J. Toxicol. Environ. Health, 2:605-618, 1976.

Tulp, M. Th.M. and Hutzinger, O. Rat metabolism of polychlorinated dibenzo-p-dioxins. Chemosphere, 9:761-768, 1978.

Vinopal, J.H. and Casida, J.E. Metabolic stability of 2,3,7,8-tetrachloro-p-dioxin in mammalian liver microsomal systems and in living mice. Arch. Environ. Contam. Toxicol., 1:122-132, 1973.

ADIPOSE TISSUE STORAGE OF POLYCHLORINATED COMPOUNDS

M.H. Bickel, W.R. Jondorf, S. Mühlebach, and
P.A. Wyss

Department of Pharmacology, University of Berne
Berne, Switzerland

ABSTRACT

Adipose tissue storage of lipophilic compounds is not simply a
matter of lipophilicity and is more complex than was commonly assumed.
DDT and related compounds are among the classical xenobiotics stored
in adipose tissue. The latter contains most of the body burden, but
the material so sequestered is also removed from potential target
organs where toxicity might become manifest. However, if the adipose
tissues are metabolized, stored material can be released. It is then
transferred into lean tissues accompanied by increased excretion via
the feces. Therefore, at the same time as the body burden overall
decreases, concentrations in certain tissues increase. This is also
the case with 2,4,5,2',4',5'-hexachlorobiphenyl (6-CB), an extremely
persistent model compound, which has been shown to be retained for
months in high constant amounts in adipose tissue when acutely
administered to rats. In rats with restricted food intake the
kinetics of 6-CB release from adipose tissue and of its redistribution and enhanced fecal excretion have been investigated. The results
may provide a model for other classes of polychlorinated compounds.

FAT STORAGE OF LIPOPHILIC XENOBIOTICS

Many drugs or their xenobiotics are not distributed evenly when
taken up by an organism. They tend to show affinities for specific
tissues and cell organelles, or body constituents such as proteins,
mucopolysaccharides, or lipids. In many cases, a xenobiotic may
become highly concentrated in some tissue or structure which can
assume a depot or storage function.

477

Storage of drugs in fat became known when the classical studies
of Brodie et al (1950, 1952) showed that a large proportion of a dose
of the lipophilic anesthetic, thiopental, would be redistributed into
adipose tissue whence it subsequently disappeared as a consequence of
its relatively rapid hepatic metabolism. At about the same time it
became clear that DDT and other organochlorine pesticides are concentrated in adipose tissue where they are also stored. Hence the
statement, repeated in textbooks ever since, that highly lipophilic
compounds are concentrated or partitioned in adipose tissue. This
kind of statement, however, turns out to be a misconception. It
overlooks the fact that thiopental is merely the most lipophilic
member in the barbiturate series; it overlooks in particular that
even more lipophilic compounds (according to the criteria of Hansch
and Leo (1979) such as imipramine and its congeners do not accumulate
in adipose tissue where their concentrations are actually minimal
(Bickel et al. 1968, 1980, 1981). Thus, fat storage of lipophilic
compounds and other factors require consideration of structural as
well as partition coefficient and lypophilicity.

FAT STROAGE OF POLYCHLORINATED COMPOUNDS

DDT and related pesticides, polychlorinated biphenyls (PCBs) and
other polychlorinated hydrocarbons are invariably highly lipophilic
and are stored in body fat. However, even with these polychlorinated
compounds it is doubtful whether lipid partition is the decisive
factor, and several aspects of their adipose tissue storage are not
clearly understood as yet.

Characteristically, the concentrations of polychlorinated compounds in adipose tissue are much greater than in other tissues and
100-1000 fold greater than in plasma. The chlorine atoms which contribute to the lipophilicity of these compounds also block metaboli-
cally vulnerable positions and thus diminish their metabolic
breakdown. The compounds thereby tend to become persistent. Their
very slow elimination, even when daily intake is quite minute, leads
to steady state concentrations (residues) which are particularly
high in adipose tissues (Figure 1).

Fat storage is thus the chief contributor to the body burden of
the polychlorinated xenobiotics. On the other hand, adipose tissue
is relatively "inert", so that material stored therin is removed from
other sites where toxic symptoms might appear, as for example, in the
nervous tissue. For this reason, fat storage can be considered as
a sequestering or portecting mechansim. However, in contrast to
eliminated material, stored material can be mobilized if the fat
compartment shrinks as happens in conditions of fasting or in some
diseased states. This can lead to unexpected toxic concentrations
in target organs as was originally suggested by Fitzhugh and Nelson
(1947). Other investigators then demonstrated that starvation of

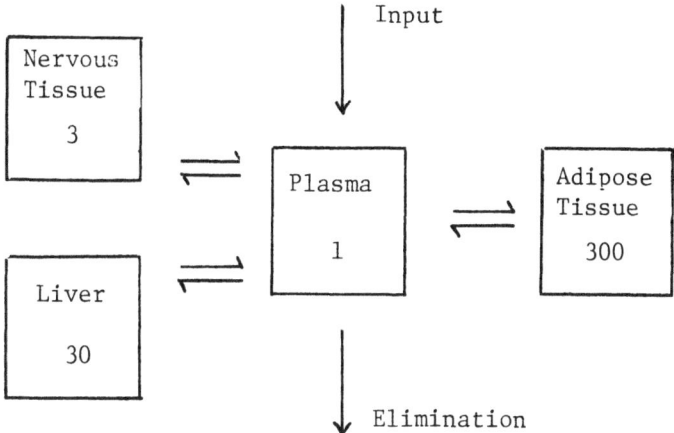

Figure 1. Pharmacokinetics of DDT and Related Compounds. Figures in Compartments are Relative Steady State Concentrations.

DDT-loaded animals leads to incresed DDT concentration in residual adipose and other tissues, but that at the same time there is in-creased elimination (Dale et al 1962, Ecobichon and Saschenbrecker 1969). Thus the risk of the appearance of toxic symptoms is increased even as the total body burden of DDT decreases. This important con-cept has become accepted even though relevant investigations are few and no details of the kinetics during starvation are known.

As is well-known, DDT and other organochlorine pesticides have been largely phased out in recent decades and residual concentrations have correspondingly decreased. At the same time the levels of en-vironmental residues of PCBs have tended to increase although the use of these industrial compounds has also been severely restricted of late. The PCB residues are not likely to dwindle rapidly because the highly chlorinated PCB isomers are also highly persistent and some are virtually unmetabolizable. The single most abundant PCB isomer in adipose tissue residues is 2,4,5,2',4',5'-hexachlorobiphenyl (6-CB).

Mathews and Anderson (1975) in their study on the effects of chlorination on the distribution and excretion of PCBs were able to show that the mono-, di-, and pentachloro isomers they had at their disposal behaved similarly to thiopental in rats with regard to re-distribution into adipose tissue and subsequent disappearance at rates determined by their metabolism. The situation was different with the hexachlorobiphenyl, 6-CB. This PCB isomer is virtually un-metabolizable in rats, and its elimination is excessively slow. The compound is slowly redistributed from other tissues into body fat which eventually contains about 75 percent of an administered dose and remains at this level over an observation period of 9 months (Mühlebach and Bickel 1981).

CHARACTERISTICS OF ADIPOSE TISSUE STORAGE

Fat storage of the extremely persistent model compound, 6-CB, shows rather unusual features which can be summarized as follows:

After administration, the redistribution into adipose tissue is extremely slow and takes weeks rather than hours.

A large proportion of an administered dose is eventually stored in the adipose tissue.

Adipose tissue storage is apparently irreversible in the sense that the amount stored is constant and not available for excretion.

Irreversibility of stroage in normally fed animals may be deceptive since the adipose tissue mass of adult rats increases considerably during the weeks and months of long-term pharmacokinetic studies. This explains why, in fact, 6-CB concentrations (in contrast to the amounts) decreased in adipose tissues over the long observation period.

In order to gain more insight into fundamentals of adipose tissue storage, rats were exposed to unphysiological conditions intended to bring about a drastic loss of adipose tissue by a chronic restriction of food intake. After a single small dose of 6-CB, time was allowed for a large proportion to be redistributed to body fat. The rats were then put on a restricted diet for 6 weeks. Experimental conditions and the gross events resulting are represented in Table 1. It can be seen that 6-CB stored in body fat is released and adipose tissue becomes virtually depleted. The released material is partly eliminated via enhanced fecal excretion and is in part redistributed to be stored in skin. Food restriction resulted in a virtually complete disappearance of adipose tissues and a considerable decrease of whole body and individual tissue weights up to the 4th week. In most tissues, 6-CB concentrations showed 2-10 fold increases up to the 4th week followed by decreases with half-lives of around 1,5 weeks. However, there was no initial concentration increase of 6-CB in adipose tissue as the restricted food intake began to melt down the adipose tissue reserves. This may have been in part due to alterations affecting the structural basis for adipose tissue storage rather than concentration or saturation effects.

In contrast to the apparent irreversibility of adipose tissue storage of 6-CB under normal conditions, fasting-induced fat mobilization reverses storage and reveals the following characteristics:

Near-total release of material stored in adipose tissue.

Table 1.   Effects of Restricted Food Intake on 6-CB
           Distribution in Rats* (percent of dose)

	Before Food Restrict.	After 6 w Food Restrict.	Controls Fed ad lib.
Adipose Tissue	47	1	80
Skin	29	47	11
Other Tissue	13	5	5
Excretion (fecal)	5	36	8

*Rats were given 2,4,5,2',4',5'-hexachlorobiphenyl (6-CB,
0.6 mg/kg, i.v.), and after 2 weeks food intake was restricted to 25 percent of the ad lib. diet for a further
6 weeks.

Initial concentration increase in tissues other than adipose
tissue.

Transfer from adipose tissue into skin and enhancement of fecal
excretion of unchanged 6-CB.

Reminiscent of the situation with DDT we find that the total body
burden of 6-CB is decreased while the concentrations in various tissues initially increase. New insights revealed by this study are
related to the fact that skin can take over the storage function
when adipose tissue is melted down. This redistribution happens at
such a rate that shrinkage of adipose tissue is not accompanied by
an initial increase of 6-CB concentration in this tissue.

As long as residues of polychlorinated compounds are found in
humans and animals on a global scale, information on their fat storage and mobilization is badly needed. The need has even increased
since the detection of polychlorinated dibenzodioxins and dibenzofurans in the environment. The observation (based on scanty information) that the latter compounds are stored in liver to a higher
degree than in adipose tissue is of little comfort in view of the
extreme toxicity of some of these contaminants. The very toxicity
of TCDF (2,3,7,8-tetrachloro-dibenzofuran) in guinea pigs was shown
to lead to fat mobilization and concomitant redistribution from
adipose tissue back to the liver as a major target organ (Decad
et al 1981).

REFERENCES

Bickel, M.H., and Gerny, R. Drug distribution as a function of
    binding competition. Experiments with the distribution
    dialysis. J Pharm Pharmacol, 32:669-674, 1980.
Bickel, M.H., and Graber, B.E. Unpublished results, 1981.
Bickel, M.H., and Weder, H.G. The total fate of a drug: Kinetics
    of distribution, excretion, and formation of 14 metabolites
    in rats treated with imipramine. Arch int Pharmacodyn, 173:
    433-463, 1968.
Brodie, B.B., Bernstein, E., and Mark, L.C. The role of body fat
    in limiting the duration of action of thiopental.
    J Pharmacol exp Therap. 105:421-426, 1952.
Brodie, B.B., Mark, L.C., Papper, E.M., Lief, P.A., Bernstein, E.,
    and Rovenstine, E.A. The fate of thiopental in man and a
    method for its estimation in biological material.
    J Pharmacol exp Therap, 98:85-96, 1950.
Dale, D.E., Gaines, T.B., and Hayes, W.J. Storage and excretion
    of DDT in starved rats. Toxicol Appl Pharmacol, 4:89-106, 1962.
Decad, G.M., Birnbaum, L.S., and Matthews, H.B. 2,3,7,8-tetrachloro-
    dibenzofuran tissue distribution and excretion in guinea pig.
    Toxicol Appl Pharmacol, 57:231-240, 1981.
Ecobichon, D.J., and Saschenbrecker, P.W. The redistribution of
    stored DDT in cockerels under the influence of food deprivation.
    Toxicol Appl Pharmacol, 15:420-432,1969.
Fitzhugh, O.G., and Nelson, A.A. The chronic oral toxicity of DDT.
    J Pharmacol exp Therap, 89:18-30, 1947.
Hansch, C., and Leo, A. Substituent constants for correlation
    analysis in chemistry and biology. New York:John Wiley, 1979.
Matthews, H.B., and Anderson, M.W. Effect of chlorination on the
    distribution and excretion of polychlorinated biphenyls.
    Drug Metab Dispos, 3:371-380, 1975.
Mühlebach, S., and Bickel, M.H. Pharmacokinetics in rats of
    2,4,5,2',4',5'-hexachlorobiphenyl, an unmetabolizable lipophilic
    model compound. Xenobiotica, 11:249-257, 1981.

STRUCTURE ELUCIDATION OF MAMMALIAN TCDD-METABOLITES

H. Poiger,* H. R. Buser**

*Institute of Toxicology
 Federal Institute of Technology and University of
  Zurich
 Zurich, Switzerland

**Federal Agricultural Research Station
  Wädenswil, Switzerland

ABSTRACT

    Biotransformation products of 2,3,7,8-tetrachlorodibenzo-p-
dioxin (TCDD) excreted in dog and rat bile were isolated and
examined by combined gas chromatography-mass spectrometry (GC-MS).
In the dog, hydroxylation of the TCDD molecule at the lateral
position with shift of a chlorine atom to the peri-position and
chlorine displacement were found to be the main routes of metabolism.
The major metabolite formed was 2-hydroxy-1,3,7,8-tetrachlorodibenzo-
p-dioxin. Dihydroxy-trichlorodibenzo-p-dioxins were also identified
as well as metabolites formed via cleavage of one or both ether bonds.
In the rat, only one metabolite has been characterized so far. It was
tentatively identified as a tetrachlorodihydroxy-diphenyl ether. At
least one additional metabolite was present which has not yet been investigated. A preliminary metabolic breakdown scheme of TCDD is
presented.

INTRODUCTION

The chemical and metabolic stability of TCDD contributes to the toxic
hazard of this compound, since it favors bioaccumulation. Metabolism
in microorganisms, if it occurs at all, has been reported to be very
slow (Matsumura and Benezet, 1973). Biodegradation in mammals has
been found to be slow in rats (Ramsey et al, 1979, Poiger and

483

Schlatter, 1979), and rapid in the Golden Syrian Hamster (Olson
et al., 1980), which is also much less sensitive to TCDD. The
biotransformation products, more polar than TCDD, are eliminated via
bile and urine in both animal species (Gasiewicz et al, this volume)
and could be separated by TLC and HPLC techniques. However, the
elucidation of the structure of any of these compounds has not yet
been reported. For our experiments the dog was chosen as experimental
animal, since it also excretes considerable amount of TCDD-metabolites in bile (Poiger et al, 1980) and can be dosed with higher
amounts of TCDD. Based on the data obtained from this animal species
the detection of corresponding compounds in the bile of the rat was
also attempted.

EXPERIMENTAL

Chemicals

     Tritiated TCDD (specific activity 41 Ci/mMol) was purified by
preparative gas chromatography (GC), using a 200x0.2cm all glass
column packed with 2 percent DC 560 on Supelcoport 80/100 mesh,
to give material with radiochemical purity of $\geq$98.5 percent.
Unlabeled TCDD was recrystallized 5 times from hot anisole. Both
chemicals were obtained from Givaudan S.A. Dübendorf, Switzerland.
The TCDD-solution used for the biological experiments was prepared by
dissolving weighed amounts of TCDD in a benzene solution of tritium
labeled TCDD, achieving a final concentration of 0.36 mg/ml (26 mCi/
mMol). GC-MS examination of this solution revealed only the presence
of TCDD and a small quantity (<1%) of penta-CDD.

Dosage

     Aliquots of the benzene solution (0.36 mg TCDD/ml) were mixed
with corn oil 1:1 and the benzene removed by evaporation under a
stream of nitrogen. In the dog experiment the total dose (5.4 mg) was
administered enterally in 4 portions of 1.8, 1.08, 1.08 and 1.44
milligrams on days 0, 2, 7 and 13, respectively. The rat received a
single dose of 100 µg of TCDD/kg.

Animal Experiments

     A one-year-old Beagle dog was cholecystectomized and a Thomas
cannula was implanted about 3 months before the experiment, during
which time the bile flow was not impaired. Immediately before
administration of TCDD, a polyethylene catheter was introduced into
the bile duct and simultaneously a small Silastic tube was inserted
into the duodenum in the distal direction via the Thomas cannula.

The polyethylene catheter was connected to a plastic bag, which was strapped on the outside of the animal. The TCDD solution was then drawn slowly into the duodenum via the Silastic tube by means of a glass syringe. During this procedure the animal was kept under slight anaesthesia. Bile was collected continuously for 2 to 4 days during which the plastic bag was changed at 1-or 2-day intervals. On days 5 and 11 both the catheter and the application tube were removed and the animals allowed to recover for a few days before insertion and application procedures were repeated as described. In the rat experiment a bile duct catheter was inserted in a 220 g female rat from the Sprague-Dawley derived ZUR: SIV-Z strain, according to a procedure described elsewhere (Bachmann and Schlatter, 1981), and the rat was left to recover for a few days. About 1 hour after dosage with TCDD, the animal was placed in a restraining cage and bile was collected for about 4 days. The animal was then allowed to recover in a normal cage for 3 days before bile was collected for another 4-day period.

Isolation of Metabolites

Ten milliliters of bile, collected from the dog after the last TCDD dose, was dried in a rotary evaporator under reduced pressure. The whole amount of bile obtained from the rat was incubated over-night with glucuronidase/arylsulfatase (Boehringer, Mannheim, Germany) at pH 5.5 and a temperature of $37^{o}C$, prior to drying. Both residues were extracted twice with 30 ml of cold ethanol. The extracts were evaporated to dryness and the residues redissolved in 50 ml of acetone. The solutions were methylated with 25 ml of methyl iodide in the presence of K_2CO_3 by refluxing overnight at a tempera-ture of $40-50^{o}C$. Then the bulk of the potassium carbonate was removed by filtration and the solvent evaporated. Twenty ml of toluene was added to the residues and insolubles were removed by fil-tration. The filtrates were then concentrated in the rotary evaporator.

For thin layer chromatography, precoated silica gel 60 plates (Merck, Germany), with a layer thickness of 0.25 mm, were used. A mixture of hexane and ethyl acetate (9:1) was used as developing solvent. The zones containing radioactive material (with the exception of the starting area) were cut out and eluted with toluene. The extracts were filtered and concentrated to a small volume. The rat bile extract was directly analyzed by GC-MS, whereas the eluate originating from the dog bile was further purified by preparative GC.

Preparative GC was performed in a Carlo Erba gas chromatograph, equipped with a splitting device and a heated, outlet pipe. A 100x0.2 cm glass column, packed with 3% SE 30 on Supelcoport 80/100 mesh was used (oven temp. $200^{o}C$). Fractions were trapped in glass tubes

filled with glass wool and the radioactive compounds were rinsed off
with hexane. The concentrated solutions were used for GC-MS analysis

GC-MS Analysis

The analysis was performed on a Finnigan 4000 quadrupole GC-MS
instrument, operating in the EI mode on 70 eV (240°C). Mass spectra
(m/z 100-450, 1.7 sec/scan) were recorded using a Finnigan 6115 data
system. The gas chromatograph was fitted with a 24 m SP 2100 fused
silica capillary column. The column conditions were: 50°C (2 min
isothermal), 20°C/min to 160°C, 5°C/min to 240°C. The He carrier gas
pressure was 0.5 at.

RESULTS

Toxicological Observations

As observed in previous experiments, oral treatment of the dog
with TCDD resulted in severe vomiting, cachexia and weight loss,
which were also the main visible symptoms after enteral application
of the dioxin. The onset of these effects was delayed about 2 hours
post-treatment. After dose 2 the animal ceased vomiting temporarily
(days 5 and 6), allowing forced feeding. Excess water consumption was
observed throughout the experiment. Four days after administration of
the last dose the animal died without any sign of recuperation.

In the rat, except for loss of weight, no visible toxic effect
was observed. It was sacrificed on day 12 of the experiment.

Excretion of Metabolites

In the dog, excretion of TCDD-derived radioactivity in the bile
reached its maximum 1 or 2 days after each dose and then decreased
quite rapidly (Figure 1). A total of about 300 µg of TCDD-derived
material (calculated from the amount of radioactivity) were excreted,
which accounted for about 5.6% of the total dose administered.
However, after doses 1 and 2 only 1.7% was excreted within 4 days,
whereas 11.1 and 8.2% were excreted within 3 days after doses 3 and
4, respectively. There was also a reduction in the bile flow from
about 100 ml/day at the beginning of the experiment to about 25 ml/day
at the end, an effect which was also observed to a minor extent in the
control animals.

In the rat, the amount of radioactivity in the bile after treatment remained relatively constant with amounts of 0.5 to 1% of the
initial dose being excreted per day.

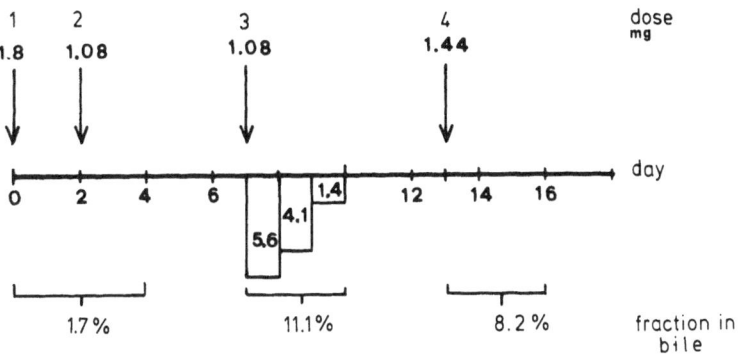

Figure 1.   Dosage of TCDD and biliary excretion of metabolites in the dog.

TLC and GC

Figure 2 shows the distribution of the radioactivity on TLC-plates. The methylated extract from the dog bile (Figure 2 a, solid line) separated into several radioactive zones ranging from r_f 0.0 to 0.9. The methylated extract from the rat bile (Figure 2 b, solid line) exhibited a different distribution pattern in that no radioactivity could be detected between r_f 0.6 and 0.9. Distribution of the other radioactive materials seems to correspond quantitatively. The material on the starting area most likely con-sists on non-methylated species. Since the majority of interfering material also was located in this area it has not yet been futher investigated.

TLC-eluates from zones II and III (Figure 2) were subjected to preparative GC. Two fractions, containing radioactive material, could be trapped from each eluate (designated IIa and IIb, and IIIa and IIIb, respectively).

GC-MS

A summary of the GC-MS data on chlorinated compounds detected in these four GC-fractions is given in Table 1. The amounts of the different compounds, as estimated from the GC-MS data, were in accordance with the amounts expected from the radioactivity measured in the samples and were between 20 and 60 ng. None of the compounds identified was detected in the bile of untreated dogs. The mass spectrum of the major component, compound II (Figure 3), with a molecular ion at m/z 350 and 4 chlorine atoms, shows a characteristic fragmentation of M^+-15 (M^+ CH_3) and M^+-43 (M^+-CH_3 -CO) which identifies this compound as a methyl ether of an hydroxylated TCDD. The strong M^+-15 ion is indicative of of a CH_3O-group in lateral (2-,3-, 7- or 8-) position (Tulp and Hutzinger 1978), suggesting that the

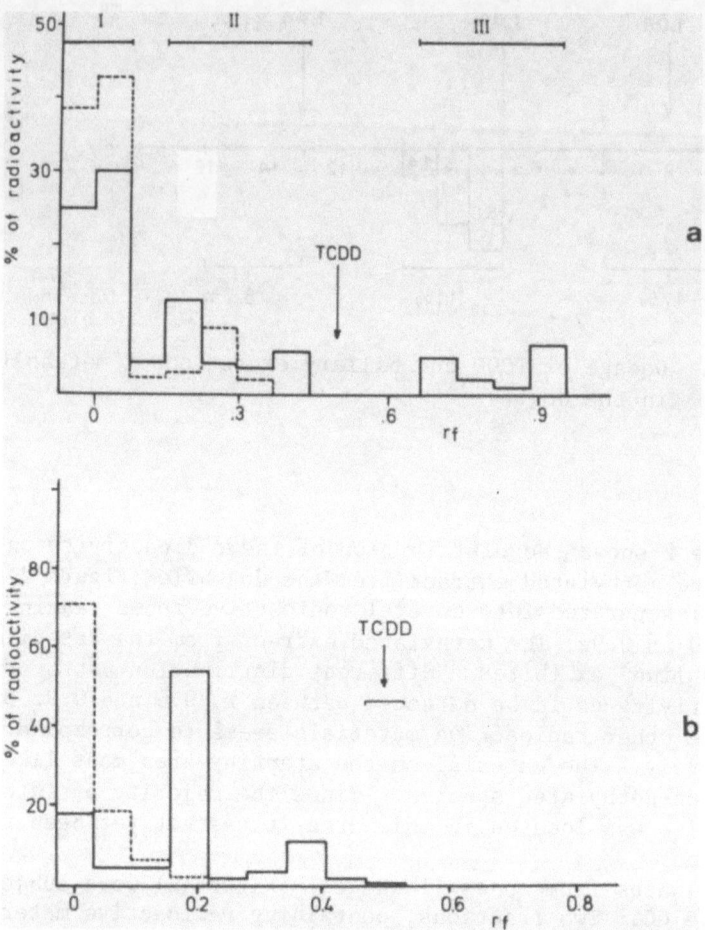

Figure 2.   Radio-TLC of methylated bile extracts (solid lines)
             from dog (a) and rat (b).   The dotted lines show
             distribution of radioactivity before methylation.

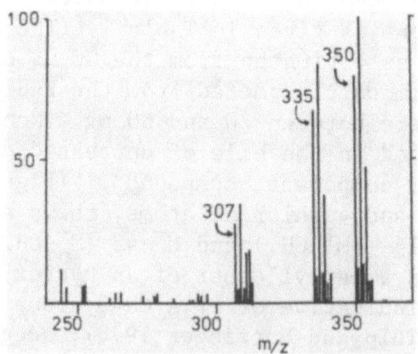

Figure 3.   Mass spectrum of compound II (dog experiment)

TABLE 1

GC-MS Data for TCDD and Methylated Metabolites (dog)

	Compound	Fraction (GC)	Elution temp.°C	M^+	No.of Cl atoms	Fragmentation
I	2378 TCDD		217.3	320	4	M^+-63; M^+-126
II	(structure)	IIIa,b	231	350	4	M^+-15; M^+-43
III	(structure)	IIIa	223.9	316	3	M^+-15; M^+-43
IVa	(structure)	IIIb	227	346	3	M^+-15; M^+-43
b	$Cl_3(OCH_3)_2$	IIIb	228.8	346	3	M^+-15; M^+-43
V	(structure)	IIb	231.8	366	4	M^+-50
VI	(structure)	IIa	134.5	206	2	M^+-15; M^+-43

compound is 2-methoxy-1,3,7,8-tetrachlorodibenzo-p-dioxin. The assignment of the methoxy group to a lateral position is supported by the fact that high temperature air oxidation of 2,3,7,8-TCDD (8 µg in a sealed quartz ampoule, heated to 620°C), followed by methylation with diazomethane led to formation of a small quantity (0.1 percent) of an isomer of compound II with an elution temperature of 232.5°C and also with a fragmentation signal at M^+-43. The absence of a strong M^+-CH$_3$ signal supports assignment of the structure of this compound as 1-methoxy-2,3,7,8-tetrachlorodibenzo-p-dioxin (methoxy group in peri-position).

Compound III (M^+=316) was identified as a trichloro-2-methoxy-dibenzo-p-dioxin (presence of a strong M^+-15 fragment). Since the same compound was also formed by high-temperature air oxidation of TCDD, assignment as 2-methoxy-3,7,8-trichloro-dibenzo-p-dioxin is strongly suggested.

Two further chlorinated compounds (IV a and b), apparently isomers, were identified as dimethoxy-trichlorodibenzo-p-dioxins. The exact positions of the substituents in these two metabolites has not yet been established.

The mass spectrum of compound V (M^+=366) did not show the characteristic fragments of laterally methoxylated dioxins (M^+-15, M^+-43). Exact mass measurements using high resolution MS indicated a composition of $C_{14}H_{10}O_3Cl_4$ (experimental mass 365.9352, exact mass 365.9385), suggesting a tetrachlorodimethoxydiphenyl ether. The fragment M^+-50 (CH_3Cl) has been found with other chlorinated 2-methoxydiphenyl ethers (Nilsson 1977).

Finally, compound VI (M^+=206) was identified as 1,2-dichloro-4,5-dimethoxybenzene based on its mass spectrum and its coelution with a synthetic reference sample.

In the isolation of the rat metabolites preparative GC of the TLC-eluates was omitted in order to prevent loss of material. Figure 4 shows the mass spectrum of the only chlorinated compound found in the TLC-eluate from the zone at r_f 0.2, where most of the radioactive material was located. The mass spectrum indicates a molecular ion at m/z 366, the presence of 4 chlorine atoms and fragmentation with formation of the M^+-46 and M^+-50 ions. The same metabolite was also found in the bile of the dog (see above) and tentatively identified as a tetrachlorodimethoxydipehnyl ether. Attempts to identify TCDD-metabolites in TLC-eluates in the region r_f 0.3-0.4 have so far failed since many interfering substances were present in these samples. Further clean-up is required.

DISCUSSION

The data on elimination TCDD metabolites in the dog bile (Figure 1) indicate increased excretion of radioactive material after

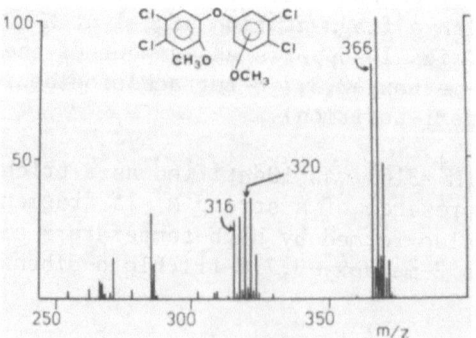

Figure 4.   Mass spectrum of rat metabolite

doses 3 and 4 as compared to doses 1 and 2. Obviously, TCDD stimu-
lates its own biotransformation by inducing the enzyme systems
responsible (MFO) in the liver. This apparently contrasts with the
elimination kinetics of the rat, where metabolites are excreted at a
much slower, but more or less constant, rate over several days. When
considering the rate of conversion of TCDD in the dog, we should note
that the intestinal uptake of the dioxin, at the mg dose level, is
incomplete and does not exceed 50 percent of the administered dose
(unpublished data from some previous experiments). This would imply
a higher conversion rate, when considering the actual fraction of the
dose absorbed. Another important difference in metabolism between the
two species is that glucuronidation of the metabolites does not occur
in the dog, whereas in the rat (and also the Syrian Hamster) the
metabolites are mainly conjugated (Gasiewicz et al, this volume).

A proposed metabolic breakdown scheme of TCDD is shown in Figure
5. It is evident that the laterally hydroxylated species can only be
formed from TCDD in a rearrangement reaction involving migration of a
chlorine substituent. This most likely arises via an NIH-shift subsequent to enzymatic formation of an arene oxide. In the dog,
metabolism of TCDD via such an arene oxide seems to be the major route
for formation of hydroxylated TCDD. In the rat, however, this route
seems to be much less important and the major metabolite is formed via
cleavage of an ether bond. It is possible that cleavage of the ether
bond involves expoxide formation at the angular carbon atom, a mechanism which is unusual in the metabolism of xenobiotics.

The toxicity of the metabolites isolated from the dog bile was
reported to be at least 100 times lower than that of the parent TCDD
(limited LD_{50} experiments, Weber et al, in press), which provides
evidence that the typical toxic symptoms are caused by the parent compound. Possibly they might be mediated by the reactive epoxide, which

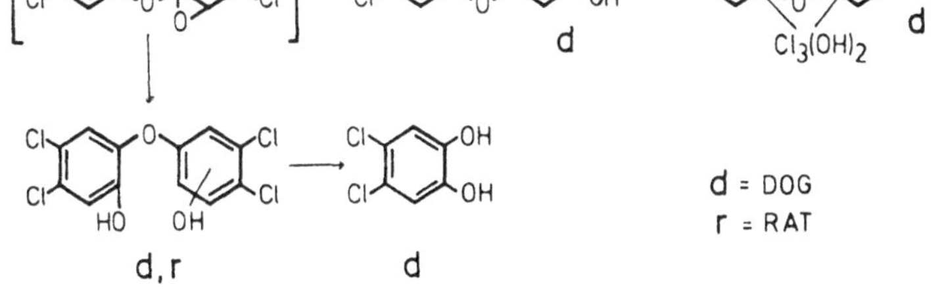

Figure 5. Proposed metabolic breakdown scheme of TCDD

could bind covalently to proteins (Guenthner et al, 1979), but this
would imply that animals metabolizing TCDD at a higher rate are more
susceptible towards TCDD, a hypothesis somewhat contradicting existing
data.

REFERENCES

Bachmann, M. and Schlatter, C.H. Metabolism of ^{14}C-emodin in the
    rat. Xenobiotica, 11(3): 217-225, 1981.
Guenthner, T. M., Fysh, J. M., and Nebert, D. W. 2,3,7,8-Tetrachloro-
    dibenzo-p-dioxin: Covalent binding of reactive intermediates
    principally in vitro. Pharmacology, 19: 12-22, 1979.
Matsumura, F., and Benezet, H. J. Studies on the bioaccumulation and
    microbial degradation of 2,3,7,8-tetrachlorodibenzo-p-dioxin.
    Environm. Health Perspect., 5: 253-258, 1973.
Nilsson, C. A. Doctoral Dissertation, University of Umea, Umea,
    Sweden, 1977.
Olson, J. R., Gasiewicz, T. A. and Neal, R. A. Tissue distribution
    excretion and metabolism of 2,3,7,8-tetrachlorodibenzo-p-dioxin
    (TCDD) in the Golden Syrian Hamster. Toxicol. Appl. Pharmac.,
    56: 78-85, 1980.
Poiger, H. and Schlatter, C.H. Biological degradation of TCDD in rats.
    Nature, 281: 706-707, 1979.
Poiger, H., Weber, H. and Schlatter, C.H. Special aspects of metabol-
    ism and kinetics of TCDD in dogs. Assessment of toxicity of
    TCDD-metabolites in guinea pigs. In: Proc. Workshop "Impact of
    Chlorinated Dioxins and Related Compounds on the Environment."
    Oct. 1980, Rome. Pergamon Press, Oxford, 1981.
Ramsey, J. C., Hefner, J. G., Karbowsky, R. J., Braun, W. H. and
    Gehring, P. J. The in vivo biotransformation of 2,3,7,8-tetra-
    chlorodibenzo-p-dioxin (TCDD) in the rat. Toxicol. Appl.
    Pharmac., 48(1): A126, (1979).
Tulp, M. Th. M. and Hutzinger, O. Identification of hydroxylated
    chlorodibenzo-p-dioxins, chlorodibenzofurans, chlorodiphenyl
    ethers and chloronaphthalenes as their methyl ethers by gas
    chromatography mass spectrometry. Biomed. Mass Spectrom., 5(3):
    224-231, 1978.
Weber, H., Poigner, H. and Schlatter, C.H. Acute oral toxicity of
    TCDD-metabolites in male guinea pigs. Submitted to Toxicol.
    Letters.

ANIMAL TOXICOLOGY

Chairman: Edward A. Smuckler
            University of California Medical
            School
            San Francisco, California   USA

ABSORPTION, DISTRIBUTION AND METABOLISM OF 2,3,7,8-

TETRACHLORODIBENZODIOXIN (TCDD) IN EXPERIMENTAL ANIMALS

T. A. Gasiewicz,* J. R. Olson,** L. H. Geiger,***
R. A. Neal****

*Environmental Health Sciences Center, University of
  Rochester School of Medicine
  Rochester, New York USA

**Department of Pharmacology and Therapeutics,
  School of Medicine
  State University of New York
  Buffalo, New York USA

***Center in Toxicology
   Vanderbilt University
   Nashville, Tennessee USA

****Chemical Industry Institute of Toxicology
    Research Triangle Park
    North Carolina USA

ABSTRACT

     Current understanding of the absorption, distribution, metabol-
ism, and excretion of 2,3,7,8-tetrachlorodibenzo-p-dioxin (TCDD) in
various mammalian species is reviewed. Previous studies on the
influence of solvents and adsorbants on dermal and intestinal absorption of TCDD suggest the vehicle for TCDD may be of importance in
assessing the relative risk of a given exposure to this compound.
In a variety of animal species the liver and adipose tissues appear
to be the major storage organs for TCDD. Interspecies differences
in TCDD distribution do not appear to be related to species differences in the acute toxic effects of this compound. Following the
administration of radiolabeled TCDD, radioactivity appears to be
eliminated from most species through a first-order process. However,
at present no clear relationship exists between the ability of a
given species to excrete TCDD and/or its metabolites and the acute
lethal toxicity of TCDD in that species. TCDD-derived radioactivity

495

was found to be largely eliminated in the feces. In hamsters and mice urinary elimination was found to be a major route of excretion. High-performance liquid chromatographic (HPLC) analyses of bile and urine from various species show that radioactivity excreted by these routes is mainly associated with metabolites of TCDD. Additional studies utilizing hamster hepatocytes suggest that TCDD is metabolized by the cytochrome P-450-containing monooxygenases. Although the chemical structures of these TCDD metabolites is unknown, additional studies suggest some of these may be glucuronide conjugates.

INTRODUCTION

As a result of its extreme toxicity and occasional presence in the environment, 2,3,7,8-tetrachlorodibenzo-p-dioxin (TCDD), commonly referred to as dioxin, has emerged as a compound for which there is great public health concern. A spectrum of toxic effects, similar to those observed with TCDD, can be produced upon exposure of animals to higher doses of other halogenated aromatic compounds. These include other polychlorinated dibenzofurans, certain poly-chlorinated biphenyl isomers and certain chlorinated azoxy- and azo-benzenes (Kimbrough, 1974; Goldstein, 1979; McConnell, 1980; McConnell and Moore, 1979). TCDD is the most potent and most ex-extensively studied toxin in this group (Schwetz, et al, 1973; McConnell et al, 1978), and thus has become a prototype for these environmentally persistent, halogenated aromatic substances.

TCDD is formed as a contaminant during the synthesis of 2,4, 5-trichlorophenol, which in turn is used to manufacture the herbicide 2,4,5-trichlorophenoxyacetic acid (2,4,5-T) and related compounds (Kimbrough, 1974; IARC, 1977). Over the last 30 years, TCDD has been implicated in a number of accidental poisonings of animals and humans (Kimbrough, 1974; Poland and Kende, 1976; Reggiani, 1978). Perhaps the most widely publicized incident occured in Seveso, Italy, in 1976, when an accident in a plant producing trichlorophenol resulted in the dissemination of in excess of 100 grams of TCDD over a wide area that included several populated communities (Firestone, 1978; Reggiani, 1978, 1980). The potential for widespread human exposure of man to low levels of TCDD has led to numerous studies attempting to es-tablish man's relative sensitivity to the toxin. Studies to date, however, have been unable to estimate the acute or chronic toxicity of TCDD in man and have not established the mechanism(s) or site(s) of action for the toxicity in experimental animals.

The extreme toxicity of some isomers of the polychlorinated dibenzodioxins has stimulated studies of their potential for wide-spread contamination of the environment and bioconcentration in certain species. TCDD appears to be a remarkably stable compound, requiring temperatures in excess of 800°C for its thermal degradation (Stehl et al, 1973). Once incorporated in the soil, TCDD appears

to be extremely resistant to degradation by microorganisms, having
a half-life of from 0.5 to > 10 years (diDomenico et al, 1980a). On
the other hand, the photolytic degradation of TCDD appears to be far
more efficient, particularly if the toxin is in the presence of a
suitable solvent which provides a hydrogen donor (diDomenico et al,
1980a); Crosby and Wong, 1976; Nestrick et al, 1980). Until recent
years, little was known about the fate of TCDD in biological systems.
In 1979, Ramsey et al, (1979) and Poiger and Schlatter (1979) provided the first report indicating that TCDD could be metabolized in
mammals. At present there are a number of reports on the pharmacokinetics of TCDD in various species of laboratory animals. This
review will examine our current understanding of the absorption,
distribution, metabolism and excretion of TCDD in various mammalian
species. The significance of the kinetic data will also be discussed in relation to the toxicity of TCDD that has been reported
for each species.

ABSORPTION OF TCDD

    In humans, exposure to TCDD occurs primarily by dermal absorp-
tion. Oral intake or inhalation are less likely routes of exposure.
However, although TCDD has extremely low volatility, inhalation of an
airborne mist, dust, or other particulate matter contaminated with
the toxin could occur. For example, in the Seveso accident, TCDD,
trichlorophenol and other materials were released into the atmosphere
and settled over an area measuring about 5 kilometers long and 700
meters wide. TCDD was found to contaminate the soil in this area at
concentrations ranging from < 0.75 to approximately 20,000 $\mu g/m^2$
(diDomenico et al, 1980b). Continuous monitoring of atmospheric
dust in the Seveso area was carried out to establish to what extent
airborne TCDD could be considered a health hazard for inhabitants
residing near heavily contaminated areas (diDomenico et al, 1980c).
In this study, TCDD levels ranged from 0.06 to 2.1 ng of TCDD/g of
airborne dust sampled. An estimated 24-hour inhalation exposure was
then calculated to be approximatley 1.4 pg of TCDD for a person who
inhaled an average of $10/m^3$ of air in 24 hours containing 0.14 ng
dust/m^3 with a TCDD concentration in the dust of 1 ppb. While the
systemic absorption of TCDD from inhaled particulates in the lung may
occur, no studies of this process have yet been reported. Nevertheless, the data cited above suggest that inhalation exposure to TCDD
may occur in heavily contaminated areas like Seveso. However, exposure by this route is unlikely to be of toxicological significance.

    Poiger and Schlatter (1980) have recently investigated the
dermal and intestinal absorption of TCDD in the rat. Environmental
exposure to TCDD usually involves exposure to another substance containing TCDD. Therefore, they examined the effect of different
formulations containing TCDD on its absorption. The results are

shown in Table 1. The liver concentration of [^{3}H]-TCDD was used as
the means of determining the uptake of the toxin. This procedure
has validity since the liver appears to be one of the major storage
sites for TCDD in the rat (Piper et al, 1973; Rose et al, 1976; Allen
et al, 1975; Van Miller et al, 1976). After oral administration of
14.7 ng of TCDD, using 50 percent ethanol as the vehicle, they found
36.7 percent of the total dose in the liver after 24 hours. Only
about half of this amount was found in the liver when the toxin was
administered orally in an aqueous mixture with soil particles. When
the orally administered TCDD was first absorbed onto activated carbon,
the absorption from the gut was almost completely prevented. The
movement of TCDD across the epidermis was also found to be highly
dependent on the nature of the formulation applied to skin. The TCDD

Table 1. Percentage of TCDD in the Liver of the Rat After
          Dermal Administration of TCDD in Various Formulations[a]

Formulation and amount administered	TCDD dose (ng)	No. of animals	Percentage of dose in the liver
Methanol, 50 µl	26	6	14.8 ± 2.6
Vaseline, 0.1 ml	26	3	1.4 ± 0.4
Polyethylene glycol 1500, 0.1 ml	350	4	9.3 ± 3.4
Polyethylene glycol 1500 +15 H$_2$O, 0.1 ml	350	4	14.1 ± 4.9
Soil/water paste, 75 mg	26	5	c. 0.05
(50 mg dry soil)	350	5	1.7 ± 0.5
	1300	3	2.2 ± 0.5
Activated carbon/water,	26	4	< 0.05
100 mg (50 mg dry)	1300	4	< 0.05

[a]Adapted from Poiger and Schlatter (1980).

level in the liver was highest when TCDD was applied in a methanol
solution (Table 1), with the dermal adsorption amounting to about
one half that observed following oral exposure. Poiger and Schlatter
(1980) also found that soil and activated carbon greatly reduced the
percutaneous uptake of TCDD (Table 1). The effect of different formulations of TCDD on the induction of chloracne on the rabbit ear was
also examined by these investigators. Threshold levels for the
induction of chloracne lesions were about 1 µg for TCDD in acetone
and about 160 µg when the toxin was absorbed onto activated carbon
before application. These studies suggest that when assessing the
relative risk of environmental exposure to TCDD, the vehicle or
formulation of the TCDD may be of greater importance than the route
of exposure. This becomes particularly significant when one considers that in the general environment TCDD is usually absorbed onto
various materials such as soil and plant surfaces.

Other kinetic studies have examined the absorption of radiolabeled TCDD following gastric intubation (oral) or intraperitonel
(i.p.) injection. In these experiments TCDD was dissolved in olive
oil or in a solution of acetone in corn oil. In studies in the rat,
from 70 to 85 percent of orally administered TCDD was found to be
absorbed (Piper et al, 1973; Rose et al, 1976; Allen et al, 1975).
Efficient uptake of TCDD from the gut was also found in the hamster,
where 73.5 percent of the oral dose was absorbed (Olson et al, 1980a).
The uptake and distribution of TCDD following i.p. injection also
appears to be very efficient. Studies in the guiena pig, rat, monkey,
mouse, and hamster suggest that essentially all of the toxin was
absorbed from the peritoneal cavity following i.p. injection
(Gasiewicz et al, 1981; Van Miller et al, 1976; Olson et al, 1980a;
Gasiewicz and Neal, 1979). In addition, a very similar time course
for the tissue distribution of i.p. and orally administered TCDD in
the hamster indicates similar rates of adsorption of the toxin
through both routes of administration (Olson et al, 1980a). Together,
these studies suggest a similar body burden of TCDD may accompany
both oral and i.p. administration of TCDD, since only 15 to 30 percent of an oral dose appears to be unabsorbed.

DISTRIBUTION OF TCDD

Tissue and Organ Distribution of TCDD

The cellular distribution of TCDD has been examined in various
species in an attempt to understand more about the pharmacokinetics
of the toxin, and perhaps to identify a target site for its toxicity.
Rose et al (1976) examined the fate of TCDD in rats following a
single oral dose of 1.0 µg [^{14}C]-TCDD/kg and following repeated oral
doses of 0.01, 0.1, or 1.0 µg [^{14}C]-TCDD/kg/day, Monday through
Friday for 7 weeks. The liver and fat contained most of the body

burden of the toxin, accounting for 50 and 10 times more [^{14}C]
activity, respectively, than any other tissue examined. As a result
of these studies, they concluded that the rate of TCDD accumulation
in the body, following single and repeated exposure, was largely
accounted for by the rate of accumulation in liver and fat. Other
studies in rats, guinea pigs, hamsters, and mice have confirmed the
selective distribution of TCDD into the liver and fat of exposed
animals.

Table 2 describes the TCDD levels found in liver and adipose
tissue of various species. At the same time, these studies also
found that the skeletal muscle, heart, testes, blood, and brain
contained the lowest concentrations of TCDD-derived radioactivity
(Van Miller et al, 1976; Olson et al, 1980a; Gasiewicz and Neal,
1979).

The monkey appears to be an exception to the general pattern of
TCDD distribution described above. A study by Van Miller et al
(1976) determined that a greater percentage of the TCDD dose was
found in the skin and muscle tissue of the monkey. At 7 days

Table 2: Distribution of Radioactivity Following
Administration of Radiolabeled TCDD

	Day Following Treatment	Liver (% Dose/g)	Liver (% Dose/liver)	Adipose (% Dose/g)
Guinea pig[a]	11	2.2 ± 0.2	21.2 ± 2.3	2.1 ± 0.2
Adult monkey[b]	7	0.09 ± 0.06	10.4 ± 6.9	0.16 ± 0.06
Infant monkey[b]	7	0.13 ± 0.07	4.5 ± 1.6	0.49 ± 0.12
Rat[b]	7	4.54 ± 0.45	43.0 ± 4.7	3.46 ± 0.21
Hamster[c]	10	4.02 ± 0.40	16.7 ± 0.6	2.03 ± 0.18

[a]2.0 µg ^{14}C-TCDD/kg, i.p. (Gasiewicz and Neal, 1979).
[b]400 µg ^{14}C-TCDD/kg, i.p. (Van Miller et al, 1976).
[c]650 µg ^{3}H-TCDD/kg, i.p. (Olson et al, 1980a).

following i.p. administration, 35.6 ± 14.4 and 22.7 ± 8.8 percent
of a dose of TCDD given to infant monkeys was found in the muscle
and skin, respectively, compared with only 4.51 ± 1.60 percent of the
dose in the liver (Van Miller et al, 1976). The data in Table 2
show that less TCDD is distributed to the liver and adipose tissue of
adult and infant monkeys as compared with other species. However, the
unusually high levels of TCDD in the skin are not unique to the
monkey. Distribution studies in the guinea pig have found high
levels of [^{14}C]-TCDD in the skin (Gasiewicz and Neal, 1979),
although these levels were not as high as observed in the monkey.

Subcellular Distribution of TCDD

The subcellular distribution of TCDD has been examined in the
liver of the rat (Allen et al, 1975), guinea pig (Gasiewicz and
Neal, 1979), and mouse (Vinopal and Casida, 1973). In all three
species, radiolabeled TCDD was found in the highest concentrations in
the microsomal fraction. In the rat up to 90 percent and in the
guinea pig and mouse approximately 50 percent of the radioactivity
present in the liver following administration of radiolabeled TCDD
was associated with the microsomal fraction. In the guinea pig and
mouse, the crude nuclear fraction had the second highest level of
TCDD-derived radioactivity followed by the mitochondrial and soluble
fractions. The high affinity of TCDD for the microsomes, particularly in the rat, may in part be due to extensive proliferation of
smooth endoplasmic reticulum (SER) in the liver of TCDD-treated
animals (Jones and Butler, 1974), with a resulting accumulation of
the highly nonpolar toxin in these lipophilic membranes. The relatively low level of TCDD in the monkey liver (Table 2) has been
suggested to be the result of a lesser degree of hepatic SER proliferation in TCDD-treated monkeys (Van Miller et al, 1976). Even
though the hepatic membranes of the monkey were found to have a
similar affinity for TCDD as those of the rat, the slower rate
proliferation and the resulting smaller quantity of SER in the monkey
liver may result in the sequestration of a lower percentage of TCDD
in this organ. As a result, the TCDD may move to other tissues, particularly those of high lipid content, such as the adipose tissue and
skin (Van Miller et al, 1976).

The relatively high levels of TCDD in the hepatic microsomal
fraction may also, in part, be due to the metabolism of the toxin at
this subcellular site. The microsomal fraction contains the cytochrome P-450 associated monooxygenases, which appear to be responsible
for at least part of the metabolism of TCDD (Beatty et al, 1978;
Guenthner et al, 1979; Olson et al, 1981). (See section on
Metabolism of TCDD.)

Chemical Nature of TCDD-Derived Radioactivity in Tissues

Analysis of the liver and adipose tissue of hamsters receiving [^{3}H]- or [^{14}C]-TCDD indicated the extractable radioactivity presents was unmetabolized TCDD (Olson et al, 1980a). These results confirm that of an earlier study which identified the radioactivity in the liver of [^{14}C]-TCDD-treated rats to be unchanged TCDD (Rose et al, 1976).

The in vivo covalent binding of [^{3}H]-TCDD to rat liver protein, ribosomal RNA, and DNA also has been reported (Poland and Glover, 1979). These investigators found that virtually all of the radio-activity present in the liver (>99.9 percent) could be extracted. The distribution of the unextractable radioactivity was: protein, 60 pmol TCDD per mol of amino acid residue; RNA, 12 pmol TCDD per mol of nucleotide residue; and DNA, 6 pmol TCDD per mol of nucleo-tide residue. The maximum covalent binding of TCDD to DNA was calculated to be about one molecule TCDD per DNA equivalent to 35 cells. This level of binding is 4 to 6 orders of magnitude lower than that of most chemical carcinogens.

In a 2-year chronic toxicity study, TCDD was found to produce an increase in the incidence of hepatocellular carcinomas and squamous cell carcinomas of the lung, hard palate/nasal turbinates, or tongue in rats (Kociba et al, 1978). However, the data reported by Poland and Glover (1979) suggest that the carcinogenic activity of TCDD may not be mediated by its covalent binding to DNA resulting in a somatic mutation.

Distribution of TCDD During Pregnancy and Postpartum

Tissue distribution studies have detected only small quantities of radioactivity from [^{14}C]-TCDD in rat fetuses following maternal exposure to the toxin (Moore et al, 1976). This study revealed that high levels of unmetabolized TCDD were excreted in milk and that each pup actually received a higher dose (μg TCDD/kg) during the first week after birth than was administered initially to the mother. These results correlate well with the finding of postnatal stimula-tion of hepatic microsomal enzymes following administration of TCDD to pregnant rats (Lucier et al, 1975). Foster mother experiments demonstrated that the postnatal inductive effects in rat pups resulted mostly from exposure of newborns to TCDD via maternal milk, although some inductive effect was found to be related solely to prenatal exposure to TCDD (Lucier et al, 1975). Both studies indicate that while TCDD crosses the placenta in the rat, exposure of the offspring occurs mainly though nursing as unmetabolized TCDD is excreted in relatively high levels in milk.

Significance of Distribution Data

In making an interspecies comparison of tissue distribution data in the literature, it should be noted that different doses of TCDD were administered and that tissue levels were analyzed at different intervals following treatment. Thus, the validity of direct comparisons of the tissue distribution data between species is questionable. With these limitations in mind, the available data suggest that tissue levels of TCDD alone appear to play little role in the interspecies differences in pathology and LD_{50} observed on administration of TCDD. For example, the hepatotoxicity observed in the rat (Jones and Butler, 1974) and mouse (Vos et al, 1974) at near LD_{50} doses of TCDD does not appear to be due solely to the relatively high hepatic levels of TCDD in these species. A similar relative concentration of TCDD (percent dose/liver) is observed in the hamster liver (Table 2) with no accompanying toxicity. The unique resistance of the hamster to the lethal effects of TCDD may allow for single exposures of 10 to 100 times more TCDD than given mice and rats, respectively. The resulting molar concentration of TCDD in the hamster liver may thus be orders of magnitude greater than in the rat and mouse, without the hamster developing signs of hepatotoxicity (Olson et al, 1980b).

ELIMINATION OF TCDD

Elimination of TCDD Following Chronic Low Level Exposure

Rose et al, (1976) described the elimination of radioactivity from rats which were administered repeated oral doses of 0.01, 0.1, or 1.0 µg [^{14}C]-TCDD/kg/day, Monday through Friday for 7 weeks. These investigators found considerable variation between individual rats, with half-life values for the elimination of TCDD-derived radioactivity ranging from 16 to 35 days. Estimates for the fraction of the dose absorbed and the rate constant for elimination over the 7-week study were obtained by fitting the body burden versus time data for each rat to a one-compartment open model exhibiting an apparent first-order rate of elimination. The whole-body half-life of [^{14}C] activity was 23.7 days. The steady-state body burden was calculated to be 21.3 percent D for rats given a daily dose of D, 5 consecutive days per week for an infinite number of weeks. By week 7 the body burden of TCDD in the body had reached 74 ± 4 percent of the ultimate steady-state value. It was calculated that 93 percent of the ultimate steady-state value would be attained if the experiment were continuted for 13 weeks. This study indicated that rats which received 0.01 µg TCDD/kg/day would not accumulate TCDD in tissues to levels which are associated with toxicity in rats. Thus it appears unlikely that TCDD would continue to accumulate

indefinitely in the tissues of rats exposed chronically to low levels
of the toxin, as has been observed with other persistent halogenated
aromatic hydrocarbons (Bickel and Muehleback, 1980).

Elimination of TCDD Following a Single Exposure

Rose et al (1976) also examined the elimination of a single oral
dose of 1.0 µg [^{14}C]-TCDD/kg in rats. The percentage of the dose
remaining in the body (body burden) as a function of time followed
apparent first-order kinetics. A mean body burden half-life of 31 +
6 days (mean + SD) was obtained from the six rats in this study. A
similar rate for the first-order elimination of TCDD was also observed
in other studies in rats, where higher doses of TCDD were administered
(Piper et al, 1973; Allen et al, 1975).

The guinea pig is the test species most sensitive to acute TCDD
toxicity (Schwetz et al, 1973; McConnell et al, 1978). Gasiewicz and
Neal (1979) reported the rate of excretion of TCDD-derived radioactivity in guinea pigs following a single i.p. dose of 0.5 µg [^{3}H]-
TCDD/kg. The cummulative excretion of [^{3}H] appeared to be linear for
23 days, resulting in an extrapolated half-life for elimination of
30.3 + 5.8 days. There was therefore no difference in the half-life
of TCDD in the guinea pig (Gasiewicz and Neal, 1979) and rat (Rose
et al, 1976), suggesting that different elimination rates could not
explain the species difference which was observed in toxicity. The
linearity of the excretion data in the guinea pig also suggested that
elimination may be a zero-order process in this species (Gasiewicz
and Neal, 1979). It should be considered, however, that the dose
which was used in these studies caused considerable toxicity and
adipose tissue mobilization. The mobilized [^{3}H]-TCDD contained in
this adipose tissue may have had considerable effect upon the observed
elimination rate.

The unusual resistance of the hamster to TCDD toxicity (Olson et
al, 1980b) also led to studies on the kinetics of TCDD elimination in
this species. The elimination of TCDD-derived radioactivity was
examined in hamsters given a single oral or i.p. treatment with [^{3}H]
or [^{14}C]-TCDD at a dose (650 µg/kg) which produced thymic atrophy but
no lethality or loss of body weight (Olson et al, 1980a). At day 35,
a total of 85 + 6.2 percent of the administered radioactivity had been
excreted (Figure 1). When the [^{3}H] body burden for each day was calculated and plotted semilograithmically (not shown) as a function of
time, the elimination appeared to be similar to the first-order
process observed earlier for the rat. The mean half-life for elimination of TCDD-derived radioactivity in the hamster was found to be
10.8 + 2.4 days (mean + SD) following the i.p. administration of
[^{14}C]-TCDD and 12.0 + 2.0 and 15.0 + 2.5 days following i.p. and p.o.
treatment with [^{3}H]-TCDD, respectively. The enhanced rate of excretion of TCDD-derived radioactivity by the hamster compared to that of

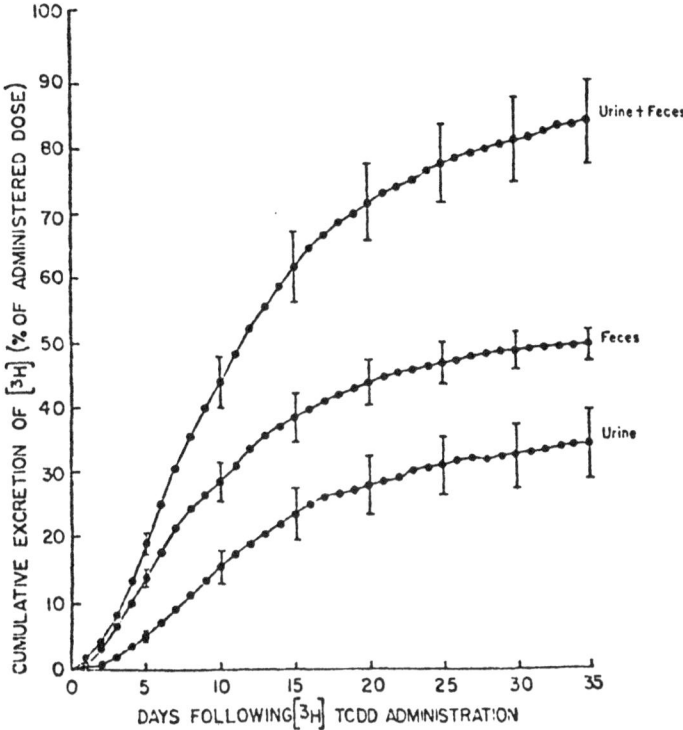

Figure 1.   The cumulative urinary and fecal elimination of [³H]
            from hamsters administered a single i.p. dose of 650 µg
            [³H]-TCDD/kg. Each data point is the percentage of the
            administered dose and represents the mean ± SD of 5
            animals (Olson et al, 1980a)

the rat and guinea pig, may contribute to, but does not appear to
explain totally the marked resistance of the hamster to TCDD toxicity.

    The excretion of TCDD in C57BL/6J and DBA/2J mice has also been
examined (Gasiewicz et al, 1981). These strains are considered to
be responsive and non-responsive, respectively, to aryl hydrocarbon
hydroxylase (AHH) induction by 3-methylcholanthrene and TCDD (Poland
and Glover, 1975). Acute, single-dose 30-day LD50 values of 620 and
132 µg TCDD/kg were found in the DBA/2J and C57BL/6J mice, respectively (Table 3). Thus, as compared to the "non-responsive " DBA/2J
strain, the "responsive" C57BL/6J mice are more sensitive to the
acute lethal as well as the inductive effect of TCDD. It was of additional interest that the LD50 value in the male B6D2F$_1$/J offspring
was approximately 300 µg/kg, or somewhat intermediate between that
of the C57BL/6J and DBA/2J parents. Intermediate responses of the
B6D2F$_1$/J offspring to TCDD have been observed for AHH induction
(Poland and Glover, 1975) as well as cleft palate production and
thymic atrophy (Poland and Glover, 1980).

Table 3.   Rates of Elimination and Toxicity of TCDD in Various Species

Species	Dose (µg/kg)	Half-life for Elimination (days)(t 1/2)	Relative % of TCDD-derived Radioactivity		LD50 (µg/kg)	Reference
			Urine	Feces		
Guinea pig	2.0 (i.p.)	30[b]	94	6	2	(Gasiewicz and Neal, 1979)
Rat	1.0 (oral)	31	>99	<1	60	(Rose et al, 1976)
Mouse:						
C57BL/6J	10.0 (i.p.)	17[b]	74	26	132	(Gasiewicz et al, 1981)
DBA/2J	10.0 (i.p.)	37[b]	70	30	620	(Gasiewicz et al, 1981)
B6D2F1/J[a]	10.0 (i.p.)	17[b]	73	27	300	(Gasiewicz et al, 1981)
Hamster	650 (i.p.)	11	59	41	>3000	(Olson et al, 1980a)
Hamster	650 (oral)	15	--	--	1157	(Olson et al, 1980a)

[a] Offspring of C57BL/6J and DBA/2J which are heterozygous at the Ah locus.
[b] Time at which 50 percent of original dose of [3H] was eliminated. This value is not to be confused with t 1/2 which is applicable to a component(s) of elimination following first-order kinetics.

Following a single intraperitoneal dose of [³H]-TCDD (10 µg/kg),
it was observed that the time at which 50 percent of the original
dose of [³H] was eliminated was approximately 17 days for the C57BL/6J
mice as compared to 37 days in the DBA/2J strain (Figure 2, Table 3).
The cumulative excretion of [³H] from [³H]-TCDD-treated B6D2F₁/J mice
was found to be remarkably similar to that from the C57BL/6J strain
(Figure 2). When the [³H]-body burden for each day was calculated
and plotted semilogarithmically (not shown), first-order elimination
kinetics was observed. However, in all cases the elimination appeared to be a function of two components, fast and slow (Table 4).
The half-time of elimination (t1/2) values for each component were
similar for the three strains of mice. However, the extrapolated
values for the relative contributions of each component to the total
excretion were markedly different. In the C57B2/6J and B6D2F₁/J mice,
approximately 35 and 65 percent of the total elimination were contributed by the fast and slow components, respectively. In the DBA/
2J mice, however, these relative contributions were approximately
12 and 88 percent, respectively (Table 4). Thus, in these strains of
mice the difference in the excretion of [³H] following [³H]-TCDD
treatment (Figure 2) appears to be due to a difference in the relative contribution of the first-order components of this elimination
rather than a marked difference in the t1/2 values for these components.

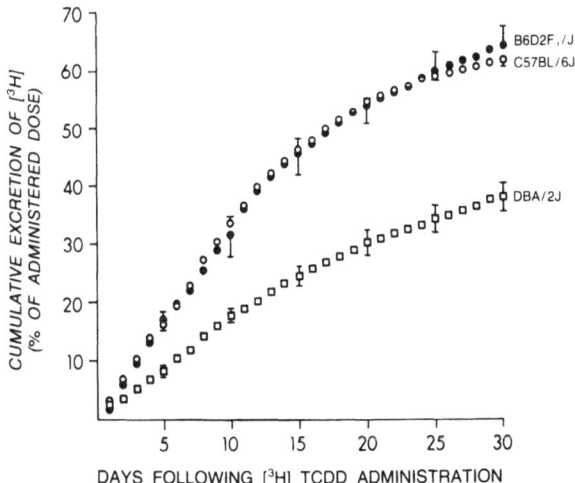

Figure 2.   The cumulative total elimination of [³H] from C57BL/6J,
             DBA/2J, and B6D2F₁/J mice administered a single i.p. dose
             of 10 µg [³H]-TCDD/kg. Each data point is the percentage
             of the original dose and represents the mean ± SD of
             three separate cages containing 4 animals per cage.

Table 4.   Estimated Half Lives of [³H] Clearance from C57BL/6J,
           DBA/2J, and B6D2F₁/J Mice Following a Single Intra-
           peritoneal Dose of 10 µg [³H]-TCDD/Kg[a]

Strain		Estimated Half-Life (t1/2)(day)	Percent Contribution To Total Elimination
C57BL/6J	Fast Component	8.5 ± 0.2	38.2 ± 2.3
	Slow Component	42.8 ± 3.7	62.8 ± 2.0
DBA/2J	Fast Component	8.9 ± 0.8	11.2 ± 2.0
	Slow Component	57.5 ± 7.8	98.2 ± 2.4
B6D2F₁/J	Fast Component	8.5 ± 1.1	35.4 ± 2.0
	Slow Component	32.7 ± 5.9	66.2 ± 2.0

Data represent the mean + S.D. of 3 determinations per time
point (see Figure 2). Each determination per time point was
performed on excreta from 4 mice.

It is possible that differences in metabolic capability of these
strains of mice may lead to differences in the metabolism of TCDD
and thus alterations in the elimination times. A dose of 10 µg
TCDD/kg has been found to give maximal or near maximal induction of
AHH activity in all three strains of mice (Poland and Glover, 1975).
To determine if the relative pattern of metabolites formed from TCDD
may be makedly different in C57BL/6J, B6D2F₁/J, and DBA/2J strains
of mice, the urine and bile from these mice treated with 10 µg [³H]-
TCDD/kg were examined by high performance liquid chromatography
(Figure 3 and 4). These data demonstrate that metabolic products of
[³H]-TCDD rather than free [³H]-TCDD are excreted in the urine and
bile from these strains of mice, and there appears to be no substantial difference in the pattern of the metabolites formed which
might explain the difference in excretion times (Figure 2).

The differences in elimination times may be in part related to
the finding that the DBA/2J mice possess approximately twice as much
total adipose tissue stores than either the C57BL/6J or B6D2F₁/J
strains (Table 5). As noted earlier the adipose tissue is a major
site of storage of TCDD. Thus, the sequestration of TCDD by the
larger adipose tissues stores in the DBA/2J strain may indeed be
responsible for the much greater contribution of the slow component
to the total elimination of [³H] as compared to the C57BL/6J and
B6D2F₁/J strains. Further pharmacokinetic analysis of the tissue
concentrations of TCDD in these mice are needed to determine the
importance of this observation. Decad et al (1981) found similar

relative differences in the half-life of elimination values of $[^{14}C]$-2,3,7,8-tetrachlorodibenzofuran and of relative adipose tissue stores in the C57BL/6J and DBA/2J strains.

These studies in mice demonstrate that other factors in addition to metabolism may influence the rate of elimination of TCDD. Furthermore, the relationship between toxicity and rate of elimination of TCDD in these strains of mice appears to be the opposite of that expected. Table 3 summarizes the LD_{50} and half-life for elimination data for TCDD in various species and three strains of mice. These data suggest no clear relationship exists between the ability of a given species to excrete TCDD and/or its metabolites and the toxicity of TCDD in that species as determined by the LD_{50} values.

Table 5. Adipose Tissue Content of Untreated Male Mice[a]

Strain	Body Weight (g)	Adipose Tissue[b]
C57BL/6J (6)	23.1 ± 1.6	5.90 ± 1.31
DBA/2J (6)	23.9 ± 1.9	11.48 ± 3.31
B6D2F$_1$/J (6)	24.3 ± 1.2	5.03 ± 1.62

[a]Results represent the mean $\pm$ S.D. of (n) determinations. Determinations of adipose tissue content were made by dissection.
[b]Percent of body weight.

Contribution of Various Routes of Elimination of TCDD

Piper et al (1973) have reported on the excretion of radioactivity in feces, urine, and expired air of rats which were administered $[^{14}C]$-TCDD. In this study, 53, 13, and 3 percent of the administered radioactivity was eliminated via the feces, urine and expired air, respectively, over a 21-day period. In another study with $[^{14}C]$-TCDD-treated rats (Rose et al, 1976), fecal excretion accounted for most if not all of the elimination of TCDD derived radioactivity, since essentially no $[^{14}C]$-activity was detected in either urine or expired air. As noted previously, lactating rats are also capable of excreting TCDD via their milk (Moore et al, 1976). Table 3 summarizes the relative percentages of the dose of radioactivity eliminated via the urine and the feces in various species administered radiolabeled-TCDD. Hamsters and mice, in contrast to guinea pigs and rats, excrete a large percentage of the eliminated radioactivity by way of the urine (approximately 41 and 26-30 percent, respectively)(Table 3).

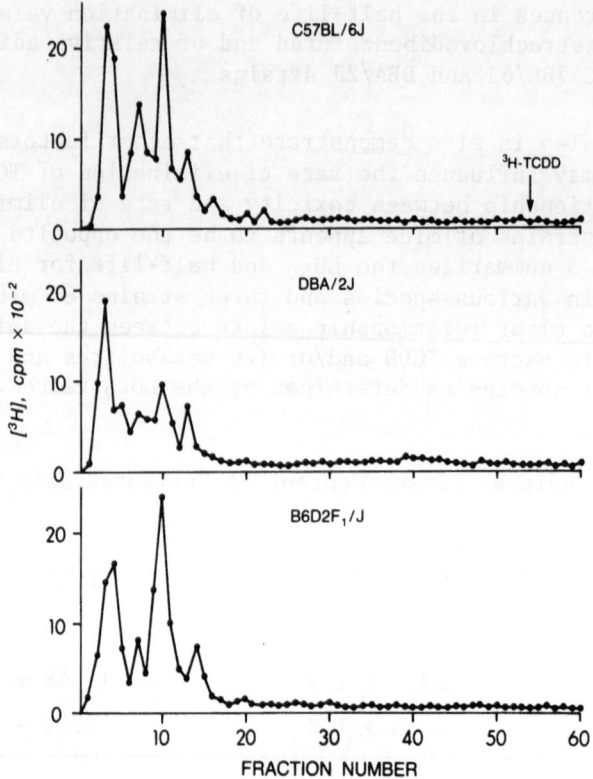

Figure 3.   High-performance liquid chromatographic separation of
            [³H]-activity contained in urine from C57BL/6J, DBA/2J,
            and B6D2F₁/J mice which were treated 7 days earlier with
            10 µg[³H]-TCDD/kg.   The Zorbax-ODS (6.2 mm x 25 cm)
            column was eluted using a 45 min. linear gradient of 0 to
            100 percent methanol in water at a flow rate of 2.2 ml/
            min/fraction.   Under these chromatographic conditions,
            [³H]-TCDD elutes in fractions 52-54.

METABOLISM OF TCDD

Chemical Nature of In Vivo Excretion Products

    Some of the earlier studies with radiolabeled TCDD found no
evidence that TCDD was metabolized either in vivo or in vitro (Fries
and Marrow 1975; Vinopal and Casida, 1973; Piper et al, 1973).
Studies on the elimination of [¹⁴C]-TCDD in rats detected the excretion of a small portion of the administered radioactivity in the
urine and expired air (Piper et al, 1976). This finding suggested
that some metabolism of TCDD may be occurring. In 1979, the excretion of unidentified metabolites of TCDD in rat bile was confirmed

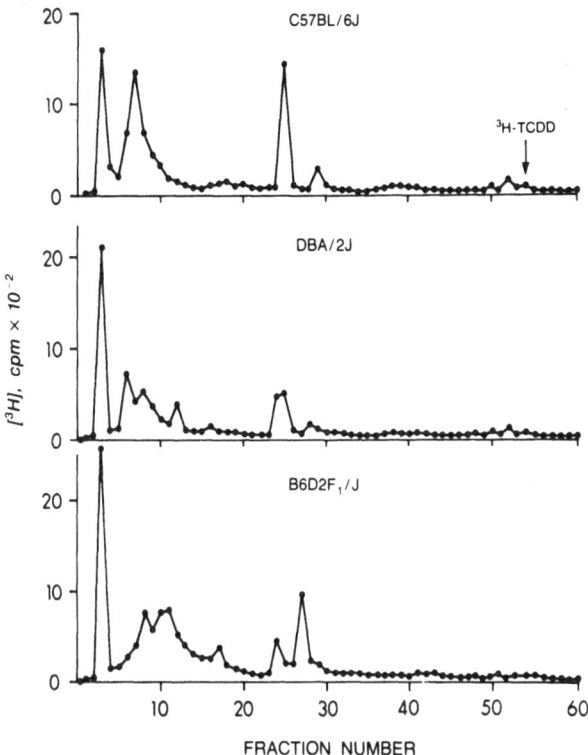

Figure 4.   High-preformance liquid chromatographic separation of
            [^{3}H]-activity contained in bile from C57BL/6J, DBA/2J,
            and B6D2F$_1$/J mice which were treated 7 days earlier
            with 10 µg [^{3}H]-TCDD/kg. The chromatographic conditions
            are given in Figure 3.

through chromatography, providing the first reports of the biological
degradation of TCDD (Ramsey et al, 1979; Poiger and Schlatter, 1979).
Both investigations suggested that only metabolites of TCDD were
eliminated in the bile of treated rats and that these metabolites
may be glucuronide conjugates.

    The chemical nature has been examined of the radioactive excre-
tion products in the urine, feces, and bile of [^{3}H]- and [^{14}C]-TCDD-
treated hamsters (Olson, et al, 1981, Neal et al, in press).   The
radioactive products analyzed by high-performance liquid chromatography (hplc) and thin-layer chromatography (tlc) and were found
to correspond to several polar metabolites of TCDD.   Figures 5 and 6
show the hplc elution profiles of [^{14}C] contained in urine and bile
of hamsters administered [^{14}C]-TCDD.   No radioactivity present in
the urine or bile eluted from the column at the position of [^{14}C]-
TCDD (fractions 46 and 47), suggesting that all of the radioactivity

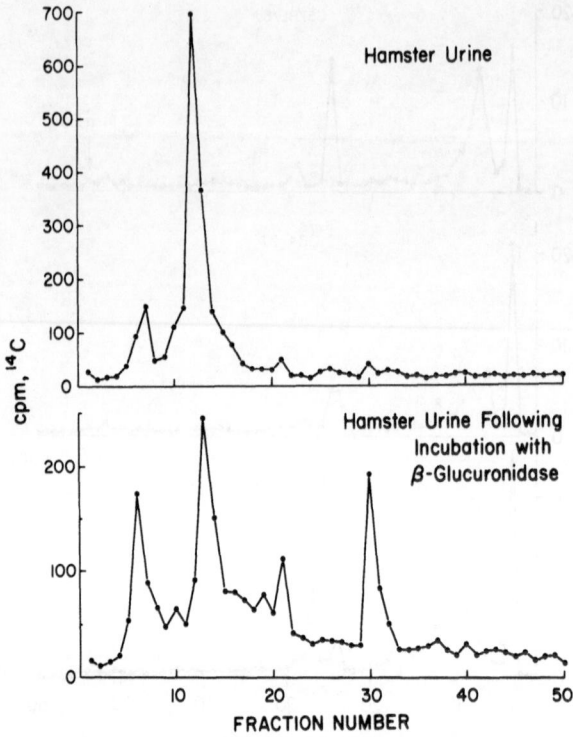

Figure 5.   High-performance liquid chromatographic separation of
            [^{14}C]-labeled products contained in urine from hamsters
            which were treated 7 days earlier with 500 µg [^{14}C]-
            TCDD/kg. The Altex Ultrasphere ODS (10 mm x 25 cm)
            column was eluted using a 30 min. linear gradient of 99
            percent water to 1 percent water in methanol at a flow
            rate of 3.0 ml/min. Fractions were collected at one min.
            intervals and the radioactivity in each fraction was
            quantitated. Under these chromatographic conditions,
            [^{14}C]-TCDD elutes in fractions 46 and 47 (Neal et al,
            in press).

was contained in metabolites of TCDD. The bottom panels of these
figures show the hplc elution profiles obtained following the incubation of urine and bile with β-glucuronidase. In these chromatograms a greater percentage of the radioactivity was observed in
the more nonpolar area (fractions 30-40) of the profile as compared
to samples which had not been treated with β-glucuronidase. No
alterations of the elution profiles were observed on chromatography
of urine and bile after incubation with aryl sulfatase. These results suggest that, in the hamster, a large portion of the biotrans-
formed TCDD was excreted in the urine and bile as glucuronide conjugates. In contrast to the TCDD-free excretion products of hamster

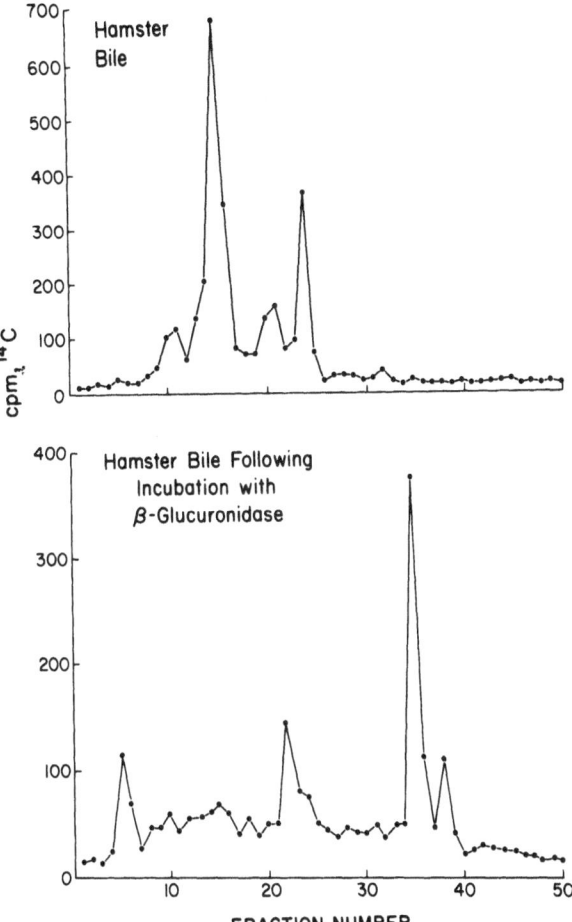

Figure 6.   High-performance liquid chromatographic separation of
            [^{14}C]-labeled products contained in bile from hamsters
            which were administered 500 μg [^{14}C]-TCDD/kg 7 days
            earlier.   The chromatographic conditions are given in
            Figure 5.   Under these chromatographic conditions, [^{14}C]-
            TCDD elutes in fractions 46 and 47 (Neal et al, in press).

urine and bile, extracts of fecal material from these animals were
found to contain unchanged TCDD as well as metabolites (Olson and
Neal, 1980).   Unchanged TCDD accounted for about 25-45 percent of
the radioactivity extracted from the feces of the hamster treated
4-9 days earlier with an i.p. dose of radiolabeled TCDD.   Thus, it
appears that a significant quantity of unchanged TCDD may enter the
intestinal lumen by a route other than the biliary pathway.

Urinary and biliary excretion products of [^{14}C]-TCDD-treated rats and guinea pigs have also been investigated (Olson and Neal, 1980). Figures 7 and 8 show the hplc elution patterns of [^{14}C] contained in the urine and bile of [^{14}C]-TCDD-treated rats and guinea pigs, respectively. None of the radioactivity eluted at the position of [^{14}C]-TCDD, suggesting that, as in the hamster and mouse, all of the radiolabeled products excreted in the urine and bile of rats and guinea pigs represent metabolites of TCDD. It is evident from Figures 3-8 that the elution times of the various radioactive metabolites in urine and bile are somewhat different in various species, suggesting some species differences in metabolism. The significance of these differences will have to await further work on the structural identification of these metabolites. The important point of these studies is that TCDD is converted to more polar metabolites prior to elimination in the urine and bile of guinea pigs, rats, mice, and hamsters (Rose et al, 1976; Olson et al, 1980a; Gasiewicz et al, 1981).

The apparently rapid elimination of TCDD metabolites upon formation suggests that the in vivo half-life for elimination of radiolabeled TCDD could provide a good estimate of the rate of TCDD metabolism in a given species. Unfortunately, this simple relationship cannot be used since, as noted earlier, the hamster eliminates significant levels of unmetabolized TCDD in the feces for a number of days following an i.p. administration of [^{14}C]-TCDD (Olson and Neal, 1980). Thus, the significance of metabolism to the half-life for elimination of TCDD in various species (Table 3) should be interpreted cautiously.

Metabolism of TCDD by Hepatic Microsomes

TCDD metabolism in in vitro mammalian systems has also been examined. In one of the earliest studies, Vinapal and Casida (1973) were unable to detect metabolism of TCDD by microsomal preparations from mouse, rat, and rabbit liver. They concluded that the chlorine atoms appear to greatly impede the metabolism of TCDD, since the nonhalogenated congener, dibenzo-p-dioxin itself, was rapidly converted to polar metabolites by liver microsomes (Vinopal and Casida, 1973). In other studies, unextractable radioactivity has been found to be bound to microsomes following the in vitro incubation of radiolabeled TCDD with rat (Nelson et al, 1977) and mouse (Guenthner et al, 1979) hepatic microsomes. NADPH was found to be a necessary cofactor for the binding of radioactivity to the microsomal protein of mouse hepatic microsomes incubated with [^{3}H]-TCDD (Guenthner et al, 1979). In addition, carbon monoxide and alpha-naphthoflavone were found to decrease, while 3-methylcholanthrene pretreatment increased the irreversible binding of TCDD-derived radioactivity to microsomal protein. Although TCDD metabolites could not be extracted from the incubations, these authors presented the binding data as

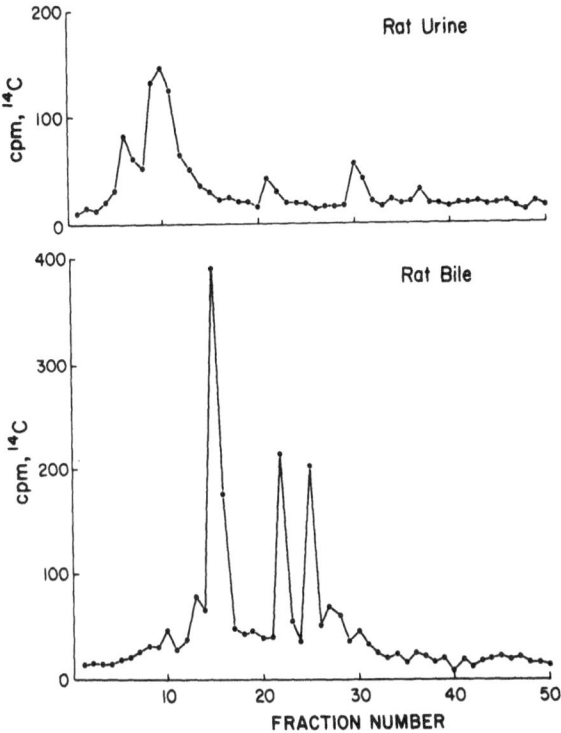

Figure 7. High-performance liquid chromatographic separation of
           [^{14}C] contained in the urine and bile of rats adminis-
           tered 500 μg [^{14}C]-TCDD/kg 7 days earlier. The chroma-
           tographic conditions are given in Figure 5. Under these
           chromatographic conditions, [^{14}C]-TCDD elutes in fractions
           46 and 47.

evidence for the metabolism of TCDD by the mouse liver cytochrome
P-450-containing monooxygenase system. A recent study has examined
the metabolism of highly purified [^{3}H]-TCDD by hamster liver microsomes (Neal et al, in press). In these experiments, [^{3}H]-TCDD was
purified by high-performance liquid chromatography since in in vitro
studies exceedingly small amounts of radioactive metabolites are
detected. These experiments confirmed the earlier studies, demonstrating that NADPH was an essential cofactor needed for irreversible
binding of TCDD-derived radioactivity to hepatic microsomal protein
(Neal et al, in press). Preliminary hplc analysis of radioactive
extracts from these incubations indicated the presence of metabolites
of TCDD although the amount of radioactivity was too low to warrant
additional experiments designed to identify the metabolites.

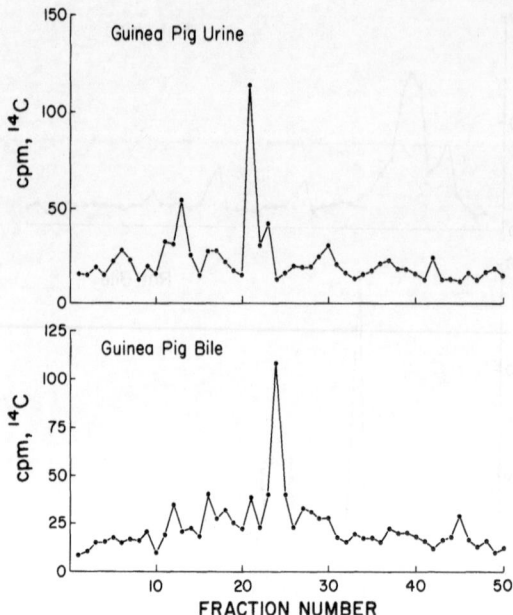

Figure 8.   High-performance liquid chromatographic separation of
[^{14}C]-activity contained in the urine and bile of guinea
pigs administered 500 μg [^{14}C]-TCDD/kg 7 days earlier.
Chromatographic conditions are given in Figure 5. Under
these chromatographic conditions, [^{14}C]-TCDD elutes in
fractions 46 and 47.

Metabolism of TCDD by Isolated Hepatocytes

    .Hepatocytes isolated from rats and hamsters have been used to
study the metabolism of [^{3}H]-TCDD (Olson et al, 1981, Neal et al,
in press). Using hamster hepatocytes, metabolites of TCDD could be
readily detected (Neal et al, in press). The high-performance liquid
chromatographic (hlpc) elution profile of radioactive products present
in the cell-free medium of an incubation of [^{3}H]-TCDD with hamster
hepatocytes is shown in Figure 9. The data indicate the presence
of at least four metabolites. The large peak at fractions 46 and 47
corresponds to unmetabolized [^{3}H]-TCDD. Radioactive peaks at fractions 10 and 24-27 correspond in retention time to metabolites found
in hamster urine (Figure 5) and bile (Figure 6), respectively. The
radioactive peak at fraction 5, in Figure 9 may represent [^{3}H]-H$_2$0.
A similar peak has been found on chromatography of the urine and bile
of [^{3}H]-TCDD-treated hamsters (Olson and Neal, 1980) and mice (Figures 3 and 4), and in the bile of [^{3}H]-TCDD-treated rats (Poiger and
Schlatter, 1979). The lower panel of Figure 9 shows the elution
profile of radiolabeled products from an aliquot of cell-free medium
which had been incubated with β-glucuronidase. As was the case with

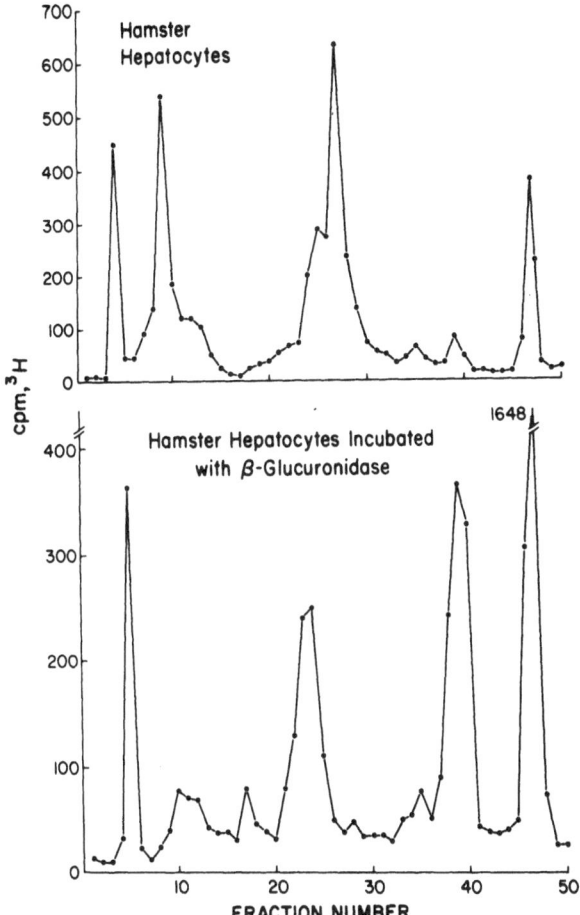

Figure 9. High-performance liquid chromatographic separation of
          [³H]-labeled products contained in the cell-free medium
          of an incubation of isolated hamster hepatocytes with
          [³H]-TCDD (2.0 µM). The chromatographic conditions are
          given in Figure 5. Under these chromatographic conditions,
          [³H]-TCDD elutes in fractions 46 and 47 (Neal et al,
          in press).

TCDD metabolites in hamster urine and bile (Figures 5 and 6), some
of the in vitro metabolites appear to be present as glucuronides.
Control incubations using boiled hepatocytes or no hepatocytes were
found to contain only [³H]-TCDD. Thus, isolated hamster hepatocytes
appear to be an effective in vitro model to examine TCDD metabolism.
In addition, they have the ability to form TCDD metabolites with
similar hplc elution times as those found in urine and bile.

Role of Cytochrome P-450 Monooxygenase Systems in TCDD Metabolism

Additional studies were performed to assess the role of the cytochrome P-450 monooxygenase systems in TCDD metabolism (Olson et al, 1981; Neal et al, in press). Rats were pretreated with a hepatic mixed-function oxidase (MFO) inducer, phenobarbital (PB) (80 mg/kg/day, i.p. x 3 days), or a single dose of TCDD (5 µg/kg, i.p.), or with the heme synthesis inhibitor $CoCl_2$ (60 mg/kg/day, s.c. x 2 days). TCDD or PB pretreatment was found to produce an increase in the rate of TCDD metabolite formation while $CoCl_2$ pretreatment reduced the rate of TCDD metabolism from that observed in the hepatocytes of untreated rats (Figure 10). Thus, TCDD has the ability to induce its own metabolism. Hepatocytes from control and PB-pretreated rats were also incubated with [^{3}H]-TCDD in the presence of SKF 525-A (0.1 mM) or metyrapone (0.5 mM). These well-known inhibitors of drug metabolism reduced the rate of TCDD metabolism without producing any change in the viability of the hepatocytes (Figure 11). Thus, the relative rate of TCDD metabolite formation in isolated rat hepatocytes appears to respond to drug-induced alterations in hepatic MFO activity. These results (Olson et al, 1981; Neal et al, in press) and the results of others (Beatty et al, 1978; Guenthner et al, 1979) suggest that TCDD is metabolized by the hepatic cytochrome P-450 monoxygenase systems.

Chemical Structure of TCDD Metabolites

At present, the chemical structure of the TCDD metabolites remains unknown. It is possible that extrahepatic metabolism of TCDD is also taking place since metabolite profiles from urine have generally been found to be different from that of bile.

Incubation with β-glucuronidase of metabolites from urine, bile, and hepatocyte incubations results in products with longer hplc retention times (Figures 5, 6 and 9), implying that one or more metabolites may be glucoronide conjugates. The presence of glucuronides suggests the formation of phenolic derivatives of TCDD. It is possible that an arene oxide of TCDD is the initial intermediate formed. This product could then be converted enzymatically or nonenzymatically to dihydrodiol or phenolic derivatives of TCDD. A number of reactive electrophilic arene oxide intermediates have been proposed to be responsible for the irreversible binding of TCDD-derived radioactivity to hepatic protein and other macromolecules (Guenthner et al, 1979).

Role of Metabolism in the Toxicity of TCDD

Although there have been no studies of the toxicity of TCDD metabolites, indirect evidence suggests that metabolism may be a detoxification process. Starting with the assumption that TCDD is

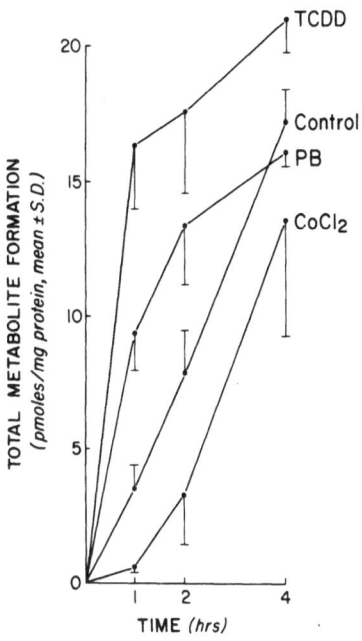

Figure 10.   The time course of TCDD metabolite formation in incuba-
              tions containing [^{3}H]-TCDD (2.0 μM) and isolated rat
              hepatocytes. The cumulative rate of metabolite formation
              in untreated controls is compared with that of rats
              pretreated with phenobarbital (PB), TCDD, or CoCl$_2$.
              Each data point represents the mean + SD of three experi-
              ments using hepatocytes isolated from different animals
              (Olson et al, 1981).

metabolized by the hepatic mixed-function oxidase (MFO) enzyme system,
one study examined the effects of altering hepatic MFO activity on
the toxicity (20-day LD$_{50}$) of TCDD in rats (Beatty et al, 1978).
These investigators used naturally occurring age and sex-related
differences in hepatic MFO activity in rats and observed that animals
with lower MFO activity also had a lower 20-day LD$_{50}$ for TCDD (see
Table 6). On the other hand, the LD$_{50}$ value for TCDD in weanling
rats was found to be increased by pretreatment with the MFO inducers,
phenobarbital and 3-methylcholanthrene. The observed relationship
between hepatic MFO activity and the LD$_{50}$ for TCDD, led to the speculation that TCDD was metabolized by the hepatic MFO enzyme system in
rats and that this metabolism resulted in the formation of a chemical
species which was less toxic than the parent compound [20]. This
study thus suggested that metabolism was a detoxification process.
Verification of this hypothesis must await the demonstration of increased excretion of TCDD metabolites in response to elevated hepatic
MFO activity, and the direct assessment of the toxicity of these
metabolites.

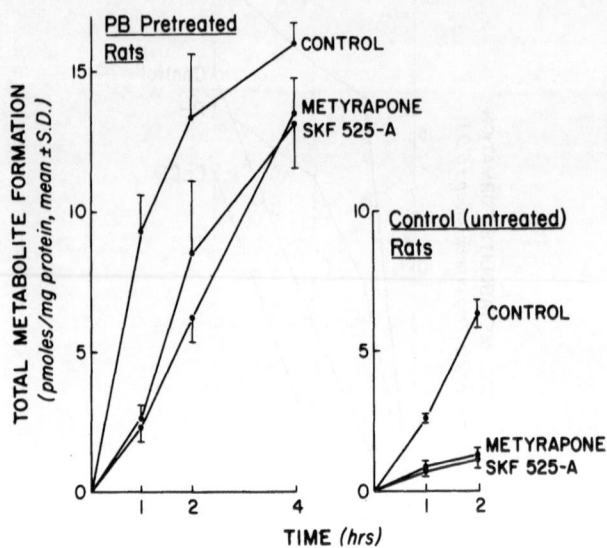

Figure 11.   Time course of TCDD metabolite formation during incuba-
             tion of [³H]-TCDD (2.0 µM) with isolated rat hepatocytes
             from control (untreated) and phenobarbital (PB)-pretreated
             rats. The incubations were performed in the presence
             and absence (control) of SKF 525-A (0.1 mM) or metyrapone
             (0.5 mM). Each data point represents the mean ± SD of
             three experiments using hepatocytes isolated from differ-
             ent animals (Olson et al, 1981).

Table 6.   Alteration of 20-Day LD_{50} Value of TCDD in Rats
            with Age, Sex and Various Pretreatments[a]

Animals and/or Pretreatment	LD_{50} (µg/kg ± S.D.)
Adult female	24.6 ± 1.4
Adult male	60.2 ± 5.6
Weanling male	
no pretreatment	25.2 ± 1.4
3-MC pretreatment[b]	44.1 ± 1.2
PB pretreatment[b]	40.9 ± 1.3

[a]From Beatty et al, 1978.
[b]Abbreviations are 3-MC, 3-methylcholanthrene; PB, phenobarbitol.

CONCLUSIONS

As shown by the recent studies of Poiger and Schlatter (1980), the nature of the formulation containing TCDD regulates, to a large extent, the absorption of the toxin following oral or dermal exposure. Combining TCDD with soil or activated carbon appears to inhibit the rate of absorption of TCDD. These and additional data suggest the formulation of TCDD is of greater importance than the route of exposure in assessing the relative risk of a given exposure to TCDD.

The liver and adipose tissue appear to be the major storage sites for TCDD in most species. Studies in rats and hamsters have indicated that the TCDD in these tissues is the parent compound rather than metabolites. The toxicological significance of the tissue distribution of TCDD is still unclear. At this time, interspecies differences in TCDD distribution do not appear to be related to species differences in the acute toxic effects observed upon exposure to TCDD.

TCDD-derived radioactivity appears to be eliminated by most species through a first-order process. The hamster eliminates TCDD at a faster rate than other species examined. The greater rate of elimination of TCDD by the hamster may in part contribute to, but not totally explain, its unusual resistance to the acute toxicity of TCDD. In addition, studies in various strains of mice show that the strain (DBA/2J) which is least susceptible to the acute lethal effects of TCDD has the slowest elimination rate. This result is opposite to . that expected from earlier studies in the rat, guinea pig and hamster, where the hamster eliminates TCDD at a significantly higher rate and is least susceptible to the acute lethal effects.

In most species, TCDD-derived radioactivity was found to be largely eliminated in the feces. In the hamster and mice, the urine was also found to be a major route of excretion. All of the TCDD-derived radioactivity excreted in the urine and bile of hamsters, guinea pigs, rats and mice appears to be in the form of metabolites of TCDD. The apparent absence of TCDD metabolites in liver and fat would suggest that once formed, the metabolites of TCDD are readily excreted. Although urine and bile appear to be free of unmetabolized TCDD, data from the hamster indicate that a significant amount of unchanged TCDD may be excreted into the intestinal lumen by some route other than bile for a number of days following treatment. In addition, the experiments with mice suggest that other factors such as the amount of adipose tissue stores may influence the total rate of elimination of TCDD and/or its metabolites. Thus, the role of metabolism in the half-life for elimination of TCDD must be interpreted with caution.

In contrast to hepatic microsomes, isolated hepatocytes are a good _in vitro_ model for study of the metabolism of TCDD. Incubation

of TCDD with hepatocytes produces metabolites of TCDD with similar
hplc elution times as those found for metabolites of TCDD excreted
in urine and bile. The relative rate of TCDD metabolite formation
in isolated rat hepatocytes was found to be increased by prior administration of inducers of the hepatic mixed-function oxidase (MFO)
enzymes. These and other studies suggest that TCDD is metabolized by
the cytochrome P-450-containing monooxygenases.

At present the chemical structures of the TCDD metabolites are
unknown. Incubation with β-glucuronidase of metabolites found in
urine, bile and hepatocyte incubations indicates some of the metabolites may be glucuronide conjugates.

REFERENCES

Allen, J.R., Van Miller, J.P. and Norback, D.H. Tissue distribution,
    excretion, and biological effects of [^{14}C] tetrachlorodibenzo-p-
    dioxin in rats. Food Cosmet. Toxicol., 13:501-505, 1975.
Beatty. P.W., Vaughn, W.K. and Neal, R.A. Effect of alteration of
    rat hepatic mixed-function (MFO) activity on the toxicity of
    2,3,7,8-tetrachlorodibenzo-p-dioxin (TCDD). Toxicol. Appl.
    Pharmacol., 45:513-519, 1978.
Bickel, M.H. and Muehleback, S. Pharmacokinetics and ecodisposition
    of polyhalogenated hydrocarbons. Drug Metab. Rev., 11:149-190,
    1980.
Crosby, D.G. and Wong, A.S. Environmental degradation of 2,3,7,8-
    tetrachlorodibenzo-p-dioxin (TCDD). Science, 195:1337-1338,1976.
Decad, G.M., Birnbaum, L.S. and Matthews, H.B. Disposition of
    tetrachlorodibenzofuran in mice. Toxicologist, 1:65, 1981.
diDomenico, A., Silano, V., Viviano, G. and Zapponi, G. Accidental
    release of 2,3,7,8-tetra-chlorodibenzo-p-dioxin (TCDD), Seveso
    Italy. V. Environmental persistence of TCDD in soil. Exotoxicol.
    Environ. Safety, 4:339-345, 1980a.
diDomenico, A., Silano, V., Viviano, G. and Zapponi, G. Accidental
    release of 2,3,7,8-tetra-chlorodibenzo-p-dioxin (TCDD), Seveso,
    Italy. II. TCDD distribution in the soil surface layer. Ibid,
    4:298-320, 1980b.
diDomenico, A., Silano, V., Viviano, G. and Zapponi, G. Accidental
    release of 2,3,7,8-tetra-chlorodibenzo-p-dioxin. VI. TCDD levels
    in atmospheric particles. Ibid., 4:346-356, 1980c.
Firestone, D. The TCDD problem: a review. Ecol. Bull., (Stockholm),
    27:39-57, 1978.
Fries, G.F. and Marrow, G.S. Retention and excretion of 2,3,7,8-
    tetrachlorodibenzo-p-dioxin by rats. J. Agr. Food Chem., 23:
    265-269, 1975.

Gasiewicz, T.A. and Neal, R.A. 2,3,7,8-Tetrachlorodibenzo-p-dioxin
    tissue distritubion, excretion, and effects on clinical chemical
    parameters in guinea pigs. Toxicol. Appl. Pharmacol., 51:329-
    339, 1979.

Gasiewicz, T.A., Gieger, L.E. and Neal, R.A. Unpublished observations,
    1981.

Goldstein, J.A. The structure-activity relationships of halogenated
    biphenyls as enzyme inducers. Ann. N.Y. Acad. Sci., 320:164-
    178, 1979.

Guenthner, T.M., Fysh, J.M. and Nebert, D.W. 2,3,7,8-Tetrachloro-
    dibenzo-p-dioxin: Covalent binding of reactive metabolic inter-
    mediates pricipally to protein in vitro. Pharmacology, 19:12-
    22, 1979.

IARC. IARC Monographs on the Evaluation of the Carcinogenic Risk of
    Chemicals to Man, i.e., Some Fumigants, the Herbicides 2,4-D and
    2,4,5-T Chlorinated Dibenzodioxins and Miscellaneous Industrial
    Chemicals. Lyon, 1977.

Jones, G. and Butler, W.H. A morphological study of the liver lesion
    induced by 2,3,7,8-tetrachlorodibenzo-p-dioxin in rats. J.
    Pathol., 112:93-97. 1974.

Kimbrough, R.D. The toxicity of polychlorinated polycyclic compounds
    and related chemicals. CRC Crit. Rev. Toxicol., 2:445-498, 1974.

Kociba, R.J., Keyes, D.G., Beyer, J.E., Carreon, R.M., Wade, C.E.,
    Dittenber, D., Kalnins, R., Frauson, L., Park, C.N., Barnard,
    S.D., Hummel, R.A. and Humiston, C.G. Results of a two-year
    chronic toxicity and oncogenicity study of 2,3,7,8-tetrachloro-
    dibenzo-p-dioxin in rats. Toxicol. Appl. Pharmacol., 46:279-
    303, 1978.

Lucier, G.W., Sonawane, B.R., McDaniel, O.S. and Hook, G.E.R.
    Postnatal stimulation of hepatic microsomal enzymes following
    administration of TCDD to pregnant rats. Chem-Biol. Interactions,
    11:15-26, 1975.

McConnell, E.E. Acute and chronic toxicity, carcinogenesis, repro-
    duction, teratogenesis, and mutagenesis in animals. In:
    Kimbrough, R.D., ed. Halogenated Biphenyls, Terphenyls,
    Naphthalenes, Dibenzodioxins and Related Products. New York:
    Elsevier/North Holland Biomedical Press, 1980. pp. 109-150.

McConnell, E.E. and Moore, J.A. Toxicopathology characteristics of
    the halogenated aromatics. Ann. N.Y. Acad. Sci., 320:138-150,
    1979.

McConnell, E.E., Moore, J.A., Haseman, J.K. and Harris, M.W. The
    comparative toxicity of chlorinated dibenzo-p-dioxins in mice
    and guinea pigs. Toxicol. Appl. Pharmacol. 44:335-356, 1978.

Moore, J.A., Harris, M.W. and Albro, P.W. Tissue distribution of
    [^{14}C] tetrachlorodibenzo-p-dioxin in pregnant and neonatal rats.
    Toxicol. Appl. Pharmacol., 37:146, 1976.

Nelson, J.O., Menzer, R.E., Kearney, P.C. and Plummer, J.R.
    2,3,7,8-Tetrachlorodibenzo-p-dioxin: In Vitro binding to rat
    liver microsomes. Bull. Environ. Cont. Toxicol., 18:9-13, 1977.
Nestrick, T.J., Lamparski, L.L. and Townsend, D.I. Identification
    of tetrachlorodibenzo-p-dioxin isomers at the 1-ng level by
    photolytic degradation and pattern recognition techniques. Anal.
    Chem., 52:1865-1874, 1980.
Olson, J.R. and Neal, R.A. Unpublished observations, 1980.
Olson, J.R., Gasiewicz, T.A. and Neal, R.A. Tissue distribution,
    excretion, and metabolism of 2,3,7,8-tetrachlorodibenzo-p-dioxin
    (TCDD) in the Golden Syrian hamster. Toxicol. Appl. Pharmacol.,
    56:78-85, 1980a.
Olson, J.R., Holscher, M.A. and Neal, R.A. Toxicity of 2,3,7,8-
    tetrachlorodibenzo-p-dioxin (TCDD) in the Golden Syrian hamster.
    Toxicol. Appl. Pharmacol., 55:67-78, 1980b.
Olson, J.R., Gudzinowicz, M. and Neal, R.A. The in vitro and in vivo
    metabolism of 2,3,7,8-tetrachlorodibenzo-p-dioxin (TCDD) in the
    rat. Toxicologist, 1:69, 1981.
Piper, W.N., Rose, J.Q. and Gehring, P.J. Excretion and tissue
    distribution of 2,3,7,8-tetrachlorodibenzo-p-dioxin in the rat.
    Environ. Health Perspect., 5:111-118, 1973.
Poiger, H. and Schlatter, C. Biological degradation of TCDD in rats.
    Nature, 281:706-707, 1979.
Poiger, H. and Schlatter, C. Influence of solvents and adsorbants
    on dermal and intestinal adsorption of TCDD. Food Cosmet.
    Toxicol., 18:477-482, 1980.
Poland, A. and Glover, E. Genetic expression of arylhydrocarbon
    hydroxylase by 2,3,7,8-tetrachlorodibenzo-p-dioxin. Evidence
    for a receptor mutation in genetically non-responsive mice.
    Mol. Pharmacol., 11:389-398, 1975.
Poland, A. and Glover, E. An estimate of the maximum in vivo
    covalent binding of 2,3,7,8-tetrachlorodibenzo-p-dioxin to rat
    liver protein, ribosomal RNA, and DNA. Cancer Res., 39:3341-
    3344, 1979.
Poland, A. and Glover, E. 2,3,7,8-Tetrachlorodibenzo-p-dioxin:
    Segregation of toxicity with the Ah locus. Mol. Pharmacol.,
    17:86-94, 1980.
Poland, A. and Kende, A. 2,3,7,8-Tetrachlorodibenzo-p-dioxin:
    Environmental contaminant and molecular probe. Fed. Proc.,
    35:2404-2411, 1976.
Ramsey, J.C., Hefner, J.G., Karbowski, R.J., Brown, W.H. and
    Gehrung, P.J. The in vivo biotransformation of 2,3,7,8-tetra-
    chlorodibenzo-p-dioxin (TCDD) in the rat. Toxicol. Appl.
    Pharmacol., 48:A162, 1979.
Reggiani, G. Medical problems raised by the TCDD contamination of
    Seveso, Italy. Arch. Toxicol., 40:161-187, 1978.
Reggiani, G. Acute human exposure to TCDD in Seveso, Italy.
    J. Toxicol. Environ. Health, 6:27-43, 1980.

Rose, J.Q., Ramsey, J.C., Mentzler, T.A., Hummel, R.A. and Gehring, P.J. The fate of 2,3,7,8-tetrachlorodibenzo-p-dioxin following single and repeated oral doses to the rat. Toxicol. Appl. Pharmacol., 36:209-226, 1976.

Schwetz, B.A., Norris, J.M., Sparschu, G.K., Rowe, V.R., Gehring, P.J. Emerson, J.L. and Gerbig, G.C. Toxicology of chlorinated dibenzo-p-dioxins. Environ. Health Perspect., 5:87-99, 1973.

Stehl, R.A., Popenfuss, R.R., Bredeweg, R.A. and Roberts, R.W. The stability of pentachlorophenol and chlorinated dioxins to sunlight, heat and combustion. In: Blair, E.A. ed. Chlorodioxins-Origin and Fate, Advances in Chemistry Series 120. Washington, D.C.: American Chemical Society, 1973. pp. 119-125.

Van Miller, J.P., Marlor, R.J. and Allen, J.R. Tissue distribution and excretion of tritiated tetrachlorodibenzo-p-dioxin in non-human primates and rats. Food Cosmet. Toxicol., 14:31-34, 1976.

Vinopal, J.H. and Casida, J.E. Metabolic stability of 2,3,7,8-tetrachlorodibenzo-p-dioxin in mammalian liver microsomal systems in living mice. Arch. Environ. Contam. and Toxicol., 1:122-132, 1973.

Vos, J.G., Moore, J. A. and Zinkl, J. G. Toxicity of 2,3,7,8-tetrachlorodibenzo-p-dioxin (TCDD) in C57BL/6 mice. Toxicol. Appl. Pharmacol., 29:229-241, 1974.

MORPHOLOGY OF LESIONS PRODUCED BY THE DIOXINS AND RELATED COMPOUNDS

Renate D. Kimbrough

Centers for Disease Control
U.S. Department of Health and Human Services
Atlanta, Georgia USA

ABSTRACT

    2,3,7,8-Tetrachlorodibenzodioxin (TCDD), other chlorinated
dibenzodioxins and related compounds, such as the chlorinated
biphenyls, chlorinated naphthalenes, brominated biphenyls, and
chlorinated dibenzofurans, affect a variety of organ systems in
different species. The organ primarily affected in rodents and
rabbits is the liver. In guinea pigs, the liver shows only mild
changes, but instead there is pronounced atrophy of the thymus and
lymphatic tissue. Chloracne-like lesions can be produced in rabbits,
subhuman primates, and the hairless mouse. Hyperkeratosis of the skin
is also prominent in cattle and horses. Changes of the nails have
been reported in humans and subhuman primates, and similar changes
have been observed in the hoofs of cattle. The gastrointestinal tract
and the gall bladder are primarily affected in subhuman primates,
while hyperplasia of the transitional epithelium of the urinary tract
can be observed in cattle, subhuman primates, and guinea pigs.

INTRODUCTION

    The chemical 2,3,7,8-tetrachlorodibenzodioxin and a number of
related halogenated aromatic compounds, such as the polychlorinated
biphenyls, other chlorinated dibenzodioxins, chlorinated dibenzofurans, chlorinated naphthalenes, and brominated biphenyls cause similar lesions in experimental and domestic animals. These
morphological alternations vary among species and vary also with dose.
In order to produce the same type of lesions, for instance, much less
of the 2,3,7,8-tetrachlorodibenzodioxin or the 2,3,7,8-tetrachloro-
dibenzofuran would have to be given than of a brominated biphenyl

527

mixture. In a number of animal species, the liver weight may be increased and the weight of the thymus, particularly in young animals, may be decreased. All of these compounds can cause weight loss, but toxic doses of the 2,3,7,8-tetrachlorodibenzodioxin in particular can cause severe emaciation. This is very prominent in guinea pigs. It has also been observed in horses which were accidentally exposed to TCDD. Similar observations were made in cows that had ingested polybrominated biphenyls and have also been reported for subhuman primates and mink. The specific organs that are preferentially affected in different species are listed in Table 1.

Table 1. List of Organs that are Preferentially Affected
in Different Species.

Species	Organs
Rat [1]	liver, thyroid
Mouse	liver
Rabbit	liver, skin
Guinea pig, rat and mouse pups	thymus, lymph nodes
Monkey	bone marrow, skin, gastro-intestinal tract
Chicken	generalized edema, vascular system
Horse	skin, liver gastrointestinal tract, hoofs
Cow	skin, hoofs, squamous meta-plasia of epithelium at different sites, urinary bladder, thyroid?

[1] Studies recently completed by the National Toxicology Program (1981) showed that Swiss Webster mice dermally exposed to TCDD at doses of 0.005 µg/female mouse and 0.001 µg/male mouse developed fibrosarcomas of the integumentary system (males: 3/42 for the vehicle, 6/28 for TCDD treated, 6/30 for TCDD + DMBA) (females: 2/41 vehicle controls, 8/27 TCDD treated, and 8/29 TCDD + DMBA treated).

THE LIVER

Depending on the concentration given and the number of doses, a variety of changes can be produced in the liver. At the lowest dose, the liver becomes enlarged and increased in weight. Microscopically, the hepatocytes are enlarged with smooth cytoplasm and eosinophil inclusions. Occasionally, pleomorphism is seen and sometimes a few mitotic figures are also noted. If these cells are examined with the electron microscope, it is found that the enlargement of the hepatocytes is due to an increase in smooth endoplasmic reticulum. The inclusions consist of concentrically arranged membranes which usually surround vacuoles (Figure 1). The vacuoles are thought to contain lipids. The mitrochondria may be enlarged and may show an increased number of abnormally arranged cristae (Figure 2). Usually, enlargement of the hepatocytes starts in the area of the central vein. At higher dose levels, the hepatocytes become vacuolated and necrosis of single liver cells is noted (Figure 3). This is particularly prominent in portal triads. There may be an increase in fibrous tissue and reticulum fibers. An increase in macrophages is noted and the Kupffer cells become more prominent. These cells often contain a brown pigment. At this stage, the liver may be darker in color. Part of the dark discoloration is caused by the presence of porphyrins within the liver. Such livers, if held under UV light, will fluoresce pink. However, in addition to the porphyrin pigment, another pigment is present. It was noted, for instance, that livers of rats fed technical pentachlorophenol, which contains chlorinated dibenzodioxins and chlorinated dibenzofurans, turned almost black (Kimbrough and Linder, 1978). It has not been determined thus far what the nature of this black pigment is. Particularly in rabbits, after a single toxic dose, the areas of necrosis can be quite extensive in the liver, and large areas of hemorrhage, as well as bile duct proliferation, are noted. If rabbits survive the toxic effects of TCDD, a repair process occurs which, after a period of weeks, results in an almost normal-appearing liver. The morphological changes which remain longest are multinucleated hepatocytes and a fibrosis surrounding the central veins. Similar observations have also been made in the rat (Jones and Butler, 1974). Horses with accidental exposure to TCDD for several weeks had even more extensive fibrosis around the central vein of the liver (Kimbrough, et al, 1977). In addition, necrosis of liver cells was noted. The fibrosis around the central vein seemed to originate in the vascular walls and at least in the horse, is reminiscent of veno-occlusive disease. Within the hepatocytes, a yellowish-green pigment was noted which probably represented bile. One horse had thrombosis of a larger blood vessel of the liver. Morphological changes in two cats which also died in this accidental poisoning episode showed primarily degeneration of liver cells in the centrolobular areas and a brown pigment in the macrophages of the liver (Kimbrough, et al, 1977). Rats fed technical pentachlorophenol over a period of about 8 months also showed lipids accumulating in many areas of the liver

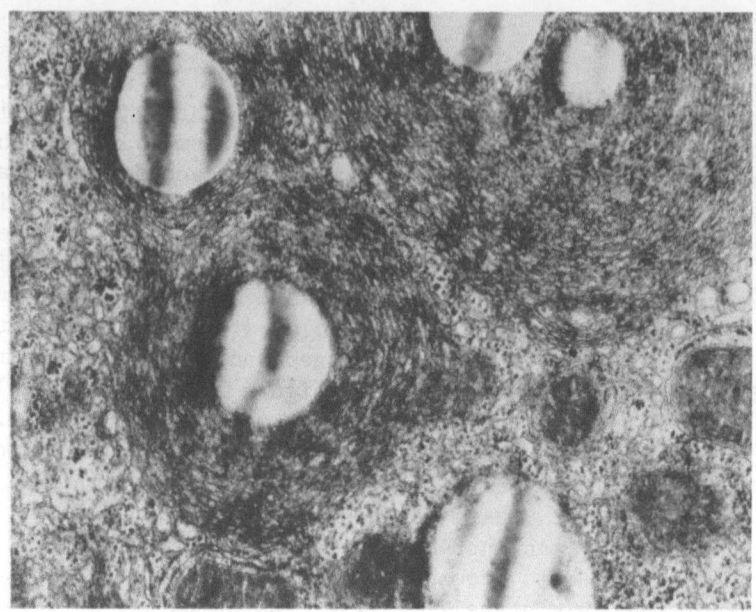

Figure 1. Electron micrograph illustrating concentrically
           arranged membranes within the cytoplasm of a
           hepatocyte. Rat was fed Aroclor 1254, a poly-
           chlorinated biphenyl (Mag X 51,300).

(Kimbrough and Linder, 1978). In addition, hepatocytes had been
replaced by fibrosis and focal inflammatory infiltrates. Multinucleated cells as well as certain amount of pleomorphism were
observed (Figure 4). Areas containing high concentrations of a very
dark pigment were also noted.

     If a compound of this type is fed continuously to rodents at dose
levels at which no systemic toxicity is noted, hepatocellular carcinomas will develop over a two-year feeding period. A variety of
lesions are observed in the livers of the dosed animals. Many of the
livers will show areas of alteration consisting of poorly circumscribed areas covering several liver lobules with either enlarged
hepatocytes or hepatoytes that appear much smaller. The cytoplasm
of these hepatocytes can either be eosinophilic or basophilic, or may
have ground-glass appearance. In addition, a number nodules may be
noted grossly in these livers. They again cover several liver lobules
and are composed of the same types of cells as those of the areas of
alteration, but they are usually well circumscribed and compress the
surrounding liver parenchyma. These nodules, which in the past used
to be referred to as hyperplasia, are now termed neoplastic nodules
(Kimbrough, et al, 1975) by a number of authors.

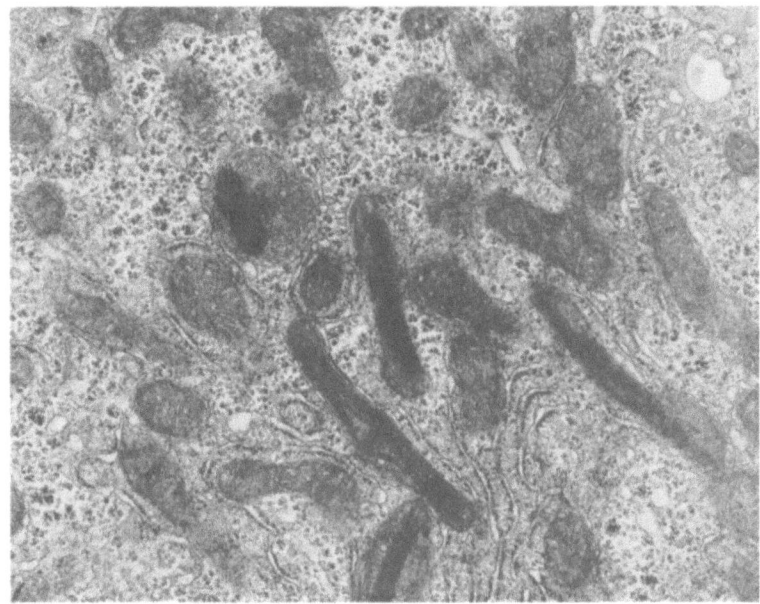

Figure 2.   Electron micrograph illustrating atypical mitochondria
with an increase in cristae. Section is from the liver
of a rat fed Aroclor 1260, a polychlorinated biphenyl
(Mag x 51,300).

In a certain proportion of these livers, hepatocellular carcinomas will also be observed. These are either well-differentiated or
they may be very undifferentiated (Kociba, et al, 1978). Most of
the studies that have demonstrated the carcinogenic potential of
chlorinated dibenzodioxins and chlorinated biphenyls have been lifetime feeding studies. Polybrominated biphenyls (PBB) are much more
persistent in the body than the other compounds just mentioned. It
is possible to produce hepatocellular carcinomas in rats with a single
dose of 1000 mg/kg PBB (Kimbrough, et al, 1981). Only with TCDD,
not with the other compounds, has it also been possible to produce
cancers of the lung, the soft and hard palate, and the nasopharynx
in rats (Kociba, et al, 1978).

In addition to the hepatocellular carcinomas, a lesion which
has been termed adenofibrosis and which consists of proliferated
glandular epithelium surrounded by either fibrous tissue or collagen
has been noted in many of the livers of rats (Figure 5) or mice fed
either chlorinated biphenyls, brominated biphenyls or related products (Kimbrough and Linder, 1974; Kimbrough, et al, 1973; and
Kimbrough, et al, 1972). The significance of this lesion is not
clear although it is usually observed concomitantly with hepato-

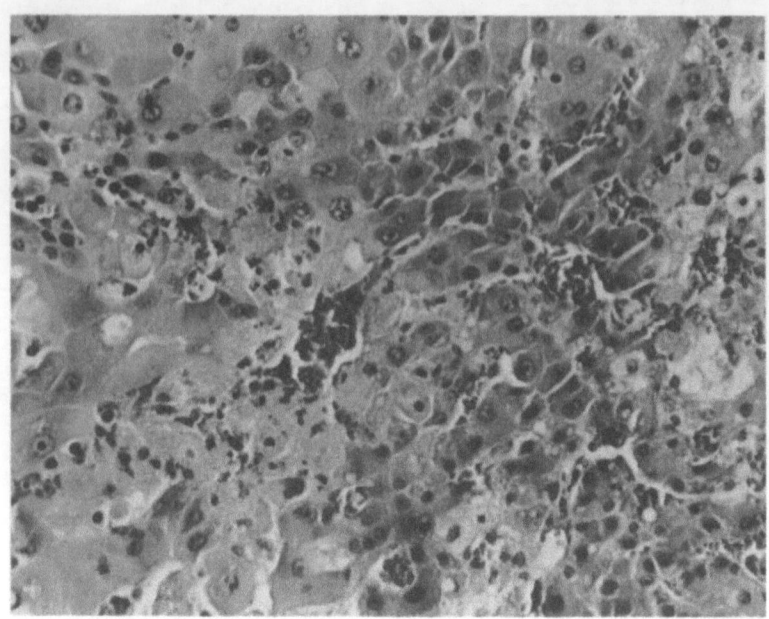

Figure 3.   Photomicrograph of section from a liver of a rabbit
            exposed dermally to TCDD. Note pleomorphism, in-
            flammatory cells, degenerated hepatocytes (HXE x 100).

cellular carcinomas. With the chlorinated biphenyls, clusters of
small epithelial cells were noted that resembled atopic pancreatic
tissue with the liver (Kimbrough, 1973). The significance of this
lesion is also not clear.

THE SKIN

    In humans and subhuman primates, a skin lesion can be produced
either by topical application of these materials or by systemic
administration. In human subjects this skin lesion is usually
referred to as chloracne. The lesion consists of elevated cysts
which give the skin a very rough appearance. At times, a hyperkeratosis can also be noted. There may be brown discoloration of the
skin. These lesions can become secondarily infected. In some
episodes, thickening of the nails has been reported in people. This
is particularly true for the recent outbreak in Taiwan where people
consumed rice oil contaminated with chlorinated dibenzofurans,
chlorinated biphenyls, terphenyls and quaterphenyls (Wong, 1981).
In the episode in Taiwan, as well as the episode in Japan, dark
discoloration of the skin was also noted. The eyes may be affected.

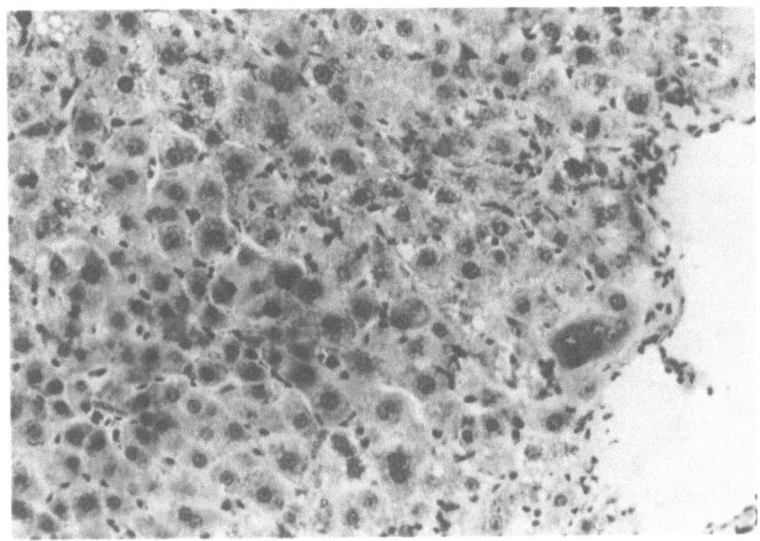

Figure 4. Photomicrograph of a section from a liver of a rat fed
technical pentachlorophenol. Note the large multinucleated liver cell, the pleomorphism and the inflammatory cells in the sinusoids (HXE x 100).

They may show swelling, and in subhuman primates, the eyelashes may
actually fall out. Involvement of the meibomian glands is usually
also noted in conjunction with eye lesions (McConnell, et al, 1978).

A lesion like chloracne can be produced by applying these types
of compounds to the inner ear of rabbits (Figure 6). The enlarged
elevated hair follicles can be noted on gross observation. Microscopically, they consist of markedly dilated hair follicles which
are filled with keratin. The sebaceous glands are partially or
completely atrophied. The epithelium surrounding the hair follicles
and also that covering the skin surface may have proliferated during
the early stage of these lesions, and in later stages the epithelium,
particularly in the hair follicles, may become atrophic. The microscopic appearance of the human skin lesions is similar to that
observed in rabbits. In humans, a secondary inflammatory reaction
is often more prominent than in rabbits, although this can also be
noted in rabbit ears. This hyperkeratotic lesion has also been
produced on the backs of rabbits and in hairless mice. In most
other experimental animals, it has not been possible to produce
such as a skin lesion. Cows will develop a very severe hyperkeratosis which consists of an increase in keratin and hair loss after
exposure to chlorinated naphthalenes and brominated biphenyls. The

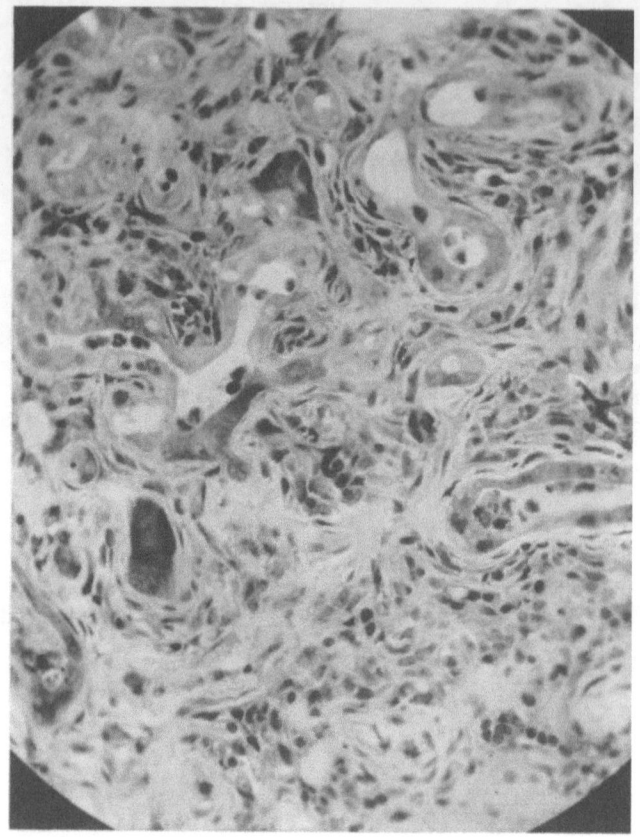

Figure 5.   Photomicrograph of a section of a liver from a rat fed
             Aroclor 1254 (PCB). Note proliferated glandular
             epithelium surrounded by fibrous tissue (methylene
             blue stain X 300).

animals that were exposed to the brominated biphenyls also showed
changes in their hoofs which might be equivalent to the changes in
the nails observed in subhuman primates and in humans. In horses
that were accidentally exposed to chlorinated dibenzodioxins, hyperkeratotic skin lesions and hair loss were also noted. The microscopic appearance of the sections of skin from these horses showed
primarily hyperkeratosis and parakeratosis.

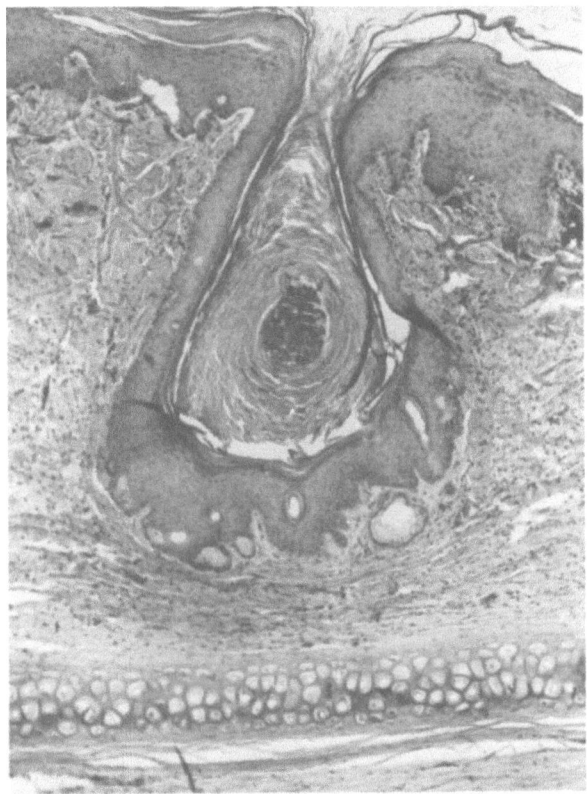

Figure 6. Photomicrograph of a section from a rabbit ear that
had been painted with TCDD. Note the proliferation
of the epithelium, dilated hair follicle filled with
keratin and atrophy of the sebaceous glands.

MORPHOLOGICAL CHANGES IN OTHER ORGANS

In a few species, particularly the guinea pig, but also rat and
mice pups, atrophy of the thymus can be produced. The decrease in
size of the thymus is almost entirely due to a loss of the cortical
lymphocytes resulting from necrosis of the lymphocytes. The clinical ramifications of this lesion are expressed by changes in the
immune system (Gupta, et al, 1973; Faith and Moore, 1977). Similarly,
the weight of the spleen and the lymphoid tissue in the gastrointestinal tract may be greatly reduced. Hypoplastic changes have also
been observed in bone marrow (Allen and Carstens, 1967). Marked
hyperplasia and inflammation, as well as the formation of mucous-
filled cysts, have been reported in the gastrointestinal tract and
the gallbladder of subhuman primates (Allen and Norback, 1977; Allen,

et al, 1974).   In the subhuman primate, large multiple papillary
neoplastic-like lesions may be seen in the glandular portion of the
stomach.   Areas of ulceration are often found within and between these
growths.   Microscopically, these lesions are characterized by hyperplasia and metaplasia.   The glands are almost entirely composed of
mucous cells.   Areas of ulceration in the glandular stomach have also
been noted in rats acutely poisoned with these compounds (Kimbrough,
et al, 1978) and in horses (Kimbrough, et al, 1977).   The urinary
tract may be affected in monkeys, guinea pigs, and cattle.   Again,
hyperplasia of the transitional epithelium is noted.   This hyperplasia
may involved the terminal portions of the collecting ducts of the
renal medulla, the renal pelvis, the ureter, and the urinary bladder
(McConnell, et al, 1979).   Occasionally, testicular atrophy has been
noted in animals showing other signs of systemic poisoning.   However,
with lower dose levels that do not result in pronounced weight loss,
this lesion is usually not observed.

An effect on the thyroid gland has been reported in rats after
exposure to polychlorinated or polybrominated biphenyls (Collins,
et al, 1977, and Sleight, et al, 1978) and in cattle after exposure
to technical pentachlorophenol (McConnell, et al, 1980).   The morphological changes consisted of smaller thyroid follicles and a decrease
in thyroxin levels.

REFERENCES

Allen, J. R., and Carstens, L. A.   Light and electron microscopic
        observations in Macacca mulatta monkeys fed toxic fat.   Am.
        J. Vet. Res., 28: 1513-1526, 1967.
Allen, J. R., Carstens, L. A., and Barsotti, D. A.   Residual effects
        of short-term low level exposure of nonhuman primates to poly-
        chlorinated biphenyls.   Toxicol. Appl. Pharmacol., 30: 440-451,
        1974.
Allen, J. R., and Norback, D. H. Carcinogenic potential of the poly-
        chlorinated biphenyls.   In: Hiatt, H. H., Watson, J. D. and
        Winsten, J. A. (Eds.), Origins of Human Cancer.   Cold Spring
        Harbor Labs., New York, 173-186, 1977.
Collins, W. T., Capen, C. C., Kasza, L., Carter, C., and Daily, R. E.
        Effect of polychlorinated biphenyl (PCB) on the thyroid gland of
        rats.   Am. J. Pathol., 89: 119-136, 1977.
Faith, R. E., and Moore, J. A.   Impairment of thymus dependent immune
        function by exposure of the developing immune sytem to 2,3,7,8-
        tetrachlorodibenzo-p-dioxin (TCDD).   J. Toxicol. Environ. Health,
        3: 451-464, 1977.
Gupta, B. N., Vos, J. G., Moore, J. A., Zinkl, J. G., and Bullock, B.
        C. Pathological effects of 2,3,7,8-tetrachlorodibenzo-p-dioxin
        in laboratory animals.   Environ. Health Perspect., 5: 125-140,
        1973.

Jones, G., and Butler, W. H. A morphological study of the liver
    lesion induced by 2,3,7,8-tetrachlorodibenzo-p-dioxin in rats.
    J. Pathol., 112: 93-97, 1974.

Kimbrough, R. D., Linder, R. E., and Gaines, T. B. Morphological
    changes in the liver of rats fed polychlorinated biphenyls. Arch.
    Environ. Health, 25: 354-364, 1972.

Kimbrough, R. D., Linder, R. E., Burse, V. W., and Jennings, R. W.,
    Adenofibrosis in the rat liver. Arch. Environ. Health, 27: 390-
    395, 1973.

Kimbrough, R. D., Brief Communication: Pancreatic-type tissue in
    livers of rats fed polychlorinated biphenyls. J. Natl. Cancer
    Inst., 51: 679-681, 1973.

Kimbrough, R. D., and Linder, R. E. Induction of adenofibrosis and
    hepatomas of the liver of Balb/cd mice by polychlorinated bi-
    phenyl (Aroclor 1254). J. Natl. Cancer Inst., 53: 547-552,
    1974.

Kimbrough, R. D., Squire, R. A., Linder, R. E., Strandberg, J. D.,
    Montali, R. J., and Burse, V. W. Introduction of liver tumors in
    Sherman strain female rats by polychlorinated biphenyl Aroclor
    1260. J. Natl. Cancer Inst., 55: 1453-1459, 1975.

Kimbrough, R. D., Carter, C. D., Liddle, J. A., Cline, R. E., and
    Phillips, P.E. Epidemiology and pathology of a tetrachloro-
    dibenzo-p-dioxin poisoning episode. Arch. Environ. Health, 32:
    77-86, 1977.

Kimbrough, R. D., and Linder, R. E. The effect of technical and pur-
    ified pentachlorophenol on the rat liver. Toxicol. Appl.
    Pharmacol., 46: 151-162, 1978.

Kimbrough, R. D., Buckley, J., Fishbein, L., Flamm, G., Kasza, L.,
    Marcus, W., Shibko, S., and Teske, R. Animal toxicology. Environ.
    Health Perspect., 24: 173-185, 1978.

Kimbrough, R. D., Groce, D. F., Korver, M. P., and Burse, V. M.
    Induction of liver tumors in female Sherman strain rats by poly-
    brominated biphenyls. J. Natl. Cancer Inst., 66: 535-542, 1981.

Kociba, R. J., Keyes, D. G., Beyer, J. E., Carreou, R. M., Wade, C.
    E., Dittenber, D. A., Kaluins, R. P., Frauson, L. E., Park, C.
    N., Barnard, S. D., Hummel, R. A., and Humiston, C. G. Results of
    a two year chronic toxicity study and oncogenicity study of 2,3,
    7,8-tetrachlorodibenzo-p-dioxin in rats. Toxicity. Appl.
    Pharmacol., 46: 279-303, 1978.

McConnell, E. E., Moore, J. A., and Dalgard, D. W. Toxicity of
    2,3,7,8-tetrachlorodibenzo-p-dioxin in Rhesus monkeys (Macacca
    mulatta) following a single oral dose. Toxicol. Appl. Pharm-
    acol., 43: 175-187, 1978.

McConnell, E. E., Hass, J. R., Altman, N., and Moore, J. A. A
    spontaneous outbreak of polychlorinated biphenyl (PCB) toxicity
    in Rhesus monkeys (Macacca mulatta). Toxicopathology Lab
    Animal Science, 29: 666-673, 1979.

McConnell, E. E., Moore, J. A., Gupta, B. N., Rakes, A. H., Luster, M. I., Goldstein, J. A., Haseman, J. D., and Parker, C. E. The chronic toxicity of technical and analytical pentachlorophenol in cattle. I. Clinicopathology Toxicol. Appl. Pharmacol., 52: 468-490, 1980.

Sleight, S. D., Mangkoewidjojo, S., Akoso, B. T., and Sanger, V. L. Polybrominated biphenyl toxicosis in rats fed an iodine-deficient, iodine-adequate, or iodine-excess diet. Environ. Health Perspect., 23: 341-346, 1978.

Wong, Chu-Kwan. PCB poisoning special issue. Clin. Med. (Taipei), 7: 1-100, 1981.

A CONSIDERATION OF THE MECHANISM OF ACTION OF 2,3,7,8-TETRACHLORO-

DIBENZO-P-DIOXIN AND RELATED HALOGENATED AROMATIC HYDROCARBONS

Alan Poland, Joyce Knutson, and Edward Glover

McArdle Laboratory for Cancer Research
Madison, Wisconsin USA

INTRODUCTION

   2,3,7,8-Tetrachlorodibenzo-p-dioxin (TCDD) is the prototype of
a group of halogenated aromatic hydrocarbons which produce a similar
and characteristic pattern of toxic and biochemical responses. The
most well-studied biochemical response to these compounds is the
induction of activity of the microsomal monooxygenase, aryl hybro-
carbon hydroxylase (AHH). The induction of AHH activity by TCDD and
congeners is mediated through their stereospecific, reversible binding
to a cytosol protein, the induction receptor. Two types of evidence
indicate that the toxic responses produced by these compounds are
also mediated through the cytosol receptor: a) the structure-activity
relationship for halogenated aromatic hydrocarbon congeners to bind
to the receptor corresponds to that for their toxic potency: and b)
several toxic responses produced by TCDD in mice, segregated with
the Ah locus, the genetic locus which determines the receptor. While
toxicity is mediated through the receptor, many tissues in various
animal species and numerous cell types in vitro, contain the receptor
and respond with the induction of AHH activity when challenged with
TCDD, but do not display toxic responses. Based on studies of the
XB cell line in culture, and the epidermal response in HRS/J mice, we
propose a model to explain tissue specificity of toxic responses. It
is proposed that in many cells, TCDD and congeners induce a limited
pleiotropic response, which consists of the induction of AHH activity
and other coordinately expressed drug metabolizing enzymes, however,
in those cells which show toxicity, an additional battery of genes is
expressed which control cell involution, division and/or differentiation. The toxicity of halogenated aromatic hydrocarbons appears to
be due to the sustained expression of a normal cellular regulatory
system, of which we were previously unaware.

2,3,7,8-Tetrachlorodibenzo-p-dioxin (TCDD) serves as the prototype of a large group of halogenated aromatic hydrocarbons (the
chlorinated dibenzo-p-dioxins, dibenzofurans, azo(xy)benzenes, biphenyls and the brominated biphenyls), which are all environmental
contaminants that share a number of properties. These compounds:
a) are all lipophilic, and most are resistant to degradation, resulting in their persistence in the environment and concentration in the
food chain; b) are approximate isostereomers, i.e., have similar
chemical structures; c) produce a similar and characteristic pattern
of toxic responses and biochemical effects; and d) appear to act by
a common mechanism. Our intent in this article is to summarize our
current understanding of the mechanism of action of TCDD and related
halogenated aromatic hydrocarbons.

ENZYME INDUCTION

Of the numerous biochemical and histologic effects produced by
the administration of TCDD and congeners, the most throughly studied
and best understood response is the induction of a battery of enzymes,
which are often called the drug metabolizing enzymes. Of this group,
the induction of the cytochrome P-450-mediated microsomal monoxygenase,
aryl hydrocarbon hydroxylase (AHH), activity, has received the greatest attention. TCDD produces a dose-related increase in AHH activity
in liver and many nonhepatic tissues in vivo (Poland et al, 1974)
and a variety of cell types in vitro (Niwa et al, 1975). The dose
of TCDD which produces one half the maximal induction of AHH activity,
the ED_{50}' in chicken embryo liver is 0.35 nmole/kg (Poland and
Glover, 1973); and similar ED_{50}'s (all about 1 nmole/kg) have been
found for this response in rat and mouse liver. Among a large series
of halogenated dibenzo-p-dioxins tested for their potency to induce
AHH activity the biologically active compounds have: a) halogen
atoms in at least three, and for maximal potency four, of the lateral
ring positions (positions 2,3,7 and 8); b) the order of potency for
substituents is Br > Cl > F; and c) at least one ring position must
be unsubstituted, i.e., the octachloro-isomer is inactive, but both
heptachloro-analogues possess activity. (The structure and ring
numbering position of the dibenzo-p-dioxin molecule and other halogenated aromatic hydrocarbons is depicted in Figure 1).

In subsequent investigations, the potency of halogenated dibenzofuran, azo(xy)benzene and biphenyl congeners to induce AHH activity
was determined in chicken embryo liver (Kende et al, 1974; Poland
et al, 1976a and b; Poland and Glover, 1977) and in rat hepatoma cell
culture (Bradlaw et al, 1980). From these studies one can generalize
the structure-activity relationship for halogenated aromatic hydrocarbons which induce AHH activity: active congeners from all chemical
classes are approximate isostereomers with halogen atoms in 3, or 4
of the lateral ring positions, and roughly fit in a rectangle 3 x 10$\overset{o}{A}$

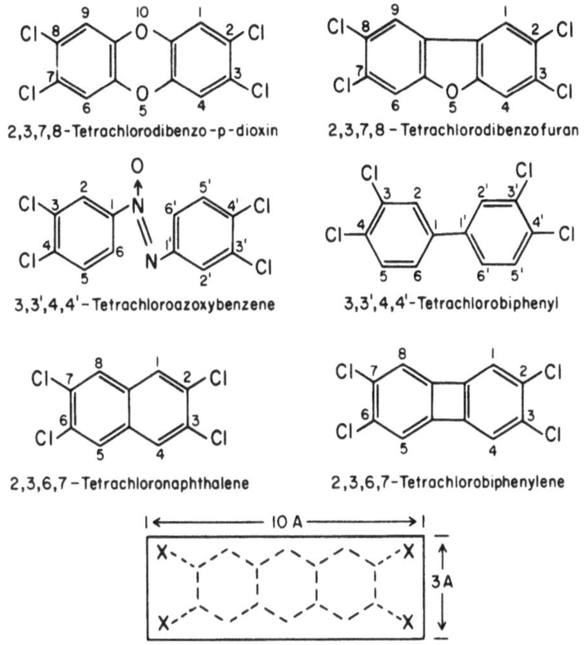

Figure 1. Representative Halogenated Aromatic Hydrocarbons. Tetrachloro-congeners of various classes of halogenated aromatic hydrocarbons demonstrate the isosterism and ring numbering systems of these compounds. The diagram at the bottom depicts the postulated recognition site on the cytosol receptor to which these isosteric compounds bind.

with halogen atoms in the corners of the rectangle. The most significant observation, is that there is excellent correspondence between the potency of halogenated aromatic hydrocarbon congeners to induce AHH activity and their toxic potency.

These observations suggested that there is a sterospecific recognitition site to which halogenated aromatic hydrocarbons bind, and this same site must mediate the induction of AHH activity and the events which result in toxicity.

INDUCTION RECEPTOR

The next step was to isolate and characterize the hypothesized recognition site or receptor for AHH induction, i.e., to look for a macromolecular species which binds ^{3}H-TCDD and which has the in vitro binding properties predicted by the in vivo biology. In the cytosol fraction of C57BL/6J mouse liver and rat liver, a macromolecular species was found which bound ^{3}H-TCDD reversibly with a high affinity

(K_D = 0.27 nM) and which approximated the ED_{50} for AHH induction ($\sim$ 1 nmole/kg) (Poland et al, 1976a). There is a small pool of these sites n = 8.4 x 10^{-14} mole/mg protein, about 8.4 pmole/gm of liver or about 5 x 10^4 molecules/cell. For a large series of halogenated dibenzo-p-dioxins, dibenzofurans, azo(xy)benzene, and biphenyl congeners, the potency of congeners to bind to the cytosol binding species, corresponded to their potency to induce AHH activity. This cytosol binding species also binds polycyclic aromatic hydrocarbons (e.g., 3-methylcholanthrene and benzo(a)pyrene), compounds which, like the halogenated aromatic hydrocarbons, induce AHH activity. Steroid hormones and thyroxine, which have receptors in the liver, do not compete with ^{3}H-TCDD for its binding species. Thus, the cytosol binding species has all the properties expected of the receptor for induction of AHH activity.

Subsequent studies have shown that the cytosol receptor mediates the nuclear uptake of TCDD (Poland et al, 1979) and that the receptor-ligand complex translocates to the nucleus (Okey et al, 1980). The receptor has been demonstrated in a variety of tissues in the rat and mouse and several cell types in culture (Carlstedt-Duke, 1979; Okey et al, 1980), all of which are inducible for AHH activity.

THE Ah LOCUS

The polycyclic aromatic hydrocarbon 3-methylcholanthrene (MC) is the classical compound which induces AHH activity. In the rat TCDD and MC induce hepatic AHH activity to the same extent and with parallel dose-response curves, but TCDD is approximately 3 x 10^4 times more potent (Poland and Glover, 1974).

In outbred and many inbred strains of mice, the administration of MC induces AHH activity and cytochrome P_1-450 in the liver and a number of extrahepatic tissues (Nebert and Gielen, 1972; Nebert et al, 1972; Thomas et al, 1972). However, certain inbred strains of mice fail to respond to the administration of MC, i.e., they show little or no induction of hepatic enzyme activity. The prototype responsive mouse strain is C57BL/6; the prototype nonresponsive strain is DBA/2. In genetic crosses and backcrosses between these two strains, the trait of responsiveness to aromatic hydrocarbons is inherited in a simple autosomal mode, and the locus controlling it is designated the Ah locus.

The potency of TCDD, relative to MC, for inducing AHH activity, prompted an examination of the effects of TCDD on mouse strains that are responsive or nonresponsive to MC. TCDD (1.2 x 10^{-7} mol/kg) induced hepatic AHH activity in all strains of mice tested, regardless of their response to MC (Poland and Glover, 1975). By a number of criteria, the hepatic cytochrome P_1-450 and enzyme activity

induced by MC or TCDD in C57BL/6 mice or by TCDD in DBA/2 mice were
the same.

Mice that are nonresponsive to MC, do respond to the more potent
stimulus TCDD, and therefore do possess the structural and regulatory genes necessary for the expression of AHH activity. It was
postulated that a) mutation in the nonresponsive mice results in a
defective recognition or receptor site, which has a diminished or
absent affinity for MC, and b) that TCDD can saturate this same receptor in nonresponsive mice because of its greater potency, and
hence presumed greater receptor affinity.

Several lines of evidence confirm this hypothesis. The dose-
response curves for TCDD induction of hepatic AHH activity for responsive strains have an ED_{50} of $\sim$ 1 nmol/kg; for nonresponsive
strains, the dose-response curve is shifted to the right with an
$ED_{50} \geq$ 10 nmol/kg. The cytosol receptor is measurable in the liver
and other tissues of responsive mice, but has a low affinity, which
is as yet unmeasurable in nonresponsive strains.

Thus, all of the available evidence support the idea that: a)
the Ah locus determines the cytosol receptor, and b) the mutation in
the Ah locus in mice nonresponsive to MC (and less sensitive to TCDD)
results in a cytosol receptor with a diminished affinity for inducing
ligands.

COORDINATE GENE EXPRESSION

The administration of MC or TCDD to laboratory animals induces
not only microsomal monooxygenase activity, but also the coordinate
expression of a number of enzyme activities, such as UDP-glucuronosyltransferase (Owens, 1977), DT-diaphorase (Beatty and Neal, 1976),
δ-aminolevulinic acid synthestase (Poland and Glover, 1973),
choline kinase (Ishidate et al, 1980), τ-aldehyde dehydrogenase
(Dietrich et al, 1978), and glutathione S-transferase (Kirsch et al,
1975). The induction of many of these enzymes is dependent on the
tissue and animal species. In inbred strains of mice, the induction
of several of these enzymes by MC or TCDD has been shown to segregate
with the Ah locus (Owens 1977; Nebert et al, 1980; Kumaki et al,
1977). Thus the cytosol receptor appears to control not only AHH
activity but also the coordinate induction of a battery of enzymes,
most of which are related to the metabolism of xenobiotics.

CONSIDERATION OF THE MECHANISM OF TOXICITY OF HALOGENATED AROMATIC
HYDROCARBONS

We now turn to a consideration of the mechanism to toxicity of
TCDD and related compounds. We will examine: a) the relationship

of the cytosol receptor to toxicity; b) the nature of the toxic responses produced by TCDD and congeners, and their tissue specificity;
c) model systems in vitro and in vivo to study toxicity; and d)
finally, a proposed model for the mechanism of action, and its implications.

STRUCTURE-ACTIVITY RELATIONSHIP

    For several classes of halogenated aromatic hydrocarbons, the
dibenzo-p-dioxins, dibenzofurans, azo(xy)benzenes and biphenyls, the
structure-activity relationship for congeners to bind to the cytosol
receptor and to induce hepatic AHH activity closely corresponds to
that observed for toxicity (for a recent review of this area see
Goldstein, 1980). This observation suggests: a) that the parent
compounds and not their metabolites produce toxicity; b) that the
toxic effects evoked by these compounds are mediated through their
binding to the cytosol receptor; and c) that the capacity of a congener to elicit one toxic effect, indicates its potential to produce
the complete syndrome.

SEGREGATION OF TOXICITY WITH THE Ah LOCUS

    As noted above, the Ah locus appears to be the structural gene
for the cytosol receptor, and among inbred strains of mice there is
a genetic polymorphism for the cytosol receptor. If the toxicity of
TCDD (and congeners) is mediated through the receptor, then one would
expect that mice with a high affinity receptor, which are sensitive
to the induction of AHH activity, would be sensitive to toxicity by
TCDD, and conversely nonresponsive strains, which have a lower affinity receptor, would be less sensitive to toxicity. This has been
found to be the case. Thymic involution, cleft palate formation in
the fetus, and hepatic porphyria produced by TCDD have all been found
to segregate with the Ah locus (Poland and Glover 1980; Jones and
Sweeney 1980).

EXAMINATION OF TOXICITY

    Animal species vary greatly in their sensitivity to halogenated
aromatic hydrocarbons. The acute oral LD_{50} of TCDD varies over a
5000 fold range in different species: LD_{50} (μg/kg)-guinea pig, 1
(Schwetz et al, 1973); chicken embryo (injected into air sac),
< 1 (Higginbotham et al, 1968); rat, 22-45 (Schwetz et al, 1973);
monkey, < 70 (McConnell et al, 1978b); rabbit, 115 (Schewtz et al,
1973); mouse, 114 (Vos et al, 1974); dog, > 300 (Schwetz et al, 1973);
bullfrog, > 500 (Beatty et al, 1976); and hamster, 5000 (Henck et al,
1981). This large variation in species sensitivity to TCDD in not

attributable to an appreciable difference in the rate of metabolism of TCDD. The guinea pig and hamster differ by more than 3 orders of magnitude in their LD_{50} for TCDD, but the whole body half-life of TCDD in these species differs by only 3-fold. For other halogenated aromatic hydrocarbons, such as 2,3,7,8-tetrachlorodibenzofuran, much larger species differences in the rate of metabolism, have been noted (Morita and Oishi, 1977; Birnbaum et al, 1980, 1981; Decad et al, 1981a and b). However, it must be noted, as in the case of TCDD, differences in metabolism and excretion of these compounds, do not account for the large differences in species susceptibility.

Animal species also differ qualitatively in the toxic effects that develop in response to halogenated aromatic hydrocarbons. Many of the toxic lesions are highly species-specific and confined to one or a few animal species (Table 1). In general, all of the halogenated aromatic hydrocarbons a) show the same order of species sensitivity and b) will elicit the same pattern of toxic responses within a given species.

Table 1. Histopathology Produced by TCDD and Congeners: Species Differences

	Monkey	Guinea Pig	Cow	Rat	Mouse	Rabbit	Chicken	Hamster
Hypoplasia, Atrophy, Necrosis								
Thymus	+[a]	+	+	+	+		+	+
Bone Marrow	+	+		+	±	+		
Testicle (seminiferous tubule)	+	+		+	+		+	
Hyperplasia, and/or Metaplasia								
Gastric Mucosa	++	0	+	0	0		0	
Intestinal Mucosa	+							++
Bladder, Renal Pelvis, Ureter	++	++	++	0	0			
Gall Bladder and Bile Duct	++		+	0	++			
Skin	++		*[b]	0	0	++		
Other								
Liver	+	0		++	+	++	+	+
Edema Formation	+	0		0	+		++	+

a) Symbols: 0, lesion not observed; +, lesion observed, number of '+' denote severity; blank, no evidence reported in literature.
b) Skin lesions in cattle are observed, but they differ from the skin lesion observed in other species.

While the scope of this review does not permit a comprehensive review of the histopathologic lesions and toxic effects produced by these compounds, it is important to summarize the lesions that comprise the toxic syndrome of the halogenated aromatic hydrocarbons.

Wasting

Following an acute lethal dose of TCDD or related compounds, all animal species have a latent period of a week or more prior to death in which they develop a reduced weight gain or weight loss accompanied by a depletion of adipose tissue (McConnell, 1980).

Lymphoid Involution

In all species studied, TCDD and congeners produce a loss of lymphoid tissue, including the spleen and lymph nodes and especially the thymus (McConnell et al, 1978b; Buu-Hoi et al, 1972). In young animals, thymic loss produced by TCDD is accompanied by suppression of the immune response. Despite considerable investigation, the precise nature of this immune suppression and the target cells affected have not been determined (Vos and Moore, 1974; Faith and Moore, 1977; Vos et al, 1980).

Fetal Wastage, Embryotoxicity and Teratogenesis

TCDD produces fetal death and resorption in rats, and cleft palate and kidney malformations in mice (Sparschu et al, 1971; Courtney and Moore, 1971).

Hepatotoxicity

TCDD and related compounds produce hepatomegaly in all species, at doses well below the lethal dose (McConnell, 1980; Kociba et al, 1979). This hyperplasia and hypertrophy of parenchymal cells, and proliferation of the smooth endoplasmic reticulum is usually accompanied by an increase in microsomal monoxygenase activity and/or the other drug metabolizing enzymes.

Other hepatic lesions vary greatly among species. In the rabbit there is widsspread necrosis (Vos and Beems 1971); in the mouse, there are focal centrilobular lesions (Vos et al, 1974), and in the guinea pig, the histopathologic changes are minimal (McConnell et al, 1978b). In the rat, the most studied species, TCDD produces lipid accumulation, pigment deposition, parenchymal cell necrosis, infiltration of inflammatory cells, fibrous proliferation in necrotic areas, some distortion of the hepatic lobules, and formation of large multinuclear parenchymal cells (Kociba et al, 1979; Gupta et al, 1973; Hinton et al, 1978; Jones and Butler, 1973).

The rat and mouse are the species most commonly employed to study toxicity, and the toxic changes described above, plus a few more (e.g., marrow hypoplasia, degeneration of the seminiferous tubules, subcutaneous edema in mice) might suffice to characterize the toxic syndrome in these species. Most of these changes involve tissue atrophy, involution and/or necrosis. However, examination of other species administered TCDD or congeners leads to a very different picture.

In the human, monkey, rabbit (ear), and hairless mouse, TCDD produces chloracne, a skin lesion characterized by hyperplasia and hyperkeratosis of the interfollicular epidermis, hyperderatosis of the hair follicle, and squamous metaplasia of the sebaceous glands which form keratinaceous comedones and cysts (Crow, 1970; Allen et al, 1977; Kimmig and Schultz, 1957; Inagami et al, 1969).

In the monkey and cow, halogenated aromatic hydrocarbons produce hyperplasia and hypertrophy of the gastric mucosa with extention of the gastric glands into the submucosa and formation of submucosal cysts (McConnell et al, 1980; Allen and Norback, 1973). Hyperplasia of the intestinal epithelium has been noted in monkeys and hamsters (Olson et al, 1980a; McConnell, 1980).

The transitional epithelium of the urinary tract (renal pelvis, ureter, and bladder) proliferates in the guinea pig, monkey, and cow (McConnell et al, 1978a and b, 1980). The extrahepatic bile ducts and/or gall bladder undergo hyperplasic hypertrophy in monkeys,cattle and horses (McConnell et al, 1978a, 1980).

This listing of toxic lesions is not intended to be exhaustive, but merely illustrative of the diversity of responses produced by halogenated aromatic hydrocarbons in different species. In Table 1 we have tried to classify these toxic responses as: a) involving hyperplasia and/or altered differentiation (metaplasia); b) involving tissue loss, hypoplasia, atrophy and/or necrosis, and c) other responses which are difficult to catergorize. Nearly all the pathologic changes affect epithelial tissues. (Involution of the lymphoid tissue, which is mesenchymal, is a notable exception, but there is no evidence that TCDD or related compounds act on lymphocytes directly (Vos et al, 1980; Knutson and Poland, 1980a). In general, it is difficult to specify the organ whose dysfunction is responsible for death in a particular species.

From this brief summary of the actions of halogenated aromatic hydrocarbons, we wish to emphasize: a) that animal species vary greatly in their sensitivity to these agents; and b) that animal species differ qualitatively in their response to these compounds, i.e., many of the tissue changes are species-specific and even lethal doses of TCDD and congeners, do not produce the same histopathologic lesions in all species.

INVESTIGATION OF TOXICITY IN VITRO

An in vitro cell culture system in which to study the toxicity of halogenated aromatic hydrocarbons would permit a more thorough characterization of the sequential events that comprise toxicity in a defined and homogeneous cell population. However, in searching for such a system, one is confronted with the in vivo observations of toxicity: a) histologic changes produced by these compounds in a given tissue often occur in only a few animal species; b) most morphologic changes develop over a period of days to weeks, and may not develop in short-term cultures; c) toxicity in some tissues consists of atrophy or necrosis, and in other tissues hyperplasia and metaplasia.

To data, over 30 cell types have been examined for their response to TCDD. These have included primary cultures, established and transformed cells from at least 6 animal species, and a variety of tissues (liver, kidney, lymphocytes, embryonal teratomas, fibroblasts) (Niwa et al, 1975; Beatty et al, 1975; Kouri et al, 1974; Knutson and Poland, 1980a and unpublished observations). A number of cell types were inducible for AHH activity, indicating the presence of the cytosol receptor. Surprisingly, there have been no observations of cell toxicity with two exceptions: a) a report by Niwa et al, (1975) that several cell lines exposed to a very high concentration of TCDD (5×10^{-6} M) for 24 hours displayed a decreased viability as measured by trypan blue uptake; and b) the XB cell-line, which is discussed below. One can draw few conclusions from these negative studies, except to speculate that: a) the true target cells were not cultured, b) cells potentially susceptible to TCDD toxicity, were cultured in an altered state of differentiation, or under conditions which prevented expression of toxicity, or c) toxicity may require the interaction of two different cell types which interact directly or through some humoral factor.

XB, a cloned mouse teratoma cell line isolated and characterized by Rheinwald and Green (1975), is the only cell line to date, in which halogenated aromatic hydrocarbons produce a dose-related characteristic "toxic" response (Knutson and Poland, 1980b). When XB cells are plated at a low density (2×10^2 cells/60 mm plate) on a feeder layer of lethally irradiated 3T3 cells, XB cells form colonies of stratified, keratinized squamous epithelium, resembling the epidermis. However, when XB cells are plated at high density (3×10^5 cells/plate) on a feeder layer, they do not spontaneously differentiate. Under these conditions of high XB density, the addition of TCDD produces a dose-related keratinization, with a maximal response at 5×10^{-11} M, which is detectable at 7 days and maximal in 10 to 13 days. If XB cells are cultured at high density in the absence of a feeder layer, but in a medium conditioned by 3T3 cells, TCDD produces terminal differentiation. Thus, TCDD acts directly on XB cells.

XB cells contain a cytosol species which binds TCDD with a high affinity, and are inducible for AHH activity by TCDD and other receptor
agonists. For over 30 halogenated aromatic hydrocarbon congeners
tested, there was an excellent correlation between the potencies of
the compounds to produce keratinization and receptor binding affinities. XB cells cultured in the absence of a feeder layer of 3T3
cells, respond to TCDD with induction of AHH activity, but do not
differentiate.

     Thus, XB cells provide the first in vitro model in which to
study halogenated aromatic hydrocarbons. TCDD and congeners act
directly on this cell type, and terminal differentiation is mediated
through the cytosol receptor. One point which we shall return to
later deserves emphasis: one can dissociate the induction of AHH
activity, from cell differentiation, i.e. 'toxicity'.

REEXAMINATION OF TOXICITY: EVOLUTION OF THE RECEPTOR-MEDIATED MODEL

     Now let us reexamine the data on the toxicity of these compounds,
in light of our proposed mechanism of action, and try to define more
clearly the elusive nature of the toxicity. Our hypothesis is that
all the toxic responses produced by TCDD and related halogenated aromatic hydrocarbons are mediated through the stereospecific and reversible binding of these compounds to the cytosol receptor, and
the sustained pleiotropic response which ensues.

     Two independent lines of evidence suggest that toxicity is medi-
ated through the cytosol receptor: a) the close correspondence between the structure-activity relationships for receptor binding and
toxicity; and b) segregation of several toxic lesions produced by
TCDD in mice with the Ah locus.

     However, many tissues in vivo and cell lines in vitro contain
the receptor and, when challenged with TCDD or congeners, respond
with the induction of AHH activity (and/or other coordinately expressed enzymes) but display no evidence or toxicity, i.e., histologic
or cytologic changes. Thus, while the receptor appears to be essential for toxicity, it is not sufficient.

     How do we account for the large quantitative difference among
animal species in their susceptibility to TCDD (and congeners), and
the qualitative differences in the response observed in certain tissues, i.e., species-specific tissue lesions? Let us review a number
of possible factors.

Pharmacokinetics

     The pharmacokinetics of TCDD, 2,3,7,8-tetrachlorodibenzofuran
and some halogenated biphenyl isomers have been investigated in

several animal species (Rose et al, 1976; Nolan et al, 1980b; Gasiewicz and Neal, 1979; Decad et al, 1981a and b; Birnbaum et al, 1980, 1981; Mizutain et al, 1977; Matthews and Tuey, 1980). For TCDD, the species differences in pharmacokinetics, rates of metabolism, and tissue localization are modest, and do not account for a significant fraction of the differences in species sensitivity (LD_{50}) to TCDD.

Receptor Affinity and Concentration

If halogenated aromatic hydrocarbons exert their toxic action through the cytosol receptor, variations in affinity of the receptor in different species, and/or variations in the concentration of the receptor in different tissues of the same species, might account for the observed differences in species sensitivity and species-specific organ responses, respectively.

The available data are limited, but do not support this view. The affinity and concentration of the cytosol receptor in the liver of the guinea pig, responsive mouse (C57BL/6J), rat, rabbit, and hamster are quite similar (varying 3- to 4-fold in concentration and affinity)(Poland and Glover, unpublished observations), while the oral LD_{50} varies 5000-fold between the hamster and guinea pig. One might argue, that comparisons of the receptor concentration in liver, are not relevant, because the critical target organ(s) is unknown.

Importance of the Induction of AHH Activity, per se

Because of the important role of microsomal monooxygenase activity in metabolizing some foreign compounds to reactive electrophilic metabolites, which covalently bind to tissue macromolecules and produce cell damage, it might be postulated that halogenated aromatic hydrocarbons, by inducing microsomal monooxygenase activity, stimulate their own metabolism to toxic metabolites. This suggestion has numerous shortcomings: a) TCDD, the prototype of this group of compounds is rather slowly metabolized, its metabolites have been isolated from bile and excreta, but never from tissues (Rose et al, 1976; Olson et al, 1980b), and the maximal estimate of covalent binding of TCDD is so small that it approaches the limits of detection (Poland and Glover, 1979; Rose et al, 1976; b) the structure-activity relationship for toxicity of the halogenated aromatic hydrocarbons is based on the parent compounds, and the chemical diversity among these various classes would presumably give rise to very different metabolites; and c) as we noted in XB cell culture, it is possible to dissociate induction of AHH activity from the morphologic ('toxic') response.

It might be argued that the stimulation of microsomal monooxygenase activity by TCDD and congeners enhances the metabolism of some vital endogenous compound(s), e.g., increased steroid degradation and lower circulating steroid hormone concentrations. This may well be true, but does not explain the spectrum of toxic responses produced by halogenated aromatic hydrocarbons. The induction of AHH activity may be viewed as a signal response, but can not be directly implicated in the mechanism of toxicity.

We have argued against the importance of various factors in determining toxicity, that is, in explaining a) the difference among species in their sensitivity to TCDD and congeners, and b) the species specificity of many of the tissue-specific lesions. How then do we explain the observation that among the many tissues and cell types which contain the cytosol receptor, only a few develop toxicity? We would like to consider a specific example, the hairless mouse, HRS/J, and then postulate a more general model.

EPIDERMAL TOXICITY IN HRS/J MICE

The most characteristic toxic effect of halogenated aromatic hydrocarbons in humans is chloracne, and a similar epidermal lesion is produced in monkeys (McConnell et al, 1978a; Allen et al, 1977), and rabbits (Kimmig and Schultz, 1957), but it is not observed in rats, mice, guinea pigs or hamsters.

Mice carrying the recessive trait of hairless (hr/hr) at a locus on chromosome 14, respond to halogenated aromatic hydrocarbons with epidermal hyperplasia, hyperkeratosis, and squamous metaplasia of the sebaceous gland with formation of keratinaceous cysts and comedones as first noted by Inagami et al, (1969). HRS/J is an inbred strain of mice segregating for the hr locus. Homozygous (hr/hr) mice are hairless, and their heterozygous littermates, (hr/+) mice, are haired. HRS/J (hr/hr) mice are identical genetically to (hr/+) mice except for one allele at the hr locus.

HRS/J mice are responsive to aromatic hydrocarbons at the Ah locus, i.e., they have a high affinity receptor. For both hr/hr and hr/+ mice: a) the affinity and concentration of the cytosol receptor is very similar (measured in the liver), and b) application of TCDD to the skin induces epidermal AHH activity with a similar time-course and dose-response curve (Knutson and Poland, unpublished observations). However, only hr/hr mice respond to TCDD with epidermal hyperplasia, hyperkeratosis and squamous metaplasia of the sebaceous glands. The failure of shaven hr/+ mice to develop chloracne in response to TCDD does not appear to be due to hair or normal hair follicles, per se, because: a) application of TCDD to the ear of hr/hr or hr/+ mice, an area devoid of hair, does not produce chloracne, and b) in the rabbit application of halogenated aromatic hydrocarbons produces epidermal hyperplasia in both the inner part of the pinna and in the haired area of the back (Vos and Beems, 1971).

Thus, hr/hr and hr/+ mice (as well as mice which are wild type, (+/+) at the hairless locus) respond to TCDD with the induction of epidermal AHH activity, but hr/hr mice express an additional pleiotropic response, which results in the hyperplastic-metaplastic changes observed in the skin. Since hr/hr and hr/+ mice differ at only one locus, hr/+ (and wild type (+/+)) mice presumably have the genes necessary for this morphologic change, but TCDD does not induce them.

An analogous situation was observed in XB cell culture. When these cells are cultured alone in fresh media, TCDD induces AHH activity; however, when XB cells are plated at a high density on a feeder layer of 3T3 cells, or in media conditioned by 3T3 cells, TCDD induces AHH activity <u>and</u> terminal differentiation. A limited response to TCDD is converted to a more extensive response, in HRS/J mice by a single allele, and in XB cells by the culture conditions.

A GENERAL MODEL OF TOXICITY

The model developed for the epidermal response in HRS/J mice, may be extended to other tissues. We propose that all tissues <u>in vivo</u> (and cell types <u>in vitro</u>) which contain the cytosol receptor respond to TCDD and related halogenated aromatic hydrocarbons with a limited pleiotropic response, i.e., the induction of AHH activity and/or other enzymes primarily involved in drug metabolism, but those tissues which show 'toxic' responses, i.e., morphologic changes such as involution, differentiation and/or proliferation, express an additional battery of genes (Figure 2). While this model may be useful in considering the toxicity of the halogenated aromatic hydrocarbons, and accounting for the species differences in specific tissue lesions, it must be emphasized that we know nothing about a) the composition of this additional battery of genes, which presumably varies with the tissue, b) the nature of the ultimate biochemical lesion, or c) even the target tissue, the dysfunction of which results in death.

Historically, the induction of the drug metabolizing enzymes by foreign chemicals is viewed as an adaptive response, the "<u>purpose</u>" of which is to increase the rate of metabolism of the foreign compound and hasten its elimination from the body. Regardless of whether one believes the cytosol receptor recognizes only foreign compounds, or responds to an endogenous physiologic ligand (e.g., a hormone), that which is regulated is the drug metabolizing enzymes. However, as we have seen, in certain tissues, the halogenated aromatic hydrocarbons acting on the receptor evoke a much broader response, which may result in cell division, differentiation and/or involution. The toxicity of these compounds appears to be due to the sustained ex-

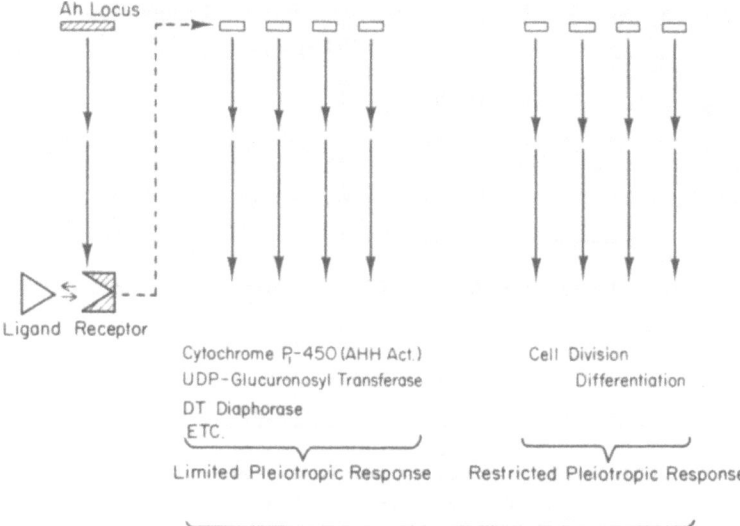

Figure 2. A Proposed Model of the Coordinate Gene Expression
           Controlled by the Ah Locus.
           The cytosol receptor, which is determined by the Ah locus,
           binds a ligand (e.g. halogenated aromatic hydrocarbon),
           and the receptor-ligand complex translocates to the nu-
           cleus. The ensuing gene activation in the skin of HRS/J
           (hr/+) mice consists primarily of the expression of en-
           zymes related to drug metabolism i.e. a limited pleio-
           tropic response. In the epidermis of HRS/J (hr/hr) mice,
           there is a larger pleiotropic response, with the addition-
           al expression of genes involved in cell division and
           differentiation. These genes are restricted in HRS/J
           (hr/+) mice. The same general model is proposed to dis-
           tinguish cells and tissues which respond with enzyme
           induction (the limited pleiotropic response) from those
           which respond with toxicity (the additional expression
           of normally restricted genes).

pression of a normal cellular regulatory system of which we were
previously unaware. In future investigation, it is hoped that we will
learn the nature and physiologic role of this regulatory system,
for only then can we hope to understand the mechanism of toxicity of
these compounds. Rational environmental policy for halogenated aromatic hydrocarbons must be based on our understanding of their
mechanism.

REFERENCES

Allen, J.R., and Norback, D.H. Polychlorinated biphenyl- and
     triphenyl-induced gastric mucosal hyperplasia in primates.
     Science, 179:498-499, 1973.
Allen, J.R., Barsotti, D.A., Van Miller, J.P., Abrahamson, L.J.,
     and Lalich, J.J. Morphological changes in monkeys consuming
     a diet containing low levels of 2,3,7,8-tetrachlorodibenzo-
     p-dioxin. Fd. Consmet. Toxicol., 15:401-410, 1977.
Beatty, P., and Neal, R.A. Induction of DT-diaphorase by 2,3,7,8-
     tetrachlorodibenzo-p-dioxin (TCDD). Biochem. Biophys. Res.
     Commun., 68:197-204, 1976.
Beatty, P.W., Holscher, M.A., and Neal, R.A. Toxicity of 2,3,7,8-
     tetrachlorodibenzo-p-dioxin in larval and adult forms of
     Rana catesbeiana. Bull. Environ. Contam. Toxicol., 5:578-581,
     1976.
Beatty, P.W., Lemack, K.J., Holscher, M.A., and Neal, R.A. Effects
     of 2,3,7,8-tetrachlorodibenzo-p-dioxin (TCDD) on mammalian
     cells in tissue culture. Toxicol. Appl. Pharmacol., 31:309-
     312, 1975.
Birnbaum, L.S., Decad, G.M., and Matthews, H.B. Disposition and
     excretion of 2,3,7,8-tetrachlorodibenzofuran in rat. Toxicol.
     Appl. Pharmacol., 55:342-352, 1980.
Birnbaum, L.S., Decad, G.M., Matthews, H.B., and McConnell, E.E.
     Fate of 2,3,7,8-tetrachlorodibenzofuran in the monkey. Toxicol.
     Appl. Pharamcol., 57:189-196, 1981.
Bradlaw, J.A., Garthoff, L.H., Hurley, N.E., and Firestone, D.
     Comparative induction of aryl hydrocarbon hydroxylase activity
     in vitro by analogues of dibenzo-p-dioxin. Fd. Cosmet. Toxicol.,
     18:627-635, 1980.
Buu-Hoi, N.P., Chanh, P.-H., Sesque, G., Azum-Gelade, M.C., and
     Saint-Ruf, G. Organs as targets of "dioxin" (2,3,7,8-tetra-
     chlorodibenzo-p-dioxin) intoxication. Naturwissenschaften,
     59:174-175, 1972.
Carlstedt-Duke, J.M.B. Tissue distribution of the receptor for
     2,3,7,8-tetrachlorodibenzo-p-dioxin in the rat. Cancer Res.,
     39:3172-3176, 1979.
Courtney, K.D., and Moore, J.A. Teratology studies with 2,4,5-
     trichlorophenoxyacetic acid and 2,3,7,8-tetrachlorodibenzo-
     p-dioxin. Toxicol. Appl. Pharmacol., 20:396-403, 1971.
Crow, K.D. Chloracne. Trans. St. Johns Hospital Dermatol. Soc.,
     56:77-99, 1970.
Decad, G.M., Birnbaum, L.S., and Matthews, H.B. 2,3,7,8-Tetrachloro-
     dibenzofuran tissue distribution and excretion in guinea pigs.
     Toxicol. Appl. Pharmacol., 57:231-240, 1981a.
Dietrich, R.A., Bludeau, P., Roger, M., and Schmuck, J. Induction
     of aldehyde dehydrogenases. Biochem. Pharmacol., 27:2343-
     2347, 1978.

Faith, R.E., and Moore, J.A. Impairment of thymus-dependent immune
    functions by exposure of the developing immune system to
    2,3,7,8-tetrachlorodibenzo-p-dioxin (TCDD). J. Toxicol. Environ.
    Health, 3:451-463, 1977.

Gasiewicz, T.A., and Neal, R.A. 2,3,7,8-Tetrachlorodibenzo-p-dioxin
    tissue distribution, excretion, and effects on clinical chemical
    parameters in guinea pigs. Toxicol. Appl. Pharmacol., 51:329-
    339, 1979.

Goldstein, J.A. Structure-activity relationships for the biochemical
    effects and relationships to toxicity. In: Kimbrough, R.D.,
    ed. Halogenated biphenyls, terphenyls, naphthalenes, dibenzo-
    dioxins, and related products. New York: Elsevier/North
    Holland, 1980. pp. 151-190.

Greenlee, W.F., and Poland, A. Nuclear uptake of 2,3,7,8-tetra-
    chlorodibenzo-p-dioxin in C57BL/6J and DBA/2J mice.
    J. Biol. Chem., 254:9814-9821, 1979.

Gupta, B.N., Vos, J.G., Moore, J.A., Zinkl, J.G., and Bullock, B.C.
    Pathologic effects of 2,3,7,8-tetrachlorodibenzo-p-dioxin in
    laboratory animals. Environ. Health Perspectives, 5:125-140,
    1973.

Henck, J.W., New, M.A., Kociba, R.J., and Rao, K.S. 2,3,7,8-Tetra-
    chlorodibenzo-p-dioxin: acute oral toxicity in hamsters.
    Toxicol. Appl. Pharmacol., 59:405-407, 1981.

Higginbotham, G.R., Huang, A., Firestone, D., Verrett, J., Ross, J.,
    and Campbell, A.D. Chemical and toxicological evaluations of
    isolated and synthetic chloro derivatives of dibenzo-p-dioxin.
    Nature. 220:702-703, 1968.

Hinton, D.E., Glaumann, H., and Trump, B.F. Studies on the cellular
    toxicity of polychlorinated biphenyls (PCBs). Effects of PCBs
    on microsomal enzymes and on synthesis and turnover of microsomal
    and cytoplasmic lipids of rat liver - a morphological and
    biochemical study. Virchows Arch. B., 27:279-306, 1978.

Inagami, K., Koga, T., Kikuchi, M., Hashimoto, M., Takahashi, H.,
    and Wada, K. Experimental study of hairless mice following
    administration of rice oil used by a "Yusho" patient.
    Fukuroka Acta. Med., 60:548-553 (In Japanese), 1969.

Ishidate, K., Tsuruoka, M., and Nakazawa, Y. Induction of choline
    kinase by polycyclic aromatic hydrocarbon carcinogens in rat
    liver. Biochem. Biophys. Res. Commun., 96:946-952, 1980.

Jones, K.G., and Sweeney, G.D. Dependence of the porphyrogenic
    effect of 2,3,7,8-tetrachlorodibenzo-p-dioxin upon inheritance
    of aryl hydrocarbon hydroxylase responsiveness. Toxicol. Appl.
    Pharmacol., 53:42-49, 1980.

Jones, G., and Butler, W.H. A morphological study of the liver
    lesion induced by 2,3,7,8-tetrachlorodibenzo-p-dioxin in rats.
    J. Pathol., 112:93-97, 1973.

Kende. A.S., Wade, J.J., Ridge, D., and Poland, A. Synthesis and
    fourier transform carbon-13 nuclear magnetic resonance
    spectroscopy of new toxic polyhalodibenzo-p-dioxins. J. Org.
    Chem., 39:931-937, 1974.

Kimmig, J., and Schultz, K.H. Chlorinated aromatic cyclic ethers as the cause of chloracne. Naturwissenschaften, 44:337-338, (In German), 1957.

Krisch, R., Fleischner, G., Kamisaka, K., and Arias, I.M. Structural and functional studies of ligandin a major renal organic anion-binding protein. J. Clin. Invest., 55:1009-1019, 1975.

Knutson, J.C., and Poland, A. 2,3,7,8-Tetrachlorodibenzo-p-dioxin: failure to demonstrate toxicity in twenty-three cultured cell types. Toxicol. Appl. Pharmacol., 54:377-383, 1980a.

Knutson, J.C., and Poland, A. Keratinizaton of mouse teratoma cell line XB produced by 2,3,7,8-tetrachlorodibenzo-p-dioxin: an in vitro model to toxicity. Cell., 22:27-36, 1980b.

Kociba, R.J., Keeler, P.A., Park, C.N., and Gehring, P.J. 2,3,7,8-tetrachlorodibenzo-p-dioxin (TCDD): Results of a 13-week oral toxicity study in rats. Toxicol. Appl. Pharmacol., 35:553-574, 1979.

Kouri, R.E., Ratrie, H., Atlas, S.A., Niwa, A., and Nebert, D.W. Aryl hydrocarbon hydroxylase induction in human lymphocyte cultures by 2,3,7,8-tetrachlorodibenzo-p-dioxin, Life Sci., 15:1585-1595, 1974.

Kumaki, K., Jensen, N.M., Shire, J.G.M., and Nebert, D.W. Genetic differences in induction of cytosol reduced NAD(L): menadione oxidoreductase and microsomal aryl hydrocarbon hydroxylase in the mouse. J. Biol. Chem., 252:157-165, 1977.

Matthews, H.B., and Tuey, D.B. The effect of chlorine position on the distribution and excretion of four hexachlorobiphenyl isomers. Toxicol. Appl. Pharmacol., 53:377-388, 1980.

McConnell, E.E. Acute and chronic toxicity, carcinogenesis, repro-duction, teratogenesis and mutagenesis in animals. In: Kimbrough, R., ed. Halogenated biphenyls, terphenyls, naphthalenes, dibenzodioxins, and related products. New York: Elsevier/North Holand Biomedical Press, 1980. pp. 109-150.

McConnell, E.E., Moore, J.A., and Dalgard, D.W. Toxicity of 2,3,7,8-tetrachlorodibenzo-p-dioxin in Rhesus monkeys (Macaca mulatta) following a single oral dose. Toxicol. Appl. Pharmacol. 43:175-187, 1978a.

McConnell, E.E., More, J.A., Haseman, J.K., and Harris, M.W. The comparative toxicity of chlorinated dibenzo-p-dioxins in mice and guinea pigs. Toxicol. Appl. Pharmacol., 44:335-356, 1978b.

McConnell, E.E., Moore, J.A., Gupta, B.N., Rakes, A.H., Luster, M.I., Goldstein, J.A., Haseman, J.K. and Parker, C.E. The chronic toxicity of technical and analytical pentachlorophenol in cattle. I. Clinicopathology. Toxicol. Appl. Pharmacol., 52:468-490, 1980.

Mizutani, T., Hidaka, K., Ohe, T., and Matsumoto. A comparative study on accumulation and elimination of tetrachlorobiphenyl isomers in mice. Bull. Environ. Contam. and Toxicol.,18:452-461, 1977.

Morita, M., and Oishi, S. Clearance and tissue distribution of
     polychlorinated dibenzofurans in mice. Bull. Environ. Contam.
     and Toxicol., 18:61-66, 1977.

Nebert, D.W., and Gielen, J.E. Genetic regulation of aryl hydrocarbon
     hydroxylase induction in the mouse. Fed. Proc., 31:1315-1327, 1972.

Nebert, D.W., Goujon, F.M., and Gielen, J.E. Aryl hydrocarbon
     hydroxylase induction by polycyclic hydrocarbons: simple
     autosomal dominant trait in the mouse. Nature New Biol., 236:
     107-110, 1972.

Nebert, D.W., Hensen, N., Perry, J., and Oka, T. Association between
     ornithine decarboxylase induction and the Ah locus in mice
     treated with polycyclic aromatic compounds. J. Biol. Chem.,
     255:6836-6842, 1980.

Niwa, A., Kumaki, K., and Nebert, D.W. Induction of aryl hydrocar-
     bon hydroxylase activity in various cell cultures by 2,3,7,8-
     tetrachlorodibenzo-p-dioxin. Mol. Pharmacol., 11:399-408, 1975.

Nolan, R.J., Smith, F.A., and Hefner, J.G. Elimination and tissue
     distribution of 2,3,7,8-tetrachlorodibenzo-p-dioxin (TCDD) in
     female guinea pigs following a single oral dose. Toxicol.
     Appl. Pharmacol., 48:A162 (Abstr.), 1979.

Okey, A.B., Bondy, G.P., Mason, M.E., Nebert, D.W., Forster-Gibson,
     C.J., Mucan, J., and Dufresne, M.J. Temperature-dependent
     cytosol-to-nucleus translocation of the Ah receptor for
     2,3,7,8-tetrachlorodibenzo-p-dioxin in continuous cell culture
     lines. J. Biol. Chem., 255:11415-11422, 1980.

Olson, J.R., Holscher, M.A., and Neal. R.A. Toxicity of 2,3,7,8-
     tetrachlorodibenzo-p-dioxin in the Golden Syrian Hamster.
     Toxicol. Appl. Pharmacol., 55:67-78, 1980a.

Olson, J.R., Gasiewicz, T.A., and Neal, R.A. Tissue distribution,
     excretion and metabolism of 2,3,7,8-tetrachlorodibenzo-p-dioxin
     (TCDD) in the Golden Syrian Hamster. Toxicol. Appl. Pharmacol.,
     56:78-85, 1980b.

Owens, I.S. Genetic regulation of UDP-glucuronosyltransferase
     induction by polycyclic aromatic compounds in mice. J. Biol.
     Chem., 252:2827-2833, 1977.

Poland, A., and Glover, E. Chlorinated dibenzo-p-dioxins: potent
     inducers of δ-aminolevulinic acid synthetase and aryl hydro-
     carbon hydroxylase. II. A study of the structure-activity
     relationship. Mol. Pharmacol., 9:736-747, 1973.

Poland, A., and Glover, E. Comparison of 2,3,7,8-tetrachlorodibenzo-
     p-dioxin, a potent inducer of aryl hydrocarbon hydroxylase with
     3-methylcholanthrene. Mol. Pharmacol., 10:349-359, 1974.

Poland, A., and Glover, E. Genetic expression of aryl hydrocarbon
     hydroxylase by 2,3,7,8-tetrachlorodibenzo-p-dioxin: evidence
     for a receptor mutation in genetically non-responsive mice.
     Mol. Pharmacol., 11:389-398, 1975.

Poland, A., and Glover, E. Chlorinated biphenyl induction of aryl
     hydrocarbon hydroxylase activity: a study of the structure-
     activity relationship. Mol. Pharmacol., 13:924-938, 1977.

Poland, A., and Glover, E. An estimate of the maximum in vivo covalent binding of 2,3,7,8-tetrachlorodibenzo-p-dioxin to rat liver protein, ribosomal RNA, and DNA. Cancer Res., 39:3341-3344, 1979.

Poland, A., and Glover, E. 2,3,7,8-tetrachlorodibenzo-p-dioxin: Segregation of toxicity with the Ah locus. Mol. Pharmacol., 17:86-94, 1980.

Poland, A., Glover, E., Robinson, J.R., and Nebert, D.W. Genetic expression of aryl hydrocarbon hydroxylase activity: induction of monooxygenase activities and cytochrome P_1-450 formation by 2,3,7,8-tetrachlorodibenzo-p-dioxin in mice genetically "non-responsive" to other aromatic hydrocarbons. J. Biol. Chem., 249:5599-5606, 1874.

Poland, A., Glover, E., and Kende, A.S. Stereospecific, high affinity binding of 2,3,7,8-tetrachlorodibenzo-p-dioxin by hepatic cytosol. J. Biol. Chem., 251:436-446, 1976a.

Poland, A., Glover, E., Kende, A.S., and DeCamp, M., Giandomenico, C.M. 3,4,3',4'-tetrachloro azoxygenzene and azobenzene: potent inducers of aryl hydrocarbon hydroxylase. Science, 194:627-30, 1976b.

Poland, A., Greenlee, W.E., and Kende, A.S. Studies on the mechanism of action of ten chlorinated dibenzo-p-dioxins and related compounds. Ann. N.Y. Acad. Sci., 320:214-230, 1979.

Rheinwald, J.G. and Green, H. Formation of a keratinizing epithelium in culture by a cloned cell line derived from a teratoma. Cell., 6:317-330, 1975.

Rose, J.Q., Ramsey, J.C., Wentzler, T.H., Hummel, R.A., and Gehring, P.J. The fate of 2,3,7,8-tetrachlorodibenzo-p-dioxin following single and repeated oral doses to the rat. Toxicol. Appl. Pharmacol., 36:209-226, 1976.

Schwetz, B.A., Norris, J.M., Sparschu, G.L., Rowe, V.K., Gehring, P.J., Emerson, J.L., and Gerbig, C.G. Toxicology of chlorinated dibenzo-p-dioxins. Environ. Health Persp., 5:87-99, 1973.

Sparschu, G.L., Dunn, F.L., and Rowe, V.K. Study of the teratogenicity of 2,3,7,8-tetrachlorodibenzo-p-dioxin in the rat. Fd. Cosmet. Toxicol., 9:405-412, 1971.

Thomas, P.E., Kouri, R.W., and Hutton, J.J. The genetics of aryl hydrocarbon hydroxylase induction in mice: a single gene difference between C57BL/6J and DBA/2J. Biochem. Genet., 6:157-168, 1972.

Vos, J.G., and Beems, R.B. Dermal toxicity studies of technical polychlorinated biphenyls and fractions thereof in rabbits. Toxicol. Appl. Pharmacol., 19:617-633, 1971.

Vos, J.G., and Moore, J.A. Suppression of cellular immunity in rats and mice by maternal treatment with 2,3,7,8-tetrachlorodibenzo-p-dioxin. Int. Arch. Allergy Appl. Immunol., 47:777-794, 1974.

Vos, J.G., Moore, J.A., and Zinkl, J.G. Toxicity of 2,3,7,8-tetrachlorodibenzo-p-dioxin (TCDD) in C57BL/6 mice. Toxicol. Appl. Pharmacol., 29:229-241, 1974.

Vos, J.G., Faith, R.E., and Luster, M.I. Immune alterations. In:
     Kimbrough, R., ed. Halogenated biphenyls, terphenyls,
     naphthalenes, dibenzodioxins, and related products. New York:
     Elsevier/North Holand Biomedical Press, 1980. pp. 241-66.

HUMAN OBSERVATIONS

Co-Chairmen: Giuseppe Reggiani
Hoffman-La Roche
Basle, Switzerland

Kenneth D. Crow
Princess Margaret Hospital
Swindon, Wiltshire U.K.

METHODOLOGY OF CLINICAL STUDIES ON EXPOSED POPULATIONS

Renate D. Kimbrough

Centers for Disease Control
U.S. Department of Health and Human Services
Atlanta, Georgia USA

ABSTRACT

Much progress has been made in the field of toxicology as it applies to safety evaluation and in the area of trace analysis for chemicals. It is now assumed by many that this technology can simply be transferred to studies of human populations in field situations. Unfortunately, many problems are encountered when this is done. The heterogeneity of the general population, the long life span of humans, the lack of information about baseline data are but a few factors. Tests to detect subclinical effects have, in many instances, not been developed to the point where they can be used as screening tools. Other tests that are used may not really be relevant for specific problems. For many such tests, it is not known how they can be used to make predictions about the health status of the individual at the time the test is done or about his or her health in the future.

INTRODUCTION

By definition, epidemiology is the study of diseases. It deals with the incidence, distribution, and control of diseases. In the past, such studies have consisted of defining disease, determining cause/effect relationships, and attempting to control factors that are casually associated with disease outbreaks. Particularly in the area of infectious diseases, much progress has been made within the past 200 years. Through epidemiological studies, we are now also beginning to understand why some cancers are present at higher rates in some population groups than in others. However, few attempts have been made to follow large groups of people prospectively to determine whether, for some reason or another, they may later develop disease.

Often it is through clinical observations that suspicions are raised about cause-effect relationships in diseases. The relationship between asbestos fibers, asbestosis, and mesothelioma were first observed and reported by clinicians. Chloracne was first reported by Herheimer in 1899 in workers who had been exposed to certain chlorinated chemicals. When the Kepone (chlordecone) episode in Virginia first came to light, it was a clinician who called the problem to the attention of the Centers for Disease Control. Similarly (Folland et al, 1978) a physician first noted that exposure to chlordimeform caused hemorrhagic cystitis in his patients. It is not known how many such episodes are not noted and, therefore, are not reported and investigated. If such occurrences are unusual, like the toxic shock syndrome (Shands et al, 1980), it is more likely that they will be reported and investigated. Sometimes it appears that evidence of a commonly occurring disease, like leukemia or other cancer is increased in specific geographic areas of the United States. Even after the obvious possible causes for such increase, such as improved reporting, special exposure to radiation and many other factors have been evaluated, no explanation for the increase can be found.

Because of these apparent spontaneous fluctuations in disease incidence, it may also happen that illnesses are erroneously associated with specific events.

These difficulties are compounded if clinical laboratory results and the results of physiological tests are compared between different study groups. This will be discussed in more detail below.

In the past 30 years, much progress has been made in the ability to detect smaller and smaller amounts of chemicals in our environment. In addition, the field of toxicology as it relates to safety testing has made enormous strides. Years ago, only the actue toxicity was determined for chemicals. Gradually, chronic toxicity testing and reproduction studies were required for many compounds. Mutagenic effects, effects on the immune response and the nervous system, including effects on behavior, became of major concern. It is now possible to demonstrate more and more subtle effects of chemicals in animals. The laboratory animals used for such testing have been made free of interfering diseases, their background incidence of cancer is known and they have been specially bred to serve specific purposes. Since it is now possible to demonstrate all of these effects in animals, it is natural to assume that such effects also occur in humans and that they should be looked for. It is the purpose of this paper to point out what the present difficulties are in undertaking such studies in humans and how this situation could perhaps be improved.

TYPES OF STUDIES

Theoretically, a variety of studies can be done to determine whether a certain event, such as exposure to a chemcial, has or will cause disease. Another way to approach the problem is to establish whether a specific disease was caused or augmented by certain events. In the past, the latter approach has been the more common. The usual tools that are used are given in Table 1. Exposure, with the exception of clinical trials of drugs, is accidental and uncontrolled in the populations that are studied. These types of studies can basically be divided into cohort studies, case-control studies and cross-sectional studies.

In such designed epidemiological studies, a number of end points can be looked for. The types of end points that are looked for depend to a great extent on what kind of chemical exposure has occurred. The greatest difficulties exist in the area of clinical-type studies, which cover biochemical changes, physiological effects, and, at times, morphological changes of tissues. It is this area of studies which I will discuss, primarily in the context of exposure of humans to chemicals.

Table 1. Methods to Assess Relationships
Between Exposures and Outcomes

---

1. Kinds of Studies:

Experimental - Control of exposure

Nonexperimental - No control of exposure
  With concurrent comparison groups
    Cohort
    Case-control
    Cross-sectional

  Without concurrent comparison groups
    Formal surveillance
      Exposed populations
      Populations with specific outcomes
    Case Clusters
    Anecdotal reports

2. Quality Control Methods.

3. Statistical Methods.

EXPOSURE

Except for occupational exposure where chemical concentrations in air and on surfaces of the work area are high (Cannon et al, 1978); Folland et al, 1978), the majority of episodes of acute poisoning in human populations have resulted from the ingestion of toxic substances (Diggory et al, 1977; Carter 1976; Cam and Nogogasyan 1963; Hayabuchi et al, 1979; Takeuchi et al, 1962).

Because of the sudden onset of illness in increased numbers of people, such acute episodes are usually recognized and their cause is identified. Because of the emergency nature of such outbreaks, preventive measures are often instituted without adequate and well-designed scientific studies which could give information on toxic dosage levels, concentrations of chemicals in blood and tissues, and their relation to abnormal clinical chemistry tests and morphological changes in tissues. Chronic health effects from continuous exposure to chemicals that cause deleterious effects are much more difficult to detect and so are delayed effects of acute short-term exposure. Basically, our knowledge in this area is limited to our experience with occupational exposure, cigarette smoking, the long-term effects of estrogenic hormones, and exposure to heavy metals. Thus, it is presently not clear whether different routes of exposure may be of more or less importance for chronic health effects.

IMPORTANCE OF THE TYPE OF EXPOSURE

Absorption by inhalation would deliver the chemicals in a very efficient way to the body since the human lungs have a surface area of about 55 m^2 and, of a given concentration of a chemical in air, much more would be taken up by the body than of the same concentration in food. However, in order for the lungs to absorb the chemical, it must be volatile or in small enough particles that it can be inhaled. Chemicals in drinking water, of course, would be ingested. They could be inhaled when water is vaporized and they could get into food through cooking. If chemicals are present in either drinking water or air and if their concentrations are known, a dose can be calculated.

Contamination of soil by chemicals is more difficult to deal with. Here it is not possible to determine with any degree of accuracy what a possible dose might be. At times, animals exposed to contaminated soil will be much more severely affected than humans exposed to the same soil (Carter, et al, 1975) or small children may acquire higher body burdens than older children and adults (Landrigan et al, 1975).

LIMITING FACTORS IN HUMAN HEALTH STUDIES

In many instances, information about the toxic effects of chemical compounds in humans is scanty or nonexistent. Even less information is available on effects produced by a combination of chemicals.
Although different species may respond differently to a chemical,
results obtained in animals studies can be used to decide what effects
should be considered in accidentally exposed human populations. A
toxic effect may, of course, be demonstrated first in humans (Folland
et al, 1978). A suspicion that illness was caused by a certain
chemical should not be disregarded even if the effect has not been
reported in animals. On the contrary, attempts should be made to
develop an animal model in which the disease can be produced.

Although experience gained in animal studies can aid in designing
human studies, the difficulties and limitations of this approach must
be realized. In animal studies, the number of animals, the dose,
age, sex, nutrition, and environment are well controlled; the incidence of background disease and the genetic makeup are known. In
humans, where uncontrolled exposure has been or is occurring, none
of these factors are controlled.

Before studies are undertaken, they must be determined to be
feasible. If very few people were exposed, a study may not yield
meaningful results. An example of this situation is one encountered
in Missouri in 1971 (Carter et al, 1975). In this expisode, several
riding arenas in Missouri were sprayed with salvage oil contaminated
with 2,3,7,8-tetrachlorodibenzodioxin, polychlorinated biphenyls, and
2,4,5-trichlorophenol. Although many animals became ill or died,
fewer than 10 persons had more than casual exposure to the contaminated arenas. The health of the individuals in this group has been
evaluated repeatedly. Because of the small number of persons and
their heterogeneity (differences in age and sex), no epidemiologic
study of this group has been made.

If large numbers of people have been exposed to a chemical or a
mixture of chemicals, a designed epidemiologic study may be feasible.
An example of such a situation is the accidental exposure of a siz-
able number of people in Michigan to polybrominated biphenyls (PBB).
In this instance, exposure occurred through the consumption of PBB-
contaminated milk, eggs, and meat (Landrigan, et al, 1979). A
similar situation exists in Triana, Alabama, where the population of
a small town (several hundred people) for many years ate locally
caught fish contaminated with polychlorinated biphenyls and high
concentrations of DDT and related materials (Kreiss, et al, 1981).
In a study we recently conducted in Triana, a cohort of 518 persons
was examined. Of these, 43.8 percent were males and 56.2 percent
were females; 36 percent were children. The ages ranged from a few
weeks to 90 years, with a median age of 27.4 years. This population
had been exposed to fish with high DDT (1,1'-(2,2,2-trichloroethenyli-

dene)-bis-4-chlorobenzene) residues. DDT blood levels ranged from
0.6 - 2820.5 ppb, and increased with age. For each age group, they
were lower in females. This illustrates that a relatively large
cohort may within itself be so diversified that for many illnesses
it would be difficult, if not impossible, to demonstrate a small
increase in a specific effect. Thus, closer examination of these
groups shows many limitations inherent in these studies. In Michigan,
the total amounts of PBB to which individuals were exposed varied a
great deal, resulting in a wide range of PBB body burdens. Furthermore, PBB is a mixture of compounds, some of which may be more toxic
than others (Robertson et al, 1981; Patterson et al, 1981). Any
measurement of total PBB body burden may not reflect differences in
exposure to specific PBB isomers. Another complicating factor in
Michigan is that this population had also been exposed to polychlorinated biphenyls (PCB). PCB may potentiate the toxic effects of
PBB. The exposure to PCB is also not uniform in this population.

       Since PBB, PCB. and DDT residues are very persistent, body
burdens do not change rapidly. Therefore, exposure levels can be
established in these populations and used as points of reference.
Since the general population is very heterogeneous, some individuals
with lower body burdens may have symptoms, but others with higher
body burdens may not (Cannon et al, 1978).

       Although in the PBB study, a cohort of 4,000 exposed people has
been established, it still may not be possible to demonstrate adverse
effects on health. One reason is the varied exposure, with a mean
PBB blood level in the cohort of 21.2 µg/l of serum and a range of
0-1900 µg/l of serum. Another is that the age at the time of exposure varied a great deal, and both males and females were exposed.
A similar situation exists in Triana, where exposure occurred over
a longer period of time than in Michigan, but the cohort is much
smaller. On the other hand, PBB is much more persistent than chlorinated aromatic compounds. Fortunately, in both situations, the
population seems relatively stable and is not greatly affected by
migration.

       In these studies, certain biases must be avoided (Sackett, 1979).
A bias may occur when the knowledge of an exposure influences the
diagnosis of or treatment for, a illness. In cohort studies, those
in the exposed group who become ill may be diagnosed and treated
earlier than those in the control group (selection bias). Continued
followup or periodic and comparable examinations of both the exposed
group and the control group can reduce this bias. Another bias may
occur when the definitions of exposure and illness are inadequate or
when the information obtained from the exposed group differs from the
information obtained from the control group (misclassification bias).
A third bias may be introduced when a factor may be involved in part,
but not all, of the relationship between an exposure and an illness
(confounding bias). This might occur in a study of the environmental

exposure to a chemical causing an effect resembling that of an unrelated illness. Here the study should be designed to eliminate this effect, or enough information should be collected to make it possible to adjust the analysis for the effect.

In the studies just mentioned, it is possible to determine body burdens of the halogenated aromatic compounds in the study populations and relate health effect findings to such body burdens. For many chemicals, this is not possible because they are rapidly metabolized and excreted, or exposure occurred many years before.

Since it is often difficult to establish exposure levels or total exposure over time in populations, health effects produced by such chemicals could be determined instead (Table 2). Unfortunately, this is usually not a good solution either unless very specific acute effects occur soom after exposure that can be casually related to the chemical exposure, such as paralysis, tremors, cholinesterase inhibition, formation of methemoglobinemia or hematuria. Chronic or delayed health effects can usually neither be detected early nor can predictions be made about their occurrence in particular individuals. Not much effort has gone into the development of tests which might detect subclinical effects in humans. Most clinical tests presently in use were developed to detect diseases, such as infections, cancer, and metabolic diseases. Over the past 30 years, testing for toxic effects of chemicals in experimental animals has become very sophisticated and standardized. Baselines have been developed in test animals for many parameters so that comparisons can be made between exposed and unexposed animals. Many of these tests cannot be done in humans; others have not been developed; and for most such tests, baseline data of normal variations in the general population are not available (Kimbrough et al, in press). For instance, most liver function tests are done on patients with a variety of illnesses rather than on normal subjects, and clinical laboratory tests are either not sufficiently standardized to detect small differences or the wrong types of liver function tests are done. Plasma lipids have now been determined in lipid research clinic population studies (Rifkind, 1980). These results and those of the HANES studies[1] will be very helpful for the interpretation of data in the future.

More sophicated tests, such as chromosome studies, are not yet part of the mainstream of medical practice. Extensive baseline data are lacking, and disagreement and confusion exists among scientists working in this area. CDC, in collaboration with the scientific community, EPA, and other Federal agencies, is developing guidelines in an attempt to promote more orderly studies and to point out pitfalls that may lead to erroneous results or misinterpretation of data (Bloom, 1981).

---

[1]Health and Nutrition Surveys (HANES) National Center for Health Statistics, 3700 East West Highway, Hyattsville, Maryland 20782

Table 2. Assessment of Health Effects in Human
          Populations After Chemical Exposures

Kinds of Measurements

1. Exposures
   Agents
   Metabolites

2. Outcomes
   Physical findings: (rashes, paralysis,
   tremor, etc.)

   Disease or disorder:
     Apparent
       Abnormal reproductive outcomes
       Growth and developmental disorders
       Behavioral or psychological disorders
       Cancer
       Other disorders (autoimmune diseases,
                         blood dyscrasias,
                         hypertension, pulmonary
                         fibrosis, allergies,
                         premature aging).

     Unapparent
       Biochemical abnormalitites (cholinesterase,
                         erythrocyte protoporphyrin,
                         urinary uroprophyrins)
       Immunologic abnormalities (e.g., decreased
                         cell mediated immune re-
                         sponse)
       Chromosomal abnormalities
       Nerve conduction abnormalities
       Other test abnormalities (pulmonary function)

3. Other Factors Mediating, Confounding, or
   Interacting with Exposures and Outcomes
   (age, nutrition, migration, occupation).

Among the pitfalls are such mundane factors as the accidental
contamination of or improper collection of specimens in the field.
Specimens may be improperly preserved or transported. Unless analytical methods are well standardized, results may not be reproducible.
If measurements, such as liver function tests, are done in a population over time, it must be assured that the tests are always done in
the same way and that a quality control system exists which would detect drifts in methodology. Such a quality control system has been
instituted for PBB determinations on the long-term followup study in
Michigan (Bruse et al, 1980).

In addition to our lack of knowledge about baseline data in the general population in many areas, we lack knowledge about how some of our findings relate to an individual's personal health. In some cases, it may be unethical to conduct tests in humans when the significance of the results, and how those results relate to health or furute illness, cannot be properly interpreted for the individual. Because of these difficulties, lack of evidence of an association between an exposure and a health effect is, in most cases, not evidence for lack of an association. On the other hand, because of bias or chance, a positive association may be erroneously noted. Thus far, the relationship between exposure and health effects has been discussed without much attention to exposure to several chemicals. When more than one exposure is involved, the existence of distinct relationships between each exposure and the illness must be verified, but the relationships among the exposures must also be examined. One exposure may confound another, explaining part or all of the relationship between the exposures and the illness. The first exposure may interact with or modify the relationship between the second exposure and the illness; in this case, the relationship the second exposure has with the illness changes with the level of the first exposure.

SUSCEPTIBILITY

The human population, as has been pointed out repeatedly, is an extremely heterogeneous group. Differences in genetic makeup may influence the toxicity of chemicals. Other factors are age, nutritional status, sex, illness, exposure to other chemicals which may have additive, potentiating, synergistic or antagonistic effects (De Brun, 1976). It is beyond the scope of this article to discuss the different parameters which may affect susceptibility in the general population (Bondy and Rosenberg, 1980). I will only cite one example. When the chlordecone workers of the now defunct Life Sciences Company were examined, it was found that some workers who had tremors had much lower chlordecone blood levels than others who did not have tremors (Cannon et al, 1978). No satisfactory explanation was ever found for this discrepancy. Many other such examples could be cited from the literature. Some such differences can be explained by differences in metabolism, but for many, no obvious reasons are evident. Further studies are needed to increase our understanding of this very important area.

CONCLUSIONS

The more subtle effects of chemicals, which can now be demonstrated in animal studies, are more difficult to include in the general armamentarium of epidemiology. This is for the following reasons: It is difficult to take highly sophisticated laboratory tests into the field and apply them to large numbers of people.

Background data about such tests are lacking in the normal general
populations. We are often unable to interpret the significance of
many such findings for human health, either immediately or for future
possible health problems in the individual. Other factors are the
enormous heterogeneity of the general population and the general
uncontrolled environment populations live in. Some of the tests used
are either nonspecific or inappropriate. Although nothing can be
done about standardizing populations, better design of studies, improved laboratory tests and development of laboratory tests applicable to field situations may improve the somewhat dismal situation
with which we are presently faced.

REFERENCES

Bloom, A.D., Ed. Guidelines for studies of human populations ex-
    posed to mutagenic and reproductive hazards. March of Dimes
    Birth Defects Foundation, 1275 Mamaroneck Ave., White Plains,
    New York 10605, 1981.
Bondy, P.K., and Rosenberg, L.E. Metabolic control and disease.
    8th Ed., W.B. Saunders Co., Philadelphia, London, Toronto, 1980.
Burse, V.W., Needham, L.L., Liddle, J.A., Bayse, D.D., and Price,
    H.A. Interlaboratory comparison for results of analyses for
    polybrominated biphenyls in human serum. J. Anal. Toxicol.,
    4:22-26, 1980.
Cam, C., and Nogogasyan, G. Acquired toxic prophyria cutanea tarda
    due to HCB. JAMA., 183:88-91, 1963.
Cannon, S.B., Veazey, J.M., Jackson, R.S., Burse, V.W., Hayes, C.,
    Straub, W.E., Landrigan, P.J., and Liddle, J.A. Epidemic Kepone
    poisoning in chemical workers. Am. J. Epidemiol, 107:529-537,
    1978.
Carter, C.D., Kimbrough, R.D., Liddle, J.A., Cline, R.E., Zack, M.M.,
    Jr., Bartel, W.E., Kiehler, R.E., and Phillips, P.E.
    Tetrachlorodibenzodioxin: an accidental poisoning episode in
    horse arenas. Science, 188:738-740, 1975.
Carter, J.J. Michigan's PBB incident: Chemical mixup leads to
    disaster. Science, 192:240-243, 1976.
DeBrun, A. Biochemical toxicology of environmental agents. Elsevier
    North Holland Biomedical Press, Amsterdam, The Netherlands,
    1976.
Diggory, M.J.P., Landrigan, P.J., Latimer, K.P., Kimbrough, R.D.,
    Liddle, J.A., Cline, R.E., and Smrek, A.L. Fatal parathion
    poisoning caused by contamination of commercial flour. Am. J.
    Epidemiol., 106:145-153, 1977.
Folland, D.S., Kimbrough, R.D., Cline, R.E., Swiggart, R.C., and
    Schaffner, W. Acute hemorrhagic cystitis. Industrial ex-
    posure to the pesticide chlordimeform. JAMA, 239:1052-1055,
    1978.

Hayabuchi, H., Yoshimura, T. and Kuratsune, M. Consumption of toxic rice oil by "Yusho" patients and its relation to the clinical response and latent period. Food Cosmet. Toxicol., 17(5):455-461, 1979.

Kimbrough, R.D., Taylor, P.R., Zack, M.M., and Heath, C.W. Studies of human populations exposed to environmental chemcials: Considerations of Love Canal. Nat. Acad. Sc., NRC, 1982, in press.

Kreiss, K., Zack, M., Kimbrough, R.D., et al. Association of blood pressure and PCB level in cross-sectional community study. JAMA, 245(24):2505-2509, 1981.

Kreiss, K., Zack, M., Kimbrough, R.D., et al. Cross-sectional study of a community with exceptional exposure to DDT. JAMA, 245(19): 1926-1930, 1981.

Landrigan, P.J., Wilcox, K.R., Jr., Silva, J.,Jr., Humphrey, H.E.B. Kauffman, C., Heath, C.W., Jr. Cohort study of Michigan residents exposed to polybrominated biphenyls: Epidemiologic and immunologic findings. Annal. NY Acad. Sci., 320:284-294, 1979.

Landrigan, P.J., Gehlback, S.W., Rosenblum, B.F., Shoults, J.M., Candelaria, R.M., Barthel, W.F., Liddle, J.A., Smrek, A.L., Staehling, N.W., and Sanders, J.F. Epidemic lead absorption near an ore smelter. The role of particulate lead. New Eng. J. Med., 292:123-129, 1975.

Patterson, D.G., Hill, R.H., Needham, L.L., Orti, D.L., Kimbrough, R.D., and Liddle, J.A. Hyperkeratosis induced by sunlight degradation products of the major polybrominated biphenyl in Firemaster. Science, 213:901-902, 1981.

Rifkind, B.M. The lipid research clinic's population studies data book. Vol. 1. The prevalence study. US DHHS, NIH Publ. No. 80-1527, 1980.

Robertson, L., Parkinson, A., and Safe, S. Induction of drug-metabolizing enzymes by fractionated commercial polybrominated biphenyls (PBBs). Toxicol. Appl. Pharmacol., 57:254-262, 1981.

Sackett, D.L. Bias in analytic research. J. Chronic. Dis., 32: 51-63, 1979.

Shands, K.N., Schmid, G.P., Dan, B.B., et al. Toxic-shock syndrome in menstruating women: Its association with tampon use and staphylococcus aureus and the clinical features in 52 cases. New Eng. J. Med., 303:1436-1442, 1980.

Tackeuchi, et al. A pathological study of Minamata disease in Japan. Acta. Neuropathol., 2:40-57, 1962.

A MORTALITY STUDY OF WORKERS EMPLOYED AT THE

MONSANTO COMPANY PLANT IN NITRO, WEST VIRGINIA

Judith A. Zack and William R. Gaffey

Monsanto Company
St. Louis, Missouri   USA

BACKGROUND

The compound 2,3,7,8-tetrachlorodibenzo-p-dioxin (TCDD) is a
highly toxic impurity that is formed in trace quantities during the
production of 2,4,5-trichlorophenoxyacetic acid (2,4,5-T). Exposure
to TCDD can cause chloracne, a skin disorder characterized by
comedones, cysts, and abscesses.[1] Outbreaks of chloracne have been
reported among workers associated with the production of 2,4,5-T and
2,4,5-T based products. Such incidents resulting from both accidental
and routine occupational exposures have been reported from several
countries.[2]

The first reported industrial accident involving exposure to
TCDD occurred in 1949 at the Monsanto Company plant in Nitro, West
Virginia. A total of 122 employees developed symptoms of chloracne
following a trichlorophenol (TCP) process accident. An undetermined
number of other employees developed symptoms of chloracne resulting
from exposure to the regular operations concerned with 2,4,5-T production over the period 1948-1969.

Between 1949 and 1953, Ashe and Suskind[3,4] examined thirty-
eight Nitro plant employees with chloracne. Twelve of these had
developed symptoms of chloracne directly following the 1949 TCP accident; twenty-six other chloracne cases had resulted from exposure to
the regular 2,4,5-T production operations. In addition to chloracne,
other signs and symptoms were observed in this group. These included
severe aches in the lower extremities, fatigue, nervousness and
irritability, loss or decrease of libido, dyspnia, and vertigo.
These findings are consistent with those that have been reported in
other industrial incidents.[5]

To examine the chronic health effects of exposure to TCDD, a
mortality study of the Nitro plant employees who had developed
symptoms of chloracne following the 1949 TCP accident was conducted.[6]
The study cohort was comprised of 121 of the 122 chloracne cases;
one female who was living as of the endpoint of the study was excluded
from analysis. At the time of the incident, it was assumed that the
symptoms were caused by exposure to unknown products of decomposition.
From today's vantage point, these symptoms suggest exposure to TCDD.
The 121-member study cohort, with a presumptive high-peak exposure
to TCDD, was followed for mortality through 1978. The entire cohort
was traced: thirty-two deaths were observed and eighty-nine persons
were confirmed as living. Analysis indicated no excess in total
mortality or in deaths from malignant neoplasms.

The study presented here examines the mortality of Nitro plant
workers who were assigned to an area of TCP or 2,4,5-T production,
with potential for exposure to TCDD. The mortality of these workers
is examined in the context of the mortality experience of the total
Nitro plant worker population.

The Monsanto Nitro plant began operations in 1922, when the
Rubber Service Laboratories purchased the plant as war surplus and
began production of chemicals and additives for the growing rubber
industry. In 1929, Monsanto Company purchased the Nitro plant from
the Rubber Service Laboratories and entered the rubber chemicals
business. Over the years, the Nitro plant has diversified to where
it now produces agricultural chemicals, paper chemicals, plasticizers,
fine chemicals, and intermediates, in addition to rubber chemicals.
The plant is situated in the Kanawha River Valley, an area containing
one of the largest concentrations of chemical production facilities
in the United States.

Of the many chemicals used over the years at the Nitro plant,
one has an established association with the occurrence of cancer in
man. Para-aminobiphenyl (PAB), used from 1941 through 1952 for use
as a rubber antioxidant and dye intermediate, was shown in 1954 by
Walpole et al.[7] to induce bladder cancer in dogs. In 1955, Melick
et al.[8] confirmed the carcinogenicity of para-aminobiphenyl to man
with the reporting of bladder tumors among workers exposed to this
chemical at two Monsanto plants. Para-aminobiphenyl was produced at
one plant and then transferred by tank car to the Nitro plant where
additional processing was carried out. The minimum duration of
exposure reported to have produced a bladder tumor is 133 days; the
latent period has ranged from 15 to 35 years.[9] An intensive screening program was instituted at Monsanto Company about 1955 to examine,
on a continuing basis, all workers exposed to this chemical. Seven
deaths from bladder cancer have occurred among Nitro plant employees
enrolled in this program. These seven deaths are included in the
present study.

Other chemicals with known health effects produced at the Nitro plant include methylparathion and carbon disulfide. The acute effects of exposure to each of these chemicals have been described[9],[10] while the chronic effects are less well understood. Besides p-aminobiphenyl, several other rubber chemicals are of potential health concern. For all of these, there is insufficient evidence to evaluate their carcinogenicity to man. Tetramethyl thiuram disulfide is considered an animal carcinogen[11] and there is some evidence, although not sufficient, that N-methyl-N,4-dinitrosaniline is also an animal carcinogen.[12] For several other rubber chemicals, there is insufficient evidence to evaluate the carcinogenicity to animals. However, these chemicals have the potential for forming nitrosamines, certain of which are known animal carcinogens.[13] Zinc dimethyl dithiocarbamate, tetramethyl thiuram disulfide and tetramethyl thiuram monosulfide have the potential to form N-nitrosodimethylamine. N-Nitrosomorpholine has been found in product samples of 2-(morpholinothio) benzothiazole and 4,4'-dithiodimorpholine (Frisone, G.J., The General Tire and Rubber Company, unpublished data).

Although several of the chemical compounds produced or used at the Nitro plant over the years have been associated with adverse health effects, no attempt has been made to relate chemical exposure to mortality with the exception of decedents exposed to the TCP or 2,4,5-T operations and potentially exposed to TCDD. As a result, the only specific hypothesis that can be tested is whether a relationship exists between potential TCDD exposure and proportional mortality, especially for malignant neoplasms. The mortality for this group is examined in addition to that of the total Nitro plant worker population.

POPULATION AND METHODS

A study cohort was developed from the Nitro plant consisting of employees active on or after January 1, 1955 with one or more years of employment on the hourly roll prior to December 31, 1977. Salaried personnel who had never worked on the hourly roll were excluded from study because many of these employees had no appreciable exposure to the plant environment and, for the most part, their exposure cannot be determined from plant records. Females and non-white males were also excluded because of their small numbers.

The cohort was assembled using government earnings reports, independent of the plant work history records. Annual earnings reports were available on a computer file from 1951 through 1977. Names and social security numbers were identified from this source. Work history records were used to supplement the earnings records. Information on race, sex, date of birth, date of hire, date of separation, and, if deceased, 2,4,5-T exposure was abstracted from these records. Ascertainment of 2,4,5-T exposure was confined to decedents only because it was too tedious to do for the entire cohort. The

information from the work history records was used to determine which
names identified from the annual earnings records met the cohort
entrance criteria. Employees identified from the earnings records
who terminated prior to 1955 were not included in this study because
work history records for all such employees were not retained prior
to this date.

Exposure to 2,4,5-T was determined by assignment to a 2,4,5-T
operation based on the work history records. 2,4,5-T exposure was
determined for all but one decedent. Employees holding a job having
plant-wide responsibilities with the potential for exposure to
2,4,5-T were, for the purposes of this study, considered to be non-
exposed.

The vital status of each member of the study cohort was determined using standard follow-up techniques and ascertained as of
December 31, 1977. Death certificates were coded by an independent
nosologist for the underlying cause of death, according to the rules
of the Eighth Revision of the International Classification of
Diseases, Adapted.[14]

Data for the total Nitro plant study population were analyzed
by the modified life-table method using the U.S. population as the
standard. With this method of analysis, the age-, race-, time-, and
cause-specific mortality rates for the U.S. general population are
applied to the person-years lived classified by age, race, and time.
A standardized mortality ratio (SMR) was calculated for 23 selected
cause of death categories. Cause-specific SMR's for 15 selected
cancer sites were calculated for subgroups of the total Nitro plant
study population defined by date of death, age at death, and year of
hire. The statistical significance of the deviation in a SMR from
100 was tested using the formula:

$$\text{standard error of SMR} = \frac{100 \ \sqrt{\text{no. observed deaths}}}{\text{no. expected deaths}}$$

If the observed SMR differed from 100 by 1.96 standard errors,
it was regarded as significant at the 5% level. A SMR was tested
for significance only when the observed number of deaths was five
or greater.

Data for those deceased were also analyzed according to 2,4,5-T
exposure using the proportional mortality method. In this case, the
expected number of deaths is calculated on the basis of proportions
of deaths observed in the U.S. general population. A proportional
mortality ratio (PMR) was calculated for 23 selected cause-of-death
categories. PMR's were calculated separately for those exposed to
2,4,5-T and for those not exposed. The statistical significance of
the deviation of a PMR from 100 was tested by calculating the 95%

confidence interval for the PMR for a given cause k according to the
formula:

$$\text{confidence interval for PMR}_k = \frac{N(\text{confidence limits for \% of observed deaths})}{\exp_k}$$

where N = number of observed deaths for all causes

Both the standardized and proportional mortality analyses were
conducted using the computer program developed by Monson.[15] The
observed mortality was compared to that of the United States white
male population for the time period of study.

RESULTS

A total of 884 men were identified for study and traced for
deaths through 1977. The entire cohort was successfully traced.
Death certificates were obtained for all deaths. Seven hundred
twenty-one (82%) were verified as living and 163 (18%) were confirmed
dead by death certificates.

Tables 1-3 characterize the total Nitro plant study population
by age at hire, year of hire, and length of employment. The distribution of the study population by age at hire indicates that the
workers were fairly young at first hire (Table 1). Seventy-two
percent of the study cohort were hired by age 30. None were hired
at age 50 or above.

Table 2 shows the distribution of the study population by year
of hire. The majority of workers (75.3%) were hired during the
period 1940-1959, while smaller percentages of workers were hired
prior to 1940 (10.8%) and after 1960 (13.9%).

An examination of the study population by length of employment
indicates a fairly even distribution over the intervals less than
10 years, 10-19 years, 20-29 years, and greater than 30 years
(Table 3).

Observed and expected deaths occurring during 1955-1977 among
the total Nitro plant study population are shown by cause in Table 4.
The SMR for all causes of death was 103 with 163 deaths observed and
158.10 expected. There were 35 deaths from malignant neoplasms
with 30.92 expected, yielding a SMR of 113. Significantly elevated
SMR's were seen for the categories of malignant neoplasms of the
genitourinary organs and of the bladder. The SMR for bladder cancer
was 989 with 9 deaths observed and 0.91 expected. This excess in
bladder cancer deaths is reflected in the elevated SMR for malignant
nioplasms of the genitourinary organs. A significantly elevated
SMR is also seen for arteriosclerotic heart disease. There were 79
deaths from this cause with 59.40 expected (SMR = 133). The SMR for

Table 1. Distribution of Total Nitro Plant Study Population
          by Age at Hire

Age at Hire	Number	Percent
<20	122	13.8
20-29	518	58.6
30-39	191	21.6
40-49	53	6.0
50+	0	0.0
Total	884	100.0

Table 2. Distribution of Total Nitro Plant Study Population
          by Year of Hire

Year of Hire	Number	Percent
Prior to 1930	21	2.4
1930-1939	74	8.4
1940-1949	361	40.8
1950-1959	305	34.5
1960-1976	123	13.9
Total	884	100.0

Table 3. Distribution of Total Nitro Plant Study Population
          by Length of Employment

Length of Employment (yrs.)	Number	Percent
<10	225	25.5
10-19	213	24.1
20-29	204	23.1
30+	242	27.4
Total	884	100.1

Table 4. Observed and Expected Number of Deaths During 1955-1977 by Cause Showing Standardized Mortality Ratios (SMR'S) for Total Nitro Plant Study Population

Cause of Death	ICDA Codes 8th Rev.	Observed	Expected	SMR
All causes of death		163	158.10	103
All malignant neoplasms	140-209	35	30.92	113
Buccal cavity and pharynx	140-149	0	1.03	0
Digestive organs and peritoneum	150-159	4	8.65	46
Stomach	151	1	1.63	61
Liver	155-156	0	0.60	0
All other digestive organs	---	3	6.42	47
Respiratory system	160-163	14	10.51	133
Lung	162-163	14	9.91	141
All other respiratory organs	---	0	0.60	0
Skin	172-173	0	0.58	0
Genitourinary organs	185-189	12	3.75	320*
Bladder	188	9	0.91	989*
All other genitourinary organs	---	3	2.84	106
Lymphatic and hematopoietic tissue	200-209	1	3.15	32
Other sites	---	4	3.25	123
Diseases of the nervous system and sense organs	320-389	0	1.28	0
Diseases of the circulatory system	390-458	92	82.59	111
Arteriosclerotic heart disease, including CHD	410-413	79	59.40	133*
All other diseases of the circulatory system	---	13	23.19	56*
Diseases of the respiratory system	460-519	6	9.13	66
Diseases of the digestive system	520-577	5	8.03	62
All other diseases	---	5	10.46	48*
External causes of death	800-998	20	15.69	127

Number at risk: 884
Person-years at risk: 13968.7
*$p < .05$

other circulatory diseases was significantly low at 56. A significant
deficit was also seen for the category of all other diseases where the
SMR was 48 with 5 deaths observed and 10.46 expected.

Trends for malignant neoplasm deaths with calendar time are shown
in Table 5. SMR's which increased consistently over the period 1955-
1977 are seen for the categories of all malignant neoplasms and malignant neoplasms of the respiratory system. The SMR for all malignant
neoplasms rose from 32 to 131. The SMR for malignant neoplasms of the
respiratory system rose from 0 to 172 reflecting a SMR for lung cancer
which increased from 0 to 181. Although based on very small numbers,
the SMR for malignant neoplasms of the digestive organs and peritoneum
consistently decreased over time from 99 to 24. The SMR for malignant
neoplasms of the genitourinary organs peaked during the period 1960-
1969. This is reflected in the SMR for bladder cancer which increased
from 0 in 1955-1959 to a peak of 1471 in 1960-1969 and decreased to
833 in 1970-1977.

The analysis of observed and expected deaths from malignant neoplasms by age at death revealed little in terms of consistent trends
with age (Table 6). For most of the cause-of-death categories, the
SMR peaked at age 45-64 rather than continuing to rise with increasing
age. Bladder cancer is one category where the SMR remained high at
age 65 and over.

The distribution of deaths from malignant neoplasms by year of
hire is shown in Table 7. No deaths from malignant neoplasms were
observed among workers hired after 1960. The SMR for all malignant
neoplasms was similar for those hired prior to 1945 and for those
hired in the period 1945-1959. The greatest difference in SMR's between those hired prior to 1945 and those hired from 1945-1959 occurs
with bladder cancer. The SMR for bladder cancer is highest for those
hired prior to 1945 with a SMR of 1111.

A subset of deaths identified from the total Nitro plant study
population was studied separately. Table 8 characterizes the decedents according to 2,4,5-T exposure. Of the 163 decedents, 58 (35.6%)
were considered to be exposed to 2,4,5-T based on their work history
records, while 104 (63.8%) were considered to be non-exposed. The
exposure of one decedent was unkown.

The results of the proportional mortality analysis by 2,4,5-T
exposure classification are presented in Table 9. The proportion of
cancer deaths among 2,4,5-T workers is lower than in the non-exposed
group (PMR: 82 vs. 122). The PMR for lung cancer deaths is slightly
higher in the exposed group (PMR: 159 vs. 117). The proportion of
deaths due to bladder cancer is higher among the decedents not exposed to 2,4,5-T. There were 7 deaths from bladder cancer among these
workers with 0.65 expected (PMR = 1077). The PMR for bladder cancer
among decedents exposed to 2,4,5-T was 909 with 2 deaths observed and

Table 5. Observed and Expected Deaths from Malignant Neoplasms During 1955-1977 by Calendar Time Showing Standardized Mortality Ratios (SMR'S) for Total Nitro Plant Study Population

| | Calendar Time | | | | | | | | |
| | 1955-1959 | | | 1960-1969 | | | 1970-1977 | | |
Cause of Death	Observed	Expected	SMR	Observed	Expected	SMR	Observed	Expected	SMR
All malignant neoplasms	1	3.13	32	13	11.81	110	21	15.98	131
Buccal cavity and pharynx	0	0.11	0	0	0.41	0	0	0.50	0
Digestive organs and peritoneum	1	1.01	99	2	3.48	57	1	4.16	24
Stomach	0	0.25	0	0	0.70	0	1	0.69	145
Liver	0	0.09	0	0	0.28	0	0	0.23	0
All other digestive organs	1	0.67	149	2	2.50	80	0	3.24	0
Respiratory system	0	0.87	0	4	3.83	104	10	5.81	172
Lung	0	0.80	0	4	3.59	111	10	5.52	181
All other respiratory organs	0	0.07	0	0	0.24	0	0	0.29	0
Skin	0	0.07	0	0	0.23	0	0	0.28	0
Genitourinary organs	0	0.33	0	5	1.37	365	7	2.05	341
Bladder	0	0.09	0	5	0.34	1471*	4	0.48	833
All other genitourinary organs	0	0.24	0	0	1.03	0	3	1.57	191
Lymphatic and hematopoietic tissue	0	0.40	0	0	1.24	0	1	1.50	67
Other sites	0	0.34	0	2	1.25	160	2	1.68	119

*p < .05

Table 6. Observed and Expected Deaths from Malignant Neoplasms During 1955-1977 by Age at Death Showing Standardized Mortality Ratios (SMR'S) for Total Nitro Plant Study Population

				Age at Death					
	<45			45-64			65+		
Cause of Death	Observed	Expected	SMR	Observed	Expected	SMR	Observed	Expected	SMR
All malignant neoplasms	2	2.39	84	21	16.85	125	12	11.69	103
Buccal cavity and pharynx	0	0.06	0	0	0.67	0	0	0.31	0
Digestive organs and peritoneum	0	0.51	0	4	4.58	87	0	3.57	0
Stomach	0	0.10	0	1	0.85	118	0	0.68	0
Liver	0	0.03	0	0	0.32	0	0	0.24	0
All other digestive organs	0	0.38	0	3	3.41	88	0	2.65	0
Respiratory system	0	0.54	0	11	6.36	173	3	3.61	83
Lung	0	0.51	0	11	5.99	184	3	3.42	88
All other respiratory organs	0	0.03	0	0	0.37	0	0	0.19	0
Skin	0	0.13	0	0	0.31	0	0	0.14	0
Genitourinary organs	1	0.20	500	3	1.43	210	8	2.12	377*
Bladder	1	0.02	5000	2	0.39	513	6	0.50	1200*
All other genitourinary organs	0	0.18	0	1	1.04	96	2	1.62	123
Lymphatic and hematopoietic tissue	0	0.55	0	1	1.60	63	0	1.00	0
Other sites	1	0.40	250	2	1.90	105	1	0.94	106

*p < .05

Table 7. Observed and Expected Deaths from Malignant Neoplasms During 1955-1977 by Year of Hire Showing Standardized Mortality Ratios (SMR'S) for Total Nitro Plant Study Population

Cause of Death	Observed	Expected	SMR	Observed	Expected	SMR	Observed	Expected	SMR
All malignant neoplasms	25	21.30	117	10	8.63	116	0	0.99	0
Buccal cavity and pharynx	0	0.70	0	0	0.31	0	0	0.03	0
Digestive organs and peritoneum	2	6.27	32	2	2.19	91	0	0.19	0
Stomach	0	1.21	0	1	0.39	256	0	0.03	0
Liver	0	0.45	0	0	0.14	0	0	0.01	0
All other digestive organs	2	4.61	43	1	1.66	60	0	0.15	0
Respiratory system	10	7.14	140	4	3.09	129	0	0.27	0
Lung	10	6.73	149	4	2.93	137	0	0.26	0
All other respiratory organs	0	0.41	0	0	0.16	0	0	0.01	0
Skin	0	0.31	0	0	0.22	0	0	0.05	0
Genitourinary organs	9	2.90	310*	3	0.76	395	0	0.09	0
Bladder	8	0.72	1111*	1	0.18	556	0	0.01	0
All other genitourinary organs	1	2.18	46	2	0.58	345	0	0.08	0
Lymphatic and hematopoietic tissue	1	1.95	51	0	1.00	0	0	0.20	0
Other sites	3	2.03	148	1	1.06	94	0	0.16	0

*p < .05

Table 8. 2,4,5-T Exposure Classification for Decedents During 1955-1977 Among Total Nitro Plant Study Population

2,4,5-T Exposure Classification	Number of Deaths	Percent
Exposed	58	35.6
Non-exposed	104	63.8
Unknown	1	0.6
Total	163	100.0

Table 9.   Observed and Expected Number of Deaths During 1955-1977 by Cause and 2,4,5-T Exposure Category Showing Proportional Mortality Ratios (PMR'S)

Cause of Death	2,4,5-T Exposure Category					
	Exposed			Non-exposed		
	Observed	Expected	PMR	Observed	Expected	PMR
All malignant neoplasms	9	10.94	82	25	20.43	122
Buccal cavity and pharynx	0	0.38	0	0	0.64	0
Digestive organs and peritoneum	0	2.80	0	3	5.74	52
Stomach	0	0.52	0	0	1.06	0
Liver	0	0.19	0	0	0.39	0
All other digestive organs	0	2.09	0	3	4.29	70
Respiratory system	6	3.78	159	8	6.81	117
Lung	6	3.57	168	8	6.42	125
All other respiratory organs	0	0.21	0	0	0.39	0
Skin	0	0.29	0	0	0.35	0
Genitourinary organs	2	0.96	208	10	2.70	370*
Bladder	2	0.22	909	7	0.65	1077*
All other genitourinary organs	0	0.74	0	3	2.05	146
Lymphatic and hematopoietic tissue	0	1.35	0	1	2.07	48
Other sites	1	1.38	72	3	2.12	142
Diseases of the nervous system and sense organs	0	0.61	0	0	0.83	0
Diseases of the circulatory system	31	26.48	117	61	55.34	110
Arteriosclerotic heart disease, including CHD	27	19.72	137	52	39.68	131*
All other diseases of the circulatory system	4	6.76	59	9	15.66	57
Diseases of the respiratory system	2	2.67	75	4	6.49	62
Diseases of the digestive system	1	3.70	27	4	4.93	81
All other diseases	3	4.31	70	2	6.70	30
External causes of death	12	9.29	129	8	9.28	86
Total number of deaths:	58	58.00		104	104.00	

*p < .05

0.22 expected. PMR's for both exposure groups are quite similar for diseases of circulatory and respiratory system. Slight differences appear in the PMR's between the two groups for diseases of the digestive system and all other diseases. However, the PMR's for both groups are quite low. The PMR for external causes of death is slightly higher in the exposed group (PMR: 129 vs. 86).

A listing of the cancer deaths among the 2,4,5-T exposed is given in Table 10. Table 11 listed the cancer deaths among the non-exposed group.

DISCUSSION

The observation made many years ago of an apparent excess in bladder cancer among Nitro plant workers was confirmed and quantified in the mortality analysis of the total Nitro plant population presented here. The SMR for bladder cancer was 989 and was the only statistically significant SMR among those for malignant neoplasms. The excess in mortality is not seen until 1960. The SMR peaked in the 1960's and declined somewhat in the 1970's. The excess also appeared to be clustered in those decedents aged 65 years and older at death and in those hired prior to 1945. This would suggest that we should see a further decline in the SMR for bladder cancer over time.

Although not statistically significant, the SMR for lung cancer appears to be elevated. The SMR increased with calendar time and was highest in those hired prior to 1945. The elevation appears to be clustered in those aged 45-64 years of age at death (SMR = 184) and does not show a gradient with age. Further analyses to evaluate trends in lung cancer deaths as they relate to occupation cannot be carried out due to limitations in the data collected in this study.

The SMR for diseases of the circulatory system was elevated at 111. This is most likely a reflection of the higher mortality from heart disease which has been observed for Charleston, West Virginia and Kanawha County, West Virginia (unpublished data, Neas LM, 1979 and Enterline PE, 1979). For the total Nitro plant study population, there was a statistically significant excess in deaths from arteriosclerotic heart disease and a deficit in deaths from other circulatory diseases. This variation in the distribution of deaths from that of the U.S. may be due to risk factor or medical care differences in the plant population and the local area. These may include differences in smoking habits, the availability and use of medical services and the specificity of diagnoses.

The proportional mortality analysis of decedents by 2,4,5-T exposure classification indicated no unusual patterns of mortality in the 2,4,5-T exposed. The proportional mortality ratio (PMR for malignant neoplasms was low (PMR = 82) in the exposed group. Lung cancer was the only site among the malignant neoplasms which was somewhat higher in the exposed group.

Table 10. Deaths Due to Malignant Neoplasms Among
Nitro Plant Workers Exposed to 2,4,5-T

Year of Birth	Year of Hire	Year of 1st Exposure	Year of Term.	Year of Death	Smoking History*	Cause of Death as Given on Death Certificate
1917	1946	1951	1972	1972	Cigarettes	Carcinoma left lung with metastases (162.1)
1911	1948	1955	1972	1972	Cigarettes	Metastatic carcinoma of the lung (162.1)
1916	1946	1959	1968	1968	Cigarettes	Bronchiogenic carcinoma of right upper lobe (162.1)
1901	1944	1956	1963	1973	Cigarettes	Carcinoma lung with metastases (162.1)
1911	1941	1948	1971	1975	Cigarettes	Bronchiogenic carcinoma with cerebral metastases (162.1)
1922	1945	1948	1972	1973	Non-smoker	Bronchiogenic carcinoma with metastases (162.1)
1923	1946	1950	1972	1972	Cigarettes	Generalized liposarcoma (171.9)
1902	1922	1948	1966	1966	Non-smoker	Metastatic carcinoma urinary bladder (188.0)**
1910	1944	1948	1968	1968	Non-smoker	Carcinoma of the urinary bladder (188.0)**

* Obtained by interview with former coworkers of decedents.
** Included on the Nitro plant PAB roster.

Table 11. Deaths Due to Malignant Neoplasms Among Nitro Plant Workers Not Exposed to 2,4,5-T

Year of Birth	Year of Hire	Year of Termin.	Year of Death	Smoking History*	Cause of Death as Given on Death Certificate
1911	1945	1966	1968	Cigarettes	Carcinoma of colon (153.8)
1904	1935	1957	1957	Cigarettes	Carcinoma of liver (157.9)
1899	1929	1959	1960	Pipe	Intraperitoneal carcinoma (158.9)
1905	1944	1970	1972	Cigarettes	Carcinoma of lung (162.1)
1909	1943	1962	1962	Cigarettes	Carcinoma of left lung (162.1)
1910	1927	1966	1970	Cigarettes	Pulmonary carcinoma (162.1)
1912	1944	1965	1965	Cigarettes	Carcinoma of apex of right lung (162.1)
1915	1937	1977	1977	Cigarettes	Carcinoma of lung (162.1)
1915	1939	1969	1970	Cigarettes	Carcinoma of lungs (162.1)
1894	1944	1960	1964	Non-smoker	Carcinoma of lung (162.1)
1912	1937	1973	1974	Cigarettes	Carcinoma of lung (162.1)
1919	1946	1963	1964	Cigarettes	Carcinoma of urinary bladder (188.0)**
1901	1933	1962	1977	Cigarettes	Carcinoma of bladder (188.0)**
1897	1941	1962	1965	Cigarettes	Carcinoma of urinary bladder (188.0)**
1898	1933	1962	1965	Smoked years ago	Carcinoma of urinary bladder (188.0)**
1905	1943	1971	1975	Cigarettes	Carcinoma of bladder (188.0)
1898	1943	1963	1970	Cigarettes	Bladder tumor (188.0)**
1888	1944	1956	1970	Unknown	Carcinoma of bladder (188.0)
1905	1933	1969	1977	Cigars	Prostatic carcinoma (185.0)
1906	1946	1968	1977	Cigarettes	Carcinoma of prostate (185.0)
1925	1945	1973	1974	Smoked years ago	Carcinoma of prostate (185.0)
1919	1943	1972	1973	Cigarettes	Hodgkin's Disease (201.0)
1901	1941	1964	1965	Cigars	Osteosarcoma arising from left arm (170.4)
1901	1943	1966	1977	Cigarettes	Carcinoma of liver & pancreas (197.8)
1922	1944	1964	1964	Cigarettes	Adenocarcinoma (199.0)

* Obtained by interview with former coworkers of decedents.
** Included on the Nitro plant PAB roster.

The PMR analysis is limited in that an assessment of the total force of mortality cannot be made. The cause-specific PMR's only approximate what an SMR analysis would have produced.[16] The PMR analysis presented here estimates the cause-specific risks associated with 2,4,5-T exposure and potential TCDD exposure.

It is interesting to compare the results of this study of Nitro plant workers potentially exposed to TCDD with the results of the study of Nitro workers involved in the 1949 TCP accident. The workers involved in that incident had presumed TCDD exposure as evidenced by chloracne. The results of the two studies are similar in that neither shows an excess in deaths from any site among malignant neoplasms.

A recent study of Ott et al.[17] found no excess in total mortality or in deaths from malignant neoplasms among workers exposed to 2,3,5-T. These workers were probably exposed to very low levels of TCDD since no cases of chloracne were observed. Other studies of workers who developed chloracne resulting from TCDD exposure have been conducted and have been reviewed.[6] At the present time, data from these various studies do not constitute corroborative evidence of a cancer risk to man for any particular cancer site.

REFERENCES

1. Greig, J.B.: The toxicology of 2,3,7,8-tetrachlorodibenzo-p-dioxin and its structural analogues. Ann. Occup. Hyg. 22: 411-420, 1979.
2. International Agency for Research on Cancer: Long-term hazards of polychlorinated dibenzodioxins and polychlorinated dibenzo-furans. IARC Internal Technical Report No. 78/001. Lyon: IARC, 1978.
3. Ashe, W.F. and Suskind, R.R.: Reports on chloracne cases, Monsanto Chemical Company, Nitro, West Virginia. Reports of the Kettering Laboratory, December 1949 and April 1950.
4. Suskind, R.R.: A clinical and environmental survey, Monsanto Chemical Company, Nitro, West Virginia. Report of the Kettering Laboratory, July 1953.
5. International Agency for Research on Cancer. IARC Monographs on the Evaluation of the Carcinogenic Risk of Chemicals to Man. Vol. 15. Some Fumigants, the Herbicides 2,4-D and 2,4,5-T, Chlorinated Dibenzodioxins and Miscellaneous Industrial Chemicals. Lyon: IARC, 1977.
6. Zack, J.A. and Suskind, R.R.: The mortality experience of workers exposed to tetrachlorodibenzodioxin in trichlorophenol process accident. J. Occup. Med. 22:11-14, 1980.
7. Walpole, A.L., Williams, M.H., and Roberts, D.C.: Tumours of urinary bladder in dogs after ingestion of 4-aminodiphenyl. Brit. J. Industr. Med. 11:105-109, 1954.

8. Melick, W.F.: First reported cases of human bladder tumors
    due to a new carcinogen - xenylamine. J. Urol. 74:760-766, 1955.
9. Melick, W.F., Naryka, J.J., and Delly, R.E.: Bladder cancer due
    to exposure to para-aminobiphenyl: a 17-year followup. J. Urol.
    106:220-226, 1974.
10. Key, M.M.: Ocuppational Diseases: A Guide to Their Recognition.
    U. S. Department of Health, Education and Welfare, Public
    Health Service, Center for Disease Control, National
    Institute for Occupational Safety and Health. DHEW (NIOSH)
    Publication No. 77-181. Washington: U.S. Government Printing
    Office, 1977.
11. International Agency for Research on Cancer. IARC Monographs on
    the Evaluation of the Carcinogenic Risk of Chemicals to Man.
    Vol. 12. Some Carbamates, Thiocarbamates, and Carbazides.
    Lyon: IARC, 1971.
12. International Agency for Research on Cancer. IARC Monographs on
    the Evaluation of Carcinogenic Risk of Chemicals to Man. Vol.
    1. Lyon: IARC, 1971.
13. Magee, P.N.,: N-nitroso compounds and related carcinogens. In:
    Searle, C.E.(ed): Chemical Carcinogens. Monograph 176.
    Washington: American Chemical Society, 1976.
14. Eighth Revision, International Classification of Diseases, Adapted
    for Use in the United States. U.S. Department of Health, Ed-
    ucation, and Welfare, Public Health Service, PHS Publication
    No. 1693. Washington: U.S. Government Printing Office, 1977.
15. Monson, R.R.,: Analysis of relative survival and proportional
    mortality. Comput. Biomed. Res. 7:325:332, 1974.
16. Decoufle, P., Thomas, T.H., and Pickle, L.W.: Comparison of the
    proportionate mortality ratio and standardization mortality
    ratio risk measures. Am. J. Epidemiol. 111:263-269, 1980.
17. Ott., G., Holder, B.B., and Olson, R.: A mortality analysis of
    employees engaged in the manufacture of 2,4,5-trichloro-
    phenoxyacetic acid. J. Occup. Med. 22:47-50, 1980.

DIOXIN AND REPRODUCTIVE EVENTS

Ralph R. Cook and Kenneth M. Bodner

Health and Environmental Sciences
Dow Chemical, USA
Midland, Michigan USA

Polychlorinated dibenzodioxins, sometimes referred to as PCDD's or simply dioxins, may be formed during the combustion of fossil fuels, the incineration of municipal wastes, or the production of chlorinated phenols.[1] Of all of the isomers of dioxins, 2,3,7,8-tetrachlorodibenzodioxin or TCDD, a low level contaminant of 2,4,5-T, trichlorophenol and its by-products, is considered the most toxic.[2]

A variety of biological effects have been indicated or hypothesized to be associated with exposures to TCDD. One set involves the potential influence of this chemical on reproduction. Animal experiments suggest adverse reproductive events may be associated with maternal exposures, but not with those of the father.[3,4,5] Reports of observational studies of humans are somewhat conflicting. The National Academy of Sciences, commissioned to study the effects of herbicides in South Vietnam, found no conclusive evidence of an association between exposure to herbicides and birth defects in humans.[6] Field and Kerr, in an ecological study in Australia, purported to have found a "linear correlation" between neural tube defects and increased use of the herbicide 2,4,5-trichlorophenoxy acetic acid (2,4,5-T), an agricultural chemical contaminated by low levels of TCDD.[7] On the other hand, Thomas reported that in Hungary, where 2,4,5-T usage had increased dramatically between 1970 and 1976, the frequency of stillbirths and congenital malformations had remained stable or decreased.[8] Smith recently published a "preliminary report of reproductive outcomes among pesticide applicators using 2,4,5-T in New Zealand".[9] No significant differences were found in the rates of congenital defects, stillbirths, and miscarriages between the chemical applicators and a control group.

There have also been dioxin-related claims of congenital malformations among offspring of United States and Vietnamese veterans

of the Vietnam conflict who were exposed to aerial herbicides spray-
ings;[10],[11] and of increased spontaneous abortions among women residing
in or near areas of Oregon sprayed with 2,4,5-T.[12] None of these
allegations have been substantiated epidemiologically and the Oregon
spontaneous abortion data, when subjected to critical review, were
found not to support the conclusion of an association with herbicide
usage.[13]

Dow Chemical USA has conducted a series of epidemiological
studies of employees potentially exposed to dioxin.[14],[15] The following is a summary of a paper by Townsend et al, which has been
submitted for publication.[16] It describes the results of an inter-
viewer-administered questionnaire survey conducted among women whose
husbands were employees of the Michigan Division of Dow Chemical
U.S.A. The purpose of the survey was to compare the proportion of
adverse outcomes between pregnancies associated with potential
paternal exposures to 2,3,7,8-TCDD or other polychlorinated dioxins,
and pregnancies with no known exposures to these chemicals.

MATERIALS AND METHODS

Selection of Study Subjects

Identification of wives whose husbands had been potentially
exposed to dioxins, and of those whose husbands had not, was accomplished via company records. Using annual personnel census lists
and individual work histories, employees who had been assigned to
chlorophenol processes for at least one month between January 1939
and December 1975 were identified. By this criterion, there were
930 male employees with potential occupational dioxin exposures.

A comparable number of controls was selected from male employees
who had no job assignments in the potential exposure areas, and who
closely matched exposed employees by date of hire. By a variety of
methods, addresses were obtained on all but 69 (3.7 percent) of the
1830 men. For the purposes of study efficiency, followup was limited
to those who lived in Midland or adjacent counties. In the exposed
group, 586 had wives potentially available for interview and 370 or
63% agreed to participate. Among the control group, 559 wives were
potentially available and 345 or 62% were interviewed.

Estimation of Exposures

Manufacture of pentachlorophenol started at the Michigan Division
of Dow Chemical, U.S.A. in 1939. Trichlorophenol production started
in 1946. In late 1963, a change was made in the trichlorophenol
process and shortly thereafter a number of employees contracted
chloracne. Operations were modified to reduce exposures and, in
1966, trichlorophenol production was transferred to a new building.

Estimation of exposures was made by an industrial hygienist who was familiar with the processes. Based on his experience, information provided by production personnel and surface contamination data, each job was assigned an exposure potential on a scale of low to high for TCDD, for all other dioxins, or for mixed exposures.

Collection and Analysis of Data

A letter explaining the purpose of the study and requesting a personal interview was mailed to the residence of each living exposed or control population member residing in Midland or adjacent counties. Appointments were made with those women willing to be interviewed. If the employee was widowed or single, or if the response indicated the wife did not wish to be interviewed, no further action was taken.

Interviews were conducted in the home by one of three trained nurse-interviewers. Appointments were arranged by geographic clusters and not by exposure status. The interviewers did not know whether they were interviewing the wives of exposed or control empolyees. At the time of the interview, the intent of the study was re-explained, each woman was advised regarding confidentiality of the information and was asked to sign an informed consent. A five-page questionnaire was used and the interview normally took less than one half-hour.

Responses dealing with health defects, congential malformations and infant deaths were coded by the Eighth Revision of the International Classification of Diseases (ICD-8).[9] Each positive response regarding congenital malformations was coded if the mother stated the condition was diagnosed prior to one year after birth or if the mother was unclear as to the time of diagnosis. Conditions identified after the first birthday were reviewed by two physicians. If either physician, blinded as to the mother's group, determined the defect could have been diagnosed during the first year of life, the condition was so coded. Since these procedures were weighted toward identifying abnormalities, the resulting data could not be meaningfully compared with state or national statistics.

For the main purposes of this study, the individual conceptus or product of conception was the observational unit, taken as independent of other conceptuses of the same pregnancy or other pregnancies of the same or different womrn. Effect of exposure to dioxins was considered to be an irreparable event; that is, any exposure of a male employee any time prior to the estimated date of each conception attributed to him was investigated as an independent variable with potential effects on the pregnancy outcome. Conception dates were estimated from the length of pregnancy and pregnancy outcome dates reported in the questionnaire. When the date of conception preceded the father's first exposure, the conceptus was assigned to the no-exposure category even though the father was in the dioxins group. This prevented a dilutional effect on the risk estimates, while simultaneously making maximum use of the available data. Two new files were thereby created, one containing only "dioxins" conceptuses and the other "non-dioxins" conceptuses.

Variables

Dependent variables included live births and the adverse outcomes of miscarriages, stillbirths, infant deaths, and live births with congenital malformations. Birth defects were analyzed as either health defects, congenital malformations, or as indicator malformations. A health defect was defined as a live birth reported by the mother to have a health problem of any nature. A congenital malformation was a report by the mother of one or more defects in the 740-759.9 range of ICD-8.[17] And an indicator malformation, one that we felt would be readily observed by the mother, was any one of the following (preceded by its ICD-8 code):

741.0 Spina Bifida

742.0 Congenital Hydrocephalus

749.0 Cleft Palate

749.1 Cleft Lip

749.2 Cleft Palate with Cleft Lip

755.0 Ploydactyly

755.1 Syndactyly, and

759.3 Down's Syndrome

Thus, indicator malformations were a subset of congenital malformations which, in turn, constitute a subset of health defects.

Independent variables consisted of four exposure categories:

TCDD Only, at any level

Any Dioxins (TCDD and others), at any level

Moderate to High TCDD Only, and

Other Dioxins Only (no TCDD).

The first two were also broken down into two duration-of-exposure subcategories: equal to or less than one year, and greater than one year total exposure.

Covariables were nine potential confounding factors or other reasons for adverse reproductive outcomes referenced in the literature: age, birth control method, complications during labor and delivery, conditions during pregnancy, medications or treatment

during pregnancy, alcohol consumption, gravidity, high risk job, and
smoking during pregnancy (Table 1). You will note from the Table,
many were broken down into subcategories. Complications during labor
and delivery were only considered a potential confounder for the
analysis of stillbirths. High risk jobs were those held by the
mother during pregnancy, such as operating room technician, x-ray
technician, bacteriologist, cosmetologist, and dental assistant.
Smoking was confined to that pregnancy, and not life-time experience.

Statistical Analysis

Statistical analyses attempted to determine whether there were
differences in the distribution of outcomes between the "exposed"
and "non-exposed" conceptuses. This was done on crude or unadjusted
outcomes, and after adjusting by stratification for the influences
of pertinent covariables. Trend analyses were also examined for the
exposures to TCDD Alone or to Dioxins Of Any Type, that is, whether
there were trends of increasing adverse pregnancy outcomes with
increasing durations of exposure.

RESULTS

Interviews were obtained from 370 wives of potentially exposed
and 345 wives of control employees (Table 2). Eleven in the first
group and 13 in the second reported no pregnancies, 3.0 and 3.8
percent, respectively. A total of 1503 conceptions occurred among
wives of the exposed and 1274 among those of the unexposed, or 4.06
and 3.69 conceptions per woman, exposed and unexposed.

As mentioned earlier, most of the analyses used the conceptus
as the observational unit. Therefore, those conceptuses conceived
by·an "exposed" employee prior to his exposure to dioxin were considered to be "non-exposed" conceptuses and placed in the control
pool for the purposes of the analyses. As illustrated diagrammati-
cally in Figure 1, the consequence of this strategy was an increase
in the number of "non-exposed" and a corresponding decrease in the
number of "exposed" conceptuses. The 370 wives of employees exposed
some time between 1939 and 1975, had a total of 1503 conceptions,
but only 737 occurred after their husbands' first exposure to dioxin.
These 737 constitute the "exposed" conceptuses. Thus, 1503 minus
737 or 766 conceptions occurred before the father was employed in
a dioxins area.

We had two options with regard to the data on the 766. We
could throw the data away, or add the experience of those 766 "pre-
exposure" conceptuses to the 1274 conceptuses that occurred among
wives of never exposed employees. We chose the latter. That gave
us 766 plus 1274 or a grand total of 2040 conceptuses in the "unexposed" category. There were 9 induced abortions among this latter
group, which reduced the number to 2031. The rest of the discussion

Table 1. Covariables

Age

<18, 18-30, >30 Years

Birth Control

None, Pills, IUD, Other

Complications During Labor and Delivery

Conditions During Pregnancy

Medications

Alcohol

$\leq$6 or >6 Drinks Per Month

Gravidity

1, 2 or 3, > 3

High Risk Job

Smoking

Table 2. Summary of Pregnancies

	Exposed	Unexposed
Interviews	370	345
No Pregnancies		
Number	11	13
Percent	3.0	3.8
Conceptions		
Number	1503	1274
Mean	4.06	3.69

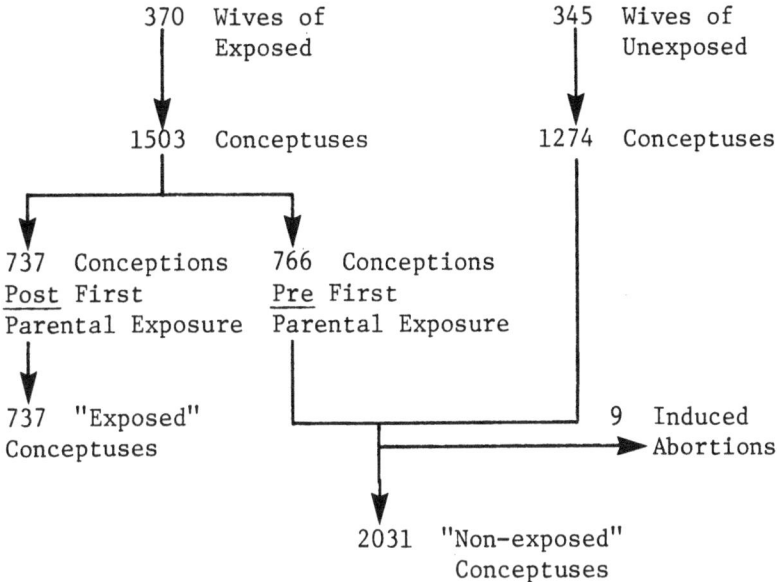

Figure 1. "Exposed" and "Non-exposed" Conceptuses

concerns the comparison of the 737 "exposed" conceptuses to the 2031
"unexposed".

Tables 3 and 4 provide crude and adjusted measures on Any Dioxin
and TCDD Only. None of the odds ratios for Any Dioxin (Table 3) were
significantly elevated, either crude or after adjustment for perti-
nant confounders. There was a notable lack of homogeneity among
some of the subtables: all conceptus deaths, spontaneous abortions,
and unfavorable outcomes.

For TCDD Only (Table 4), none of the odds ratios were signifi-
cantly elevated. Notice that some of the odds ratios, All Conceptus
Deaths and Infant Deaths, increased slightly after adjustment.
Notice also that the subtables, for the same three dependent vari-
ables - All Conceptus Deaths, Spontaneous Abortions, and Unfavorable
Outcomes, are marginally non-homogeneous.

The covariable-specific subtables for these exposure-outcome
combinations that showed evidence of non-homogeneity were examined
for significant associations, some 123 in number. Seven such associations were found; five with odds ratios greater than one, and two
with odds ratios less than one. Recurrent, in the subtables with
elevated odds ratios, was a group of 9 TCDD-exposed conceptuses who
contributed 3 spontaneous abortions and only 6 live births to the
analyses. Mothers of these conceptuses characteristically were
18-30 years of age, gravida 4+, were not using birth control methods

Table 3.   Crude and Adjusted Odds Ratios, P Values and
           Homogeneity of Subtables, All Outcomes, Any Dioxin

Outcome	Crude		Adjusted		
	Or	P	Or	P	Homogenity (P)
All Conceptus Deaths	1.14	.31	1.03	.44	<.01
Stillbirths	1.27	.26	1.06	.49	.20
Spontaneous Abortions	1.12	.23	1.03	.44	.02
Infant Deaths	0.64	.15	0.63	.15	.61
Health Defects	0.93	.37	0.85	.20	.57
Congenital Malformations	0.96	.48	0.85	.62	.62
All Unfavorable Outcomes	1.03	.42	0.97	.41	.05

Table 4.   Crude and Adjusted Odds Ratios, P Values, and
           Homogenity of Subtables, All Outcomes, TCDD Only

Outcome	Crude		Adjusted		
	Or	P	Or	P	Homogenity (P)
All Conceptus Deaths	0.99	.93	1.02	.49	.09
Stillbirths	1.03	.54	0.97	.56	.37
Spontaneous Abortions	0.98	.49	0.96	.45	.06
Infant Deaths	0.74	.32	0.82	.41	.42
Health Defects	1.00	.54	0.93	.40	.60
Congenital Malformations	1.18	.30	1.08	.43	.86
All Unfavorable Outcomes	0.98	.47	0.98	.47	.06

at conception, reported no illnesses during pregnancy, did not work
at high risk jobs, took medications during pregnancy, and had fewer
than seven drinks per week. Since TCDD Only was a subset of Any
Dioxin, and Spontaneous Abortions was a subset of All Conceptus
Deaths, which, in turn, was a subset of All Unfavorable Outcomes,
it is probably not surprising that this one covariable-specific
group produced such a ripple effect. I should note that a shift of
even a single covariable eliminated statistical significance and,
in some cases, changed the direction of the association.

Not shown are the results for the categories moderate-to-high
TCDD and other dioxins above. Their numbers were quite small, but
none of the results were statistically significant; each, of course,
was a subset of at least one of the categories shown. Crude and
adjusted trend analyses based on length-of-exposure categories were
also non-significant.

DISCUSSION

In summary, there were no statistically significant differences
in crude or adjusted risk estimates for any of the pregnancy outcomes
studied. Nor were there significant trands overall with increased
duration of exposure.

The observed lack of association of adverse pregnancy outcomes
with differences in paternal exposure to dioxins may have a variety
of explanations. There may have been problems related to selection,
misclassification, confounding or power that prevented us from pick-
ing up true differences in the two groups.[18,19] These are, of course,
problems with any epidemiological study and are discussed in the
referenced report.[16]

In addition, there may have been a problem with recall bias.
That is, those among the wives of the exposed, having been informed
of the intent of the study prior to the interview, may have ruminated
about adverse outcomes and reported more, relatively, than did the
controls. This would have produced an artificial elevation in some
or all of the risk estimates.

It is also possible that the results reflect the true state of
nature; that is, there is no risk to human reproduction caused by
paternal exposures to dioxins. This explanation is consistent with
most of the animal and human data we have to date.

In conclusion, it is our interpretation that, under the conditions of this study, there were no biologically meaningful associations between adverse pregnancy outcomes and paternal dioxin
exposures.

REFERENCES

1. Bumb, R.R., Crummett, W.B., Cutie, S.S., et al: A source of
       chlorinated dioxins. Science; 210:385-390, 1980.
2. Schwetz, B.A., Norris, J.M., Sparschu, G.L., et al: Toxicology
       of chlorinated dibenzo-p-dioxins. Environ Health Perspec;
       5:87-99, 1973.
3. Khera, K.S., Ruddick, J.A.: Ploychlorodibenzo-p-dioxins:
       Perinatal effects and the dominant lethal test in Wistar rats.
       Chlorodioxins - Origin and Fate. In: Blair, E.H. Advances
       in Chemistry, Series 120, American Chemistry Society, Wash,D.C.
       1973.

4. Murray, F.J., Smith, F.A., Nitschke, K.D., et al: Three-gener-
    ation reproduction study of rats given 2,3,7,8-tetrachloro-
    dibenzo-p-dioxin (TCDD) in the diet. Tox and Applied Pharm;
    50:241-252, 1979.
5. Lamb, J.C., Moore, J.A., Marks, T.A.: Evaluation of 2,4-dichloro-
    phenoxy-acetic acid (2,4-D), 2,4,5-trichlorophenoxyacetic acid
    (2,4,5-T), and 2,3,7,8-tetrachlorodibenzo-p-dioxin (TCDD)
    toxicity in C57BL/6 mice. Reproduction and fertility in
    treated male mice and evaluation of congenital malformations
    in their offspring. National Toxicology Program Manuscript
    NTP-80-44, 1980.
6. National Academy of Science (NAS): The effects of herbicides in
    South Vietnam. NAS Committee on the Effects of Herbicides in
    Vietnam, Division of Biological Sciences. Assembly of Life
    Sciences, National Research Council, NAS, Washington, D.C., 1974.
7. Field, B., Kerr, C.: (Letter to the Editor) Herbicide use and
    incidence of neural-tube defects. Lancet 1341-42, 1979.
8. Thomas, H.F.: (Letter to the Editor) 2,4,5-T use and congenital
    malformation rates in Hungary. Lancet 214-215, July 1980.
9. Smith, A.H., Matheson, D.P., Fisher, D.O., Chapman, C.J. Pre-
    liminary report of reproductive outcomes among pesticide
    applicators using 2,4,5-T. New Zealand Med Journal 93:177-9,
    1981.
10. Bogen, G.: (Letters) Symptoms in Vietnam veterans exposed to
    Agent Orange. JAMA 242:2391, 1979.
11. Tung, T.T., Anh, T.K., Tuyen, B.Q., et al: Clinical effects of
    massive and continuous utilization of defoliants on civilians.
    Etudes Vietnamiennes 9:57-87, 1971.
12. U.S. Environmental Protection Agency: Report of assessment of a
    Field investigation of six-year spontaneous abortion rates of
    three Oregon areas in relation to forest 2,4,5-T spray prac-
    tices. Prepared by Epidemiologic Studies Program, Human
    Effects Monitoring Branch, Benefits and Field Studies Division,
    Washington, D.C., 1979 (Unpublished)
13. Wagner, S.L., Witt, J.M., Norris, L.A., et al: A scientific
    critique of the Alsea II study and report. Manuscript from
    Environmental Health Sciences Center, Oregon State University,
    Corvallis, October 25, 1979.
14. Ott, M.G., Holder, B.B., Olson, R.D.: A mortality analysis of
    employees engaged in the manufacture of 2,4,5-trichloro-
    phenoxyacetic acid. JOM 22:47-50, 1980.
15. Cook, R.R., Townsend, J.C., Ott, M.G., Silverstein, L.G.:
    Mortality Experience of employees exposed to 2,3,7,8-tetra-
    chlorodibenzo-p-dioxin (TCDD). JOM 22:530-532, 1980.
16. Townsend, J.C., Bodner, K.M., VanPeenen, P.F.D., Olson, R.D.,
    Cook, R.R.: Survey of reproductive events of wives of employees
    exposed to chlorinated dioxins. (Submitted for publication
    1981)

17. International Classification of Diseases: Namual of the Inter-
     national Statistical Classification of Diseases, Injuries and
     Causes of Death. Eighth Revision. World Health Organization,
     Geneva, 1967.
18. MacMahon, B., Pugh, T.F.: Epidemiology. Principles and Methods.
     Little, Brown and Company, Boston, 1970.
19. Cook, R.R., Bond, G.G.: Epdiemiology: Basic Concepts. Presented
     at American Chemical Society Meeting, 27 August 1981 (In press).

SIGNIFICANCE OF CUTANEOUS LESIONS IN THE SYMPTOMATOLOGY OF

EXPOSURE TO DIOXINS AND OTHER CHLORACNEGENS

K. D. Crow

Princess Margaret Hospital
Swindon
Wiltshire, United Kingdom

ABSTRACT

After a brief review of the subject of chloracnegenic toxicity, this paper examines two proposals. Firstly that the acnegenic potential of all chloroacnegens correlates perfectly with their systemic toxicity and, secondly, that with almost no exceptions chloracne still appears to be the most sensitive marker of poisoning by chloracnegens in the human subject. It is as such a marker that it is of great importance. Various alternative markers are examined and rejected in favor of chloracne. It is, of course, emphasized that the failure of chloracne to appear after apparent poisoning rules out short and medium term effects but cannot entirely preclude the potential existence of long term low dose effects, although none have so far been proven statistically to exist in over 30 years of follow-up.

INTRODUCTION

The cutaneous lesions associated with chloracne may briefly be summarized as follows:

Chloracne	Palmar and Plantar hyperhidrosis
Ophthalmic	Phrynoderma
Pigmentation	Hypertrichosis
Porphyric changes	

In view of the fact that they are non-specific, we can dispose straight away with all these changes except for two. First and foremost there is chloracne, an eruption which is unique in being the skin's only consistent response to poisoning with chloracnegenic

605

chemicals; and secondly, there are the peculiar changes which may take place in the meibomian gland of the eyelid. These glands, with the eyelashes, form modified pilosebaceous units in every way analgous to those in the skin. Therefore we might well expect them to react in exactly the same way to chloracnegenic toxins--and so they do, with the formation of keratinous cysts and disappearance of the meibomian gland. Recent work in experimental animals suggests that the cerumen glands in the auditory canal, again modified sebaceous glands, appear to undergo squamous metaplasia in exactly the same way. (McConnell, 1981). This change is so specific to chloracnegenic poisoning that I propose to call these eyelid changes "opthalmic chloracne" in order to distinguish it from non-specific irritation, conjunctivitis, and other things of that sort.

We can define a dioxin-type chloracnegen very roughly as a mol-ecule with two adjacent aromatic ring structures, which has planarity, symmetry with lateral adjacent halogens, lack of steric hindrance, and the ability to fit into a box-like configuration of some $10 \times 3 \, \text{A}^{\circ}$. However, recent work by McKinney suggests that this concept needs mod-ification in order to accommodate toxic and non-toxic halogenated aromatics which do not accord with the above definition (McKinney and McConnell, 1980). For example the 2,3,7,8-isomer of Tetrachlornaph-thalene fits the above criteria perfectly but is non-toxic and non-acnegenic, (Crow, 1970) whereas 2,3,7-Tribromonaphthalene, which does not fit, is highly toxic and acnegenic. His theory of electronic polarisation has been described elsewhere, but the importance of this finding as far as I am concerned here, lies in the fact that, despite the changing definition, acnegenicity and toxicity still appear to go hand in hand.

Now, if chloracne is to retain its significance as a vital marker of poisoning by chloracnegens, we must be able to make an unequivocal diagnosis. Clinical features alone, even to an expert, may be insuf-ficient, but, taken in conjunction with other factors such as dis-tribution, age, clustering, and contact with known chloracnegens or other chemicals with the specific molecular characteristics, should be sufficient to make a positive diagnosis. Where there is doubt, and particularly if some hitherto unknown toxin seems responsible, histological examination of the skin is essential. This is because the conversion of the pilo-sebaceous follicle in skin or on the edge of the eyelid, together with an early underlying disappearance of the sebaceous glands, lack of inflammatory cells, and a diminished bac-terial flora, (Cunliffe et al, 1975) would seem to be quite specific. Indeed, since the highly typical keratinous cysts are only formed by the squamous metaplasia of infundibular and sebaceous cells, the presence (partial or total) of the one without the absence of the other is an incompatibility. Of all these changes, the squamous metaplasia and thus disappearance of the sebaceous gland, is probably unique to these poisons. It has been shown to occur in human subjects and appropriate experimental animals with every known chloracnegen.

Furthermore, in other animal species such as cattle and mice, changes
of a similar but lesser nature are observed, and while hyperkeratosis
is more pronounced and sebaceous gland metaplasia less so, it seems
highly likely that this is but a species variant of the same pathological process of chloracne (McConnell, 1981).

Toxicologically, all chloracnegens, and particularly the
chlorinated dioxins, induce certain effects which appear to be common
to all experimental animal species so far investigated and to be
strictly equivalent. These are firstly their binding by the Cytosolic
induction receptor for the Ah locus (Poland and Glover, 1979); secondly, their ability to induce thymic atrophy; thirdly, their fetoxicity
and lastly, their acnegenicity. Only enzyme induction and acnegenicity, however, have been demonstrated with every known chloracnegen.
Furthermore, it is only these two effects (enzyme induction and
acnegenicity) which can readily be measured in live human subjects
and there has so far been no recorded incidence of an acute death due
to a chloracnegen where a full autopsy was performed. In fact, few
deaths have ever been recorded in the human subject within a few weeks
of chloracnegenic poisoning; where later autopsies have been carried
out, as in Yusho victims, the thymus has not so far been mentioned.
(Higuchi, 1976a). Furthermore, fetotoxicity and abortion have been
so far recorded only among Yusho mothers, poisoned in Japan with PCBs
and PCDFs, but its incidence was very low compared to the incidence
of cutaneous and ophthalmic chloracne (Higuchi, 1976b), a fact which
I should like to emphasize most strongly. Chlorodioxins should not be
considered in isolation since it is accepted that all chloracnegens
can and should be equated toxicologically.

CHLORACNEGENICITY EQUALS GENERAL TOXICITY

There is little doubt that chloracnegenicity and general toxicity of all relevant systems, are strictly equivalent in any one
species. This point has additional importance when we consider that
the converse must also be true (Karamaschi et al, 1981) (Kaufman and
Djerassi, 1979). For example, human and animal data have proved that
Tetrachlornaphthalene is non-acnegenic and also non-toxic (Crow, 1970)
(despite its perfect symmetry and planarity) whereas the penta-and
hexachlornaphthalenes are highly toxic and very acnegenic (Shelly
and Kligman, 1957). Let us look than at some of the evidence supporting this view. Most of the relevant animal work has consisted of
LD50 and similar experiments as well as the induction of chloracneform
changes on the rabbit ear. The latter is a relatively crude assay,
however, and therefore of limited value. However, where the general
toxicity to experimental animals has been evaluated, there has been
fairly close correlation with chloracne. Among the known chloracnegens accurate toxicological data on tetrachloroazobenzene (TCAB)
and azoxybenzene (TCAOB) has been lacking partly due to the failure
of some workers to realize that azo-bonds are destroyed by the

microbial flora of the mammalian gut and thus oral toxicity studies
have very limited value. Skin absorption experiments have been
largely ignored. However, as far as it goes, the toxicity of TCAB
isomers parallels their acnegenicity, as has similarly been reported
by many other workers with respect to PCBs, PCDDs, PCDFs and chloronaphthalenes (Huff et al, 1980) (Poland, 1978) (McConnell et al, 1978)
(Moore et al, 1979) (Karamaschi et al, 1981).

It is generally accepted that the induction of aryl hydrocarbon
hydroxylase (AHH) correlates with all other toxic parameters including cutaneous and ophthalmic chloracne (Huff et al, 1980). Slight
discrepancies have not altered the general correctness of this observation. The determination of AHH induction has the advantages of
simplicity and accuracy as well as the total absence of sex, strain
and other differences which make animal work more difficult. Furthermore, correlation holds for both TCAB and TCAOB and not only for
acnegenicity, but also lymphoid hypoplasia (Joint NIEHS/IARC Working
Group Report, 1978) (Sundstrom, 1980). The coorelative data form a
most powerful body of experimental evidence supporting the view that a
quantitative estimation of chloracnegenicity, whether cutaneous or
ophthalmic, is of enormous value in determining the probable degree or
exposure and general poisoning. These experimental data are reinforc-
ed by human findings. For example, fairly severe exposure to the dust
of crude pentachlorophenol (PCP) and concentrated hexachlorodioxin-
containing residues produced in one factory only very mild chloracne
and no evidence of systemic toxicity (Crow, 1976). Similarly, much
the same sort of exposure to 3,4,3',4'-TCAB, encountered during the
manufacture of 3,4,-dichloroaniline in another chemical plant, resulted in very mild chloracne and once again no evidence of systemic
toxicity (Crow, 1977).

Where chemical contact has been gross, the worst known examples
being during the clean-up following explosions during 2,4,5-trichloro-
phenol (TCP) manufacture, as at Nitro, West Virginia, in 1949
(Suskind, 1978), BASF in Ludwigschafen in 1953 (Goldmann, 1973), and
Grenoble in 1968 (Dugois et al, 1968), chloracne was generally severe,
and many cases had some signs or symptoms of laboratory abnormalities
indicating systemic toxicity. However, one of the most outstanding
examples of correlation between chloracnegenicity and general toxicity
was the Seveso accident where, with the exception of some heavily
exposed children, the chlorance was mild or very mild and acute toxicity was not detected. (Report of the Italian Parliamentary Commission
of Enquiry, 1978).

Conversely, in the severe Japanese Yusho poisoning due to
accidental ingestion of PCBs and PCDFs, it was reported that a classification of chloracne both cutaneous and ophthalmic into Grades 1-4
showed a correlation between the severity of the chloracne and systemic toxicity (Okumura and Katsuki, 1969) (Holmstedt, 1980).

SENSITIVITY OF CHLORACNE AS A MARKER

There is little doubt that the most important contribution which chloracne, both cutaneous and ophthalmic, has to make to this subject is its unique function as the most sensitive single marker we have, indicating exposure to chloracnegenic toxins and particularly chloro-dioxins (Joint NIEHS/IARC Working Group, 1978) (Kaufman and Djerassi, 1979).

Let us examine some of the evidence supporting this proposal. Of 78 workers heavily exposed to TCDD during 2,4,5-TCP manufacture, (Jirasek, 1973) and his colleagues found only one case without chlor-acne, but only 55 of the 78 showed symptoms and signs or laboratory findings suggesting systemic intoxication. In an explosion at a 2,4,5-TCP plant in West Virginia in 1949 (Zack and Suskind, 1980) chloracne was found to be the most consistent marker of poisoning. The same was true of the Coalite explosion in 1968 during the manufac-ture of the same chemical. Seventy-nine workers developed chloracne, but there were very few who had clinical or laboratory evidence of systemic poisoning (May, 1973). Similarly, after a similar explosion in Holland in 1963 chloracne again was almost the only abnormal find-ing (Joint NIEHS/IARC Working Group, 1978).

This sensitivity is even more striking where there have been moderate or heavy exposures to weaker chloracnegens such as in the two factory incidents I mentioned: the heavy exposure to hexachlorodoxins during PCP manufacture; and the lighter exposure to TCAB during man-ufacture of 3,4-dichloroaniline. The minimal nature of the chloracne produced in these incidents was nevertheless the sole means whereby exposure to, and absorption of, chloracnegenic toxins was revealed (Crow, 1979). But, the strongest evidence in support of the sugges-tion that chloracne is the most sensitive marker we have, comes from the Seveso accident. There were 187 cases of chloracne, mild indeed in most cases and unaccompanied by any obvious signs, symptoms or laboratory findings of systemic toxicity (Karamaschi et al, 1981). The indicated sensitivity of the human subject to chloracne led the Joint NIEHS/IARC Working Group (1978) to recommend that an Interna-tional Chloracne Register be compiled because chloracne was a classic way of defining precisely a cohort known to have been exposed to these toxins, (Huff et al, 1980).

Finally, again, the Japanese PCB and PCDF poisonings of persons of both sexes differed from every other recorded incident of poisoning by chloracnegens, since the route of absorption of the toxins was largely by ingestion, skin contact and inhalation - usually the major routes - being essentially excluded. The absence of chloracne, either cutaneous or ophthalmic, in roughly 10% of those cases who had symp-toms, signs and laboratory abnormalities was unique until the 1979 Taiwan poisoning (Wong et al, 1981) which was the exact counterpart of the earlier Japanese Yusho disaster and showed a similar incidence

of symptomatic cases without cutaneous or ophthalmic chloracne. In
every other recorded outbreak of chloracnegenic poisoning, not only
with dioxins, chloracne has been found in almost 100% of those known
to have been poisoned.

The usefulness of chloracne as a marker is underlined if we
consider some other highly sensitive markers of chloracnegenic poisoning and their advantages and disadvantages. Firstly aryl hydrocarbon
hydroxylase induction is possibly even more sensitive, but not so far
available as a simple clinical screen. Serum triglycerides have not
always been raised where chloracne has been mild but very obvious
(Crow, 1977). Fetotoxicity may be mentioned only to be dismissed,
whereas examination of the thymus requires an autopsy.

It should not necessarily be assumed that long term effects may
not yet become evident, even in the absence of chloracne, since every
case of chloracnegenic poisoning must, because of skin absorption,
have a degree, however small, of systemic toxicity. However, follow-up
examinations of admittedly small cohorts now cover a span of 32 years
and, so far, only malignancy in the form of soft tissue sarcomata has
appeared. A cause/effect relationship is possible because of the
extreme rareness of this type of tumor but has not been proved
(Honchar and Halperin, 1981) (Zack and Suskind, 1980) (Cook et al,
1980) (Moses and Selikoff, 1981).

Therefore, it is fair to say that, leaving aside soft tissue
sarcomata at the moment, the suggestion that any other long-term
effects are ascribable to chloracnegen exposure, cannot be reasonably
sustained in the absence of a clear history of diagnosis of chloracne.
Rare exceptions to any such pronouncement in human subjects there must
be, but these exceptions, I believe, simply go to prove the general
rule.

REFERENCES

Cook, R. R., Townsend, J. C., Ott, M. G., and Silverstein, L. G.
     (1980) Mortality experience of employees exposed to 2,3,7,8-
     tetrachlorodibenzo-p-dioxin (TCDD) Jour. Occup. Med., 22: 530-
     532.
Crow, K. D., (1970) Chloracne: A critical review including a
     comparison of two series of cases of acne from chlornapthalene
     and pitch fumes. Trans. St. Johns Hosp. Derm. Soc. -56: 79-99.
Crow, K. D., (1976) Personal observations.
Crow, K. D., (1977) Personal observations.
Crow, K. D., (1979) Personal observations.
Cunliffe, W. J., Williams, M., Edwards, J. C., Williams, S., Holland,
     K. T., Robert, C. D., Holmes, R. L., Williamson, D. and Palmer,
     W. C., (1975) An explanation for chloracne - an industrial
     hazard. Acta. Dermato. Vener. (Stockholm) 55:211-214.

Dugois, P., Amblard, P., Aimard, M., and Deshors, G. (1968) Acne chlorique collective et accidentelle d'un type noveau. Bull. Soc. Franc. Derm. Syph., 76: 260, and Lyon Med. 219: 203.

Goldman, P. J., (1973) Severe acute chloracne, a mass intoxication by 2,3,6,7 tetrachlorodibenzo-dioxin. Hautarzt. 24: 149-152.

Higuchi, K., (1976a) PCB Poisoning and Pollution. pp. 70-77. Academic Press (London).

Higuchi, K., (1976b) PCB Poisoning and Pollution. pp. 19-21. Academic Press (London).

Holmstedt, B., (1980) Prolegomena to Seveso. Arch. Toxicol. 44: 211-230.

Honchar, P. A., Halperin, W. E., (1981) 2,4,5-Trichlorophenol and soft tissue sarcoma. Lancet I: 268-269.

Huff, J. E., Moore, J. A., Saracci, R., and Tomatis, L., (1980) Long term hazards of polychlorinated dibenzo-dioxins and polychlorinated dibenzo-furans. Environ. Health Pers., 36: 221-240.

Jirasek, L., (1973) Acne chlorina, porphyria cutanea tarda and other manifestations of general intoxication during the manufacture of herbicides. Part I. Ceskoslovenska Dermatologie 48: 306-317.

Joint NIEHS/IARC Working Group Report on long term hazards of polychlorinated dibenzodioxins and polychlorinated dibenzofurans. IARC. Lyons, 1978.

Karamaschi, F., Del Corno, G., Favaratti, C., Guambelluca, S. E., Montesarchio, E., and Fara, G. M., (1981) Chloracne following environmental contamination by TCDD in Seveso, Italy. International Journal of Epidemiology 10: 135-143.

Kaufman, G., and Djerassi, L. S., (1979) Toxicity of polybrominated naphthalene in man -- the clinical syndrome. Proceedings of the International Conference on Critial Current Issues in Environmental Health Hazards, Tel Aviv, Israel March 4-7, 1979.

McConnell, E. E., (1981) Personal Communication.

McConnell, E. E., Moore, J. A., and Dalgard, D. W. (1978) Toxicity of tetrachlorodibenzo-p-dioxin in Rhesus monkeys (Macaca Mulatta) following a single oral dose. Toxicol. Appl. Pharmacol. 43: 175-187.

McKinney, J. D. and McConnell, E. E., (1980) Structural Specificity and the Dioxin Receptor Workshop on the impact of chlorinated dioxins and related compounds on the environment. Instituto Superiore Di Sanita, Rome, October 22-24 (In press).

May, G. (1973) Chloracne from the accidental production of tetrachlorodibenzo-dioxins. Brit. Journ. Indust. Med. 30: 283.

Moore, J. A., McConnell, E. E., Dalgard, D. W. and Harris, M. W. (1979) Comparative toxicology of 3-halogenated dibenzo-furans in guinea pigs, mice and Rhesus monkeys. Annals N. Y. Acad. Sci. 320: 151-163.

Moses, M. and Selikoff, I.J., (1981) Soft tissue sarcomas, phenoxy herbicides and chlorinated phenols. Lancet I: 1370.

Okumura, M., and Katsuki, S. (1969) A clinical study of oil disease (Chlorinated biphenyl poisoning) particularly the internal medical signs. Fukuoka Acta. Med. 60: 440-446.

Poland, A. Joint NIEHS/IARC Working Group. Report on long term
    hazards of polychlorinated dibenzo-dioxins and polychlorinated
    dibenzo-furans. p. 39. IARC. Lyons, June 1978.
Poland, A, and Glover, E. (1979) 2,3,7,8 Tetrachlorodibenzo-p-dioxin:
    segregation of toxicity with the Ah locus. Mol. Pharmacol. 17:
    86-94.
Report of the Italian Parliamentary Commission of Enquiry into the
    Escape of Toxic Substances which took place on 10 July 1976 in
    the Icmesa Factory and into the potential health risks to the
    Environment, deriving from industrial activity. Rome, July 1978.
Shelley, W. B., and Klingman, A. M. (1957) Experimental production of
    acne by penta and hexachlornapthalenes. Arch. Derm. 75: 689.
Sundström, G. (1980) 3,3',4,4' Tetrachloroazobenzene and 3,3', 4, 4',
    tetrachloroazoxybenzene, chloroacneigens and potent enzyme
    inducers -- an overview. Workshop on the impact of chlorinated
    dioxins and related compounds on the environment. Instituto
    Superiore di Sanita. Rome, October 22-24 (In press).
Suskind, R. R. (1978) Chloracne and associated health problems in
    the manufacture of 2,4,5-T: Report to the Joint Conferences,
    National Institute of Environmental Health Sciences, Inter-
    national Agency for Research on Cancer. Lyon, France, 11
    January.
Wong, K. C., Hwang, M. Y., (1981) PCB Poisoning, Special Issue.
    Clinical Medicine. (Taipei) 8: 83-88 (Chinese).
Zack, J., and Suskind, R. R. (1980) The mortality experience of work-
    ers exposed to tetrachlorodibenzo-p-dioxin in a trichlorophenol
    process accident. Journ. Occup. Med. 22:11-14.

SOFT TISSUE SARCOMAS: CLUES AND CAUTION

Ralph R. Cook

The Dow Chemical Company
Health and Environmental Sciences
Midland, Michigan   USA

Hardell and Eriksson[1] have made a significant contribution to the scientific literature, focusing attention on the need for further research on soft tissue sarcomas. Indeed, as they pointed out in their letter[1], such research is already underway throughout the world.

I would like to caution, however, against overinterpretation of the current data. They provide us with more sophisticated questions. They do not provide any definitive answers. The current reports are too incomplete, both individually and in aggregate, to permit formulation of a clear picture of the possible associations between TCDD and soft tissue sarcomas. By analogy, we are in the process of fitting together the pieces of a large jigsaw puzzle. Some of the pieces on the table may be superfluous, false leads that waste time and resources. Other pieces are, and maybe always will be, missing; but with each new bit of research, we are building a more complete picture, determining whether or not TCDD has major human implications and, if it does, at what level and for which disease.

During my formal training in epidemiology, I was taught to exercise a healthy skepticism toward the results of any epidemiological study, whether the research be mine or that of another. To do otherwise would be to ignore the vagaries of this observational science that are introduced by selection, misclassification, confounding and chance. Nothing over the subsequent years has induced me to change this mind set. My objective in this brief paper is to pass along some of that lesson.

First let me address the outcome, soft tissue sarcomas. Hajdu pointed out in a recent review article that soft tissue sarcoma is a generic term for a heterogeneous group of lesions that include more

than 100 tumors.[2] And there are at least 50 reactive lesions that
resemble soft tissue neoplasms. The nomenclature of these growths
is in flux; and the tumors are so rare that few pathologists have
been able to achieve a high degree of expertise and consistency in
the diagnosis of specific neoplasms.[3] To data, no single pathologist
has reviewed all the histopathological specimens. To date, we do not
know whether we are dealing with a single predominant clinical entity
or a multitude of diseases. The former would argue for cause and
effects. The latter would argue against it. One of the reasons vinyl
chloride is accepted as a human carcinogen is the specificity of its
effect. At high levels, it produces a single type of soft tissue
sarcoma: angiosarcoma. It does not produce a general increase of all
soft tissue sarcomas.

     Second, let me comment on the exposures. The common exposure
link between the various reports of soft tissue sarcomas in occupational groups is 2,3,7,8-tetrachlorodibenzodioxin, one of the most
toxic chemicals known to man. But is it the real common link? The
rigor of exposure documentation varies from study to study. In some,
based on detailed technical knowledge of the chemical process,
analyses of reaction contaminants, and clinical evidence of excessive
group exposure, e.g., chloracne, there is strong evidence that
2,3,7,8-TCDD is at least one of the common exposure links. In other
reports, the evidence is more presumptive, and in some, totally
lacking.

     Epidemiologists from Monsanto and Dow have reported on four soft
tissue sarcomas cases occurring in their occupational cohorts exposed
to high levels of 2,3,7,8-TCDD.[4,5,6] The reports were later reviewed
by Honchar and Halperin.[7] Of the four cases, three are deceased and
one is still living.[8] Table 1 gives an update on the four. There are
two new bits of information: Case number 3 also had 2,4,5-T process
exposures and also had chloracne.

     Notice the patterns of year of birth, process, chloracne, and
cigarettes. Are some of these important, or are they "red herrings"?
We don't know, yet. These questions will have to be explored in
future epidemiological studies. Parenthetically, I would like to
point out there have been no cases of soft tissue sarcoma in a cohort
of Dow 2,4,5-T employees, none of whom had chloracne.[9] Nor has there
been any soft tissue sarcoma cases reported in a BASF cohort.[10]
Members of the BASF cohort were exposed to 2,3,7,8-TCDD during a 1953
accident in a trichlorophenol production unit or during cleanup and
repair following the accident. Exposures were heavy. Many developed
severe chloracne, and at least one died as a consequence of overex-
posure.[11]

     The work of Hardell et al uses a different type of methodology,
the case-control approach; and reports on different patterns and
levels of exposure.[12,13] In their first study, they reported a 5- to

Table 1   Summary Data of Soft Tissue Sarcoma Cases From
          Three Occupational Cohorts Exposed To 2,3,7,8-TCDD

Case Number	1	2	3	4
Year of Birth	1920	1921	1923	1921
Year of Hire	1946	1950	1946	1951
First Exposure	1948	1964	1950	1951
Year of Death	1978	1975	1972	NA
Company	Monsanto	Dow	Monsanto	Dow
Process	TCP	TCP	2,4,5-T TCP	TCP
Chloracne	+	±	+	+
Cigarettes	+	+	+	+

6-fold increased risk for those presumptively exposed to phenoxyacetic
acids, predominately 2,4,5-T, and chlorophenols.[12] I say presumptively
because the exposure information was subjectively reported by the
study participants and/or next of kin. Objective records were incomplete and difficult to interpret. In the second study, they also
estimated the risk associated with exposure to phenoxy acids not contaminated with TCDD and found the same magnitude of risk.[13]

I personally find this confusing. The justification for combining chlorophenols and 2,4,5-T is the common contaminant, TCDD.[13] If
that be the case, then phenoxyacetic acids without this contaminant
shouldn't demonstrate an increased risk.

In both of their studies, Hardell et al first administered mail
questionnaires and then followed up with telephone interviews on a
selected subset of subjects. While the questionnaire was designed
with the idea of masking the intent of the investigation, the telephone inquiry zeroed in specifically on the exposures of interest:
phenoxyacetic acids and chlorophinols.

Interestingly, in their second study the frequency of the various exposures among both cases and controls were reported and most
of the risk estimates were elevated. As indicated in Table 2, the
risk estimates are broken down by whether they were achieved with
data from the questionnaire alone or telephone supplement. Note that
only one, exposure to sodium chlorate, is not elevated. Notice also
that nicotine has a risk of 6.1. Unfortunately, we do not know the

Table 2   Risk Estimates By Agent By Collection Method
In Eriksson et al Case-Control Study

Agent	Questionnaire Alone	Telephone Supplement
Phenoxy Acids and Chlorophenols	--	5.1
Phenoxy Acids Only	--	6.8
Chlorophenols Only	--	3.3
Phenoxy Acids Minus 2,4,5-T and Chlorophenols	--	4.2
Asbestos	1.3	--
Glass Fiber	2.1	--
Power Saws	1.5	--
Smoking	1.1	--
Other Pesticides	2.3	--
Amitrol	2.0	--
Bromofos	2.0	--
DDT	1.3	--
Dinoseb	3.0	--
Fenitrothion	2.0	--
Mercury Seed Rx	1.9	--
Lindane	2.0	--
Na Chlorate	1.0	--
Nicotine	6.1	--
Others	2.1	--

Adapted from: Eriksson, M. Berg, N.O., Hardell, L., Moller, T., and Axelsson, O., 1979. Case-Control Study on Malignant Mesenchymal Tumors of the Soft Parts and Exposure to Chemical Stustances. Lakartidninger 76:3872-75. (Translated for EPA by Literature Research Co., Annandale, Va.)

risk estimates on the agent of concern were derived from the questionnaire alone, nor do we know what the risk estimates would have been on the other agents if they had been probed in the same depth as were exposures to phenoxies and chlorophenols.

There are at least four alternate interpretations of this table. One, exposure to phenoxyacetic acids, chlorophenols and their contaminants may cause soft tissue sarcomas and the elevated risks of all other agents are unimportant. Two, there was recall bias. Those with disease ruminated about their previous exposures and overreported relative to the nondiseased controls. Three, there were confounding variables. That is, other causes of the sarcomas were not adequately controlled in design or analysis. Or, four, phenoxies and chlorophenols are only indirectly associated with the disease. They and all of the agents listed are surrogates for particular types of occupations in which the true cause is found. We will not know which of the above is or are correct without further research.

In summary, important information continues to be accumulated. For those charged with making administrative decisions from these data, the speed by which they are collected must be agonizingly slow. For those exposed to these agents and frightened by the premature declarations of cause and effect, the apparent indecision of the scientific community must be frustrating. Nonetheless, we must continue methodically to collect the data, analyze them, subject our reports to peer review, refine our hypotheses, and retest them on independent groups of subjects. Only then will we know whether 2,4,5-T, or chlorophenols, or 2,3,7,8-TCDD cause specific types of diseases. And, if they do, at what levels and under what conditions.

REFERENCES

1. Hardell, L. and Eriksson, M. Soft-Tissue Sarcomas, Phenoxy Herbicides, and Chlorinated Phenols. The Lancet, p. 250, August 1, 1981.
2. Hajdu, S.I. Soft Tissue Sarcomas: Classification and Natural History. CA-A Cancer Journal for Clinicians 31(5):271-280, Sept/Oct, 1981.
3. Patterson, W.B., Rubin, P., Editor; Bakemeier, R.F., Associate Ed. Chapter XIX. Soft Tissue Sarcoma. In: Clinical Oncology for Medical Students and Physicians. Fifth Edition. The University of Rochester School of Medicine and Dentistry. American Cancer Society, pp. 210-217, 1978.
4. Zack, J.A., Suskind, R.S. The Mortality Experience of Workers Exposed to Tetrachlorodibenzodioxin in a Trichlorophenol Process Accident. J. Occup. Med. 22:11-20, 1980.
5. Cook, R.R., Townsend, J.C., Ott, M.D. Mortality Experience of Employees Exposed to 2,3,7,8-Tetrachlorodibenzo-p-dioxin (TCDD). J. Occup. Med. 22:250-532, 1980.

6. Zack, J.A. and Gaffey, W. A Mortality Study of Workers Employed
    at the Monsanto Chemical Plant in Nitro, West Virginia. 1980.
    Unpublished.

7. Honchar, P.A. and Halperin, W.E. 2,4,5-T, Trichlorophenol, and
    Soft Tissue Sarcoma. Letter to the Editor, The Lancet, 268-269,
    1981.

8. Cook, R.R. Dioxin, Chloracne, and Soft Tissue Sarcoma. Letter
    to the Editor, The Lancet, 618-619, 1981.

9. Ott, M.G., Holder, B.B. and Olson, R.D. A Mortality Analysis of
    Employees Engaged in the Manufacture of 2,4,5-Trichlorophenoxy-
    acetic Acid. J. Occup. Med. 22:47-50, 1980.

10. Thiess, A.M. and Frentzel-Beyme, R. Mortality Study of Persons
    Exposed to Dioxin After an Accident Which Occurred in the BASF
    on 13 November 1953. Paper presented at MEDICHEM Congress V,
    San Francisco, September 5-9, 1977.

11. Goldman, P.J. Severe Acute Chloracne. A Mass Intoxication by
    2,3,7,8-Tetrachlorobenzodioxin. Der Hautarzt 24:149-152, 1973.
    Translated for EPA by Literature Research Company, Annandale,
    Virginia.

12. Hardell, L. and Sandstrom, A. Case-Control Study: Soft Tissue
    Sarcoma and Exposure to Phenoxyacetic Acids or Chlorophenols.
    Br. J. Cancer. 39:711-717, 1979.

13. Eriksson, M., Berg, N.O., Hardell, L., Moller, T. and Axelson,O.
    Case-Control Study on Malignant Mesenchymal Tumors of the Soft
    Parts and Exposure to Chemical Substances. Lakartidningen 76:
    3872-3875, 1979.

RISK ASSESSMENT

Chairman: Allan P. Gray
Dynamac Corporation
Rockville, Maryland USA

RISK ASSESSMENT: INTRODUCTORY REMARKS

Allan P. Gray

Dynamac Corporation
Enviro Control Division
Rockville, Maryland   USA

This section considers risk assessment methodologies and attempts to evaluate the feasibility of applying these techniques to the chlorinated dibenzo-p-dioxins in general and to TCDD in particular.

Faced with this task, one quickly realizes the complexity of the problem and the number of imponderables involved in any attempt to derive a meaningful risk assessment of a substance such as TCDD that: a) is extremely toxic, in fact considered one of the most toxic substances known; and b) is not itself a product but a trace contaminant of, potentially, many products. Some questions that need to be addressed in considering the feasibility of developing valid risk assessments for TCDD are:

Can the wide variations in sensitivity of different species to TCDD be accommodated? (The oral LD_{50} in the guinea pig is approximately 1/50th of that in the rat, which animal is in turn apparently much more sensitive to the lethal effects of TCDD than man).

Can realistic estimates of exposure levels be made given current stringent control procedures which continue to reduce the degree of contamination of products to ever lower, nearly undetectable levels?

Is TCDD a true, genotoxic carcinogen or is carcinogenicity nongenetic in origin and a secondary consequence of a toxic insult and tissue damage?

In the Symposium, the Risk Assessment Session took the form of a panel discussion. Following brief, formal presentations by each

621

panelist, there was a general, open discussion. The general discussion is not included in this Proceedings volume but only the formal presentations of four of the six panelists. These are:

Mr. L. Mark Wine, attorney with the law firm of Kirkland and Ellis, discusses the scientific-legal interface and the difficulties scientists face in providing the kinds of definitive technical judgments demanded by the legal process.

Dr. Joseph Rodricks, of Clement Associates and formerly of the FDA, discusses general principles that need to be considered in developing a risk assessment;

Dr. Lester Lave, of the Brookings Institute, discusses risk assessment methodologies and their application to TCDD;

Ms. Judith Hushon, of the Dynamac Corporation, presents a risk assessment of TCDD;

Dr. Roy Albert, of the Institute of Environmental Medicine, New York University Medical Center, and Dr. Patricia Honchar, of the National Institute of Occupational Safety and Health, elected not to submit their papers for publication.

ASSESSING RISK IN LEGAL PROCEEDINGS: HOW TO DEAL WITH SCIENTIFIC

UNCERTAINTY IN TOXIC SUBSTANCES RISK ASSESSMENTS

L. Mark Wine

Kirkland & Ellis
1776 K Street, NW
Washington, D.C. USA

ABSTRACT

Scientists and lawyers will continue to work together in assessing the risk of toxic substances, including the dioxins, in the
future. Scientists are uncomfortable in the legal system, in part
because of the inherent conflict between how scientists deal with
uncertainty and the lawyers' need for concrete answers. Scientists
must apply their professional judgment beyond what is specifically
proved by the scientific data. Some specific suggestions and examples
of common risk assessment pitfalls are given. Related aspects regarding assessing the risk of 2,3,7,8-TCDD are also mentioned.

INTRODUCTION

The legal aspects of risk assessments and dioxins are sufficiently important and complex to warrant considerably more attention
than can be devoted to it in this proceedings. Thus, I cannot give
you very many answers about how to deal in future legal proceedings
with dioxin-related issues. Instead, I will focus here on what I
believe to be the fundamental problem of any risk assessment for any
toxic substance -- how to translate the inherent conservatism of
scientists in dealing with uncertainty into the legal system's need
for concrete answers. This is the fundamental problem which con-
fronts lawyers concerned with issues dependent on answers to scientific questions.

I want to focus on how scientific risk assessment can best meet
the needs of the legal system. After briefly reviewing the legal
setting where risk assessments of dioxins are most likely to be

623

found, I will discuss some general principles that apply to all toxic
substance risk assessments. I will then comment briefly on risk
assessments for 2,3,7,8-TCDD.

THE LEGAL SETTING AND THE SCIENTIST'S ROLE

Most scientists are inherently uncomfortable when thrust into
the strange, seemingly Byzantine legal system. The role of a scientist in presenting risk assessment in the legal process is heavily
influenced by the type of legal proceeding involved. The legal system
is quite complex. It includes activities as diverse as internal
government analysis, scientific advisory board hearings, and jury
trials.

The scientist's role in legal proceedings involving risk assessment is influenced not only by the legal setting but also by the
nature of the testimony. Some scientists will be called upon to do
nothing more than present the results of their own research, a
relatively easy task. More often, a scientist will be called on to
evaluate the research of others as well and, consequently, either
support or criticize the work of peers. This too is not very difficult, particulatly when compared to the most difficult role of all--
presenting a comprehensive assessment of risks in the face of
incomplete data.

The last situation is most relevant to this discussion, and I
will emphasize it. My comments are directed primarily to government
regulation, for that is where most risk assessing is found today.
Nevertheless, most of my observations are applicable to product
liability and other litigation as well.

LEGAL RISK ASSESSMENTS--GENERAL CONSIDERATIONS

Scientists participating in a legal risk assessment proceeding
face several problems in dealing with uncertainty and in communicating
complex data to often scientifically naive lawyers and decisionmakers.
After explaining the problems, I will offer several specific suggestions which can make risk assessments more useful in the legal context.

The inherent conflict between the scientific process' method of
dealing with uncertainty and the legal system's need for answers is
at the heart of the discomfort scientists find when thrust into the
legal arena. Over the decades, the scientific method has evolved a
simple approach to uncertainty. Studies are designed, undertaken
and reported in a logical fashion. Where questions remain, new
studies are designed and the logical, step-by-step approach continues.

This is a long, often evolutionary process, with few revolutionary break-throughs. Research results are subject to established and accepted systems of quality control, principally peer-review publication. Competent scientists report their results and carefully set forth all of the caveats which apply to their data.

Once a certain experiment is completed, additional studies are planned and the range of uncertainty is thereby reduced. Scientists understand all of this and have reason to be proud of the internal discipline the scientific method has established. The credibility that science still enjoys today is due in substantial part to the professional ethics inherent in that system.

But the scientific method breaks down in the law. Lawyers cannot deal with uncertainty very well. The need for an answer overrides the search for absolute truth. Perhaps the extreme example is the jury system. Although juries are charged with "finding the truth," the system if far from perfect. Complicated decisions are made by well-meaning, but unqualified jurors. But decisions get made nevertheless. In the legal system there is no such thing as absolute "truth," but truth only in the context of what the ultimate fact-finder decrees.

When confronted by scientific uncertainty, lawyers will not settle for more studies. Lawyers will insist that scientists go beyond the specific studies, and apply their years of professional experience in making informed and thoughtful scientific judgments from the available data. But, they must necessarily go beyond the hard data.

For many scientists, the temptation when called by a lawyer, will be to decline to participate or to limit their role. Yet a scientist must respond. The legal system needs all the competent scientific advice that it can get. Unfortunately if the competent, knowledgeable scientists abdicate in any given area, the vacuum will be filled either by the unqualified scientist or the non-scientist.

What lawyers need is the best scientific judgment. This means something more than pure speculation, but often something less than proven scientific fact.

A legally useful risk assessment has several important elements:

First, the scientist must carefully evaluate the quality of the available data, either discarding or discounting those data that fail to meet normally accepted scientific standards. Non peer-reviewed data in particular should be viewed with suspicion. Qualitative differences among researchers exist, and should be recognized. Scientists do disagree, and such disagreements should be frankly acknowledged.

Second, the scientist must consider the breadth of the available data and how consistent they are. Replicated studies by qualified, independent researchers giving similar results should increase confidence. Information on possible mechanisms of toxicity is important.

Third, the scientist must consider carefully the available human data, particularly data helping to define the relative sensitivity of humans as compared to animals. Too often, the available human data are ignored or their significance minimized because they are not as easy to work with as are the animal data, with their carefully controlled experimental results.

Fourth, the scientist must resist the temptation to be too conservative in his or her interpretation. Conservatism has its place. But, if one starts to examine all of the conservative assumptions that go into some risk assessments, one realizes that the end result is highly biased. For example, "worst-case" exposure estimates are too often substituted for careful use of available data. To be sure, assumptions must be made, but they can be made as realistic as possible by using all of the available data and a lot of common sense.

Fifth, the scientist must identify his/her assumptions. Of course, assumptions are always needed in risk assessments for toxic substances, but too often they are hidden under a mass of data. The resulting risk assessment will be reported without the scientific limitations inherent in the assumptions if those limitations are not explicitly stated and emphasized.

Sixth, the scientist must effectively communicate the relative certainty he/she have in the final analysis. For example, one should state risks in ranges or order of magnitude concepts rather than precise numbers. If one must make "worst-case" assumptions, one should be sure to let that be known. But, even more important, one should try to make some judgments about a more realistic assessment.

Seventh, the scientist must minimize technical jargon. Analysis and judgments should be communicated concisely and clearly. The scientist's audience will include many without the benefit of technical expertise.

RISK ASSESSMENT FOR DIOXINS

As I stated at the outset, I do not intend to participate in the scientific discussion of the risk assessment of the dioxins, but as a prelude to what follows let me present a few thoughts of one lawyer who has spent several years dealing with the scientific data concerning 2,3,7,8-TCDD.

Few chemical substances have been studied as extensively as has 2,3,7,8-TCDD. Chronic and acute toxicity, mutagenicity, reproductive effects, teratogenicity, immunosuppression, and pharmacokinetics have all been studied extensively in animals. Numerous studies of human exposure also are available.

Not only are there many excellent studies available, but many of them have been replicated by first-rate researchers with remarkably consistent results. The studies go far beyond the screening-type bioassays usually found, and include careful studies examining the mechanism of toxicity.

Similarly, TCDD's formation and environmental behavior have been studied extensively. Considerable environmental monitoring has also been completed.

But, to my mind, most important of all are the available human data. Particularly important are the data from general population exposures at Seveso which show humans are not specially sensitive. The soft-tissue sarcoma data from the United States cohorts also merit attention.

This large body of data is remarkably consistent and supplies strong evidence for use in a risk assessment. Because the data are so extensive, many of the typical problems of risk assessment are limited with TCDD. Because so much data are available, fewer assumptions need to be made. As compared to most toxic substances, the unanswered questions on 2,3,7,8-TCDD are few. Yet uncertainty still remains.

For the scientists participating in this Panel discussion, I have one additional word of advice. A good comprehensive risk assessment takes considerable analysis and time to develop. While scientists cannot yet provide the ultimate risk assessment they can help light the way for others to follow where more careful reflection and thought are possible.

CONCLUSION

The challenge to those who follow is to use all of the available data--not just a scrap here or there--and to weave it into a coherent comprehensive and easily understandable, assessment of risk.

But make no mistake about it, some uncertainty will prevail. The challenge remains not only to use the data sensibly and thought-fully, but also to communicate to others the level of uncertainty and how it compares with information used to assess other compounds.

SOME GENERAL OBSERVATIONS ON THE SAFETY OF PCDDs

J. V. Rodricks

Environ Corporation
Washington, D. C. USA

I shall direct my remarks towards the problem of protecting the general public from the polychlorinated dibenzo-p-dioxins (PCDDs). In particular, I shall give my view of the types of data and analysis necessary to decide whether regulation or some other form of control is appropriate, and, if so, the extent of control that should be sought. My interest is not in the specialized problem of accidental exposure, but rather in the problem of the introduction of PCDDs into the environment through the continued use of phenoxy herbicides, through combustion, and through the disposal of chemical wastes.

Let me begin by describing the general methodology customarily used by the regulatory agencies and other institutions to treat hazard and exposure data for purposes of establishing standards. Ordinarily, data derived from studies in experimental animals provide the basis for such standard-setting activities. Certainly, reliance on such data is essential for newly introduced substances. For substances to which humans have already been exposed, it might seem desirable to rely on epidemiological data. Unfortunately, such data usually have serious limitations for use in setting standards for purposes of protecting the general population. If epidemiological studies have revealed an association between a disease and exposure to a toxic substance, they frequently lack adequate dose-response data. If epidemiological studies have revealed no detectable association between exposure to a substance and human disease, they are frequently found to be insufficiently sensitive to detect risks that, while small, might still be of significance to the general population.

So even for substances to which the general population has been exposed for many years, it is usually necessary to rely upon animal data. This is inherently an uncomfortable scientific position, but it is necessary. I might note that to permit human exposure to a substance to continue without imposing limits constitutes a decision that the exposure is without significant risk, so that if such a decision is made, it should be a conscious one, based on the best available scientific information.

Thus, for substances such as food additives, pesticides used on food crops, and drinking-water contaminants, standard-setting for purposes of protecting the general population relies heavily on data obtained in experimental animals. The usual methods for arriving at acceptable exposure levels for these substances can be simply summarized:

1. Data on animal toxicity are obtained under a variety of exposure conditions, and the adverse effects occurring at the lowest exposure level (usually in chronic studies) are identified. A level of exposure producing no observable effect is then identified. This "no observed effect level" (NOEL) is divided by a safety factor, usually 100 for data reflecting chronic exposure, to arrive at an acceptable human exposure level. This NOEL-safety factor approach was introduced in the 1950s and is still used. The use of a safety factor is based on the notion that humans might be more sensitive to the effects of an agent and that the range of susceptibilities in the human population is expected to be much wider than it is in the animal population. However, the specific safety factors used have no scientific basis. Further, although it is not claimed that the level of exposure found acceptable under this method is without risk, no attempt is made to ascertain the level of risk that is being tolerated.

2. If a substance is found to be carcinogenic, a conceptually different approach is taken. An attempt is made explicitly to estimate risk under the conditions of exposure of the general population. A decision is then made concerning the level of risk that is tolerable in the general population.

3. For some substances, factors such as the cost of reducing exposure, the benefits of the substance, and the availability of control technology are also considered, and these factors may result in modification of the tolerable exposure level obtained based on analysis of the health data.

If we now consider the problems of environmental contaminants such as PCDDs (and let me assume for a moment that contamination is sufficiently widespread so that the general population is exposed), I think we find these schemes for standard-setting to be problematic,

especially the one used to treat nonneoplastic effects. The most
sensitive nonneoplastic response to a PCDD is seen in the reproductive system. Specifically, 2,3,7,8-TCDD has been found to affect
reproductive performance in rats at doses of 10 and 100 ng/kg/day
(Murray et al, 1979); the NOEL for this effect was reported to be a
ng/kg/day.[1] Schantz et al, (1979) have reported reproductive failure
in monkeys exposed for 7 months prior to mating and during pregnancy
at a dose of 1.7 ng/kh/day of TCDD; no lower dose was used in this
study, thus a NOEL was not established. It would not be surprising
if the NOEL for reproductive effects of TCDD in monkeys were found to
be lower than 1 ng/kh/day.

     However, if it is tentatively assumed that 1 ng/kg/day is the
NOEL for TCDD in a test animal, we are then left with the decision
of how far below this level we should go to protect the health of
the general population. We have no scientific basis for making this
decision. We only have custom to guide us--that is, we divide by 100
and pray. The selection of a safety factor is said by some to be a
matter of "scientific judgment;" while there appears to be some
scientific basis for the use of safety factors, it is difficult to
ascertain what "judgment" could possibly inform the selection of
specific factors. The method we have for treating nonneoplastic forms
of toxicity is, to my view, inherently unsatisfactory and new methods
need to be developed, based on the concept now applied to carcinogens
that the available dose-response data should be used to make an
explicit estimate of risk. I realize that there are enormous difficulties that must be overcome before we can arrive at suitable
methods, but I think it is time to begin to move away from the concept
that toxicologists can decide what is "safe" by simply selecting arbitrary "safety factors."

     But, let me resort to this custom for purposes of treating the
TCDD data. If you accept that a safety factor of 100 is a useful
guide, the use of the NOEL of 1 ng/kg/day yields a tentative tolerable daily TCDD intake in humans of 10 pg/kg.

     It is now necessary to compare this exposure level to actual
exposure in the general population; we find, however, that insufficient data exist to permit this comparison. It is clear that the
question of TCDD exposure in the general population is unanswered and
needs to be answered before any intelligent decision can be made
about risk and needed controls. The tentative tolerable daily TCDD
intake of 10 pg/kg can be used to guide analytical chemists in that
it provides the basis for establishing a target detection limit for
their assays. Thus, if it is assumed that contaminated fish are the
most likely source of TCDD exposure in the general population, we can
estimate the level of TCDD detection that should be achieved for
analytical methods used for the surveillance of fish. For example,

----

[1]Reanalysis of these data suggests that this dose may not be a
  NOEL (Nisbet and Paxton 1981).

a person weighing 50 kg consuming an average of 25 g of fish each day
would be exposed at the tentative tolerable dose of 10 pg/kg if the
fish were contaminated at a concentration of 20 ng/kg (20 ppt). Thus,
an exhaustive sampling of the fish supply using methods capable of
detecting TCDD in the range of 10-20 ppt should provide highly useful
information about the magnitude of the margin-of-safety in the general
population. Such a survey seems to be the most important priority for
data gathering on TCDD. Analytical methods are already available to
accomplish this task.

I have not considered the carcinogenicity of TCDD. I shall note,
however, that there is relatively strong evidence that TCDD is not a
directly acting carcinogen. It can be argued that for such a carcinogen, the method of using a NOEL coupled with a safety factor can
provide as much protection as this method of standard-setting provides
for other forms of toxicity. It is beyond the scope of this paper to
discuss the merits of this argument, but if it is accepted, then consideration of the carcinogenicity data developed by Kociba et al,
(1978) will lead to the same estimate of a tentative tolerable level
of intake that was derived on the basis of reproductive toxicity. I
am sure we shall return to this subject in the discussion because it
is by no means as simple as I have just made it sound.

Thus, if you accept the methodology I have described, it provides
a useful guide to the analytical chemist and for the development of
exposure information. I reemphasize that the collection of adequate
exposure data should be the next order of business.

Questions of the adequacy of the margin-of-safety for the general
population (all my estimates are tentative), the need to establish
TCDD limits in the various affected media, the relative effectiveness
of various control strategies (e.g., is it more effective to ban the
use of phenoxy herbicides or to prohibit incineration of chlorophenols), and the cost of various forms of regulation or control can
only be considered after the missing piece of the risk formula--
exposure--is added. This is not to say that we should refrain from
other forms of information gathering or avoid control actions in
specific areas where gross contamination is apparent.

We certainly know enough about TCDD to pursue vigorously specific
instances of food, air, or water contamination in which human exposure
is likely to approach the range of clearly identifiable risks.

I shall close with an observation concerning the extraordinary
promotional and microsomal enzyme-inducing properties of TCDD. In
view of these properties, it is not unreasonable to speculate that a
TCDD carcinogenesis bioassay in which exposure to other carcinogens is
minimized is likely to have lessened the risk of this substance. I am
not sure what should be made of this speculation, but it seems to me
worthy of further discussion in the scientific community.

REFERENCES

Kociba, R. J., Keyes, D. G., Beyer, J. E., Carreon, R. M., Wade, C.
    E., Dittenber, D. A., Kalnis, R. P., Frauson, L. E., Park, C. N.,
    Barnard, S. D., Hummel, R. A., and Humiston, C. B. Results of a
    two-year chronic toxicity and oncogenicity study of 2,3,7,8-
    tetrachlorodibenzo-p-dioxin in rats. Toxicol. Appl. Pharmacol.
    46:279-303, 1978.

Murray, F. J., Smith, F. A., Nitschke, K. D., Humiston, C. B., Kociba,
    R. J., and Schwetz, B. A. Three-generation reproduction study
    of rats given 2,3,7,8-tetrachlorodibenzo-p-dioxin in the diet.
    Toxicol. Appl. Pharmacol. 50:241-252, 1979.
Nisbet, I. C. T., and Paxton, M. B., Personal communication. Clement
    Associates, Washington, D. C. 1981.
Schantz, S. L., Barsotti, D. A., and Allen, J. R., Toxicological
    effects produced in nonhuman primates chronically exposed to
    fifty parts per trillion 2,3,7,8-tetrachlodibenzo-p-dioxin
    (TCDD). Toxicol. Appl. Pharmacol. 48:A180, 1979. (Abstract)

RISK ASSESSMENT FOR REGULATION OF DIOXIN (TCDD)

Lester B. Lave

The Brookings Institute
Washington, D.C.   USA

INTRODUCTION

People are concerned about health problems, particularly risks
of cancer, and reproductive problems. Whether judged by contributions
to the American Cancer Society, appropriations to the National Cancer
Institute, and stories in the media, these concerns are important.
At the same time, few people have even the most rudimentary idea of
the nature of the risks, their magnitude, and of what can be done to
lower risk. Consequently, people seem to demand simplistic solutions
to these complicated, emotion-packed problems.

An example of a simplistic solution is the Delaney Clause in the
Food, Drug, and Cosmetic Act that forbids FDA to allow any substance
to be added to food which is found to cause cancer in animals or man.
The FDA opposed this amendment as being both unnecessary and too con-
straining. Subsequent history has shown the FDA to be correct. The
Delaney Clause has been used only a few times since the general FDA
statutes aimed at adulterated food and toxic substances are almost
invariably sufficient. Where FDA has been painted into a corner, as
with saccharin, Congress decided that the Delaney Clause was not in
the public interest and forbid the banning of saccharin. The point
is that emotional subjects can easily lead to simplistic legislation
that is adverse to the public interest.

I am a devoteé of risk analysis and take every opportunity to
urge the use of quantitative risk analysis (QRA) in formulating both
private and regulatory responses to exposure to toxic substances.
However, scientists tend to play Marie Antoinette to a worried public,
asserting that QRA embodies the best scientific knowledge and that a
risk of 10^{-7} is so small as to be negligible is akin to "letting

them eat cake." Regulations must be accepted by the general public
and the goals and methods at least vaguely understood. If the public
generally believes that a death from cancer is many times worse than
death from other causes, regulations must embody this value. If the
general public, or an articulate group claiming to represent them,
believe that accidents are many times more likely than scientists
estimate, stonewalling will be dysfunctional.

The United States is a democracy. Views that are strongly held
by a large segment of the population must be accounted for in the
legislative and regulatory process. QRA will improve regulatory decisions only insofar as it embodies public values and earns the
confidence of the public. Industry's objective should not be to
co-opt regulators; they will both go down in flames together. Rather,
it must be to convince the interested scientists and public that the
answers of QRA can be believed and that the regulatory decisions
reflect the best available scientific knowledge and public values.

I have just gone through an exercise at the Nuclear Regulatory
Commission which examined the desirability of setting quantitative
risk goals. Such goals would be desirable in that they would give
more consistent criteria to be used in design and licensing and would
better inform the public about goals and licensing criteria. Quantitative risk levels were proposed, such as 10^{-5} and 10^{-7}, and extensive
justifications were written as to when these risk goals were plausible
in comparison with other risks. The problem is that quantitative risk
goals must be acceptable to the general public. Until someone suc-
ceeds in convincing them that these levels are acceptable, the debate
remains one among a cloistered group and the outcome will not make
regulatory decisions more acceptable.

UNCERTAINTIES IN TCDD RISK ANALYSIS

My knowledge of dioxin is limited. However, it is apparent that
important aspects of the problem are being ignored. In an otherwise
praiseworthy review of 2,4,5-T, Silvex, and TCDD, Perry J. Gehring
simply asserts that the multistage, linearized dose-response relationship is conservative, and implicitly assumes that the available
bioassay data can be accepted, that there will be few or no sensitive
individuals, and that accidents involving significantly greater exposure will not occur. Furthermore, there is the possibility of
interactions of TCDD with other substances that might vastly enhance
the effect of either.

Certainly, the linearized, multistage model is a conservative
one, but it is not the most conservative one. In the absence of
knowledge of the mechanisms by which TCDD causes responses, none of
the standard models are more plausible than their competitors. The
better part of wisdom is to derive estimates using a range of

relevant models and to set out this range of estimates without asserting more confidence in any estimate than current knowledge would
support. By using only a single conservative model, undue emphasis
is given to what is a single point in a large range.

Bioassays have numerous problems, from experimental design to
implementation. Often the numbers of animals are inadequate to draw
confident conclusions about the toxic nature of the substance. These
data must be treated with a great deal of skepticism, whether the
outcomes are positive or negative.

Humans are born with vastly different genetic inheritances.
These inherent sensitivities can be heightened by exposure to various
substances, nutritional history, etc. Generally, the population
exhibits a vast range of susceptibilities to any substance. This
range must be accounted for in the risk analysis.

All human activities involve accidents. It is safe to forecast
that some members of the work force and general public will be exposed
to large doses of any widely used toxic substance. Again, the risk
analysis must account for this exposure.

Finally, substances may interact chemically and biologically to
enhance or suppress effects. More has to be learned about such interactions in practice.

REGULATING TCDD

Regulators differ from scientists in that they must make timely
decisions. Putting off regulating in order to gather more evidence
is a decision. Thus, I regard the above comments on risk assessment
as being guides to doing this task better in the future. More practical, short-term suggestions for risk analysis of TCDD are needed to
guide regulation.

I am unwilling to bear any risk without an offsetting benefit.
Thus, the first step in making an intelligent regulatory decision on
TCDD is to specify and, if possible, quantify the benefits of TCDD,
for example, using herbicides containing TCDD. These benefits must
be relative to the next best herbicide, including the manufacturing
costs, relative costs of application, and effectiveness. The risks
must also be relative to the next best herbicide. Is any herbicide
desirable? If the answer is yes, the questions become which herbicide ought to be used for what purposes and under what conditions.
Comparing the teratogenic risks of TCDD with the risks of coffee is
not very helpful (unless coffee is used as a herbicide). Comparative
risk assessment can be immensely helpful if the risks that are being
compared are similar.

The risk analysis requires several steps. The first is a careful review of data from epidemiology and toxicology. What are the results and how much confidence can be placed in each? In extrapolating from animal bioassays, a range of plausible assumptions should be employed and the resulting range of risk estimates reported. In practice, it seems unlikely that much will be known about especially sensitive individuals or the interactions of TCDD with other environmental substances.

Finally, I would advise regulators to adopt an approach where gaining knowledge will tend to decrease the actual level of risk. That is, regulators should tend to make conservative assumptions in the face of ignorance. As the ignorance is dispelled, accurate assumptions can be substituted for the conservative ones, resulting generally in a slight decrease in risk levels. This framework gives manufacturers a strong incentive to conduct further research so as to lessen uncertainty.

RISK ASSESSMENT FOR 2,3,7,8-TETRACHLORODIBENZO-P-DIOXIN (TCDD)

Janice D. Longstreth and Judith M. Hushon

Dynamac Corporation
Enviro Control Division
Rockville, Maryland USA

ABSTRACT

This report presents efforts to develop a quantitative risk
assessment for 2,3,7,8-tetrachlorodizenzo-p-dioxin (TCDD), a contaminant of phenoxyherbicides such as 2,4,5-T and silvex. Based
on the available exposure information and assuming a contamination
level of 0.01 ppm TCDD in silvex and 2,4,5-T, it is estimated that
the general population may be exposed to a daily TCDD dose of
about 0.007 pg/kg. Subpopulations consuming only TCDD-contaminated beef or dairy products are estimated to receive doses of
0.44 or 1.11 pg/kg/day, respectively. The doses estimated for
occupational exposure range from a low of 0.001 pg/kg/day (flaggers) to a high of 2.18 pg/kg/day (rights-of-way hand applicators).
Because TCDD induces tumors in rodents, high-to-low dose extrapolations using the probit, logit, Weibull, multi-hit, and multistage models were performed in order to determine a range of virtually safe doses (VSDs) associated with several levels of human
risk. An unusual pattern of predictions resulted from these analyses, with the multistage model, normally the most conservative,
being the least conservative. Based on this result and the uncertainty as to the mechanism of TCDD's carcinogenic effect, it was
concluded that the use of these models to estimate human risk is
inappropriate. This conclusion, in conjunction with a lack of
evidence that TCDD is a genotoxic carcinogen, led to an estimate
of human risk from TCDD based on an acceptable daily intake (ADI)
of 1 pg/kg/day. At the assumed level of TCDD contamination in
2,4,5-T and silvex, two populations receive estimated doses that
slightly exceed this ADI and hence may be at risk: individuals
eating only contaminated dairy products and pesticide applicators
hand spraying rights-of-way.

INTRODUCTION AND APPROACH

This report is designed to present, evaluate, and utilize the data available for a risk assessment of 2,3,7,8-tetrachlorodibenzo-p-dioxin (TCDD). Using a variety of risk assessment methods and models, the findings discussed here integrate information on exposure from occupational and dietary sources and toxicity information from laboratory animal tests into a quantitative risk assessment.

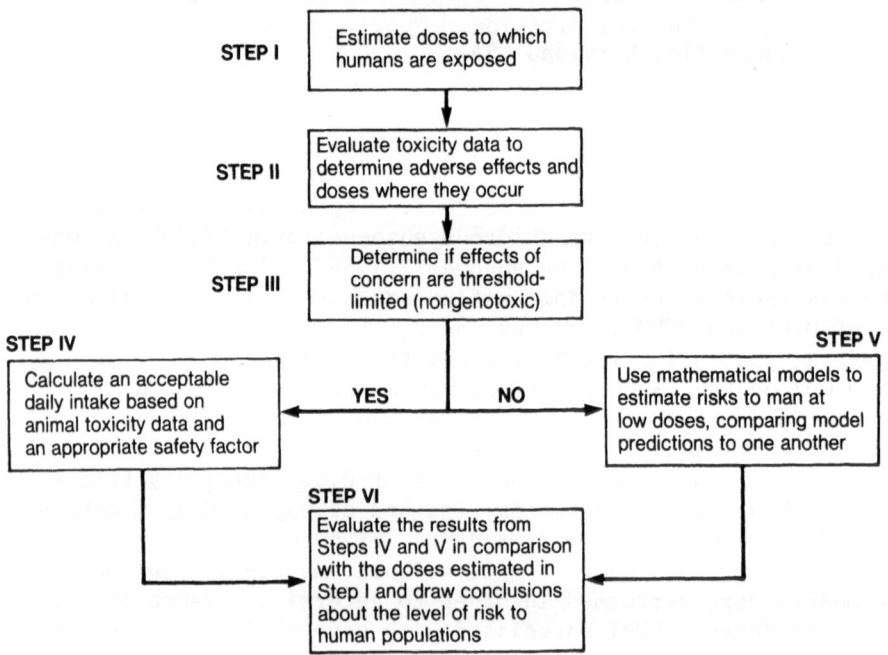

Figure 1.   Decision Tree for Estimating Human Risk from Exposure to Toxic Chemicals

Because all chemicals pose some risks under some conditions, the purpose of this risk assessment is not to find the exposure level at which no risk exists, but simply to identify the risks at any given exposure level. Our approach to this assessment can be expressed as the series of steps shown in Figure 1.

Estimation of Exposure

TCDD occurs as a contaminant in two widely used phenoxy herbicides: 2,4,5-trichlorophenoxyacetic acid (2,4,5-T) and 2,4,5-
trichlorophenoxypropionic acid (silvex). Both pesticides use
2,4,5-trichlorophenol (2,4,5-TCP) as a starting material, the
2,4,5-TCP being made by hydrolysis of tetrachlorobenzene with sodium hydroxide in ethylene glycol. TCDD is formed as a side product
of the process and its concentration in current formulations of
2,4,5-T and silvex ranges between 0.01 ppm (Gehring 1980) and 0.1
ppm (USEPA 1981).

EPA, as part of its continuing evaluation of substances registered under FIFRA, published an in-depth assessment of exposure
to 2,4,5-T, silvex, and TCDD for pesticide applicators, farm
workers, and the general population (USEPA 1981). Both inhalation
and dietary routes of exposure were considered. However, EPA did
not consider exposure from incineration of chlorine-containing
hydrocarbons (Bumb et al 1980). For exposure information, this
assessment relies on the EPA data.

Based on the 2,4,5-T exposure data for applicators and farm
workers, the use pattern information for silvex and 2,4,5-T, and
assuming that uptake of TCDD occurs at the same rate as 2,4,5-T
the lifetime daily dose of TCDD received by these workers can be
estimated as shown in Table 1 (details of the calculation are
presented in Appendix A). The far right column of Table 1 gives a
ranking of the estimated doses. It is interesting to note that
seven out of the ten highest doses are received by applicators
spraying rights-of-way, indicating that this group will probably
be at greatest risk. Table 2 gives the EPA estimates of the doses
received by the general population and special subgroups from
dietary exposure to TCDD. In Table 3, the information in Tables 1
and 2 is combined to estimate the total dose received by certain
categories of pesticide applicators and farm workers who could be
exposed both through their diets and their occupations.[1]

Health Effects in Animals

TCDD is a highly toxic substance. As shown in Figure 2, reported oral LD_{50} values for TCDD range from a low value of 0.6 µg/

----

[1]It should be noted that the data from the EPA exposure assessment (USEPA 1981) were appropriately caveated and the authors
went to great length to identify and explain their assumptions.
Where these limitations apply directly to this risk assessment,
they are reiterated; however, those limitations relating to
different scenarios, e.g., an increased use of 2,4,5-T and silvex,
are not repeated.

Table 1. Estimated Doses of TCDD Received by Pesticide Applicators and Farmworkers During Work[a]

Use Pattern	Exposed Group	Est. No. Exposed	Average Daily Dose[b] pg/kg/day	Rank by Size of Dose
FORESTRY				
Aerial	Pilots	73	9.04×10^{-2}	14
	Mixer/Loaders	73–145	1.50	5
	Flaggers	ND	7.24×10^{-2}	16
	Supervisors	ND	9.64×10^{-2}	12
Ground Broadcast				
Tractor Mistblower	Mixer/Loader	90–180	2.89×10^{-1}	10
	Tractor/operator/ worker	90	9.41×10^{-2}	13
	Supervisor	ND	8.68×10^{-2}	15
Backpack Sprayer	Applicators	300	5.06×10^{-1}	8
	Mixer/Supervisor	ND	3.01×10^{-2}	18
RANGE AND PASTURE[c]				
Aerial	Pilots	130	1.66×10^{-2}	20
	Mixer/Loaders	130–260	8.57×10^{-2}	16
	Flaggers	800	1.37×10^{-3}	22
Ground Backpack	Applicators	20,000	1.77×10^{-2}	19
RICE[c]				
Aerial	Pilots	307	2.63×10^{-3}	21
	Mixer/Loader	307	3.94×10^{-2}	17
	Flaggers	6,500–9,500	3.29×10^{-5}	23

Use Pattern	Exposed Group	Est. No. Exposed	Average Daily Dose[b] pg/kg/day	Rank by Size of Dose
RIGHTS-OF-WAY[c]				
Aerial	Pilots	25	7.24×10^{-1}	7
	Mixer/Loaders	25–50	2.89	1
Ground				
Selective	Applicators (hand) Basal	1,380	2.53	2
Cut Stump	Applicators (hand)	60	7.99×10^{-1}	6
Mixed Brush	Applicators (hand)	270	1.57	4
	Truck boom Applicators	178	9.94×10^{-2}	11
Railroad	Crew of Four	114	5.25×10^{-1}	9
Electric Power	Applicators (hand)	400	1.6	3

[a]Based on exposure data taken from USEPA (1981).
[b]Details of the calculation are given in the appendix. dose estimates are based on assumed contamination of 0.01 ppm (Gehring 1980). At 0.1 ppm, doses would go up 10-fold, i.e., 10^{-2} would become 10^{-1}.
[c]This information was extrapolated from the information on forestry workers by assuming that the exposure concentration would increase proportionally to the application rate.

Table 2. Estimated Doses of TCDD Received by the General Population and Certain Local
Populations from Contaminated Beef and Dairy Products

Exposed Group	Dose	
	(pg/day)[a]	(pg/kg/day)[b]
General population	0.5	7×10^{-3}
Local populations that consume:		
only contaminated beef	31	0.44
only contaminated dairy products	78	1.11
only contaminated beef and dairy products	117	1.67

[a]Adjusted for an application period once/5 years as indicated in USEPA (1981).
[b]Adjusted using a 70-kg human.

Table 3. Estimated Doses from Dietary and Occupational Exposure

Exposed Population	Weighted Mean Occupational Dose[a] (pg/kg/day)	Total Dose (occupational plus dietary) (pg/kg/day)			
		Consuming An Average Diet	Consuming Only Contaminated Beef	Consuming Only Contaminated Dairy	Consuming Only Contaminated Beef and Dairy
Pilots	0.05	0.06	0.49	1.16	1.72
Mixer/Loaders	0.11–0.35	0.12–0.36	0.55–0.79	1.21–1.46	1.78–2.02
Flaggers[b]	$(1.37-1.79) \times 10^{-4}$	$(7.14-7.18) \times 10^{-3}$	0.44	1.11	1.67
Backpack Applicators	0.03	0.04	0.47	1.14	1.70
Hand Applicators[c]	2.18	2.19	2.62	3.29	3.85

[a] Weighted according to the occupational population size, see Appendix A for details.
[b] Excluding the forestry flaggers, as no population size was available.
[c] This information pertains exclusively to those individuals treating rights-of-way.

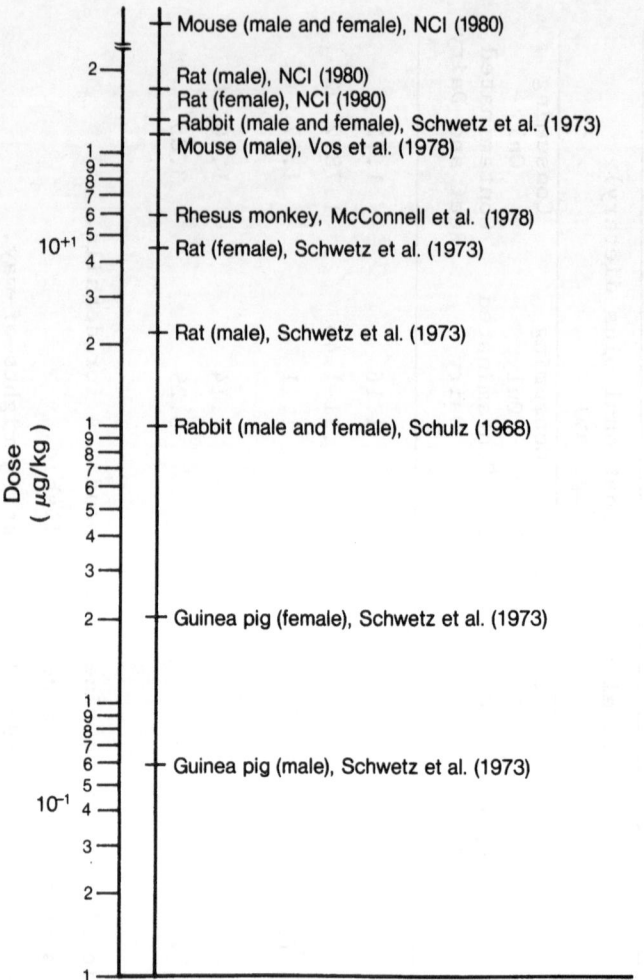

Figure 2.   Acute Lethality of TCDD (Oral LD$_{50}$)

kg for male guinea pigs (Schwetz et al, 1973) to a high value of
greater than 200 μg/kg for male and female B6C3F1 mice (NCI 1980).
As displayed in Figure 3, reported adverse effects of subchronic
administration of TCDD include:

   O   depressed lymphocyte responsiveness to tetanus toxin in
       guinea pigs (Zinkl et al 1973)

   O   a variety of reproductive effects including abortions,
       still births, and failure to conceive in monkeys (Schantz
       et al 1979); decreased litter size, fertility, growth and
       viability in F$_1$ and F$_2$ rats (Murray et al 1979); terato-
       genicity and fetotoxicity in rats (Sparschu et al 1971);
       and cleft palate in mice (Smith et al 1976).

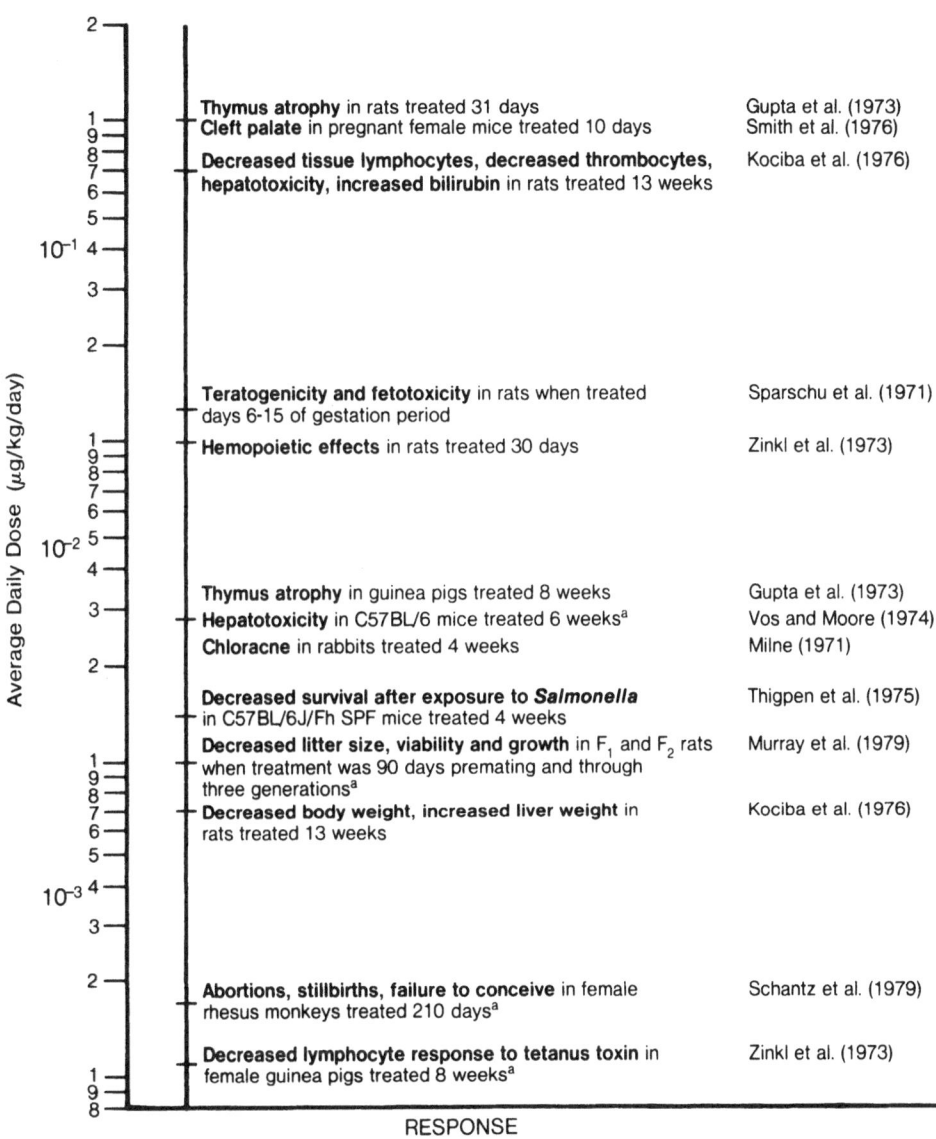

^aDoses corrected to indicate average daily dose.

Figure 3. Subchronic Toxicity of TCDD

O decreased resistance to <u>Salmonella</u> infections in mice
   (Thigpen et al 1975)

O decreased body weight and increased liver size in rats
   (Kociba et al 1976)

O thymus atrophy in guinea pigs and rats (Gupta et al 1973)

O hepatotoxicity in rats and mice (Kociba et al 1976, Vos
   and Moore 1974)

O chloracne in rabbits (Milne 1971).

The toxicity of TCDD has also been examined in a number of
chronic studies using rodents (Kociba et al 1978, Van Miller et
al 1977, Toth 1979, and NCI 1980). Carcinogenesis was the endpoint of concern in these studies, and all found either statistically significant increases in tumorigenesis (Kociba et al 1978,
Van Miller et al 1977, and NCI 1980) or suggestive evidence that a
carcinogenic process was taking place (Toth et al 1979).

In the Kociba et al (1978) study, male and female Sprague
Dawley rats were fed diets providing 0.1, 0.01 and 0.001 μg/kg/day
for two years. Histopathological examination of the tissues from
these rats showed that at the highest dose, there were statistically significant increases in hepatocellular carcinomas and
squamous cell carcinomas of the hard palate and/or the nasal tur-
binate in both males and females, and in squamous cell carcinomas
of the tongue in males. The USEPA (1981) in reviewing these data
obtained an outside histopathological review which found essentially the same target organ pattern, but a slightly higher total
numbers of tumors. The data from that review for the toal numbers
of statistically significant tumors appear in Table 4.

In the NCI (1980) bioassay, 50 Osborne Mendel rats and 50
B6C3F1 mice of each sex were gavaged with TCDD suspended in a
vehicle of 9:1 corn oil: acetone 2 days a week for 104 weeks.
For rats and male mice, the doses given were 0.01, 0.05 and 0.5
μg/kg/wk; for female mice they were 0.04, 0.2 and 2.0 μg/kg/week.
Using the Fisher exact P test to evaluate the results of the
histopathological examinations, NCI found a dose-related, significantly increased incidence relative to controls of follicular cell
adenomas or carcinomas of the thyroid in male rats in all three
test groups. In addition, subcutaneous tissue fibromas were significantly increased in the high dose male rats. In high dose
female rats, hepatocellular carcinomas and neoplastic nodules,
subcutaneous tissue fibrosarcomas, and adrenal cortical adenomas
all showed statistically significant increases.

Table 4.   Synopsis of the Kociba and NCI Data
on Tumors in Animals Given TCDD[a]

Squire Review (USEPA 1981) of Kociba Study (1978)

Dose Level (mg/kg/day)	0	$1x10^{-6}$	$1x10^{-5}$	$1x10^{-4}$
Male rats (tr/n)[b]	0/77	2/44	1/49	9/44
Female rats (tr/n)	16/86	8/50	27/50	34/47

NCI Study (1980)

Dose Level 1 (mg/kg/day)	0	$1.43x10^{-6}$	$7.14x10^{-6}$	$7.14x10^{-5}$
Male rats (tr/n)	1/69	5/48	8/50	11/50
Female rats	5/75	1/49	3/50	14/49
Male mice	15/73	12/49	13/49	27/50

Dose Level 2 (mg/kg/day)	0	$5.71x10^{-5}$	$2.86x10^{-5}$	$2.86x10^{-4}$
Female mice	22/74	20/50	19/48	31/47

[a]Data taken from EPA (1981).
[b]Number of animals with tumors over total number of animals.

In high dose male mice, significant increases in hepatocellular carcinomas and neoplastic nodules were observed.   In high dose
female mice, statistically significant increases in hepatocellular
adenomas and carcinomas, fibrosarcomas, histiocytic lymphomas,
thyroid follicular cell adenomas and cortical adenomas or carcinomas were observed.   A summary of the information indicating total
numbers of statistically significant tumors appears in Table 4.[2]

---

[2]The numbers given in Table 4 are for tumors judged significant
on the basis of the Fisher exact P test, and are those that were
evaluated by EPA in their risk assessment of TCDD (USEPA 1981).
The NCI, however, uses the Cochran-Armitage test for trends and
concluded that only the occurrence of the following tumors was related to the administration of TCDD: thyroid tumors in male rats,
liver tumors in female rats, liver tumors in male and female mice,
and thyroid tumors and possibly histiocytic leukemias in female
mice.   When the Cochran-Armitage test was applied to the Kociba
et al (1978) study, all sites judged significant by Fisher's
exact P were also significant by this second test.

Evidence of Genotoxicity

There is some indication that TCDD may be a direct-acting frame-shift mutagen (Hussain et al 1972, Seiler 1973); however, the data are conflicting (McCann, as cited in USEPA 1981, Nebert et al 1976) and need to be resolved. In mammalian mutagenesis test systems, TCDD has generally given negative results (Khera and Ruddick 1973, Green et al 1977). However, when given twice weekly for 13 weeks at a dose of 2 µg/kg, TCDD caused some chromosomal aberrations in rat bone marrow cells. There is no indication, however, that TCDD caused mutations in those individuals (mothers and fetuses) who were exposed during the July 1976 Seveso, Italy, factory accident (Reggiani 1977 as cited in USEPA 1981, Tuchman-Duplesis 1977 as cited in USEPA 1981).

ASSESSMENT OF HUMAN RISK

As noted in NAS (1980), there have historically been two principal approaches to estimating "acceptable" levels of exposure to various agents. One approach involves calculating an "Acceptable Daily Intake" (ADI) by applying a safety factor to doses of the agent that did not produce an observed effect in animals (No observed adverse effects level, NOAEL). The second approach relies on the use of mathematical models to extrapolate experimental dose-response information from the high levels at which such information is obtained to the low levels likely to be encountered by human populations. This latter approach generally assumes that there is no threshold for the response being extra-polated. It was originally used for radiation damage but has more recently found widespread use for estimating the risks due to chemical carcinogenesis. As a general rule, in estimating the human risks posed by an agent, if there is evidence of a carcino-genic effect, the high-to-low dose extrapolation models are used. However, they theoretically should be applied only to carcinogenic effects that result from genotoxicity of the agent, e.g., alkyla-tion of DNA.

In this risk assessment, because of uncertainty as to the carcinogenic mechanism of TCDD, the animal carcinogenicity bio-assay data have been evaluated using a variety of high-to-low dose extrapolation models.

Estimation of Human Virtually Safe Doses (VSDs) Based on Various Mathematical Models

In this report, the derivation of human VSDs from animal data on carcinogenesis is treated as a two-step process. The animal data to be modeled are identified and extrapolated to the desired

probability (risk) levels and the resulting animal VSDs are then
converted from animal to man based on a surface area conversion
factor. A number of mathematical models exist for the first step
of high-to-low dose extrapolation, but although many of them
adequately fit the data in the observable range, they often yield
widely divergent results upon extrapolation to low doses (Krewski
and Van Ryzin 1981). Thus, the choice of a model becomes largely
a matter of judgment. This judgment should include, however, an
evaluation of the data set to be extrapolated and any information
on the biological behavior of the compound under study, as well as
an assessment of the statistical fit of the data to the model.
This last factor should perhaps be given less weight as it is an
assessment of random error which, given the number of assumptions
and approximations involved in choosing a model and the data to
fit to it, may contribute little to the overall error in the
extrapolation. Indeed, because of these considerations and also
because a comparison between models can provide additional insight
into an evaluation of the risks posed by a particular compound,
this report compares and contrasts the VSDs predicted by five
different models: namely the probit (P), logit (L), Weibull (W),
multi-hit (MH) and multistage (MS) models. A brief description
of these models follows.

The probit and logit models are both based on the notion that
the response of a given population reflects the distribution of
the individual tolerances of the animals in that population
(Krewski and Van Ryzin 1981). The probit model specifically
assumes that the distribution of the logarithms of these tolerances
is Gaussian (normal) against the logarithm of the dose, whereas the
logit model assumes that these tolerances show a logistic distribution (NAS 1980). The Weibull and the multi-hit models are both
generalizations of the one-hit model which assumes that a positive
result occurs after a single receptor has been exposed to a single
effective dose unit of a substance (Hoel et al 1975). The Weibull
differs from the one-hit in that it does not assume independent
equal action for incremental dose units. The multi-hit model
differs from the one-hit in that it assumes that a series of hits
are required at each receptor in order to cause a positive result.
The multistage model assumes that carcinogenesis and other irreversible self-replicating adverse effects result from the occurrence of a number of different random events with the age-specific
rate of occurrence of each event being linearly related to dose
(Krewski and Van Ryzin 1981).

All of the models assume no threshold, but not all are linear
at low doses. The probit model is inherently sublinear at low
doses and generally leads to relatively low risk estimates in that
region.. The logit, Weibull, and multi-hit models are linear only
when the dose-response curves are concave at moderate and high

doses. Thus, these models generally predict higher risks at low
doses than the probit model. The multistage model provides for
dose-response curves that are both linear at low doses and convex
at moderate doses. Krewski and Van Ryzin (1981), in examining the
predictions of these models for a number of compounds, found that
the VSD predictions at a given risk level can generally be ranked
as follows:

$$probit > multi-hit > logit > Weibull > multistage$$

In their theoretical form, all these models assume that the background response rate is zero. Often, however, this is not the
case and one has to deal with a spontaneous incidence rate. This
background rate can mechanistically be thought to occur either independently of or additively to the response of interest. As
noted by Krewski and Van Ryzin (1981), while the choice of the way
to model the background is somewhat crucial, the biological or
statistical criteria for making this choice are often elusive. In
this report, independent background is assumed.

When a compound such as TCDD induces tumors at multiple sites
and in two species, a choice of which data to fit to these
mathematical models must be made. As pointed out by the Food
Safety Council (1980), "the VSD for risk for each organ alone may
be higher than the VSD for the combined risk; and when the individual responses are considered equally undesirable, the combined
risk should also be determined." This reasoning presumably led
the USEPA (1981) to fit the multistage model to a data set derived
from the Squire review that included a total increased incidence
of all statistically significant tumors in female rats from the
Kociba et al study (1978). The choice of the Kociba female rat
data was made on the basis that these data, when fitted to the
multistage model, gave the greatest estimated slope for the human
dose-response curve, q_1*. In this assessment, the USEPA (1981)
analysis is taken a step further and these data are fitted to
four additional models: the logit, the probit, the Weibull, and
the multi-hit.[3] The VSDs[4] that result from fitting the Squire
(USEPA 1981) data to these models are given in Figure 4. The results of the chi square goodness-of-fit tests were barely adequate

---

[3]We are indebted to Dr. Daniel Krewski for providing us with Risk
81, the computer program for these four models, and to Dr. Kenny
Crump for Global 79, the program for the multistage model.
[4]It should be noted that the data points plotted in Figure 4 are
the best point estimates rather than the upper confidence limits
(UCL) used by the Carcinogenesis Assessment Group (USEPA 1981).
Plots of the UCLs did not change the pattern of the model predictions but simply lowered the VSDs still further.

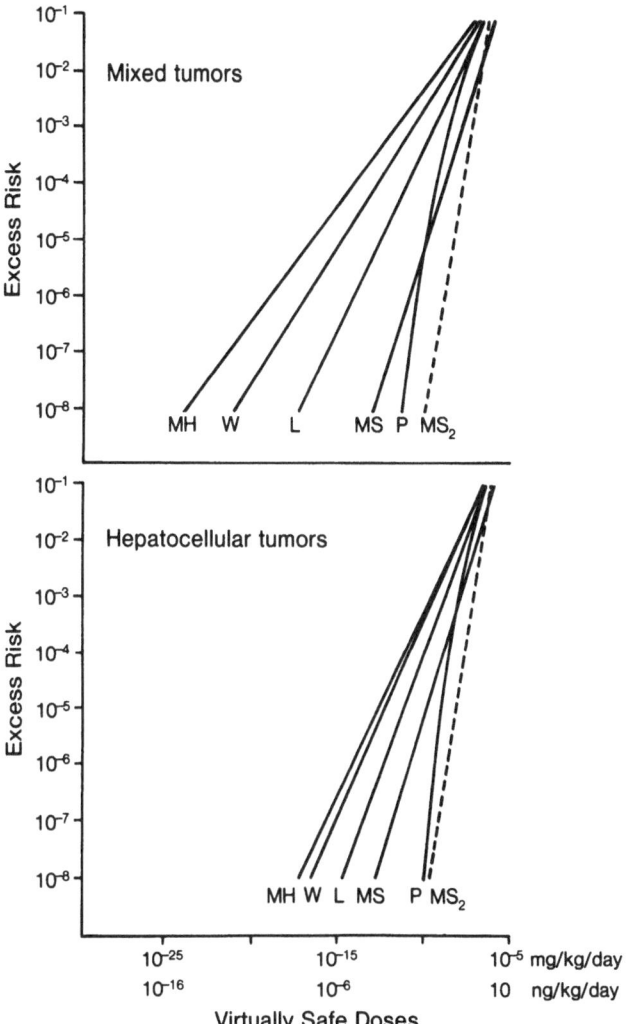

Figure 4. Model Predictions

for the probit (p = 0.05), the logit (p = 0.04) and the Weibull
(p = 0.02) models and poor for the multi-hit (p = 0.01) and the
multistage (p<0.01). However, an adequate fit (p = .44) was obtained for the multistate model when the top dose was dropped and
the model refitted to the lower two doses (USEPA 1981). The VSD
predictions from that fit are presented as the dashed line in
Figure 4 which is identified as MS_2.

   In evaluating the predictions from these models, it was noted
that the multistage model, generally considered to be the most
conservative (Krewski and Van Ryzin 1981) was, in this instance,

less conservative and was the least conservative when a good fit
was obtained. In evaluating this information, several explanations come to mind. First, it may be inappropriate to combine
tumor information from several sites. Conceivably the tumorigenic
processes at these sites could be different so that either the
number of "hits" or "stages" required differs, or the distribution
of tolerances for tumors at each site differs. Is such is the
case, then a poor fit or an unexpected pattern might result.

To test this hypothesis, a data set from a single site was
selected and fitted to the five models. The data set selected was
that for hepatocellular tumors (USEPA 1981) and the selection was
based on the fact that this site showed the highest carcinogenic
potency, β, according to the one-hit model. This exercise resulted
in a better fit for the probit ($p = 0.12$), the logit ($p = 0.08$),
and the Weibull ($p = 0.05$) models, an adequate fit for the multi-
hit ($p = 0.04$) model, but still a poor fit for the multistage ($p =
0.01$) model.[5] As shown in Figure 4, however, it did not change
the order of the predictions by these models, but only reduced the
VSDs.

A second possible explanation of the unusual pattern of predictions was that the dose metameter used was incorrect, i.e.,
that the effective dose of TCDD somehow differed from the administered dose. Using these five models, Krewski and Van Ryzin (1981)
found essentially the same pattern for vinyl chloride observed
here for TCDD, whereas 14 other compounds gave the expected pattern. For vinyl chloride, the difference was attributed to the
concave nature of the dose-response curve and this was found to be
correctable by using pharmacokinetic information to derive an
effective dose that reflected the effects of saturated metabolism
on the administered dose. At the present time, the applicability
of this particular hypothesis to TCDD cannot be examined. The
mechanism by which TCDD exerts its carcinogenic effect is still
unknown.

Although its behavior (carcinogenic in more than one species
and tumor production at multiple sites) is classically that of a
compound that is metabolized to a proximate electrophilic carcinogen that binds to DNA (Food Safety Council 1980), there is no
evidence that this is in fact how TCDD works. Indeed, TCDD shares
none of the other properties of well-known electrophilic carcinogens (activity in short-term tests for carcinogenicity and genotoxicity) suggesting that if the effective dose differs from the
administered dose, it is probably not due to saturated metabolism.
It is possible, however, that saturation of some other process may
be occurring. TCDD is known to form complexes with the arylhydro-

---

[5]Until the top dose level is dropped.

carbon hydroxylase (Ah) receptor which then interacts with the
genome to activate a number of genes (Okey 1981). Formation of such
a complex is possibly subject to saturation and should this complex
play a role in the neoplastic process, then its effective concentration may not be reflected by changes in the administered dose.

   If, for example, some of the TCDD is bound to animal tissue,
and slowly released, the release rate may be the rate determining
step for toxicity. Higher doses might than result in a greater
amount of binding with release at the same rate, but over a longer
period of time resulting in increased toxic effects due essentially
to the same level of exposure but over a longer time interval.
This phenomenon would not be consistent with the normal risk extrapolation models but would be consistent with the behavior of TCDD.
The data of Seefeld and Peterson (1981) for groups of rats fed
single doses of 25 or 50 μg of TCDD might be interpreted to support this conclusion. In both groups, deaths were first observed
about 2 weeks after treatment. Ultimately 75 percent of the high
dose group died while only 25 percent of the 25 μg group died.
Seefeld and Peterson (1981) would say the higher dose (longer exposure) resulted in a longer reduced nutrition period resulting in
increased deaths.

   Another possible explanation for the exceptional behavior of
TCDD in these models is that the mechanism by which TCDD induces
tumors has a threshold and that these non-threshold models are inappropriate.

   More research is needed to determine the pharmacodynamics of
TCDD metabolism. Only in this way can it be determined whether a
genotoxic effect does exist, but is being masked by some sort of a
threshold-producing reaction.

   The somewhat unusual behavior of TCDD with these five models
makes the choice among them difficult. Based on the information
on the various compounds examined by Krewski and Van Ryzin (1981),
it seems likely that there is some missing factor that precludes
the valid application of these models to the high-to-low dose
extrapolation for TCDD. Because of this, and because at the present time TCDD is not a proven genotoxic compound, this risk assessment evaluates the hazard to humans on the basis of the ADI.

Calculation of an Acceptable Daily Intake (ADI)

   For all phenomena that are threshold limited, no adverse
effects should result from exposure to doses that are below the
threshold level. In order to transpose information from animal to
man, however, various agencies have taken the approach of applying
a "safety factor" to animal data in order to estimate an ADI for

humans. (For a review of this subject see NAS 1980.) This safety
factor generally is large enough to account for interspecies variation in sensitivity as well as individual variation within a single
species. Although no single safety factor has emerged as a universal
standard for all chemicals, a 100-fold factor has been commonly used
when a clear no-effect level has been established in a sensitive test
animal. If the reported data are questionable or only a lowest effect level is reported, a safety factor of 1,000 is used.

In reviewing the literature, the lowest no observable adverse
effect level (NOAEL) reported was a dose of 1 ng/kg/day given orally
for 2 years to rats (Kociba et al, 1978). However, as indicated in
Figure 3, Zinkl et al (1973) noted that guinea pigs exposed for 4
weeks to this same dose showed a decreased lymphocyte count in response to tetanus toxin, and a slightly higher dose (1.7 ng/kg/day)
was associated with abortions, stillbirths and failure to conceive
in female rhesus monkeys (Schantz et al, 1979). Because of these
latter observations, this report bases its ADI on the guinea pig data
reporting a minimal effect level of 1 ng/kg/day. Applying a one-
thousand-fold safety factor to this dose, an ADI of 1 pg/kg/day is
calculated.

Conclusions

A comparison of the calculated ADI of 1 pg/kg/day with the
levels estimated in Table 3, indicates that the doses received by
two populations exceed the ADI and thus these populations may be
at risk from the adverse effects of TCDD. These are hand applicators treating rights-of-way with the pesticides 2,4,5-T and silvex,
and individuals whose dairy products come only from herds grazing
on pastures and range lands treated with 2,4,5-T or silvex. The
doses received by these two groups, however, are only 2- to 3-fold
higher than the ADI and given the number of conservative approximations and assumptions that were required to estimate exposure
for these groups, it may well be that the actual dose received is
somewhat lower.[6,7]

---

[6] The data in Table 3 were based on the USEPA (1981) exposure
assessment which has been criticized by Braun (1980) for methodology errors in the estimation of the rate of absorption of
2,4,5-T for pesticide applicators. Accordingly, Braun (1980) recalculated the TCDD doses for these occupational populations.
These data are presented in Appendix B. At a contamination level
of 0.01 ppm TCDD, these data substantiate the conclusion that
certain categories of rights-of-way applicators receive daily
doses slightly in excess of the ADI.

[7] The doses estimated for individuals consuming only contaminated
dairy products assumed that the residue levels in beef cattle adipose tissues could be extrapolated directly to residues in butterfat. Firestone et al (1979) has shown, however, that this is not

the case for Holsteins where the level of dioxins in adipose
tissue was approximately twice that found in butterfat.

It must also be pointed out that not all of the assumptions
made in this assessment were the most conservative. In particular,
the choice of a contamination level of 0.01 ppm (Gehring 1981)
TCDD in 2,4,5-T may be from 4-fold (Lavy et al 1979, EPA 1981) to
10-fold (EPA 1981) too low. Were this the case, then the hand
applicators could be receiving a dose 8 to 20 times more than the
ADI. Data on the actual dose received by applicators who treat
rights-of-way are necessary in order to resolve this particular
point.

REFERENCES

Braun, W.H. Direct Testimony, FIFA Docket Nos. 415 et al. Exhibit
     902, 1980.
Bumb, R.R., Crummett, W.B., Cutie, S.S., Glendhill, J.R., Hummel,
     R.H., Kagel, R.O., Lamparski, L.L., Luoma, E.V., Miller, D.L.,
     Nestrick, T.J., Shadoff, L.A., Stehl, R.H., and Woods, J.S.
     Trace chemistries of fire: a scource of chlorinated dioxins.
     Science, 210:385, 1980.
Firestone, D., Clower, M., Jr., Barsetti, A.P., Teske, R.H., Long,
     P.E. Polychlorodibenzo-p-dioxin and pentachlorophenol residues
     in milk and blood of cows fed technical pentachlorophenal. J
     Agric Food Chem, 27:6, pp. 1171-1177, 1979.
Food Safety Council. Quantitative Risk Assessment, Chapter 11. In:
     Proposed System for Food Safety Assessment. Final Report of
     the Scientific Committee of the Food Safety Council, pp. 137-
     160, 1980.
Gehring, P.J. Direct Testimony, FIFRA Docket Nos. 415 et al,
     Exhibit 912, 1980.
Green, S., Moreland, F., and Sheu, C. Cytogenic effect of 2,3,7,8-
     tetrachlorodibenzo-p-dioxin on rat bone marrow cells. U.S.
     Food and Drug Admin., Washington, D.C. FDA By-Lines, 7:292,
     1977.
Gupta, B.N., Vos, J.G., Moore, J.A., Zinkle, J.G., and Bullock, B.C.
     Pathological effects of TCDD in laboratory animals. Environ
     Health Perspect, 5:125, 1973.
Hoel, D.G., Gaylor, D.W., Kirschstein, R.L., Saffiotti, V., and
     Schneiderman, M.A. Estimation of risks of irreversible de-
     layed toxicity. J. Toxicol and Environ Health, 1:133, 1975.
Hussain, S., Ehrenberg, L., Lofroth, G., and Gejvall, T. Mutagenic
     effects of TCDD on bacterial systems. Ambio, 1:32, 1972.
Khera, K.S., and Ruddick, J.A. Polychlorodibenzo-p-dioxins: pre-
     natal effects and the dominant lethal test in Wistar rats.
     In: Blair, E.D. (ed.), Chlorodioxins - Origins and Fate. Adv
     Chem Ser, No. 120. Am. Chem. Soc., Washington, D.C., p. 70,
     1973.

Kociba, R.J., Kesler, P.A., Park, C.N., and Gehring, P.J. 2,3,7,8-Tetrachlorodibenzo-p-dioxin results of a 13-week oral toxicity study in rats. Toxicol Appl Pharmacol, 35:553, 1976.

Kociba, R.J., Keyes, D.G., Beyer, J.E., Carreon, R.M., Wade, C.E., Dittenber, D.A., Kalnins, R.P., Frauson, L.E., Park, C.N., Barnard, S.D., Hummel, R.A., and Humiston, C.G. Results of a two-year chronic toxicity and oncogenicity study of 2,3,4,8-tetrachlorodibenzo-p-dioxin in rats. Toxicol Appl Pharmacol, 46:279, 1978.

Krewski, D., and Van Ryzin, J. Dose response models for quantal response toxicity data. In: Sorgo, M.C., Dawson, D., Rao, J.N.K., and Salch, E., eds. Current Topics in Probability and Statistics, North Holland, Amsterdam, 1981.

McConnell, E.E., Moore, J.A. and Dolgard, D.W. Toxicity of TCDD in Rhesus monkey (Macaca mulatta) following a single oral dose. Toxicol Appl Pharmacol, 43:175, 1978.

Milne, M.H. Formation of 2,3,7,8-tetrachlorodizenzo-p-dioxin by thermal decomposition of sodium 2,4,5-trichlorophenate. Nature, 232:395, 1971, as cited in Ambient Water Quality Criteria for 2,3,7,8-tetrachlorodizenzo-p-dioxin, USEPA, 1980, EPA 440/5-80-072.

Murray, F.J., Smith, F.A., Nitschke, K.D., Humiston, C.G., Kociba, R.J., and Schwetz, B.A. Three-generation reproduction study of rats given 2,3,7,8-tetrachlorodizenzo-p-dioxin (TCDD) in the diet. Toxicol Appl Pharmacol, 50:241, 1979.

National Academy of Sciences. Problems of Risk Estimation. In: Drinking Water and Health, Vol. 3, pp. 25-65, 1980.

National Cancer Institute. Bioassay of 2,3,7,8-tetrachlorodibenzo-p-dioxin for possible carcinogenicity. (Gavage study.) U.S. Department of Health and Human Services Pub. No. 80-1765, 1980.

Nebert, D., Thorgeirsson, S., and Felton, J. Genetic differences in mutagenesis, carcinogenesis, and drug toxicity. In: DeSerres, F., Fouts, J., Bend, J., and Philpot, R. (eds.), In Vitro Metabolic Activation in Mutagenesis Testing. Elsevier/North Holland Biomedical Press, Amsterdam, pp. 105-124, 1976.

Ramsey, J.C., Lavy, T.L., and Braun, W.H. Exposure of forest workers to 2,4,5-T; calculated dose levels. Unpublished report, 1979.

Schantz, S.L., Barsott, D.A., and Allen, J.R. Toxicological effects produced in non-human primates exposed to fifty parts per trillion 2,3,7,8-tetrachlorodibenzo-p-dioxin (TCDD). Toxicol Appl Pharmacol, 48:A180, 1979.

Schultz, K.H. Clinical picture and etiology of chloracne Arbeitis-Medizen Sogialmedizin Arbeitshugiene 3:25, 1968.

Schwetz, B.A., Norris, J.N., Sparshu, G.L., Rowe, V.K., Gehring, P.J., Emerson, J.L., and Gerbig, C.G. Toxicity of chlorinated dibenzo-p-dioxins. Environ Health Perspect, 5:87, 1973.

Seefeld, M.D. and Peterson, R.E. Mechanisms for TCDD Toxicity in the Rat. International Symposium on Chlorinated Dioxins and Related Compounds, Arlington, Va, 1981.

Seiler, J.P. A survey on the mutagenicity of various pesticides. Experientia, 20:622, 1973.

Smith, F.A., Schwetz, B.A., and Nitschke, K.D. Teratogenicity of TCDD in CF-1 mice. Toxicol Appl Pharmacol, 38:517, 1976.

Sparschu, G.L., Dunn, F.L., and Rowe, V.K. Study of the teratogenicity of TCDD in the rat. Food Cosmet Toxicol, 9:405, 1971.

Thigpen, J., Faith, R., McConnell, E., and Moore, J. Increased susceptibility to bacterial infection as a sequela of exposure to 2,3,7,8-tetrachlorodizenzo-p-dioxin. Inf Imm, 12:1319, 1975.

Toth, K., Somfai-Relle, K., Sugar, J., and Bence, J. Carcinogenicity of herbicide 2,4,5-trichloro-phenoxyethanol-containing dioxin and of pure dioxin in Swiss mice. Nature, 278:548, 1979.

USEPA. Risk assessment on (2,4,5-trichlorophenoxy) acetic acid. (2,4,5-T), (2,4,5-trichlorophenoxy) propionic acid (silvex), 2,3,7,8-tetrachlorodibenzo-p-dioxin (TCDD). Carcinogen Assessment Group. Office of Health and Environmental Assessment EPA-600/6-81-003, 1981.

VanMiller, J.P., Lalich, J.J., and Allen, J.R. Increased incidence of neoplasms in rats exposed to low levels of TCDD. Chemosphere, 6:625, 1977.

Vos, J.G., and Moore, J. Suppression of cellular immunity in rats and mice by maternal treatment with 2,3,7,8-tetrachlorodibenzo-p-dioxin. Int Arch Allerg Appl Immunol, 47:777, 1974.

Vos, J.G., Kreeftenberg, J.G., Engel, H.W.B., Minderhoud, A. and Van Noorle Jansen, L.M. Studies on 2,3,7,8-tetrachlorodibenzo-p-dioxin induced immune suppression and decreased resistance to infection: endotoxin hypersensitivity, serum zink concentrations and effects of thymosin treatment. Toxicology, 9:75, 1978.

Zinkl, J., Vos, J.G., Moore, J.A., and Gupta, B.N. Hematologic and clinical chemistry effects of 2,3,7,8-tetrachlorodibenzo-p-dioxin in laboratory animals. Environ Health Perspect, 5:111, 1973.

ACKNOWLEDGEMENTS

Our thanks to D. McEwan for his help with the programming, Dr. D. Kreski, M. Samuels and W. Perry for discussions on the models and their interpretations, and Ms. S. Goetz and N. Winning for typing.

APPENDIX A

The doses used in this assessment were derived from the exposure information given in Table 1, p. 7, Appendix F of the USEPA
(1981) Quantitative Risk Assessment of 2,4,5-T, Silvex and TCDD.
Based on data from Ramsey et al (1979), EPA estimated that forestry
pilots were exposed to 2,4,5-T at an hourly rate of 0.015 mg/kg.
The USEPA (1981) further estimated that pilots were exposed at this
level approximately 200 hrs/year. Assuming that 2,4,5-T is contaminated with 0.01 ppm TCDD, an average daily dose for forestry
pilots can be calculated as follows:

$$\frac{0.015 \text{ mg/kg/hr X 200 hrs/yr X } 10^{-8} \text{ mg TCDD/mg 2,4,5-T}}{365 \text{ days/yr}}$$

$$= 8.22 \times 10^{-11} \text{ mg/kg/day}$$

$$= 8.22 \times 10^{-2} \text{ pg/kg/day}$$

In order to account for the TCDD received via the use of
silvex, the use rate information in Table 5, p. 14 of Appendix F
(EPA 1981) was used. For example, on forests, the rate of silvex
to 2,4,5-T use was 1:10. Thus, the dose of TCDD received from use
of both 2,4,5-T and silvex would be approximately 110 percent of
that received from 2,4,5-T use alone:

$$1.1 \times 8.22 \times 10^{-2} \text{ pg/kg/day} = 9.04 \times 10^{-2} \text{ pg/kg/day}$$

Note that the above dose is the average daily dose per year
received by pilots during those years that they spray 2,4,5-T and
silvex. If that period is only 10 years out of an approximate life-
span of 70 years, then their lifetime daily dose would be seven-
fold less (9.04×10^{-2} pg/kg/day $\div$ 7 = 1.29×10^{-2} pg/kg/day).
This last dose would be appropriate for comparison to the VSDs
estimated from various models derived from chronic (i.e., lifetime)
animal studies. However, since the final comparison was made to an
ADI that was based on subchronic toxicity data, the 9.02×10^{-2}
pg/kg/day daily dose was judged to provide the more appropriate
comparison.

Calculation of Weighted Mean Occupational Doses

In order to reduce the exposure data to more manageable proportions, the occupational population was divided into five categories: pilots, mixer/loaders, flaggers, backpack applicators and
hand applicators. A sample weighted mean average daily dose for
pilots can be calculated as follows.

For each use pattern (forestry, range and pasture, etc.) the
number of pilots exposed is multiplied by the average daily dose

and the resulting doses for all use patterns are summed and then divided by the total number of pilots:

Use	Est. No. Exposed	Average Daily Dose	Population Dose
Forestry	73	9.04×10^{-2}	659.92×10^{-2}
Range	130	1.66×10^{-2}	215.8×10^{-2}
Rice	307	0.26×10^{-2}	79.8×10^{-2}
Rights-of-way	25	72.4×10^{-2}	1810.00
	TOTAL 535		TOTAL 2765.5

$$2765.5 \times 10^{-2} \text{ pg/kg/day} \div 535 = 5.17 \times 10^{-2}$$
$$= 0.05 \text{ pg/kg/day}$$

APPENDIX B

The data in Table B-1 are derived from the direct testimony of Werner H. Braun (1980). The doses calculated are the daily doses received by workers on those days that they are actually applying pesticides. As such, they are not directly comparable to the doses estimated from the USEPA (1981) exposure assessment. These latter doses are the average daily doses received over a work year. Thus, if an applicator only worked 50 days per year, his daily dose (based on the USEPA 1981 information) would differ from the average daily dose calculated by a factor of 365/50 or 7.3. We feel that the criticisms by Braun of the USEPA (1981) methodology are valid and that his doses more accurately reflect applicator exposures. They still support the conclusion, however, that certain categories of rights-of-way applicators receive a dose slightly in excess of the ADI and thus that monitoring information on these workers is needed.

Table B-1. TCDD EXPOSURES

Applicator Category		ASSUMPTION 1: 0.04 ppm TCDD Content [1]		ASSUMPTION 2: 0.01 ppm TCDD Content [2]	
		Daily Dose (pg/kg)[3]	Average Daily Lifetime Dose (fg/kg) [4]	Daily Dose (pg/kg)[3]	Average Daily Lifetime Dose (fg/kg) [4]
FORESTRY					
Aerial:					
Pilots	1	1.1	48.	.28	12.
	2	1.1	30.	.28	7.5
Mixers	1	2.1	23.	.53	5.8
	2	2.1	6.8	.53	1.7
Mechanics	1	.12	1.3	.03	.33
	2	.12	.39	.03	0.1
Flaggers	1	.12	1.3	.03	.33
	2	.12	.39	.03	0.1
Ground Broadcast:					
a) Tractor Mistblower					
Driver	1	2.2	51.	.55	13.
	2	2.2	15.	.55	3.8
Mixer	1	2.8	65.	0.7	16.
	2	2.8	19.	0.7	4.8
Supervisor	1	.88	20.	.22	5.0
	2	.88	6.1	.22	1.5
b) Backpack Sprayers					
Applicator	1	2.5	96.	.63	24.
	2	2.5	29.	.63	7.3
Mixer	1	.64	25.	.16	6.3
	2	.64	7.4	.16	1.9

Table B-1 (Cont.)

Applicator Category		Daily Dose (pg/kg)[3]	Average Daily Life-time Dose (fg/kg)[4]	Daily Dose (pg/kg)[3]	Average Daily Life-time Dose (fg/kg)[4]
		ASSUMPTION 1: 0.04 ppm TCDD Content[1]		ASSUMPTION 2: 0.01 ppm TCDD Content[2]	
Range					
Aerial:					
Pilots	1	.56	15.	.14	3.8
	2	.56	9.6	.14	2.4
Mixers	1	2.6	18.	.65	4.5
	2	2.6	5.4	.65	1.4
Mechanics	1	.16	1.1	.04	.28
	2	.16	.34	.04	.09
Flaggers		.08	.16	.02	.04
Ground:					
Backpack					
Applicators	1	.92	3.6	.23	0.9
	2	.92	1.1	.23	.28
RICE					
Aerial:					
Pilots	1	.56	25.	.14	6.3
	2	.56	15.	.14	3.8
Mixers	1	3.4	37.	.85	9.3
	2	3.4	11.	.85	2.8
Mechanics	1	.16	1.7	.04	.43
	2	.16	.54	.04	.14
Flaggers		.08	.01	.02	.003
R-O-W					
Aerial:					
Pilots	1	4.5	375.	1.1	94.
	2	4.5	234.	1.1	59.
Mixers	1	3.3	69.	.83	17.
	2	3.3	21.	.83	5.3
Mechanics	1	.16	3.3	.04	.83
	2	.16	1.0	.04	.25

(Continued)

Table B-1 (Cont.)

Applicator Category		ASSUMPTION 1: 0.04 ppm TCDD Content [1]		ASSUMPTION 2: 0.01 ppm TCDD Content [2]	
		Daily Dose (pg/kg) [3]	Average Daily Lifetime Dose (fg/kg) [4]	Daily Dose (pg/kg) [3]	Average Daily Lifetime Dose (fg/kg) [4]
Ground:					
a) Selective Basal					
Applicators	1	6.2	299.	1.6	75.
	2	6.2	90.	1.6	23.
Supervisor	1	.44	21.	.11	5.3
	2	.44	6.4	.11	1.6
b) Cut Stump:					
Applicators	1	6.2	299.	1.6	75.
	2	6.2	90.	1.6	23.
c) Mixed Brush:					
Hand					
Applicator	1	9.3	251.	2.3	63.
	2	9.3	75.	2.3	19.
Truck					
Applicators	1	.92	25.	.23	6.3
	2	.92	7.4	.23	1.9
d) Railroad:					
Crewmen	1	5.6	143.	1.4	36.
	2	5.6	43.	1.4	11.
e) Electric Power					
Applicators	1	5.6	151.	1.4	38.
	2	5.6	45.	1.4	11.
Foreman	1	.16	3.2	.04	0.8
	2	.16	.96	.04	.24

Source: Braun (1980).

[1] The daily dose and average daily lifetime dose are from Appendix B, in Braun (1980) columns E and H, respectively.

[2] I calculated the daily dose and average daily lifetime dose by reducing the doses under assumption 1 by a factor of 4, reflecting the reduction from 0.04 to 0.01 ppm TCDD in 2,4,5-T.

[3] Picograms (10^{-12} grams) per day.

[4] Femtograms (10^{-15} grams) per day.

LABORATORY SAFETY AND WASTE MANAGEMENT

Chairman: Alvin L. Young
           Veterans Administration
           Washington, D. C. USA

AN OVERVIEW OF LABORATORY AND WASTE MANAGEMENT GUIDELINES

FOR TOXIC CHLORINATED DIBENZO-P-DIOXINS AND DIBENZOFURANS

A. L. Young

Office of Environmental Medicine
Veterans Administration
Washington, DC, USA

ABSTRACT

Numerous analytical and toxicological laboratories are currently conducting or proposing to conduct experiments with toxic polychlorinated dibenzo-p-dioxins (PCDDs) and dibenzofurans (PCDFs). It is important that these laboratories have adequate safety and waste management procedures. For example, the facility should be appropriately designed for handling hazardous materials in a variety of sample matrices. Protocols should be written and evaluated for the safe handling of analytical standards and contaminated samples. The laboratory should have a written and instituted medical surveillance program for all laboratory personnel. Special care should be given to the handling, storing, shipping, and disposal of laboratory wastes.

INTRODUCTION

How realistic are the health hazards associate with the laboratory handling and disposal of toxic chlorinated dioxins, e.g., 2,3,7,8-tetrachlorodibenzo-p-dioxin (TCDD)? An episode involving the deliberate synthesis of TCDD was reported in 1975 by Oliver[8]. This involved three young (male) scientists working with pure TCDD in the laboratory. Two of the men were exposed to the dioxin while attempting to synthesize it by heating trichlorophenol in an alkaline solution in the presence of a catalyst or by heating prepared potassium trichlorophenate in a closed system. Both men wore overalls and plastic gloves and allegedly took the utmost care to avoid inhalation or skin contact. The third man was a colleague of the other men and had

667

been working with the diluted dioxin standards they had prepared.
His work also had been done with the utmost caution and with
special care to avoid personal contamination. Three clinical
features were common to the three men, namely, chloracne,
hyperpigmentation, and hypercholesteremia (increased levels of
cholesterol). The lesson learned from this episode was that a
sophisticated laboratory safety program is essential.

In 1972 Jensen[5] reported on two cases of chloracne in
employees of an outside contractor that had been working on a
piece of equipment exposed (but thought to have been decontaminated)
to TCDD 3 years earlier in an industrial explosion in Derbyshire,
England. A young son of one of these employees also developed
chloracne. The presumed source of the child's contamination was
the father's working clothes. The lessons learned from this
episode were a) its difficult to clean-up TCDD once the workplace
becomes contaminated; b) proper waste (used equipment) management
is important; and c) contaminatd clothing should be carefully
handled.

Despite the experiences reported by Oliver[8] and Jensen[5],
it should be noted that since 1970 hundreds of studies, both
analytical and toxicological, have been conducted safely with
2,3,7,8-TCDD. Moreover, Young[13] in 1980 reported on three
scientists that had routinely handled 2,3,7,8-TCDD contaminated
environmental samples since 1973. These scientists had spent
thousands of hours collecting, handling, extracting, and analyzing
biological and environmental samples contaminated with parts-per-
billion (ppb) levels of TCDD. In April 1979, the three scientists
served as subjects in a study of 2,3,7,8-TCDD in human adipose
tissue. The analysis of subcutaneous fat from the abdominal wall
(surgically removed under local anesthesia) revealed that two of
the three scientists had detectable levels of TCDD; however, these
values were never more than 4 parts-per-trillion above the limit
of detection (2 parts-per-trillion). The health of these
scientists at the time of the dioxin analysis was good. The
binding of TCDD to environmental matrices may have reduced the
extent of exposure. Thus, the handling of TCDD in environmental
samples may not constitute the same hazards associated with the
handling of "analytical standards". The prudent scientist will
exercise care in handling all samples, regardless of source.

The guidelines for working in the analytical or biological
laboratory with toxic chlorinatd dibenzo-p-dioxins and dibenzo-
furans can be separated into six areas: a) Planning, Common
Sense and Public Relations; b) Laboratory Design; c) Safe Handling
Techniques; d) Monitoring Laboratory Contamination; e) Medical
Surveillance; and f) Disposal of Laboratory Wastes.

PLANNING, COMMON SENSE, AND PUBLIC RELATIONS

When a researcher selects the area of investigation,
considerable attention is given to the requirement for instrumentation and/or laboratory animals. Frequently, the investigator
overlooks until late in the formulation of plans the requirements
associated with procuring, safe handling, storing, and disposing
of the target chemical and its associated wastes. Indeed many
laboratories and laboratory personnel are simply not equipped to
work with the toxic dibenzo-p-dioxins and dibenzofurans. When
selecting these chemicals as the target for investigation, the
researcher must carefully determine the studies to be conducted,
the quantity of chemical required (micrograms to milligrams),
where can it be obtained or synthesized and how can it be shipped
to the laboratory, how will it be administered to the test
animals, what are the health implications for the laboratory
workers, and how can the contaminated wastes generated from the
experiment be disposed. The point I am making, is that the
scientist must carefully weigh the decision to proceed with
research on toxic dioxins and furans. Once the decision is made
to proceed, laboratory guidelines must be formulated.

Laboratory guidelines for the safe handling of toxic
chemicals are of little value unless they are exercised with
common sense. In a recent (1981) publication on handling
hazardous chemicals, the National Research Council[6] concluded
1) laboratory safety can be achieved only by the exercise of
judgement by informed, responsible individuals; 2) an essential
part of the development of scientists is that they learn to work
with and to accept the responsibility for appropriate use of
hazardous substances; 3) good laboratoary practice requires
mandatory safety rules; 4) a good laboratory safety program must
always be based on participation of both laboratory administration
and students and/or employees; and 5) the individual worker must
learn to think about possible hazards and seek information and
advice before beginning any experiment.

The Laboratory Director has the ultimate responsibility for
laboratory safety. The Director must insure that an effective
institutional safety program is in place. In addition, the
Director must establish a public relations program. I have found
that when the entire institution is aware of the important
research that is being conducted within that facility, they will
be more supportive. Hileman et al[4] noted that one method to
enchance public relations was to prepare a video tape that
summarized the precautions taken in the handling of PCDDs and
PCDFs. This video tape was widely shown throughout the laboratory
to those individuals who did not have a direct involvement in the
PCDD and PCDF program but who wanted to know what steps were taken
to protect them from any potential exposure. Public relations,

however, involves dealing with more than facility, department, or company personnel -- it involves dealing with the community outside the laboratory.

When laboratories deal with toxic chemicals, especially chemicals that have gained as much attention in the mass media as the dioxins, they must be prepard to deal with reporters and inquiries from the press. The Director of the laboratory, principal investigators, and staff should determine a course of action in the event a press inquiry is received. At a minimum, the response should be a statement that includes the nature of the research, its importance, the name of the investigator and area of expertise, description of the containment facilities and safety programs, and the funding agency. The statement should provide a clearly and simply written summary that can help the reporter to prepare an accurate story.

LABORATORY DESIGN

Outterson and Hickman[9] point out that the objectives of designing a laboratory specifically for working with hazardous materials are to protect the workers, prevent any environmental insult, and to provide for the scientific integrity of the experiments being conducted. They recommended that the three major considerations in constructing a hazardous materials laboratory were engineering design, administrative controls, and personal protective equipment.

The National Research Council of Canada[7] noted that laboratory design should encompass a method to control access to the laboratory and that the laboratory area be divided into at least three areas, thus allowing for separation of low and high hazard work areas. The low hazard work areas include areas for sample preparation, extraction, and clean-up; and chemical clean-up and activation; glassware washing and testing; and chemical, solvents, and waste storage. The high hazard work area should house the analytical instrumentation and the area where the samples are concentrated and the samples and the standards are analyzed. Both the low and high hazard work areas must be situated away from high traffic flow. The areas should be clearly posted, continually secured (i.e., a locking system without a master lock), and accessed only be authorized personnel.

Laboratory design also includes the basic layout of the ventilation, utilities, and provision for waste disposal[9]. A hazardous materials laboratory must have an excellent venting system for hoods, glove boxes, and analytical instruments. The venting system should have a means of monitoring airflow. It should also contain a filter unit system which allows maintenance and replacement of the units without exposing personnel or contaminating the environment. The vacuum and water lines should be designed to prevent hazardous contaminants from entering the central water or drainage system.

SAFE HANDLING TECHNIQUES

Futrell[3] noted that the exquisite toxicity of the 2,3,7,8-TCDD isomer and its particular physical and chemical properties dictated special handling procedures. The low volatility and very high stability of the compound dictate that it will not migrate if spilled and that solvents must be used to remove and/or transport the material. Futrell[3] suggests that the major toxic hazard associated with TCDD in the laboratory is the use of calibration standards to develop and confirm analytical procedures. Consequently, storage and handling of standards, spiked samples, and regents are primary concerns in managing personnel safety.

Beck[2] has provided an excellent "list" of safety precautions applicable to laboratories working with toxic PCDDs and PCDFs. Methods for handling, transporting, and storing of toxic chemicals are recommended. Beck[2] emphasized that for residue analyses TCDD solutions should principally be used so that the solid pure substance does not have to be handled. In general the concentration of the stock solutions should be below 1 mg/mL, that of the working solution 1-100 ng/mL. For analytical work not more than 150 ng TCDD should be used per single test so that major contamination can be avoided should an accident occur. Futrell[3] has recommended the use of the less toxic 1,2,3,4-TCDD isomer as a calibration standard and for developing and demonstrating the efficiency of the analytical procedure. Because the number of calibration runs is so high in trace analysis work, the substitution of the less toxic isomer decreases the potential or exposure by at least tenfold[3].

MONITORING LABORATORY CONTAMINATION

Laboratories that handle PCDDs and PCDFs must continually monitor for laboratory contamination. Contamination occurs primarily because of incorrect handling of samples and standards, improper venting of analytical instruments, or poor hosuekeeping procedures. It is surprising how often the latter item is overlooked. All laboratory areas must be kept neat and clean; in no instance should dry sweeping or dry mopping be permitted. Wet mopping with disposable sponge mop is recommended to clean floor areas while benches and hood areas should be cleaned with wet (water or solvent) disposable towels.

The major reason for a contaminate monitoring program has been
amply described by Hileman et al[4]. They point out the prime
reason for such a program is to confirm the belief that adequate
safety and handling precautions are being taken -- it is proof
that the system works. Each worker knows the extent to which his
laboratory technique conforms to safe handling practices, but he
may not know explicitly about the technique of others. Thus the
monitoring program gives all the personnel involved the confidence
in their own and others abilities to safely work with these
hazardous compounds[4].

MEDICAL SURVEILLANCE

Futrell[3] has noted that the most significant safety element
in working with the toxic chlorinated dibenzo-p-dioxins is the
personnel who are involved in the program. He recommended that
only Ph.D. level personnel and/or analysts with several years of
laboratory experience be assigned to dioxin projects and that only
a limited number (e.g., four) are assigned at anyone time.
Furthermore, all laboratory personnel must be thoroughly briefed
on the hazards involved and the physical and chemical properties
of TCDD. Lastly, in order to make a safety program work, the
individual worker must be cooperative, noncomplacent, and totally
honest. Appropriate health recommendations must be followed.
Anderson[1] pointed out that if a laboratory worker ignores a
developing, potential health hazard, he endangers not only himself
but his fellow workers as well. The laboratory worker must play
an active role in his own health and safety.

The first step that an institution or facility should take in
establishing a medical surveillance program is to conduct a risk
assessment. The more specific the assessment of hazards, the
better the medical surveillance. In addition to hazard
recognition, all of the individuals, (electricians, janitors,
instrument repair personnel, etc.) having contact with a
laboratory must be identified and considered in the risk
assessment[1]. Access to the laboratory may in some cases be
limited to individuals participating in a segment of the medical
surveillance program.

Once a risk assessment is completed, and the "at risk"
personnel are identified, then an occupational health program can
be tailored to the laboratory program. Wolfe and Lathrop[12]
described three components to a comprehensive surveillance system,
namely, environmental monitoring, medical evaluations, and
tracking of personnel. The medical evaluation should consist of a
standardized but comprehensive physical examination that
incorporates extensive biochemical analyses. Wolfe and
Lathrop[12] provided details for just such an examination.

Likewise, the National Research Council of Canada[7] has outlined a complete medical examination for all personnel who work in a hazardous contaminants area. Tracking of personnel over time is a vital component of a proper surveillance system, and is intended to identify chronic sequalae of toxic occupational exposures[12]. Efforts to conduct long-term follow-up studies of current and former employees exposed to dioxins and furans will be required if we are to answer the scientific questions related to these chemicals and human health.

DISPOSAL OF LABORATORY WASTES

Indiscriminate disposal of laboratory wastes containing PCDDs or PCDFs is unacceptable and is being curtailed by a combination of local, state, and federal regulations. It is important that an institutional safety plan provide for the regular disposal of contaminated wastes. Unfortunately, as Taft et al[11] have noted, much of the dioxin-contaminated waste being generated by laboratories is currently being stored on-site. However, because these laboratories generate small volumes having a low level of contamination, alternate methods for dealing with these wastes maybe practical. Taft et al[11] describe seven methods directly applicable to the laboratory disposal of laboratory generated wastes contaminated with TCDD and related materials. The most appropriate methods appear to incineration and photolysis. Indeed Staub and Tsang[10] have concluded that practical incinerator environments are within the present technological capability.

CONCLUSION

Laboratories that work with the toxic PCDDs and PCDFs require safety precautions that are not ordinarily encountered elsewhere. Our understanding of environmental and health problems related to these chemicals has increased steadily and this is now being reflected in improvements in the design of and working conditions in research laboratories.

REFERENCES

1. Anderson, J.H., Jr. 1980. Medical aspects of occupational health in a laboratory setting. Chapter 9, p. 281-297. In: A.A. Fuscaldo, B.J. Erlick, and B. Hindman (eds.). Laboratory Safety: Theory and Practice. Academic Press, New York.

2. Beck, H. 1982. Safety precautions for the handling of TCDD. These Proceedings.

3. Futrell, J.H. 1982. Safe handling of toxic chemicals:
    perspective of a bench chemist and laboratory manager. These
    Proceedings.

4. Hileman, F.D., T. Mazer, and D.E. Kirk. 1982. Program for
    monitoring potential contamination in the laboratory following
    the handling and analyses of chlorinated dibenzo-p-dioxins and
    dibenzofurans. These Proceedings.

5. Jensen, N.E. 1972. Chloracne: three cases. Proc. R. Soc. Med.
    65(8):697-688.

6. National Research Council. 1981. Prudent Practices for
    Handling Hazardous Chemicals in Laboratories. Committee on
    Hazardous Substances in the Laboratory. National Academy
    Press, Washington, D.C. 291 p.

7. National Research Council of Canada. 1981. Polychlorinated
    Dibenzo-p-dioxins - Limitation to the Current Analytical
    Techniques. Associate Committee on Scientific Criteria for
    Environmental Quality. Publication NRCC No. 18576 of the
    Environmental Secretariat, Ottawa, Canada.

8. Oliver, R.M. 1975. Toxic effects of 2,3,7,8-tetrachloro-
    dibenzo-1,4-dioxin in laboratory workers. Brit. J. Ind. Med.
    32(1): 49-53.

9. Outterson, G.G., and C.H. Hickman. 1982. Laboratory design
    consideration for the handling and analysis of chlorinated
    dioxins and related compounds. These Proceedings.

10. Shaub, W.M. and W. Tsang. 1982. Physical and chemical
    properties of dioxins in relation to their disposal. These
    Proceedings.

11. Taft, L.G., P.R. Betltz, and B.C. Garrett. 1982. Laboratory
    handling and disposal of chlorinated dioxin wastes. These
    Proceedings.

12. Wolfe, W.H. and G.D. Lathrop. 1982. A medical surveillance
    program for scientist exposed to dioxins and furans. These
    Proceedings.

13. Young, A.L. 1980. Direct Testimony -- Environmental Fate of
    2,3,7,8-tetrachlorodibenzo-p-dioxin and 2,4,5-trichloro-
    phenoxyacetic acid. Administrative Hearings on 2,4,5-T and
    Silvex. Environmental Protection Agency, 401 M Street, S.W.
    Washington, DC, July 22, 1980.

LABORATORY DESIGN CONSIDERATIONS FOR THE HANDLING AND ANALYSES OF CHLORINATED DIOXINS AND RELATED COMPOUNDS

G. G. Outterson and C. H. Hickman

Battle Columbus Laboratories
Columbus, Ohio, USA

ABSTRACT

The objectives of designing a laboratory specifically for working with hazardous materials are to protect the workers, prevent any environmental insult and to provide for the scientific integrity of the experiments being conducted.

Guidelines are offered for the design or retrofitting of such a laboratory. Three major considerations are described including engineering design, administrative controls and personal protective equipment. Examples showing implementation of the guidelines are given.

DESIGN CONSIDERATIONS FOR A HAZARDOUS MATERIALS LABORATORY

There are three main objectives in designing a facility for working with hazardous materials such as 2,3,7,8-tetrachloro-dibenzo-p-dioxin. These are to protect the worker, prevent environmental insult and to provide for the scientific integrity of the experiments. First and foremost, the facility must be designed with the people who work in it in mind. The layout of the facility, the ventilation provided, the personal protective equipment and the actual experimental procedures must be designed with their safety in mind. It is also important that control over hazardous materials is not lost and that the material can be accounted for at all times. Finally, it is important that the facility design not interfere with the scientific experiments to be carried out. The three objectives can be achieved by

675

considering the following general considerations; engineering
design, administrative control, and personal protective
equipment. Engineering design includes the basic layout of the
facility, ventilation, utilities, and provision for waste
disposal. The administrative control might include operating
procedures, emergency procedures, maintenance, and external
reviews. Personal protective equipment will include such items as
gloves, lab coats, coveralls, aprons, and respirators.

ENGINEERING DESIGN

    When designing a facility on the drawing board there is a
great deal of flexibility in actually laying out the laboratory.
Unfortunately, most of us are far more familiar with retrofitting
an existing laboratory. In that case, the layout of the hoods and
other major equipment may be fixed or may be dictated by existing
permanent features. The first thing to decide is what are going
to be the uses of the facility. For example, in an analytical
facility the traffic patterns are likely to be different than
those of a synthetic facility or a facility to be used for
toxicological or animal studies. If varied activities are
contemplated, sensitive analytical tasks should be separated from
the high background inherent in some synthesis work. The primary
work stations for a analytical facility may be hoods and
analytical instruments. By locating these stations with
forethought, personnel movement can be minimized. Similarly, for
a synthesis laboratory the preparations, purification and
analytical work-up operations may be conducted in different hoods.
These hoods should be located as much as possible to facilitate
work and transfer of the hazardous materials between these major
work areas. One specific feature that has been adopted at
Battelle is the use of passthroughs between hoods. Two hoods have
been arranged side-by-side and an opening has been cut between the
hoods within the hood. A refrigerator may be attached to one side
of the hood. Any vapors that collect within the refrigerator are
safely captured and removed by the hood when the door is opened.
It is also possible to attach, for example, a small glove box to
another side of a hood to allow installing an analytical balance.
This glove box can be fitted with a simple ducting to provide a
low level of ventilation sufficient to exhaust any hazardous
materials that might be present, but at the same time, at a low
enough velocity to permit accurate weighing.

    Such a box should not be a makeshift arrangement. The
installation must be designed with a clear understanding of
ventilation principles. If it is necessry to use a booster
blower, care must be taken to ensure that ordinary ductwork is
never pressurized with dangerously contaminated air. This may
require interlocks to ensure that the booster blower does not work

without reduced pressure throughout the exhaust ductwork. A
visual and audible alarm should be provided on any hood where
failure to ventilate can create a dangerous situation.

There probably has never been a laboratory designed that had
too much bench space. With that in mind, there are two approaches
to providing bench space. One is to have an island or a series of
islands of benches either projecting from the wall or free
standing. The island approach can provide more bench space in a
given square foot area, but frequently, fewer wall cabinets are
installed. The other approach is to have the benches against the
walls. Other considerations for the physical layout are such
items as safety showers, and emergency lighting. Trickle charged
battery powered lights are available which can provide emergency
lighting so that in the event of a power blackout, the scientists
will be able to transfer the hazardous material into a safe
condition, either by storage or destruction, and safely exit the
laboratory.

A final consideration may be security. For materials, as
toxic and frightening to the public as some of the chlorinated
dioxins, it may be important to provide physical protection of the
materials by locking doors or perhaps locking the hoods or secure
containers where the materials are stored. Access to toxic
chlorinated dioxins can then be limited to those individuals who
are familiar with the hazards of working with the materials.

Guidelines for the design of a laboratory should include the
subject of waste disposal. This subject will be discussed in a
technical manner later in these proceedings, but, from a general
point of view, there are three types of effluents requiring
consideration: air, liquid, and solid wastes. Each of these three
needs to be addressed separately. All air exhausted from the
labortory may be passed through a High Efficiency Particulate Air
Filter. This filter is designed to collect any aerosols or
particles that may be swept out of the hood. The performance
standard for these filters is 99.95% efficiency when challenged by
a 0.3 micrometer particle size aerosol. Rough filters that are
are used upstream to extend their service life. For hazardous
materials that are volatile, charcoal filters are available.
These are considerably more difficult to use and much more
expensive than particulate filters. Either type can be provided
in a bag-in, bag-out system. In these systems when a filter is to
be changed, a hatch is opened and the filter can be pulled out
into a bag. The bag can then be sealed and a new filter can be
installed reducing the exposure of the person changing the filter
to any hazardous material that might be trapped or absorbed on the
filters. A key problem with all filters, is how often do they
need to be changed. High Efficiency Particulate Air Filters are

usually changed when the pressure drop across the filter increases
(due to clogging) so as to reduce the operation of the hood. In
the simplest case no automatic controls are provided and the air
flow rate changes continuously throughout the service life of the
filters. The blower is designed to provide an adequate flow when
the filter's useful life is over. With a new filter air flow may
be much greater. Such changes may or may not be tolerable from
safety and operations standpoints, but can make ventilation system
balancing virtually impossible.

Balancing to maintain reduced pressure in the labaratory, is
generally considered necessary for work with highly toxic
materials. A much better system is to provide a pneumatically-
operated automatic control damper. When the filters are new, the
damper provides a substantial pressure drop in series with the
filters. As filters gradually clog in service, the pressure drop
is monitored and the damper opens automatically to maintain the
same pressure drop. Thus, the volume flow rate will be
essentially unchanged until the economic life of the filters is
over, at which time, the damper will be fully open and the flow
rate will begin to drop slowly as filters clog further.

For charcoal filters, there are no generally accepted methods
of predicting or determing the end of the filter's useful service
life. Several approaches have been tried, including sampling
tubes, and challenges with gases that are not readily absorbed,
such as halogenated hydrocarbons. As a practical matter, charcoal
filters often are changed on a regular basis such as quarterly or
annually.

Liquid wastes are often reduced by evaporation and filtered
through an activated charcoal treatment system. One disadvantage
of water treatment systems are that they generate a solid waste,
namely, the charcoal. Charcoal is classified a hazardous material
under U.S. Department of Transportation shipping regulations. The
charcoal can be collected along with other solid wastes, and when
facilities are available the bulk may be incinerated. When the
amount of wastes are small the hazardous materials may be
decontaminated through chemical reaction.

Most of the ventilation in a laboratory is provided by the
hood or hoods. Several factors affect the design of the hood:
face velocity, work space and periodic decontamination. In the
past it was often thought that if a little was good, a lot more
would be much better. There is increasing recognition among
industrial hygienists that in the case of face velocity, this is
not true. A face velocity somewhere between 100 and maybe 150
linear feet per minute average across the face of the hood is
probably ideal from the capture standpoint. In view of high
energy costs, emphasis has been given recently to studies showing

that 40 to 60 feet per minute may be entirely adequate for materials of relatively low toxicity. Body movement past the hood face creates cross-drafts that interfere with the face velocity, so a traffic pattern around the hood face requires a higher hood flow rate. At some higher velocity it is clear that exposure may actually be increased by the effects or turbulance around the hood face and around the worker himself.

The work space in the hood should be large enough for the work being conducted. Extraneous equipment should be removed in order to maximize the available workspace. Accumulated chemical storage interferes with hood function. A ventilated cabinet can be provided for storage and may be connected to an exhaust system that runs continuously.

Decontamination requirements can also be preplanned. Periodically, the entire surfaces of the inside of the hood may be rinsed or wiped in order to remove any build-up of contamination. The permanent equipment, paint, and the material of construction must be suitable for whatever decontamination is necessary. Suitability includes both the cleanability (non-porous nature) of the surface and the ability to withstand the decontamination chemicals that are used. The most expensive but perhaps the most flexible approach is to use a stainless steel hood. Other approaches are to make sure that any paint to be used in the hood is compatible with any solvent to be used in the washdown procedure. In some cases strippable coatings are used, and are peeled off for disposal after some period of service.

ADMINISTRATIVE CONTROLS

Administrative controls include those actions taken by operators and managers for operating a facility safely. They involve insuring that the workers have adequate training, creating and implementing general rules for experiments to be conducted, and making detailed plans for all experiments. The importance of detailed planning cannot be over-emphasized when working with hazardous materials. Although the physical manipulations may be very similar to normal operations, the margin for error may be far narrower.

Another area of administrative control, never obvious until it fails, is planned preventive maintenance and inspection. In the case of ventilation equipment, routine maintenance is essential and the alternative is to accept failures in service. Critical hoods at Battelle are given preventive maintenance weekly. Maintenance mechanics begin their workday earlier than research staff, and a given critical hood is serviced on the same day each week before the official research workday begins. This arrangement minimizes opportunity for miscommunication and work interruption. A final administrative control which often is quite valuable is external review. Most organizations should have some hazard review procedures established. If not, it may well be worthwhile to invite a colleague or perhaps someone from your Safety Department. Very often ideas may be generated by disinterested parties that might contribute greatly to the overall safety of your operation.

PERSONAL PROTECTIVE EQUIPMENT

Personal protective equipment for use in laboratories handling toxic materials generally include gloves, clothing, and respirators. Aprons, labcoats or coveralls will provide some dermal and clothing protection to laboratory workers. Some materials, such as disposable synthetic coveralls, will prevent solid and in some cases liquid penetration. However, these offer little protection against volatile hazardous materials. Gloves should be changed regularly since most materials will eventually penetrate gloves. Whenever a solvent or liquid comes in contact with glove material it will start to penetrate, even if the challenge liquid is removed. The American Society of Testing Materials provide tests for determining the permeation and penetration rates of various chemicals through glove materials and these tests should be considered in the selection of gloves.

Air purifying respirators are available as half masks or full face respirators. The latter provide splash and particle protection for the eyes, and also can generally provide better respiratory protection by virtue of more stable position and smoother facial contours in the seal area. Interchangeable cartridges are available including a combination type that incorporates both a charcoal bed and a high efficiency (radionuclide) filter. A greater level of protection is available by use of positive pressure respirators, which require (among other things) an air line to each worker and reliable source of high-quality breathing air.

Respirators in general, can be considered a mixed blessing. Their effective use requires training, fit testing, maintenance, and inspection. Conceivable, false confidence (from use of ineffective respirators) could have adverse effects on safe technique. Respirator use can presumably increase the frequency of hand-face contact, with attendant increased exposure potential. All respirator uses should be guided by the Occupational Safety and Health Administration (OSHA) and American National Standards Institute (ANSI) standards.

SAFE HANDLING OF TOXIC CHEMICALS:

PERSPECTIVES OF A BENCH CHEMIST AND LABORATORY MANAGER

J. H. Futrell

Department of Chemistry
University of Utah
Salt Lake City, Utah USA

ABSTRACT

In a university environment we have evolved a number of working rules for handling and analysis of toxic samples: (1) secure storage and isolation of samples during workup; (2) use of less toxic surrogate compounds for calibration and testing of analytical scheme; (3) dedicated instruments, work area and personnel; (4) limited number of competent, knowledgeable, motivated personnel doing actual analyses; (5) redundant containment, verified by wipe tests; (6) medical surveillance; (7) close supervision and, (8) careful disposal of contaminated waste. We have also found it advantageous to have a toxicologist working with the analytical team and an analyst working with the toxicologists. With close collaboration and good communication of these groups serious exposure incidents have been avoided and a safe working environment has been maintained.

INTRODUCTION

This paper describes the procedures devised for safe handling of the exquisitely toxic compound, 2,3,7,8-tetrachlorodibenzo-p-dioxin (TCDD). Initial work in our laboratory involved the development of analytical methodology based upon high resolution capillary column gas chromatography - high resolution mass spectrometry (GC-MS) for the analysis of tissue samples. The analytical methodology has been described in a review article[1] and will not be discussed further in this report. Rather I shall discuss the sample handling protocol developed for dealing with this highly toxic substance in a university laboratory/research environment.

In addition to serving as Principal Investigator for phase I
development of analytical methods for TCDD samples the author was
Associate Director (Chemical Analysis) and later Director of the
Flammability Research Center (FRC) of the University of Utah.
Both the Toxicology and Analytical Groups of the laboratory were
involved in generating and handling toxic substances. For this
reason the FRC facility was located off campus, with general
laboratory practices, including safety, resembling industrial
laboratory rather than teaching laboratory practices. One task of
FRC was the analysis of soil samples which potentially contained
high levels (ppm) of TCDD. Only modest changes in the general
analytical protocol were required to ensure containment of TCDD.

Except for the somewhat different environment - controlled
access versus open access - the procedures employed were
essentially identical and only one description will be given.
Table 1 is an outline of the special precautions taken for TCDD
analyses. The discussion will follow the same format as the
outline. With the exception of the nature of special physical
examinations and surveillance for chloracne, the outline applies
equally well for other highly toxic materials.

PROCEDURES

In all of our work with toxic or potentially toxic materials
we have followed what maybe termed the "buddy system" for both the
Toxicology and Analytical groups. This means that an observer is
always present during the active phase of an experiment so that
prompt corrective action may be taken in the event an accident
occurs. Any potential exposure because of breakage, spillage,
leaks or carelessness is noted and pointed out to both the
individual and a supervisor.

For general work with toxic substances generated by combustion
we have used a "mouse spectrometer" to pinpoint conditions for
maximal generation and identification of toxicants[2,3]. More
generally, laboratory procedures for animal exposure studies,
product analysis, and on-line sampling were reviewed by a team of
supervisors, which included a medical doctor, an analytical
chemist, a toxicologist and an engineer, prior to initiating a
study with new materials and later while the work was in progress.

For TCDD the equisite toxicity of the 2,3,7,8-TCDD isomer and
its particular physical and chemical properties dictated special
handling procedures. In particular the low volatility and very

Table 1.   Outline of Special Safety Precautions for 2,3,7,8-TCDD
           Analyses

A. Samples

    1. Secure Storage and Disposal
    2. Minimum Handling
        a. Calibration of Instruments Using 1,2,3,4-TCDD Isomer
        b. Minimum Number of Replicates of Spiked Samples
        c. Prompt Return of Samples to Storage
    3. Redundant Containment
    4. Waste Retrieval and Disposal

B. Analysis Area

    1. Dedicated Negative Pressure Hood
    2. Sealed Vials
    3. Stainless Steel Tray Carrier
    4. Clean Working Area
    5. Rapid Air Change
    6. Wipe tests

C. Analytical Instruments

    1. Dedicated to TCDD Analyses
    2. Charcoal Traps on Exhaust
    3. Pumps Vented to High Velocity Hoods
    4. Tools, Instruments, Syringes, Forceps, Etc.-Strictly
        Segregated
    5. Wipe Tests

D. Personnel

    1. Limited Number
    2. Highly Trained
    3. Highly Motivated
    4. Highly Cautious
    5. Protective Clothing
    6. Special Medical Examinations

high stability of the compound dictate that it will not migrate if
spilled and that solvents must be used to remove and/or transport
the material. The tenacity with which it is absorbed on surfaces
at low concentration requires that isotopically labelled TCDD be
added during the extraction and cleanup procedures and that the

final analysis be based on the ratio of unlabelled (native) to
labelled material (in the case of trace analysis). This is a
variation of the isotopic dilution technique in which the amount
of labelled TCDD added substantially exceeds the amount present in
the analyte. Consequently, the major toxic hazard is associated
with the labelled material added as carrier and with calibration
standards used to develop and confirm the analytical procedure.
Therefore the control of these reagents, rather than the samples
being analyzed, is the major hazard. Consequently, storage and
handling of spiked samples and reagents are primary concerns from
the viewpoint of personnel safety.

Sample Handling

     Secure handling and storage of: (1) samples from the time that
labelled TCDD carrier is added, (2) calibration reagents, and (3)
concentrates containing the TCDD fraction are of paramount
importance. In a controlled-access laboratory such as the FRC,
where analysis of toxic substances is a routine procedure, all
that we required was carefully labelled storage containers in a
refrigerator dedicated to toxic materials storage. All containers,
traps, syringes, etc., were labelled "TCDD" in prominent letters.
This signified to all laboratory personnel the highly toxic nature
of the samples and that they were "off-limits" for everyone except
those directly involved with TCDD analyses.

     In the Chemistry Department a more positive security procedure
was employed because of the continual flux of students through the
building. Standard "Hazardous Chemical in Use" signs on the doors
were amplified by hand-lettered signs which proclaimed that highly
toxic materials were stored in the respective rooms which were
also locked at all times when no personnel were present. A
special refrigerator was labelled "TCDD" and marked with a (home-
made) skull and cross-bones symbol. It was kept locked at all
times except when samples were being removed or added. Keys were
issued only to the analyst and observer. A spare key was kept in
a locked drawer inside my office.

     Very early in the project we decided to use 1,2,3,4-tetchloro-
dibenzo-p-dioxin as a calibration standard and for developing and
demonstrating the efficiency of the analytical procedure. Synthesized in our laboratory by standard methods, the 1,2,3,4-TCDD
isomer is less toxic by a factor of $10,000^4$. Its mass spectrum
and retention time by packed column gas chromatography are indis-
tinguishable from the 2,3,7,8-TCDD isomer. Only by high resolution
capillary column GC or by negative ion chemical ionization[5] is
it possible to distinguish the two compounds. Because the physical
properties are essentially identical and the 1,2,3,4-TCDD isomer
can be followed through a separations/analysis scheme with the
same results, we substituted this compound for all our preliminary

work with methods development and instrument calibration. Only
for the final runs, for which the isotopic peaks of TCDD in actual
samples were references to ^{37}Cl-TCDD, were the instruments and
personnel involved in handling the 2,3,7,8-TCDD isomer. Because
the number of calibration runs is so high in trace analysis work,
the substitution of the less toxic isomer decreased our potential
for exposure by at least tenfold.

As indicated in Table 1, it is also important to realize that
running samples in triplicate are usually sufficient for trace
analysis work; e.g., at the parts-per-trillion range. There is a
temptation to run additional samples because of the large
experimental uncertainty in results obtained at or near the
detection limit of the analytical methods. However, under these
conditions the precision (not accuracy) increases only with the
square root of the number of repetitions. Consequently a lot of
samples must be run - not atypically consuming all the available
sample - for only a modest increase in precision. Considering the
current status of inter-laboratory reproducibility of TCDD
analyses[6] there is limited justification for running large
number of duplicate samples.

Redundant containment of samples is also considered important.
For transporting samples, e.g., between laboratories, careful
packaging of a sealed container within a sealed container and with
with adequate absorbent packing materials, is considered essential.
In the laboratory sample preparation area absorbent paper inside a
tray surrounded by more absorbent paper on a bench top or with a
raised lip or within a hood is considered an optimum arrangement.

Only a small number of samples are carried from the preparation
area to the analysis area at a time, and they are carried in sealed
vials in a special container in a tray lined with absorbent paper.
Some extracts and all calibration standards were packaged in crimp-
top rubber-septum-sealed glass vials carried in aluminum racks
fabricated to provide a close (but not tight) fit. Some extracts
were in sealed (fused) glass tubes which required that the tip be
broken to access. In such cases the tubes were individually
frozen with liquid nitrogen and resealed immediately after
analysis. They were also transported in specially-fabricated
aluminum blocks.

All solid waste was retained in special containers for
ultimate disposal by the sponsoring agency. Samples known or
suspected to be contaminated with 50 ppb or greater levels of TCDD
were collected in a single, appropriately labelled (High Level
TCDD Waste) 5 gallon can and sealed for disposal. One role for
the observer was to watch for stray drops spilled during syringe
or pipette transfer of any solution containing ppm level TCDD.
Wipes, paper from the sample containment tray, rubber gloves and

charcoal traps from the GC effluent line (discussed later) were
all added to this container. Solvents used in cleaning syringes,
tools, ion sources, etc., potentially containing TCDD were
evaporated to dryness (at room temperature) in glass beakers in a
special hood. These beakers were than added to the high level
waste drum. Other solid wastes (e.g., those materials used but not
in contact with high levels of TCDD) were sealed in double plastic
bags for routine disposal as chemical wastes.

Analysis Area

In our work the analysis area consisted of two rooms for
sample preparation and analysis, respectively, plus a third room
used for sample storage. The sample preparation area was considered the most critical and for both our locations we selected
a relatively small room containing a laminar flow hood with a
stainless steel working surface and rim to prevent spills from
leaking to the floor. The floor was painted with epoxy paint over
concrete to resist absorption of spilled solutions. Not only the
hood, but also the entire room was dedicated to TCDD work for
periods of months at a time when we were actively engaged in TCDD
analyses.

All transport of samples was by means of absorbent paper-lined
stainless steel trays, as described above. To the maximum extent
possible these procedures were also applied in the sample
preparation area. All sample weighing, processing, and transfer
steps were carried-out over absorbent paper which was periodically
removed and added to the high level waste can for disposal. A
clean working area, with a fresh supply of absorbent paper, was
maintained at all times. All spatulas, syringes, pipettes, etc.,
used in TCDD sample preparation were carefully segregated and
cleaned thoroughly before being returned (ultimately) to general
laboratory use. Broken items were disposed of as high level waste.

During 6 years of work with TCDD the only spill occurred
inside the hood. A 10-gram soil sample containing about 4.3 ng of
2,3,7,8-TCDD and 12 mL of alcoholic KOH was spilled and largely
absorbed by the layer of absorbent paper covering the working
space within the hood. The paper was immediately stripped off and
the hood working surface was cleaned with chloroform-soaked paper
towels, followed by dry towels, followed by acetone-soaked towels.
These towels plus the rubber gloves worn by the analyst were added
to the high level waste drum.

Wipe tests were taken periodically using hexane-soaked
Kimwipes which were later extracted with hexane in a Soxhlet
apparatus for 1 hour and concentrated for GC-MS analysis. The
hood area, floor near the hood, working surface near the GC and
GC-MS and computer and MS console areas were all checked for TCDD

contamination at the conclusion of sample work-up cycles. No
evidence for contamination was found at the level of about 10^{-9}g/
1000 cm^2 of surface samples. Despite the negative wipe test
results, the hood, instrument work areas and other work surfaces
potentially exposed to TCDD were thoroughly cleaned with
chloroform-soaked paper towels before releasing these instruments
and work areas for other purposes. Similarly all tools, spatuals,
syringes, etc., were washed three times in chloroform and acetone
before they were returned to the general inventory of laboratory
supplies.

As a parenthetical note it should be added that both
laboratory facilities are equipped with rapid change, once-
through air conditioning and heating systems which replace the air
about once every 3 minutes in the laboratory areas. Although less
important for TCDD than for many other toxicants, this is
considered highly desirable for laboratories dealing with
hazardous chemicals.

Analytical Instruments

As mentioned above, it is highly desirable that the
laboratories used for TCDD analysis be closed to other traffic.
Similarly, analytical instruments should be dedicated only to TCDD
analysis. For many, perhaps most laboratories, this will not be
possible for extended periods of time. Consequently the GC and
GC-MS instruments should be reserved for blocks of time in which
other analyses are excluded. It is also important that instrument
maintenance personnel be aware of possible contamination with
TCDD.

Diffusion pump oil and forepump oil should be considered as
hazardous chemical waste after TCDD has been run in an instrument.
The possibility that valves, sampling ports, GC columns and
transfer lines may be contaminated should be pointed out to other
users of the instrumentation. A charcoal trap can be added
conveniently to the GC detector/splitter effluent line to trap any
high boiling point material such as TCDD. The pumps should be
vented to hoods for general safety reasons.

Wipe tests similar to those described above are advantageously
applied in order to certify that instruments are not contaminated.
We have used them, for example, to show that ion sources and the
vacuum housing and pump lines adjacent to the source were free of
contamination after several months of TCDD analyses.

An unanswered question to date is the fate of TCDD injected
into the mass spectrometer for analysis. The wipe tests described
above have shown it does not deposit at the first cold surface
after leaving the ion source. This was validated in a definitive

way at the Dow Chemical Laboratories, Midland, Michigan, using ^{14}C-labelled TCDD. The ion source, vacuum housing and adjacent pumping line were "counted" and shown to be at background level[7]. This is a surprising result which suggests the accommodation coefficient of stainless steel surfaces for TCDD may be significantly less than unity. A likely sump for TCDD is the diffusion pump. However, the high operating temperatures of modern pumps and the concomitant self-cleaning action of repeated distillation suggests TCDD may be transferred to the forepump. This is also a likely respository if it goes through the vacuum system as an aerosol. In any case, a survey undertaken for writing this paper of several laboratories involved for many years in TCDD analysis provided no definite information on its ultimate fate in the mass spectrometer. Provisionally, we suggest that both the diffusion pump oil and forepump oil be treated as if they were contaminated with TCDD. Personnel involved with MS maintenance should also be informed of possible contamination and an upper limit estimate should be made of the total amount of TCDD analyzed with a particular instrument which might still be present in the apparatus.

Personnel

Probably the most significant safety element is the personnel who are involved. At the outset we decided that students and technicians should not be involved in TCDD analysis. Only Ph.D. level chemists and/or analysts with several years of laboratory experience were assigned to these projects. A very limited number (maximum of four in our experience) were assigned at any one time, and all were thoroughly briefed on the hazards involved and the physical and chemical properties of TCDD.

Surgical rubber gloves and conventional laboratory coats and safety glases were worn as protective clothing. The involatility of the compound suggested to us that other protective devices would be of marginal value and might even introduce additional hazards. The buddy system of having an informed observer for all operations involving dangerous quantities of the toxic 2,3,7,8-TCDD isomer was strictly observed.

All workers potentially exposed to TCDD were provided complete physicial examinations at 6-month intervals by the University Medical Center. Special tests deemed appropriate for monitoring health effect of TCDD exposure are discussed elsewhere[8], many of which were included in our personnel surveillance program. In addition, color photographs were taken weekly of the principal analysts to document the absence of any evidence of chloracne during the most active phase of our work with TCDD. Later we substituted before and after photographs as a more reasonable protocol when TCDD analyses became an intermittent activity for the laboratory.

CONCLUSIONS

The laboratory environment, mix of personnel, and distribution of responsibilities of our facility are probably atypical and have also changed significantly over the 6-year period in which we were engaged in the analysis of samples containing TCDD. Accordingly, each laboratory manager will find it necessary to adapt the recommended procedures to his particular situation. The key points are, in my opinion, (1) the prime responsibility for personal safety resides with the analyst himself and (2) all personnel who must enter a laboratory containing TCDD must be briefed on the hazards involved. The analogy with radioisotope safety is a good one to use in describing the problem, with the added complication that monitoring is much more difficult.

Thru adherence to the principles outlined in Table 1 and discussed in this paper we have experienced no unfortunate incidents of personnel exposure to these equisitely toxic agents. It is also a pleasure to acknowledge the significant contributions of our two technical monitors, Dr. Ralph Ross, formerly of the EPA, and now at USDA and Major Alvin Young, Ph.D., USAF, who carefully briefed us on the biological hazards of TCDD at the outset of our work with this chemical.

REFERENCES

1. N. H. Mahle and L. A. Shadoff, The Mass Spectrometry of Chlorinated Dibenzo-p-dioxins, in: "Biomedical Mass Spectrometry" G. Waller, ed., In Press.
2. K. J. Voorhees, I. N. Einhorn, F. D. Hileman, and L. H. Wojcik, The identification of a highly toxic bicyclophosphate ester in the combusion products of a fire-retarded urethane foam, J. Polymer Sci., 13:293 (1975).
3. J. H. Petajan, K. J. Voorhees, S. C. Packham, R. C. Baldwin, I. N. Einhorn, M. L. Grunnett, B. G. Dinger, and M. M. Birky, Extreme toxicity from combustion products of a fire retarded - polyurethane foam, Science, 187:742 (1975).
4. Discussed by K. S. Khera and J. A. Ruddick, Polychlorodibenzo-p-dioxins: perinatal effects and the dominant lethal test in wistar rats in: "Chlorodioxins-Orgin and Fate," E. H. Blair, ed., Adv. in Chem. Series, 120:70 (1973) and E. H. McConnell, J. A. Moore, J. K. Haseman and M. W. Harris, Toxicol. Appl. Pharmacol. 44:335 (1978).
5. D. F. Hunt, T. M. Harvey and J. W. Russell, Oxygen as a reagent gas for the analysis of 2,3,7,8-tetrachlorodibenzo-p-dioxin by negative ion chemical ionization mass spectometry, J. Chem. Soc., Chem. Commun., 152 (1975).

6.   M. L. Gross, Interlaboratory validation in dioxin analysis,
     presented at International Symposium on Chlorinated Dioxins
     and Related Compounds These <u>Proceedings</u>.
7.   L. A. Shadoff, Personal Communication, Analytical
     Laboratories, 574 Building, Dow Chemical Laboratories, U.S.A.,
     Midland, Michigan 48640. Also described briefly in reference
     1.
8.   G. D. Lathrop and W. H. Wolfe, A medical surveillance program
     for scientists exposed to dioxins and furans. These
     <u>Proceedings</u>.

SAFETY PRECAUTIONS FOR THE HANDLING OF TCDD

H. Beck

Bundesgesundheitsamt (Federal Health Office)
Berlin, Federal Republic of Germany

ABSTRACT

During the past 80 years there have been numerous reports on a severe form of acne and related symptoms associated with exposure to TCDD and/or structurally similar aromatic chlorine compounds. Many of these reports dealt with the exposure of laboratory staff and workers in the chemical industry. The need thus arises to prescribe particularly stringent safety precautions for laboratory experiments with these extremely toxic substances. At the same time a critical evaluation of the necessity of these experiments should be carried out.

In 1976, a list of safety precautions for the handling of TCDD were prepared by the Federal Health Office; these are presented in detail. Numerous technical details as well as possibilities for improvisation are discussed. Moreover, simple solutions for the safe disposal of contaminated solids and liquids are described. Finally, the necessity of medical check-ups for laboratory staff who handle these compounds is emphasized.

INTRODUCTION

As early as 1899 - many years before the discovery of TCDD - a German physician described a severe form of acne that he observed in workers in plants producing chlorine by electrolysis. He termed this disease "chloracne" (Herxheimer, 1899). Since that time, many reports on accidents indicate that chloracne and related symptoms have been observed in over 1,000 chemical plant

691

workers in plants and laboratories. These include many cases
with severe outcomes.

The recognition and confirmation of the structure of TCDD by
the German industrial chemist, W. Sandermann in 1957, was
connected with a severe case of chloracne in one of his assistants
(Sandermann, 1957). This incident resulted also in the first
systematic investigation of the extreme toxicity of this compound
(Kimmig and Schulz, 1957). Even in recent years, scientists in
the laboratory have developed chloracne by failing to avoid
exposure to TCDD (Oliver, 1975). These events underline the
necessity of special safety precautions for the handling of TCDD,
particularly now since a large number of investigations are being
conducted to analyze the effects of the Seveso accident.

The Seveso accident was also the reason that the Federal
Health Office in Berlin drew up a list of safety precautions for
the handling of TCDD. A large number of analytical laboratories
were analyzing food and other samples for TCDD contamination and
had to prepare analytical standards of TCDD as well. The
application of these precautions should not be limited to TCDD,
but should also apply to the handling of other toxic compounds
such as the hexa- and octa-chlorinated dibenzodioxins,
dibenzofurans, naphthalenes, azobenzenes and others. These
precautions were originally elaborated on for analytical
laboratories but many of them are also applicable to toxicology
laboratories.

1. Before starting an experiment involving TCDD the
    following question should be asked: "Is it necessary
    to carry out this test"? Even if our knowledge on this
    and related substances is still incomplete, we should
    not, just for that reason, give our consent to such an
    experiment. There is a series of examples where we could
    prevent the testing of samples for TCDD residues by a
    critical risk/benefit evaluation.

2. The next problem applies to the choice of personnel. The
    following criteria should be taken into consideration:

    a. Only well-trained workers with sufficient experience
        and manual skill should be chosen.

    b. The workers should be reliable and responsible, so
        that they will not panic in the event of accidents.
        Furthermore, the slightest incident should be report-
        ed to the supervisor.

c. Workers whose state of health is not entirely satis-
    factory as well as women who might become pregnant
    should be excluded.

d. The workers should decide on their own whether they
    are willing to work with these dangerous substances
    and the special conditions that are required.

3. Anyone having contact with operations of this type should
    be fully informed of the particularly high toxicity of
    TCDD and the potential risk involved. This includes
    janitorial and glassware cleaning staffs, incinerator
    operators, etc. The safety precautions recommended for
    TCDD largely correspond to those for the handling of
    radioactive material in the microcurie range. It should
    be noted, however, that unlike contamination with radio-
    active material which is easily recognized by monitoring,
    contamination by TCDD is far more difficult to detect.

4. Operations involving TCDD should be conducted, if possible,
    in separate rooms, clearly labelled and accessible to auth-
    orized staff only. Otherwise, the working area which
    might become contaminated by such operations, should be
    defined and marked, e.g., adhesive tape in conspicuous
    colors. In such situations, however, there should be at
    least a lockable hood under which proper operations are
    performed. If available, the hood material should have a
    smooth surface, e.g., stainless steel. If during any
    operation TCDD may become dispersed as particulate matter,
    the hood must also be provided with a suitable filter
    (e.g. activated carbon) which is easy to exchange.

5. Outlets of gas chromatographs and vacuum pumps of the
    mass spectrometry equipment should be vented into
    activated carbon and connected to a hood. In most cases,
    attention should be paid to the fact that the TCDD will
    condense in the hoses leading to vacuum pumps. Consequent-
    ly, cleaning operations on such equipment require special
    care.

    Large-scale equipment of the type mentioned will ofte be n
    impossible to accommodate within the restricted access
    area. Only in rare cases, will a mass spectrometer be
    assigned to the exclusive purpose of TCDD analysis. How-
    ever, since equipment that has been used for this purpose
    should be considered "contaminated" and labelled accord-
    ingly, access to such apparatus must be limited and staff
    should be informed of the potential risks involved.

6.   When a TCDD spill occurs during analytic operations,
      contamination must be localized. "Double receptacles"
      should always be used. Particularly suited to this task
      are stainless steel throughs lined with filter paper or
      plastic-backed absorbent paper. Volumetric flasks and
      similar receptacles holding TCDD solutions should always
      be place inside beakers, particularly if placed inside
      the refrigerator. When there is no refrigerator
      specially designated for this purpose, solutions should
      additionally be placed inside a larger receptacle (e.g.
      an exsiccator). For transportation of such solutions
      within the laboratory building (e.g. to the mass spectro-
      meter room), break-resistant containers should be used.

7.   When working with TCDD, disposable protective gloves
      should be used. This also applies to injections into
      the gas chromatograph or HPLC equipment. Laboratory
      coats, buttoned at the back should be used. These coats
      as well as towels should be washed separately. Contam-
      inated coats and, if necessary, other garments should be
      taken off immediately and treated like solid wastes. In
      Germany, the use of protective glasses during laboratory
      experiments is a requirement under general laboratory
      guidelines.

8.   For residue analyses TCDD solutions should principally
      be used so that the solid pure substance does not have
      to be handled. If possible, this should hold true for
      other tests as well. In general, the concentration of
      the stock solutions should be below 1 mg/ml, that of the
      working solutions 1-100 ng/ml. For analytical work not
      more than 150 ng TCDD should be used per single test run
      so that major contamination can be avoided should an
      accident occur· (e.g. broken glass). In Germany, the
      transport of standard solutions still represents an
      awkward problem. Consignments of TCDD have had to be
      taken personally from the supplying laboratory since
      there is no official institution distributing stock
      solutions.

9.   TCDD in dissolved form is decomposed when irradiated with
      UV light (including sunlight) (Crosby et al, 1971;
      Plimmer et al, 1973). Therefore it is necessary to
      protect standard and sample solutions from such
      irradiation. Furthermore, the concentration of the
      standard solutions should be checked from time to time.

10.   Balance calculations should be performed for each test
      in order to establish a recovery rate. This makes it

indicating damage by TCDD has not been defined. Since this problem is still unresolved it could not be considered in the studies at hand. Contributions on this subject are urgently needed.

It is hoped that the discussion of safety precautions will contribute to a better protection of the health of all scientific workers and technical staff working with such toxic compounds.

REFERENCES

Crosby, D. G., Wong, A. S., Plimmer, J. R., and Woolson, E. A., 1974, Photodecomposition of chlorinated dibenzo-p-dioxins, Science., 173:748-749, 1971.
Herxheimer, K., 1899, Uber Chlorakne, Munch. Med. Woch, 46.1:278.
Kimmig, J. and Schulz, K. H., 1957, Occupational acne (so-called chloracne) due to chlorinated aromatic cyclic ethers, Dermatologia, 115(4):540-546.
Oliver, R. M., 1975, Toxic effects of 2,3,7,8-tetrachlorodibenzo-1,4-dioxin in laboratory workers, Brit. J. Ind. Med., 32(1):49-53.
Plimmer, J. R., Klingebiel, U. I., Crosby, D. G., and Wong, A. S. 1973, Photochemistry of dibenzo-p-dioxins, Adv. Chem. Ser., 120: 44-54.
Sandermann, W., Stockmann, H., and Casten, R., 1957, Uber die pyrolyse des pentachlorphenols, Chem. Ber., 90-690-692.
Stehl, R. H., Papenfuss, R. R., Bredeweg, R. A., and Roberts, P. W., 1973, The stability of pentachlorophenol and chlorinated dioxins to sunlight, heat and combustion, Adv. Chem Ser. 120:119-125.

PROGRAM FOR MONITORING POTENTIAL CONTAMINATION IN THE
LABORATORY FOLLOWING THE HANDLING AND ANALYSES OF
CHLORINATED DIBENZO-P-DIOXINS AND DIBENZOFURANS

F. D. Hileman, T. Mazer and D. E. Kirk

Monsanto Research Corporation
Dayton, Ohio   USA

ABSTRACT

A program of safety wiping has been established to monitor
the workplace for polychlorinated dibenzo-p-dioxins and
polychlorinated dibenzofurans. The procedure uses isooctaine
wetted wipes to obtain the samples followed by a simple alumina
column cleanup procedure and gas chromatographic-mass
spectrometric analysis of the samples. A quality control program
has been set up as a check on the effectiveness of the procedure.
Detection limits of five nanograms per wipe are routinely obtained
for the tetrachlorodibenzo-p-dioxins and tetrachlorodibenzofurans.

INTRODUCTION

As a major chemical manufacturing company, Monsanto has
established a facility for the trace analysis of contaminants
which may be present in products and waste streams (Throdahl,
1980). Particular attention has been given to the analysis of
polychlorianted dibenzo-p-dioxins (PCDDs) and polychlorinated
dibenzofurans (PCDFs). Performing these analyses necessarily
involves handling analytical standards which contain PCDDs and
PCDFs. The extreme toxicity of certain of these compounds creates
a potential health hazard for personnel involved in the analyses.

To facilitate the safe handling analysis of these compounds
Monsanto has established isolated sample storage, workup and
analysis facilities. These include a Chemical Safety Laboratory
where standards are stored and dilute standard solutions are

697

prepared. An isolated Sample Preparation Laboratory has also been
provided which is used solely for the workup of samples which
could contain PCDDs or PCDFs. Finally, an isolated Sample
Analysis Facility has been established in which the instrumentation has been specifically designated for the analysis of
these chlorinated compounds.

Access to all of these facilities is strictly limited. The
only people who may use these laboratories are those whose work is
directly related to the analysis of PCDDs or PCDFs. These
individuals participate in a health monitoring program involving
annual physical examinations and they are thoroughly indoctrinated
into the safe handling of these compounds. This indoctrination
includes the viewing of a special video tape that summarizes all
the precautions that are taken in the handling of PCDDs and PCDFs.
In addition, this video tape is shown to any other interested
individuals throughout the laboratory who do not have a direct
involvement in the PCDD and PCDF program but who want to know what
steps are being taken to protect them from any potential exposure.

In view of the precautions being taken, one might question why it
is necessary to have a contaminate monitoring program. The prime
reason for such a program is that it confirms the belief that
adequate precautions are being taken. In other words, it is proof
that the system works. This is particularly necessary when
dealing with a multiplicity of workers involved in the PCDD and
PCDF projects. Each worker knows the extent to which his
laboratory technique conforms to safe handling practices, but he
may not know explicitly about the technique of others. Monsanto's
monitoring program gives all involved the confidence in their own
and others abilities to safely work with these hazardous
compounds.

PROCEDURES

When the monitoring program was established, several criteria
were set up to guide its development. First, the program was to
be comprehensive in monitoring all PCDDS and PCDFs, including
monochloro-through octachloro-substituted compounds. Second, high
sensitivity was desired to ensure good detection limits. Third,
high specificity was deemed essential to avoid false positives
which could raise undue alarm and cause work delays. Fourth, the
sample preparation procedure was to be rapid and easily adapted to
handling ten to fifteen samples at a time. At the same time
automation on the analysis was desirable so that excessive
manpower was not spent in analyzing the samples. These criteria
have for the most part been met in the following monitoring
program.

Sample Collection

Due to the nature of the samples being analyzed and the nonvolatile nature of most PCDDs and PCDFs the sample collection procedure made use of wipes. Whatman No. 40 ashless filter paper was moistened with isooctane (note, all solvents were Burdick and Jackson "Distilled in Glass" grade, Muskegon, Michigan). The technician carrying out the wiping wore disposable gloves which were changed between each wipe to prevent cross contamination. Typically a 100 cm^2 surface area was used for each wipe. Surfaces to be wiped were designated by the individuals carrying out the PCDD and PCDF analyses since these individuals were most likely to know which surfaces might be to be contaminated. In addition, areas which should not be contaminated (e.g., telephone receivers, door knobs, water cooler handles) were also wiped to allow an assessment of potential contamination outside the isolated laboratories. After completing the wiping, the samples were placed into 100 mL wide-mouthed bottles with a Teflon-lined cap and saved for subsequent processing.

To better assess the data that was generated during the wipe study a quality assurance/quality control (QA/QC) program was adopted. Two spike samples and one blank were required for every ten wipe samples that were taken. The spikes were prepared by spiking a 12.5 cm^2 piece of aluminum foil with a standard containing mono- through octachloro-dibenzo-p-dioxin and mono- through octachloro-dibenzofuran. Spiking levels were typically two times and twenty times the lower detection limit that could be achieved. The solvent was then allowed to evaporate, and the foil was wiped in a manner identical to the other wipes. The blank sample simply involved wiping a piece of unspiked aluminum foil.

Sample Preparation

Forty milliliters of petroleum ether were added to the bottle containing the wipe and the bottle and contents were shaken for fifteen minutes. The petroleum either was then transferred to a disposable 250 mL bottle. Two additional 10 mL petroleum ether rinses of the bottle containing the wipe were made combined with the original 40 mL of petroleum either. Fifty millilters of a 1N KOH solution were then added to the petroleum either and the mixture shaken for five minutes. The aqueous layer was removed and 50 mL of organic free water was added to and shaken with the petroleum ether.

Again the aqueous layer was removed and the peteroleum ether was dried over 2 grams(g) of anhydrous sodium sulfate. The petroleum ether was then transferred to a centrifuge tube with two additional 10 mL rinses of the sodium sulfate. It was then concentrated to 0.5 mL under a gentle stream of purified nitrogen.

This concentrate was then processed on a chromatographic
column packed with 4 g of Woelm Basic Alumina (ICN Bichemicals,
Cleveland, Ohio) which had been activated at 500°C for 24 hours.
The column was topped with 2 g of anhydrous sodium sulfate. The
concentrate was applied to the column and the column eluted with
10 mL of a 2% dichloromethane/hexane (V/V) solution. The eluent
was discarded. A second elution was then made with 15 mL of 50%
dichloromethane/hexane (V/V). Three hundred microliters of
dodecane (Aldrich Chemical, Milwaukee, Wiconsin) were then added
to the 50% dichloromethane/hexane eluent which was concentrated to
300 μL with gentle heating (70°C) under a stream of purified
nitrogen. The sample was transferred to a 2 mL vial with a
compression seal lid suitable for use with an automatic liquid
sampler.

Sample Analysis

Sample analysis was carried out on a Hewlett-Packard 5985B
GC-MS system. The chromatography was performed using a 30-meter
SE-54 fused silica capillary column (J and W Scientific, Rancho
Cordova, California). The gas chromatograph was operated at an
initial temperature of 190°C with a hold of one minute followed by
a temperature program of 10°C/minute to 275°C. Helium was used as
a carrier gas at a head pressure of 7 psi. The mass spectrometer
was operated in a batch mode in which a liquid sampler
automatically injected the sample with the mass spectrometer
operating in a selected ion mode detecting ions characteristic of
mono- through octachloro-dibenzo-p-dioxin and mono- through
octachloro-dibenzofuran.

Hardcopy output of the analytical results was automatically
produced as chromatographic traces for each characteristic ion.
It was then possible to visually examine the data to identify any
apparent positive results which could then be reanalyzed to
confirm or deny the presence of PCDDs or PCDFs on the wipe
samples.

RESULTS AND DISCUSSION

The methodology for this safety monitoring procedure is based
on a GC-MS system as the analysis device. To many this may seem
an expensive way of carrying out these analyses compared with
other techniques such as electron capture gas chromatography.
However, the inherent specificity combined with sensitivity make
GC-MS the anslysis method of choice, especially since the system
being used is so easily automated. This automation allows the
system to analyze these samples during non-work hours and thus
these wipe analyses do not greatly interfere with other analyses.

The recovery results and method detection limits for a typical analysis of PCDD and PCDF spiked wipes are given in Table 1. From this table it can be observed that the recoveries for monochloro- and dichloro-dibenzo-p-dioxins and dibenzofurans are low. This results from the loss of these compounds from the aluminum foil during the evaporation of the solvent. This is illustrated for the PCDDs in Table 2 in which the spiking solution was deposited directly onto the isooctane moistened wipe. The wipe was then processed in the normal manner as described in the Procedure Section. The results given in Table 2 show that the recoveries are significantly improved for the monochloro-and dichloro- compounds. However, the information provided in the QA/QC data in Table 1, in which the spiking and wiping was done in a manner which very closely simulates the actual taking of samples, is far more valuable. Table 1 illustrates that wiping is an effective method for assesssing the presence of trichloro- through octachloro-substituted dibenzo-p-dioxins and dibenzofurans. If monochloro- and dichloro- substituted compounds are to be detected at low levels then an alternative sampling procedure such as air sampling should be used.

The importance of the wiping solvent is illustrated in Table 3. For these samples the aluminum foil pieces were spiked with PCDDs and the solvent allowed to evaporate. Both dry and water-moistened filter papers were then used to wipe the foils. These procedures were evaluated because they have been used by inspection teams visiting our facilities and we wanted to establish the effectiveness of these procedures in providing reasonable results. From Table 3 it can be observed that the use of water-wetted wipes is a very ineffective way of collecting PCDDs on a wipe. The use of the dry wipe and the mechanical gathering of the PCDDs onto the wipe actually yields higher recoveries than the water-wetted wipes. Note also that the spike levels for Table 3 have been raised approximately one order of magnitude from those used in Tables 1 and 2 in order to have a sufficient amount of sample to be detected by the GC-MS system.

In Tables 1 and 2 the method detection limits have been given for the analysis of each chlorine class of PCDD and PCDF. It is possible to reduce these limits by decreasing the final volume of dodecane solvent in the sample preparation step. This has not been done since the 300 μL sample volume is very compatible with the use of the automatic liquid sample, which must have sufficient sample volume to repeatedly rinse the syringe and then inject a sample. Since one criterion of the wipe methodology has been to keep the analyses automated, the detection limits have thus been set by convenience rather than for ultimate sensitivity. Another reason for the 300 μL final volume is that this volume is easily reproduced during the solvent blow-down step. Internal standards are not used and therefore all quantitations are based on external

Table 1. Analysis of Wipe Samples for Chlorinated Dibenzo-p-dioxins and Chlorinated Dibenzofuran

Sample	Mono	Di	Tri	Tetra	Penta	Hexa	Hepta	Octa
			Chlorinated Dibenzo-p-dioxins, ng/wipe					
Spike Level	19.7	8.5	10	9	18.8	31.8	18.2	38.0
Wipe – Spiked	4.7	5.1	8	10	15.8	30	20	38
Recovery	24%	60%	80%	111%	84%	94%	110%	74%
Spike Level	197	85	100	90	188	318	182	380
Wipe – Spiked	41	43	65	62	128	221	135	270
Recovery	21%	51%	65%	69%	68%	69%	74%	71%
Blank	ND	ND	ND	ND	ND	ND	ND	ND
Method Detection Limit	10	5	5	4	5	5	5	5
			Chlorinated Dibenzofurans, ng/wipe					
Spike Level	9.0	16.0	17.3	11.2	25.9	44.1	9.5	27.4
Wipe – Spiked	ND	ND	6.1	6.3	20	34	6.5	20
Recovery	–	–	35%	56%	77%	77%	68%	73%
Spike Level	90	160	173	112	259	441	95	274
Wipe – Spiked	27	77	106	82	175	293	54	178
Recovery	30%	48%	61%	73%	68%	66%	57%	65%
Blank	ND	ND	ND	ND	ND	ND	ND	ND
Method Detection Limit	25	25	6	5	5	5	5	5

Table 2. Analysis of Wipe Samples Spiked Directly Onto the Wipe With Chlorinated Dibenzo-p-dioxins

Quantity Found, ng/wipe

	Mono	Di	Tri	Tetra	Penta	Hexa	Hepta	Octa
Direct Low Spike Level	20	8	10	9	19	32	18	38
Wipe – Found	14	6.6	10.5	9.5	21	26	15	35
Recovery	69%	83%	105%	108%	101%	80%	84%	92%
Direct High Spike Level	196	85	100	90	188	318	182	380
Wipe – Found	123	59	72	61	139	254	149	270
Recovery	63%	69%	72%	68%	74%	80%	82%	71%
Blank	ND	ND	ND	ND	ND	ND	ND	ND
Method Detection Limit	4	4	3	3	4	4	4	4

Table 3. Analysis of Wipe Samples Using Both Wet (Water) and Dry Wipes for the Presence of Chlorinated Dibenzo-p-dioxins

Quantity Found, ng/wipe

	Mono	Di	Tri	Tetra	Penta	Hexa	Hepta	Octa
Aluminum Foil, Low								
Spike Level	98	42	50	45	94	159	91	190
Wipe (dry) - Found	34	14	20	19	39	105	37	72
Recovery	35%	33%	40%	42%	42%	66%	41%	38%
Wipe (wet) - Found	2	ND	1	1	2	3	ND	4
Recovery	2%	0%	2%	2%	2%	2%	0%	2%
Aluminum Foil, High								
Spike Level	982	425	500	450	938	1590	910	1900
Wipe (dry) - Found	265	157	225	189	591	970	482	893
Recovery	27%	37%	45%	42%	63%	61%	53%	47%
Wipe (wet) - Found	10	8	14	13	30	48	18	40
Recovery	1%	2%	3%	3%	3%	3%	2%	2%
Blank	ND	ND	ND	ND	ND	ND	ND	ND
Instrument Detection								
Limit	2	2	1	1	2	3	3	4

standards which makes the reproducibility of the final sample volume more critical. The reason that no internal standards are used is that we also analyze for the presence of the internal standards (either $^{13}Cl_4$-2,3,7,8-tetrachlorodibenzo-p-dioxin or $^{37}Cl_4$-2,3,7,8-tetrachlorodibenzofuran) on the wipes. This is a particularly important analysis since these isotopically labeled compounds are the most widely used analytical standards in our laboratory, being added as an internal standard to almost all samples.

If a PCDD or PCDF is found on a wipe sample, an immediate cleanup of the contaminated area is carried out. The immediate area and surrounding areas are rewiped and the resulting wipes are analyzed for the presence of PCDDs or PCDFs. In certain cases particular areas have been found always to be contaminated with detectable levels of PCDDs and PCDFs. The injection ports on liquid chromatographs used for sample cleanup are an excellent example. For these cases the area in question is covered as much as possible with a disposable cover (i.e., aluminum foil) which is regularly changed. The area is also posted with signs warning personnel that it is a potential contamination site.

CONCLUSION

A program for monitoring PCDDs and PCDFs in the workplace has been described. The procedures combine an effective quality assurance/quality control program with automated analyses to provide a reasonable assessment of the effectiveness of the analyses as well as the cleanliness of the laboratory. The program achieves reasonable detection limits and specificity for the highly toxic PCDDs and PCDFs while not being so burdensome that cost and time prevent it from being used on a regular basis.

REFERENCES

Throdahl, M.C., Practical Implications of Risk Management in Industry, Presentation to the New York Academy of Sciences, New York, New York, March 18, 1980.

A MEDICAL SURVEILLANCE PROGRAM FOR SCIENTISTS

EXPOSED TO DIOXINS AND FURANS

William H. Wolfe and George D. Lathrop

Epidemiology Division
USAF School of Aerospace Medicine
Brooks AFB, San Antonio, Texas USA

ABSTRACT

The rationale for medical surveillance of individuals exposed to toxic substances and the components and objectives of a comprehensive surveillance program is discussed. The applicability of these principles and concepts to toxic chlorinated dibenzo-p-dioxins (dioxins) and chlorinated dibenzo-p-furans (furans) is then considered, emphasizing the broad range of biomedical effects suspected to be caused by these substances. A format for the medical evaluation of individuals occupationally exposed to these chemicals is presented and selected medical examination procedures are discussed. The importance of a comprehensive unified program of surveillance involving medical evaluation, industrial hygiene techniques, epidemiologic analysis, and long-term population tracking is presented.

INTRODUCTION

Environmental surveillance programs in industries were initially provided by employers on a voluntary basis, and subsequently, these programs came to be required by Federal and State regulatory agencies such as Occupational Safety and Health Administration in the 1970s. Industrial hygiene programs have traditionally been employed in industrial settings, but they have been far less evident in medical and research laboratory settings. However in dealing with laboratory personnel working with toxic chlorinated dibenzo-p-dioxins (dioxins) and chlorinated dibenzo-p-furans (furans) and similar compounds, a comprehensive medical surveillance system as well as a classical

industrial hygiene program is necessary. Medical-legal
considerations have also contributed to the emphasis placed on
medical surveillance in industry. However, even elegant hygiene
programs may not be enough to answer medical-legal questions in
the absence of a scientific perspective. It was only a few years
ago that a group of United States Government workers received
large compensation settlements for nonspecific and subclinical
abnormalities on liver biopsy even without evidence of a casual
link between their condition and occupational exposures. Thus,
while concerns for the current and long-term health and safety of
employees and compliance with legal and regulatory requirements
are the primary motivating forces behind such programs, the
scientific uncertainties surrounding the dioxins and furans
increase the importance of effective broad-based medical
surveillance efforts.

A comprehensive medical surveillance system for scientists
and laboratory workers exposed to dioxins and similar chemicals
should meet several key objectives. Subclinical, acute,
subacute, and long-term, adverse health effects from attributable
exposures can be identified by surveillance, depending on the
elegance or invasiveness of the examination. However, it is
imperative that medical endpoints from the full spectrum of
illness caused by the toxic agents be included in the program.
The medical evaluation process will also provide a basis for the
meaningful analysis of data pertaining to the exposure to these
potentially toxic chemicals. A systematic approach to medical
surveillance will allow comparable data from many sources to be
combined. Occasionally, causal relationships between illness and
exposures can be defined from observations of suboptimal sample
sizes when case clustering of unusual conditions occurs, but such
revelations are relatively uncommon. Thus, a larger number of
subjects with the same or comparable exposures should be
evaluated in comparable ways. This should provide an analysis
that overcomes many of the deficiencies of small sample size and
nonstandardized procedures encountered in much of the current
medical literature, and complex issues surrounding the dioxins
and furans may be clarified.

There are three components to a comprehensive surveillance
system: environmental monitoring; medical evaluation; and
tracking of personnel. Classical industrial hygiene approaches
to toxic substances in the workplace have been used by large
industries for many years, and improved analytical chemistry
technology has enhanced the capability of the industrial
hygienist to detect smaller and small concentrations of chemicals
in the workplace. These capabilities have resulted in a tendency
to reduce exposure to meet the levels of detection, especially in
exposures to the more toxic substances. Environmental monitoring
coupled with threshold limit values and action levels, are useful

tools. However, programs based solely on these principles remain limited to the field of physical chemistry; exposure standards are frequently arbitrarily assigned, and there is no satisfactory mechanism to extrapolate such data directly to human populations. This problem assumes an even greater degree of importance as more females enter the work force, and we become concerned about more direct fetotoxic and teratogenic effects. In the United States, threshold limit values, emergency exposure limits, and other standards are generally based upon "no effect levels" in healthy male workers. The applicability of these standards to a growing proportion of female workers is unknown at this time.

Comprehensive medical evaluations conducted on a periodic basis are of major importance in developing an effective surveillance program. The medical evaluations can be sharply focused for personnel exposed to many chemical and physical agents; unfortunately, we do not have this luxury when dealing with the dioxins and furans. Attempts to identify a specific chronic condition of syndrome arising from exposures to these compounds have been unsuccessful, except for chloracne.[1,2] While we are able to consistently demonstrate a large group of acute symptoms and physical findings following exposure to these chemicals (Table 1), the question of whether significant chronic effects are produced in humans is a controversial issue which many of us must deal with on a daily basis.[3,8] Reviews of physical chemistry data, animal toxicity data, human exposure case reports, and epidemiologic studies have been relatively unsuccessful in identifying objective medical endpoints for the chronic effects of exposure.[4,5,7,9] The list of known or suspected acute and subacute effects following TCDD exposure is somewhat overwhelming, and many of the endpoints are highly subjective and extremely difficult to evaluate.[6,7] While chloracne appears to be a consistent chronic effect of moderate to heavy exposure, the implications of this condition on the long-term health of exposed workers is unknown.[9] At best, one can develop a list of potential large organ systems which should be carefully evaluated during preemployment and periodic physical examination. The following target organ systems/areas can be identified: (1) Dermatologic; (2) Hepatic; (3) Neoplastic; (4) Neuropsychiatric; (5) Endocrine/reproductive; (6) Renal; (7) Immunologic; and (8) Hemopoietic.

Table 1. The Acute/Subacute "Symptom Complex" Following TCDD Exposure (as Derived from Reviews of Both Human and Animal Literature)

Dermatologic: Chlorance, Porphyria, Hyperpigmentation, Hirsutism, alopecia   (continued)

Table 1. The Acute/Subacute "Symptom Complex" Following TCDD
         Exposure (as Derived from Reviews of Both Human and
         Animal Literature) - Cont.

Hepatic: Increased cholesterol and triglycerides, abnormalities
          in liver function tests, hepatomegaly

Neuropsychiatric: Peripheral neuropathy, asthenic syndrome

Endocrine/Reproductive: Adnormalities of glucose metabolism,
                          hypothyroidism, decreased fertility,
                          and increased spontaneous abortion rates

Renal: Proteinuria, decreased renal output, tubular degeneration,
        glomerular degeneration, renal glucosuria

Immunologic: Decreased resistance to infection, thymic atrophy

Helopoietic: Decreased lymphocyte counts, anemia, thrombocytopenia

Cardiac: Bradycardia, trachycardia, atrial fibrillation

Gastrointestinal: Nausea, vomiting, diarrhea, gastritis, abdominal
                   pain

     Ideally one, would like to have a sensitive and specific
examination or laboratory procedure to detect these chemicals or
their effects in human tissues. Unfortunately, there is a lack
of clearly defined endpoints in the scientific literature, and,
other than chloracne, distinct clinical syndromes or unique
effects indicative of chronic illness have not been identified.[1,9]
The signs and symptoms currently attributed to exposure are
confounded by age and other causes, and the effect, if present,
may be lost in common symptoms from other causes of disease
(unlike vaginal adeno-carcinoma from diethyl stilbesterol
ingestion and angiosarcoma of the liver from vinyl chloride
exposure). Currently, research efforts are being directed toward
the development of methods to identify specific signs or symptoms
of dioxin/furan exposure and differentiate them from background
effects normally seen in any similar nonexposed population.
Improvements in biochemical technology and the conduct of
epidemiologic studies with a sufficient number of study subjects
and statistical power may clarify this issue. In the absence of
sensitive and specific indicators of exposure, a comprehensive
examination format must be developed around the target organ
systems. As can be seen in Table 2, much of the proposed medical
evaluation involves biochemical analysis.

Table 2. Format for a Physical Examination Following TCDD
Exposure

Organ System	Procedure
General:	General Examination
Dermatologic:	Dermatologic examination, delta-aminolevulinic acid
Hepatic:	Cholesterol/HDL cholesterol, triglycerides, SGOT/SGPT, Gamma Fluteryl Transpeptidase, Aklaline Phosphatase, LDH, Urine Porphyrins, Urine Porphobilinogen
Neuropsychiatric:	Complete Neurological Examination *MNPI *Psychological Performance Tests *Nerve Conduction Velocities CPK
Endocrine/Reproductive:	Fasting Blood Glucose Two Hour Post Prandial Glucose Thyroid Profile Semen Analysis: count, motility, volume, percentage abnormal Careful Reproductive History of Subject and Spouse
Renal:	Urinalysis BUN Creatinineom vinyl chloride
Immunologic:	Protein Electrophoresis *Immunoelectrophoresis *Quantitative Immunoglobulins *Monila Skin Tests
	*If subject reports an abnormal history or review of systems or is at highest risk of unprotected exposure

The clinical laboratory evaluation can also range from the
simple to the complex, but the individual tests should be
carefully selected to cover as many of the suggested signs and

symptoms as possible. Evaluation of hepatic, renal, and
endocrine function, red and white blood cell production and
cholesterol/triglyceride metabolism should definitely be included
in any program. A dermatologic examination coupled with
urine-porphyrin and/or prophobilonogen testing will help evaluate
the suggested effects of the dioxin and furans on the skin. An
evaluation of reproductive function should also be conducted and
should include a careful fertility history of both the patient
and spouse or spouses. The presence of birth defects in
offspring should also be determined. The potential for the
planar dioxin/furan molecule to intercalate in DNA to produce
frame-shift mutations is particularly worrisome, and could be a
mechanism for paternally-transmitted birth defects. Immunological
status can most easily be evaluated by careful history-taking
with emphasis on susceptibility to infections. Various skin
sensitivity tests can also be used, but frequently will not
demonstrate abnormalities until well after a history of frequent
infectious disease is evident. This extensive biochemical
battery is included in the physical examination in an attempt to
identify conditions in their subclinical stages when intervention
has the best chance of success. Several of the other examination
procedures such as nerve conduction velocities, neurological/
psychological performance testing, semen analysis, and the more
extensive immunologic testing can be reserved for those
individuals reporting an abnormal history, having abnormal
examination findings, and for in-depth research studies.

Other tests such as testicular biopsies, porphyrin metabolite
patterns, mitogenic responses of lymphocytes, and T and B cell
lympocyte determinations have been suggested by some investigators.
However, these procedures may be desirable on a case by case
basis, but are not generally suitable for large scale screening
efforts at this time. In any attempt to evaluate the potential
adverse effects of dioxin/furan exposure, substantial effort must
be put forth to enhance the quality of the collected data.
Examinations and history-taking protocols should be prepared and
followed precisely. These efforts will insure that all workers
receive the medical care they deserve and valuable medical and
scientific data will be collected.

While the examination should be comprehensive in order to
evaluate the wide range of possible or suggested effects of
exposure, it should also be practical. Tests and procedures
should be of an accepted technology and should be subjected to
rigorous quality control standards to enhance their validity.
Some procedures are primarily of research interest and should be
subjected to more detailed evaluation before inclusion in a
reasonable screening program. With current technology, tissue
biopsies for dioxins and furans are quite sensitive (parts-per-
trillion) but not very specific. It is quite difficult to

differentiate between the isomeric forms. Dioxins and furans
from sources such as commercial or industrial incinerators can be
mistaken for chemicals from occupational exposures. The tests
are quite difficult to perform on human tissue, requiring an
extensive amount of specimen purification. If a tissue biopsy is
positive there is no assurance that occupational exposure had
indeed occurred or that adverse health effects would follow.
Conversely, if a biopsy is negative, it would not rule out
occupational exposure and the subsequent development of adverse
health effects. We do know that the body can accumulate these
chemicals and selectively store them in adipose tissue, but the
dynamics of this process are unclear, and the duration of storage
is unknown. Of course, a careful medical/occupational health
history should be obtained at each examination. As presented in
Table 3, the history should also be comprehensive, and attention
should be directed to repoductive, recreational and leisure
activities, and other sources of confounding exposures.
Confouding exposures are exposures to chemical and physical
agents not in the workplace which could cause adverse medical
effects. Examples of confounding exposure are shown in Table 4.

Table 3. Questionnaire Outline for Occupational Health Data

---

1. Demographic Data: Name, Social Security Number,
                             Residence, Phone Number, Date of
                             Birth, Race, Educational History

2. Occupational History: Detailed Work History

                             Specific Questioning about Occupation/
                             Jobs of Special Interest

                             Specific Questions about Exposure to
                             Chemical/Physical Agents of Special
                             Interest

3. Recreational and Leisure Activities Involving Hazardous
   Agents or Exposures

4. Marital History

5. Reproductive History of Subject and Spouse (or Spouses)

6. Medical Review of Systems, Keyed to those Organ Systems of
   Primary Interest

7. Detailed Smoking and Alcohol Use Histories

Table 4. Potentially Confounding Exposures

TOXIC CHEMICAL AND PHYSICAL AGENTS IN THE MODERN ENVIRONMENT:

Pesticides   Therapeutic Drugs

Cleaning Agents   "Recreational" Drugs

Industrial Chemicals   DDT, PCB, Other Dioxins

Alcohol   Radiation

OTHER DISEASES:

Viral

Bacterial

Parasitic

Tracking of personnel over time is also a vital component of a proper surveillance system, and is intended to identify chronic sequalae of toxic occupational exposures. If the initial data base is haphazardly maintained, or if individuals become "lost" with the passage of time, it becomes virtually impossible to determine whether chronic adverse health effects have occurred as a result of occupational exposures to chemical, physical, or other stresses of employment. Long-term follow-up of the health status of current and former employees exposed to toxic substances will permit epidemiologic studies based on person-years of observation. These efforts require a coordinated review of multiple record keeping systems, but is an extremely valuable and vital tool if we are to answer the scientific questions posed by the dioxin/furan issues in a correct manner.

The optimal medical surveillance program for dioxin and furan exposed workers, as well as for workers exposed to other toxic substance (Figure 1) requires a unified approach based on periodic medical evaluation, industrial hygiene, epidemiologic analysis, and population tracking. These elements should be applied in a coordinated manner, which can be adapted to meet the changing direction of toxiologic concern. Each element of this approach has inherent limitations: medical surveillance programs suffer from the lack of precision in the examinations and clinical laboratory procedures; classical industrial hygiene

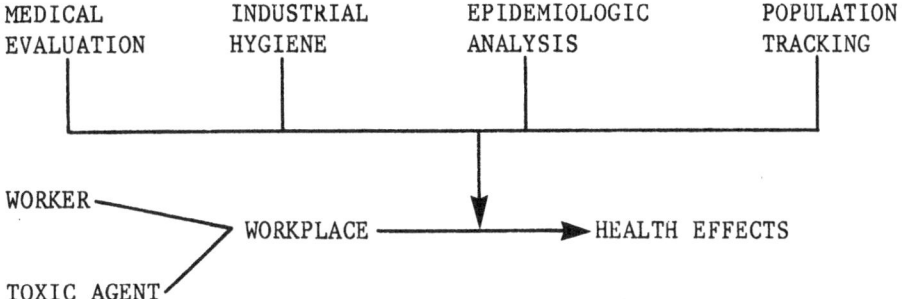

Fig. 1. A Medical Surveillance Model for Occupational Exposure
         to Toxic Agents

efforts often have little direct relationship to human populations
and disease; epidemiologic studies usually involve relatively
small numbers of subjects, exposed to confounding substances,
with a poor expectation of quick study results; and long-term
personnel tracking is extremely difficult to accomplish.
However, despite the weaknesses of each element, they complement
one another when used in a coordinated program, and many of the
weaknesses are overcome. A carefully structured program of
environmental monitoring, medical surveillance, long-term
tracking, and epidemiologic analysis focused by the use of
predictive toxicology will provide the best health service to the
employees. It will identify workers at risk of significant
exposure, will detect toxic effects at the earliest stage of
pathogenesis, and will hopefully provide quality data to help
clarify the key scientific and social issues we all face.

REFERENCES

1. K. D. Crow, Chloracne, <u>Trans. St. John Hosp. Derm. Soc.</u>
    56:79-99 (1970).
2. R. D. Kimbrough, ed., "Halogenated Biphenyls, Terphenyls,
    Naphthalenes, Dibenzodioxins and Related Products."
    Amsterdam, The Netherlands 406 p. Elsevier/North-Holland
    Biomedical Press, Amsterdam (1980).
3. E. Homberger, G. Reggiani, J. Sambeth, and H. K. Wipf, The
    Seveso Accident: Its Nature, Extent and Consequences,
    <u>Ann. Occup. Hyg.</u>, 22:327-370 (1979).
4. L. Jirasek, J. Kalensky, and K. Kubec, Acne chlorina and
    porphyria cutanea tarda during the manufacture of herbicides,
    Part I, <u>Czech. Dermatol.</u>, 48(5):306-317 (1973).

5. L. Jirasek, J. Kalensky, K. Kubec, J. Pazderova, and E. Lucas,
    Acne chlorina, porphyria cutanea tarda and other
    manifestations of general intoxication during the manufacture
    of herbicides, Part 2, Czech. Dermatol., 49(3):145-157 (1974).
6. R. M. Oliver, Toxic Effect of 2,3,7,8-Tetrachloro-dibenzo-1,4-
    dioxin in Laboratory Workers, Br. J. Ind. Med. 32:46-53
    (1975).
7. A. Poland, W. F. Greenlee, and A. S. Kende, Studies on the
    mechanism of action of the chlorinated dibenzo-p-dioxins and
    related compounds, N.Y. Acad. Sci. 320:214-230 (1979).
8. G. Regianni, Acute Human Exposure to TCDD in Seveso, Italy,
    J. Toxicol. Environ. Health 6:27-43 (1980).
9. A.L. Young, J. A. Calcagni, C. E. Thalken and J. W. Tremblay,
    The Toxicology, Environmental Fate, and Human Risk of
    Herbicide Orange and Its Associated Dioxin, Technical Report
    OEHL-TR-78-92, USAF Occupational and Environmental Health
    Laboratory, Brooks AFB, Texas, 247p (1978).

LABORATORY HANDLING AND DISPOSAL OF CHLORINATED DIOXIN WASTES

L. G. Taft, P. R. Beltz, and B. C. Garrett

Battelle, Columbus Laboratories
Columbus, Ohio, USA

ABSTRACT

The handling and disposal of waste materials containing toxic chlorinated dibenzo-p-dioxins has become a major problem facing laboratories currently conducting or considering conducting research with TCDD and related chemicals. Much of the dioxin-contaminated waste being generated by these laboratories is currently being stored on-site. The small volume of this waste and the low level of contamination present make laboratory methods of degradation or destruction an alternative option worthy of consideration for many laboratories. Each of the methods or techniques which have been demonstrated to result in the chemical destruction or deactivation of TCDD will be discussed briefly with regard to its use for disposal of laboratory wastes contaminated with these materials. In addition to these directly applicable methods, other techniques which appear to have potential for laboratory-scale dioxin disposal will be identified and discussed briefly. These methods should be evaluated by laboratories handling dioxin-related materials as alternatives to the centralized collection and storage of dioxin-contaminated wastes.

INTRODUCTION

There has been a great deal of research conducted on the chlorinated dioxins since their discovery as contaminants of herbicide formulations in 1970. The intensity of this research dramatically increased after the explosion and subsequent widespread contamination that occurred at Seveso, Italy in 1976. As a result of this research, a great deal is known about the chemical,

physical, and biological characteristics of 2,3,7,8-tetrachloro-
dibenzo-p-dioxin (TCDD) and, to a more limited extent, other
chlorinated dioxin materials. There has been considerable study
of methods to dispose of commerical chemicals and production
wastes which contain these compounds. Much of this research was
conducted in support of the disposition of Herbicide Orange stocks
by the United States Air Force and also in support of
decontamination at Seveso, Italy.

A number of techniques have been developed which have
demonstrated the chemical destruction of the TCDD molecule. In
some cases, the TCDD molecule is known to undergo complete or
partial dechlorination, forming the less chlorinated or totally
dechlorinated dibenzo-p-dioxin. These materials are much less
toxic than TCDD and present much less of a disposal problem. In
most cases, however, although the disappearance of TCDD has been
demonstrated, the resulting chemical species have not been
identified. Each of the methodologies or techniques which have
resulted in the destruction of TCDD will be discussed briefly with
regard to its use for laboratory-scale disposal of wastes
contaminated with chlorinated dioxin materials. In addition,
other techniques which appear to have a potential for the
laboratory-scale disposal of dioxin materials will be identified
and briefly discussed. These disposal methods represent potential
alternatives to the centralized collection and storage of
dioxin-contaminated wastes.

HANDLING OF DIOXIN WASTE

Although chlorinated dioxins are extremely hazardous
materials, they have been handled safely in laboratory situations
for many years. In general, wastes contaminated with these
materials can be effectively handled using precautions commonly
taken when using microcurie amounts of radioactive compounds (Dow
Chemical Company, 1970).

Commonly suggested procedures for handling dioxin waste
closely parallel those suggested for handling the actual materials
themselves; 1) the use of protective equipment, 2) attention to
personal hygiene, 3) total confinement of materials, and 4) prompt
and proper disposal of wastes. Waste materials generated from the
laboratory-handling of chlorinated dioxin materials might include:
1) disposable equipment and materials such as labcoats, gloves,
paper wipes, safety glasses, and bench top mats; 2) excess stock
and working solutions; 3) solid and liquid absorbents used in
various types of traps; and 4) solutions and other solid materials
resulting from cleaning of contaminated surfaces and glassware.
Typically, these liquid and solid wastes contain very small
amounts of dioxin-related materials, perhaps in the nanogram

through microgram range. With the current unwillingness of
virtually all haulers and disposers of normal laboratory waste to
accept wastes containing chlorinated dioxins, much of this waste
is currently being stored. At our laboratory in Columbus, Ohio,
waste liquids contaminated with low levels of dioxins are
currently evaporated in disposable containers and the residue
stored as dioxin-contaminated solid waste.

DISPOSAL OF DIOXIN WASTE

Methods with Direct Applicability

There are several methods which have been used successfully
to achieve the destruction of TCCD. These include incineration,
photolysis, ozonoloysis, soil biodegradation, radiolysis,
chloroiodide treatment, and ruthenium tetroxide treatment.

Incineration

Incineration is the high temperature reaction of material
with oxygen, which results in the degradation of the chemical
compounds. At present, it is a widely used, highly developed,
large-scale, commercial disposal technology which has been
successfuly used for the disposal of TCDD-containing wastes
(Ackerman et al, 1978). In the laboratory, it has been used to
demonstrate the destruction of several chlorinated hydrocarbon
pesticides as well as TCDD. Conditions currently considered
adequate for the safe disposal and complete destruction of TCDD
and most other toxic chlorinated hydrocarbons are a temperature of
at least 1,000°C (1832°F) with a residence of 2 seconds (Wilkinson
et at, 1978).

There are a number of specific laboratory procedures which
have been used for the thermal destruction of highly toxic
chlorinated pesticides which could possibly be adapted for the
small-scale laboratory incineration of TCDD wastes. These include
differential thermal analysis, dry combustion, ashing in a muffle
furnace (Kennedy et al, 1969). However, the best approach appears
to be the quartz tube apparatus (Duval and Rubey, 1976). A small
laboratory-scale quartz tube apparatus was used by these authors
to study the thermal decomposition of kepone and related
pesticides. An important element of this apparatus is the folded
quartz tube which is contained in a Lindberg tube furnace. Stehl
et al, (1973) used a similar apparatus in conjunction with a
Sargent tube heater to decompose milligram quantities of TCDD.
Temperatures of 800°C (1472°F) and residence times of 21 to 50
seconds were achieved with this apparatus. As shown in Table 1,
TCDD readily decomposes above a temperature of 800°C.

Table 1. Decomposition of 2,3,7,8-TCDD
Using A Quartz Tube Apparatus

Temperature		Percent Decomposition at retention time of	
C	F	21 seconds	50 seconds
500	(932)	39	42
600	(1,112)	40	59
700	(1,292)	50	53
800	(1,472)	99.5	--

Source: Stehl et al, 1973

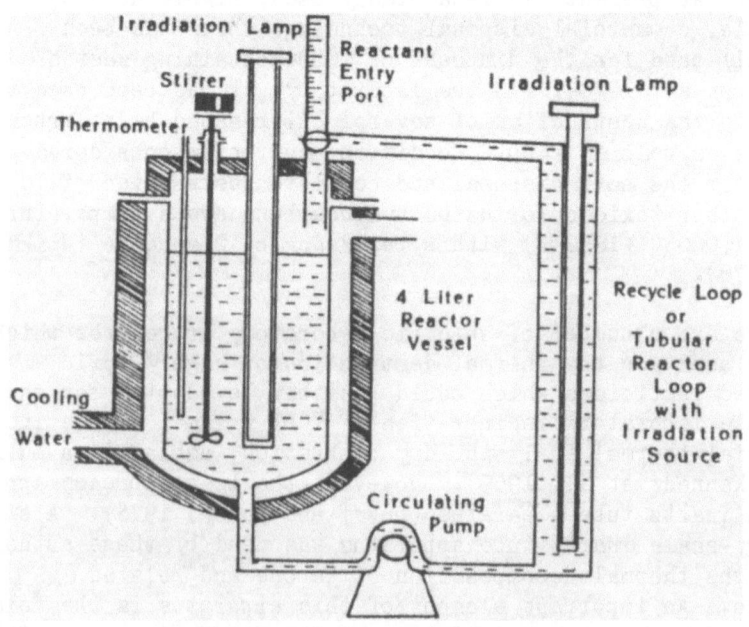

Fig. 1. Reactor design for disposal of waste chemicals.

Source: Piver, 1980

Photolysis

Photolysis is another bench-scale technology which has often been used for decomposition of highly toxic chemicals including TCDD. The laboratory apparatus required to achieve the photo-decomposition of these compounds is relatively simple and readily available (Crosby and Li, 1969; Watkins, 1974; and Calvert and Pitts, 1969). The major components of a simple photoreactor used by Crosby et al, (1969) are the reaction chamber, fabricated from borosilicate glass, and the fluorescent ultraviolet light source. Many similar reactors have been fabricated in the laboratory using fluorescent UV light sources as well as low pressure and high pressure mercury vapor lamps. The principal design features of a general-purpose photolytic reactor proposed by Piver (1980) for the deactivation and disposal of small quantities of experimental chemicals are illustrated in Figure 1. From the extensive experimental study of TCDD photodegradation reported in the literature, it should be rather straightforward to modify such a reactor to accomplish the routine photodegradation of liquid laboratory-generated TCDD wastes. TCDD has been shown to be photochemically unstable to UV radiation when dissolved in a wide variety of solvents and solvent systems. Generally, TCDD decomposes in less than 24 hours in the presence of UV light and a suitable solvent. Half-life values for TCDD under these conditions are known to vary from as long as 6 hours (Crosby and Wong, 1977) to as short as 40 minutes (Stehl et al, 1973).

Ozonolysis

Ozonolysis, like incineration and photolysis, is another bench-scale technique which has been successfully used to achieve the chemical destruction of TCDD. The basic principle of ozonolysis involves the chemical oxidation of the substrate by bubbling ozone through a liquid solution (Wilkinson et al, 1978). The required apparatus is quite simple and should be easily assembled in the laboratory. It involves, basically, an oxygen supply, an ozone generator, a reactor vessel containing a gas diffuser, and perhaps some additional form of agitation. Ozonolysis has recently been used experimentally by Cavolloni and Zecca, (1977) to achieve a 97 percent destruction of TCDD by bubbling ozone through a carbon tetrachloride-water solution for 50 hours. The method reportedly depends upon the oxidation of the aerosol formed by carbon tetrachloride and water.

Soil Biodegradation

Soil biodegradation is another method of disposal appropriate for small quantities of dioxin-contaminated materials (Wilkinson et al, 1978). It has been both recommended and utilized as a method of dioxin disposal (Kearney et al, 1973; and Young et al,

1976). Soil biodegradation depends upon microbial activity associated with the organic fraction of soil to degrade the organic molecule to less toxic products. The degradation process depends upon good chemical-soil contact, which is generally achieved by soil mixing of some form. Soil incorporation to a depth of 15 cm has been suggested as a means of achieving adequate mixing and to avoid aerial movement of soil-borne particles away from the site of application. The degradation of TCDD has been observed in several experiments. In comparison with other chemicals, particularly pesticides, TCDD is known to be persistent in soils. Its degree of persistence is similar to many of the chlorinated hydrocarbon pesticides. Measurement of TCDD residues after 350 days of incubation at 28°C showed that about 40 percent of the original 100-ppm application was degraded in a sandy soil as well as a silty clay loam soil (Kearney et al, 1972; and Kearney et al, 1973). Additional research conducted by the United States Air Force showed that under specific soil and climatic conditions, an 80 percent reduction in the soil concentration of TCDD was achieved over a period of about 2 years (Young et al, 1976).

Radiolysis

Radiolysis is an additional method which has been used experimentally to degrade the chlorodibenzo-p-dioxins. Radiolysis is essentially an extension of the photolyic method which uses energy absorbed from gamma radiation to partially degrade the dioxin molecule. As with photolysis, the gamma radiation does not destroy the dioxin ring system but only removes the chlorine substituents. The reaction product, being a mixture of less chlorinated dioxins, exhibits a much lower degree of toxicity and hazard potential. Commercial experimental irradiators are available which can be used directly to decompose chlorinated dioxin materials in solution. Such equipment has been used to achieve the reductive dechlorination of TCDD as well as its octachloro analog OCDD (Octachlorodibenzo-p-dioxin). In one such study by Buser (1976), the radiolysis of OCDD was carried out on 10 mL of a 25-ppm solution (hexane-benzene) using a ^{60}Co source at a dose level of 1.4 mega rad per hour. Significant dechlorination was achieved after a 16-hour exposure. In a similar investigation by Fanelli et al, (1978) an experimental irradiator containing a 10.0 Ci ^{60}Co source was used to study the effect of gamma ray irradiation of TCDD dissolved in dioxane, acetone, and ethanol. In this study, the concentration of a 100-ppm TCDD ethanol solution was reduced by 97 percent after a 30-hour exposure to 1.0 mega rad per hour.

Chloroiodide Treatment

The treatment with chloroiodide is another laboratory method which has been used to achieve the decomposition of TCDD. The method reportedly involves the cleavage of two ether bonds of the dibenzo-p-dioxin molecule. This work, recently reported by Botre' et al, (1979), involves the use of two specific chloroiodides, alkyldimethybenzylammonium chloroiodide and 1-hexadecylpyridinium chloroiodide generated from quarternary ammonium salt surfactants, to catalytically degrade the TCDD molecule. Degradation rates/for TCDD of 71 and 92 percent, respectively, were observed with these two chloroiodides 72 hours after treatment. Degradation products included phenols, chlorophenols, and phenoxychlorophenols.

Ruthenium Tetroxide Treatment

The experimental degradation of TCDD has very recently been demonstrated by treatment with ruthenium tetroxide (Ayres, 1981). Ruthenium is one of the platinum group of metals and its tetroxide is a powerful oxidizing agent which reacts with a wide range of organic substances including those aromatic compounds which do not react with permanganate. It can be handled safely in the laboratory as an aqueous solution or in certain non-nucleophilic organic solvents. The treatment of a carbon tetrachloride solution containing 70 µg/mL of TCDD with about 10 equivalents of ruthenium tetroxide at 70°C resulted in a half-life of less than 15 minutes. Similar results were obtained by adding the oxidant to a refluxing carbon tetrachloride solution of OCDD. Ayres recommended the use of ruthenium tetroxide for the decomtamination of laboratory glassware and apparatus and the periodic decontamination of industrial reactors to prevent the buildup of polychlorinated dibenzo-p-dioxin residues.

METHODS WITH POTENTIAL APPLICABILITY

There are other methods which have been used to dispose of small quantities of hazardous chemicals that, although they have not been used for TCDD destruction, may have a potential applicability to the disposal of dioxin wastes. The most promising of these include microwave plasma destruction, molten salt combustion, ozone/photolytic destruction, micropit disposal, and catalytic dichlorination.

Microwave Plasma Destruction

The microwave plasma destruction process was developed by
Lockheed's Palo Alto Research Laboratory as a destruction method
for hazardous materials. This process, as it relates to the
destruction of pesticides and hazardous wastes, has recently been
reviewed (Wilkinson et al, 1978; and Bailin et al, 1978). The
process uses the energy of microwave discharges to promote
chemical reactions which result in the chemical decomposition of
organic materials. Laboratory-scale models of the apparatus have
been used to achieve near total decomposition of several
organophosphate and chlorinated hydrocarbon pesticides as well as
PCBS (polychlorinated biphenyls) (Bailin et al, 1975; and Bailin
and Hertzler, 1976). The microwave plasma apparatus can be
constructed in the laboratory from standard commercially available
hardware and electronic components. Details of construction,
operation, and application can be found in the above references.

Molten Salt Combustion

The technology of molten salt combustion has been developed
over the past 20 years by the Atomics International Division of
Rockwell International Corporation. Its application to the
destruction of hazardous wastes and the construction details of a
bench-scale molten salt combustor capable of disposing of 250 to
1,000 grams per hour of waste is discussed by Yosim et al, (1978).
The concept of molten salt combustion of hazardous wastes involves
the injection of air and waste beneath the surface of molten salt.
The molten salt is sodium or potassium carbonate containing 10
percent sodium sulphate kept at 800-1000°C. These temperatures
are within the range required for the thermal destruction of TCDD.
This apparatus has been used to decompose chlorinated hydrocarbons
and is capable of handling waste materials in solid form.

Ozone/Photolytic Destruction

This technology has been developed by Houston Research, Inc.
as a method of destroying or detoxifying hazardous chemicals in
solution. The technology is basically a combination of two
technologies previously discussed, ozonoloysis and photolysis.
The process technology, which has been reviewed by Wilkinson et
al, (1978), involves the activation of the organic molecule to a
highly reactive state by UV light, coupled with its subsequent
oxidation by ozone. It is particularly attractive as a potential
method of TCDD destruction because of the susceptibility of TCDD
to decomposition of both ozone and UV light. The apparatus is
relatively simple and can easily be assembled in the laboratory.

It is essentially that described for ozonolysis with the addition
of a mercury vapor lamp. Two bench-scale reactors of 10- and
21-liter volumes have been constructed and tested (Mauk et al,
1976). Details of construction and operation are discussed in the
above references.

Micropit Disposal

The micropit disposal method was developed at Iowa State
University and is an outgrowth of a procedure used to dispose of
excess pesticides (Rogers and Allen, 1978). It involves the
placement of wastes in a relatively small, confined system
(disposal pit), which facilitates biodegradation by direct
microbial action. The construction details are shown in Wilkinson
(1978) and include a partially submerged, 113-L polyethylene
barrel into which a layer of rock and a layer of soil is placed.
The two layers are hydraulically connected by a vertically
standing, perforated clay drainage tile. Inside the barrel and on
top of the layer of soil is placed, a perforated metal basket-
containing a layer of gravel. Liquid pesticide wastes are placed
into the metal basket and the system is continuously aerated.
Microbial seeding and nutrient addition can be accomplish as
necessary. The method may have potential application for the
disposal of TCDD-containing laboratory wastes since microbial
degradation of TCDD has been shown to occur to a limited degree
(Matsumura and Benezet, 1973).

Catalytic Dechlorination

Catalytic dechlorination is a simple procedure of chemical
treatment which involves the action of a catalyst to redutively
dechlorinate an organic compound. A number of catalysts have been
used to achieve the dechlorination of chloro-aromatic compounds.
These include Raney nickel, palladium on charcoal, and nickel
boride. Of these, nickel boride has been studied the most
extensively. The nickel boride catalyst is prepared in situ by
the reaction of sodium borohydride with an alcohol solution of
nickel chloride. A series of studies involving the treatment of
several persistent chlorinated insecticides (e.g., DDT) by nickel
boride has resulted in the reductive dechlorination of these
compounds in about 30 minutes at room temperature (Dennis and
Cooper, 1975; Dennis and Cooper, 1976; Dennis and Cooper, 1977;
and Cooper and Dennis, 1978). This technique has been cited as a
potential disposal method for small quanitites of highly
chlorinated pesticides as well as TCDD itself (Cooper and Dennis,
1978; and Esposito et al, 1980).

CONCLUSION

Wastes from a laboratory conducting research with chlorinated dibenzo-p-dioxin materials might include disposable equipment and materials, excess stock and working solutions, solid and liquid absorbents and cleaning solutions, all of which may be contaminated with low levels of a very toxic chemical. In general, such wastes require handling procedures commonly used for highly toxic or radioacitve materials. The disposal of these wastes, which are generated in small amounts and which typically contain small amounts of contamination, have become a major problem facing laboratories working experimentally with TCDD and related chemicals. The unavailability of normal laboratory waste disposal channels has resulted in the storage of these wastes in the laboratory.

There are seven methods which may be directly applicable to the laboratory disposal of laboratory generated wastes contaminated with TCDD and related materials. These include incineration, photolysis, ozonolysis, soil biodegradation, radiolysis, chloroiodide treatment, and ruthenium tetroxide treatment. Although the seven methods should be considered as possible candidates for further laboratory evaluation, the most appropriate methods for further laboratory development appear to be incineration and photolysis. Each of these two methods has been studied extensively for the molecular destruction of highly toxic chlorinated materials, and the required apparatus has been adequately characterized and is very amenable to laboratory fabrication and operation. The soil biodegradation method, although operationally a very simple approach, is based on less direct and critical experimentation and suffers from the additional disadvantage that long time periods would be required for reasonably complete molecular destruction of TCDD contamination. The remaining methods, ozonolyisis, radiolysis and treatment with chloroiodide or ruthenium tetroxide, although promising, are based only on very limited, experimental data.

In addition, there are five small-scale chemical disposal methods which, based on their use for the non-TCDD related chlorinated pesticide disposal, may have a potential applicability to the disposal of TCDD-contaminated laboratory wastes. These methods include microwave plasma destruction, molten salt combustion, ozone/photolytic destruction, micropit disposal and catalytic dechlorination. The most appropriate methods for primary evaluation in TCDD degradation studies appear to be molten salt combustion and ozone/photolytic destruction.

Based on present data and the current state of knowledge, it is feasible for laboratories presently working with chlorinated dioxin materials to accomplish the degradation of TCDD in contaminated wastes within the dioxin laboratory. The above laboratory disposal methods represent potential alternatives to the centralized collection and storage of dioxin wastes. Because of the potential for human exposure and the current feasibility of laboratory methods capable of TCDD destruction, the centralized collection and storage of TCDD wastes is no longer necessary, and indeed, can no longer be justified.

REFERENCES

Ackerman, D.G., H.J. Fisher, R.J. Johnson, R.F. Maddalone, B.J. Mathews, E.L. Moon, K.H. Scheyes, C.C. Shih, and R.F. Tobias. At-Sea incineration of Herbicide Orange onboard the M/T Vulcanus. EPA-600/2-78-086, April 1978.

Ayres, D.C. Destruction of polychlorinated dibenzo-p-dioxins. Nature, 290: 323-324, 1981.

Bailin, L.J., B.L. Hertzler, and D.A. Oberacker. Detoxification of pesticides and hazardous wastes by the microwave plasma process. In: Kennedy, M.V., ed. Disposal and Decontamination of Pesticides. ACS Symposium Series 73. Washington, D.C.: American Chemical Society, 1978. pp. 49-72.

Bailin, L.J. and B.L. Hertzler. Development of microwave plasma detoxification process for hazardous wastes, Phase I. Lockheed Missiles and Space Company, Inc., NTIS PB-268 526/161, 1976.

Bailin, L.J., M.E. Sibert, L.A. Jonas, and A.T. Bell. Microwave decomposition of toxic vapor simulants. Environ. Sci. Technol., 9(3): 254-258, 1975.

Botre', C., A. Memoli, and F. Alhaique. On the degradation of 2,3,7,8-tetrachlorodibenzo-para dioxin (TCDD) by means of a new class of chloriodides. Environ. Sci. Technol., 13(2): 228-231, 1979.

Buser, H.R. Preparation of quantitative standard mixtures of polychlorinated dibenzo-p-dioxins and dibenzofurans by ultraviolet and gamma irradiation of the octochloro compounds. Journal of Chromatography, 129: 303-307, 1976.

Calvert, J.G. and J.N. Pitts. Photochemistry, New York: John Wiley and Sons, 1969.

Cavolloni, L., and L. Zecca. La decomposizione del TCDD mediante ozono. Medicina Termale e Climatologia 34: 73-74, 1977.

Cooper, J.J., and W.H. Dennis, Jr. Catalytic dechlorination of organochlorine compounds IV: mass spectral identification of DDT and heptachlor products. Chemosphere, 4: 299-305, 1978.

Crosby, D.G., and A.S. Wong. Environmental degradation of
    2,3,7,8-tetrachlorodibenzo-p-dioxin (TCDD). Science, 195:
    1337-1338, 1977.
Crosby, D.G., and M.Y. Li. Herbicide photodecomposition. In:
    Kearney, P.C. and Kaufman, D.D., ed. Degradation of
    Herbicides, New York: Marcel Dekker, Inc., 1969. pp. 321-363.
Dennis, W.J., Jr., and W.H. Cooper. Catalytic dechlorination of
    organochlorine compounds III: lindane. Bull. Environ.
    Contamin. and Toxicol., 18(1): 57-59, 1977.
Dennis, W.J., Jr., and W.H. Cooper. Catalytic dechlorination of
    organochlorine compounds II: heptachlor and chlordane.
    Bull. Environ. Contamin. and Toxicol., 16(4): 425-430, 1976.
Dennis, W.J., Jr., and W.H. Cooper. Catalytic dechlorination of
    organochlorine compounds I: DDT. Bull. Environ. Contamin.
    and Toxicol., 14(6): 738-744, 1975.
Dow Chemical Company. Summary of safe handling of
    2,3,7,8-tetrachlorodibenzo-p-dioxin in the laboratory.
    Biochemical Research Laboratory, Midland, Michigan. 1970.
Duvall, D.S., and W.A. Rubey. Laboratory evaluation of high
    temperature destruction of kepone and related pesticides.
    Tech. Rep. UDRI-TR-76-21. University of Dayton Research
    Institute. EPA-600/2-76-299, 1976.
Esposito, M.P., T.O. Tierman, and F.E. Dryden. Dioxins.
    EPA-600/2-80-197, 1980.
Fanelli, R.C., Chiabrando, M. Salmona, S. Garattini, and P.G.
    Caldera. Degradation of 2,3,7,8-tetrachlorodibenzo-p-dioxin
    in organic solvents by gamma ray irradiation. Experentia,
    34(9): 1126-1127, 1978.
Kearney, P.C., A.R. Isensee, C.S. Helling, E.A. Woolson, and J.R.
    Plimmer. Environmental significance of chlorodioxins. In:
    Blair, E.H., ed. Chlorodioxins-Origin and Fate, Advances in
    Chemistry, Series 120. Washington, D.C.: American Chemical
    Society, 1973. pp 105-111.
Kearney, P.C., E.A. Woolson, A.R. Isensee, and C.S. Helling. TCDD
    in the environment: sources, fate, and decontamination.
    Environ. Health Perspect., 5: 273-277, 1973.
Kearney, P.C., E.A. Woolson, C.P. Ellington. Persistence and
    metabolism of chlorodioxins in soils. Environ. Sci.
    Technol., 6(12): 1017-1019, 1972.
Kennedy, M.V., B.J. Stojanovic, and F.L. Shuman, Jr. Chemical and
    thermal methods for disposal of pesticides. Residue Rev.,
    29: 89, 1969.
Matsumura, F., and H.J. Benezet. Studies on the bio-accumulation
    and microbial degradation of 2,3,7,8-tetrachlorodibenzo-p-
    dioxin. Environ. Health Perspect., 5: 253-258, 1973.

Mauk, C.E., H.W. Prengle, Jr., and J.E. Payne. Oxidation of
    pesticides by ozone and ultraviolet light. U.S. Army
    mobility Equipment Research and Development Command, AD-A028
    306/9ST, 1967.

Piver, W.T. Deactivation and disposal methods for small
    quantities of experimental chemicals. In: Walters, D.B.,
    ed. Safe Handling of Chemical Carcinogens, Mutagens,
    Teratogens and Highly Toxic Substances. Vol. II. Ann Arbor:
    Ann Arbor Science, 1980. pp. 555-572.

Rogers, C.J., and R. Allen. Developing technology for
    detoxification of pesticides and other hazardous materials.
    In: Kennedy, M.V. ed. Disposal and Decontamination of
    Pesticides, ACS Symposium Series 73. Washington, D.C.:
    American Chemical Society, 1978. pp. 10-117.

Stehl, R.H., R.R. Papenfuss, R.A. Bredeweg, and R.W. Roberts. The
    stability of pentachlorophenol and chlorinated dioxins to
    sunlight, heat and combustion. In: Blair, E.H., ed.
    Chlorodioxins - Origin and Fate. Advances in Chem. Ser. 120.
    Washington, D.C.: American Chemical Society, 1973. pp.
    119-125.

Watkins, D.A.M. Implications of the photochemical decomposition
    of pesticides. Chemistry and Industry, 5: 185-190, 1974.

Wilkinson, R.R., G.L. Kelso, and F.C. Hopkins. State-of-the-art
    report: pesticide disposal research. EPA-600/2-78-183,
    1978.

Yosim, S.J., K.M. Barclay, and E.F. Grantham. Destruction of
    pesticides and pesticide containers of molten salt
    combustion. In: Kennedy, M.V., ed. Disposal and
    Decontamination of Pesticides. ACS Symposium Series 73.
    Washington, D.C.: American Chemical Society, 1978. pp.
    118-130.

Young, A.L., C.E. Thalken, E.L. Arnold, J.M. Cupello, and L.G.
    Cockerham. Fate of 2,3,7,8-tetrachlorodibenzo-p-dioxin
    (TCDD) in the environment: summary and decontamination
    recommendations. USAFA-TR-76-18, USAF Academy, Colorado,
    1976.

PHYSICAL AND CHEMICAL PROPERTIES OF DIOXINS

IN RELATION TO THE THEIR DISPOSAL

W. M. Shaub and W. Tsang

Chemical Kinetics Division
Center for Chemical Physics
National Bureau of Standards
Washington, D.C. USA

ABSTRACT

The physical and chemical properties of polychlorinated dibenzo-p-dioxins have been considered in relation to prospects for their formation and destruction in incinerator environments. Detailed equilibrium and chemical kinetic considerations have been used in performing qualitative assessments. It is concluded that there are no apparent thermodynamic barriers to their destruction and that kinetic control is a dominating factor in practical incinerator environments. This analysis as well as a consideration of some existing experimental data are used to suggest some useful guidelines and to indicate research which should be carried out in the future regarding dioxin disposal.

INTRODUCTION

Although there is inadequate data to explicitly describe the complete destruction or formation mechanisms of polychlorinated dibenzo-p-dioxins (PCDDs) under thermal stress, enough information is known or can be inferred such that important characteristic features of these processes can be elucidated. We focus attention upon this class of chemicals in view of the fact that there have recently been reported several instances in which dioxins have been associated with gaseous and particulate emissions of incinerators. Two extensive reviews are available (Harris et al., 1981; Esposito et al., 1980). There has been a recent suggestion by some scientists (Bumb et al., 1980) that PCDDs may be ubiquitous as a consequence of trace chemical processes that occur during normal combustion. This hypothesis has been questioned (Kriebel, 1981). It appears at the present

time that this issue has not been resolved. What is known is that
several dioxins, particularly 2,3,7,8-tetrachlorodibenzo-p-dioxin,
are extremely toxic substances (Poland and Kende, 1976). It is
therefore desirable to determine what if any conditions for incineration are likely to minimize formation and maximize destruction of
these species.

      This paper will briefly review the molecular processes which
are responsible for destruction of a large polyatomic molecule. We
next discuss the limitations imposed by equilibrium thermodynamics.
This will be followed by an assessment of kinetic control factors.
Some considerations regarding the role of non-gas phase and/or
heterogeneous processes will be presented. These factors will be
utilized to interpret some of the experimental research that has recently been reported in the literature. Recommendations shall be
presented regarding future research as well as practical guidelines
pertinent to incinerator operation and possibly risk assessment.

      The work presented here is in response to a congressional man-
date, and as directed under Section 5002 of the Resource Conservation Recovery Act of 1876 (RCRA), and through support of the NBS
Office of Recycled Materials, the DOE Office of Energy for Municipal
Waste, and the U.S. Air Force Environics Division, Tyndall AFB.

REACTION MECHANISMS

      The kinetic processes which control the decomposition of a com-
plex organic molecule such as PCDD in a practical incinerator environment may be described by the specification of reaction mechanisms. These reaction mechanisms in turn are composed of a series
of elementary processes. The rates of these elementary steps in
combination with appropriate computer programs can describe the temporal evolution of the chemical species during the combustion process
in an incinerator. Regardless of the detailed mechanism, the fate
of the molecule is either a stepwise breakdown to smaller, often more
oxygenated species or alternatively a transformation in concert with
other species to a higher molecular weight material (e.g., soot formation).

      It is appropriate to first consider those kinetic processes
which are operative in the gas phase. Historically, it is these
processes which have been studied most and are best understood.
Furthermore, we shall show that an understanding of the general features which characterize gas phase decomposition processes provides
in turn, insight into the nature of reactions operative in other
phases. As we have discussed previously (Tsang and Shaub, 1981),
the important elementary processes operative in the gas phase in an
incinerator environment may be categorized as either unimolecular
or bimolecular reaction steps. The former are represented by reactions of the type

$$A \xrightarrow{ k_u } P$$

where A represents a complex organic molecule and P denotes a product (or products) which may not necessarily be a (or several) stable species. The rate of destruction of A by this unimolecular reaction step is $dA/dt = -k_u A$. These reactions for the present purposes can be classified as those involving either bond rupture or complex fragmentation. Biomolecular reactions may either involve addition or metathesis.

$$A + R \xrightarrow{ k_b } P + (C)$$

where A and P are previously defined before, R denotes a species capable of undergoing reaction with A, and C is the other product in metathesis reactions. The rate of destruction of A by bimolecular steps is

$$dA/dt = -(\sum_i k_{bi} R_i) A$$

where the summation over the subscript, i, denotes the fact that in a practical incinerator environment, there may be many reactive species capable of undergoing chemical reaction with A.

THERMODYNAMICS CONSTRAINTS

In order to completely describe the gas phase decomposition of a large organic molecule, it would be necessary to specify exactly all of the reaction steps in a mechanism. In practice, this has not been done to date even for so simple a molecule as methane. Approximations are necessary. In particular, thermodynamics imposes important constraints on the system and in view of the lack of knowledge of detailed mechanism, the first step must be an investigation of the consequences if equilibrium is attained.

A complete thermodynamic characterization of the equilibrium distribution of species present in a gas phase mixture is possible providing appropriate hydrodynamic variables, e.g., temperature and pressure, thermodynamic data for all pertinent species, the initial fuel/oxidizer equivalence ratios (a measure of the amount of fuel and oxidizer available) are specified. Up to the present time, the fundamental limitation to complete equilibrium characterization of many systems is due to the lack of thermodynamic data for numerous large organic molecules. This has been particularly true in the case of aromatic and chlorinated hazardous waste chemicals such as benzenes, phenols, biphenyls, biphenyl ethers, dibenzofurans and dibenzo-p-dioxins. Recently we have used estimation techniques (Shaub, 1981a, 1981b; Tsang and Shaub, 1981) to construct a thermodynamic

Table 1. Gas Phase Heat of Formation for Chlorinated Hazardous
          Waste Compounds

Class	ΔH_f^o (g, 298 K) kcal·mole^{-1}
Polychlorinated biphenyls	+35.93 → + 0.07
Polychlorinated biphenyl ethers	+10.3 → - 54.25
Polychlorinated dibenzofurans	+18.26 → - 48.12
Polychlorinated dibenzo-p-dioxins	-22.7 → - 100.2

data base for these molecules. Table 1 for example, summarizes
ranges of gas phase heats of formation estimated for polychlorinated
biphenyls (PCBs), polychlorinated biphenyl ethers (PCBEs), polychlorinated dibenzofurans (PCDFs), and polychlorinated dibenzo-p-
dioxins (PCDDs). In general, heats of formation decrease with increased degree of chlorination and number of oxygen atoms present
in the molecule. Variations within each class are attributable to
nearest neighbor interactions when using our estimation techniques.
The extent to which these estimates are reliable has been discussed
previously (Shaub, 1981a). These estimated heats of formation as
well as estimates of heat capacities and entropies (Shaub, 1981b)
can be utilized to construct the necessary thermodynamic data input
for equilibrium computations. A typical data sheet for a chlorinated dioxin is presented in Table 2. We have used this information
together with pre-existing thermodynamic data for smaller polyatomic
as well as atomic and diatomic species (Stull and Prophet, 1971; Cox
and Pilcher, 1970; and Stull and Westrum, 1969) to simulate equilibrium chemical thermodynamic distributions of polychlorinated
aromatic species for a range of specifications of initial fuel/
oxidizer equivalence ratios, temperatures and pressures characteristic of a practical incinerator environment. We are able to conclude that in terms of equilibrium thermodynamics, at temperatures
in excess of 500°C, destruction of polychlorinated aromatics is essentially complete. That is, there is no single compound with any
special thermodynamic stability characteristics. Within any particular class of chlorinated compounds, e.g., dioxins, there is a trend
toward greater stability with increased chlorination if there are
no other sources of hydrogen available in the waste feedstock. It
is often found in practice that temperatures much in excess of 500°C
are required to effect complete destruction. For example, Duvall
and Rubey (1977) report that commercial PCBs and some dioxins undergo decomposition in air between 640-740°C at a residence time of one
second. A 99.995% destruction efficiency is reported for some PCBs
at a temperature of 1000°C and the same residence time. This is an
indication that at least at the lower temperatures, (T ≳ 900°C) reactions may be kinetically controlled. It is important to recognize,
however, that while complete gas phase thermodynamic equilibrium may

Table 2. Thermodynamic Data Sheet

Octachlorodibenzo-p-dioxin $C_{12}O_1Cl_8$ (Ideal gas mol. wt. 459.744 state)

T°K	C_p°	S_T°	$-(G_T^\circ-H_0^\circ)/T$	$H_T^\circ-H_0^\circ$	ΔH_f°	ΔG_f°	Log K_p
	cal mol⁻¹ K⁻¹				kcal mol⁻¹		
200	54.54	111.60	81.45	6.03	-100.72	-71.94	78.602
298	70.86	136.58	95.58	12.23	-100.21	-57.93	42.454
300	71.12	137.02	95.84	12.36	-100.20	-57.65	41.989
400	83.45	159.25	108.97	20.11	-99.44	-43.57	23.799
500	92.81	178.93	121.03	28.95	-98.57	-29.70	12.979
600	99.88	196.50	132.17	38.60	-97.68	-16.01	5.829
700	105.20	212.32	142.51	48.86	-96.79	-2.46	0.768
800	109.23	226.64	152.15	59.60	-95.87	10.95	-2.992
900	112.32	239.69	161.16	70.68	-95.06	24.26	-5.890
1000	114.72	251.66	169.62	82.04	-94.17	37.46	-8.186
1100	116.61	262.68	177.59	93.61	-93.78	50.58	-10.047
1200	118.11	272.90	185.11	105.35	-92.39	63.62	-11.584
1300	119.33	282.40	192.23	117.22	-91.59	76.59	-12.874
1400	120.32	291.28	198.99	129.20	-90.83	89.50	-13.969
1500	121.14	299.61	205.43	141.28	-90.09	102.34	-14.908

not be attained in an incinerator environment, a biomolecularly controlled local (pseudo) thermodynamic equilibrium may be expected to
develop in practical incinerator environments for small inorganic
radicals, particularly those associated with hydrogen/oxygen reactions (e.g., H, O, OH, H_2O) by virtue of the fact that they undergo
fast biomolecular reactions with each other. This means that to a
good approximation, equilibrium thermodynamic calculations should
serve adequately to "track" concentration of these small inorganic
species with variations of temperature and pressure over time-scales
that are characteristic of an incineration process.

Gas Phase Kinetic Control Factors

It is evident that gas phase processes for destruction of chlorinated dioxins in practical incinerators are kinetically controlled.
The destruction of chlorinated dioxins may then be described by the
relationship:

$$-dA/dt = k_u A + (\Sigma_i k_{bi} R_i)A.$$

It is instructive to consider first the situation in which destruction is governed predominantly by a unimolecular mode of decomposition. It can be shown that under this condition, and subject to a
requirement of 99.99% destruction efficiency, the minimum transit
time required for achievement of this extent of decomposition is
related to the unimolecular rate constant by the expression:

$$t(99.99\%) = 9.212 \, k_u^{-1}$$

Similarly, for 99.9999% destruction efficiency:

$$t(99.9999\%) \approx 1.5 t(99.99\%),$$

The unimolecular rate constant, k_u, may be written as (pressure dependence is assumed negligible under conditions of incineration)

$$k_u = A_\infty \exp(-E_\infty/RT).$$

Above, R has the value 1.987 cal/(mol·deg K). Values for the pre-
exponential factor (units of sec^{-1}) and the activation energy (units
of cal/mol) may be conveniently estimated for a dioxin molecule undergoing decomposition via simple bond rupture - a process which
sets an upper limit range upon the decomposition time. A reasonable range for the value of the pre-exponential factor is $\approx 10^{15.5} \pm$
0.5 sec^{-1} (Tsang and Shaub, 1981). Substituent chlorine atom
effects (increasing degree of chlorination) should not be expected
to result in A_∞ values outside of this spectrum of values. The activation energy for unimolecular decomposition of a dioxin molecule
via simple bond rupture should have a value approximately equal to

the bond dissociation energy of the weakest bond. Carbon–carbon bond rupture in the aromatic rings is ruled out as being prohibitive due to resonance stabilization in the aromatic rings. The C–H bonds and C–Cl bonds in benzene and benzyl chloride have dissociation energies of 111 kcal·mol^{-1} and 95 kcal·mol^{-1} respectively. An estimate of the bond dissociation energy of the CO bond in dioxins is obtained as follows. The heat of formation of $C_6H_5O\cdot$ (phenoxy radical) is ΔH_f^o (g, 298) = 11.9 kcal·mol^{-1} as determined from the reported heat of phenol, ΔH_f^o (g, 298) = –23.0 kcal·mol^{-1} (Cox and Pilcher, 1970) the bond dissociation energy of the C_6H_5O–H bond, 87 kcal·mol^{-1} (Colussi et al., 1977), and the heat of formation of atomic hydrogen, ΔH_f^o (g, 298) = 52.1 kcal·mol^{-1} (Stull and Prophet, 1971). An estimate of the C–O bond dissociation energy in dioxins is that it should be comparable to the C–O bond dissociation energy in diphenyl ether. Since the heat of formation of the phenyl radical is ΔH_f^o (g, 298) = 78.5 kcal·mol^{-1} (Benson, 1976), and the heat of formation of diphenyl ether is ΔH_f^o (g, 298) = 11.94 kcal·mol^{-1} (Cox and Pilcher, 1970), it follows that the bond dissociation energy of the C–O bond in diphenyl ether is:

$$BDE \approx \Delta H_f^o \ (g, \ 298)_{C_6H_5\cdot} + \Delta H_f^o \ (g, \ 298)_{C_6H_5O\cdot} \ -$$

$$\Delta H_f^o \ (g, \ 298)_{C_{12}H_{10}O} \approx 78.5 \ kcal\cdot mol^{-1}$$

It is unlikely that there is any significant resonance energy associated with the central ring of dioxin. In fact, an examination of the central rings in dibenzopyran and 9,10-dihydroanthracene, using thermochemical estimates (Benson, 1976) and reported values for the heats of formation of these compounds (Cox and Pilcher, 1970), leads to the conclusions that there is no significant resonance strengthening of the bonds to the central ring bridge atoms in these compounds and further, that there is a slight central ring destabilization of about 1–4 kcal mol^{-1} due to ring strain. Thus, by analogy, it is reasonable to assume that it is unlikely that the bond dissociation energy of the C–O bond in dioxin exceeds 80 kcal mol^{-1}. In fact, it may be lower if partial resonance stabilization of the resulting phenoxy-like radical structure occurs during the bond-breaking process.

It follows from the above discussion that in the case of unimolecular dissociation of dioxins, a reasonable estimate for the unimolecular rate constant is

$$k_u \approx 10^{15.5} \pm \exp(-80000/RT) \ sec^{-1}.$$

This value for k_u can be utilized in the expression for t(99.99%) to determine the time required for efficient destruction at various temperatures. In Table 3 we present a summary of these times for $A_\infty = 10^{15}$ and also $A_\infty = 10^{16}$ to indicate the effect of the variation of this parameter. Clearly, effective removal of dioxins over short

Table 3. 99.99% Dioxin Decomposition Times, Gas Phases

$k_u = 10^{+15} \; e^{-80000/RT}$				$k_u = 10^{+16} \; e^{-80000/RT}$
$\tau =$	°C	°K		$\tau =$
2 billion years	227	500		200 million years
6 years	477	750		0.6 years
46.5 minutes	727	1000		4.65 minutes
1 second	977	1250		1/10 second
1/2 second	1000	1273		5/100 second
4 milliseconds	1227	1500		400 microseconds
5 microseconds	1727	2000		1/2 microsecond

periods of time can only occur at elevated temperatures, under conditions of unimolecular kinetic control.

The contribution of bimolecular reactions to the destruction of
dioxins may be considered as follows. As mentioned previously, the
significance of the existence of local thermodynamic equilibrium in
an incinerator environment is that we can, via computer simulation,
determine the concentrations of small inorganic radicals. We have
concluded from our equilibrium thermodynamics analysis (Tsang and
Shaub, 1981) that with respect to bimolecular removal, the major reactive species in the case of fuel rich combustion are hydrogen
atoms, and in the case of stoichiometric combustion or excess air
combustion, hydroxyl radicals (OH) and to a lesser extent, oxygen
atoms. Larger radicals, at elevated temperatures characteristic of
an incinerator combustion environment are generally present at lower
concentrations due to instability. In most cases they are also less
reactive. Consider stoichiometric or excess air combustion of dioxins. We have determined from our previous analysis of kinetic
rate constant data for the rates of attack of OH upon organic substrates that, at incinerator conditions, bimolecular rate constants
will be of the order of $k_b \sim 10^9 \pm 1 \; liter \cdot mol^{-1} \cdot sec^{-1}$ (Tsang and
Shaub, 1981). We have estimated the concentration of OH radicals as
a function of temperature under the assumption of local thermodynamic
equilibrium of small inorganic radicals. The validity of this assumption has been demonstrated (Ernst et al., 1978; Muller et al.,
1980; and Creighton, 1980). It is therefore possible to infer that
a resident time of about 1 second and a temperature of the order of
1230 K (957 C) would be required to insure 99.99% destruction efficiency of dioxins via attack of OH. Thus, it can be seen by comparison with Table 3, both unimolecular and bimolecular reactions
may contribute significantly to dioxin removal at the temperatures
where 99.99% removal occurs over timescales typical of incinerator
operation.

As we have mentioned previously, OH attack upon unsaturates tends to be electrophilic in nature, leading us to expect a lowering of the rates of bimolecular attack upon chlorinated dioxins and other compounds, such as chlorinated biphenyls, with increasing degrees of chlorination. Thus, effluent emissions from incinerators should tend to contain predominantly more highly chlorinated dioxins if in fact they do contain dioxins at all. In the case of PCBs, this effect should be more marked as C-Cl bonds are the weakest bonds in these molecules and therefore effects associated with bimolecular attack should be significantly enhanced as, under these circumstances (i.e., high bond energies), unimolecular reactions would be less predominant. These predictions are supported by experimental results reported in recent literature (Harrison et al., 1980; Duvall and Rubey, 1977). In the case of fuel-rich systems, H atom concentrations tend to be lower than the corresponding OH concentrations in stoichiometric or excess air combustion. This means that since rate constants for H atom attack at incineration temperatures are comparable to k_b for OH attack, unimolecular decomposition of dioxins may tend to be more important in fuel-rich combustion.

It is useful to make another observation regarding the incineration of hazardous materials such as dioxins. A distinction should be made between mean furnace operating temperatures which may tend to be low relative to destruction of dioxins (750 C is not atypical in municipal incinerators), and the much higher temperatures which are obtained in the flame regions of the furance. Thermal profiles in the region of flames are commonly in the range 1250 C to 1750 C over a distance in excess of 15-20 cm with peak temperatures (1750 C and higher) extending over distances of 5-10 millimeters or more (McDonald 1980; Bechtel, 1980; Switzer et al., 1980; Oran et al., 1980; and Dibble et al., 1980). In a practical incinerator environment, these distances may be greatly extended. As is evident from Table 3, it is these high temperature regions which are primarily responsible for destruction of dioxins and other similar hazardous waste stream materials. At flow velocities of 10-50 ft·s^{-1}, 99.99% (or even 99.9999%) gas phase destruction is to be expected if all the dioxin material is successfully heated to these temperatures. Thus while average furnace temperatures may be lower than what is required for dioxin disposal, local flame temperatures, if felt by all molecules in the system, can provide the impetus for successful decomposition.

With regard to the formation of chlorinated dioxins in the gas phase once temperatures for destruction of chlorinated dioxins are achieved, the same driving forces which are available to destroy dioxins will also lead to decomposition of dioxin precursors such as chlorinated phenols. Additionally, in a municipal incinerator, conconcentrations of hazardous wastes tend to be low relative to the bulk of the waste stream. Formation of dioxins in the gas phase would require a minimum occurrence of a bimolecular step, character-

ized by a very slow rate relative to the overall process of decomposition, due to low values associated with bimolecular rate constants for highly sterically hindered reactions (Gardiner, 1972).
There are reports of a few dioxin precursors (predioxins - e.g.,
polychlorinated 2-phenoxyphenols) apparently directly producing high
yields of chlorinated dioxins under mild (ca. 275 C) thermal stress
(Esposito et al., 1980; Nilsson et al., 1974). It should be noted
that there is no indication that these materials are especially
prevalent in municipal waste streams. In addition, there has been
no demonstration that conversion of these predioxins to dioxins occurs in the gas phase. Such a reaction would be expected to be
highly sterically hindered and possibly retarded by resonance stabilization of an intermediate phenoxy radical structure.

The previous discussion has assumed a model in which all hazardous waste materials such as chlorinated dioxins are in a homoge-
nous gas phase mixture, i.e., pre-mixed burning, and that further,
all of the gas is successfully elevated to the high flame temperatures required for successful decomposition. In practical incineration this is not the case. Fuel and air are not pre-mixed, although
a highly efficient spray injection liquid hazardous waste incinerator may approach this ideal. When improper mixing occurs, extremely
fuel-rich or fuel-lean pockets of gas may pass through some parts of
the incinerator and may possibly escape from the furnace area particularly if loading is excessive. If the mixture is exceedingly
fuel-rich or fuel-lean, the gas pockets may not support ignition,
and consequently material destruction may not occur due to formation of local cold spots or lack of reactive species. Additionally,
extremely fuel-rich combustion may promote undesirable condensation
reactions and lead to soot formation which can perturb energy transport to and from regions in the incinerator in addition to other undesirable effects. Recently, liquid injection incinerators were
utilized onboard the M/T incinerator ship to destroy organochlorine
wastes containing 2,3,7,8-tetrachlorodibenzo-p-dioxin (TCDD) as an
impurity to the extent of an average 2 ppm (Ackerman et al., 1978).
The flame temperature in the liquid injection incinerators was maintained at all times above 1250 C during the incineration of organochlorine wastes. Residence times normally ranged from 0.5 to 2 seconds. Combined TCDD destruction efficiencies of >99.93% were reported for three burns. An efficiency >99.99% was reported for one
burn. It is clear from these results that the predicted efficient
destruction of dioxins in the gas phase is realizable in practical
excess air incineration. Slight departures from expected theoretical efficiencies may have been due to some of the problems discussed
previously.

Non Gas Phase Effects

In the last few years, there have been several reports of dioxin emissions from full-scale combustion facilities (Olie et al.,

1977; Buser and Bosshardt, 1978; Buser et al., 1978; and Eiceman et al., 1979, 1980). In general, emissions from municipal waste incinerators have been consistently lower than those reported for industrial waste incinerator facilities (Harris et al., 1980). The majority of reports of dioxin emissions have been concerned with analysis of fly ash samples, particularly those collected from electrostatic precipitators. Although there are problems with the analytical methodology, these observations make necessary the consideration of non gas phase.

In a typical incinerator equipped with an electrostatic precipitator, the principal areas for loss of solid which are of concern are the following:

a) Fly ash and other particulates ≤10 μm in diameter may pass through the settling chambers and centrifugal separators into the electrostatic precipitator. Here, assuming efficient operation, particles as small as 0.1 μm to 0.01 μm will be collected. This material is highly oxidized, with carbon or soot contents (organic fraction) commonly as low as 5% by weight and ranging up to 40% by weight (Domino, 1979) in poorly designed or inefficiently operated incinerators. (It is possible that real time optical probing for soot deposited on fly ash may prove to be a useful indicator of efficient incinerator operation). Glassy particulates may be harder to trap due to the difficulty associated with ionization of this material. Fly ash in hoppers from electrostatic precipitators is subject to management by containment. However, due to the size of this particular matter great care in handling may be essential, particularly with dry fly ash, to avoid raising a cloud of this material.

b) Particles smaller in size than 0.01 μm will readily escape through an electrostatic precipitator and pass out the stacks. This fly ash is highly oxidized and has been thoroughly thermally stressed. This material is not subject to management by containment and represents a direct emission into the environment. Dioxin transport, if it occurs to an appreciable extent on this material could be particularly serious as deposition efficiencies for inhaled particles of this size range strongly favor pulmonary and tracheobronchial regions as opposed to nasopharyngeal regions of the respiratory tract (Natusch and Wallace, 1974). In this respect, it is also particularly important to control soot emissions (organics) as for example, benzo[a]pyrene, a common soot constituent, has been found to predominate in small pulmonary depositing particles (Natusch and Wallace, 1974).

The inorganic content of fly ash is present as metal oxides.
The relative amounts of the metal oxides present in fly ash vary
widely from one sample to another in different incinerators. On the
average SiO_2 is the principal inorganic oxide ($\approx 55\%$). Fly ash from
municipal incinerators tends to also have significantly higher
ratios (Domino, 1979; Cheremisinoff and Morresi, 1976; and Hardesty
and Pahl, 1979) of SiO_2/Al_2O_3 and CaO/Fe_2O_3 than fly ash from lignite
coal. The Al_2O_3 fraction in municipal incinerator fly ash is often
low due to aluminum can recycling. Generally, fly ash will be basic
when there are high amounts of calcium and/or aluminum and low
amounts of irons and/or aluminum. Quartz (SiO_2) and many types of
glass also tend to be basic. That is, there may be many hydroxyl
(OH) groups on the surface of these materials. The ash fusion tem-
peraturs (Domino, 1979) of most municipal incinerator fly ash are
comparable to those reported for fly ash from various coals (Hardesty
and Pohl, 1979), and typically range from about 2000 F (1093 C,
1366 K) to about 2500 K (1371 C, 1644 K). Thermal conductivities of
fly ash tend to be somewhat larger than those for most coals.

The above information about fly ash is helpful towards development of an understanding of the conditions under which one might expect dioxins to be present in this material. That this is so, many
be seen from the discussion presented below. At one atmosphere pressure and a temperature of 600 C, energy transport from the surface
of a 1000 μm diameter particle of coal to its center takes place during a time-scale of the order of one second. Under similar conditions, a 100 μm coal particle will be heated during a time-scale of
∿10 millisecond (Russel et al., 1979). These figures are approximate time-scales, but nonetheless they serve to bring out the point
that sub-micron size coal particles will heat to the core in times
much shorter than a millisecond. Since fly ash particles have similar thermal conductivities, it is reasonable to expect that during
incineration, submicron size particles should be uniformly heated to
high temperatures on time-scales that are short relative to transit
times through high temperature zones. Fusion, which is known to
commonly occur, may further promote heating of particle cores if the
particles are transformed to a liquid state due to convective transport of heat and oxidants. Under these conditions, dioxins present
inside sub-micron size fly ash particles may be subject to the same
unimolecular processes as in the gas phase. Additionally, oxidative
attack of hydroxyl or other small radicals on particles may also
contribute towards decomposition. For example, it has recently been
demonstrated that OH radicals rapidly attack the surface of carbonaceous materials even at room temperatures to produce approximately
equal amounts of CO and CO_2 (Mulcahy and Yound, 1975). Since the
surface of fly ash in incinerators may be basic, surface bound hydroxyl radicals may also play a role. Other surface effects may
also be operative. The implication of this discussion is that, providing these particles are subjected to the same high temperature
oxidative conditions as is the gas in an incinerator, high destruc-

tion efficiencies are to be expected. In fact, the completeness of
oxidation of inorganic material in sub-micron size fly ash may imply
that dioxin and other organic component levels in these particles
should be lower than in the gas phase under conditions of fuel-lean,
high temperature incineration.

Ash which collects in hoppers in the region of the electrostatic precipitor may commonly range in size up to about 10 μm. The
uncertainties regarding time-scales for heating could be as large as
an order of magnitude. This does not affect the nature of our analysis regarding sub-micron sized particles. However, as regards particles with diameters on the order of 10 mμ, if, due to uncertainties, particle heating time-scales were actually higher by an order
of magnitude, the time to heat the core of these particles could be
a significant fraction of the particle transit times through the
highest temperature zones. In this respect more experimental work
in determining actual particle heating times would be very helpful.
As can be determined from reports of fly ash collected from some
electrostatic precipitators, levels of dioxins have been found to be
in the ppb range. Note that if the input waste stream contained ppm
levels of dioxins, or of material that may be converted to dioxins
(to the extent of ppm dioxin levels during time-scales similar to
those in incinerators), a 99.99% combustion destruction efficiency
due to rapid particle heat up ·relative to transit times would imply
ppt to ppb levels of dioxins will be present in electrostatic precipitator fly ash consistent with what has been reported in some instances. This would imply electrostatic precipitator fly ash has
probably been uniformly heated. Whether or not this analysis has
validity in an actual incinerator requires statistically representative information about input feed streams that has not presently
been reported. This information would also provide in part, a useful guide in determining the effectiveness of varying furnace flame
temperatures and residence times.

Recently, formation of polychlorinated dibenzofurans (PCDFs) at
levels approaching 1% has been reported from the pyrolysis of polychlorinated biphenyls (PCBs) in sealed quartz mini-ampoules at temperatures of 550-850 C (Buser et al., 1978b). Molar ratios of PCB
to oxygen in an ampoule were varied from 1:7.5 to about 1:75. The
maximum.amounts of PCDFs reported were found at 550 C. At 850 C
levels found are reported as <0.01%. The important point associated
with this work, although not brought out by these authors, is the
fact that at the lower temperatures (550 C) where highest formations
occurred, their results clearly show the reaction to be self-inhibiting. That is, decreasing the amount of PCBs in a quartz ampoule by
a factor of ten actually results in more than a six-fold increase in
decomposition. Since gas phase reactions involving oxygen are not
important at these temperatures, the self-inhibiting nature of this
reaction clearly rules out any chain branching mechanisms and suggests that the reactions are taking place on the surface of the
quartz ampoules.

    Synthesis of chlorinated dioxins and PCDFs have also been re-
ported under similar conditions in quartz ampoules from chlorinated
benzenes (Buser, 1979) and chlorophenates (Rappe and Marklund, 1978).
It is highly likely that again, surface effects are also operative
here as no kinetically plausible gas phase processes are conceivable
at these temperatures for formation of these compounds. Additionally, it should be noted that there is a long history of problems
due to surface effects in quartz and glass vessels that have been
encountered in the study of gas phase reactions and halogen-contain-
ing organic compounds. As quartz tends to be basic, it is possible
that hydroxyl groups on the surface of the quartz ampoules may be a
reactant source for promoting formation of dioxins from PCBs and
other precursors. As mentioned previously, fly ash may also tend to
be basic and usually has a high SiO_2 content. Thus, in a manner
similar to quartz, fly ash may possibly act as a promoter for surface reactions leading to formation of dioxins and similar materials
in an incinerator environment. The relative surface of glass ampoules compared to that of fly ash in an incinerator has not been
established. Thus the validity of extrapolating the latter results
is uncertain. However, the observation of similar distribution patterns in chromatograms of dioxins obtained from municipal incincera-
tor, and controlled experiments involving chlorophenates in quartz
ampoules (Buser et al., 1978) and on leaves or wood (Rappe et al.,
1978) supports this suggestion. It is important to note that in a
study of production of PCDFs from specific PCBs that was carried out
in quartz ampoules (Buser et al., 1978b), the PCBs were destroyed
with an efficiency >99.99% at a temperature of 850 C. Since PCBs
are kinetically more stable than dioxins, comparable destruction
efficiencies are to be expected for the latter. Thus, while fly
ash may possibly promote dioxins formation at lower temperatures
around 550 C, high destruction efficiencies still appear to be possible if the fly ash is subsequently subjected to elevated temperatures for times comparable to those required for gas phase destruction at elevated temperatures, providing the ash is uniformly
stressed. This latter point is important, since if dioxines ore
precursors are trapped inside particles and the cores of particles
are not thermally stressed, destruction efficiencies may be lower
than expected, due to a thermal insulating effect.

CONCLUSIONS AND SUGGESTION FOR FUTURE RESEARCH

    Below, we enumerate a number of tentative conclusions which may
be drawn from the above discussions, in the hope that they will provide some useful guidance to enable persons concerned with the disposal of dioxins and similar materials to understand to some extent
what is possible via incineration. It seems reasonable to use to
expect that 99.99% or higher destruction efficiencies are well within
the capabilities of modern municipal waste and hazardous waste incinerators. Our basic conclusion is that destruction efficiencies
are not likely to be limited by thermodynamic or kinetic control

factors at high enough temperatures and sufficient residence times. However, in light of our discussions above, the following points should be considered:

a) Mixing is very important and should be monitored closely, especially to avoid cold spots which may result in insufficient thermal stressing.

b) As particle size increases, core heating time-scales may be lengthened. This must be considered an important factor in terms of particle transit times through the furnace and in fly ash disposal.

c) A careful, statistically-representative, characterization of the input waste feed stream is necessary to properly further assess actual destruction efficiencies.

d) Providing stack (uncontained) emissions of dioxins and related materials are reduced to acceptable levels, incineration of dioxins should in principle be a manageable problem since ash in hoppers is <u>contained</u> ash and therefore subject to further treatment if needed.

As regards future research, it would be most helpful if physical properties of various fly ash, e.g., thermal conductivities, fusion temperature, alkalinity, surface reactivity, porosity, etc., were more thoroughly determined. Rate processes of chemical reactions are readily subject to scientific scrutiny, and more carefully designed experiments are needed in the future, particularly with regard to kinetics of reactions involving materials such as dioxins and precursors on various surfaces. It should also be noted that since dioxins and similar chemical substances may be found in some fly ash, the prior history of disposal practices of fly ash would dictate that there may exist a need to investigate the chemical, photochemical, and biological <u>fate</u> of these substances as they may exist in some ash dump sites. Certainly, some monitoring of historical sites would be instructive.

REFERENCES

Ackerman, D. G., Fisher, H. J., Johnson, R. J., Maddalone, R. F., Matthews, B. J., Moon, E. L., Scheyer, K. H., Shih, C. C., Tobias, R. F., and Venezia, R. A., <u>At Sea Incineration of Herbicide Orange Onborad the M/T VULCANUS</u>, Washington, D.C., U.S. Environmental Agency, EPA-600/2-78-086, April, 1978.
Bechtel, J. H., Laser Probes of Premixed Laminar Methane-Air Flames and Comparison with Theory, in: <u>Laser Probes for Combustion Chemistry</u>, D. R. Crosley, ed., Washington, D.C., American Chemical Society (1980), pp. 850102.

Benson, S. W., Thermochemical Kinetics, John Wiley and Sons, New York (1976).

Bumb, R. R., Crummett, W. B., Cutie, S. S., Gledhill, J. R., Hummell, R. H. Kagel, R. O., Lamparskii, L. L., Luoma, E. V., Miller, D. L., Nestricle, T. J., Shadoff, L. A., Stehl, R. H., and Woods, J. S., Trace Chemistries of Fire: A Source of Chlorinated Dioxins, Science, 210:385-390 (1980).

Buser, H. R., Formation of Polychlorinated Dibenzofurans (PCDFs) and Dibenzo-p-Dioxino (PCDDs) from the Pyrolysis of Chlorobenzenes, Chemosphere, 8:415-424 (1979).

Buser, H. R., Bosshardt, H.-P., and Rappe, C., Identification of Polychlorinated Dibenzo-p-dioxin Isomers Found in Fly Ash, Chemosphere, 7:165-172 (1978).

Buser, H. R., Bosshardt, H.-P., and Rappe, C., Formation of Polychlorinated Dibenzofurans (PCDFs) from the Pyrolysis of PCGs. Chemosphere, 7:109-119 (1978).

Buser, H. R., and Bosshardt, H.-P., Polychlorierte Dibenzo-p-dioxine, Dibenzofurane und Benzole in der Asche kommunaler und industrieller Verbrennungslagen, Mitt Gebiete Lebenson Hys., 69:191-199 (1978).

Chermisinoff, P. N., and Moresi, A. C., Energy from Solid Wastes, Marcel Dekker, Inc., New York (1976), pp. 59.

Colussi, A. J., Zabil, F., and Benson, S. W., Int. J. Chem. Kin., 9:161-170 (1977).

Cox, J. D., and Pilcher, G., Thermochemistry of Organic and Organometallic Compounds, Academic Press, New York (1970).

Creighton, J. R., Rate of Methane Oxidation Controlled by Free Radicals, in: Laser Probes for Combustion Chemistry, D. E. Crosley, ed., American Chemical Society, Washington, D.C. (1980), pp. 357-363.

Domino, F. A., ed., Energy from Solid Waste-Recent Developments, Noyes Date Corp., Park Ridge, N. J. (1979), pp. 301.

Duvall, D. S., and Rubey, W. A., Laboratory Evaluation of High Temperature Destruction of Polychlorinated Biphenyls and Related Compounds, Washington, D.C., EPA-600/2-77-228 (1977).

Eiceman, G. A., Clement, R. E., and Karasek, F. W., Analysis of Fly Ash from Municipal Incinerators for Trace Organic Compounds, Anal. Chem., 51:2343-2350 (1979).

Eiceman, G. A., Vlau, A. C., and Karasek, F. W., Ultrasonic Extraction of Polychlorinated Dibenzo-p-dioxins and Other Organic Compounds from Fly Ash from Municipal Incinerators, Anal. Chem., 52:1492-1496 (1980).

Ernst, J., Wagner, H. Gg., and Zellner, R. A., Combined Flash Photolysis Shock-Tube Study of the Absolute Rate Constants for Reactions of the Hydroxyl Radical with CH_4 and CF_3H around 1300 K, Ber. Bunsenges Phys. Chem., 82:313-409 (1978).

Esposito, M. P., Tiernan, T. O., and Dryden, F. E., Dioxins, Washington, D.C., EPA-600/2-80-197 (1980).

Gardiner, W. C., Jr., Rates and Mechanisms of Chemical Reactions, W. A. Benjamin, Inc., Menlo Park, California (1972), pp. 76: 89-90.

Hardesty, D. R., and Pohl, J. H., The Combustion of Pulverized Coals –
    An Assessment of Research Needs, in: 10th Materials Research
    Symposium on Characterization of High Temperature Vapors and
    Gases, J. W. Hastie, ed., Washington, D. C., NBS Spec. Pub.
    561, Vol. II (1979), pp. 1407–1449.
Harris, J. C., Anderson, R. C., Goodwin, B. E., and Rechsteiner,
    C. E., Dioxin Emissions from Combustion Sources: A Review of
    the Current State of Knowledge, final report to ASME, New York
    (1980).
Kriebel, D., Science, 213:1060 (1981).
McDonald, J. R., Laser Probes for Combustion Applications, in:
    Laser Probes for Combustion Chemistry, D. R. Crosley, ed., Am-
    erican Chemical Society, Washington, D.C. (1980), pp. 19–58.
Mulcahy, M. F. R., and Young, B. C., The Reaction of OH Radicals
    with Carbon at 298 K, Carbon, 13:115–121 (1975).
Muller, C. H., III, Schofield, K., and Steinberg, M., Laser-Induced
    Fluorescence: A Powerful Tool for the Study of Flame Chemistry,
    in: Laser Probes for Combustion Chemistry, D. R. Crosley, ed.,
    American Chemical Society, Washington, D.C. (1980), pp. 103–130.
Natusch, D. F. S., and Wallace, J. R., Urban Aerosol Toxicity: The
    Influence of Particle Size, Science, 186:695–699 (1974).
Nilsson, C.-A., Anderson, K., Rappe, C., and Westermark, S.-O.,
    Chromatographic Evidence for the Formation of Chlorodioxins
    from Chloro-2-Phenoxyphenols, J. Chromatog., 96:137–147 (1974).
Olie, K., Vermeulen, P. L., and Hutzinger, O., Chlorodibenzo-p-
    dioxins and Chlorodibenzofurans are Trace Components of Fly Ash
    and Flue Gas of Some Municipal Incinerators in the Netherlands,
    Chemosphere, 6:455–459 (1977).
Poland, A., and Kende, A., 2,3,7,8-Tetrachlorodibenzo-p-dioxins:
    Environmental Contaminant and Molecular Probe, Fed. Proc., Fed.
    Am. Soc. Exp. Biol., 35(12):2404–2410 (1976).
Rappe, C., Marklund, S., Buser, H. R., and Bosshardt, H. P., Forma-
    tion of Polychlorinated Dibenzo-p-Dioxins (PCDDs) and Dibenzo-
    furans (PCDFs) by Burning or Heating Chlorophenates, Chemosphere,
    7:269–281 (1978).
Russel, W. B., Saville, D. A., and Greene, M. I., A Model for Short
    Residence Time Hydropyrolysis of Single Coal Particles, AICHE,
    25:65–80 (1979).
Shaub, W. M., Procedure for Estimating the Heats of Formation of
    Aromatic Compounds: Chlorinated Benzenes, Phenols, and Dioxins,
    to be submitted to Thermochimica Acta.
Shaub, W. M., Estimated Thermodynamic Functions for Some Chlorinated
    Benzenes, Phenols, and Dioxins, to be submitted to Thermo-
    chimica Acta.
Stull, D. R., and Prophet, H., JANAF Thermochemical Tables, Washing-
    ton, D.C., U.S. Government Printing Office NSRDS-NBS 37 (1971).
Stull, D. R., Westrum, E. F., Jr., and Sinke, G. C., Chemical Ther-
    modynamics of Organic Compounds, John Wiley and Sons, New York
    (1969).

Switzer, G. L., Roquemore, W. M., Bradley, R. P., Schreiber, P. W., and Roh, W. B., CARS Measurements in Simulated Practical Combustion Environments, in: <u>Laser Probes for Combustion Chemistry</u>, D. R. Crosley, ed., American Chemical Society, Washington, D.C. (1980), pp. 303-311.

Tsang, W., and Shaub, W. M., Chemical Processes in the Incineration of Hazardous Materials, Paper No. 15, 182nd National Meeting, American Chemical Society, Division of Environmental Chemistry, New York, August 23-28 1981.

THE DESIGN, IMPLEMENTATION, AND EVALUATION
OF THE INDUSTRIAL HYGIENE PROGRAM USED
DURING THE DISPOSAL OF HERBICIDE ORANGE

James W. Tremblay

Lockwood, Andrews, & Newnam, Inc.
San Antonio
Texas, USA

ABSTRACT

During the Summer of 1977 the United States Air Force (USAF) disposed of 8.4 million liters of Herbicide Orange, a 50/50 mixture of 2,4-D and 2,4,5-T. The herbicide contained approximately 23 kilograms of 2,3,7,8-tetrachlorodibenzo-p-dioxin (TCDD). Disposal was accomplished by high-temperature incineration at sea aboard the incinerator ship, M/T Vulcanus. Under provisions of United States Environmental Protection Agency (EPA) permits, the USAF was required to conduct comprehensive quality control, environmental and industrial hygiene monitoring. The disposal operations were accomplished in compliance with all EPA permit requirements. This paper focuses on the industrial hygiene workplace air sampling for 2,4-D, 2,4,5-T, and TCDD. The sampling collection methods, equipment, and materials as well as sample handling, processing, and analytical techniques are described. Results of the industrial hygiene sampling program are reviewed. Noted levels for 2,4-D and 2,4,5-T were well below permissible exposure levels of 10 mg/m^3. TCDD was not detected in any industrial hygiene air samples, with lower limit of detection on the order of 30 ng/m^3. It is concluded that similar sampling and analysis regimens may be used for workplace monitoring of 2,4-D, 2,4,5-T, and TCDD. Needed research to simplify TCDD air sampling is suggested.

INTRODUCTION

During the summer of 1977, the United States Air Force (USAF) disposed of 8.4 million liters of Herbicide Orange by high temperature incineration at sea aboard a specially designed ship. The land-based transfer and loading operations and the at-sea incineration operations were conducted under U.S. Environmental Protection Agency (EPA) special permits. Among other things, the EPA permits required the Air Force to conduct industrial hygiene and ambient air monitoring for the concentrations of 2,4-D, 2,4,5-T, and TCDD.

The objective of this paper is to briefly summarize the historical background leading to the 1977 disposal operations, to describe the land-based and shipboard operations for transfer of the Herbicide Orange to the incinerator ship, and to relate the sample collection and analytical procedures and results of the industrial hygiene monitoring. The techniques for sample collection as successfully used during the 1977 disposal operations are evaluated in light of current practice.

HISTORICAL BACKGROUND

In April 1970, the Secretaries of Agriculture; Health, Education and Welfare; and the Interior jointly announced the suspension of certain uses of 2,4,5-T. These suspensions resulted from published studies indicating that 2,4,5-T was a teratogen. Subsequent studies revealed that the teratogenic effects had resulted from a toxic contaminant in the 2,4,5-T, identified as 2,3,7,8-tetrachloro-dibenzo-p-dioxin (TCDD). Subsequently, the Department of Defense suspended the use of Herbicide Orange. At the time of the suspension, the Air Force had an inventory of 5.2 million liters of Herbicide Orange in South Vietnam and 3.2 million liters at the Naval Construction Battalion Center (NCBC), Gulfport, Mississippi. In September, 1971, the Department of Defense directed that the Herbicide Orange in South Vietnam be returned to the United States and that the entire 8.4 million liters be disposed of in an environmentally safe and efficient manner. The 5.3 million liters were moved from South Vietnam to Johnston Island in the North Pacific Ocean for storage in April, 1972. The average concentration of TCDD in the Herbicide Orange was about 2 parts-per-million, and the total amount of TCDD in the entire Herbicide Orange stock was approximately 23 kilograms.

Various techniques of destruction and recovery of the herbicide were investigated by the Air Force from 1971 to 1974.

In December, 1974, the USAF filed a final environmental impact statement (Anonymous, 1974) with the President's Council on Environmental Quality on the disposition of Herbicide Orange by destruction aboard a specially designed incineration vessel in a remote area of the Pacific Ocean, west of Johnston Island.

Between February, 1975, and February, 1977, the Air Force evaluated a private proposal to reprocess the Herbicide Orange by selectively removing the TCDD using activated coconut charcoal. The reprocessing method was found to be feasible, but a satisfactory method for disposing of the TCDD-ladened carbon could not be demonstrated.

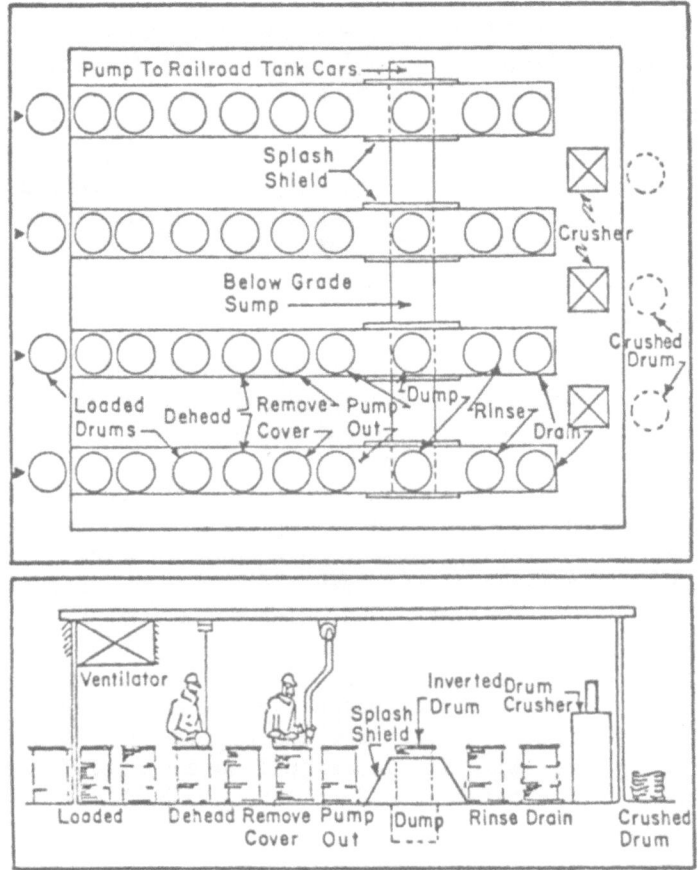

Fig. 1. Dedrumming Facility, Gulfport, Mississippi.

As a result of the public hearing held on April 8, 1977, the EPA issued a research permit to the USAF and Ocean Combustion Services, B.V. (OCS) (Anonymous, 1977). This permit authorized the transfer, loading the transport of the Herbicide Orange from the Naval Construction Battalion Center, Gulfport, Mississippi, to a designated site in the North Pacific Ocean for the purpose of at-sea incineration in accordance with the provisions of the Marine Protection, Research and Sanctuaries Act of 1972 as amended. The vessel contracted for the at-sea incineration was the Dutch-owned ship, M/T Vulcanus, a ship registered in Singapore and previously used in the North Atlantic Ocean and the Gulf of Mexico to destory chlorinated hydrocarbon wastes (Wastler et al., 1975). A total of three herbicide loadings were required to incinerate the total stocks of Herbicide Orange along with the diesel fuel rinsings: one loading from Gulfport, and two loadings from Johnston Island.

DESCRIPTION OF LAND-BASED OPERATIONS

The operations at both storage sites were similar in many ways At both sites, the 55-gallon (208 liters) drums of Herbicide Orange were transported short distances from their storage location to a centralized facility. The herbicide was drained from the drums and transferred to the M/T Vulcanus. Following emptying, the drums were rinsed with diesel fuel and subsequently crushed. The rinsing from empty drum cleaning was combined with the herbicide and transferred to the ship for later incineration at sea.

NCBC, Gulfport, Mississippi

The centralized dedrumming facility at the NCBC was a temporary, enclosed facility measuring approximately 11 meters by 11 meters with an interior ceiling height of approximately 3 meters (Fig. 1). A ventilation System capable of providing approximately 57 air changes per hour was equipped with in-line activated charcoal filters in the exhaust ventilation system to reduce vapor emissions to the outside air. Within this enclosed facility were four identical processing lines. Each line consisted of a self-closing entry door to admit full drums; a roller conveyor along which drums were moved in an upright position; a position equipped with a heavy duty electrically operated deheading cutter; a suction wand to remove the greatest portion of the herbicide from a deheaded drum; a spray device beneath the conveyor over which the deheaded and emptied drum was inverted and rinsed with 7.5 liters of diesel fuel; a commercial, heavy-duty crusher; and a self-closing exit door through which the crushed drums were passed.

Once each drum was deheaded, the contents were removed by the sunction wand, leaving approximately 11 liters of liquid in the drum. The drum was then manually inverted, and the remaining herbicide was collected in an open trough beneath the conveyor. Each drum was permitted to drain into the same trough for a minimum period of five minutes after which it was sprayed with 7.5 liters of diesel fuel, allowed to drain while still inverted for a minimum of two minutes, and then crushed end-to-end to approximately one-third its original volume. The rinsed and crushed drum was then passed through the exit door and stacked with all other crushed drums.

The liquid herbicide from the suction wands and the herbicide and diesel fuel rinsing from the below-grade, open trough were pumped to 37,800 liter capacity rail tank cars. Air displaced from the tank cars during filling was filtered through an activated charcoal filter. The rail cars were moved along a rail spur approximately 3.2 kilometers to a dockside location where the herbicide was transferred to the incinerator ship, M/T Vulcanus. Displaced air from the ship's cargo tanks was also filtered through activated chatcoal.

A total of 15,470 drums of Herbicide Orange was processed in this fashion at the NCBC between May 24, 1977, and June 10, 1977. Two 8-hour shifts of approximately 55 men each accomplished the dedrumming/transfer operations. These men were all USAF officers/ technicians from Kelly AFB, Texas; Hill AFB, Utah; Robins AFB, Georgia; Tinker AFB, Oklahoma; and McClellan AFB, California. All workers were provided daily changes of freshly laundered work clothes, and men working within the dedrum facility were provided protective clothing including cartridge respirators, face shields, rubber aprons and rubber gloves. With only a few exceptions, the men rotated through all jobs involved in the dedrumming/transfer operations. All personnel were given pre-operational and post-operational physical examinations consisting of a complete medical history, complete neurological examination and the following laboratory procedures: 1) complete hemoglobin, including hematocrit and platelet count; 2) prothombin time; 3) serum lipids; 4) serum glutamic pyruvate transaminase (SGOT); 5) serum glutamic pyruvate transaminase (SGPT); 6) serum bilirubin; 7) blood glucose; 8) complete urinalysis; and 9) a chest x-ray.

Johnston Island

The centralized dedrum facility at Johnston Island was a temporary, open facility measuring approximately 9 meters by 27 meters, consisting of a concrete pad, roof and moveable canvas walls to exclude rain (Fig. 2). This open facility was located adjacent to the Herbicide Orange storage site on the northwest end of Johnston Island. Nearly constant east winds ranging from 4.8 to 9.5 meters per second provided natural ventilation and carried released vapors away from occupied areas. Two processing lines consisting of fabricated metal racks and open troughs were located in the west two-thirds of the facility. The east one-third contained pumps and drivethrough for for fuel trucks that were used to transport the dedrummed herbicide to the M/T Vulcanus. Full drums of herbicide were transported to the dedrum facility in sets of four using forklifts equipped with specially designed clamps. The drums were placed on the inclined metal racks in four groups of 12 drums each. Each set of 12 drums was handled independently by the dedrumming crew. Once a set of 12 drums was on the rack and the forklifts had withdrawn, a crew member would punch a vent hole near the top of each inclined drum to allow the crew's supervisory personnel to check the contents. Any drums found to contain other than Herbicide Orange were removed from the line and held for further testing. Three or more closely spaced holes were then punched in the bottom of each drum and the contents allowed to drain into the open troughs. Once the herbicide had stopped flowing from the drums, they were allowed to drain for a five-minute period after which the interior of each drum was rinsed twice with a total of 7.5 liters of diesel fuel. The diesel fuel rinsing drained into the open troughts, combining with the herbicide. After the 12 drums in each set had drained for a minimum of two minutes, they

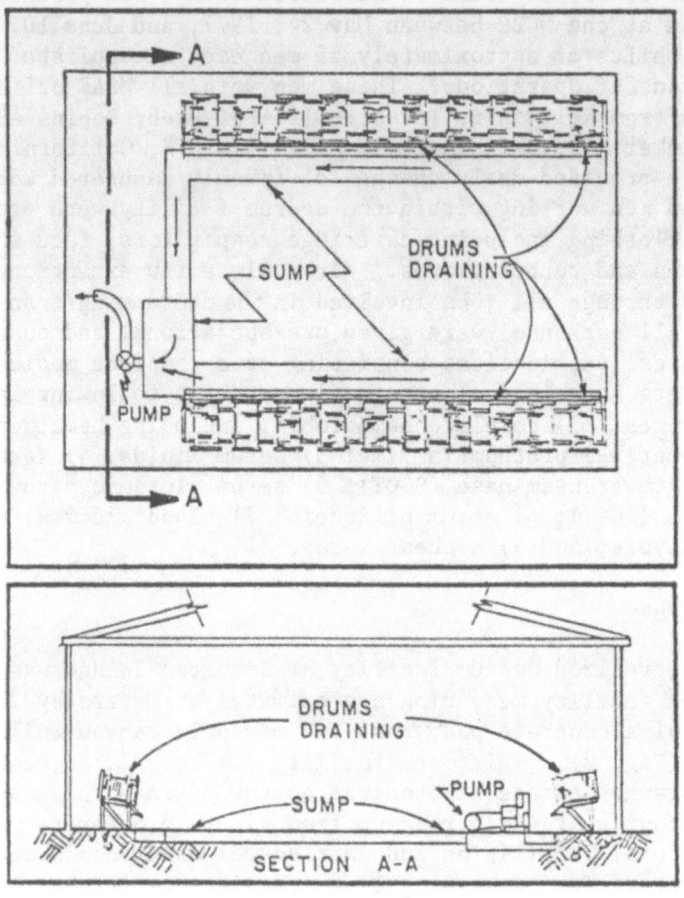

Fig. 2. Dedrumming Facility, Johnston Island.

were transported to a nearby drum crusher which consisted of a large
weight suspended between two vertical I-beams. One drum at a time
was crushed along its longitudinal axis. When approximately 30
drums had been crushed, they were removed, banded and stacked together near the crusher. The liquid herbicide and diesel fuel rinsing from the drums flowed into the two open troughs to a below-grade
sump. The material was pumped from this sump into modified fuel
tankers and transported in 11,340 liter lots to dockside where the
material was pumped aboard the M/T Vulcanus.

A total of 24,795 drums of herbicide was processed in this
fashion between July 27, 1977, and August 23, 1977. Two 10-hour
shifts of approximately 50 men each were used. The workers were
civilian employees of a contractor engaged to perform the dedrumming
operations. USAF officers monitored all operations. As at NCBC,

all workers were provided daily changes of freshly laudered work clothes, and men working within the dedrum facility were provided protective clothing consisting of cartridge respirators, face shields, rubber aprons, rubber gloves and boots. Unlike at NCBC, men on each crew remained in the same job through the dedrumming/transfer operations. A requirement of employment was pre- and post-operational physical examinations similar to those given the workers at the NCBC.

LAND-BASED AND SHIPBOARD AIR MONITORING PROGRAMS

Both environmental and occupational monitoring was accomplished at each land site and aboard the M/T Vulcanus. This section outlines only the industrial hygiene programs. Essentially, the same equipment, methods, and procedures were used. The only significant difference between the two land-based operations was that all sampling at the NCBC site was accomplished by members of the US Air Force Occupational and Environmental Health Laboratory (USAF OEHL), Brooks AFB, Texas, while sampling at the Johnston Island site was conducted by Battelle Columbus Laboratories (BCL), Columbus, Ohio, under contract to the USAF. Shipboard sampling was accomplished by TRW, Inc., and USAF OEHL. In general, the industrial hygiene sampling program at each of the land sites consisted of daily air samples within the dedrum facilities with rapid analysis (approximately 24-hour turn-a-round time) for 2,4-D and 2,4,5-T. Samples collected for analysis of TCDD were analyzed after-the-fact. Pre-operational and post-operational background sampling was also accomplished. Environmental monitoring programs are reported elsewhere (Doan 1979, and Thomas et al., 1978). Shipboard industrial hygiene sampling is described later.

Monitoring Equipment and Procedures

Two different methods were employed for industrial hygiene sampling for 2,4-D, 2,4,5-T, and TCDD. These procedures had been developed and field tested by the USAF OEHL.

Sampling for 2,4-D and 2,4,5-T

Sampling for 2,4-D and 2,4,5-T was accomplished utilizing Chromosorb 102 as an absorption medium, a granular polymer well suited for collection of chlorinated hydrodarbon vapors in air (Thomas and Seiber, 1974). The polymer was packed in micropipet tubes which were then wrapped in new aluminum foil and stored in rubber stoppered test tubes. The sampling appratus consisted of a Mine Safety Appliance Model G Personnel Sampling Pump (Fig. 3). The Chromosorb 102 tubes were connected to the pumps with Tygon or latex rubber tubing. A flow rate of 0.50 liters/minute (L/min) for periods ranging from five to ten hours was used, yielding an air sample volume of approximately 150 to 300 liters. This sampling time corresponded to the length of approximately one-half shift and was expected to

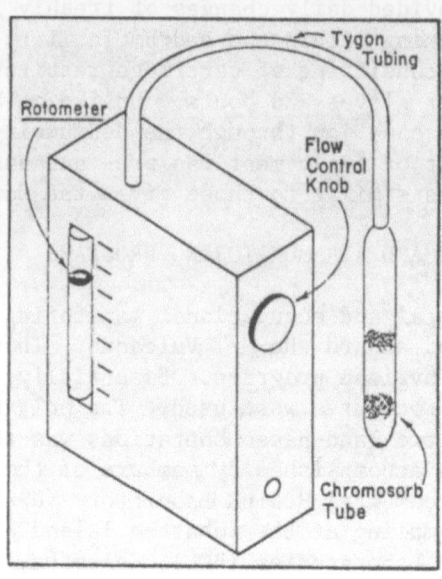

Fig. 3. MSA Model G battery operated personnel air sampler.

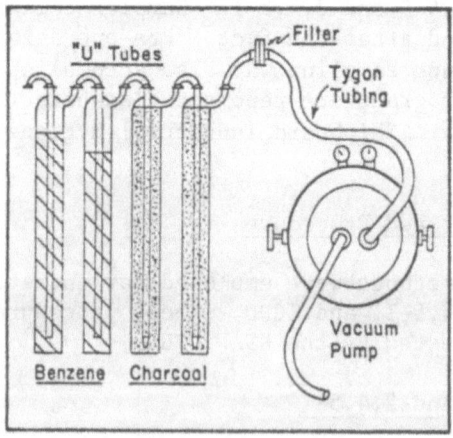

Fig. 4. USAF benzene impinger train used in sampling for TCDD.

yield sufficient adsorption to permit ready analysis. Sampling rates
were checked hourly with a calibrated rotameter to insure that 0.50
L/min flow rate was maintained. Where possible, the pumps were maintained on constant "high" charge by providing connections to available 110-vole power supply. When the Chromosorb 102 tubes were removed from the field for lab analysis, the individual tubes were
wrapped in aluminum foil and contained in rubber stoppered test tubes.

Sampling for TCDD

Air sampling for TCDD was accomplished using benzene as a collection medium. The Chromosorb 102 medium had not been evaluated
for adsorption and subsequent TCDD analyses. The benzene sampling
apparatus consisted of a train of four Greenberg-Smith impingers
(Fig. 4). The first two impingers were fritted and each contained
approximately 350 ml of benzene. The third and fourth impingers
were modified by removal of the fritts and contained activated carbon
to adsorb vaporized benzene. The two benzene impingers were wrapped
with aluminum foil providing a light barrier that would prevent any
photo-decomposition of the TCDD collected in the sample. Following
the four impingers, an in-line paper filter was attached with Tygon
tubing to prevent carbon particles from entering the Millipore pump.
The pumps were operated directly from 110-volt AC power, and the flow
rate was one L/min. The duration of sampling ranged from three to
five hours, yielding an air sample volume from 180 to 300 liters.
Flow rates were checked hourly using a calibrated rotameter, and
total volume of air sampled was calculated from these hourly flow
rates. The maximum running time of five hours was dictated by ambient temperatures ranging from 22 to 33°C and the saturation limitations of the carbon to adsorb the benzene vapors. Samples were removed from the sampling sites with impinger trains intact in special
wooden holders. The benzene was drained into new brown glass jars
in a "clean" laboratory area. The impinger glassware was rinsed with
benzene into the sample container to collect any materials adhering
to the impinger walls. All impinger glassware was rinsed three times
with acetone and once with benzene prior to reuse in the field.

Analytical Procedures and Methodologies

The analytical procedures and methodologies used throughout
this project (Project PACER HO) were developed, refined, tested and
repeatedly used throughout the variety of field studies conducted by
USAF Occupational and field studies conducted by the USAF Occupational and Environmental Health Laboratory over a five-year period
from 1972 to 1977.

Industrial hygiene samples collected during the PACER HO operations were analyzed at one of four different laboratories. At Gulfport, Mississippi, samples collected for 2,4-D/2,4,5-T were analyzed
by the U.S. Department of Agriculture, Gulf Southwest Laboratory,
Gulfport, Mississippi. At Johnston Atoll, the 2,4-D/2,4,5-T analyses
were accomplished on site by Battelle Columbus Laboratories, Columbus,
Ohio. Supplementary 2,4-D/2,4,5-T analyses were accomplished by the
US Air Force Occupational and Environmental Health Laboratory, Brooks
Air Force Base, Texas. Analysis of all samples for TCDD were ac-
complished by the Brehm Laboratory, Weight-State University, Dayton,
Ohio.

Analytical Methodologies for 2,4-D and 2,4,5-T

The sample preparation and analytical procedure for 2,4-D and 2,4,5-T used by the three laboratories were essentially the same. Following is a brief description of the preparation and analytical techniques used for 2,4-D/2,4,5-T analysis of Chromosorb 102 air samples (Thomas et al., 1978; Doan, 1979):

- Chromosorb 102, 60/80 mesh, Johns-Mansville Corp.

- Remove adsorbent and glass wool plug from the collector tube and place in an alundum Soxhlet thimble.

- Add 150 ml of hexane to the 250 ml Soxhlet extractor flask and extract adsorbent for 1 h (50 cycles).

- Concentrate extract to 1 ml and make up to 4 ml with iso-octane for gas chromatography.

Analytical Methodologies for TCDD

The sample preparation and analytical procedures for the TCDD analysis were as follows (Erk et al., 1978):

- Combine the impinger train benzene collection medium in amber bottles.

- Concentrate 250 ml of impinger benzene liquid to dryness using a stream of nitrogen and gentle heating (50°C).

- Transfer the residue to a 2 ml Reacti-Vial (Supelco).

- Using 0.5 ml of benzene (Nanograde), dissolve the residue just prior to analysis.

The Varian 2740 Gas Chromatograph-Extranuclear Quadrupole Mass Spectrometer System (GS-MS) was used for the analysis. Operating parameters for the GS-MS system were as shown in Table 1.

LAND-BASED AND SHIPBOARD MONITORING RESULTS

Detailed results of environmental and occupational monitoring at both sites have been reported by Thomas (1978) and Doan (1979). This section outlines only the industrial hygiene monitoring results for each site. All availabile data have indicated that there were no adverse environmental impacts on air, water, or land resources at either land site or at the designated ocean burn site as a result of land-based dedrumming, transfer operations, and at-sea incineration operations.

Table 1. GC-MS Operating Parameters for the Analysis of TCDD

GC-Column:	1/4" O.D. glass packed with 3% OV-3 on Chromosorb W(HP)
GC-Carrier gas:	Helium - 65 ml/min
GC-Temperatures:	Injector - 220°C
	Column - 265°C
	Transfer line - 290°C
MS-Ionizing voltage:	23.5 eV
MS-Mass monitored:	m/e 322

Naval Construction Battalion Center (NCBC), Gulfport, Mississippi

The results of the industrial hygiene monitoring programs at the NCBC are summarized below:

The industrial hygiene air sampling results for 2,4-D and 2,4,5-T, and TCDD are presented in Table 2. Five operational industrial hygiene samples were collected during each shift from the four corners within the enclosed dedrum facility. Four of these samples were for 2,4-D/2,4,5-T using Chromosorb 102, while the fifth sample was a benzene impinger in one corner of the facility collected for TCDD analysis. The Chromosorb 102 and benzene impinger samplers were placed in low traffic areas near the four corners of the enclosed facility to prevent interference with work activity within the facility. As shown in Table 2, vapor concentrations of the n-butyl esters of 2,4-D and 2,4,5-T ranged from 7.76-141.15 µg/m³, respectively. The uniformity of concentrations of herbicide respectively. The uniformity of concentrations of herbicide vapors within the dedrum facility is demonstrated by the absence of variability of 2,4-D/2,4,5-T data among the four sampling locations. All noted levels were well below the time weighted average Threshold Limit Value (TLV) of 10,000 µg/m³ for either 2,4-D or 2,4,5-T as adopted by the American Conference of Governmental Industrial Hygienists (ACGIH). No TCDD was detected in any of the 27 benzene impinger samples. The minimum detectable concentrations for TCDD ranged from 22.4 to 35.9 ng/m³.

Johnston Island

The results of the industrial hygiene monitoring programs at Johnston Island are summarized below. There were two distinct loading operations during the Johnston Island phase of the project. The first dedrum/transfer (first loading) operation was conducted from July 27, 1977, to August 5, 1977, and the second loading from August 17, 1977, to August 23, 1977. The industrial hygiene sampling of

Table 2. Results of Industrial Hygiene Air Samples Collected Inside the Dedrumming Facility Project PACER HO, NCBC, Gulfport, Mississippi, May 24–June 10, 1977

Parameter	Sample location dedrum facility Naval Construction Battalion Center (NCBC)			
	SE Corner	NE Corner	SW Corner	NW Corner
No. of samples	28	28	14	14
NBE[a]2,4-D (µg/m³):				
Range	8.7-141.15	7.86-136.35	7.76-134.9	15.18-105.11
Std. Dev.	31.45	34.55	36.25	27.01
Mean	52.99	53.72	54.58	51.5
NBE[a]2,4,5-T (µg/m³):				
Range	5.52-65.11	5.70-76.36	3.01-79.62	7.59-51.31
Std. Dev.	14.98	18.57	21.02	12.79
Mean	26.40	29.93	32.39	25.93
TCDD:				
No. of samples	27	0	0	0
Mean	ND[b]	-	-	-

[a]NBE is normal-butyl ester.
[b]ND is non-detectable at minimum detectable concentrations that ranged from <'22.4 to '33.9 ng/m³.
Note: The time-weighted Threshold Limit Value for either 2,4-D or 2,4,5-T is 10,000 µg/m³ (see text.

Table 3.   Results of Industrial Hygiene Air Samples Collected Inside
           the Dedrumming Facility, Project PACER HO, Johnston Island
           First Loading July 27-August 5, 1977

Parameter	Sample location Dedrum Facility Johnston Island, first loading		
	SW Corner	NW Corner	E Wall
No. of samples	3	3	3
NBE[a]2,4-D (µg/m³):			
Range	12.8-16.0	4.79-18.33	0.50-2.58
Std. Dev.	1.77	7.30	1.37
Mean	14.84	9.99	1.03
NBE[a]2,4,5-T (µg/m³):			
Range	6.92-8.84	2.26-8.28	–
Std. Dev.	1.05	3.24	–
Mean	8.12	4.58	–
TCDD:			
No. of samples	4	0	0
Mean	ND[b]	–	–

[a]NBE is normal butyl ester.
[b]ND is non-detectable at minimum detectable concentrations that
ranged from <'8.06 to <'13.89 ng/m³.
Note: The time-weighted Threshold Limit Value for either 2,4-D or
       2,4,5-T is 10,000 µg/m³ (see text).

the Johnston Island operations differed from the sampling at the
NCBC. The facility was larger and open to natural ventilation, and
the dedrum operations were far different as described earlier. Because of these and other factors, the industrial hygiene sampling
program was modified to include true "breathing zone" samples for
2,4-D and 2,4,5-T from selected worker positions. In general, there
were three worker positions evaluated using the Chromosorb 102 tubes.
These positions were selected after an analysis of all positions revealed that these worker locations represented the greatest possibility of receiving a significant exposure. The first was the position occupied by those workers who punched the vent holes in each
drum. When the vent holes were punched, internal pressure in many
drums was released; and there was a possibility of elevated exposures to workers who punched the several drain holes in each drum,
and the third position was the operator of the sump pump. In the
latter two cases, these workers were close to open troughs of flow-

Table 4. Results of Industrial Hygiene Air Samples Collected Inside
          the Dedrumming Facility, Project PACER HO, Johnston Island,
          Second Loading August 17–August 23, 1977

Parameter	Sample location Dedrum Facility Johnston Island, second loading	
	SW Corner	NW Corner
No. of samples	1	1
NBE[a]2,4-D (μg/m^3):	18.78	6.60
NBE[a]2,4,5-T (μg/m^3):	7.35	2.27
TCDD:		
No. of samples	5	0
Mean	ND[b]	–

[a]NBE is normal butyl ester.
[b]ND is non-detectable at minimum detectable concentrations that
ranged from <6.64 to <23.41 ng/m^3.

ing Herbicide Orange. In addition to these "breathing zone" samples, air samples within the dedrum facility were also collected for
2,4-D, 2,4,5-T, and TCDD. Tables 3, 4, and 5 present the results of
these sampling programs.

    The levels noted within the dedrum facility at Johnston Island
were on the order of two to five times lower than those noted at the
NCBC, Gulfport, Mississippi. These lower concentrations probably
resulted, from much greater dilution by natural ventilation of the
open facility at Johnston Island. The noted levels of 2,4-D and
2,4,5-T were well below the ACGIH TLV of 10,000 μg/m^3. No TCDD was
detected in any of the samples analyzed.

Description of Shipboard Industrial Hygiene
Sampling Operations

    The M/T Vulcanus, converted in 1972 from a bulk carrier, was
designed to carry approximately 3500 cubic meters of liquid wastes.
Two high temperature incinerators are installed aft. Depending upon
the characteristics of a given waste, the ship can incinerate up to
25.0 metric tons per hour. Normal incinerator operating flame temperature in 1500°C; and nominal incinerator residence time is 1.0
second (Fig. 5).

Table 5. Results of Industrial Hygiene "Breathing Zone" Samples
          Collected Inside the Dedrumming Facility, Project PACER
          HO, Johnston Island

Vent parameter	Sample locations Dedrum Facility, Johnston Island (see text)		
	Vent punchers	Drain punchers	Pump operator
	First loading (July 27–August 5, 1977)		
No. of samples	8	10	5
NBE[a]2,4-D (μg/m^3):			
Range	2.14–30.8	7.64–19.18	6.11–26.78
Std. Dev.	8.35	5.73	8.18
Mean	17.92	19.18	14.36
NBE[a]2,4,5-T (μg/m^3):			
Range	0.57–16.1	3.79–13.6	2.43–11.48
Std. Dev.	4.52	2.95	3.61
Mean	8.70	9.54	6.32
	Second loading (August 17–23, 1977)		
No. of samples	12	7	0
NBE[a]2,4-D (μg/m^3):			
Range	8.38–40.28	15.96–38.0	–
Std. Dev.	10.47	8.53	–
Mean	23.20	23.04	–
NBE[a]2,4,5-T (μg/m^3):			
Range	6.49–22.22	8.82–22.53	–
Std. Dev.	6.06	5.20	–
Mean	13.21	13.68	–

[a]NBE is normal butyl ester.
Note: The time-weighted Threshold Limit Value for either 2,4-D or
       2,4,5-T is 10,000 μg/m^3 (see text).

      During the Herbicide Orange disposal operations, the ship con-
ducted three burns. Average rate of incineration was 15 metric tons
per hour. Flame temperatures ranged from 1375°C to 1576°C. Results
of incinerator stack sampling revealed that the TCDD destruction
efficiency exceed 99.87% during each of the three burns. The destruction efficiencies for 2,4-D and 2,4,5-T exceeded 99.999% as reported by Ackerman (1978).

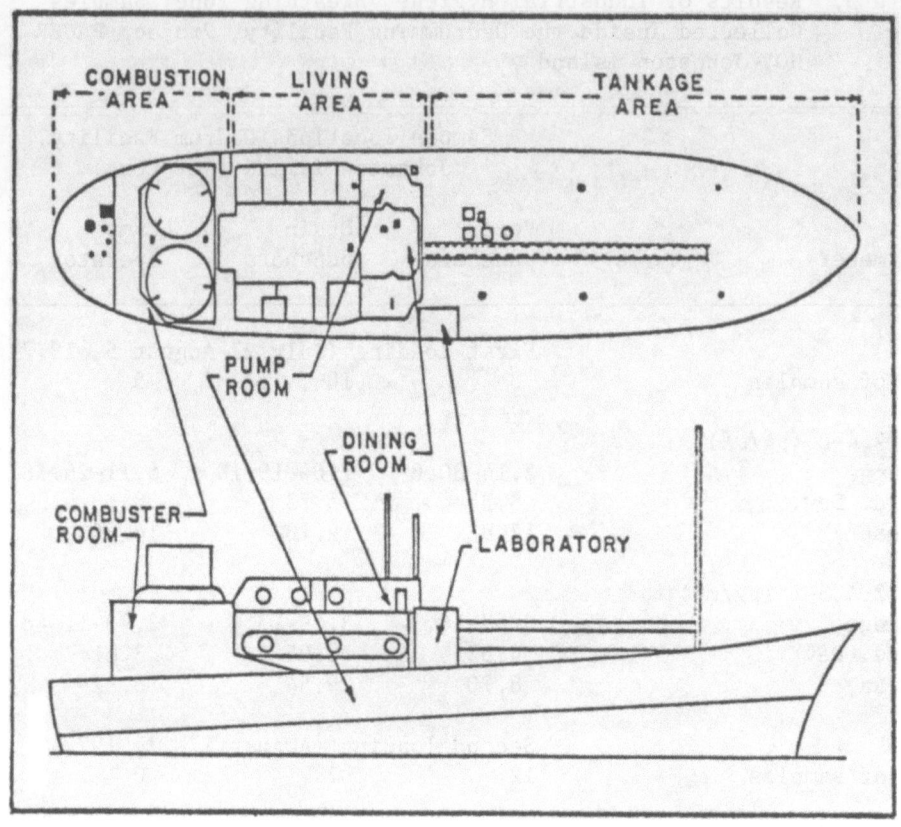

Fig. 5.   The M/T Vulcanus and the Industrial Hygiene Sampling Scheme.

As was the case for land-based operations, workplace areas
aboard the ship were sampled to determine airborne concentrations of
Herbicide Orange components. Air sampling was conducted in four
areas during the three burns: the incinerator combustion room where
incinerator control operators worked; the pump room where all wastes
were pumped; the portable laboratory where all environmental and
safety samples were analyzed; and the ship's dining room.

Air samples aboard the ship in these four areas were collected
using the same basic techniques used during the land-base monitoring.
Because of the potential hazards of handling benzene aboard the ship,
ambient air samples were collected only for the 2,4-D and 2,4,5-T
using the MSA Model G samplers and Chromosorb 102 tubes. Results of
the shipboard workspace air sampling for the three burns are presented in Tables 6, 7, and 8.

Table 6. Results of Industrial Hygiene Air Sampling (first burn)[b] Collected Aboard the M/T Vulcanus, Project PACER HO, off Johnston Island, North Pacific Ocean, 1977

Parameter	Combustor room	Pump room	Dining room	Laboratory
Ambient temperature	93°C	32°C	24°C	23°C
No. of samples	6	8	7	7
NBE[a] 2,4-D ($\mu g/m^3$):				
Range	0.22–5.1	0.42–62.2	<0.22–1.03	0.16–<0.58
Std. Dev.	2.01	21.2	0.35	0.17
Mean	1.68	9.73	<0.55	<0.47
NBE[a] 2,4,5-T ($\mu g/m^3$):				
Range	0.06–2.50	0.1–10.8	<0.08–1.29	<0.08–0.29
Std. Dev.	0.95	3.58	0.07	0.09
Mean	<0.95	2.09	<0.12	<0.22

[a]NBE is normal-butyl ester.
[b]First burn — 11 days, from July 13 to July 25, 1977.
Note: The time-weighted Threshold Limit Value for either 2,4-D or 2,4,5-T is 10,000 $\mu g/m^3$.

Table 7. Results of Industrial Hygiene Air Sampling (second burn)[b] Collected Aboard the M/T Vulcanus, Project PACER HO, off Johnston Island, North Pacific Ocean, 1977

Parameter	Combustor room	Pump room	Dining room	Laboratory
Ambient temperature	93°C	32°C	24°C	23°C
No. of samples	5	5	3	3
NBE[a]2,4-D ($\mu g/m^3$):				
Range	1.73-68.3	0.70-7.31	0.56-1.37	<0.28-<0.29
Std. Dev.	27.8	8.39	1.00	0.006
Mean	29.18	12.60	0.54	<0.28
NBE[a]2,4,5-T ($\mu g/m^3$):				
Range	0.84-38.2	0.22-10.3	0.15-0.48	<0.14-0.15
Std. Dev.	15.57	3.80	0.18	0.006
Mean	16.19	4.61	0.8	<0.14

[a]NBE is normal-butyl ester.
[b]First burn - 9 days, from August 7 to August 16, 1977.
Note: The time-weighted Treshold Limit Value for either 2,4-D or 2,4,5-T is 10,000 $\mu g/m^3$.

Table 8. Results of Industrial Hygiene Air Sampling (third burn)[b]
          Collected Aboard the M/T Vulcanus, Project PACER HO, off
          Johnston Island, North Pacific Ocean, 1977

Parameter	Combustor room	Dining room
Ambient temperature	93°C	24°C
No. of samples	5	7
NBE[a]2,4-D ($\mu g/m^3$):		
Range	9.97-114	0.53-1.52
Std. Dev.	38.31	0.432
Mean	65.07	1.034
NBE[a]2,4,5-T ($\mu g/m^3$):		
Range	6.09-59.6	0.15-0.43
Std. Dev.	20.08	1.28
Mean	33.01	$\leq$0.268

[a]NBE is normal-butyl ester.
[b]Third burn - 10 days, from August 24 to September 3, 1977.
Note: The time-weighted Threshold Limit Value for either 2,4-D or
       2,4,5-T is 10,000 $\mu g/m^3$.

Burn No. 1 was conducted between July 13 and July 25, 1977.
Shipboard air samples were collected during this period. The results of this sampling are shown in Table 6. A total of 28 Chromosorb 102 samples were analyzed for the n-butyl esters of 2,4,-D and
2,4,5-T. As was anticipated, very low levels were noted in the
dining room and laboratory. Relatively higher levels were found in
the combustor and pump rooms. In all cases, the noted levels of
2,4-D and 2,4,5-T were well below the ACGIH TLV of 10,000 $\mu g/m^3$.

Similar results were noted from sampling these locations during
the second (August 7 to August 16, 1977) and third (August 24 to
September 3, 1977) burns (see Tables 7 and 8, respectively). As
would be expected, the airborne concentrations noted on the second
and third burns reflected increasing concentrations. In all cases,
however, the levels were well below the TLV.

SUMMARY AND CONCLUSIONS

The USAF accomplished industrial hygiene and ambient air sampling of all land-based dedrumming/transfer operations and shipboard operations of Project PACER HO, the USAF project to dispose of
8.4 million liters of Herbicide Orange.

The results of these sampling programs revealed that under the worst case noted the levels of 2,4-D and 2,4,5-T vapors were well below the Threshold Limit Value of 10,000 $\mu g/m^3$. The noted levels were at least two orders of magnitude below the TLVs. TCDD was not detected in any air samples.

Approximately 200 personnel carried out the dedrumming activities at the NCBC, Gulfport, Mississippi, and at Johnston Island. Comparisons of available pre- and post-operational medical examinations of military personnel involved revealed no apparent physical effects as a result of these activities.

The sampling and analytical procedures used during these operations may be employed in similar activities.

Research is indicated to determine the efficacy and efficiency of air sampling for dioxins and dibenzofurans using a solid sorbent collection medium such as Chromosorb 102.

Sampling programs such as those described in non-industrial environments should be employed as a means of verifying the proper operations and functioning of engineering control systems (hoods, booths, etc.).

REFERENCES

D. G. Ackerman, H. J. Fisher, R. J. Johnson, R. F. Maddalone, B. J. Mathews, E. L. Moon, K. H. Scheyer, C. C. Shin, and R. F. Tobias, in: "At-Sea Incineration of Herbicide Orange Onboard the M/T Vulcanus," Environmental Protection Technology Series EPA-600/2-78-086. Office of Research and Development, U.S. Environmental Protection Agency, Research Triangle Part, North Caroline (1978), 263 p.

Anonymous, in: "Disposition of Orange Herbicide by Incineration," Final Environmental State, Department of the Air Force, Washington, D.C. (1974), 737 p.

Anonymous, in: "Marine Protection, Research and Sanctuaries Act (Ocean Dumping) Research Permit Number 770DHOOIR," U.S. Environmental Protection Agency, Washington, D.C. (1977), 15 p.

T. R. Doane, in: "Land Based Environmental Monitoring at the Naval Construction Battalion Center, Gulfport, Mississippi Disposal of Herbicide Orange," Technical Report of the U.S. Air Force Occupational and Environmental Health Laboratory, Brooks Air Force Base, Texas (1979), 83 p.

S. D. Erk, M. L. Taylor, and T. O. Tiernan, in: "Environmental Monitoring in Conjunction with Incineration of Herbicide Orange at Sea," Activities of the Brehm Laboratory, Wright State University, Dayton, Ohio, Presentation to the 1978 National Conference and Exhibition on Control of Hazardous Material Spills, Miami, Florida (1978), 31 p.

T. C. Thomas and J. N. Seiber, in: "Chromosorb 102, an Efficient
    Medium for Trapping Pesticides from Air," <u>Bull. Environ. Contam.
    Toxicol.</u>, 12(1):17-25 (1974).

T. A. Wastler, C. A. Offutt, C. K. Fitzsimmons, and P. E. Des Rosiers,
    in: "Disposal of Organochlorine Wastes by Incineration at Sea,"
    Environmental Protection Technology Series EPA-430/9-75-014,
    Office of Water and Hazardous Materials, Environmental Protec-
    tion Agency, Washington, D.C. (1975), 233 p.

PANEL REPORTS

Chairman: Otto Hutzinger
            University of Amsterdam
            Amsterdam, The Netherlands

One key activity during the International Symposium on Dioxins involved discussions that were conducted by pre-selected members of Blue Ribbon Panels, one for each of the technical sessions. Each panel met one or more times during the four-day symposium and was charged with summarizing scientifically supported knowledge, drawing conclusions supported by the data, listing unresolved issues, and recommending future research direction.

The following are reports developed by the Blue Ribbon Panels.

ANALYTICAL CHEMISTRY

Chairman: Harold MacLeod, Ph.D.

Panel Members

Rudolph-Han Buser, Ph.D.	Robert L. Harless, Ph.D.
Warren B. Crummett, Ph.D.	T.J. Nestrick, M.S.
Aubrey Dupuy, Ph.D.	Lewis Shadoff, Ph.D.
David Firestone, Ph.D.	Thomas Tiernan, Ph.D.
Michael Gross, Ph.D.	Halle Tosine, Ph.D.

The panel, a group of experienced analysts representing agencies from North America and Europe, examined past achievements, the status quo, and future developments.

Progress in the development of analytical methodology for the determination of chlorinated dibenzodioxins (CDDs) and dibenzofurans (CDFs) in products and environmental samples has been extensive and dramatic during the last decade. Thus, the limit of detection for the tetrachlorodibenzo-p-dioxins (T_4CDD) in products has been lowered from one part per million in 1969 to 1 part per billion in 1980. Similarly, the limit of detection for 2,3,7,8-T_4CDD in environmental samples has developed to a part per trillion in 1978 from 50 parts per billion in 1970. Furthermore, the ability to separate a specific isomer in a particular isomer group from all of its isomers and other congeners has advanced from the ability to separate 2,3,7,8-T_4CDD from only two of its isomers in 1974 to an ability to separate all of the 22 T_4CDD isomers in 1978. Likewise, all 10 isomers of H_6CDD have been separated as have the two isomers of H_7CDD.

Such rapid development of highly sensitive methodology suitable for the determination of specific compounds among large numbers of isomers in a series of homologous compounds, as well as a vast number of other related compounds, sets new standards for progress in analytical science. It was achieved by the continuous investment in the finest manpower and equipment, both operating near their optimum potential. Leadership and cooperation by industry, academic, and government agencies were required to accomplish the goal. In no small measure this International Symposium has significantly contributed to this rapid progress by convening viable working groups

in which deliberations and free exchange of ideas has taken place.
Continuation of this forum in future years would be expected to maintain this momentum.

Emphasis was given to documenting analytical problems for environmental and commercial sample matrices and the ability of methods
of analysis to produce reliable data. There was unanimous agreement
that reliability of the method to provide sound data for dioxin residues in various matrices was absolutely necessary. Otherwise, conclusions reached by other disciplines using these data would be in
jeopardy if not completely erroneous. It is axiomatic that analytical chemistry serves as the basis on which all our numbers are being
generated, for frequently the users of data are inclined to accept
without question their reliability when making decisions.

Analysts by nature are individualistic with strong opinions on
how to approach an analytical problem. Frequently the end result is
a myriad of methods designed to reach the same goal: a procedure
capable of generating a number that truly reflects a residue's status
in its native or foreign environment. Thus we are presented with a
large variety of methods from which to choose. There are some who
feel strongly that an effort to standardize these procedures should
be made in one form or another.

It is the concensus of the panel that methodology should not
be rigidly standardized. To the contrary, laboratories should be
free to develop their own approaches to an analytical requirement.
However, it is important that the laboratories attain a high level
of analytical proficiency through experience and a quality assurance
program. It is also desirable that the labroatories monitor performance via participation in interlaboratory check sample programs.
The panel recognizes that the first tentative steps have been taken
to develop interlaboratory quality assurance programs, e.g., the
dioxin implementation plan, the Canada/U.S. round robin check sample
for fish, and the exchange of other samples between various laboratories in Europe, the United States and Canada. However, these efforts should be strengthened and financially supported by all interested parties and agencies.

Currently, analysts are working at very low levels whose significance on human health is not understood absolutely. At present,
agencies in the United States and Canada have issued advisories to
jurisdictions consuming Great Lakes fish which expressed concern
about ingestion of fish containing more than 10 (New York State),
20 (Health and Welfare Canada), and 25 (Food and Drug Administration)
ppt of 2,3,4,7-TCDD. It is important that available methodology be
demonstrated to be capable of determining reliability of 2,3,7,8-TCDD
at these levels. Another concern directly related to this same situa-

tion is the analysts' need for guidance from other disciplines on what level of residue a method would be expected to quantitate in each matrix. In the interest of practicality, this need should no longer be avoided but confronted by all concerned. It is suggested that on recognition of a potential residue problem by any interested party, one of the first prerequisites in planning an adequate response is involvement of the analyst to establish practical quantitation levels.

Primary standards for all individual dioxin congeners and their derivatives are a crucial problem needing immediate attention. They are required for further expansion and validation of present methodology. Common concerns dictate that on an international basis, a cooperative approach be taken and a suitable repository be set up for acquiring and handling these highly toxic compounds and related substances.

Methodology for the determination of dioxins and their derivatives, especially TCDD, is on the leading edge of analytical capabilities, requiring a high degree of analyst expertise to detect and confirm the identity of low ppt levels of some congeners. This presents problems associated with different matrices such as biological, effluents and particulates, waste disposal areas, etc. It has been observed that limits of detection at levels of 0.5 to 10 or 15 ppt may differ widely in their value between samples of the same species, e.g., fish, soil, etc. Thus, there is need of further researching, development and validation of methods, particularly in the areas of waste disposal dumps, incinerators, and commercial products such as fuels for reciprocating engines, power plants, etc.

A serious drawback with present methodology is its complexity and resultant low volume sample output. This has made it difficult to accumulate sufficient data on a matrice's residue status that is statistically significant. An exception is probably the fish surveys. Research should be directed to developing procedures and techniques that would conceivably maintain accuracy and/or precision with high volume output. Such areas as bioassay, MS/MS techniques, etc., require immediate attention.

ENVIRONMENTAL CHEMISTRY

Chairman: Philip C. Kearney, Ph.D.

Panel Members
Donald Barnes, Ph.D.   David L. Stalling, Ph.D.
Fumio Matsumura, Ph.D.   Donald I. Townsend, Ph.D.
Christoffer Rappe, Ph.D.   H.K. Wipf, Ph.D.

We need to understand the photodecomposition of PCDDs and PCDFs on particulate matter and/in aerosols.

While pioneering work has been done on photodecomposition of certain PCDDs in solution and nonreactive surfaces, little is known about how these results would predict photodecomposition of PCDDs and PCDFs adsorbed on to particles; specifically, flyash and soil.

The answers to this question would help us to understand the significance and persistence of PCDDs and PCDFs introduced into the environment through a) combustion and b) leachates from dumpsites.

We need to understand the bioavailability of PCDDs and PCDFs on particulate matter.

A series of reports have raised perplexing questions: Why does 2,3,7,8-TCDD predominate in fish samples? And what is the biological significance of inhalation of PCDD/PCDF-contaminated flyash?

In order to resolve such questions, additional studies are needed. (Poiger and Schlatter have already conducted an experiment in this area. EPA and Dow are designing a cooperative experiment on fish.)

The results are needed to address the question which follows on the heels of each new report on discovery of PCDDs and PCDFs on particles; that is, "So what?"

How do PCDD and PCDF distributions in soils vary with time?

There has been some discussion about an "aging effect" in which PCDDs and PCDFs become progressively harder to extract as time goes by. These reports have not included all of the PCDDs and PCDFs, however. This information would be helpful in determining the eventual fate of the different PCDDs and PCDFs in the environment and in directing our efforts to those species of greatest concern.

In order to understand the environmental behavior of PCDDs and PCDFs in the environment, its behavior in the soil must be studied in detail. Significant changes can probably be spotted and rationalized by systematic studies of the variation of fingerprint profiles of PCDDs and PCDFs with time. It should be possible to distinguish between biotic and abiotic changes. General rules for the prediction of the behavior of individual isomers may be deduced.

We should make every effort to gather information on PCDFs as well as PCDDs in our experiments.

Because the toxicological responses of organisms are quite similar for PCDD and PCDF exposures, survey analyses for these contaminants should include both classes. Minimally, profiles (concentration estimates for each degree of chlorination for PCDD and PCDF congeners) should be determined. Also patterns for the various 2,3,7,8 substituted congeners should be characterized. Isomer specific analyses are most useful and could provide data that permit classification of pollution sources. Pattern recognition techniques can readily be utilized as the dimension of M-space is fixed by the number of isomers of each class. In this manner classification of pollution sources could be done quantitatively and similarities among patterns of PCDDs and PCDFs in samples more readily recognized.

What are the PCDD and PCDF profiles on flyash and related emissions from many different combustion and other sources, and can we use these in conjunction with pattern recognition techniques to identify sources?

Attempts have been made to relate PCDDs and PCDF profiles observed in the environment to those associated with various sources. The general tactic has been to relate the relative amounts of various isomers or isomer groups in source and environmental samples. Because of the large number of isomers, and the large number of variables in the experiments, far more data are necessary to perform all but the simplest studies. Nevertheless, this approach continues to be a promising technique for future work and should be expanded to include a larger field of study as well as more specific isomer analysis as analytical techniques become available.

What is the significance of evaporation of PCDDs and PCDFs from soil and water surfaces into the atmosphere compared to combustion sources?

At this stage there is no hard data indicating the magnitude of evaporation of these chemicals. Since they are known to be relative labile against photochemical reactions but stable in soil, evaporation is expected to greatly influence the overall persistency figure for them in the environment. Also, it is important to know the factors influencing the rate of evaporation from soil, the most important parameter being the effect of water which is known to aid evaporation of DDT and other pollutants from the soil surfaces.

Explain the mechanism for PCDDs degradation in the aquatic ecosystem which gives the shift in isomer patterns observed in the environment and in biological samples.

Laboratory experiments are needed to establish the factors responsible for the markedly disproportioned PCDD and PCDF residues observed in aquatic biota and sediments. PCDF residues in fish are primarily 2,3,7,8-tetra- and penta-chloro congeners while sediments are highly enriched with hepta- and octachloro PCDFs (and OCDD). Similar residue profiles are observed for PCDDs. Rapid metabolism of isomers having vicinal hydrogens would most readily explain the occurrence of these residue patterns in biological organisms. Similar patterns were observed in the Yusho patients. Does adsorption of the hepta- and octa- CDDs and CDFs on carbonaceous particulate matter in the sediments explain the enrichment of the more highly chlorinated PCDDs and PCDFs in sediments? Is photochemistry involved?

What are the ambient levels of PCDDs and PCDFs in the environment (soil)?

An increasing number of reports show the occurrence of PCDDs and PCDFs in various parts of the environment. In most cases the amounts reported are low. It is difficult to put these into a perspective without knowing what (if any) background level we are generally experiencing. For theoretical reasons and from a number of studies, we can expect that PCDD and PCDF levels, if present, will be below the level of detection in air and water. Therefore, the question needs to be focused on the ambient levels in the soils -- in the city, in the country, etc.

Not enough is known about the ambient levels of PCDDs and PCDFs in the environment (soil). There are indications that this type of contamination is widespread. Systematic distribution studies involving individual isomers or fingerprint patterns are necessary to get a general picture on a large scale (urban vs rural areas in the U.S. and other countries). On a smaller scale, similar studies may be used to locate point sources and distinguish them from a general background due to a number of nonidentifiable sources.

Is uptake the limiting factor in microbial degradation?

The rate of degradation of TCDD in soil and aquatic sediment is known to be slow. One possibility is that this type of chemical is not readily picked up by microorganisms in soil, and therefore is not subjected to microbial degradative forces. There are two pieces of supporting evidence: First, Poiger and Schlatter have shown that TCDD incorporated in soil/water paste was not readily picked up by the experimental animal, while that dissolved in methanol was. Second, Matsumura has shown that the use of ethyl acetate or DMSO as a solvent-carrier for TCDD stimulated the process of TCDD degradation in a culture of microorganisms. This point could be very important in assessing the future potentials for designing microbial degradation systems for these pollutants.

Should PCDD and PCDF surveys be made for the human food chain?

Because complex mixtures of PCDFs and PCDDs have been found in samples of fish from the Great Lakes and other fresh water ecosystems, residue measurements are needed to evaluate these and other known human dietary exposure sources (such as gelatine). For example, a potential risk group may be people who consume large amounts of therapeutic drugs in gelatine capsules. PCDF have not been measured in nursing mothers but levels may eventually be found because of the existence of PCDFs in PCBs and chlorinated phenols.

What is the relative importance of factors that contribute to emissions for PCDDs and PCDFs; e.g., feed material, temperature, residence time, etc.

The paper by Shaub and Tsang contributed valuable insights into this question. Their work needs to be extended and complemented by analysis of feedstocks and deliberations of engineers on the topics of incinerator design and retrofitting.

What are the PCDD and PCDF emissions from burning of PCP-treated wood?

A number of laboratory and theoretical studies have identified burning of PCP-treated wood as a major source of PCDD entrance

into the environment.   These predictions need to be tested by experimentation.

*What is the environmental fate of PCDDs and PCDFs that are
found in bottom ash of certain municipal waste combustors?*

International studies by several investigators have reported
finding PCDDs and PCDFs in the bottom ash of municipal wasts combustors.   The eventual fate of these materials need to be determined.

ENVIRONMENTAL TOXICOLOGY

Chairman: Eugene E. Kenaga, Ph.D.

Panel Members
Lorris G. Cockerham, Ph.D.   J. Russell Roberts, Ph.D.
Michael Gilbertson   Charles E. Thalken, D.V.M.
Donald D. Harrison   Alvin L. Young, Ph.D.

Recent findings of up to 100's of ppt of various isomers of polychlorinated dibenzo-p-dioxins and dibenzofurans in fish in the Great
Lakes and some rivers have raised the question of the toxicity of such
chemicals. Representative chemicals which are found are the 2,3,7,8-
tetrachloro, hexachloro, and octachloro derivatives. The fact that
they have been found in fish, birds, and sediment establishes some
degree of persistence. The nature of such chemicals leads one to
believe they bioconcentrate from water to some degree. This is
supported by the fact that when found they are usually detected in
fish and not in water. Except for TCDD, there is little, if any, data
on the toxicity of these chemicals to fish or wildlife.

Basic data needed for hazard evaluation is almost completely
missing aside from data on 2,3,7,8-TCDD. Examples of needed data are
water solubility, octanol-water partition coefficient, vapor pressure,
sediment partitioning, experimental bioconcentration factors, acute
toxicity (with delayed observation) on rats and fish, at a minimum.
Depending on the results of this preliminary set of data, some idea
of the toxicity to reproduction of some key representative organisms
may be needed, especially if monitoring data continue to show concentrations of such chemicals to be increasing in the environment.

It is recommended that key tests be done on only a limited number
of dioxin and benzofuran derivatives and a very limited number of
organisms on a "need to know" basis. The specific problems should be
addressed after this data becomes available and not automatically run
the entire checklist of test methods and species which could be run.
For chemicals such as certain benzofurans which are already present in
fish, water, or sediments in minute amounts, a controlled fish reproduction test in situ could be accomplished such as in artificially
isolated lagoons of the Great Lakes. Such a test would be a product
of the effect of the total load of toxicants (i.e., DDE, PCB's, PCDD's

and PCDF's) which is not accomplished in laboratory tests. This is
a field test problem which needs the special attention of fisheries
biologists.

A number of predictive models, partition coefficients, and
regression equations are available which could be used. However,
since the majority of these compounds are likely to have octanol-water
partition coefficients (Kow) values in excess of one million, the
validation needs to be extended with particular attention to bioaccumulation and the possibility of anomolies with high molecular weight
compounds.

The correlating equations are based upon either Kow or water
solubility. With the exception of 2,3,7,8-TCDD, data on these parameters are unavailable, the relative Kow values of the various groups
of homologues should be established to provide a better Kow base for
predictions. Additionally, the isomeric specificity at highly chlorinated PCDD's and PCDF's for Kow relationships should be established.

The extent to which dioxins other than 2,3,7,8-TCDD are degraded
in vertebrates is not sufficiently understood, albeit analogies can be
drawn with the chlorinated benzofurans and polychlorinated biphenyls
which suggest that such processes should reflect a high degree of
structural specificity. Without an understanding of these patterns,
it is difficult to determine with confidence the exposure encountered
by an organism simply on the basis of the residues in its tissues.
Thus, the bioaccumulation patterns of various highly chlorinated
dibenzodioxins and dibenzofurans should be established for vertebrates.

Because a complex mixture of dibenzodioxins and dibenzofurans are
known to have entered from a variety of sources, there is a need to
monitor the levels and effects in the environment.

There is first a need to investigate the extent of contamination
on a geographic basis for a variety of compounds to determine how it
relates to potential sources and to identify hotspots where environmental monitoring is needed. In areas where elevated levels are
found, there is a need for selection of appropriate biological monitoring species, both to investigate the incidence of effects among the
population, and to document the trends in the level of contamination
over time.

Because dioxin analyses are so costly, locations for monitoring
should be selected taking into account the likely sources that may
contaminate the ecosystem.

Some other toxicological problems which come up with all chemicals including the dioxins and related chemicals, are the number of
species which need to be tested for representation of various phyla,

classes, or families of species. It has been noted that there are similar ranges (several orders of magnitude) in sensitivity, for example, between species of fish as there may be between fish and aquatic invertebrate species. Therefore, a representative of 2 to 3 species is probably adequate except for evaluation of specific organisms. Some data on surrogate species extrapolation exist in the literature.

Another problem is field validation of laboratory data, or vice versa. It cannot be expected that they can ever duplicate each other. However, if the data from such experiments come within one order of magnitude, it can be considered to be of limited confirmation and some consolation, but, also unfortunately, possibly fortuitous. Caution should be exercised. More examples of good laboratory simulation of field tests need to be developed.

From the data presented at this conference, it is evident that certain concentrations of TCDD in soil, sediments and water are related to no-effect levels in animals. We should now be making practical use of this information.

BIOCHEMISTRY AND METABOLISM

Chairman: Stephen Safe, Ph.D.

Panel Members
Marcel H. Bickel, Ph.D.   Allan B. Okey, Ph.D.
M. Gillner, Ph.D.   Richard Peterson, Ph.D.
H.B. Matthews, Ph.D.   George D. Sweeney, M.D., Ph.D.
Daniel W. Nebert, M.D., Ph.D.

A.   METABOLISM OF DIOXINS AND RELATED COMPOUNDS

Properties	PCDDs	PCDFs	PCBs	PCNs
1. Rate of metabolism decreases with increasing chlorine substitution	+	+	++	+
2. Hydroxylation site specificity	+	+	+	-
3. Phenolic (or diol) compounds are the major metabolites and have been characterized	+	+	++	+
4. Binding to macromolecules observed	+	-	++	-
5. Biohydroxylation catalyzed by monooxygenases	+	+	++	+
6. Metabolites are generally less toxic than their hydrocarbon precursors and the process (metabolism) results in detoxication	+	-	+	-

+ = some data; ++ = considerable data; - = no data.

Recommendations For Future Work

1. Further characterization of all the 2,3,7,8-TCDD metabolites.

2. Comparative metabolic studies in diverse animal species.

3. Toxicity testing of 2,3,7,8-TCDD metabolites (presumably using synthetic compounds).

4. Additional metabolic studies on other relevant PCDDs and related compounds.

5. Do the metabolites bind to the receptor (and are some toxic)? (Test the bile metabolites in the binding assay.)

6. Is TCDD-protein covalent binding important in toxicity?

B.   PHARMACOKINETICS

Properties	PCDDs	PCDFs	PCBs	PCNs
1. Highly dependent on species	+	+	++	-
2. Regulation of residues by the size of the fat reservoirs	-	-	+	-
3. Marked effect of structure on fat retention (e.g., ortho substitution - PCBs)	-	-	+	-
4. Increasing chlorine content of the halohydrocarbons results in increased long-term retention in adipose tissue except for highly chlorinated compounds > Cl_7.	-	-	+	-
5. Readily metabolized congeners are rapidly removed from tissues	+(?)	+	++	+

- = very little data; + = some data; ++ = considerable data.

Recommendations

It is clear that the pharmacokinetics of dioxins and related compounds is dependent on numerous factors (lipophilicity, molecular volume, shape, etc.) and is species dependent. Appropriate mathematical models should be developed to explain the myraid of results.

C.   BIOCHEMISTRY AND MECHANISM OF ACTION

Properties	PCDDs	PCDFs	PCBs	PCNs
1. Marked effect of structure on activity (most active chemicals are isosteric with 2,3,7,8-TCDD)	++	+	++	-
2. Correlation of AHH induction with toxicity	++	+	+	-
3. Correlation of avidity of receptor binding and toxicity	++	+	+	-
4. Segregation of activity with Ah locus	++	+	+	-

DIOXINS AND RELATED COMPOUNDS--UNRESOLVED PROBLEMS

1.   Confirmation that binding to the receptor is required for the biologic and toxic effects.

2.   Determination of the structural factors which facilitate the ligand-receptor binding.

3.   The mechanism of the interaction between the ligand-receptor and DNA and the related controls.

4.   Is the induction of $P-450_c$ responsible for any of the toxic responses?

5.   There is a need for the development of more in vitro assays.

6.   Several anomalies must be resolved or explained, e.g.:

    (a)   Why is the guinea pig more susceptible to TCDD toxicity than the hamster even though their receptor levels are comparable?

      (b) Is there any evidence that ligand-receptor interactions are different in tissues of different species (e.g., hamster, rat, guinea pig)?

7. More studies are required on the levels of the human receptor and the structure of the mammalian <u>Ah</u> receptor.

8. Are there epigenetic mechanism of toxicity and are dioxins and related compounds active as carcinogens or promoters?

9. Porphyria

      (a) The utility of altered porphyrin patterns as a tool for diagnosing toxicity in mammals should be determined

      (b) The protective roles/mechanism of iron depletion and antioxidants should be elucidated.

10. Wasting Syndrome--What is the mechanism and is the set-point hypothesis valid?

ANIMAL TOXICOLOGY

Chairman: Edward Smuckler, M.D., Ph.D.

Panel Members
Renate D. Kimbrough, M.D.
Richard J. Kociba, D.V.M., Ph.D.
Robert A. Neal, Ph.D.
Alan Poland, M.D.

Toxicology is the science of adverse (toxic) reactions. Animal toxicology generally focuses on the responses of vertebrates, although studies of invertebrates are not unknown. In large part, however, our information base is derived from studies in rodents and monkeys. The following report is directed at summarizing some of the known and unknowns in the animal toxicology of a number of polyhalogenated aromatics, including the polyhalogenated dibenzodioxins, dibenzofurans and azoxybenzenes.

The wealth of data available suggests that there are several target systems which are sensitive to these classes of compounds, including epithelial structures (especially skin), liver, thymus, and the fetus. Reports of toxic effects in other systems seem to be less consistent. More work is needed, however, in separating those responses that are degenerative from those that are restorative/proliferative/adaptive on/or those resulting from host response to other organ system changes.

Critical evaluation of physiologic changes in the animal as a whole have been less thorough; the available data appear to be conflicting. Particularly important is the question of reversibility of adverse effects. Is the thymus lesion caused by TCDD really reversible as suggested? What effect does age have on the development of this response? What is the long term effect of TCDD treatment and recovery on immunological competence?

For TCDD, absorption, distribution, metabolism and excretion data are very different in different species; the overall impression, however, is that TCDD is poorly metabolized. On the basis of studies in this area as well as the fact that there are similarities of effects among species, the general impression is that for TCDD at least, metabolism is not a requisite for adverse effects. The relationship of the parent TCDD or its metabolic products to the observed tissue changes must be investigated more thoroughly, particularly for the compound 2, 3, 7, 8-TCDD.

There are a number of problems which must be confronted in order to analyze actions and/or effects of the polyhalogenated dibenzodioxins and dibenzofurans. First, there has been no consistent analysis of the effect of an agent on both structure and function; the available data on injury to the animal are largely descriptive. The supposition has been made, particularly with regard to analysis of visceral modification that PBB's, PCB's and TCDD's result in similar acute and chronic morphological changes in parenchymal organs. Without a detailed comparison, this seems an unwarranted supposition. In order to justify it, a concerted effort must be made to demonstrate similarities (and differences) in structural and biochemical changes. Secondly, studies on the mechanisms of action of these compounds must be conducted in order to differentiate degenerative from proliferative or restorative changes. Thirdly, more time must be devoted to analyzing death and its associated phenomena in the whole organism rather than in individual systems or compartments. And finally, a more detailed comparison of in vivo and in vitro metabolic data must be made to determine how well in vitro tests will predict the outcome(s) of in vivo exposure.

There are a number of key observations that can be used to identify important areas to be examined further. For instance, the cutaneous response to TCDD appears to be a proliferative one for the epithelium thickens and hyperkeratis is prominent. This response should be studied in more detail to determine if the skin lesions represent increased cellular proliferation, increased cellular longevity or increased cellular synthetic activity. Such a study should also include an analysis of the rate of gene product formation e.g., keratin. It is also important to know if TCDD is sequestered in the skin cells. If Dr. Poland's findings are correct and applicable to skin, then nuclear localization of TCDD should be demonstrable by autoradiography.

Another problem which requires further study, is the temporal quantitation of hepatocellular response to TCDD. Studies to address this problem must include analysis of both structural and functional cellular change. It may be possible in such studies to separate out responses or effects which modify phenotype, e.g., cause proliferation, from those which are purely destructive and cause cell death and degeneration. In these studies, the cellular and subcellular localization of TCDD in hepatocytes should be sought and Dr. Poland's hypothesis should be tested by seeking gene product modification.

The carcinogenic potential of TCDD is not defined. Does it act as an initiator or a promoter (or both)? Clearly, TCDD acts as a promoter in the Peraino two-step liver system and just as clearly, administration of TCDD in food results in liver neoplasia, but the definitive experiment showing TCDD as an initiator has not yet been performed. Dr. Pitot's experiments, but this time using TCDD as an initiator, should be repeated.

It has been suggested that TCDD intoxication, at least in some species, results in vascular change and associated hemorragic phenomena. However, changes in clotting function or vascular permeability have not been analyzed. "Chick edema" disease could be caused either by decreased osmotic pressure or by reduced regulation of vascular permeability. The difference in mechanism should lend itself to testing.

TCDD (and other halogenated hydrocarbons) can potentially intercalate in membranes as well as mucleic acids. Is it possible that these compounds modify membrane properties? If so, then cytochemical studies should provide some interesting answers.

The striking thymic involution seen following TCDD intoxication is not a property of TCDD alone but is also seen with other toxins and may be independent of the endrocrine axis. More research is required to determine what is unique about thymocytes and their associated parenchymal cells that makes them so sensitive to toxic insults.

The ultimate goal of animal toxicology studies is to serve as indicators or predictors of human hazard. To reach this goal, however, some validation will be required that adverse effects observed in animals really are predictive of human reactions. This validation can only be achieved through a careful assessment of intravitam and post mortem structural and functional changes in humans whose exposure to known levels of these compounds can be documented.

HUMAN OBSERVATIONS

Co-Chairman: Kenneth D. Crow, M.D.
Co-Chairman: Giuseppe Reggiani, M.D.

Panel Members
Donald F. Austin, M.D.   Renate D. Kimbrough, M.D.
Ralph R. Cook, M.D., M.P.H.   George D. Lathrop, M.D., Ph.D.
William R. Gaffey, Ph.D.   Barclay M. Shepard, M.D.
Alastair Hay, Ph.D.   Allan H. Smith, M.D.
Patricia A. Honchar, Ph.D.

    The panel considered as its first priority the identification
of controversial issues. The first was determining exposure. The
assessments of high, medium or low exposure (above background) are
all relative terms that vary quite widely in the literature, and
this makes comparison between studies very difficult.

    The skin lesion, chloracne, is an important indicator of ex-
posure to TCDD and other known chloracnegens and is probably the
most sensitive indicator of exposure.

    As far as human health problems resulting from TCDD exposure
are concerned, the panel has reviewed teratogenicity, fetotoxicity,
cardiovascular disease, neurotoxicity, chromosome aberrations,
hepatic prophyria, and carcinogenicity.

    Teratogenicity and fetotoxicity are controversial because TCDD
is both a fetotoxin and a teratogen in certain laboratory animals.
Information indicating that TCDD is fetotoxic or teratogenic in
humans is lacking.

    The Alsea II study concluded that 2,4,5-T contaminated with
TCDD is fetotoxic. (The herbicide was claimed to have been the
cause of an increase in spontaneous abortions in the Alsea region
of the state of Oregon). This study, however, has been severely
criticized on methodological grounds and many remain skeptical about
its findings. In addition to this, there is evidence from other
studies which challenges the suggestion that TCDD is fetotoxic in
humans. There is no evidence that TCDD is a teratogen in humans.

Elevations in blood lipids have occurred in workers exposed to TCDD. The most consistent observation is an elevation in triglyceride levels. Elevated cholesterol levels have been observed in some cases. The significance of these findings is not clear. There is no evidence that exposure to TCDD is associated with an increase of cardiovascular disease.

Peripheral neurotoxicity has been observed in workers exposed to high levels to TCDD, PCBs and pentachlorophenols. There are no objective clinical signs of central nervous system damage in individuals exposed to chloracnegens; however, many have complained of a wide range of subjective symptoms in severe poisoning. Some of these symptoms have also been reported in cases of low exposure and the significance of this is not understood.

Chromosome aberrations have not been observed in people exposed to TCDD. However, there have been reports of chromosome abnormalities in workers using pesticides. The significance of such findings do not permit any predicitions to be made about the health risk to the individual.

Two severe industrial poisonings due to TCDD have produced a number of cases of classic hepatic porphyria. Recent work indicates that a particular porphyria can be indentified which may be specific to chloracnegen poisoning. This work should be pursued at every opportunity.

In connection with carcinogenicity, the panel reviewed 4 Swedish scientific papers on the carcinogenicity of the phenoxy acids and the chlorophenols. These studies report a positive association between exposure to phenoxy acids and chlorophenols and soft tissue sarcomas and lymphomas, but not colon cancer. These findings must be replicated in other areas and by different methods before a cause and effect relationship between phenoxy acids and soft tissue sarcomas or lymphomas can be concluded.

Four cases of soft tissue sarcoma have been observed in two epidemiological studies of workers exposed to 2,4,5-trichlorophenol at Dow Chemical and Monsanto. In all these cases, the men were exposed to high concentrations of TCDD and in every case there was diagnosed, or suspected, chloracne.

A study is being conducted in New Zealand comparing occupational exposures between soft tissue sarcoma cases and patients with other forms of cancer identified from the National Cancer Registry. A preliminary analysis utilizing occupational data in the Cancer Registry relating to the time of registration does not reveal any occupational differences between the two groups.

A cohort of herbicide applicators in Finland exposed to phenoxy herbicides for at least two weeks between 1955 and 1971 have been monitored since 1972. The incidence of tumours in this group is no higher than would be expected. Thus far there have been 20 cancer deaths (4 less than would be expected) in this group, but no cases of soft tissue sarcoma have been observed.

LABORATORY SAFETY AND WASTE MANAGEMENT

Chairman: Alvin L. Young, Ph.D.

Panel Members
Hans Beck, Ph.D.   Walter M. Shaub, Ph.D.
Jean H. Futrell, Ph.D.   Lee G. Taft, Ph.D.
Fred D. Hileman, Ph.D.   James W. Tremblay, P.E.
D.R. Hilker   Wing Tsang, Ph.D.
George G. Outterson, Ph.D.   William Wolfe, M.D., M.P.H.

Health Surveillance Program

     A health surveillance program should consist of a standardized
but comprehensive physical examination that incorporates extensive
biochemical analyses. Preemployment examinations must be conducted
and a careful medical/occupational health history should be obtained
at each examination. Frequency of examinations should be based on
degree of exposure risks (rosters of laboratory personnel, past and
present, should be maintained for future tracking purposes).

     A major need is the development of methodologies to interpret
fluctuations in biochemical tests as a function of exposure. The
current recommendation for an individual having an abnormal test
result is to repeat the examination. If the result is still abnormal,
attempt to exclude other potential causes (e.g., alcohol), assess
other clinical indicators and subclinical signs (e.g., nerve conduction) and evaluate the individual's risk status in the laboratory.

     Determinations need to be made as to whether informed consent
documents are adequate to protect management and female employees
from problems arising from teratogenic and fetotoxic effects.

Decontamination Methods

     For laboratory spills or contamination, no standard methods of
decontamination are available. An evaluation of different decontamination methods should be conducted.

Monitoring Residues in the Laboratory

Although wipe techniques are routinely used for monitoring residues in the laboratory, there is a <u>critical</u> need for determining "safe limits". Thus, a contamination standard is needed in the United States and other countries for surfaces and it should encompass dioxins, furans, and PCB's. For example, in Seveso, the Italian government accepted a safety standard for TCDD of 0.75 micrograms/m^2 inside family dwellings.

Non-sophisticated monitoring programs (i.e., one that incorporates automated procedures) are more likely to be used in a laboratory. Compliance with complex, labor intensive programs requiring hands-on cleanup and injection is generally sub-optimal.

Use of Protective Clothing

Data are needed on the effectiveness of various types of gloves and coveralls to provide personal protection. Moreover, a method is needed to "clean" contaminated personnel as a consequence of spills or barriers provided by protective clothing. The present recommendation is to thoroughly wash with soap and water.

Laboratory Design and Equipment

Equipment should be properly installed and maintained. Data are needed, however, on the fate of chlorinated dioxins, Furans, and related compounds in the internals of sophisticated laboratory equipment (e.g., the GC-MS).

Floor surfaces should be made of non-porous material and areas of the laboratory should be designated "high" or "low" risk; the integrity of these areas must be maintained.

Methods for Disposal of Laboratory Wastes

Data are needed on "practical" but safe means of disposing of laboratory wastes (both liquid and solid). In laboratories handling animal wastes or biological tissue, different procedures may be needed for storing and disposing of contaminated material.

Public Relations

Audiovisual material may be an effective means of educating laboratory, support, and non-laboratory personnel to the Safety procedures for working with hazardous material as well as to the analytical or toxicological project. Information exchange is needed between laboratories regarding safety programs.

Labelling of Environmental Samples

Frequently, personnel handle the same sample differently when analyzing for dioxins versus inorganic contaminants. There must be a consistency in handling samples.

AUTHOR INDEX